Immunology Methods Manual
Volume 2

IMMUNOLOGY METHODS MANUAL
The Comprehensive Sourcebook of Techniques

Volume 2

Edited by

Ivan Lefkovits

Basel Institute for Immunology

Grenzacherstrasse 487

Postfach

CH-4005 Basel

Switzerland

ACADEMIC PRESS

Harcourt Brace & Company, Publishers

San Diego London Boston New York Sydney Tokyo Toronto

Academic Press, Inc.
525 B Street, Suite 1900, San Diego, California 92101-4495, USA
http://www.apnet.com

Academic Press Limited
24-28 Oval Road, London NW1 7DX, UK
http://www.hbuk.co.uk/ap/

ISBN 0–12–442712–X

Library of Congress Cataloging-in-Publication Data

A catalogue record for this book is available from the British Library

Typeset by Unwin Bros Ltd., Old Woking, Surrey
Printed in Great Britain by Unwin Bros Ltd., Old Woking, Surrey

97 98 99 00 01 02 EB 9 8 7 6 5 4 3 2 1

Cover picture by Louis Du Pasquier: Chromosome spreads made with *Xenopus* lymphoid tumor cells.

Preface

The manual that you now hold in your hands exists because many of us – scientists from this institute and elsewhere – were convinced that this type of book is needed. It is different from other methodology books in several ways. Firstly, it covers a wide spectrum of know-how in immunology. Secondly, it is project-oriented rather than protocol-oriented, with the emphasis on describing the environment in which each method was developed and is used. We hope that immunologists, developmental biologists, students and advanced researchers, and workers in clinical and diagnostic laboratories, will find in it what they need for their daily work and for developing long-term research strategies.The book is intended for scientists who are at the forefront of immunology and for those who will join the avant-garde by using these methods. One of the aims of the book is to provoke dialogue between the authors and the method users. The hard copy of the book provides an excellent starting point for this and the CD-ROM version additionally enables the user to copy-paste, reshuffle and print out those methods performed daily. The next logical step, which will lead to a truly interactive world, is when user groups and authors are connected via the Internet.

The manual contains advanced methods as well as descriptions of basic protocols. It will be used not only by those who study the immune system of man or mouse but also by those who intend to apply the methods to other species, for instance chicken, frog, fish or sheep. It also describes in detail the reasons why certain species are more attractive than others for solving a certain problem.

This book is large and versatile and more than 300 authors have contributed to it. Who are these people? How were they recruited? Two decades ago when I prepared the first volume of *Immunological Methods*, the world of immunology was still so small that we all knew each other. Since then, numerous subdisciplines of immunology have evolved and the revolutionary development of the field resulted in a situation whereby the section of immunology that any one scientist intimately knows is becoming increasingly smaller.

Twenty years ago it was possible to find 'complete methodological know-how' within one good institute. Today this is no longer the case. The field has become so vast and the information so plentiful that scientists have lost their universality, and now even the collective wisdom of an institute cannot encompass all available knowledge. We have to have access to the skills of other scientists and other institutes worldwide. Information in bulk is indeed available within our global village, whereas sorted and comprehensive information is still a rare commodity.

When preparing the preliminary list of the fields I intended to cover in this book, I recognised humbly that for many research areas I did not know the experts. Thus, I decided to create an editorial structure which would allow me to delegate the recruiting of the contributors. This became feasible by assembling a group of Section Editors. It was satisfying when the people I contacted said 'Yes' to the project. We agreed that the Section Editors would be free to choose their contributors and have as many of them as they felt they needed. In some instances they chose contributors from their own laboratory, in other instances they asked for contributions from outside. Clearly, I only had a limited idea as to who could best write up the most important and up-to-date methods in cold-blooded vertebrates, sheep or chickens. However, once Louis Du Pasquier, Wayne Hein and Olli Vainio agreed to assemble the relevant methodology, the task became straightforward. I was quite surprised by the blunt reaction of some contributors to the *Xenopus* methodology, who stated that they would not have written the contributions had they not been asked to do so by Louis Du Pasquier. I am sure that the sheep methodology will be read by many more research workers than just those working on sheep – it will be a section that serves as a source of inspiration for many researchers. The chicken methodology provides a good contrast. The methods show that much can be learned from studying these species and much can be adapted for use in other species.

Many molecular biological methods relevant for studying the immune system are included. By no means could, or should, this methods book compete with Maniatis (Molecular Cloning, CSH 1989). However, it is certain that those scientists studying rearrangement of immunoglobulin genes, or those wishing to study the expression of chimeric recombinant molecules in myeloma cell systems, or maybe antibody-based fusion proteins, antibody fragments and engineered molecules in mammalian cells, will find the best approach for their problem in this book.

The lymphocyte remains the central theme of the methods book; because the source of lymphocytes is not only the spleen, lymph node or peripheral blood, also considered are resident pulmonary lymphocytes or lymphocytes from skin, gut, liver, mammary gland and uterus. Cells other than lymphocytes (e.g., mast cells or follicular dendritic cells) are also dealt with in detail.

Methods for establishing models of autoimmunity are described: whether in arthritis, experimental allergic encephalomyelitis, insulin-dependent diabetes mellitus, myasthenia gravis, thyroiditis, uveoretinitis, systemic lupus erythematosus or any other autoimmune disease.

The section on studying HIV (Section 29) is of great interest and relevance. This is reflected in the description of methods for HIV growth, measurement and neutralization, HIV-specific cytotoxic T lymphocyte analyses, preparation and use of vaccinia virus vectors for HIV proteins, among others.

Various applications of confocal microscopy are of upcoming interest. Included are methods for co-localization, signaling observed with intracellular dyes, and single-cell calcium release measured on the flow cytometer.

Turning our attention back to lymphocytes, various culture techniques, lymphocyte migration tests, adhesion assays, and techniques for cytokine preparations and tests can all be found in this manual.

Viruses and cytotoxic T cell responses are dealt with by several contributors (vesicular stomatitis virus, vaccinia virus, lymphocytic choriomeningitis virus, etc.) Apoptosis has become a methodologically rich study area worked on in several sections (cell lines resistant to apoptosis, staining of apoptotic lymphocytes).

It would be inconceivable to omit methodology on B and T cell lineages: detection of multipotent hematopoietic cells in pre-liver embryos, development of B lineage cells from uncommitted progenitors *in vitro*, and growth of human B cell precursors in bone marrow stromal cell-dependent cultures.

Across all the approaches, the use of PCR technology with many applications is provided; one is the procedure to establish gene expression patterns of single cells.

It was not so long ago that the MHC and peptide interactions became an easy target for precise study. Many methods are given, whether for the analysis of MHC class II – binding motifs with bacteriophage peptide display libraries, peptide-binding assays or for antagonist analysis for MHC class II. Additional themes such as HLA typing and the analysis of heat shock proteins are also included.

Clinically useful antibody assays are reproduced; whether we have in mind systemic rheumatic diseases, reactive arthritis or T cell responses in rheumatology, the methods for such studies are described. A strong section on mucosal immunity (interaction of the immune system with the components of intestinal flora) is presented. Methodology for experiments on gnotobiotic animals is shown in detail.

Some further sections of the manual were planned but could not be completed by the deadline. I was thus faced with a pragmatic choice: either to wait for the remaining few sections or to publish a timely book to which additional sections could be appended later via a CD-ROM update. I opted for the latter approach and I therefore hope to see the project develop from a printed manual to a 'living organism' with social interactions. It is anticipated that CD-ROM updates will be made available at intervals to keep the manual à jour with latest developments.

During the time the manual was in press, the table of contents was put on the Internet. This means of communication will be utilized at some stage in the future as we plan to set up an Internet Bulletin Board. One aim of the Bulletin Board is to enable the exchange of information between authors and users, as well as among the users themselves. Another proposal is to list the suppliers mentioned in the book and, where possible, establish direct links to their catalogues.

In the 1970s we witnessed the revolutionary development of new tools and new techologies, and clearly this process of innovation did not diminish. On the contrary, new tools, methods and approaches keep emerging, and this work hopefully is indicative of the fact that our fingers are on the pulse of development. This manual presents methods that are 'alive', that are used and that reflect the state of the art in experimental immunology.

The principle of a particular method might be conceived as an elegant solution to a problem. The actual bench-work method, however, has to evolve step-by-step from a clumsy prototype to a perfected procedure. Initially, inefficient procedures are applied. Later, when the methods are perfected, it is often surprising that the initial approach worked at all. For example, for the analysis of V gene rearrangements, standard molecular biology tools exist today. In the early 1970s André Traunecker, Christine Brack and Susumu Tonegawa spent day after day analyzing electron micrographs of DNA strands to identify which gene segments joined which. Indeed, there are many examples of tedious approaches which have become easy-to-use tools. I hope the reader will find many of them in this book.

I wish to thank all the contributors for willingly sharing their methods with the scientific community. Furthermore, I would like to thank them for being amenable to making necessary changes to their contributions. I would like to express my sincere gratitude to the Section Editors who put a great deal of effort into engaging the best people in the field to contribute, and I also wish to thank the Advisory Board whose suggestions throughout the preparation of this manual were invaluable. My special acknowledgment goes to one member of the Advisory Board who sadly is no longer among us, Georges Köhler, who surveyed the list of Section Editors and made important suggestions. When the list was complete he commented, 'With this crew, it will fly high'.

Preparing a book like this evolves not only as a scientific endeavor but also as a social one. Many of the Section Editors are former members of our institute and I generally know who is who and who is where. They often became busy or important (or both), but they still agreed to assemble a section. A good friend of mine, Herman Waldmann, declined to act as Section Editor with the words, 'Ivan, what do you prefer, to gain a section or to lose a friend?'. Herman was preparing to move from Cambridge to Oxford, but he suggested that John Adair take his place; he produced a section that is a valuable contribution to this methods manual.

To sum up, I would like to quote from one of the last conversations I had with Niels Jerne just before he became terminally ill (he died on 7th October 1994). When I explained this project and showed him the list of Section Editors he said, 'It reminds me of the horizontal structure of our institute: assembling good people, giving them freedom and giving them the full responsibility to do their best with it'.

IVAN LEFKOVITS
FEBRUARY 1996

Editor's Note

These lines are being written when the publication date of the manual is just around the corner: the page proofs have been checked and the printing press is about to 'rotate'. When I consider all the editorial work I have ever done, this project, admittedly, was the toughest, but also the most interesting and rewarding.

As with any project, this one has also had its ups and downs, but when viewed as a whole and considering its scope, it went rather smoothly. In the Preface I expressed my appreciation to several people or groups of people without mentioning the pleasant and efficient cooperation I had with Academic Press. Tessa Picknett, Emma White, Tamsin Cousins, Shammima Cowan, Anne Doris, Simon Haggis, Ed Pentz, Helen Knapp and many young and enthusiastic members of staff at Academic Press made my editorial task a pleasure.

Well, I mentioned ups and downs: the preparation of the CD-ROM was a straightforward task, apart from the section editors being asked to prepare hypertext links. Precise instructions were issued and sent out. Instantly, the telephone rang and rang, e-mails accumulated, and letters arrived saying that it was not clear what should be done. One section editor commented ironically: 'Did I miss something in my primary education? I do not know what do you mean by hypertext links, anchors, starts and ends'. Nevertheless, even this frustrating intermezzo had its happy end – the CD-ROM has all the features to be expected from a modern, high-quality electronic product.

We all know that a giga-byte is a lot of memory. But still it is amazing that when all the 2500 pages were stored on the CD-ROM there was plenty of free 'space'. We therefore made a last-minute decision to put part of a video on the immune system (prepared by André Traunecker and Stefan Meyer on the occasion of the 25th anniversary of the Basel Institute for Immunology) onto the remaining space of the CD-ROM. This video is very good and has excellent special effects (it could hardly be kept secret that the same software was used as for Steven Spielberg's *Jurassic Park*). The inclusion of part of this video in the CD-ROM will serve as a test: if this hybrid product is successful, an updated version of the CD-ROM could contain animated film sequences related to various techniques. Thus, if the user clicks on a figure, a short film sequence would show the technique.

In conclusion I wish to emphasize another important feature of the methods manual. At some point in the future readers will have the chance to establish contact with the section editors and the contributors; in addition, reader-to-reader contacts will be possible. Two platforms for these interactions will be set up. The 'passive' one will be to view the web 'home page' where various updates and comments will be entered regularly (http://www.hbuk.co.uk/ap/books/lefkovits/). From the home page there will be a link to a discussion group, and this will form the 'active' part of the interaction. At the moment, i.e. August 1996, it is at the planning stage, but the reader is encouraged to browse through the home page occasionally to see how it develops.

I am confident that this methods manual will live up to all the users' expectations, and I am sure that many readers will contribute their know-how to the next edition.

IVAN LEFKOVITS
AUGUST 1996

Contents

Contents

Contents

Contents

Cumulative Contents for Volumes 1–4

Cumulative Contents

Editors

EDITOR
Dr Ivan Lefkovits, Basel Institute for
Immunology, Postfach, Grenzacherstrasse
487, CH-4005 Basel, Switzerland
Tel: +41 61 605 1259
Fax: +41 61 605 1364
E-mail: lefkovits@bii.ch

EDITORIAL ASSISTANT
Leslie Nicklin, Basel Institute for
Immunology, Postfach, Grenzacherstrasse
487, CH-4005 Basel, Switzerland
Tel: +41 61 605 1336
Fax: +41 61 605 1222
E-mail: nicklin@bii.ch

ADVISORY EDITORS
Dr Louis Du Pasquier, Basel Institute for
Immunology, Grenzacherstrasse 487,
CH-4005 Basel, Switzerland
Tel: +4161 605 1269
Fax: +4161 605 1222
E-mail: dupasquier@dial.eunet.ch

Dr Georges Köhler (Deceased),
Max-Planck-Institut für Immunobiologie,
79108 Freiburg, Stubeweg 51, Germany

Dr Fritz Melchers, Basel Institute for
Immunology, Postfach, Grenzacherstrasse
487, CH-4005 Basel, Switzerland
Tel: +41 61 605 1238
Fax: +41 61 605 1300

Dr Charles M Steinberg, Basel Institute for
Immunology, Grenzacherstrasse 487,
Postfach, CH-4005 Basel, Switzerland
Tel: +4161 605 1111
Fax: +4161 605 1222
E-mail: steinberg@dial.eunet.ch

Dr Harald von Boehmer, Faculté de
Médicine Necker, INSERM Unité 373,
156, Rue de Vaugirard,
F-75730 Paris Cedex 15, France
Tel: +331 40 60 53 81
Fax: +331 40 61 55 90

SECTION EDITORS
Dr John Adair, Axis Genetics plc, Babraham,
Cambridge CB2 4AX, UK
Tel: +44 1223 837611
Fax: +44 1223 837604

Dr Luciano Adorini, Roche Milano Ricerche,
Via Olgettina 58, I–20132 Milan, Italy
Tel: +39 2288 4816
Fax: +39 2215 3208
E-mail: adorinil@dibit.hsr.it

Dr Andrei Augustin, National Jewish Center
for Immunology and Respiratory Medicine,
1400 Jackson Street, Denver, CO 80206, USA
Tel: +1 303 398 1318
Fax: +1 303 398 1396
E-mail: augustina@njc.org

Dr Robbert Benner, Department of
Immunology, Erasmus University Rotterdam,
Postbus 1738, 3000 DR Rotterdam,
The Netherlands
Tel: + 31 10 408 7191
Fax: +31 10 436 7601

Dr Stuart Berger, The Wellesley Hospital
Research Institute, 160 Wellesley Street,
Toronto, Ontario M4Y 1J3, Canada
Tel: +1 416 926 5148
Fax: +1 416 926 5109
E-mail: @whri.on.ca

Dr Louis Du Pasquier, Basel Institute for
Immunology, Grenzacherstrasse 487,
CH-4005 Basel, Switzerland
Tel: +4161 605 1269
Fax: +4161 605 1222
E-mail: dupasquier@dial.eunet.ch

Editors

Dr Jose A Garcia Sanz, Basel Institute for Immunology, Postfach, Grenzacherstrasse 487, CH-4005 Basel, Switzerland
Tel: +41 61 605 1303
Fax: +41 61 605 1364
E-mail: garcia@bii.ch

Dr Philip Griebel, Veterinary Infectious Disease Organization, 120 Veterinary Road, University of Saskatchewan, Saskatoon, Saskatchewan S7N 5E3, Canada
Tel: +1 306 966 7478
Fax: +1 306 966 7478
E-mail: griebelp@sask.usask.ca

Dr Klaus-Ulrich Hartmann, Institute for Experimental Immunology, Philipps University Marburg, Deutschhausstrasse 1, D-35507 Marburg, Germany
Tel: +49 6421 284 051
Fax: +49 6421 288 925
E-mail: hartmank@mailer.uni-marburg.de

Dr Maarten P Hazenberg, Department of Immunology, Erasmus University, Rotterdam, Postbus 1738, 3000 DR Rotterdam, The Netherlands
Tel: +31 10 408 7191
Fax: +31 10 436 7601

Dr Wayne Hein, Basel Institute for Immunology, Postfach, Grenzacherstrasse 487, CH-4005 Basel, Switzerland
Tel: + 41 61 605 1352
Fax: +41 61 605 1364
E-mail: hein@bii.ch

Dr Hans Hengartner, Institut für Exp. Immunologie, Universitätsspital Zürich, Sternwartstrasse 12, CH-8091 Zürich, Switzerland
Tel: + 41 1255 2989
Fax: +41 1255 4420
E-mail: HHENG@usz.unizh.ch

Dr Julia L Hurwitz, Department of Immunology, St Jude Children's Research Hospital,
332 N Lauderdale, Memphis, TN 38101, USA
Tel: +1 901 495 2464
Fax: +1 901 495 3107
E-mail: julia.hurwitz@stjude.org

Dr Beat Imhof, Basel Institute for Immunology, Postfach, Grenzacherstrasse 487, CH-4005 Basel, Switzerland
Tel: + 41 61 605 1270
Fax: +41 61 605 1364
E-mail: imhof@bii.ch

Dr Manfred Kopf, Basel Institute for Immunology, Grenzacherstrasse 487, CH-4005 Basel, Switzerland
Tel: + 41 61 605 1319
Fax: +41 61 605 1364
E-mail: kopf@bii.ch

Dr Marie H. Kosco Vilbois, Geneva Biomedical Research Institute, Glaxo Wellcome R & D SA, Case postale 674, CH-1228 Plan-les-Ouates/Geneva, Switzerland
Tel: +41 22 706 9666
Fax: +41 22 794 6965
E-mail: MHKV10664@ggr.co.uk

Dr Pieter J M Leenen, Erasmus University, Department of Immunology, PO BOX 1738, 3000 DR Rotterdam, The Netherlands
Tel: +31 10 408 7181/8188
Fax: +31 10 436 7601
E-mail: leenen@immu.fgg.eur.nl

Dr Carlos Martinez A, Department of Immunology and Oncology, Centro Nacional de Biotecnologia/CSIC, UAM Campus de Cantoblanco, E-28049 Madrid, Spain
Tel: + 341 585 4544
Fax: +341 372 0493
E-mail: cmartineza@samba.cnb.uam.es

Dr Fiona McConnell, Basel Institute for Immunology, Postfach, Grenzacherstrasse 487, CH-4005 Basel, Switzerland
Tel: + 4161 605 1312
Fax: +4161 605 1364

Dr Roald Nezlin, Department of Immunology, The Weizmann Institute of Science, 76100 Rehovot, Israel
Tel: +972 8 934 3652
Fax: +972 8 946 6966
E-mail: linezlin@weizmann.weizmann.ac.il

Dr Christopher J Paige, The Wellesley
Hospital Research Institute, 160 Wellesley
Street East, Toronto, Ontario M4Y 1J3,
Canada
Tel: +1 416 926 7751
Fax: +1 416 926 5109
E-mail: paige@whri.on.ca

Dr José Quintans, Pathology Department
Box 414, University of Chicago, 5841 S
Maryland Avenue, Chicago, IL 60637, USA
Tel: +1 312 702 9206
Fax: +1 312 702 3701
E-mail: j-quintans@uchicago.edu

Dr Hans-Georg Rammensee, Institut für
Zellbiologie der Universität Tübingen,
Abteilung Immunologie, Auf der Morgenstelle
15, (Verfügungsgebäude), D-72070 Tübingen,
Germany
Tel: +49 7071 29 87628; +49 7071 29 80991
Fax: +49 7071 29 5653

Dr Antonius Rolink, Basel Institute for
Immunology, Grenzacherstrasse 487,
CH-4005 Basel, Switzerland
Tel: +4161 605 1265
Fax: +4161 605 1364
E-mail: rolink@bii.ch

Dr Gek Kee Sim, Immunology Lab,
1825 Sharp Point Drive, Fort Collins,
CO 80525, USA
Tel: +1 970 224 5387
Fax: +1 970 493 7333
E-mail: sim@ez0.ezlink.com

Dr Jaroslav Sterzl, Institute of Microbiology,
Academy of Sciences of the Czech Rep.,
Videnska 1083, CZ 14200 Prague 4-Krc,
Czech Republic
Tel: +422 475 2363
Fax: +422 472 1143
E-mail: immuno@biomed.cas.cz

Dr Hans-Jürgen Thiesen, Institut für
Immunologie, Schillingallee 70,
D-18055 Rostock, Germany
Tel: +49 381 494 5871
Fax: + 49 381 494 5882
E-mail: hans-juergen.thiesen@uni-rostock.de

Dr Helena Tlaskalová-Hogenová, Institute of
Microbiology, Academy of Sciences of the
Czech Rep., Videnskà 1083, CZ-14200
Prague 4-Krc, Czech Republic
Tel: +422 475 2345
Fax: +422 472 1143
E-mail: tlaskalo@biomed.cas.cz

Dr Auli Toivanen, Department of Medicine,
Turku University, Kiinamyllynkatu 4-8,
FIN-20520 Turku, Finland
Tel: +358 2 611 611
Fax: +358 2 611 030
E-mail: auli.toivanen@utu.fi

Dr Paavo Toivanen, Department of Medical
Microbiology, Turku University,
Kiinamyllynkatu 13, FIN-20520 Turku, Finland
Tel: +358 2 333 7426
Fax: +358 2 233 0008
E-mail: paavo.toivanen@utu.fi

Dr André Traunecker, Basel Institute for
Immunology, 487 Grenzacherstrasse,
CH-4058 Basel, Switzerland
Tel: +41 61 605 1287
Fax: +41 61 605 1364
E-mail: traunecker@bii.ch

Dr Olli Vainio, Department of Medical
Microbiology, Turku University,
Kiinamyllnkatu 13, FIN-20520 Turku, Finland
Tel: +358 2 333 7400
Fax: +358 2 233 0008
E-mail: olli.vainio@utu.fi

Dr Jacques JM van Dongen, Department of
Immunology, Erasmus University Rotterdam,
PO Box 1738,
3000 DR Rotterdam, The Netherlands
Tel: +31 10 408 8094
Fax: +31 10 436 7601
E-mail: vandongen@immu.fgg.eur.nl

Dr Siegfried Weiss, Gesellschaft für
Biotechnologische Forschung mbH,
Mascheroder Weg 1, D-38124 Braunschweig,
Germany
Tel: +49 531 6181 230
Fax: +49 531 6181 292
E-mail: siw@gbf-braunschweig.de

Editors

Dr Hartmut Wekerle, Max Planck Institut für
Psychiatrie, Am Klopferspitz 18a,
D-82152 Martinsried, Germany
Tel: +4989 8578 3551
Fax: +4989 8578 3790

Dr Michael V Wiles, Max-Planck Institut für
Molekulare Genetik, Ihnestrasse 73,
D-14195 Berlin, Germany
Tel: +49 30 8413 1350
Fax: +49 30 8413 1380
E-mail: wiles@mpimg-berlin-dahlem.mpg.de

Dr Gillian E Wu, Department of Immunology,
Medical Science Building, University of
Toronto, Toronto, Ontario M5S 1AB, Canada
Tel: +1 416 926 5149
Fax: +1 416 926 5109
E-mail: wu@immune.med.utoronto.ca

Dr Rolf Zinkernagel, Institut für Exp.
Immunologie, Universitätsspital Zürich,
Schmelzbergstrasse 12, CH-8091 Zürich,
Switzerland
Tel: +411 255 2386
Fax: +411 255 4420
E-mail: zink@pathol.unizh.ch

Contributors

Dr Garry I Abelev, Laboratory of
Immunochemistry, Cancer Research Center,
Russian Academy of Medical Sciences,
115478 Moscow, Russia
Tel: +7 095 323 5910
Fax: +7 095 324 1205

Dr Nevin J Abernethy, Genesis Research
and Development, Corporation Ltd,
PO Box 50, Auckland, New Zealand
Tel: +64 9 373 5600
Fax: +64 9 373 2189
E-mail: n.abernethy@genesis.co.nz

Dr John Adair, Axis Genetics Plc, Babraham,
Cambridge CB2 4AX, UK
Tel: +44 1223 837 611
Fax: +44 1223 837 604

Dr Mark A Adkison, Department of Medicine
and Epidemiology, School of Veterinary
Medicine, University of California, Davis, CA
95616, USA
Tel: +1 916 752 9318
Fax: +1 916 752 0414
E-mail: maadkison@ucdavis.edu

Dr Luciano Adorini, Roche Milano Ricerche,
Via Olgettina 58, I-20132 Milan, Italy
Tel: +39 2 288 4816
Fax: +39 2 215 3203
E-mail: adorinil@dibit.hsr.it

Dr Ann Ager, Laboratory for Cellular
Immunology, National Institute for Medical
Research, The Ridgeway, Mill Hill, London
NW7 1AA, UK
Tel: +44 181 959 36 66
Fax: +44 181 913 85 29
E-mail: a-ager@nimr.mrc.ac.uk

Dr Vincent Aguirre, The Center for Blood
Research Inc., Harvard Medical School,
200 Longwood Ave., Boston, MA 02115, USA
Tel: +1 617 278 3245
Fax: +1 61 278 3131
E-mail: aguirre@cbrv1.med.harvard.edu

Dr Jeff Alexander, Cytel Corporation,
3525 John Hopkins Court, San Diego,
CA 92121, USA
Tel: +1 619 552 3000
Fax: +1 619 552 8801

Dr Arna E Andrews, Centre for Animal
Biotechnology, University of Melbourne,
Parkville, 3052 Victoria, Australia
Tel: +61 3 9344 8001
Fax: +61 3 9347 4083
E-mail: AEA@rubens.its.unimelb.edu.au

Dr Jon R Appel, Torrey Pines Institute for
Molecular Studies, 3550 General Atomics Ct.,
San Diego, CA 92121, USA
Tel: +1 619 455 3847
Fax: +1 619 455 3804
E-mail: jonappel@tpims.org

Dr K Armour, Immunology Division,
Cambridge University, Department of
Pathology, Tennis Court Road, Cambridge
CB2 1QP, UK
Tel: +44 1223 333 702
Fax: +44 1223 333 875

Dr T Petteri Arstila, Department of Medical
Microbiology, Turku University,
Kiinamyllynkatu 13, FIN-20520 Turku, Finland
Tel: +358 2 333 71
Fax: +358 2 233 0008

Dr Jean-Pierre Aubry, Geneva Biomedical
Research Institute, Glaxo Wellcome
R & D SA, Case postale 674, CH-1228
Plan-les-Ouates/Geneva, Switzerland
Tel: +41 22 706 9666
Fax: +41 22 794 6965

Contributors

Dr Andrei A Augustin, National Jewish Center for Immunology & Respiratory Medicine, 1400 Jackson Street, Denver CO 80206, USA
Tel: +1 303 398 1318
Fax: +1 303 398 1396
E-mail: augustina@njc.org

Dr Martin F Bachmann, Ontario Cancer Institute, 610 University Avenue, Toronto, Ontario M5G 2M9, Canada
Tel: +1 416 946 2000 ext 5471
Fax: +1 416 946 2086
E-mail: bachmann@oci.utoronto.ca

Dr I Begara, Moredum Research Institute, 408 Gilmerton Road, Edinburgh EH17 7JH, Scotland, UK
Tel: +44 131 664 3262
Fax: +44 131 664 8001
E-mail: begai@mri.sari.ac.uk

Dr Anke Bender, Max-Planck-Institut für Psychiatrie, Abteilung Neuroimmunologie, Am Klopferspitz 18a, D-82152 Martinsried, Germany
Tel: +49 89 8578 3581
Fax: +49 89 8578 3790

Dr Robbert Benner, Department of Immunology, Erasmus University Rotterdam, Postbus 1738, 3000 DR Rotterdam, The Netherlands
Tel: +31 10 408 7191
Fax: +31 10 436 7601

Dr Laurent A Bentolila, Wellesley Hospital, Research Institute, 160 Wellesley Street E, Toronto, Ontario M4Y 1J3, Canada
Tel: +1 416 926 5151
Fax: +1 416 926 5109
E-mail: kotel@pasteur.fr

Dr Stuart A Berger, The Wellesley Hospital Research Institute, 160 Wellesley Street, Toronto, Ontario M4Y 1J3, Canada
Tel: +1 416 926 5148
Fax: +1 416 926 5109
E-mail: berger@whri.on.ca

Dr Antonio Bernad, Department of Immunology & Oncology, Centro Nacional de Biotecnología/CSIC, UAM-Campus de Cantoblanco, E-28049 Madrid, Spain
Tel: +34 1 585 4530
Fax: +34 1 372 0493
E-mail: abernad@samba.cnb.uam.es

Dr Andre TJ Bianchi, DLO Institute for Animal Science and Health, Department of Mammalian Virology, POB 65, 8200 AB Lelystad, The Netherlands
Tel: +31 320 23 8238
Fax: +31 320 23 8668
E-mail: A.T.J.Bianchi@id.dlo.nl

Dr Wim JA Boersma, DLO Institute for Animal Science and Health, POB 65, 8200 AB Lelystad, The Netherlands
Tel: +31 320 23 8098
Fax: +31 320 23 8050
E-mail: W.Boersma@id.dlo.nl

Dr Jean-Yves Bonnefoy, Geneva Biomedical Research Institute, Glaxo Wellcome R & D SA, Case postale 674, CH-1228 Plan-les-Ouates/Geneva, Switzerland
Tel: +41 22 706 9666
Fax: +41 22 794 6965

Dr Thierry Boon, Ludwig Institute for Cancer Research, UCL 7459, 74 avenue Hippocrate, B-1200 Brussels, Belgium
Tel: +32 2 764 7580
Fax: +32 2 764 7590
E-mail: boon@licr.ucl.ac.be

Dr Willi Born, National Jewish Center for Immunology and Respiratory Medicine, 1400 Jackson Street, Denver CO 80206, USA
Tel: +1 303 398 1087
Fax: +1 303 398 1396
E-mail: bornw@njc.org

Dr Paul Borron, Department of Biochemistry, Lawson Research Centre, University of Western Ontario, London, Ontario N6A 3K7, Canada
Tel: +1 519 661 3074

Dr Lisardo Boscá, Instituto de Bioquímica, CSIC, Universidad Complutense, E-28040 Madrid, Spain
Tel: +34 1 394 1853
Fax: +34 1 394 1782
E-mail: boscal@eu.cmax.sim.ucm.es

Dr Richard L Boyd, Department of Pathology and Immunology, Monash Medical School, Commercial Road, Prahran 3181, Victoria, Australia
Tel: +61 3 9276 2738
Fax: +61 3 9529 6484
E-mail: boyd@cobra.path.monash.edu.au

Dr Monika Bradl, Max-Planck-Institut für Psychiatrie, Am Klopferspitz 18a, D-82152 Martinsried, Germany
Tel: +49 89 8578 3578
Fax: +49 89 8578 3790

Dr Andrew M. Bray, Chiron Mimotopes Pty Ltd, 11 Duerdin Street, Clayton Victoria 3168, Australia
Tel: +61 3 9565 1111
Fax: +61 3 9565 1199
E-mail: Andrew_Bray@cc.chiron.com

Dr Vincent Brichard, Ludwig Institute for Cancer Research, UCL 7459, 74 avenue Hippocrate, B-1200 Brussels, Belgium
Fax: +32 2 762 9405
E-mail: brichard@licr.ucl.ac.be

Dr CG Duncan Brown, Centre for Tropical Veterinary Medicine, The University of Edinburgh, Easter Bush, Roslin Midlothian EH25 9RG, Scotland, UK
Tel: +44 131 650 6287
Fax: +44 131 650 6287
E-mail: PAULINE.McMANUS@ed.ac.uk

Dr Raymond Bujdoso, Department of Clinical Veterinary Medicine, University of Cambridge, Madingley Road, Cambridge CB3 0ES, UK
Tel: +44 1223 337733
Fax: +44 1223 337610

Dr A Burny, Université Libre de Bruxelles, Laboratoire de chimie biologique, 67, rue des Chevaux, B-1640 Rhôde St-Genèse, Belgium
Tel: +32 2 650 9824
Fax: +32 2 650 9839

Dr Søren Buus, Institute for Medical Microbiology and Immunology, Panum Inst. 18.3.22, Blegdamsvej 3, DK-2200 Copenhagen N, Denmark
Tel: +45 35 32 7885
Fax: +45 35 32 7853
E-mail: S.Buus@sb.immi.ku.dk

Dr F Carr, Biovation Ltd, Marischal College, Broad Street, Aberdeen AB9 1AS, UK
Tel: +44 1224 273 197
Fax: +44 1224 273 198

Dr Ana C Carrera, Department of Immunology & Oncology, Centro Nacional de Biotecnología/CSIC, UAM-Campus de Cantoblanco, E-28049 Madrid, Spain
Tel: +34 1 585 4537
Fax: +34 1 372 0493
E-mail: acarrera@samba.cnb.uam.es

Dr Patrick Caspers, F Hoffmann-La Roche Ltd, PRPG, Bldg 66, Rm 611, CH-4070 Basel, Switzerland
Tel: +41 61 688 62 56
Fax: +41 61 688 14 48
E-mail: patrick.caspers@roche.com

Dr Rachel Chap, Department of Chemical Immunology, The Weizmann Institute of Science, 76100 Rehovot, Israel
Tel: +972 8 934 2985
Fax: +972 8 934 4141

Dr Jacques Charlemagne, Laboratoire d'Immunologie Comparée, Université Pierre et Marie Curie, CNRS URA 1135, Boîte 29, F-75252 Paris Cedex 05, France
Tel: +33 1 44 27 34 16
Fax: +33 1 44 27 34 09

Dr Chen-lo H Chen, Wallace Tumor Institute 378, University of Alabama at Birmingham, 824 6th Avenue South, Birmingham, AL 35294-3300, USA
Tel: +1 205 934 3370
Fax: +1 205 934 1875
E-mail: chen@ms.ccc.uab.edu

Contributors

Dr Steven J Chmura, Pathology Department Box 414, University of Chicago, 5841 S Maryland Avenue, Chicago IL 60637, USA
Tel: +1 312 702 9206
Fax: +1 312 702 3701

Dr Isabelle Chrétien, Basel Institute for Immunology, Postfach, Grenzacherstrasse 487, CH-4005 Basel, Switzerland
Tel: +41 61 605 1269
Fax: +41 61 605 1364
E-mail: chrétien@bii.ch

Dr Eric Claassen, DLO Institute for Animal Science and Health, POB 65, 8200 AB Lelystad, The Netherlands
Tel: +31 320 23 8006
Fax: +31 320 23 8008
E-mail: H.J.H.M.CLAASSEN@ID.DLO.NL

Dr Y Cleuter, Department of Molecular Biology, Université Libre de Bruxelles, 67 rue des Chevaux, B-1640 Rhôde St-Genèse, Belgium
Tel: +32 2 650 9826
Fax: +32 2 650 9839

Dr Nicholas Cohen, University of Rochester, Department of Microbiology and Immunology, Box 672, 601 Elmwood Avenue, Rochester, NY 14642, USA
Tel: +1 716 275 3402
Fax: +1 716 473 9573
E-mail: ncohn@medinfo.rochester.edu

Dr Ian G Colditz, CSIRO, Division of Animal Health, Armidale 2350, Australia

Dr DDS Collie, Department of Veterinary Clinical Studies, The University of Edinburgh, Veterinary Field Station, Easter Bush, Midlothian EH25 9RG, UK
Tel: +44 131 650 1000
Fax: +44 131 650 6599

Dr W Marieke Comans-Bitter, Department of Immunology, Erasmus University Rotterdam, POB 1738, 3000 DR Rotterdam, The Netherlands
Tel: +31 10 408 8090
Fax: +31 10 436 7601
E-mail: vandongen@immu.fgg.eur.nl

Dr Alison Connor, Wellesley Hospital Research Institute, 160 Wellesley Street E, Toronto, Ontario M4Y 1J3, Canada
Tel: +1 416 926 5151
Fax: +1 416 926 5109
E-mail: connor@immune.med.utoronto.ca

Dr Anne Cooke, University of Cambridge, Department of Pathology, Immunology Division, Tennis Court Road, Cambridge CB2 1QP, UK
Tel: +44 1223 33 3907
Fax: +44 1223 33 3914
E-mail: ac@mole.biol.cam.ac.uk

Dr Catherine Corbel, Institut d'Embryologie Cellulaire et Moléculaire du CNRS et du Collège du France, 49 bis, avenue de la Belle Gabrielle, F-94736 Nogent-sur-Marne Cedex, France
Tel: +33 1 48 73 60 90
Fax: +33 1 48 73 43 77
E-mail: corbel@infobiogen.fr

Dr Pierre Coulie, Université Catholique de Louvain, Cellular Genetics Unit, UCL 7459, 74 avenue Hippocrate, B-1200 Brussels, Belgium
Tel: +32 2 764 7599
Fax: +32 2 762 9405
E-mail: coulie@licr.ucl.ac.be

Dr Bozena Cukrowska, Institute of Microbiology, Academy of Sciences of the Czech Republic, Videnskà 1083, CZ-142 00 Prague 4 Krc, Czech Republic
Tel: +422 475 2368
Fax: +422 472 1143
E-mail: immuno@biomed.cas.cz

Dr Ana Cumano, Unité de Biologie
Moléculaire du Gène, Institut Pasteur,
25 Rue du Dr. Roux, F-75724 Paris Cedex 15,
France
Tel: +33 1 40 61 30 56
Fax: +33 1 45 68 85 48
E-mail: cumano@pasteur.fr

Mr Jean-Pierre Dangy, Basel Institute for
Immunology, Postfach,
Grenzacherstrasse 487, CH-4005 Basel,
Switzerland
Tel: +41 61 605 1270
Fax: +41 61 605 1364
E-mail: dangy@bii.ch

Dr Ayub Darji, Gesellschaft für
Biotechnologische Forschung mbH,
Mascheroder Weg 1,
D-38124 Braunschweig, Germany
Tel: +49 531 6181 253
Fax: +49 531 6181 292
E-mail: ADA@gbf-braunschweig.de

Dr Natalie J Davidson, DNAX Research
Institute, 901 California Avenue,
Palo Alto CA 94304, USA
Tel: +1 415 496 1124
Fax: +1 415 496 1200
E-mail: davidson@dnax.com

Dr Glyn Dawson, Department of Pediatrics,
University of Chicago, 5841 S Maryland
Avenue, Chicago IL 60637, USA
Tel: +1 312 702 9206
Fax: +1 312 702 3701

Dr Mark de Boer, Pangenetics BV,
E van Calcarstraat 30, 1963 DG Heemskerk,
The Netherlands
Tel: +31 251 241 356
Fax: +31 251 250 719
E-mail: mdb_pg@euronet.nl

Dr Tjerk WA de Bruin, Department of
Medicine, GO2.228, University Hospital
Utrecht, Heidelberglaan 100,
3584 CX Utrecht, The Netherlands
Tel: +31 30 250 7399
Fax: +31 30 251 1893
E-mail: t.w.a.bruin@digd.azu.nl

Dr Arnaud de Guerra, Laboratoire
d'Immunologie Comparée, Université Pierre et
Marie Curie, CNRS URA 1135, Boîte 29,
F-75252 Paris Cedex 05, France
Tel: +331 4427 4427
Fax: +331 4427 3866

Dr John de Kruif, Department of
Immunology, University Hospital Utrecht,
Heidelberglaan 100, 3584 CX Utrecht,
The Netherlands
Tel: +31 30 250 7590 or 250 7678
Fax: +31 30 251 7107
E-mail: j.deKruif@lab.azu.nl

Dr José Luis de la Pompa, The Ontario
Cancer Institute, Departments of Medical
Biophysics and Immunology, 610 University
Avenue, c/c Dr TW Mak, Toronto, Ontario
M5G 2M9, Canada
Tel: +1 416 204 5309
Fax: +1 416 204 5320
E-mail: jdelapom@amgen.com

Dr Etienne De Plaen, Ludwig Institute for
Cancer Research, UCL 7459, 74 avenue
Hippocrate, B-1200 Brussels, Belgium
Tel: +32 2 764 7479
Fax: +32 2 762 9405
E-mail: deplaen@licr.ucl.ac.be

Dr Sylvie Degermann, Basel Institute for
Immunology, Postfach,
Grenzacherstrasse 487, CH-4005 Basel,
Switzerland
Tel: +41 61 605 1249
Fax: +41 61 605 1364
E-mail: degermann@bii.ch

Dr Sylvie Delassus, Unité de Biologie
Moléculaire du Gène, Institut Pasteur,
25 Rue du Dr Roux, F-75724 Paris Cedex 15,
France
Tel: +33 1 40 61 30 56
Fax: +33 1 45 68 85 48
E-mail: sylvix@pasteur.fr

Dr Su-jun Deng, Burroughs Wellcome Co.,
3030 Cornwallis Road,
Research Triangle Park NC 27709, USA

Contributors

Dr Soléenne Desravines, New York University, Department of Biology, 1009 Main Bldg, 100 Washington Square East, New York, NY 10003, USA

Mr Mark Dessing, Basel Institute for Immunology, Postfach, Grenzacherstrasse 487, CH-4005 Basel, Switzerland
Tel: +41 61 605 1369
Fax: +41 61 605 1364
E-mail: dessing@bii.ch

Dr Christine D Dijkstra, Department of Cell Biology and Immunology, Vrije Universiteit, Van der Boechorststraat 7, 1081 BT Amsterdam, The Netherlands
Tel: +31 20 444 8080
Fax: +31 20 444 8081
E-mail: cd.dijkstra.cell@med.vu.nl

Dr Bonnie N Dittel, Section of Immunobiology, Yale University School of Medicine, 310 Cedar Street, New Haven, CT 06510 8023, USA
Tel: +1 203 785 9386
Fax: +1 203 737 1764
E-mail: dittel@sprynet.com

Dr Klaus Dornmair, Max-Planck-Institut für Psychiatrie, Am Klopferspitz 18a, D-82152 Martinsried, Germany
Tel: +49 89 8578 3566
Fax: +49 89 8578 3790
E-mail: dornmair@genmic.biochem.mpg.de

Dr Kenneth Dorshkind, Division of Biomedical Sciences, University of California, Riverside CA 92521 0121 USA
Tel: +1 909 787 4534
Fax: +1 909 787 2193
E-mail: dorsh@ucrac1.ucr

Dr Louis Droogmans, Department of Molecular Biology, Université Libre de Bruxelles, 67 rue des Chevaux, B-1640 Rhôde St-Genèse, Belgium
Tel: +32 2 650 9816
Fax: +32 2 650 9839
E-mail: LOUISD@DBM.ULB.AC.BE

Dr Philippe Druet, Hôpital Purpan, Pavillon Lefebvre, Place du Docteur Baylac, F-31059 Toulouse Cedex, France
Tel: +33 61 77 92 92
Fax: +33 61 77 92 91
E-mail: Philippe Druet@purpan.inserm.fr

Dr Louis Du Pasquier, Basel Institute for Immunology, Postfach, Grenzacherstrasse 487, CH-4005 Basel, Switzerland
Tel: +41 61 605 1269
Fax: +41 61 605 1222
E-mail: dupasquier@dial.eunet.ch

Ms Lisbeth Dudler, Basel Institute for Immunology, Postfach, Grenzacherstrasse 487, CH-4005 Basel, Switzerland
Tel: +41 61 605 1247
Fax: +41 61 605 1364

Dr Dominique Dunon, CNRS URA 1135, Université Pierre et Marie Curie, Bât. C-30 7ème étage, 9 Quai Saint-Bernard, F-75005 Paris, France
Tel: +33 1 44 27 34 36
Fax: +33 1 44 27 34 45
E-mail: dunon@ccr.jussien.fr

Dr Herman N Eisen, Center for Cancer Research, MIT, Bldg E17 128, Cambridge MA 02139 4307, USA
Tel: +1 617 253 6406
Fax: +1 617 258 8728
E-mail: hneisen@mit.edu

Dr Adelheid Elbe, Division of Immunology, Allergy and Infectious Diseases, Department of Dermatology, University of Vienna Medical School, VIRCC, Brunner Strasse 59, A-1235 Vienna, Austria
Tel: +43 1 86 364 465
Fax: +43 1 86 634 339
E-mail: adelheid.elbe@univie.ac.at

Dr Victor H Engelhard, Department of Microbiology and Beirne Carter Center for Immunology Research, University of Virginia, Charlottesville VA 22901, USA
Tel: +1 804 924 2423
Fax: +1 804 924 1221
E-mail: vhe@virginia.edu

Dr Zelig Eshhar, Department of Immunology, The Weizmann Institute of Science, 76100 Rehovot, Israel
Tel: +972 8 934 3965
Fax: +972 8 934 4141
E-mail: lieshhar@weizmann.weizmann.ac.il

Dr Kirsten Falk, Department of Biochemistry and Molecular Biology, Harvard University, 7 Divinity Avenue, Cambridge, MA 02138-2092, USA
Tel: +1 617 495 5612
Fax: +1 617 496 8351

Ms Christine S Falk, Institut für Immunologie, Universität München, Goethestrasse 31, D-80336 München, Germany
Tel: +49 89 5996 684
Fax: +49 89 5160 2236
E-mail: moritz@ifi.med.uni-muenchen.de

Dr Liam Fanning, University College Cork, Department of Medicine, c/o Clinical Sciences Building, Cork University Hospital, Wilton, Cork, Ireland
Tel: +353 2190300
Fax: +353 21345300

Dr Julien S Fellah, Laboratoire d'Immunologie Comparée, Université Pierre et Marie Curie, CNRS URA 1135, Boîte 29, F-75252 Paris Cedex 05, France
Tel: +331 4427 4427
Fax: +331 4427 3866

Dr M Fernie, Rowett Research Services, Greenburn Road, Bucksburn, Aberdeen AB21 9SB, UK
Tel: +44 1224 716 226
Fax: +44 1224 716 225

Dr S Forster, Celsis International Plc, Cambridge Science Park, Milton Road, Cambridge CB4 4FX, UK
Tel: +44 1223 426 008
Fax: +44 1223 426 003

Dr Carolyn A Foster, Sandoz Forschungsinstitut, Department of Dermatology, A-1235 Vienna, Austria
Tel: +43 1 866 34 685
Fax: +43 1 866 34 354
E-mail: Foster_C@A1.WIENV1.Sandoz.com

Dr JM Rodriguez Frade, Department of Immunology and Oncology, Centro Nacional de Biotecnología/CSIC, UAM-Campus de Cantoblanco, E-28049 Madrid, Spain
Tel: +34 1 585 4533
Fax: +34 1 372 0493
E-mail: jmmellado@samba.cnb.uam.es

Dr Thierry Francey, Institute of Veterinary Virology, University of Bern, Langgass-Strasse 122, CH-3012 Bern, Switzerland

Dr Ronald Frank, Gesellschaft für Biotechnologische Forschung mbH, Molekulare Erkennung, Mascherorder Weg 1, D-38124 Braunschweig, Germany
Tel: +49 531 6181 720
Fax: +49 531 6181 795
E-mail: frank@gbf-braunschweig.de

Dr Sara Fuchs, Department of Immunology, Weizmann Institute of Science, 76100 Rehovot, Israel
Tel: +972 8 934 2618
Fax: +972 8 934 4141
E-mail: lifuchs@wiccmail.weizmann.ac.il

Dr J Victor Garcia, Department of Virology and Molecular Biology, St Jude Children's Research Hospital, 332 N Lauderdale, Memphis, TN 38101, USA
Tel: +1 901 495 2611
Fax: +1 901 523 2622
E-mail: victor.garcia@stjude.org

Dr Jose A Garcia-Sanz, Basel Institute for Immunology, Postfach, Grenzacherstrasse 487, CH-4005 Basel, Switzerland
Tel: +41 61 605 1303
Fax: +41 61 605 1364
E-mail: garcia@bii.ch

Dr M Eric Gershwin, Division of Rheumatology/Allergy and Clinical Immunology, TB 192, University of California, Davis, Davis CA 95616, USA
Tel: +1 916 752 2884
Fax: +1 916 752 4669
E-mail: megershwin@ucdavis.edu

Contributors

Dr Igal Gery, Laboratory of Immunology, National Eye Institute, NIH, Bldg 10, Rm 10N208, Bethesda, MD 20892 1858, USA
Tel: +1 301 496 4159
Fax: +1 301 402 0485
E-mail: igery@helix.nih.gov

Dr Paolo Ghia, Division of Hematologic Malignancies, Room D-738, Dana Farber Cancer Institute, 44 Binney Street, Boston, MA 02115 6084, USA
Tel: +1 617 632 5119
Fax: +1 617 632 5167
E-mail:
paolo_ghia@macmailgw.dfci.harvard.edu

Dr Roland H Gisler, Basel Institute for Immunology, Postfach, Grenzacherstrasse 487, CH-4005 Basel, Switzerland
Tel: +41 61 605 1240
Fax: +41 61 605 1382

Dr Alexander R Gottschalk, Pathology Department, Box 414, University of Chicago, 5841 S. Maryland Ave, Chicago IL 60637, USA

Dr Thomas Graf, EMBL, Postfach 10.2209, Meyerhofstrasse 1, D-69117 Heidelberg, Germany
Tel: +49 6221 387 410
Fax: +49 6221 387 516
E-mail: graf@embl-heidelberg.de

Dr Ulf Grawunder, Department of Pathology, Washington University, School of Medicine, 660 S Euclid Avenue, Campus Box 8118, St. Louis, MO 63110, USA
Tel: +1 314 362 4207
Fax: +1 314 362 4096
E-mail: ulf@pathology.wustl.edu

Dr Bernard S Green, Department of Pharmaceutical Chemistry, The Hebrew University Faculty of Medicine, 91120 Jerusalem, Israel
Tel: +972 2 758 310
Fax: +972 8 548 335
E-mail: csgreen@weizmann.weizmann.ac.il

Dr Howard M Grey, La Jolla Institute for Allergy and Immunology, 10355 Science Center Drive, San Diego CA 92121, USA
Tel: +1 619 558 3500
Fax: +1 619 558 3525

Dr Philip Griebel, Veterinary Infectious Disease Organization, 120 Veterinary Road, University of Saskatchewan, Saskatoon, Saskatchewan S7N 5E3, Canada.
Tel: +1 306 966 7478
Fax: +1 306 966 7478
E-mail: griebelp@sask.usask.ca

Dr Gillian M Griffiths, MRC Laboratory of Molecular and Cell Biology, University College London, Gower Street, London WC1E 6BT, UK
Tel: +44 171 380 7806
Fax: +44 171 380 7805
E-mail: dmcbgil@ucl.ac.uk

Dr Luigi ME Grimaldi, Neuroimmunology Unit, DIBIT - S Raffaele Scientific Institute, Via Olgettina 58, I-20132 Milan, Italy
Tel: +39 2 2643 2791
Fax: +39 2 2643 4855
E-mail: grimall@dibit.hsr.it

Dr Kees Groeneveld, Department of Immunology, Erasmus University Rotterdam, POB 1738, 3000 DR Rotterdam, The Netherlands
Tel: +31 10 408 8087
Fax: +31 10 436 7601
E-mail: kgroeneveld@immu.fgg.eur.nl

Dr Monique Grommé, Division of Cellular Biochemistry, The Netherlands Cancer Institute, Plesmanlaan 121, 1066 CX Amsterdam, The Netherlands
Tel: +31 20 512 1976
Fax: +31 20 512 1989
E-mail: MONQGROM@NKI.NL

Dr Françoise Guillet, Laboratoire d'Immunologie Comparée, Université Pierre et Marie Curie, CNRS URA 1135, Boîte 29, F-75252 Paris Cedex 05, France
Tel: +331 4427 4427
Fax: +331 4427 3866

Dr José-Carlos Gutiérrez-Ramos,
The Center for Blood Research Inc., Office
254, Laboratory 259-260, 200 Longwood Avenue, Boston, MA 02115, USA
Tel: +1 617 278 3240
Fax: +1 617 278 3131
E-mail: gutierrez@cbrv1.med.harvard.edu

Dr Dirk Haasner, Global Project Management, F Hoffmann-La Roche Ltd,
Bldg 52, Rm 1308, CH-4070 Basel,
Switzerland
Tel: +41 61 688 6071
Fax: +41 61 688 9369

Dr David M Haig, Moredun Research Institute, 408 Gilmerton Road, Edinburgh EH17
7JH, UK
Tel: +44 131 664 3262
Fax: +44 131 664 8001
E-mail: haigd@mri.sari.ac.uk

Dr Karel Hala, Institute for General and Exp.
Pathology, University of Innsbruck, Fritz-Pregl-Strasse 3, A-6020 Innsbruck, Austria
Tel: +43 512 507 3103
Fax: +43 512 507 2867

Dr Alf Hamann, Department of Immunology,
Med Klinik, Universitätskrankenhaus Eppendorf, Martinistr. 52, D-20246 Hamburg,
Germany
Tel: +49 40 4717 3613
Fax: +49 40 4717 4243
E-mail: hamann@uke.uni-hamburg.de

Dr A Hamilton, Biocure Ltd, Balgownie Technology Centre, Campus 3, Aberdeen Science
and Technology Park, Aberdeen AB22 8GW,
UK
Tel: +44 1224 703 649
Fax: +44 1224 823 039

Mr Philippe Hammel, Basel Institute for
Immunology, Postfach, Grenzacherstrasse
487, CH-4005 Basel, Switzerland
Tel: +41 61 605 1270
Fax: +41 61 605 1364

Dr Juergen Hammer, Roche Milan Ricerche,
Via Olgettina 58, I-20132 Milan, Italy
Tel: +39 2 288 4803
Fax: +39 2 215 3203
E-mail: 101750.3060@compuserve.com

Dr W Harris, University of Aberdeen, Department of Cell and Molecular Biology, Institute
of Medical Sciences, University of Aberdeen,
Aberdeen AB25 2ZD, UK
Tel: +44 1224 273 185
Fax: +44 1224 273 144
E-mail: w.j.harris@abdn.ac.uk

Dr Leonard C Harrison, The Walter and Eliza
Hall, Institute of Medical Research, PO Royal
Melbourne Hospital, Victoria 3050, Australia
Tel: +61 3 9345 2555
Fax: +61 3 9347 0852
E-mail: harrison@wehi.edu.au

Dr Klaus-Ulrich Hartmann, Institute of
Experimental Immunology, Philipps University
Marburg, Deutschhausstrasse 1,
D-35037 Marburg, Germany
Tel: +49 6421 284 051
Fax: +49 6421 288 925
E-mail: hartmank@mailer.uni-marburg.de

Dr John B Hay, Sunnybrook Health Science
Centre, Room S 234, University of Toronto,
2075 Bayview Avenue, North York, Ontario
M4N 3M5, Canada
Tel: +1 416 480 5707
Fax: +1 416 480 5737

Dr Maarten P Hazenberg, Department of
Immunology, Erasmus University Rotterdam,
POB 1738, 3000 DR Rotterdam,
The Netherlands
Tel: +31 10 408 7191
Fax: +31 10 436 7601

Dr Wayne Hein, Basel Institute for Immunology, Postfach, Grenzacherstrasse 487, CH-4005 Basel, Switzerland
Tel: +41 61 605 1352
Fax: +41 61 605 1364
E-mail: hein@bii.ch

Contributors

Dr Ronald C Hendrickson, Biological Mass Spectrometry Laboratory, University of Virginia, McCormick Road, Charlottesville, VA 22901, USA
Tel: +1 804 924 7994
Fax: +1 804 982 2781
E-mail: rch6k@virginia.edu

Dr Hans Hengartner, Institut für Exp. Immunologie, Universitätsspital Zürich, Sternwartstrasse 12, CH-8091 Zürich, Switzerland
Tel: +41 1 255 2989
Fax: +41 1 255 4420
E-mail: HHENG@usz.unizh.ch

Dr Kent Heyborne, Perinatal Resource Center, Swedish Medical Center, 501 East Hampden Avenue, Englewood CO 80110, USA
Tel: +1 303 788 8550
Fax: +1 303 788 8554
E-mail: heyborne@msn.com

Dr Fumiya Hirayama, Medical University of South Carolina, Ralph H Johnson Veterans Affairs, Medical Center,
Charleston SC 29401-5799, USA
Tel: +1 803 577 5011
Fax:+1 803 722 7741

Dr Robert S Hodges, Protein Engineering Network of Centres of Excellence, Department of Biochemistry, University of Alberta, Edmonton AB T6G 2H7, Canada
Tel: +1 403 492 2758
Fax: +1 403 492 0095

Dr Maarten A Hoijer, Department of Immunology, Erasmus University Rotterdam, POB 1738, 3000 DR Rotterdam,
The Netherlands
Tel: +31 10 408 7800
Fax: +31 10 436 7601
E-mail: Hoijer@immu.fgg.eur.nl

Dr Arne Holm, Kernisk Institutt, Landbohojskolen, Thorvaldsenvej 40, DK-1871 Frederiksberg C, Denmark
Tel: +45 35 282 429
Fax: +45 35 282 089
E-mail: ah@pepsyn.chem.kvl.dk

Dr Sonsoles Hortelano, Instituto de Bioquimica, CSIC, Universidad Complutense, E-28040 Madrid, Spain
Tel: +34 1 394 1853
Fax: +34 1 394 1782

Dr John D Horton, University of Durham, Department of Biological Sciences, Science Labs, South Road, Durham DH1 3LE, UK
Tel: +44 191 374 3359
Fax: +44 191 374 2417
E-mail: J.D.Horton@durham.ac.uk

Dr Richard A Houghten, Torrey Pines Institute for Molecular Studies, 3550 General Atomics Ct., San Diego CA 92121, USA
Tel: +1 619 455 3803
Fax: +1 619 455 3804
E-mail: houghten@tpims.org

Dr Ellen Hsu, New York University, Department of Biology, 1009 Main Bldg, 100 Washington Square East, New York, NY 10003, USA
Tel: +1 212 998 8200
Fax: +1 212 995 4015
E-mail: HSU@ACF2.NYU.EDU

Dr Thomas F Huff, Department of Microbiology and Immunology, Virginia Commonwealth University, 1101 East Marshall Street, Box 980678, Richmond VA 23298-0678, USA
Tel: +1 804 828 2317
Fax: +1 804 828 9946
E-mail: huff@gems.vcu.edu

Dr Donald F Hunt, Department of Chemistry and Pathology, University of Virginia, McCormick Road, Charlottesville VA 22901, USA
Tel: +1 804 924 3610
Fax: +1 804 296 3159
E-mail: dfh@virginia.edu

Dr Julia L Hurwitz, Department of Immunology, St Jude Children's Research Hospital, 332 N Lauderdale, Memphis, TN 38101, USA
Tel: +1 901 495 2464
Fax: +1 901 495 3107
E-mail: julia.hurwitz@stjude.org

Dr Devon Husband, Protein Engineering Network of Centres of Excellence, Department of Biochemistry, University of Alberta, Edmonton AB T6G 2H7, Canada
Tel: +1 403 492 2758
Fax: +1 403 492 0095

Ms Patricia R Hutchings, University of Cambridge, Department of Pathology, Immunology Division, Tennis Court Road, Cambridge CB2 1QP, UK
Tel: +44 1223 337 733 ext. 3917
Fax: +44 1223 333 694

Dr Antonio Iglesias, Max-Planck-Institut für Psychiatrie, Abteilung Neuroimmunologie, Am Klopferspitz 18a, D-82152 Martinsried, Germany
Tel: +49 89 8578 3591
Fax: +49 89 8578 3790

Dr Beat A Imhof, Basel Institute for Immunology, Postfach, Grenzacherstrasse 487, CH-4005 Basel, Switzerland
Tel: +41 61 605 1270
Fax: +41 61 605 1364
E-mail: imhof@bii.ch

Dr Elisabeth A Innes, Moredun Research Institute, 408 Gilmerton Road, Edinburgh EH17 7JH, UK
Tel: +44 131 664 3262
Fax: +44 131 664 8001
E-mail: innel@mri.sari.ac.uk

Dr Shozo Izui, Department of Pathology, CMU, 1 rue Michel Servet, CH-1211 Geneva, Switzerland
Tel: +41 22 702 5741
Fax: +41 22 702 5746

Dr Birgit Jaspert, Institute for Experimental Immunology, Philipps University Marburg, Deutschhausstrasse 1, D-35037 Marburg, Germany
Tel: +49 6421 284 051
Fax: +49 6421 288 925
E-mail: hartmank@mailer.uni-marburg.de

Dr Dieter E Jenne, Max-Planck-Institut für Psychiatrie, Abteilung Neuroimmunologie, Am Klopferspitz 18a, D-82152 Martinsried, Germany
Tel: +49 89 8578 3588
Fax: +49 89 8578 3790
E-mail: jenne@vms.biochem.mpg.de

Dr Suzan HM Jeurissen, Institute for Animal Science and Health (ID-DLO), POB 65, 8200 AB Lelystad, The Netherlands
Tel: +31 320 23 8238
Fax: +31 320 23 8050
E-mail: S.H.M.Jeurissen@id.dlo.nl

Dr Gui-quan Jia, The Center for Blood Research Inc, Harvard Medical School, 200 Longwood Avenue, Boston, MA 02115, USA
Tel: +1 617 278 3248
Fax: +1 617 278 3131
E-mail: jia@cbrv1.med.harvard.edu

Dr Petra Jonas, Department of Immunology, I Med Klinik, Universitätskrankenhaus Eppendorf, Martinistrasse 52, D-20246 Hamburg, Germany
Tel: +49 40 4717 6787
Fax: +49 40 4717 4243

Dr Michael H Julius, Department of Immunology, Medical Science Bldg, University of Toronto, Toronto, Ontario M5S 1A8, Canada
Tel: +1 416 978 6382
Fax: +1 416 978 1938

Dr Thomas W Jungi, Institute of Veterinary Virology, University of Bern, Langgass-Strasse 122, CH-3012 Bern, Switzerland
Tel: +41 31 631 2502
Fax: +41 31 631 2534
E-mail: jungi@ivv.unibe.ch

Dr Stephen L Kaattari, School of Marine Science, Virginia Institute of Marine Science, College of William and Mary, Gloucester Point, VA 23062 1346, USA
Tel: +1 804 642 7000
Fax: +1 804 642 7186
E-mail: kaattari@vims.edu

Contributors

Dr ER Karamova, Laboratory of Immuno-chemistry, Cancer Research Center, Russian Academy of Medical Sciences, 115478 Moscow, Russia
Tel: +7 095 323 5810
Fax: +7 095 324 1205

Dr Barbara L Kee, Department of Biology, University of California, 9500 Gilman Drive, 0366 La Jolla, CA 92093, USA
Tel: +1 619 534 8797
Fax: +1 619 534 7550
E-mail: bkee@jeeves.ucsd.edu

Dr Laurie J Kennedy, Immunology Research Group, Faculty of Medicine, University of Calgary, 3330 Hospital Drive NW, Calgary, Alberta T2N 1N4, Canada
Tel: +1 403 220 4551
Fax: +1 403 270 7410
E-mail: lkennedy@acs.ucalgary.ca

Dr Fabienne Kerfourn, Laboratoire d'Immu-nologie Comparée, Université Pierre et Marie Curie, CNRS URA 1135, Boîte 29, F-75252 Paris Cedex 05, France
Tel: +331 4427 4427
Fax: +331 4427 3866

Dr R Kettmann, Faculty of Agronomy, B-5030 Gembloux, Belgium
Tel: +32 81 622 156
Fax: +32 81 613 888
E-mail: bimaphya@fsaga.ac.be

Dr Se-Ho Kim, Department of Chemical Immunology, The Weizmann Institute of Science, 76100 Rehovot, Israel
Tel: +972 8 934 2391
Fax: +972 8 934 4141
E-mail: liseho@wiccmail.weizmann.ac.il

Dr S King, Biovation Ltd, Marischal College, Broad Street, Aberdeen AB9 1AS, UK
Tel: +44 1224 273 197
Fax: +44 1224 273 198

Dr D King, Celltech Therapeutics Ltd, 216 Bath Road, Slough, Berks SL1 4EN, UK
Tel: +44 1753 534 655
Fax: +44 1753 537 108

Dr W E F Klinkert, Max-Planck-Institut für Psychiatrie, Abteilung Neuroimmunologie, Am Klopferspitz 18a, D-82152 Martinsried, Germany
Tel: +49 89 8578 3583
Fax: +49 89 8578 3790
E-mail: klinkert@alf.biochem.mpg.de

Dr Guus Koch, DLO Institute for Animal Science and Health, Department of Avian Virology, POB 658200 AB Lelystad, The Netherlands
Tel: +31 320 23 8238
Fax: +31 320 23 8668
E-mail: G.Koch@id.dlo.nl

Ms Anne Koniski, University of Rochester, Department of Pediatrics, Hematology and Oncology, Box 777, 601 Elmwood Avenue, Rochester, NY 14642 USA
Tel: +1 716 275 2981
Fax: +1 716 273 1039

Dr Manfred Kopf, Basel Institute for Immunology, Postfach, Grenzacherstrasse 487, CH-4005 Basel, Switzerland
Tel: +41 61 605 1319
Fax: +41 61 605 1364
E-mail: kopf@bii.ch

Dr Marie H Kosco-Vilbois, Geneva Biomedical Research Institute, Glaxo Wellcome R & D SA, Case postale 674, CH-1228 Plan-les-Ouates/Geneva, Switzerland
Tel: +41 22 706 9666
Fax: +41 22 794 6965
E-mail: MHKV10664@ggr.co.uk

Dr Philippe Kourilsky, Unité de Biologie Moléculaire du Gène, Institut Pasteur, 25 rue du Dr Roux, F-75724 Paris Cedex 15, France
Tel: +33 1 45 68 85 46
Fax: +33 1 45 68 85 48

Dr Frantisek Kováru, Department of Physiology, University of Veterinary and Pharmaceutical Sciences, Palackeho 1 3, CZ-612 42 Brno, Czech Republic
Tel: +42 5 4132 1107
Fax: +42 5 748 841

Dr Hana Kozakova, Department of Immunology and Gnotobiology, Institute of Microbiology, Academy of Sciences of the Czech Republic, CZ-54922 Novy Hrádek, Czech Republic
Tel: +42 443 957 22
Fax: +42 443 957 79
E-mail: immuno@mbu.anet.cz

Dr Alexander T Kozhich, Laboratory of Immunology, National Eye Institute, NIH, Bldg 10, Rm 10N208, Bethesda, MD 20892, USA
Tel: +1 301 496 4159
Fax: +1 301 402 0485
E-mail: akozhich@helix.nih.gov

Dr Georg Kraal, Department of Cell Biology and Immunology, Vrije Universiteit, Van der Boechorststraat 7, 1081 BT Amsterdam, The Netherlands
Tel: +31 20 444 8080
Fax: +31 20 444 8081
E-mail: g.kraal@med.vu.nl

Dr Leonor Kremer, Department of Immunology and Oncology, Centro Nacional de Biotecnología/CSIC, UAM-Campus de Cantoblanco, E-28049 Madrid, Spain
Tel: +34 1 585 4533
Fax: +34 1 372 0493
E-mail: lkremer@samba.cnb.uam.es

Dr Guido Kroemer, CNRS-UPR420, 19 rue Guy Môquet, BP 8, F-94801 Villejuif, France
Tel: +33 1 49 58 35 13
Fax: +33 1 49 58 35 09

Dr Jon D Laman, Division of Immunological and Infectious Diseases, TNO Prevention and Health (TNO-PG), PO Box 2215, 2301 CE Leiden, The Netherlands
Tel: +31 71 518 1542
Fax: +31 71 518 1901
E-mail: JD.Laman@PG.TNO.NL

Dr Kenneth S Landreth, Department of Microbiology and Immunology, The Mary Babb Randolph Cancer Center, WV University Health Science Center, Morgantown, WV 26506, USA
Tel: +1 304 293 4067
Fax: +1 304 293 2134
E-mail: KLANDRET@WVU.EDU

Dr Peter Lane, Basel Institute for Immunology, Postfach, Grenzacherstrasse 487, CH-4005 Basel, Switzerland
Tel: +41 61 605 13 12
Fax: +41 61 605 13 64
E-mail: lane@bii.ch

Dr Anton W Langerak, Department of Immunology, Erasmus University Rotterdam, POB 1738, 3000 DR Rotterdam, The Netherlands
Tel: +31 10 408 1111
Fax: +31 10 436 7601

Dr Chris S Lantz, Department of Pathology, Research North, Beth Israel Hospital, Harvard Medical School, 330 Brookline Avenue, Boston, MA 02215, USA
Tel: +1 617 667 2422
Fax: +1 617 667 8210

Dr Olli Lassila, Turku Immunology Center, Department of Medical Microbiology, Turku University, Kiinamyllynkatu 13, FIN-20520 Turku, Finland
Tel: +358 2 333 7411
Fax: +358 2 233 0008
E-mail: olli.lassila@utu.fi

Dr Tucker W LeBien, Department of Laboratory Medicine and Pathology, University of Minnesota Medical School, 420 Delaware Street SE, Minneapolis, MN 55455-0315, USA
Tel: +1 612 626 1422
Fax: +1 612 624 2400
E-mail: lebie001@maroon.tc.umn.edu

Dr Birgit Ledermann, Sandoz Pharma Ltd, Bau 386, Raum 444, Lichtstrasse, CH-4002 Basel, Switzerland
Tel: +41 61 324 5395
Fax: +41 61 324 2990
E-mail: Birgit.ledermann@gwa.sandoz.com

Contributors

Dr Marlies Leenaars, Department of Immunology, Erasmus University Rotterdam, Postbus 1738, 3000 DR Rotterdam,
The Netherlands
Tel: +31 10 408 8142
Fax: +31 10 436 7601

Dr Pieter JM Leenen, Department of Immunology, Erasmus University Rotterdam, POB 1738, 3000 DR Rotterdam, The Netherlands
Tel: +31 10 408 7181
Fax: +31 10 436 7601
E-mail: leenen@immu.fgg.eur.nl

Dr Gertz Likhtenshtein, Department of Chemistry, Ben-Gurion University of the Negev, POB 653, Beer-Sheva 84105, Israel
Tel: +972 7 472 187
Fax: +972 7 274 359
E-mail: gertz@bgu.ac.il

Dr Werner Lindenmaier, Gesellschaft für Biotechnologische Forschung mbH, Mascheroder Weg 1, D-38124 Braunschweig,
Germany
Tel: +49 531 6181 203
Fax: +49 531 6181 202
E-mail: wli@gbf-braunschweig.de

Dr Ariel B Lindner, Department of Chemical Immunology, The Weizmann Institute of Science, 76100 Rehovot, Israel
Tel: +972 8 934 2985
Fax: +972 8 934 4141
E-mail: bnariel@wiccmail.weizmann.ac.il

Dr Christopher Linington, Max-Planck-Institut für Psychiatrie, Am Klopferspitz 18a, D-82152 Martinsried, Germany
Fax: +49 89 8578 3790

Dr Alexandra Livingstone, Department of Biology, Imperial College of Science, Technology and Medicine, Prince Consort Road, London SW7 2BB, UK
Tel: +44 171 594 5416
Fax: +44 171 584 9075
E-mail: a.m.livingstone@ic.ac.uk

Dr Ton Logtenberg, Department of Immunology, University Hospital Utrecht, Heidelberglaan 100, 3584 CX Utrecht, The Netherlands
Tel: +31 30 250 6578
Fax: +31 30 251 1893
E-mail: t.logtenberg@lab.azu.nl

Dr John W Lowenthal, CSIRO DAH, Australian Animal Health Laboratory,
PO Bag 24, Geelong, Victoria 3220, Australia
Tel: +61 52 275 000
Fax: +61 52 275 555
E-mail: j.lowenthal@aahl.dah.csiro.au

Dr L Lujan, Departamento Patologia Animal, Facultad Veterinaria, Miguel Servet 177, 50013 Zaragoza, Spain
Tel: +34 76 76 16 09
Fax: +34 76 76 16 08
E-mail: llujan@mvet.unizar.es

Dr Christophe Lurquin, Ludwig Institute for Cancer Research, UCL 7459, 74 avenue Hippocrate, B-1200 Brussels, Belgium
Tel: +32 2 764 7476 / 32 2 764 7450
Fax: +32 2 762 9405 / 32 2 764 7590
E-mail: lurquin@licr.ucl.ac.be

Dr C Roger MacKenzie, Institute for Biological Sciences, National Research Council of Canada, Ottawa, Ontario K1A 0R6, Canada
Tel: +1 613 990 0833
Fax: +1 613 941 1327
E-mail: mackenzie@biologysx.lan.nrc.ca

Ms Barbara Maget, Institut für Immunologie, Universität München, Goethestrasse 31, D-80336 München, Germany
Tel: +49 89 5996 684
Fax: +49 89 5160 2236
E-mail: moritz@ifi.med.uni-muenchen.de

Dr Tak W Mak, The Ontario Cancer Institute, Departments of Medical Biophysics and Immunology, 610 University Avenue, Room 8 712, Toronto, Ontario M5G 2M9, Canada
Tel: +1 416 204 2236
Fax: +1 416 204 5300
E-mail: tmak@oci.utoronto.ca

Dr Leos Mandel, Department of Immunology and Gnotobiology, Institute of Microbiology, Academy of Sciences of the Czech Republic, CZ-54922 Novy Hrádek, Czech Republic
Tel: +42 443 957 22
Fax: +42 443 957 79
E-mail: immuno@mbu.anet.cz

Mr Santos Mañes, Department of Immunology and Oncology, Centro Nacional de Biotecnología/CSIC, UAM-Campus de Cantoblanco, E-28049 Madrid, Spain
Tel: +34 1 585 4533
Fax: +34 1 372 0493
E-mail: smanes@samba.cnb.uam.es

Dr Philippe Marchetti, CNRS-UPR420, 19 rue Guy Môquet, BP 8, F-94801 Villejuif, France
Tel: +33 1 49 58 35 13
Fax: +33 1 49 58 35 09

Ms Anne Marcuz, Basel Institute for Immunology, Postfach, Grenzacherstrasse 487, CH-4005 Basel, Switzerland
Tel: +41 61 605 1269
Fax: +41 61 605 1364

Dr Elisabeth Märker-Hermann, I Med Klinik und Poliklinik, Johannes Gutenberg Universität, Langenbeckstrasse 1, D-55101 Mainz, Germany
Tel: +49 6131 171
Fax: +49 6131 222 332
E-mail: un65rd@genius.embnet.dkfz-heidelberg.de

Dr Aaron J Marshall, The Wellesley Hospital Research Institute, 160 Wellesley Street E, Toronto, Ontario M4Y 1J3, Canada
Tel: +1 416 926 5151
Fax: +1 416 926 5109
E-mail: aaron@immune.med.utoronto.ca

Dr Helene M Martin, Centre for Animal Biotechnology, University of Melbourne, Parkville 3052 Victoria, Australia
Tel: +61 3 9344 7348
Fax: +61 3 9347 4083
E-mail: hmm@rubens.its.unimelb.edu.au

Dr Carlos Martinez-A, Department of Immunology and Oncology, Centro Nacional de Biotecnología/CSIC, UAM-Campus de Cantoblanco, E-28049 Madrid, Spain
Tel: +34 1 585 4544
Fax: +34 1 372 0493
E-mail: cmartineza@samba.cnb.uam.es

Dr Gianvito Martino, Neuroimmunology Unit - DIBIT, Department of Neurology, University of Milano, S. Raffaele Scientific Institute, Via Olgettina 58, I-20132 Milan, Italy
Tel: +39 2 2643 4853
Fax: +39 2 2643 4855
E-mail: marting@dibit.hsr.it

Dr Donald W Mason, MRC Cellular Immunology Unit, Sir William Dunn School of Pathology, South Parks Road, Oxford OX1 3RE, UK
Tel: +44 1865 27 55 94
Fax: +44 1865 27 55 91
E-mail: dmason@molbiol.ox.ac.uk

Dr Patrick Matthias, Friedrich Miescher Institute, POB 2543, Mattenstrasse 22, CH-4002 Basel, Switzerland
Tel: +41 61 697 6661
Fax: +41 61 697 9386
E-mail: matthias@fmi.ch

Dr Fiona McConnell, Basel Institute for Immunology, Postfach, Grenzacherstrasse 487, CH-4005 Basel, Switzerland
Tel: +41 61 605 1312
Fax: +41 61 605 1364

Dr Peter J McCullagh, Division of Clinical Sciences, John Curtin School of Medical Research, The Australian National University, PO Box 334, Canberra ACT 2601, Australia
Tel: +61 6 249 2550
Fax: +61 6 249 0413

Mr Ian McDermott, The Wellesley Hospital Research Institute, 160 Wellesley East, Toronto, Ontario M4Y 1J3, Canada
Tel: +1 416 926 4833
Fax: +1 416 926 5109
E-mail: ian@whri.on.ca

Contributors

Dr D McGregor, Rowett Research Services, Greenburn Road, Bucksburn, Aberdeen AB21 9SB, UK
Tel: +44 1224 716 226
Fax: +44 1224 716 225
E-mail: dpm@rri.sari.ac.uk

Dr Kelly McNagny, EMBL, Postfach 10.2209, Meyerhofstrasse 1, D-69117 Heidelberg, Germany
Tel: +49 6221 387 417
Fax: +49 6221 387 516
E-mail: mcnagny@embl-heidelberg.de

Dr Fritz Melchers, Basel Institute for Immunology, Postfach, Grenzacherstrasse 487, CH-4005 Basel, Switzerland
Tel: +41 61 605 1238
Fax: +41 61 605 1300

Dr Mario Mellado, Department of Immunology and Oncology, Centro Nacional de Biotecnología/CSIC, UAM-Campus de Cantoblanco, E-28049 Madrid, Spain
Tel: +34 1 585 4533
Fax: +34 1 372 0493
E-mail: jmmellado@samba.cnb.uam.es

Dr Rob H Meloen, DLO Institute for Animal Science and Health, Laboratory for Molecular Recognition, POB 65, 8200 AB Lelystad, The Netherlands
Tel: +31 320 23 8136
Fax: +31 320 23 8050
E-mail: R.H.Meloen@id.dlo.nl

Dr Paul A Monach, Pathology Department MC 6079, University of Chicago, 5841 S Maryland Avenue, Chicago IL 60637, USA
Tel: +1 312 702 1260
Fax: +1 312 702 3778
E-mail: pamonach@midway.uchicago.edu

Dr M Moss, Aberdeen and NE Scotland Blood Transfusion Service, Regional Transfusion Centre, Foresterhill Road, Aberdeen AB25 2ZW, UK
Tel: +44 1224 685 685
Fax: +44 1224 695 351

Dr Ernst W Müllner, Institute of Molecular Biology, Vienna Biocenter, University of Vienna, Dr. Bohr-Gasse 9, A-1030 Vienna, Austria
Tel: +43 1 795 15 26 22
Fax: +43 1 795 15 29 01
E-mail: em@univie.ac.at

Dr Saran A Narang, Institute for Biological Sciences, National Research Council of Canada, Ottawa, Ontario K1A 0R6, Canada
Tel: +1 613 990 3247
Fax: +1 613 941 1327
E-mail: narang@biologysx.lan.nrc.ca

Dr Andrew D Nash, Centre for Animal Biotechnology, University of Melbourne, Parkville 3052 Victoria, Australia
Tel: +61 3 9344 7348
Fax: +61 3 9347 4083
E-mail: adn@rubens.its.unimelb.edu.au

Dr Hovav Nechushtan, Department of Biochemistry, Hebrew University Hadassah Medical School, POB 12272, Jerusalem 91120, Israel
Tel: +972 2 758 288
Fax: +972 2 411 663

Dr Jacques J Neefjes, Division of Cellular Biochemistry, The Netherlands Cancer Institute, Plesmanlaan 121, 1066 CX Amsterdam, The Netherlands
Tel: +31 20 512 1977
Fax: +31 20 512 1989
E-mail: JNEEFJES@NKI.NL

Dr Anne Neisig, Division of Cellular Biochemistry, The Netherlands Cancer Institute, Plesmanlaan 121, 1066 CX Amsterdam, The Netherlands
Tel: +31 20 512 1977
Fax: +31 20 512 1989

Dr Harald Neumann, Max-Planck-Institut für Psychiatrie, Am Klopferspitz 18a, D-82152 Martinsried, Germany
Tel: +49 89 8578 3560
Fax: +49 89 8578 3790

Dr Roald Nezlin, Department of Immunology, The Weizmann Institute of Science, 76100 Rehovot, Israel
Tel: +972 8 934 3652
Fax: +972 8 946 6966
E-mail: linezlin@weizmann.weizmann.ac.il

Dr Aliki-Anna Nichogiannopoulou, The Center for Blood Research Inc, Harvard Medical School, 200 Longwood Avenue, Boston, MA 02115, USA
Tel: +1 617 278 3245
Fax: +1 617 278 3131
E-mail: nichogianopo@cbrv1.med.harvard.edu

Dr Kirsten Niebuhr, Gesellschaft für Biotechnologische Forschung mbH, Zellbiologie und Immunologie, Mascheroder Weg 1, D-38124 Braunschweig, Germany
Tel: +49 531 6181 441
Fax: +49 531 6181 411
E-mail: kin@gbf-braunschweig.de

Dr Lars Nitschke, Max-Planck-Institut für Immunbiologie, Postfach 1169, D-79108 Freiburg, Germany
Tel: +49 761 5108 221
Fax: +49 761 5108 330
E-mail: Nitschke@immunbio.mpg.de

Dr Rebecca L O'Brien, National Jewish Center for Immunology and Respiratory Medicine, 1400 Jackson Street, Denver CO 80206, USA
Tel: +1 303 398 1158
Fax: +1 303 398 1396
E-mail: obrienr@njc.org

Dr Tania D Obranovich, Davies Ryan De Boos, 1 Little Collins Street, Melbourne, Victoria 3000, Australia
Tel: +61 3 9254 2888
Fax: +61 3 9254 2880
E-mail: tobranovich@davies.com.au

Dr Makio Ogawa, Medical University of South Carolina, Ralph H Johnson Veterans Affairs, Medical Center, Charleston SC 29401 5799, USA
Tel: +1 803 577 5011 Ext. 7372
Fax: +1 803 722 7741

Dr Søren Østergaard, Kernisk Institutt, Landbohojskolen, Thorvaldsenvej 40, DK-1871 Frederiksberg C, Denmark
Tel: +45 35 282 429
Fax: +45 35 282 089

Dr Lars Østergaard Pedersen, Institute for Medical Microbiology and Immunology, Panum Institute 18.3.22, Blegdamsvej 3, DK-2200 Copenhagen N, Denmark
Tel: +45 35 32 7900
Fax: +45 31 35 6310

Dr Randall J Owens, Frederick Research Center, Southern Research Institute, 431 Aviation Way, Frederick MD 21701, USA
Tel: +1 301 694 3232
Fax: +1 301 694 7223
E-mail: frc-owens@sri.org

Dr Christopher J Paige, The Wellesley Hospital Research Institute, 160 Wellesley Street East, Toronto, Ontario M4Y1J3, Canada
Tel: +1 416 926 7751
Fax: +1 416 926 5109
E-mail: paige@whri.on.ca

Dr Eeva Palojoki, Department of Medical Microbiology, Turku University, Kiinamyllynkatu 13, FIN-20520 Turku, Finland
Tel: +358 2 333 71
Fax: +358 2 233 0008

Dr Nicole M Parish, University of Cambridge, Department of Pathology, Immunology Division, Tennis Court Road, Cambridge CB2 1QP, UK
Tel: +44 1223 33 3917
Fax: +44 1223 33 3694
E-mail: np@mole.bio.cam.ac.uk

Dr Sylvie Partula, Laboratoire d'Immunologie Comparée, Université Pierre et Marie Curie, CNRS URA 1135, Boîte 29, F-75252 Paris Cedex 05, France
Tel: +331 4427 4427
Fax: +331 4427 3866

Dr Sulabha S Pathak, Department of Immunology, Erasmus University Rotterdam, Postbus 1738, 3000 DR Rotterdam, The Netherlands

Contributors

Dr Lucette Pelletier, Hôpital de Purpan, Université Paul Sabatier, Place du Docteur Baylac, F-31059 Toulouse Cedex, France
Tel: +33 61 77 92 95/90
Fax: +33 61 77 92 91
E-mail: Lucette.Pelletier@purpan.inserm.fr

Dr Jacqueline LMH Pennycook, Wellesley Hospital Research Institute, 160 Wellesley Street East, Toronto, Ontario M4Y 1J3, Canada
Tel: +1 416 926 5151
Fax: +1 416 926 5109
E-mail: jacq@immune.med.utoronto.ca

Dr Ernst Peterhans, Insitute of Veterinary Virology, University of Bern, Langgass-Strasse 122, CH-3012 Bern, Switzerland
Tel: +41 31 631 2413
Fax: +41 31 631 2534
E-mail: peterhans@ivv.unibe.ch

Dr Stephan Petrasch, Medizinische Universitätsklinik, Knappschaftskrankenhaus, In der Schornau 23 25, D-44892 Bochum, Germany
Tel: +49 234 299 3400
Fax: +49 234 299 3409

Dr Luca Piali, Theodor-Kocher Institut, Freiestrasse 1, CH-3012 Bern, Switzerland
Tel: +41 31 631 4161
Fax: +41 31 631 3799
E-mail: PIALI@tki.unibe.ch

Dr Clemencia Pinilla, Torrey Pines Institute for Molecular Studies, 3550 General Atomics Ct., San Diego CA 92121, USA
Tel: +1 619 455 3843
Fax: +1 619 455 3804
E-mail: pinilla@tpims.org

Dr Jiri Plachy, Institute of Molecular Genetics, Academy of Sciences of the Czech Republic, Flemingovo n 2, CZ-166 37 Prague 6, Czech Republic
Tel: +422 243 10 234
Fax: +422 243 10 955

Dr Philippe Poussier, The Wellesley Hospital Research Institute, 160 Wellesley Street E, Toronto, Ontario M4Y 1J3, Canada
Tel: +1 416 926 5059
Fax: +1 416 926 5109
E-mail: poussier@whri.on.ca

Dr Christine Power, Geneva Biomedical Research Institute, Glaxo Wellcome R & D SA, Case postale 674, CH-1228 Plan-les-Ouates/Geneva, Switzerland
Tel: +41 22 706 9666
Fax: +41 22 794 6965

Ms Julie AR Pribyl, Department of Laboratory Medicine and Pathology, University of Minnesota Medical School, 420 Delaware Street SE, Minneapolis, MN 55455-0315, USA
Tel: +1 612 626 4839
Fax: +1 612 624 2400
E-mail: rehma001@maroon.tc.umn.edu

Dr Gabriele Proetzel, Boehringer Mannheim, Therapeutics, Nonnenwald 2, D-82372 Penzberg, Germany
Tel: +49 8856 60 30 06
Fax: +49 8856 60 32 01

Dr Wouter C Puijk, DLO Institute for Animal Science and Health, POB 65, 8200 AB Lelystad, The Netherlands
Tel: +31 320 23 8135
Fax: +31 320 23 8050
E-mail: W.C.Puijk@id.dlo.nl

Dr José Quintans, Pathology Department Box 414, University of Chicago, 5841 S Maryland Avenue, Chicago IL 60637, USA
Tel: +1 312 702 9206
Fax: +1 312 702 3701
E-mail: j-quintans@uchicago.edu

Dr Sheela Ramanathan, The Wellesley Hospital Research Institute, 160 Wellesley Street E, Toronto, Ontario M4Y 1J3, Canada
Tel: +1 416 967 3671
Fax: +1 416 926 5109
E-mail: sheela@whri.on.ca

Dr Hans-Georg Rammensee, Institut für Zellbiologie der Universität Tübingen, Abteilung Immunologie, Auf der Morgenstelle 15, (Verfügungsgebäude), D-72070 Tübingen, Germany
Tel: +49 7071 29 80991 / 49 7071 29 87628
Fax: +49 7071 29 5653

Dr Michael JH Ratcliffe, Department of Microbiology and Immunology, McGill University, 3775 University Street, Montreal, Quebec H3A 2B4, Canada
Tel: +1 514 398 3934
Fax: +1 514 398 7052
E-mail: MRATCLIFFE@NEXUS.MICROIMM. MCGILL.CA

Dr Ehud Razin, Department of Biochemistry, Hebrew University Hadassah Medical School, POB 12272, Jerusalem 91120, Israel
Tel: +972 2 758 288
Fax: +972 2 411 663
E-mail: ERazin@ MD2.HUJI.AC.IL

Dr Christopher Reardon, Department of Dermatology, B-153, University of Colorado, Health Sciences Center, Denver, CO 80262, USA
Tel: +1 303 270 5045
Fax: +1 303 270 8272
E-mail: chris.reardon@uchsc.edu

Dr Zuzana Rehakova, Department of Immunology and Gnotobiology, Institute of Microbiology, Academy of Sciences of the Czech Republic, CZ-54922 Novy Hrádek, Czech Republic
Tel: +42 443 957 22
Fax: +42 443 957 79
E-mail: immuno@mbu.anet.cz

Dr Jean-Christophe Renauld, Ludwig Institute for Cancer Research, UCL 7459, 74 avenue Hippocrate, B-1200 Brussels, Belgium
Tel: +32 2 764 7464
Fax: +32 2 762 9405
E-mail: renauld@licr.ucl.ac.be

Dr John D Reynolds, Immunology Research Group, Faculty of Medicine, University of Calgary, 3330 Hospital Drive NW, Calgary, Alberta T2N 1N4, Canada
Tel: +1 403 220 4552
Fax: +1 403 270 7410
E-mail: reynolds@acs.ucalgary.ca

Dr Stuart Rison, Geneva Biomedical Research Institute, Glaxo Wellcome R & D SA, Case postale 674, CH-1228 Plan-les-Ouates/Geneva, Switzerland
Tel: +41 22 706 9666
Fax: +41 22 794 6965

Dr Christina E Roark, Department of Immunology, University of Colorado, Health Sciences Center, Denver, CO 80206, USA
Tel: +1 303 398 1097
Fax: +1 303 398 1225
E-mail: kotzinlab@njc.org

Dr Jacques Robert, Department of Microbiology and Immunology Medical Center, University of Rochester, Rochester NY 14642, USA
Tel: +1 716 275 3412
Fax: +1 716 473 9573

Dr Stuart J Rodda, Chiron Mimotopes Pty Ltd, 11 Duerdin Street, Clayton Victoria 3168, Australia
Tel: +61 3 9565 1111
Fax: +61 3 9565 1199
E-mail: Stuart_Rodda@cc.chiron.com

Dr Antonius Rolink, Basel Institute for Immunology, Postfach, Grenzacherstrasse 487, CH-4005 Basel, Switzerland
Tel: +41 61 605 1265
Fax: +41 61 605 1364

Dr Louise A Rollins-Smith, Vanderbilt University, Departments of Microbiology and Immunology and of Pediatrics, Nashville, TN 37232, USA
Tel: +1 615 322 4397
Fax: +1 615 322 4399
E-mail: louise.rollins-smith@mcmail.vanderbilt.edu

Contributors

Dr Thibaud Roman, Laboratoire d'Immunologie Comparée, Université Pierre et Marie Curie, CNRS URA 1135, Boîte 29, F-75252 Paris Cedex 05, France
Tel: +33 1 44 27 44 27
Fax: +33 1 44 27 38 66

Dr David R Rose, Department of Medical Biophysics, University of Toronto, Ontario Cancer Institute, 610 University Avenue, Toronto, Ontario M5G 2M9, Canada
Tel: +1 416 946 2970
Fax: +1 416 946 6529
E-mail: drose@oci.utoronto.ca

Dr Olaf Rötzsch, Department of Biochemistry, Harvard University, 7 Divinity Avenue, Cambridge, MA 02138 2092, USA
Tel: +1 617 495 5612
Fax: +1 617 496 8351

Dr Kenneth H Roux, Department of Biological Science, The Florida State University, Tallahassee FL 32306 3050, USA
Tel: +1 904 644 5037
Fax: +1 904 644 0481

Dr Patricia Ruiz, Max-Delbrück Center for Molecular Medicine, Robert-Rösslestrasse 10, D-13125 Berlin-Buch, Germany
Tel: +49 30 9406 3581
Fax: +49 30 9406 3832
E-mail: ruiz@orion.rz.mdc-berlin.de

Dr Johanneke G H Ruseler-van Embden, Department of Immunology, Erasmus University Rotterdam, POB 1738, 3000 DR Rotterdam, The Netherlands
Tel: +31 10 408 7796
Fax: +31 10 436 7601

Dr Kevin W Ryan, Department of Immunology, St Jude Children's Research Hospital, 332 N Lauderdale, Memphis, TN 38101, USA
Tel: +1 901 495 3411
Fax: +1 901 523 2622
E-mail: kevin.ryan@stjude.org

Dr Huub FJ Savelkoul, Department of Immunology, Erasmus University Rotterdam, Postbus 1738, 3000 DR Rotterdam, The Netherlands
Tel: +31 10 408 8142
Fax: +31 10 436 7601
E-mail: Savelkoul@immu.fgg.eur.nl

Dr Wim MM Schaaper, DLO Institute for Animal Science and Health, POB 65, 8200 AB Lelystad, The Netherlands
Tel: +31 320 23 8133
Fax: +31 320 23 8050
E-mail: W.M.M.Schaaper@id.dlo.nl

Dr Dolores J Schendel, Institut für Immunologie, Universität München, Goethestrasse 31, D-80336 München, Germany
Tel: +49 89 5996 667
Fax: +49 89 5160 2236
E-mail: schendel@ifi.med.uni-muenchen.de

Dr Jörn Schmitz, Division of Viral Pathogenesis, Department of Medicine, Beth Israel Hospital, 330 Brookline Avenue, Boston MA 02215, USA
Tel: +1 617 667 5206
Fax: +1 617 667 8210
E-mail: jschmitz@bih.harvard.edu

Dr David H Schwartz, Center for Immunization Research, The Johns Hopkins University, Hampton House, Rm 125, 624 N Broadway, Baltimore MD 21205 1901, USA
Tel: +1 410 955 1622
Fax: +1 410 955 2791
E-mail: DSCHWART@PHNET.SPH.JHU.Edu

Dr Christian Seiser, Institute of Molecular Biology, Vienna Biocenter, University of Vienna, Dr Bohr-Gasse 9, A-1030 Vienna, Austria
Tel: +43 1 795 15 26 30
Fax: +43 1 795 15 29 01
E-mail: cs@univie.ac.at

Dr Heng Fong Seow, Macfarlane Burnet
Centre for Medical Research, PO Box 254,
Yarra Bend Rd, Fairfield Victoria 3078,
Australia
Tel: +61 3 9282 2239
Fax: +61 3 9689 2125 / 61 3 9282 2100
E-mail: heng@burnet.mbcmr.unimelb.edu.au

Dr Alessandro Sette, Cytel Corporation,
3525 John Hopkins Court, San Diego,
CA 92121, USA
Tel: +1 619 552 3000
Fax: +1 619 552 8801

Dr Jeffrey Shabanowitz, Biological Mass
Spectrometry Laboratory, University of
Virginia, McCormick Road,
Charlottesville VA 22901, USA
Tel: +1 804 924 7994
Fax: +1 804 982 2781
E-mail: js4c@faraday.clas.virginia.edu

Ms Nisha Shah, Department of Laboratory
Medicine and Pathology, University of
Minnesota Medical School, 420 Delaware
Street SE, Minneapolis, MN 55455 0315, USA
Tel: +1 612 626 4839
Fax: +1 612 624 2400
E-mail: shahx002@maroon.tc.umn.edu

Dr David A Shapiro, School of Marine
Sciences, Virginia Institute of Marine Science,
College of William and Mary,
Gloucester Point, VA 23062, USA
Tel: +1 804 642 7243
Fax: +1 804 642 7186
E-mail: shapirod@vims.edu

Dr Sadhana Sharma, Allelix
Biopharmaceuticals Inc., 6850 Goreway Drive,
Mississauga, Ontario K4V 1V7, Canada
Tel: +1 905 677 0831
Fax: +1 905 677 9595

Dr Ann-Bin Shyu, Department of
Biochemistry and Molecular Biology,
University of Texas Medical School, 6431
Fannin Street, Houston TX 77030, USA
Tel: +1 713 792 5398
Fax: +1 713 794 4150
E-mail: abshyu@utmmg.med.uth.tmc.edu

Dr Chris Siatskas, Department of Pathology
and Immunology, Monash Medical School,
Commercial Road, Prahran 3181, Victoria,
Australia
Tel: +61 3 9276 3223
Fax: +61 3 9529 6484
E-mail: chris@cobra.path.monash.edu.au

Dr John Sidney, Cytel Corporation,
3525 John Hopkins Court, San Diego,
CA 92121, USA
Tel: +1 619 552 3000
Fax: +1 619 552 8801

Dr Gek-Kee Sim, Immunology Laboratory,
1825 Sharp Point Drive,
Fort Collins CO 80525, USA
Tel: +1 970 224 5387
Fax: +1 970 493 7333
E-mail: sim@ez0.ezlink.com

Mr Steve Simmonds, MRC Cellular
Immunology Unit, Sir William Dunn School of
Pathology, South Parks Road,
Oxford OX1 3RE, UK
Tel: +44 1865 27 56 00
Fax: +44 1865 27 55 91
E-mail: simmonds@molbiol.ox.ac.uk

Dr Francesco Sinigaglia, Roche Milano
Ricerche, Via Olgettina 58,
I-20132 Milano, Italy
Tel: +39 2 288 4805
Fax: +39 2 215 3203
E-mail: Francesco.Sinigaglia@roche.com

Dr Jiri Sinkora, Department of Immunology
and Gnotobiology, Institute of Microbiology,
Academy of Sciences of the Czech Republic,
CZ-54922 Novy Hrádek, Czech Republic
Tel: +42 443 957 22
Fax: +42 443 957 79
E-mail: immuno@mbu.anet.cz

Dr Marek Sinkora, Department of
Immunology and Gnotobiology, Institute of
Microbiology, Academy of Sciences of the
Czech Republic, CZ-54922 Novy Hrádek,
Czech Republic
Tel: +42 443 957 22
Fax: +42 443 957 79
E-mail: immuno@mbu.anet.cz

Contributors

Dr Jonathan CA Skipper, University of Oxford, Molecular Immunology Group of Molecular Medicine, John Radcliffe Hospital, Oxford OX3 9DU, UK
Tel: +44 1865 222 329
Fax: +44 1865 222 502
E-mail: jskipper@molbiol.ox.ac.uk

Dr Craig L Slingluff Jr, Department of Surgery, Box 181, University of Virginia, Charlottesville, VA 22901, USA
Tel: +1 804 982 5474
Fax: +1 804 924 1221
E-mail: cls8h@avery.med.virginia.edu

Dr Richard M Smith, Cambridge University, Department of Pathology, Tennis Court Road, Cambridge CB2 1QP, UK
Tel: +44 1223 33 3702
Fax: +44 1223 33 3875

Ms Tammie Smith Benzinger, Pathology Department MC 6079, University of Chicago, 5841 S Maryland Avenue, Chicago IL 60637, USA
Tel: +1 312 702 1260
Fax: +1 312 702 3778

Dr Miriam C Souroujon, Department of Immunology, Weizmann Institute of Science, 76100 Rehovot, Israel
Tel: +972 8 934 4141
Fax: +972 8 934 4141
Email: lisouro@weizmann.weizmann.ac.il

Dr Igor Splichal, Department of Immunology and Gnotobiology, Institute of Microbiology, Academy of Sciences of the Czech Republic, CZ-54922 Novy Hrádek, Czech Republic
Tel: +42 443 957 22
Fax: +42 443 957 79
E-mail: immuno@mbu.anet.cz

Dr RV Srinivas, Department of Infectious Diseases, St Jude Children's Research Hospital, 332 N Lauderdale, Memphis, TN 38101, USA
Tel: +1 901 495 2359
Fax: +1 901 521 1311
E-mail: rv.srinivas@stjude.org

Dr Pramod K Srivastava, Department of Biological Sciences, Fordham University, Larkin Hall, Bronx, NY 10458, USA
Tel: +1 718 817 3669
Fax: +1 718 817 3691
E-mail: psrivastava@murray.fordham.edu

Dr Ulrich Steinhoff, Max-Planck-Institut für Infektionsbiologie, Monbijoustrasse 2, D-10117 Berlin, Germany
Tel: +49 30 2802 6389
Fax: +49 30 2802 6212

Dr Renáta Stepánková, Department of Immunology and Gnotobiology, Institute of Microbiology, Academy of Sciences of the Czech Republic, CZ-54922 Novy Hrádek, Czech Republic
Tel: +42 443 957 22
Fax: +42 443 957 79
E-mail: immuno@mbu.anet.cz

Dr Jaroslav Sterzl, Institute of Microbiology, Academy of Sciences of the Czech Republic, Videnskà 1083, CZ-142 00 Prague 4 - Krc, Czech Republic
Tel: +422 475 2363
Fax: +422 472 1143
E-mail: immuno@biomed.cas.cz

Dr Stefan Stevanovíc, Universität Tübingen, Institut für Zellbiologie, Abteilung Immunologie, Auf der Morgenstelle 15, D-72070 Tübingen, Germany
Tel: +49 7071 298 7645
Fax: +49 7071 295 653
E-mail: stefan.stevanovic@uni-tuebingen.de

Dr J Steven, Rowett Research Services, Greenburn Road, Bucksburn, Aberdeen AB21 9SB, UK
Tel: +44 1224 716 226
Fax: +44 1224 716 225

Dr Martin Stieger, F Hoffmann-La Roche Ltd, PRPG, Bldg 66, Rm 609A, CH-4070 Basel, Switzerland
Tel: +41 61 688 76 84
Fax: +41 61 688 14 48
E-mail: martin.stieger@roche.com

Dr Georg Stingl, Division of Immunology, Allergy and Infectious Diseases, Department of Dermatology, University of Vienna Medical School, A-1090 Vienna, Austria
Tel: +43 1 403 6933
Fax: +43 1 403 1900
E-mail: georg.stingl@akh-wien.ac.at

Dr Brigitta Stockinger, Department of Molecular Immunology, National Institute for Medical Research, The Ridgeway, Mill Hill, London NW7 1AA, UK
Tel: +44 181 959 3666, ext. 2190
Fax: +44 181 913 8531
E-mail: b-stocki@nimr.mrc.ac.uk

Dr Anette Stryhn, Institute for Medical Microbiology and Immunology, Panum Institute, 18.3.22, Blegdamsvej 3, DK-2200 Copenhagen N, Denmark
Tel: +45 35 32 7883
Fax: +45 35 32 7851
E-mail: A.Stryhn@sb.immi.ku.dk

Dr Yuri Sykulev, Center for Cancer Research, MIT, Bldg E17 128, Cambridge, MA 02139 4307, USA
Tel: +1 617 253 6404
Fax: +1 617 258 8728 or 6172
E-mail: ysykulev@mit.edu

Dr Jean-Pierre Szikora, Ludwig Institute for Cancer Research, UCL 7459, 74 avenue Hippocrate, B-1200 Brussels, Belgium
Tel: +32 2 764 7500
Fax: +32 2 762 9405
E-mail: szikora@licr.ucl.ac.be

Dr Dan S Tawfik, Centre for Protein Engineering, MRC Centre, Hills Road, Cambridge CB2 2QH, UK
Tel: +44 1223 402 127
Fax: +44 1223 402 140
E-mail: dst@macpost.mrc.lmb.cam.ac.uk

Dr P Tempest, Cambridge Antibody Technology, The Science Park, Melbourn, Cambs SG8 6JJ, UK
Tel: +44 1763 263 233
Fax: +44 1763 263 413

Dr Hans-Jürgen Thiesen, Institut für Immunologie, Schillingallee 70, D-18055 Rostock, Germany
Tel: +49 381 494 5871
Fax: +49 381 494 5882
E-mail: hans-juergen.thiesen@uni-rostock.de

Dr G John M Tibbe, Department of Immunology, Erasmus University Rotterdam, Postbus 1738, 3000 DR Rotterdam, The Netherlands
Tel: +31 10 408 8090
Fax: +31 10 436 7601
E-mail: Tibbe@immu.fgg.eur.nl

Dr Helena Tlaskalová-Hogenová, Institute of Microbiology, Academy of Sciences of the Czech Republic, Videnskà 1083, CZ-142 00 Prague 4 - Krc, Czech Republic
Tel: +422 475 2345
Fax: +422 472 1143
E-mail: tlaskalo@biomed.cas.cz

Dr Paavo Toivanen, Department of Medical Microbiology, Turku University, Kiinamyllynkatu 13, FIN-20520 Turku, Finland
Tel: +358 2 333 7426
Fax: +358 2 233 0008
E-mail: paavo.toivanen@utu.fi

Dr Auli Toivanen, Department of Medicine, Turku University, Kiinamyllynkatu 4 8, FIN-20520 Turku, Finland
Tel: +358 2 611 611
Fax: +358 2 2611 030
E-mail: auli.toivanen@utu.fi

Mr André Traunecker, Basel Institute for Immunology, Postfach, Grenzacherstrasse 487, CH-4005 Basel, Switzerland
Tel: +41 61 605 12 87
Fax: +41 61 605 13 64
E-mail: traunecker@bii.ch

Dr Ilja Trebichavsky, Department of Immunology and Gnotobiology, Institute of Microbiology, Academy of Sciences of the Czech Republic, CZ-54922 Novy Hrádek, Czech Republic
Tel: +42 443 957 22
Fax: +42 443 957 79
E-mail: immuno@mbu.anet.cz

Contributors

Dr Clive A Tregaskes, Institute for Animal Health, Compton, Newbury, Berkshire RG20 7NN, UK
Tel: +44 1635 578 411 ext. 2645
Fax: +44 1635 577 263
Email: CLIVE.TREGASKES@BBSRC.AC.UK

Dr Sylvie Trembleau, Roche Milano Ricerche, Via Olgettina 58, I-20132 Milan, Italy
Tel: +39 2 288 4803
Fax: +39 2 215 3203

Dr Gordon Tribbick, Chiron Mimotopes Pty Ltd, 11 Duerdin Street, Clayton Victoria 3168, Australia
Tel: +61 3 9565 1111
Fax: +61 3 9565 1199
E-mail: Gordon_Tribbick@cc.chiron.com

Dr Talip Tümkaya, Department of Immunology, Erasmus University Rotterdam, POB 1738, 3000 DR Rotterdam, The Netherlands
Tel: +31 10 408 1111
Fax: +31 10 436 7601
E-mail: tumkaya@immu.fgg.eur.nl

Dr Alexander Turchin, New York University, Department of Biology, 1009 Main Bldg, 100 Washington Square East, New York, NY 10003, USA
Tel: +1 410 327 2167
Fax: +1 212 995 4015
E-mail: aturchin@welchlink.welch.jhu.edu

Dr A Turner, Delta Biotechnology Ltd, Castle Court, Castle Boulevard, Nottingham NG7 1SB, UK
Tel: +44 1159 553 355
Fax: +44 1159 551 299

Dr Jaakko Uksila, Department of Medical Microbiology, Turku University, Kiinamyllynkatu 13, FIN-20520 Turku, Finland
Tel: +358 2 333 71
Fax: +358 2 233 0008
E-mail: jaakko.uksila@utu.fi

Dr Olli Vainio, Department of Medical Microbiology, Turku University, Kiinamyllynkatu 13, FIN-20520 Turku, Finland
Tel: +358 2 333 7400
Fax: +358 2 233 0008
E-mail: olli.vainio@utu.fi

Dr Salvatore Valitutti, Basel Institute for Immunology, Postfach, Grenzacherstrasse 487, CH-4005 Basel, Switzerland
Tel: +41 61 605 1394
Fax: +41 61 605 1364
E-mail: valitutti@bii.ch

Dr Joop P van de Merwe, Department of Immunology, Erasmus University Rotterdam, POB 1738, 3000 DR Rotterdam, The Netherlands
Tel: +31 10 408 7801
Fax: +31 10 436 7601
E-mail: vandemerwe@immu.fgg.eur.nl

Dr Judy van de Water, Division of Rheumatology/Allergy and Clinical Immunology, TB 192, University of California, Davis, CA 95616, USA
Tel: +1 916 752 2154
Fax: +1 916 752 4669
E-mail: javandewater@ucdavis.edu

Dr René van den Beemd, Department of Immunology, Erasmus University Rotterdam, POB 1738, 3000 DR Rotterdam, The Netherlands
Tel: +31 10 408 1111
Fax: +31 10 436 7601
E-mail: beemd@immu.fgg.eur.nl

Dr A Van den Broeke, Department of Molecular Biology, Université Libre de Bruxelles, 67 rue des Chevaux, B-1640 Rhôde St-Genèse, Belgium
Tel: +32 2 650 9826
Fax: +32 2 650 9839
E-mail: AVDB@DBM.ULB.AC.BE

Dr Pierre van der Bruggen, Ludwig Institute for Cancer Research, UCL 7459, 74 avenue Hippocrate, B-1200 Brussels, Belgium
Tel: +32 2 764 7431
Fax: +32 2 762 9405
E-mail: vanderbrugge@licr.ucl.ac.be

Dr Jessica van der Heijden, Department of Immunology, University Hospital Utrecht, Heidelberglaan 100, 3584 CX Utrecht, The Netherlands
Tel: +31 30 250 7590 or 250 7580
Fax: +31 30 251 7107
E-mail: j.vanderHeijden@lab.azu.nl

Dr Ruurd van der Zee, Utrecht University, Veterinary Faculty, Department of Infectious Diseases and Immunology, PO Box 80.165, 3508 TD Utrecht, The Netherlands
Fax: +31 30 2533 555

Dr Jacques JM van Dongen, Department of Immunology, Erasmus University Rotterdam, POB 1738, 3000 DR Rotterdam, The Netherlands
Tel: +31 10 408 8094
Fax: +31 10 436 7601
E-mail: vandongen@immu.fgg.eur.nl

Dr Willem van Eden, Utrecht University, Veterinary Faculty, Department of Infectious Diseases and Immunology, PO Box 80.165, 3508 TD Utrecht, The Netherlands
Tel: +31 30 2534 358
Fax: +31 30 2533 555
E-mail: w.eden@vetmic.dgk.ruu.nl

Dr Leo MC van Lieshout, Department of Immunology, Erasmus University Rotterdam, POB 1738, 3000 DR Rotterdam, The Netherlands
Tel: +31 10 408 7796
Fax: +31 10 436 7601

Dr Adri Van Oudenaren, Department of Immunology, Erasmus University Rotterdam, Postbus 1738, 3000 DR Rotterdam, The Netherlands
Tel: +31 10 408 8141
Fax: +31 10 436 7601
E-mail: VanOudenaren@immu.fgg.eur.nl

Dr Aline Van Pel, Ludwig Institute for Cancer Research, UCL 7459, 74 avenue Hippocrate, B-1200 Brussels, Belgium
Tel: +32 2 764 7575
Fax: +32 2 762 9405
E-mail: vanpel@licr.ucl.ac.be

Dr Florencio Varas, Unidad de Biología Molecular y Celular, CIEMAT, Avda Complutense 22, E-28040 Madrid, Spain
Tel: +34 1 346 6484
Fax: +34 1 346 6393
E-mail: varas@ciemat.es

Dr Dietmar Vestweber, Institut für Zellbiologie, ZMBE, Mendelstrasse 11, D-48149 Münster, Germany
Tel: +49 251 838 617
Fax: +49 251 838 616
E-mail: vestweb@uni-muenster.de

Dr Markku Viander, Department of Medical Microbiology, Turku University, Kiinamyllynkatu 13, FIN-20520 Turku, Finland
Tel: +358 2 333 71
Fax: +358 2 233 0008
E-mail: markku.viander@utu.fi

Ms Barbara P Vistica, Laboratory of Immunology, National Eye Institute, NIH, Bldg 10, Rm 10N210, 10 Center Dr MSC 1858, Bethesda, MD 20892 1858, USA
Tel: +1 301 496 4159
Fax: +1 301 402 0485
E-mail: bvistica@POserver-b.nih.gov

Dr TP Wallace, Peptide Therapeutics Ltd, 321 Cambridge Science Park, Milton Road, Cambridge CB4 4WG, UK
Tel: +44 1223 423 333
Fax: +44 1223 423 111

Dr Hans-Joachim Wallny, Biotest Pharma GmbH, Landsteiner Strasse 5, D-63303 Dreieich, Germany
Tel: +49 6103 801 345
Fax: +49 6103 801 781

Dr Rudolf Wank, Institut für Immunologie, Universität München, Goethestrasse 31, D-80336 München, Germany
Tel: +49 89 5996 667
Fax: +49 89 5160 2236
E-mail: wank@ifi.med.uni-muenchen.de

Dr NJ Watt, Moredum Research Institute, 408 Gilmerton Road, Edinburgh EH17 7JH, UK
Tel: +44 131 664 3262
Fax: +44 131 664 8001

Contributors

Dr Jürgen Wehland, Gesellschaft für Biotechnologische Forschung mbH, Mascheroder Weg 1, D-38124 Braunschweig, Germany
Tel: +49 531 6181 415
Fax: +49 531 6181 444
E-mail: jwe@gbf-braunschweig.de

Dr N Weir, Celltech Therapeutics Ltd, 216 Bath Road, Slough, Berkshire SL1 4EN, UK
Tel: +44 1753 534 655
Fax: +44 1753 537 108

Dr Siegfried Weiss, Gesellschaft für Biotechnologische Forschung mbH, Mascheroder Weg 1, D-38124 Braunschweig, Germany
Tel: +49 531 6181 230
Fax: +49 531 6181 292
E-mail: siw@gbf-braunschweig.de

Dr Hartmut Wekerle, Max-Planck-Institut für Psychiatrie, Am Klopferspitz 18a, D-82152 Martinsried, Germany
Tel: +49 89 8578 3551
Fax: +49 89 8578 3790

Dr P White, Biovation Ltd, Marischal College, Broad Street, Aberdeen AB9 1AS, UK
Tel: +44 1224 273 197
Fax: +44 1224 273 198

Dr Georg Wick, Institute for General and Experimental Pathology, University of Innsbruck, Fritz-Pregl-Strasse 3, A-6020 Innsbruck, Austria
Tel: +43 512 507 3100
Fax: +43 512 507 2867
E-mail: GEORG.WICK@UIBK.AC.AT

Dr Douglas A Wiesner, Department of Pediatrics, University of Chicago, 5841 S Maryland Avenue, Chicago, IL 60637, USA
Tel: +1 312 702 9206
Fax: +1 312 702 3701

Dr Michael V Wiles, Max-Planck-Institut für Molekulare Genetik, Ihnestrasse 73, D-14195 Berlin, Germany
Tel: +49 30 8413 1350
Fax: +49 30 8413 1380
E-mail: wiles@mpimg-berlin-dahlem.mpg.de

Dr Thomas H Winkler, Basel Institute for Immunology, Postfach, Grenzacherstrasse 487, CH-4005 Basel, Switzerland
Tel: +41 61 605 1272
Fax: +41 61 605 1364
E-mail: winkler@bii.ch

Dr Thomas Wirth, MSZ - Institut für Med. Strahlenkunde u. Zellforschung, Universität Würzburg Versbacher Strasse 5, D-97078 Würzburg, Germany

Dr Thomas Wölfel, I Medizinische Klinik der Johannes Gutenberg-Universität, Langenbeckstrasse 1, D-55101 Mainz, Germany
Tel: +49 6131 3383
Fax: +49 6131 17 3364
E-mail: woelfel@mzdmza.zdv.uni-mainz.de

Dr Paul R Wood, CSIRO, Division of Animal Health, Animal Health Research Laboratory, Parkville 3052, Victoria, Australia
Tel: +61 3 9342 9700
Fax: +61 3 9342 9830
E-mail: PAUL.WOOD@MEL-DAH.CSIRO.AU

Dr David C Wraith, Department of Pathology and Microbiology, School of Medical Sciences, University of Bristol, University Walk, Bristol BS8 1TD, UK
Tel: +44 117 928 7883/7581
Fax: +44 117 928 7896
E-mail: D.C.WRAITH@bris.ac.uk

Dr Gillian E Wu, Department of Immunology, Medical Sciences Bldg, University of Toronto, Toronto, Ontario M5S 1AB, Canada
Tel: +1 416 926 5149
Fax: +1 416 926 5109
E-mail: wu@immune.med.utoronto.ca

Dr Alan J Young, Basel Institute for Immunology, Postfach, Grenzacherstrasse 487, CH-4005 Basel, Switzerland
Tel: +41 61 605 1352
Fax: +41 61 605 1364
E-mail: young@bii.ch

Dr John R Young, Institute for Animal Health, Compton, Newbury, Berkshire RG20 7NN, UK
Tel: +44 1635 577 267
Fax: +44 1635 577 263
E-mail: JOHN.YOUNG@BBSRC.AC.UK

Dr N Martin Young, Institute for Biological Sciences, National Research Council of Canada, Ottawa, Ontario K1A 0R6, Canada
Tel: +1 613 990 0855
Fax: +1 613 941 1327
E-mail: young@biologysx.lan.nrc.ca

Dr Calvin CK Yu, Wellesley Hospital, Research Institute, 160 Wellesley Street E, Toronto, Ontario M4Y 1J3, Canada
Tel: +1 416 926 5151
Fax: +1 416 926 5109
E-mail: calvin@immune.med.utoronto.ca

Dr Naoufal Zamzami, CNRS-UPR420, 19 rue Guy Môquet, BP 8, F-94801 Villejuif, France
Tel: +33 1 49 58 35 13
Fax: +33 1 49 58 35 09
E-mail: zamzami@infobiogen.fr

Dr Rolf M Zinkernagel, Institut für Exp. Immunologie, Universitätsspital Zürich, Schmelzbergstrasse 12, CH-8091 Zürich, Switzerland
Tel: +41 1 255 2386
Fax: +41 1 255 4420
E-mail: zink@pathol.unizh.ch

Section 9

MHC Ligands and Peptide Binding

Section Editor
Hans-Georg Rammensee

LIST OF CHAPTERS

Introduction

9.1

Hans-Georg Rammensee

Abteilung Immunologie, Institut für Zellbiologie, University of Tübingen, Tübingen, Germany

As in all sciences, development of new methods serves to enhance progress in immunology. The last 10 years have brought us a number of new techniques, or the perfection of methods already known. It is the purpose of this section to provide detailed information on new or newly improved techniques in the field for those working on the functional analysis of MHC molecules and other molecules involved in antigen processing, as well as those working in applied immunology. The methods covered center around the analysis of peptides associated with MHC class I and class II molecules. Three of the contributions deal with the analysis of natural peptide ligands of MHC molecules. Chapter 9.2 describes the methods for immunoprecipitating MHC molecules, dissociating the peptides by acid treatment, and separating them by reversed-phase HPLC. The following two chapters deal with the structural identification of the eluted peptides. Chapter 9.3 concentrates on sequencing of ligands by Edman degradation, with special emphasis on pool sequencing, which has been recognized as a powerful tool for the identification of allele-specific peptide ligand motifs for both MHC classes. Chapter 9.4 describes the use of tandem mass spectrometry for sequencing of MHC ligands in minute amounts.

The following three chapters cover the use of peptide binding assays for characterization of MHC peptide specificity. The concept of this approach is several years older than that of natural ligand analysis. However, it is only recently that peptide binding assays have been perfected. Just a few years ago, rather confusing results based on peptide binding to MHC molecules were reported, owing to inadequate methods and lack of insight. The insight meant here, the structural basis of MHC–peptide interaction, is the result of detailed high-resolution crystallography of defined MHC–peptide complexes, as pioneered by D. Wiley and colleagues. This area — crystallography of MHC molecules — as well as computer modeling, is not covered in this section.

Chapters 9.8 and 9.9 deal with techniques which have a very important bearing on applied immunology, that is, on vaccine development and on the design of new approaches to T cell-mediated immunotherapy. Whereas the culture of $CD4^+$ human T cells is established in many laboratories, the work with antigen-specific human $CD8^+$ CTL appears to be more complicated. For this reason, chapter 9.8 is an expert treatise on the handling of human $CD8^+$ CTL. In many cases, the genes encoding CTL epitopes are unknown. This is true for most of the peptides recognized by peptide-directed allo-reactive CTL, and also for many of the CTL directed against infectious agents like viruses or intracellular bacteria or parasites, and still for most of the tumor-associated antigens. Pioneering work by T. Boon and colleagues has provided methods of identifying such genes, as described in chapter 9.9.

The peptides we find associated with MHC molecules of normal, malignant, or infected cells are not only defined by the peptide binding specificity of the MHC molecule. In addition, the entire processing machinery cutting out the peptides from proteins is believed to imprint its specificity on the identity of

ligands. Nevertheless, we are far from understanding the details of this process, although we do know some of the molecules involved, such as the peptide transporter TAP, and other molecules most likely or potentially involved, such as the proteasomes and heat shock proteins. Thus, the last chapters deal with techniques used to analyze some of these molecules: Assays for analyzing specificity and function of TAP, and finally, methods for the functional study of heat shock proteins as putative chaperones in antigen processing.

We hope that in this section we have covered those methods which will be in high demand in the years to come.

Isolation of MHC-restricted peptides by TFA extraction

9.2

Kirsten Falk
Olaf Rötzschke

Department of Biochemistry and Molecular Biology, Harvard University, Cambridge, Massachusetts, USA

TABLE OF CONTENTS

Immunology Methods Manual
ISBN 0–12–442712–X

Abstract

MHC-restricted peptides are proteolytically generated protein fragments, which are non-covalently associated to proteins of the major histocompatibility gene complex (MHC class I or class II molecules). The peptide–MHC complexes are located on the cell surface of antigen presenting cells and represent the target structure of the T cell receptor. Owing to their short length of about 8 to 25 amino acids, MHC-restricted peptides represent linear T cell determinants which can be isolated under relatively drastic and denaturing conditions. One approach is the extraction of peptides with acidic solvents, such as trifluoroacetic acid (TFA). Briefly, peptides are first extracted from cells, tissues, or purified MHC molecules by treatment with aqueous TFA, followed by a size exclusion separation. The low-molecular-weight material is then subjected to reversed-phase high-performance liquid chromatography (HPLC) and the peptide-containing HPLC fractions can subsequently be analyzed for the presence of T cell antigens or can be used for amino acid sequencing experiments.

Introduction

Proteins contain the antigenic determinants for antibodies as well as for the T cell receptor (TCR). But in contrast to antibodies, the TCR does not directly interact with the native protein: the protein first has to undergo extensive proteolytic degradation. This 'processing' leads to the generation of a set of relatively short peptide fragments, which contains the actual T cell antigen. However, this peptide antigen is not antigenic *per se*. To acquire biological activity it has to bind to receptor-like molecules encoded by the major histocompatibility gene complex (MHC). The noncovalent (hormone/receptor-like) complexes of peptide and MHC molecules represent the specific target structure of the TCR.

The MHC gene complex includes two different sets of genes encoding for MHC molecules: MHC class I and class II molecules (Klein 1979). MHC class I molecules consist of a heavy chain associated with a relatively small protein (β_2-microglobulin), while class II molecules consist of an α- and a β-chain. Both classes of molecule function as peptide receptor molecules, but are part of two different branches of the immune defense system (reviewed by Germain and Margulies (1993)). CD8$^+$ T cells are restricted to MHC class I molecules. They are usually cytotoxic and their main function is the elimination of infected or transformed cells. This protection covers basically all tissues, so that MHC class I molecules are expressed by almost all somatic cells. The ligand supply for MHC class I molecules is mainly provided by the 'endogenous' processing pathway. This pathway appears to be designed to produce peptide samples of most proteins expressed within a cell, so that it allows a representative display of the intracellular protein content on the cell surface. MHC class II-restricted CD4$^+$ T cells, in contrast, fulfill mainly immune-regulatory functions; they respond by a release of interleukins. The distribution of MHC class II molecules is restricted to only a few specialized antigen presenting cells (APC) such as B cells, dendritic cells, or macrophages, all capable of endocytosis. Apparently, the major pathway for the supply of MHC class II ligands, the 'exogenous' processing pathway, has evolved to provide peptide samples from extracellular protein sources. However, most of the ligands found on class II molecules actually derive from endogenous proteins, though mainly from compartments involved in the process of endocytosis.

Although MHC class I and class II molecules are composed of different subunits, their overall structure is very similar (Bjorkman et al 1987; Brown et al 1993). The peptide binding site of both classes of MHC molecules is shaped like a groove. It is formed by a β-pleated sheet which is flanked by two α-helical regions and forces the peptide ligand to adopt an almost fully extended conformation (Madden et al 1991; Stern et al 1994). In the case of class I molecules, the groove is closed at both ends, allowing only very limited length variations of the ligands. Usually, they are only between 8 and 10 amino acids long (Falk et al 1991b; Jardetzky et al 1991). MHC class II ligands, in contrast, range from 12 to 25 amino acids (Rudensky et al 1991; Chicz et al 1992; Hunt et al 1992). They extrude the groove at both ends and truncated versions of the same peptide are often found. The tight association between MHC molecule and peptide-ligand is accomplished by a number of noncovalent peptide–MHC interactions, mainly hydrogen-bond formations with the backbone of the peptide, but also by other interactions involving some of the ligand's side-chains. The latter in particular impose strong steric or chemical restrictions, so that only a small subset of the ligand pool actually binds to a given MHC molecule.

During evolution, at least two events have effected the MHC gene complex to overcome this bias (reviewed by Klein and Figueroa 1986). First, the number of genes encoding for MHC molecules has been enhanced by gene duplication; humans, for example, express three different MHC class I (HLA-A, -B, -C) and three different class II molecules (HLA-DR, -DP, -DQ). Second, MHC molecules are highly polymorphic and most of the polymorphic residues cluster within the peptide binding groove. On the level of the population, therefore, more than a hundred different allelic forms of MHC molecules exist, all having distinct binding characteristics. These binding requirements are also reflected in the primary structure of the peptides: essentially all ligands presented by a given MHC molecule share a characteristic amino acid pattern. The existence of these patterns (or 'peptide motifs') is particularly evident in the case of MHC class I ligands (Falk et al 1991b). Mostly, they contain two invariant amino acid positions, which interact with specific pockets of the MHC molecule. One of these 'anchor' amino acids is always located at the peptide's C-terminus. The side-chain is usually hydrophobic (aliphatic or aromatic) and, in some cases, positively charged. The other anchor is often at position 2, but the chemical character of the anchor side chain is highly dependent on the allelic form of the MHC class I molecule.

Similar motifs have also been identified for the ligands of MHC class II (Hammer et al 1992; Falk et al 1994). Owing to the length variation, their MHC-binding region is not always located at the same absolute position, but, as a common feature, appears always to start with a hydrophobic aliphatic or aromatic anchor amino acid. However, compared to MHC class I ligands, the motifs are more complex in general and more similar to each other. As a consequence, MHC class II molecules display a much higher degree of 'promiscuity', meaning that the same ligand is presented by a number of different allelic forms of MHC class II molecules (Sinigaglia et al 1988; Panina-Bordignon et al 1989).

Besides binding, selective processing events also influence the structure of MHC ligands. For example, the limited number of allowed C-terminal residues appears not to be dependent on MHC class I binding specificity alone (Rötzschke and Falk 1991). It is likely also to be a consequence of a matching specificity of the whole processing machinery, which probably involves proteases, transporters, and chaperones as well as the MHC molecules themselves. In the case of class II ligands, similar effects might cause the over-representation of ligands having a Pro residue one amino acid away from the N-terminus (Kropshofer et al 1993; Falk et al 1994). x-Ray crystallographic data suggest that this residue is located outside the class II binding cleft (Stern et al 1994), so that, for example, proteolytic 'trimming' of the N-terminal ligand residues might cause this phenomenon.

MHC molecules present probably several thousands of different self-peptides. T cells responding against these peptides are usually eliminated during their maturation within the thymus. A T cell antigen, therefore, under normal circumstances has to originate from a 'foreign' source, such as a virus-protein. Other examples of pathogens known to provide MHC-restricted peptide antigens are myco-bacteria or *Plasmodium*. Tumor-specific pep-tide-antigens represent another important group of MHC-restricted T cell antigens. In contrast to pathogen-derived antigens, they originate from tumor-specific self-proteins. Tumor-specific peptides must either cover a part of the protein which has undergone a mutation, or derive from a protein which is significantly overexpressed in the transformed tissue. The two other cases in which autoanti-gens trigger T cell responses are peptides associated with autoimmune diseases and transplantation antigens. While in the first case a break of self-tolerance has to occur, the antigenicity of the latter ones is simply due to the artificial exposure of donor-specific pep-tide–MHC complexes to the T cells of the recipient. Mismatches between the haplotype of the donor and the recipient are very likely and either the different allelic forms of donor's

MHC molecules alone, the allele-specific set of self-peptides presented in the context of these MHC molecules ('allo-peptides'), or peptide antigens which derive from donor-specific or polymorphic proteins ('minor histocompatibility antigens') induce the T cell response.

A relatively large number of MHC-restricted peptide antigens has already been isolated. Their detection is relatively easy, since the *in vitro* T cell assays used for this purpose are very sensitive and selective. The identification of their amino acid sequence is much more complicated. The typically low copy number of these antigens and the complex composition of the MHC ligand pool make their sequence determination usually a very difficult task. In contrast, the sequencing of abundant self-peptides or the identification of ligand motifs by 'pool sequencing' is relatively easy. The term 'pool sequencing' refers to the multiple sequence analysis of the combined ligand pool, which allows the identification of the specific amino acid patterns characterizing the ligand motif (see Chapter 9.3). These motifs have been identified for the ligands of MHC class I as well as of class II molecules and can be used for the prediction of peptide antigens (Rötzschke et al 1991c; Pamer et al 1992).

Isolation of MHC-restricted peptides

MHC-restricted peptides can be extracted by the use of acidic solvents, such as aqueous trifluoroacetic acid (TFA). TFA combines several advantages: first, it is a well-established peptide solvent and is widely used for peptide separations by high-performance liquid chromatography (HPLC). Second, it denatures proteins and solubilizes peptides by providing a low-pH environment as well as by its chaotropic ability. Third, it is volatile, so that it is easily removable from the peptide isolate. In

addition to an HPLC device, only standard laboratory equipment is required for the peptide extraction. The extraction can be performed either directly from cells or tissues (Rötzschke et al 1990a) or from purified MHC molecules (Falk et al 1991b). The protocols for both methods are listed below.

Direct extraction of cells or tissues

The direct extraction of cells or tissues is a very quick approach and requires relatively low amounts of starting material (Fig. 9.2.1). The method contains only three steps: peptide extraction with TFA, gel filtration, and RP-HPLC separation. The extract obtained is relatively crude, so that sequencing of peptides succeeds only in exceptional cases (Udaka et al 1992). Nevertheless, the quality and quantity of the extracted material are usually sufficient for detection of natural T cell antigens with *in vitro* T cell assays.

The yield of a peptide antigen extracted directly from cells or tissue is usually only a few nanograms. Despite the low amount, the loss during the preparation is very low; for two peptide antigens the recovery has been estimated to range between 25% and 100% (Falk et al 1991a). The relatively high yield is probably due to the fact that the large amount of nonrelevant material in the extract prevents losses by nonspecific adsorption. The collec-

tion of peptide antigens isolated by this method includes 'classically' restricted viral (Rötzschke et al 1990b; Del Val et al 1991), mycobacterial (Pamer et al 1992), tumor (Wallny et al 1992; Wölfel et al 1994), minor H- (Rötzschke et al 1990a; Sekimata et al 1992), and allo-antigens (Rötzschke et al 1991b; Udaka et al 1992) as well as antigens extracted from so-called 'non-classical' MHC molecules (Aldrich et al 1994). In addition to MHC-associated peptides, this approach also allows the isolation of putative precursor peptides or peptides derived from other pathways (Falk et al. 1990). The discrimination between MHC-associated and other (for example, half-processed) forms of the peptide antigen is relatively easy. Fully-processed peptide antigens can usually only be extracted from cells which express the restricting MHC molecule (Falk et al 1990; Griem et al 1991). Therefore, as a control, cells should also be extracted which express only the antigen but not the restricting MHC molecule (see Fig. 9.2.8).

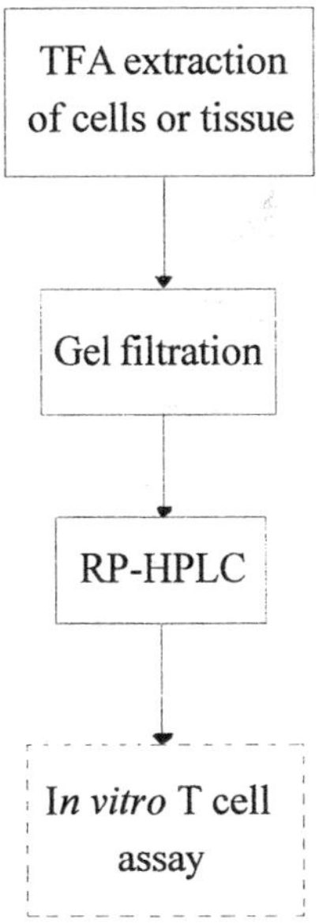

TFA extraction

Figure 9.2.1 Direct peptide extraction of cells or tissues with TFA (see text).

Starting material

0.5–1.0 g cells or tissue (e.g. 2–4 mouse spleens)

Solvents

0.1% (v/v) aqueous trifluoroacetic acid (TFA)

10% (v/v) TFA

Equipment

Cell homogenizer (Dounce)

Cell disrupter, ultrasound sonifier (micro tip)

Ultracentrifuge

250 ml round-bottom flask

Lyophilizer

1. Transfer the starting material together with 10 ml 0.1% TFA into the Dounce homogenizer. Homogenize the material with a couple of strokes using a piston with narrow gap. Remove the homogenate, wash the homogenizer with 5 ml 0.1% TFA and combine the washing with the homogenate. Check the pH and adjust to pH 2 with 10% TFA if necessary.
2. Sonicate the suspension on ice with 2×10 ultrasound bursts. Stir the sonicated solution for 30 min on ice, spin 30 min at 160,000g and transfer the supernatant into a 250 ml round-bottom flask.
3. Transfer the pellet with 7 ml 0.1% TFA into the Dounce homogenizer and homogenize the suspension. Wash the homogenizer with 3.5 ml 0.1% TFA and repeat step 2.
4. Combine the supernatants in a 250 ml round-bottom flask, freeze them with liquid nitrogen, and lyophilize overnight.
    Note: Do not spread the extract over the entire wall of the flask, otherwise redissolution of the lyophilized material might become difficult.

Gel filtration

Solvents

0.1% (v/v) TFA

Equipment

Homogenizer (Dounce)

Peristaltic pump

UV detector (280 nm)

250 ml round-bottom flask

Lyophilizer

Sephadex G25 (Pharmacia)

Separation parameters

- Bed volume: ~25 $\times$ 1.6 cm
- Eluent: 0.1% (v/v) TFA
- Flow rate: 1.6 ml min^{-1}

1. Prepare the column by resuspending the column material directly in 0.1% TFA. Let the material swell for 1 h and fill the column by adding the entire material at once to avoid discontinuities of the bed.
    Note: Store the column in water containing 0.02% azide. Before using the column again, reequilibrate with 3 column volumes of 0.1% TFA.
2. Add a total volume of 1.5 ml 0.1% TFA to the lyophilized material. Resuspend in a 2 ml Dounce homogenizer. Transfer the suspension into an Eppendorf cup and shake 5–10 min. Spin the sample for 10 min at top speed.
3. Transfer the supernatant into a separate tube and repeat step 2 with the pellet. Combine both supernatants and spin again 10 min at top speed.
4. Apply the cleared supernatant carefully on top of the column bed and elute with 0.1% TFA. Monitor the separation with a UV detector (Fig. 9.2.2).
5. Collect all fractions that elute beyond the exclusion volume in a 250 ml round-bottom flask. Lyophilize the material as described under 'TFA Extraction' step 4.

Preparative reversed-phase HPLC

Special care must be taken if a synthetic peptide analogue of the antigen already exists in the laboratory, which might have been

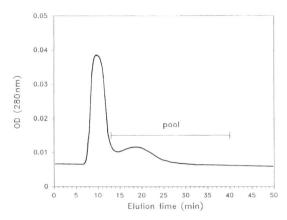

Figure 9.2.2 Elution profile of the gel filtration separation. The entire separation takes about 50 min. The major peak represents the high-molecular-weight material, which elutes within the exclusion volume of the column. The elution of the low-molecular-weight material starts after approximately 13 min. Fractions to combine are indicated. Note that the pool is extended to fractions which elute well behind the low-molecular-weight peak to avoid a loss of peptidic material caused by retardation.

separated on your HPLC-device. In that case one should keep in mind that the *in vitro* T cell assays are very sensitive; at least MHC class I-restricted T cells recognize their antigens down to the lower nanomolar or even picomolar range. Often trace amounts of a synthetic peptide antigen that was separated a couple of runs before the TFA extract has been injected can still easily be detected in the T cell assay and interfere with the recognition of the extracted natural antigen. To avoid false positive results, one should make sure that the HPLC system is 'clean' *before* injecting the TFA extract; this can be done by testing the fractions of a blank run with the antigen-specific T cell in an *in vitro* assay (see 'Trouble Shooting').

Solvents

0.1% (v/v) TFA

Equipment

2 ml homogenizer (Dounce)

Preparative HPLC device

Speedvac concentrator

HPLC materials and parameters

- Flow rate: $1\,\mathrm{ml\,min^{-1}}$
- Fraction size: 1 min
- Fraction range: 15–40 min
- Column: analytical reversed-phase column (e.g., Superpac Pep-S, Pharmacia: $4.6 \times 250\,\mathrm{mm}$, C2/C18, particle size $5\,\mu\mathrm{m}$)
- Sample loop: 1 ml
- Solvent A: 0.1% TFA in H_2O
- Solvent B: 0.1% TFA in acetonitrile
- *Gradient*:

min	0	5	40	45	50
%A	100	100	40	40	100
%B	0	0	60	60	0

1. Preequilibrate the column with one or two blank runs until the baseline is 'clean' and stable.
   Note: Perform at least one blank run between two separations. Large and hydrophobic material in particular can stick on the column and might interfere with the next run.
2. Add a total volume of 1 ml 0.1% TFA to the lyophilized material. Resuspend in a 2 ml Dounce homogenizer and shake the sample afterwards for 5–10 min.
3. Spin the sample 10 min at top speed and inject the clear supernatant.
4. Monitor the separation with a UV detector at 220 nm and collect the fractions (Fig. 9.2.3).
5. Store the fractions at –70°C or dry them directly in a speedvac

concentrator without additional heating. The dried fractions can be stored at –20 or –70°C.

TFA extraction of purified MHC molecules

The protocol for the extraction of peptides from purified MHC molecules is relatively complex (Fig. 9.2.4). The complete method consists of 6 or 7 steps and requires in addition to 5–40 g of cellular starting material several milligrams of an antibody specific for the MHC molecule. The antibody can either be produced *in vivo* as ascites or *in vitro* as hybridoma supernatant, but for the sake of higher purity of the preparation, the latter method should be preferred.

However, the protocol starts with the isolation of the antibody by a protein A or protein G column, followed by coupling of the purified antibody to Sepharose beads. The antibody-coated beads are used in the next step for the immunoprecipitation of MHC molecules, which have previously been solubilized in a detergent lysate. MHC-associated peptides are then directly extracted from the beads with TFA (a prior elution of the bound MHC molecules is not necessary) and separated on a RP-HPLC column.

The total yield of peptides extracted from purified MHC molecules typically reaches the nanomole range and the purity of the obtained peptide fractions is relatively high. It is the method of choice if sequencing is the aim of the experiment. A relatively large number of MHC class I- or class II-restricted self-peptides have already been sequenced from material obtained by this or similar methods (MHC class

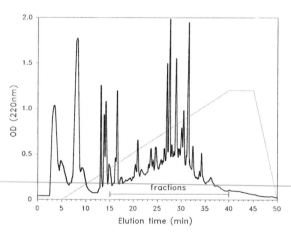

Figure 9.2.3 Elution profile of HPLC separation of a mouse spleen extract. The solid line represents the UV absorption of the eluted material, the dashed line the gradient of the eluent (see text). The range of fractions taken is indicated. The HPLC chromatogram is very complex and, usually, does not show any signals that derive from MHC-restricted peptides or even that represent T cell antigens. Although basically all peaks of the chromatogram, with regard to MHC-restricted antigens, represent nonrelevant material, it is usually characteristic for extracts derived from a particular cell line or even for tissue extracts of a particular mouse strain. It is therefore important for judging the reproducibility and for the quality control of the preparation.

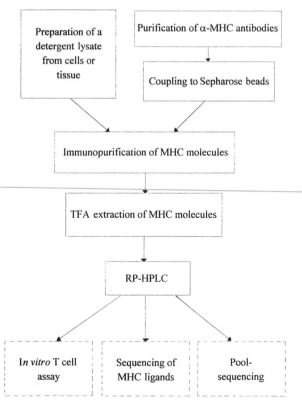

Figure 9.2.4 Peptide extraction of purified MHC molecules with TFA (see text).

I ligands are reviewed by Falk and Rötzschke (1993) and Engelhard (1994); MHC class II ligands are reviewed by Chicz et al (1993), Malcherek et al (1993), and Falk et al (1994)). The combined ligand pool can also be subjected to pool sequencing. Ligand motifs identified by this approach cover a broad range of different allelic forms of 'classical' MHC class I molecules (reviewed by Falk and Rötzschke 1993) and of MHC class II molecules (Malcherek et al 1993; Falk et al 1994) as well as one 'non-classical' MHC class I molecule (Rötzschke et al 1993). Of course, the peptides extracted from MHC molecules with TFA or other acidic solvents can also be tested in *in vitro* assays. The detection of several natural T cell antigens has already been reported (Rötzschke et al 1991a; de Bueger et al 1993) and some of them could be even sequenced by Edman degradation (Nelson et al 1992; Vignali et al 1993). But, again, the sequence determination of T cell antigens is still not a routine task and even the application of sophisticated sequencing equipment and technique such as tandem mass spectrometry succeeds only exceptionally (Cox et al 1994).

Purification of antibodies from hybridoma supernatants

Starting material

2–10 liters hybridoma supernatant

Chemicals and solutions

PBS

$(NH_4)_2SO_4$

10% (w/v) NaN_3

1 M $NaHCO_3$

0.1 M acetic acid/0.15 M NaCl

0.1 M $NaHCO_3$/0.15 M NaCl

0.1 M and 0.02 M NaH_2PO_4

(Adjust pH to 8.0 for protein A; adjust pH to 7.0 for protein G.)

Equipment

Dialysis tubings

Protein A or protein G column material (e.g., protein A- or protein G-Sepharose FF, Pharmacia)

UV detector (280 nm)

1. Transfer the hybridoma supernatant into the coldroom and add $(NH_4)_2SO_4$ with vigorous stirring until the solution reaches 50% saturation (314 g per 1 liter of supernatant). Add the $(NH_4)_2SO_4$ in small portions (100–250 g) and wait until all of the already added $(NH_4)_2SO_4$ has been dissolved before adding the next portion. Continue to stir for an additional 2 h. Stop stirring and leave the solution in the coldroom for 1–2 days so that the proteins precipitate gravitationally.

   Note: The purification of antibodies by protein A or protein G can also be performed directly with the untreated hybridoma supernatant, but the yield is higher if the antibodies are concentrated by $(NH_4)_2SO_4$-precipitation.

2. Suck off most of the supernatant with a pump and resuspend the precipitate in the remaining solution (~1/10 of the original volume). Spin down the precipitate (10 min, 10,000*g*) and discard the supernatant. Add 10–50 ml of 20 mM phosphate buffer directly on to the pellet in the centrifugation bottle and leave it in the coldroom for 1 h.

   Note: The pH of the phosphate buffer depends on whether the purification is done with protein A (pH 8.0) or protein G (pH 7.0). Refer to the manufacturer's manual as to which column material is required for your antibody (species, subtype).

3. Combine the redissolved pellets and add 20 mM phosphate buffer up to 1/10 of the original volume of the hybridoma supernatant. Stir for 30 min

and clear the solution by spinning for 10 min 10,000*g*. If the supernatant is not clear yet, filter through a prefilter followed by a 0.45 μm filter. Add NaN_3 to a final concentration of 0.02%.

4. Apply the solution on a protein A or protein G column (bed volume 2–10 ml; flow rate 50 ml h^{-1}). The column should have been previously equilibrated with at least 5 bed volumes of 0.1 M phosphate buffer pH 8.0 (protein A) or pH 7.0 (protein G).

   Note: If ascites is used instead of hybridoma supernatant, dilute the ascites 1:5 with phosphate buffer before applying it to the column and continue as described in the following steps.

5. Wash with 0.1 M phosphate buffer until the baseline is stable. Elute with 0.1 M acetic acid–0.15 M NaCl into tubes which already contain 0.4 ml 1 M $NaHCO_3$ to neutralize the acid (fraction size ~10 ml). Monitor the washing and the elution of the column with a UV detector (280 nm). Repeat steps 5 and 6 until no more antibody can be isolated (see Fig. 9.2.5).

6. Dialyze the antibody-containing eluate overnight at 4°C in 0.1 M $NaHCO_3$–0.15 M NaCl. Determine the yield in a protein assay (e.g., Bradford). The yield usually ranges between 5 and 20 mg per liter of hybridoma supernatant.

   Note: Store the antibodies at −20 or −70°C.

Coupling of the antibodies to Sepharose beads

Solvents

1 mM HCl

0.1 M glycine pH 8.0

10% (w/v) NaN_3

Material

Antibody solution in 0.1 M $NaHCO_3$–0.15 M NaCl (0.2–2 mg ml^{-1})

CNBr-Sepharose, Pharmacia

Shaking device, end-over-end

Typically, between 5 and 40 mg of antibody is used for one coupling experiment. In parallel to the antibody-coated beads, glycine-blocked beads should also be prepared. They will later be used during the immunoprecipitation of the MHC molecules as material for a precolumn and will also be subjected to a mock extraction.

1. Transfer the dry Sepharose beads into a tube. Use 375 mg of beads per 10 mg antibody. Also transfer the same amount of beads into a second tube for preparation of glycine-coated beads.
2. Add 40 ml of 1 mM HCl to each tube and place the tubes on ice. Shake them occasionally and let the beads swell for 30 min.

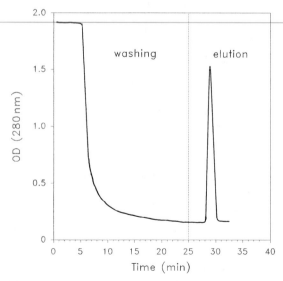

Figure 9.2.5 Elution profile of the antibody purification by protein A or protein G column (see text).

3. Spin down the beads (2 min, 200*g*, *no brake*) and discard the supernatant.
4. Add the antibody solution (0.2–2.0 mg ml^{-1}) or 20 ml of 0.1 M glycine pH 8.0, respectively. Shake slowly end over end for 3 h at room temperature. Save a few μl of the antibody solution to determine the coupling efficiency.

   Note: The coupling efficiency increases with the concentration of the antibody solution, but relatively dilute solutions also give reasonable results.
5. Spin down the antibody-coated beads slowly (2 min, 200*g*, no brake) and remove the supernatant. Determine the coupling efficiency by measuring the protein concentration of the original antibody solution and of the solution after coupling in a photometric protein assay (e.g., Bradford). Calculate the efficiency using the following formula:

   Efficiency (%) =

   $$\frac{\text{OD after coupling}}{\text{OD before coupling}} \times 100$$

   Note: The coupling efficiency is usually above 70%.
6. Add 20 ml 0.1 M glycine pH 8.0 to the antibody-coated beads to block any remaining CNBr groups. Shake end over end for 1 h.
7. Spin down the antibody-coated beads and the glycine-coated beads (2 min, 200*g*, no brake). Wash them 3 times with 20 ml PBS. Prepare a 1:5 slurry with PBS–0.02% NaN$_3$ and store the beads at 4°C.

   Note: The antibody-coated beads can be stored for several months.

Preparation of a detergent lysate

Cellular material

4–40 × 10^9 cells or up to 50 g tissue

Equipment

Homogenizer (only for tissue)

0.45 μm filter (low protein binding)

Solvents and chemicals

Lysis buffer (PBS containing 1% (v/v) Nonidet P-40, containing protease inhibitors, e.g. 0.1 mM phenylmethylsulfonyl fluoride (PMSF) (stock solution in ethanol), 2 μg ml^{-1} aprotinine, 2 μg ml^{-1} leupeptine, 0.0002% (w/v) pepstatin (stock solution in DMSO))

1. Transfer the cells or the homogenized tissue together with ice-cold lysis buffer into a beaker. The ratio of the cell pellet to lysate volume should range between 1/5–1/10. Stir for 45 min on ice.

   Note: If the volume of the lysate exceeds 1 liter it is more convenient to isolate the membranes by differential centrifugation before preparing the lysate (Gorga et al 1984).
2. Spin the lysate for 5 min at 800*g* to remove nuclei and cell debris. Transfer the supernatant into ultracentrifugation tubes.
3. Spin for 45 min at 120,000*g* (4°C). Remove the tubes carefully from the rotor. If lipids are floating on the surface of the lysate, suck them off with a water pump.

   Note: The removal of lipids should be done carefully, otherwise they will stick to the column in the subsequent affinity purification.
4. Combine the supernatant and filter the lysate through a 0.45 μm filter. The

lysate should either be used immediately for the isolation of MHC molecules or stored at –70°C.

Affinity purification of MHC molecules

Solvents and chemicals

PBS–1.0% Nonidet P-40

PBS–0.1% Nonidet P-40

PBS

Equipment

Antibody-coated Sepharose beads and glycine-coated beads.

Small columns, tubing, and clamps

Peristaltic pump

1. Transfer the slurry of the antibody-coated beads and the glycine-coated beads into small columns, so that the bed volume ranges between 0.5 and 2 ml ($\equiv$5–20 mg antibody). The entire affinity purification should be performed in the coldroom.
2. Wash and equilibrate the columns with 5–10 bed-volumes of PBS–1% Nonidet P-40 (flow rate 1 ml min^{-1}) and connect them in series. Place at least one column filled with glycine-coated beads at the top of the series, so that it functions as a precolumn.
   Note: The MHC preparations are purer the more columns there are in front of the column filled with the specific antibody. If more than one MHC molecule is going to be purified, place the most important column at the end of the series.
3. Apply the clear lysate to the columns with a flow rate of 0.5 ml min^{-1}.
   Note: If the lysate appears turbid, filter it again before column

application, otherwise it will stick to the columns.

4. Disconnect the columns and wash each column carefully (also the one filled with glycine-coated beads) first with 10 ml PBS–1% NP-40, then with 20 ml PBS–0.1% NP-40 (flow rate 0.5 ml min^{-1}).
   Note: Since the lysate can still be used later for the purification of other MHC molecules, refreeze it at –70°C.
5. Wash each column finally with 3 ml PBS to remove most of the detergent. Let the column drain dry and with the peristaltic pump suck most of the buffer out of the column bed.

TFA extraction of purified MHC molecules

Solvents

0.1% TFA

10% (v/v) TFA

Equipment

2 ml Eppendorf cups

1. Transfer the beads with 1.5 ml 0.1% TFA into 2 ml Eppendorf cups. Readjust the pH to 2 with a few μl 10% TFA if necessary.
2. Shake the tubes for 10 min. Spin down the beads (15 s, 14,000 rpm) and collect the supernatant in another vial.
3. Extract the beads again with 2 × 750 μl 0.1% TFA as described in step 2 and combine the extracts.
4. Spin the extract again (5 min, 14,000 rpm) to remove remaining beads.
   Note: Either inject the extract immediately into the HPLC column or store it below –20°C.

HPLC separation of the TFA extract

The TFA extract can be separated either on a standard analytical RP column or on a micro-bore column. The use of the latter leads to better resolution and at least a 10-fold higher sensitivity with regard to the UV detection of eluted peptides. Both systems can be used, but if the extract is prepared for separation on a microbore column, it is necessary to remove the high-molecular-weight material (mainly antibodies, MHC molecules and β_2-microglobulin) to avoid overloading the column. The protocols for both separations are described below (see also section 'Preparative Reversed-phase HPLC' earlier if the TFA extraction is performed to isolate natural peptide antigens).

HPLC separation with a standard analytical RP column

Solvents

0.1% (v/v) TFA

Equipment

Preparative HPLC device

Speedvac concentrator

HPLC materials and parameters

- Flow rate: 1 ml min^{-1}
- Fraction size: 1 min
- Fraction range: 15–40 min
- Column: analytical reversed-phase column (e.g. Superpac Pep-S, Pharmacia: 4.6 × 250 mm, C2/C18, particle size 5 μm)

(a)

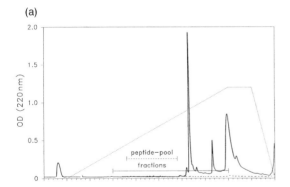

(b)

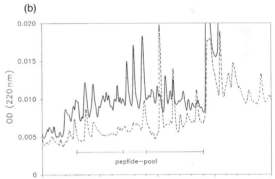

Figure 9.2.6 Elution profile of an HPLC separation on a standard analytical RP column. (a) entire range of the separation; (b) close-up of the range in which peptides are detected. The solid line refers to the separation of material extracted from α-MHC class I-coated beads, the dashed line in (b) to the separation of a mock extraction from glycine-blocked beads. The range of fractions to be tested in *in vitro* T cell assays and of fractions to be combined for pool sequence analysis are indicated. This figure shows a typical HPLC chromatogram as obtained by the separation of MHC class I extracts. The significant peaks between 30 and 33 min derive from β_2-microglobulin and the light chain of the antibody; MHC molecules and the heavy chain of the antibody elute between 35 and 45 min. The amount of the extracted peptides is generally not sufficient to produce signals; they usually do not exceed 0.02 OD. By comparing the profile with that derived from the extract of the glycine-coated beads, the range in which peptides elute can be identified as a slight bulge of the baseline (b). The alignment of both profiles is also necessary to discriminate fractions containing nonspecifically adsorbed material from fractions containing 'abundant' peptides. For example, the two signals detected between 22 and 23 min probably refer to two MHC ligands, while the third significant peak, eluting between 23 and 24 min is caused by nonspecifically adsorbed material, since it is also detected in the mock extract. If peptide-containing fractions are to be combined for pool sequence analysis, these fractions should be excluded from the pool.

- Sample loop: 0.1 ml
- Solvent A: 0.1% TFA in H_2O
- Solvent B: 0.1% TFA in acetonitrile
- Gradient:

min	0	5	40	45	50
%A	100	100	40	40	100
%B	0	0	60	60	0

1. Reduce the volume of the extract (3 ml) in a speedvac concentrator (without additional heating) to 1 ml by successively adding the extract into the same cup.
2. Preequilibrate the column with 0.1% TFA and perform a blank run (see also 'Preparative Reversed-phase HPLC').
3. Spin the extract for 5 min at 14,000 rpm.
4. Inject the sample, monitor the separation with a UV detector (220 nm) and collect the fractions (see Fig. 9.2.6).
5. Store the fractions at –70°C or dry them directly in a speedvac concentrator without additional heating. The dried fractions can be stored at –20°C.

HPLC separation with a microbore RP column

Solvents

0.1% TFA

Materials

Preparative HPLC device suitable for microbore separations (e.g., Smart-system, Pharmacia)

Centricon-10 filters, Amicon

Speedvac concentrator

HPLC materials and parameters

- Flow rate: 0.5 ml min⁻¹

- Fraction size: 0.5 min
- Fraction range: 15–50 min
- Column: microbore reversed-phase column (e.g., µRPC, Pharmacia: 2.1 × 100 mm, C2/C18)
- Sample loop: 0.2 ml
- Solvent A: 0.1% TFA in H_2O
- Solvent B: 0.1% TFA in acetonitrile
- *Gradient*:

min	0	5	25	45	55	60	65
%A	100	100	80	60	40	40	100
%B	0	0	20	40	60	60	0

1. Transfer the extract into Centricon-10 filter.
2. Spin at 4°C at 5000 rpm. Collect the eluate and wash the filter twice with 250 µl 0.1% TFA.
3. Combine the eluates and concentrate them down to a volume of 200 µl in a

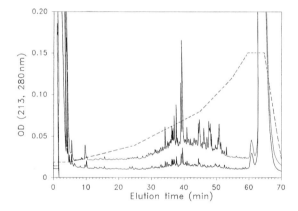

Figure 9.2.7 Elution profile of an MHC class II extract, separated on a microbore RP column (the upper profile is detected at 213 nm, the lower one at 280 nm). Compared to the separation on a standard column, the chromatogram is much more detailed. The range of eluted peptides extends from 20 to 55 min and most of the peaks actually derive from MHC ligands. In this example, the signals of some of the ligands exceed 0.1 OD, but even peptides detected with less then 0.02 OD can still be sequenced. Particularly if the separation is performed to obtain material for peptide sequencing, it is also advised to use a fraction collector with automatic peak recognition to achieve a fractionation that is adequate for the superior separation of a microbore column.

speedvac concentrator (no additional heating) by successively adding the eluate into the same cup.

4. Spin the extract 5 min at 14,000 rpm before injecting the sample.
5. Monitor the separation with a UV

detector and collect the fractions (see Fig. 9.2.7).

6. Store the fractions at −70°C or dry them immediately in the speedvac concentrator.

Subsequent analysis of the extracted material

The HPLC fractions obtained by both methods, the direct extraction of cells and the extraction of purified MHC molecules, can be used for the detection of T cell antigens in *in vitro* assays. If the sequence of the antigen is already known or can be predicted, the identity of the extracted antigen can be confirmed by coelution experiments with a synthetic peptide analogue. The respective protocols as well as a brief description of the sample preparation for sequencing experiments are listed below.

Detection of T cell antigens with *in vitro* assays

Most of the antigens identified so far have been extracted from MHC class I molecules. Reports of the isolation of naturally processed MHC class II restricted antigens are relatively rare, though, in general, the TFA extraction method should be applicable. One reason might be, that MHC class I-restricted T cells usually respond more sensitively than MHC class II-restricted T cells. However, if the peptide antigen is MHC class I-restricted, the antigen-specific T cell is cytotoxic and the fractions can be tested in a ^{51}Cr-release assay as listed below. The assay for the detection of MHC class II-restricted antigens depends on the

nature of the T cell response. If the T cell is cytotoxic the same assay might be used as the one described for class I-restricted T cells. Otherwise, the fractions have to be tested in proliferation assays. However, in any of these cases, the HPLC fractions should be prepared and handled basically in the same way and also the amount of material tested per well should be in the same range as described.

Solutions

PBS

Medium

Na^{51}CrO$_4$ in PBS

Materials

96-well round-bottom plates

CO$_2$-incubator

γ-Counter

Cells

Target cells expressing the restricting MHC molecule

Antigen-specific cytotoxic T cells

In particular, the relatively crude extracts obtained by direct extraction of cells or tissue might reveal some HPLC fractions which are nonspecifically toxic to the target cells. The HPLC fractions, obtained from extracts of purified MHC molecules, in contrast, are much purer and, usually, do not contain any toxic fractions. However, a control experiment to identify toxic fractions by incubating the HPLC fractions with ^{51}Cr-labeled target cells alone (without addition of the antigen-specific T cells) should be done in any case.

1. If the HPLC fractions have not been dried yet, dry them in a speedvac concentrator without additional heating.
2. Add 200–500 μl of sterile PBS to the dried HPLC fractions and shake for 10 min.
3. Label 10^6–10^7 target cells by incubating them with 0.2 mCi Na^{51}CrO$_4$ in PBS in a volume of 50–100 μl for 45 min at 37°C. Wash the labeled cells 3 times with 5 ml medium and adjust to 10^5 cells ml^{-1}.
4. Transfer 100 μl/well of the labeled target cells into a 96-well plate. Add 50 μl of HPLC fraction and incubate for 90 min in a CO_2 incubator. Incubate at least 6 wells with PBS instead of HPLC fraction. These will be used to determine the spontaneous and the maximal ^{51}Cr release.

   Note: If the peptide antigen is HLA-A2.1-restricted, the sensitivity of the assay can be enhanced by incubating the target cells with the MA2.1 antibody prior to peptide loading (Bodmer et al 1989; Wölfel et al 1994).

5. Add 50 μl of T cells (specific recognition) or medium alone (toxicity control) to each well. The effector/target ratio should range between 1:1 and 20:1. Add 50 μl of medium or detergent (e.g. 0.01% NP-40 in PBS) instead of the T cells to the wells used for the determination of the spontaneous and maximal ^{51}Cr release, respectively. Incubate for 6 h at 37°C in the CO_2-incubator.

Note: Often a shorter incubation time is used. However, in our hands, an incubation period of 6 h has given the best results.

6. Harvest 100 μl of supernatant of each well using a multichannel pipette. Count the samples in a γ-counter.
7. The specific ^{51}Cr release is calculated from

Specific release (%) =

$$\frac{cpm\ (sample) - cpm\ (spont)}{cpm\ (max) - cpm\ (spont)} \times 100$$

where max is maximum and spont is spontaneous.

Note: See Fig. 9.2.8.

Rechromatographic analysis of extracted antigens and synthetic peptide analogues

If the sequence of the antigen is known or can be predicted, the identity of an extracted antigen can be verified by comparing the elution behavior of the extracted antigen with that of a synthetic peptide analogue. This can be accomplished by separating both under varying elution conditions on a preparative HPLC column and determining their retention times by testing the fractions in an *in vitro* assay. The following aspects should be considered when planning these experiments:

1. Separate *first* the extracted antigen and *then* the synthetic peptides. If large amounts of synthetic peptide antigen are injected into an HPLC column, trace amounts tend to elute in subsequent separations ('memory effect'). They would interfere with the detection of the natural antigen in the *in vitro* assay.

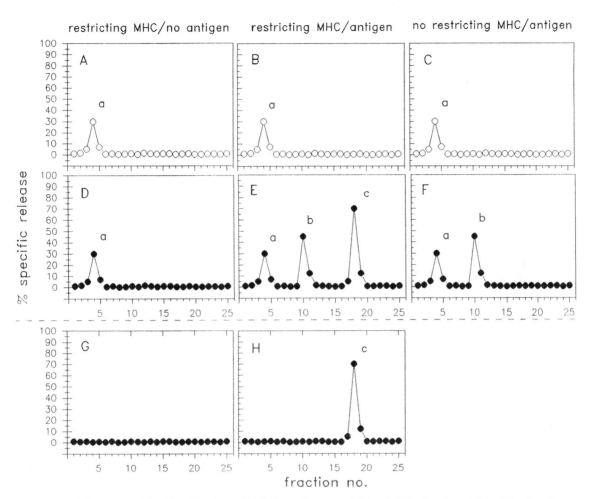

Figure 9.2.8 Example of the identification of HPLC fractions containing MHC class I-restricted T cell antigens by *in vitro* T cell assay. The figure shows the screening of fractions derived from a direct TFA extraction of cells (A–F) or from an extraction of purified MHC molecules (G,H). The extracts are prepared from cells which express the restricting MHC molecule but do not express the antigenic protein (A, D, G), from cells expressing both the restricting MHC molecule as well as the antigen (B, E, H), and from cells which express the antigenic protein but do not express the restricting MHC molecule (C, F). The range of fractions tested refers to the locations indicated in Figs 9.2.3 and 9.2.6, respectively. Fractions are tested on ^{51}Cr-labeled target cells with antigen-specific T cells (solid circles) or, as a control, with medium alone (open circles). In this example, the testing of fractions obtained by the direct extraction of cells which express the antigen as well as the restricting MHC molecule (E), revealed three fractions which trigger a ^{51}Cr release (a, b, c). One of these fractions (a) contains some toxic material, since it induces a nonspecific ^{51}Cr release even if no T cells are added (A, B, C). The other two peaks (b, c) indicate fractions containing antigens which actually derive from the antigenic protein; they are not found in the TFA extract from cells which lack this protein (D). Fraction (b) contains a peptide which represents a putative precursor or derives from a nonrelevant pathway, while the fully processed peptide antigen is detected in fraction (c). In contrast to the putative precursor peptide, the fully processed antigen is not found in the extract derived from antigen-expressing cells which lack the restricting MHC molecule (F; Falk et al 1990). Vice versa, the extraction of purified MHC molecules (H) reveals only the fully processed MHC-associated peptide antigen (c), but not the putative precursor (b).

2. Reduce the slope of the gradient. This enhances the resolution of the separation and allows more accurate fractionation, so that the determination of the retention time is more precise.
3. Use at least two different eluents. For example, the TFA might be replaced by heptafluorobutyric acid, and also phosphate buffers of various pH might be used to provide different elution conditions (Falk et al 1991a).
4. Sequence identity of the extracted and the synthetic peptides can only be assumed if both elute with precisely identical retention times. At least in our hands, the retention time of synthetic peptides is slightly affected by the amount of peptide subjected to HPLC separation. It is therefore advised, to perform two subsequent separations with the synthetic antigen: in the first separation, only a few nanograms should be injected, still sufficient to be detected in the T cell assay. In the second separation, a few micrograms can be used, just enough to produce a peak in the HPLC profile.

Sample preparation for sequencing experiments

The sample preparation for the sequencing of individual ligands is very simple, since it basically involves only drying the HPLC fractions in the speedvac concentrator and passing them to the sequencing facility. Generally, peptide-containing fractions should not be collected or dried in glass vessels, since their use can lead to losses by nonspecific adsorption on the surface of the glass. For the preparation of samples for pool sequencing experiments, only fractions should be combined that do not show significant peaks in the respective HPLC profile (See Fig. 9.2.6b). Otherwise, the multiple sequence analysis might reveal, for example, the sequence of an abundant self-peptide or β_2-microglobulin instead of the ligand consensus motif.

Trouble shooting

HPLC problems

If the HPLC system is 'contaminated' with a synthetic peptide antigen, the antigen can only be removed under drastic conditions. A change of the column is usually not sufficient, since the main source for this 'memory-effect' are peptides which are trapped or adsorbed somewhere within the injection system. The following protocol can be used for cleaning the entire injection system of the HPLC setup and, if necessary, also of the column. However, before using this cleaning protocol, consult your equipment and column manufacturer to deter-

mine whether this protocol may damage your equipment.

1. Disconnect the column from the HPLC system.
2. Inject a total of about 2 ml of 1 N NaOH in several portions and elute it through the injection system and the tubing which is usually connected to the inlet of the column.
3. Inject 2 ml of 6 M urea–50% acetic acid as described in step 2.
4. Repeat steps 2 and 3 twice.
5. If the column is also to be cleaned, reconnect the column, set the flow

rate to 25% of the regular rate and inject 50–200 µl 6 M urea–50% acetic acid several times and elute it over the column.

6. Perform two blank runs by injecting only solvent A (usually 0.1% TFA)

7. Test the fractions of the last blank run in a T cell assay. Repeat the protocol if the system is not yet clean.

been hydrolyzed during swelling or during the coupling reaction: check pH of the solutions used for the swelling (0.1 M HCl, pH should be 4.0) and for the coupling (0.1 M NaHCO$_3$, pH 8–9)

Preparation of a detergent lysate

Lysate cannot be filtered because it blocks the filter

- Lipids or other cell debris have not been removed properly after centrifugation: clear the lysate supernatant again by ultracentrifugation.

Purification of antibodies from hybridoma supernatants

No or very little amount of antibody eluted

- Subtype or species-origin of the antibody does not match the column material (Protein A or G): check the manufacturer's manual.
- Only a small subpopulation of the B cell hybridoma still produces the antibody: test the titer of the hybridoma supernatant; subclone the hybridoma cells if the production is low.

Large amounts of nonspecific antibodies eluted

- Serum supplement of the medium already contained antibodies: check by applying a sample of the serum on the protein A or protein G column.

Coupling of the antibodies to Sepharose beads

Low coupling efficiency of the antibodies to the CNBr–Sepharose beads

- Active groups of the Sepharose have

Detection of T cell antigens with *in vitro* assays

Only toxic fractions detected in the ^{51}Cr-assay

- Too much sample added to the assay: titrate the toxic fractions in the assay

No antigenic fractions detected

- Relative amount of the extracted peptide antigen is too low: repeat the assay with a larger amount of extracted material or try to extract the peptide from purified MHC molecules by using a higher amount of cellular starting material (see 'TFA Extraction of Purified MHC molecules').

More then one antigenic fraction detected

- Some of the fractions contain precursor peptides or peptides derived from other processing pathways: prepare a control extraction from a cell expressing the antigenic protein but lacking the restricting MHC molecule (See Fig. 9.2.8).

A fraction recognized by a specific T cell appears after each HPLC separation (also after a blank-run)
- HPLC system is 'contaminated' with a

synthetic peptide: clean the HPLC (see above).

References

Aldrich CJ, Delloux A, Woods AS, Cotter PJ, Soloski NJ, Forman J (1994) Identification of a Tap-dependent leader peptide recognized by allo-reactive T cells specific for a class Ib antigen. Cell 79: 649–658.

Bjorkman PJ, Saper MA, Samraoui B, Bennet WS, Strominger JL, Wiley DC (1987) Structure of the human class I histocompatibility antigen HLA-A2. Nature 329: 506–512.

Bodmer H, Ogg G, Gotch F, McMichael A (1989) Anti-HLA-A2 antibody-enhancement of peptide association with HLA-A2 as detected by cytotoxic T lymphocytes. Nature 342: 443–446.

Brown JH, Jardetzky TS, Gorga JC, et al (1993) Three-dimensional structure of the human class II histocompatibility antigen HLA-DR. Nature 364: 33–39.

Chicz RM, Urban RG, Lane WS, et al (1992) Predominant naturally processed peptides bound to HLA-DR1 are derived from MHC-related molecules and are heterogenous in size. Nature 358: 764–768.

Chicz RM, Urban RG, Gorga JC, Vignali DAA, Lane WS, Strominger JL (1993) Specificity and promiscuity among naturally processed peptides bound to HLA-DR alleles. J Exp Med 178: 27–47.

Cox AL, Skipper J, Chen Y, et al (1994) Identification of a peptide recognized by five melanoma-specific human cytotoxic T cells. Science 264: 716–719.

de Bueger M, Verrek F, Blokland E, et al (1993) Isolation of a HLA-A2.1 extracted human minor histocompatibility peptide. Eur J Immunol 23: 614–618.

Del Val M, Schlicht H-J, Ruppert T, Reddehase MJ, Koszinowski UH (1991) Efficient processing of an antigenic sequence for its presentation by MHC class I molecules depends on its neighboring residues in the protein. Cell 66: 1145–1153.

Engelhard VH (1994) Structure of peptides associated with MHC class I molecules. Curr Opin Immunol 6: 13–23.

Falk K, Rötzschke O (1993) Consensus-motifs and peptidel ligands of MHC class I molecules. Semin Immunol 5: 81–94.

Falk K, Rötzschke O, Rammensee H-G (1990) Cellular peptide composition governed by major histocompatibility complex class I molecules. Nature 348: 248–251.

Falk K, Rötzschke O, Deres K, Metzger J, Jung G, Rammensee H-G (1991a) Identification of naturally processed viral nonapeptides allows their quantification in infected cells and suggests an allele-specific T cell epitope forecast. J Exp Med 174: 425–434.

Falk K, Rötzschke O, Stevanović S, Jung G, Rammensee H-G (1991b) Allele-specific motifs revealed by sequencing of self-peptides eluted from MHC molecules. Nature 351: 290–296.

Falk K, Rötzschke O, Stevanović S, Jung G, Rammensee H-G (1994) Pool-sequencing of natural HLA-DR, DQ, and DP ligands reveals detailed peptide motifs, constraints of processing, and general rules. Immunogenetics 39: 230–242.

Germain RN, Margulies DH (1993) The biochemistry and cell biology of antigen processing and presentation. Annu Rev Immunol 11: 403–450.

Gorga JC, Foran J, Burakoff SL, Strominger JL (1984) Use of the HLA-DR antigens incorporated into liposomes to generate HLA-DR specific cytotoxic T lymphocytes. Methods Enzymol 108: 607–613.

Griem P, Wallny HJ, Falk K, et al (1991) Uneven tissue distribution of minor histocompatibility proteins versus peptides is caused by MHC expression. Cell 65: 633–640.

Hammer J, Takacs B, Sinigagla F (1992) Identification of a motif for HLA DR1 binding peptides using M13 display libraries. J Exp Med 176: 1007–1013.

Hunt DF, Michel H, Dickinson TA, et al (1992) Peptides presented to the immune system by the murine class II major histocompatibility complex molecule I-A^d. Science 256: 1817–1820.

Jardetzky TS, Lane WS, Robinson RA, Madden DR, Wiley DC (1991) Identification of peptides bound to the class I MHC molecule HLA-B27. Nature 353: 326–329.

Klein J (1979) The major histocompatibility complex of the mouse. Science 203: 516–521.

Klein J, Figueroa F (1986) Evolution of the major histocompatibility complex. Crit Rev Immunol 6: 295–387.

Kropshofer H, Max H, Halder T, Kalbus M, Müller CA, Kalbacher H (1993) Self-peptides from four HLA-DR alleles share hydrophobic anchor residues near the NH$_2$-terminal including proline as a stop signal for trimming. J Immunol 151: 4732–4742.

Madden DR, Gorga JC, Strominger JL, Wiley DC (1991) The structure of HLA-B27 reveals nonamer self-peptides bound in extended conformation. Nature 353: 321–325.

Malcherek G, Falk K, Rötzschke O, et al (1993) Natural ligand motifs of two HLA molecules associated with myasthenia gravis. Int Immunol 5: 1229–1237.

Nelson CA, Roof RW, McCourt DW, Unanue ER (1992) Identification of the naturally processed form of hen egg white lysozyme bound to the murine major histocompatibility complex class II molecule I-A^k. Proc Natl Acad Sci USA 89: 7380–7383.

Pamer EG, Harty JT, Bevan MJ (1992) Precise prediction of a dominant class I MHC-restricted epitope of *Listeria monocytogenes*. Nature 353: 852–855.

Panina-Bordignon P, Tan A, Termijtelen A, Demotz S, Corradin G, Lanzavecchia A (1989) Universally immunogenic T cell epitopes: promiscuous binding to human MHC class II and promiscuous recognition by T cells. Eur J Immunol 19: 2237–2242.

Rötzschke O, Falk K (1991) Naturally occurring peptide antigens derived from the MHC class I-restricted processing pathway. Immunol Today 12: 447–455.

Rötzschke O, Falk K, Wallny HJ, Faath S, Rammensee H-G (1990a) Characterization of naturally occurring minor histocompatibility peptides including H-4 and H-Y. Science 249: 1587–1589.

Rötzschke O, Falk K, Deres K, et al (1990b) Isolation and analysis of naturally processed viral peptides as recognized by cytotoxic T cells. Nature 348: 252–254.

Rötzschke O, Falk K, Faath S, Rammensee H-G (1991) On the nature of peptides involved in alloreactivity. J Exp Med 174: 1059–1071.

Rötzschke O, Falk K, Stevanović S, Jung G, Walden P, Rammensee H-G (1991b) Exact prediction of a natural T cell epitope. Eur J Immunol 21: 2891–2894.

Rötzschke O, Falk K, Stevanović S, et al (1993) Qa-2 molecules are peptide receptors of higher stringency than ordinary class I molecules. Nature 361: 642–644.

Rudensky AY, Preston-Hurlburt P, Hong S-C, Barlow A, Janeway CA Jr (1991) Sequence analysis of peptides bound to MHC class II molecules. Nature 353: 622–627.

Sekimata M, Griem P, Egawa K, Rammensee H-G, Takiguchi M (1992) Isolation of human minor histocompatibility peptides. Int Immunol 4: 301–304.

Sinigaglia E, Guttinger M, Kilgus J, et al (1988) A malaria T-cell epitope recognized in association with most mouse and human MHC class II molecules. Nature 336: 778–780.

Stern LJ, Brown JH, Jardetzky TS, Wiley DC (1994) Crystal structure of the human class II MHC protein HLA-DR1 complexed with an influenza virus peptide. Nature 368: 215–221.

Udaka K, Tsomides T, Eisen HN (1992) A naturally occurring peptide recognized by alloreactive CD8$^+$ cytotoxic T lymphocytes in association with class I MHC protein. Cell 69: 989–998.

Vignali DAA, Urban RG, Chicz RM, Strominger JL (1993) Minute quantities of a single immunodominant foreign epitope are presented as large sets by major histocompatibility complex class II molecules. Eur J Immunol 23: 1602–1607.

Wallny HJ, Deres K, Faath S, et al (1992) Identification and quantification of a naturally presented peptide as recognized by cytotoxic T lymphocytes specific for an immunogenic tumor variant. Int Immunol 4: 1085–1090.

Wölfel T, Schneider J, Meyer zum Büschenfelde K-H, Rammensee H-G, Rötzschke O, Falk K (1994) Isolation of naturally processed peptides recognized by cytotoxic T lymphocytes (OTL) on human melanoma cells in association with HLA-A2.1. Int J Cancer 57: 413–418.

Isolation of MHC-restricted peptides

Multiple sequence analysis of MHC ligands

9.3

Stefan Stevanović

Deutsches Krebsforschungszentrum, Heidelberg, Germany

TABLE OF CONTENTS

Immunology Methods Manual
ISBN 0–12–442712–X

Abstract

The quickest way to gain comprehensive information on natural ligands bound by MHC class I or II molecules is sequence analysis of the whole peptide pool eluted from those presenting proteins. This chapter introduces multiple sequence analysis as a means of revealing most of the secrets of allele-specific peptide presentation.

Introduction

Characterization of MHC ligands faces two obstacles: On the one hand, the amounts of individual ligands are very low, so for successful analysis several billions of cells are needed. On the other hand, peptides bound to one type of MHC molecule show very similar properties, and it is tedious work to separate thousands of different ligands. Both problems are overcome by pool sequencing. Analysis of the complete mixture of MHC-extracted peptides by automated Edman degradation reveals characteristics common to most peptides within the pool, usually without problems concerning the size of the sample. Compared to analyzing individual natural ligands, only one sequencing experiment is needed to provide comprehensive information. Compared to binding assays with amino acid replacements in synthetic peptides, the results of natural ligand pool sequencing include features of processing and transport of MHC-presented peptides.

Materials

For multiple sequence analysis of MHC ligands, only very few, very expensive devices are needed: a protein sequencer able to analyze peptides at least in the low picomole range, and access to a computer executing programs suited for performing (even multiple) searches in protein or nucleotide databases. The most widely used protein sequencers are provided by Applied Biosystems, Beckman Instruments, and Hewlett Packard (addresses are listed below); and the software package of the GCG (Genetics Computer Group, Devereux et al 1984) is indispensable for database searching. All data presented in this report were generated using Applied Biosystems 476A or 477A protein sequencers (standard chemicals and protocols) and the HUSAR program for database screening, developed in the German Cancer Research Center and based upon the GCG software package.

Suppliers

Applied Biosystems Division of Perkin Elmer, 850 Lincoln Center Drive, Foster City, CA 94404, USA

Beckman Instruments, Inc., 2500 Harbor Boulevard, E-26-C, Fullerton, CA 92634-3100, USA

Hewlett-Packard Company, 1601 California Avenue, Palo Alto, CA 94304, USA

Multiple sequence analysis: Opportunities and limitations

When pool sequencing of MHC I ligands was first reported (Falk et al 1991), the details of multiple sequence analysis had not yet been elaborated. MHC I-extracted peptide pools were then judged by the appearance and disappearance of amino acids at certain positions within the peptides making up the pool. Meanwhile, detailed descriptions of multiple sequence analysis (e.g., Stevanović and Jung 1993; Metzger et al 1995) facilitated the exact quantification of every amino acid detected within a sequencing experiment. Multiple sequence analysis takes care of

- different recoveries of phenylthiohydantoin (PTH)-amino acids under the conditions of automated Edman degradation (compensated by the use of correction factors);
- background effects, by analyzing control pools and individual peptides;
- sample loss during sequencing (calculated as 'repetitive yield', which is the part of the sample left after one cycle of sequencing);
- lag effects, observed as carry-over of amino acid signals in subsequent cycles;
- efficient recognition of signals by an optimal setting of thresholds.

Automated Edman degradation, however, still has its limitations. Cysteine is not detectable without a modification procedure, and tryptophan is very unstable and therefore also often undetectable. All other amino acids appear with different but rather stable recoveries, allowing the use of correction factors. The determination of correction factors has been described in detail previously (Metzger et al 1994); they differ between individual sequencers and may change slightly after several months because different batches of chemicals or HPLC columns and gradients are used for the separation of PTH-amino acids. Table 9.3.1 shows the correction factors applied to the experiments described in this chapter.

The lag effect represents a major problem in the analysis of peptide pools. Tailing of signals in subsequent cycles makes it difficult to identify positions where certain amino acids are not allowed, and weak signals might be hidden under the lag of strong signals. Lag effects cannot be corrected, because they vary between different amino acids, different amounts of sample present, and different positions within the sequencing experiment.

Identifying the C-terminal amino acid of a peptide or a peptide pool is a challenging task in sequence analysis. Owing to problems of sample fixation (when just one amino acid is left upon the carrying matrix in the reaction chamber), signals of those amino acids are much lower than expected. Nevertheless, in most cases of individual peptide sequencing and of MHC I ligand pool sequencing the C-terminal site can be successfully analyzed. In MHC II ligand pools, consisting of peptides of different lengths, no clear-cut C-terminal end can be seen.

Table 9.3.1 Recoveries and correction factors of 19 PTH-amino acids (AA) after automated Edman degradation in two protein sequencers.

477A			476A		
PTH-AA	Recovery (average) (%)	Correction factor	PTH-AA	Recovery (average) (%)	Correction factor
A	71.2	1.40	A	90.8	1.10
D	39.1	2.56	D	71.6	1.40
E	47.2	2.12	E	57.3	1.75
F	93.2	1.07	F	97.5	1.03
G	64.7	1.55	G	83.1	1.20
H	11.6	8.62	H	43.6	2.29
I	99.9	1.00	I	97.7	1.02
K	89.6	1.12	K	78.9	1.27
L	94.6	1.06	L	96.2	1.04
M	102.0	0.98	M	70.4	1.42
N	46.2	2.16	N	63.7	1.57
P	66.8	1.50	P	85.4	1.17
Q	44.0	2.27	Q	73.2	1.37
R	21.1	4.74	R	63.3	1.58
S	24.7	4.05	S	26.9	3.72
T	47.1	2.12	T	39.0	2.56
V	95.7	1.04	V	96.3	1.04
W	21.4	4.67	W	21.2	4.72
Y	69.7	1.43	Y	76.2	1.31

Multiple sequence analysis: Practice

The most comprehensive account of sequencing results is an amino acid table, listing the picomole amounts of every PTH-amino acid in each cycle of Edman degradation. But beware! Automated calibration/calculation procedures of the sequencer's software have to be checked very carefully for correctness. In most HPLC chromatograms, the calculated baseline used for quantification of PTH-amino acids does not fit exactly with the real baseline. In the low picomole range in particular, such errors may strongly enhance or reduce the true amount of amino acids. Thus, the quantification of every single peak has to be confirmed by the operator.

Once the amino acid table is perfected, the signals have to be highlighted. Signal definition may use the rate of increase in picomoles of an amino acid compared to the previous cycle, or the highest amount present in one distinct cycle. Thus, signals are easily identified when analyzing individual peptides, since the two criteria usually coincide. Multiple sequence analysis considers only the increase in amount compared to the two preceding cycles for signal definition. In pool sequencing experiments, signals are classified as follows:

- **Strong**, if there is an increase in amount of more than 100% compared to the lesser amount of the two previous cycles.
- **Intermediate**, if the increase is more than 50%.
- **Weak**, if the increase is between 20% and 50%.

By considering only the rate of increase, the use of correction factors can be avoided. To avoid mistakes, however, the absolute amounts must be considered too. If there is, for example, an increase of alanine from 0.5 to 1.2 pmol, this effect is probably caused by background noise. If serine (a low responder with recoveries of about 25% compared to alanine) shows the same increase, this might be significant!

A particular problem in pool sequence analysis is posed by plateaus, where amino acids are detected in rather high amounts over several sequencing cycles. The signal definition does not work in these cases, and only the first position of such a plateau would be highlighted. For this reason, a rule was established for post-signal positions which proved very helpful in the analysis of peptide pools: Positions following a signal are defined as a signal of the same quality if the amount is at least 90% of the signal's amount ('plateau rule'). For an example, see the aspartate in positions 3–5 of the B*39011 ligand pool (Table 9.3.2).

For a first survey it may be useful to plot the pmole profiles of all amino acids as in Figs 9.3.1 and 9.3.2. Such diagrams make it easy to detect minima. maxima, and plateaus. The details of how to determine a peptide motif or how to call three sequences out of one sequencing experiment are described in the paragraphs below.

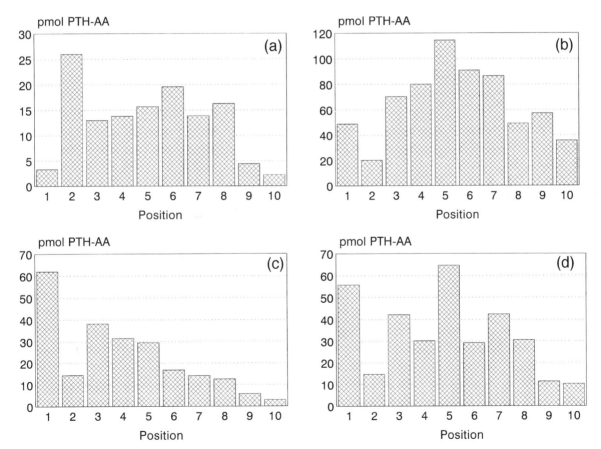

Figure 9.3.1 Multiple sequence analysis of B*39011-extracted peptides. The pmole profiles of arginine (a), leucine (b), serine (c), and tyrosine (d) represent examples for anchor (a, b) and nonanchor (c, d) residues. Note the low amounts in position 2 (b, c, d) according to the anchor rule (see text), and the decrease in amount from position 8 to 9 (a, c, d) compared to the increase in position 9 (b).

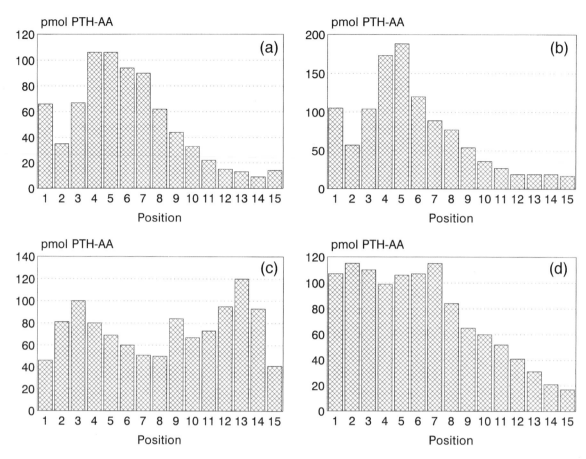

Figure 9.3.2 Multiple sequence analysis of DRB1*0405-extracted peptides. The pmole profiles of phenylalanine (a) and tyrosine (b) demonstrate the clustering of signals for the first anchor, P1, in positions 4/5 of the sequencing experiment. Aspartic acid (c) appears as a cluster in positions 12/13, corresponding to anchor position P9, while leucine (d) shows no preference for any position.

Pool sequencing of MHC I ligands

In Table 9.3.2, pool sequencing data of HLA-B*39011-extracted peptides are shown. Note that the highlighting of signals is not strictly according to the general rules described above. For MHC I ligand pools, there is an additional point of signal definition, called the 'anchor rule'. This says that anchor positions are disregarded for the rate-of-increase determina-tion. In our example, cycle 2 represents an anchor position of B*39011 ligands, occupied by histidine and arginine. Most other amino acids show a drastic decrease in cycle 2, as can be seen in Fig. 9.3.1 (see also anchor definition below). Taking the pmole values in cycle 2 for comparison, almost every amino acid in cycle 3 had to be defined as a signal. By

Table 9.3.2 Multiple sequence analysis of the natural ligand pool eluted from HLA-B*39011 molecules.

Cycle	I Ile	V Val	L Leu	F Phe	M Met	A Ala	G Gly	T Thr	S Ser	Y Tyr	P Pro	H His	Q Gln	N Asn	E Glu	D Asp	K Lys	R Arg
1	76.3	43.3	48.9	68.7	117.0	111.3	46.8	118.3	62.0	55.8	6.2	0.1	182.1	54.7	89.3	0.0	11.3	3.3
2	27.9	19.5	20.1	26.7	13.9	46.8	28.1	16.1	14.4	14.7	**15.7**	**27.6**	119.3	9.2	29.9	1.9	13.2	**26.0**
3	95.5	59.9	70.4	91.8	54.9	181.2	48.7	28.4	38.2	42.2	10.2	**24.2**	61.9	5.9	30.3	82.3	2.5	13.0
4	82.3	66.5	80.0	59.6	42.0	160.5	145.3	26.9	31.5	30.2	**114.9**	**24.1**	72.0	6.7	**90.1**	90.6	41.7	13.8
5	116.6	191.4	114.4	52.7	32.6	117.0	119.6	64.6	29.5	64.7	86.8	19.6	54.9	7.4	71.9	82.2	26.7	15.7
6	175.0	172.0	90.8	38.5	28.9	77.9	60.4	29.2	16.8	29.2	56.4	13.9	38.9	11.8	49.2	59.0	18.8	19.6
7	112.5	132.0	86.4	37.5	23.7	53.4	38.5	25.4	14.2	42.4	32.2	11.3	29.3	17.1	40.3	32.3	9.6	13.9
8	56.1	104.7	49.2	15.7	20.3	45.6	21.8	24.1	12.6	30.6	17.7	0.3	24.9	8.0	44.1	26.9	15.1	16.3
9	25.6	38.3	57.2	7.3	7.5	18.5	11.3	9.5	5.8	11.4	6.9	1.3	9.0	2.8	17.7	14.3	2.7	4.4
10	13.9	19.2	35.6	4.0	4.5	10.7	7.8	6.6	3.0	10.3	4.2	0.0	3.8	1.3	8.4	8.6	0.7	2.2

Table 9.3.3 The most prominent signals of the HLA-B*39011 ligand pool.

AA (cycles compared	Rate of increase	AA$^{pos.}$	pmol (uncorrected)	AA$^{pos.}$	pmol (corrected)
H (2→1)	276.0	V^5	191.4	A^3	253.7
D (3→2)	43.3	A^3	181.2	S^1	251.1
K (4→3)	16.7	I^6	175.0	T^1	250.8
P (4→3)	11.3	V^6	172.0	H^2	237.9
R (2→1)	7.9	A^4	160.5	D^4	231.9
V (5→3)	3.2	G^4	145.3	G^4	225.2
G (4→3)	3.0	I^5	116.6	A^4	224.7
E (4→3)	3.0	P^4	114.9	D^3	210.7

applying the anchor rule, the amounts of PTH-amino acids in position 3 are compared to the corresponding amounts in position 1, and only seven signals are left in position 3.

To get an overview of the most prominent signals, the highest rates of increase and the highest pmole amounts are listed (Table 9.3.3). In order not to miss amino acids yielding low recoveries in Edman degradation, the list has to comprise corrected data (correction factors: histidine 8.62; proline 1.50, valine 1.04; see Table 9.3.1).

For the determination of a common motif of MHC I-extracted peptides, the signals are classified as follows:

- **Anchor** amino acids always cause strong signals (i.e., highlighted in boldface and underlined). In addition, they represent a small set of closely related amino acids predominating in positions where most other amino acids show a drop in the pmole amount. In our example, cycle 2 is such an anchor position: the basic amino acids histidine and arginine show strong increases (accompanied only by a strong signal of proline). The pmole values of most other amino acids collapse in cycle 2; for seven amino acids cycle 2 represents the absolute minimum between positions 1 and 8. This increase/decrease pattern is essential for the evaluation of anchor positions, otherwise asparagine as a signal standing alone in position 7 could easily be misinterpreted as an anchor residue.

However, the characteristic collapse in the pmole profile observed in anchor positions is not seen in position 7.

- **Auxiliary** anchors are defined in positions with several related amino acids predominating, but without a drastic decrease of other amino acids, and with the accompaniment of unrelated ones. In our example, position 6 shows a clear preference for isoleucine and valine; both show strong signals, incorporating by far the highest amounts detected in this position. These two hydrophobic amino acids are not anchor residues because they are accompanied by signals of asparagine and arginine, but, more importantly, position 6 represents a minimum for none of the other amino acids.

- **Preferred** residues are identified by outstanding signals in positions where many other amino acids are also present. For B*39041 ligands, P^4, V^5, or A^3 are preferred residues in the respective positions.

Leucine in position 9 (boxed) does not fulfill any of the requirements necessary for signal definition. On the other hand, the slight increase of leucine has to be compared to the drastic breakdown of all other amino acids, which is typical for anchor positions (also seen in Fig. 9.3.1). This observation, together with the knowledge of many anchors in C-terminal positions of MHC I ligands, leads to the inclusion of L^9 into the motif.

The B*39011 ligand motif

Position	1 2 3 4 5 6 7 8 9
Anchor (bold) or auxiliary	**R** I **L**
anchor residues	**H** V
Preferred residues	P A D D N N K
	D G V R
	P T
	K Y
	E E

This example of multiple sequence analysis of B*39011 ligands was chosen because of several features:

- The anchor position 2 is occupied by amino acids with rather low recoveries in Edman degradation — histidine and arginine. Nevertheless, those anchor amino acids are easily identified by the use of correction factors and by watching the pmole profile of all amino acids.
- The C-terminal anchor would have been missed by considering only signals without an analysis of the pmole profiles.
- Glutamine shows the highest pmole amounts present in positions 1 and 2. This observation, however, must not be misinterpreted. Although glutamine values are initially very high; they are steadily decreasing and there is not one signal within this column. Therefore, glutamine is not considered significant throughout the experiment.

Pool sequencing of MHC II ligands

In contrast to MHC I ligands, class II-bound peptides are not of uniform length. This is true for the C-terminal and the N-terminal parts. The existence of truncation variants with different numbers of amino acids between the N-terminus and the first anchor residue is reflected in Edman degradation by a clustering of signals. Because the distance between N-terminal residue and anchor position one (P1) is indeed rather similar in the majority of MHC II ligands, the signal clusters usually extend over two to three sequencing cycles. Most DR1 ligands, for example, carry the amino acid which serves as the first anchor (Rammensee et al 1995) in absolute position 4 or 5 (six of nine natural DR1 ligands reported so far use position 5, two use position 4 as P1 anchor. P1 anchor residues are indicated in bold):

```
V G S D W R F L R G Y H Q Y A
V G S D W R F L R G Y H Q Y A Y D G
V G S D W R F L R G Y H Q Y
  G S D W R F L R G Y H Q Y A
    S D W R F L R G Y H Q Y A
I P A D L R I I S A N G C G
  R V E Y H F L S P Y V S P K E S P
Y K H T L N Q I D S V K V W P R R P T
A I L E F R A M A Q F S R K T D
```

Analyzing signal clusters is the specific feature of MHC II ligand pool sequencing, but signal clusters are advantageous in another way. Since the frequency of individual ligands is quite different between MHC I and MHC II receptors, some predominant peptides in class II ligand preparations may distort the results. While individual MHC I ligands very rarely make up more than 5% of the whole peptide amount, some special class II-associated peptides are presented by more than 20% of the corresponding MHC II molecules. Those sequences, however, are identified by their sharp signals in single sequencing positions, and therefore can be distinguished from the pool's signals.

Table 9.3.4 shows the amino acid table of DRB1*0405 ligand pool sequencing (T. Friede et al, 1996); only strong and intermediate signals are highlighted. Most striking is the appearance of many signal clusters in positions 1–8 while only few signal clusters are observed in later positions. Regarding first signals beyond position 8, only three clusters are found: asparagine$^{9-10}$, glutamate$^{11-13}$, and arginine$^{8-9}$, the latter belonging to a cluster extending over positions 2–9. Looking more closely, the glutamate cluster is accompanied by an aspartate signal pair (weak signal in position 12, intermediate in 13). Out of the many signals in positions 1–8, those representing signal pairs of comparable height are considered to be more significant (such as phenylalanine$^{4-5}$, tryptophan$^{4-5}$, glutamate$^{2-3}$) than those extended over more than three cycles (such as histidine$^{2-5}$, glutamine$^{2-5}$, or arginine$^{2-9}$). Once again, the pmole profile (Fig. 9.3.2) shows very clearly the preference for aromatic residues in positions 4/5 and for acidic residues in positions 12/13, while leucine represents an example of amino acids without preference for a distinct position. Comparing the properties of the amino acids in terms of possible interactions with the MHC residues, the essence of the sequencing results is as follows:

- Positions 2/3: a very high proline signal, probably as a result of processing events (Falk et al 1994), and detected in most MHC II ligand pools
- Positions 2/3: preference for polar residues, especially glutamate and aspartate
- Positions 4/5: predominance of the aromatic residues phenylalanine and tyrosine
- Positions 5–8: aliphatic residues are preferred, with slight differences between isoleucine (positions 5–7), methionine,

alanine (both positions 6–7), and valine (very high amounts in positions 7–8)
- Positions 9/10: small, hydrophilic residues like asparagine, serine, threonine, glycine, and aspartate
- Positions 12/13: preference for the negatively charged amino acids aspartate and glutamate.

Single signals did not influence analysis of this pool. Before the 'natural ligand motif of HLA-DRB1*0405-presented peptides' is defined, some of the pool's features must be discussed. Leucine shows no increase in amount over positions 1–7, therefore escaping the signal definition. The rather high amount of leucine, however, could lead to the assumption that this residue is favored in each of these positions. On the other hand, leucine (like alanine), belongs to the most common amino acids, so that considerable amounts have to be expected only for statistical reasons. In summary, leucine may be present in positions 1–7 of DRB1*0405 ligands, but it does not contribute to allele-specific binding. The same is true for arginine, which, as a special problem in automated Edman degradation, shows a very high lag-effect (i.e., tailing in subsequent cycles). Thus, the steadily increasing amounts of the arginine signal cluster in positions 2–9 are caused by a cumulative effect, leading to the same summary as for leucine.

To define the ligand motif of any MHC class II molecule, the first anchor is set to relative position 1 (P1, in our case representing cycles 4/5 of Edman degradation), and all other anchors are referred to this position.

The DRB1*0405 ligand motif

Relative position	–3	–2	–1	1	2	3	4	5	6	7	8	9	10
Anchors (bold) or preferred residues				**Y** **F**			**V** **A** I M		**N** **S** T G		**E** **D**		

It is important to recognize that MHC II ligand motif definition, if based only upon one experimental method, is a risky undertaking. Although pool sequencing is able to reveal all the characteristics of an MHC II ligand mixture,

Table 9.3.4 Multiple sequence analysis of the natural ligand pool eluted from HLA-DRB1*0405 molecules.

Cycle	I Ile	V Val	L Leu	F Phe	W Trp	M Met	A Ala	G Gly	T Thr	S Ser	Y Tyr	P Pro	H His	Q Gln	N Asn	E Glu	D Asp	K Lys	R Arg
1	85.5	106.1	106.5	66.1	4.3	13.4	110.6	91.0	35.0	34.0	104.5	13.1	1.9	12.4	12.5	55.8	46.0	40.1	12.1
2	41.2	85.3	114.7	35.4	5.1	11.6	141.9	73.0	29.4	36.9	57.3	**275.4**	**14.4**	**33.8**	**31.5**	**97.5**	**81.3**	48.8	**30.8**
3	46.4	73.2	109.7	**66.8**	4.5	**30.6**	104.1	72.8	31.7	29.3	**103.5**	**184.7**	**17.8**	**47.5**	**38.3**	**96.6**	**100.4**	52.5	**39.5**
4	54.8	71.5	98.9	**105.7**	6.2	**23.8**	63.6	62.9	39.2	28.2	**173.0**	99.2	**18.6**	**68.0**	28.6	85.4	79.8	50.8	**42.7**
5	**90.1**	83.0	106.3	**106.1**	5.1	21.5	61.5	43.1	43.0	19.0	**187.5**	70.1	**20.0**	**78.3**	29.6	66.3	68.9	58.0	**45.6**
6	**97.4**	**117.4**	106.8	93.8	5.8	**38.5**	86.2	32.0	36.0	27.7	120.0	56.7	18.4	69.3	29.9	71.3	59.7	55.2	**53.6**
7	**95.0**	**159.7**	114.5	89.5	5.0	**40.8**	**119.2**	29.0	30.2	23.8	88.7	53.5	15.5	44.4	33.8	46.3	50.8	42.0	**57.4**
8	77.4	**199.0**	84.4	61.8	2.5	26.9	81.7	30.2	29.0	23.5	77.1	61.7	10.3	30.2	32.6	28.0	49.6	36.7	**58.8**
9	57.2	164.6	64.9	43.8	**4.5**	29.3	71.0	35.6	37.7	28.1	53.9	64.8	**19.4**	39.9	**57.3**	28.6	**83.9**	28.3	**63.4**
10	38.5	133.6	60.0	33.4	**4.2**	17.3	76.4	39.7	35.4	29.5	35.8	62.6	12.2	32.8	**72.5**	22.5	67.3	26.0	49.3
11	32.7	133.9	51.9	21.6	3.8	22.0	86.5	37.5	29.5	22.5	27.2	64.4	9.5	40.0	51.4	**32.9**	73.3	27.9	48.3
12	28.0	91.0	41.1	14.7	2.4	13.6	58.3	27.8	19.8	16.8	18.7	59.9	8.3	39.8	36.9	**48.6**	95.3	25.1	49.2
13	23.0	64.9	30.6	12.5	2.7	10.5	42.5	29.4	17.7	17.7	18.8	49.8	9.1	36.8	24.5	**45.8**	**120.3**	26.7	56.3
14	16.8	40.7	20.7	8.6	1.0	6.3	34.9	23.1	14.8	15.9	16.9	33.2	7.2	34.1	28.6	34.3	92.5	24.3	38.6
15	14.5	30.2	16.9	**13.6**	**3.5**	4.7	23.4	xxx	xxx	xxx	13.2	23.5	3.2	13.0	xxx	16.9	40.7	17.9	xxx
16	9.1	20.3	10.2	6.9	1.8	3.4	20.9	21.6	11.1	10.1	13.1	16.9	3.4	14.2	23.1	19.8	50.0	18.2	18.7

analysis of individual ligands and their alignment to the motif, results of peptide binding assays, and molecular modeling according to the MHC crystal structure are important additional tools which complete the picture. On the other hand, determination of class II motifs without pool sequencing is extremely labor-intensive task, typically relying on a large number of synthetic peptides and amino acid replacements.

Sequence analysis of heterogenous HPLC fractions

Usually, the last step in MHC ligand isolation is HPLC separation. Although HPLC systems are able to separate more than hundred peptides during a single run, the very similar properties of MHC ligands (class I ligands as well as class II ligands) prevent purification of MHC-bound peptides to homogeneity. Very often, multiple peptides are hidden under a single peak. A combination of careful sequence analysis and multiple database searching makes it possible to identify two or even three peptide sequences simultaneously.

For the analysis of individual sequences, it is advisable to use the amount of increase instead of the rate of increase for signal definition. This is because the background is more stable than in complex peptide pools, and the amount of each peptide present in the mixture can more easily be estimated.

As an example, Table 9.3.5 shows the amino acid table obtained after sequencing of HPLC peak of an MHC II (HLA-DRw52b, K. Falk et al unpublished data) ligand preparation. In this experiment, signals were highlighted if increases ≥ 1 pmol were seen, or if the absolute amount was ≥ 5 pmol, with the plateau rule applied. The presence of three signals in most cycles indicated a mixture of three peptides; the signal strength indicated amounts of 5–10 pmol for each of the peptides. Because there were two or four signals in the first four cycles, all signals from position 5 to position 11

were chosen for multiple searches using the FINDPATTERNS program: the combination (I,RS)(Y,L,N)(A,K,N)(R,A)(A,E,I)(E,G,K)(V,Y,K) was compared with two large protein databases, SWISS-PROT and PIR. While 78 sequence stretches of the SWISS-PROT database (12 of human origin) matched positions 5–11 of the analyzed peptide mixture, 174 PIR entries (49 of human proteins) matched. After comparing the signals of positions 1–4 and 12–15 to the database hits, exactly three sequences were left. The only signals not assigned to any protein sequence were proline[1], proline[2], and tyrosine[5]. Further analysis completed the data:

- RNFERNKAIKVI (EBV-induced receptor, 254–265) ~12 pmol, repetitive yield 97.5%.
- VTRYIYNREEYARF (MHC II β, 59–72) ~5 pmol, repetitive yield 97.8%.
- KVHGSLARAGKVRGQ... (ribosomal protein S30, 1→...) ~9 pmol, repetitive yield 95.6%.

The last sequence represents the N-terminal part of a ribosomal protein. Such DNA- or RNA-binding proteins are sometimes copurified with MHC ligands, probably because their basic nature allows a rather strong interaction with the antibody used for immunoprecipitation.

Table 9.3.5 Sequence analysis of the peptides hidden under one HPLC-peak of an MHC II ligand preparation.

Cycle	I Ile	V Val	L Leu	F Phe	A Ala	G Gly	T Thr	S Ser	Y Tyr	P Pro	H His	Q Gln	N Asn	E Glu	K Lys	R Arg
1	2.06	**8.52**	2.45	1.38	3.07	4.56	1.91	1.96	0.86	**7.12**	0.00	1.58	0.72	1.33	**6.21**	**7.27**
2	1.18	**9.89**	1.42	1.48	3.33	3.57	**3.38**	0.86	0.43	**5.11**	0.29	1.40	**6.11**	0.94	1.23	1.33
3	1.80	2.82	1.56	**11.83**	2.39	3.42	1.19	0.55	0.86	3.46	**3.10**	2.06	2.06	1.08	1.46	**2.70**
4	2.66	2.09	1.16	2.22	0.86	2.80	0.00	0.00	**3.19**	2.89	0.11	0.94	0.42	**2.39**	0.95	0.04
5	**6.81**	1.12	1.64	2.27	1.42	2.39	0.94	**1.82**	**3.28**	1.05	0.74	1.16	1.16	1.65	0.65	**5.62**
6	2.90	1.59	**8.49**	1.28	1.31	2.05	0.49	0.00	**5.73**	0.92	0.49	1.14	**7.36**	1.18	1.03	2.69
7	1.39	1.52	1.93	0.94	**7.10**	1.77	0.36	0.50	1.92	1.20	0.57	0.92	**7.28**	0.86	**8.11**	2.12
8	0.67	1.23	1.16	0.64	**10.01**	1.53	0.49	0.00	0.88	0.97	0.49	0.73	4.49	0.61	1.70	**5.00**
9	**7.97**	1.75	0.91	0.82	**8.67**	1.52	0.36	0.00	0.67	1.45	0.37	0.65	3.33	**2.91**	1.16	3.36
10	2.42	1.98	1.06	0.57	2.77	**5.48**	0.19	0.00	0.73	2.04	0.27	0.77	1.83	**2.89**	**6.46**	1.77
11	1.18	**4.97**	1.04	0.55	1.88	2.43	0.23	0.00	**3.12**	1.95	0.37	1.05	1.17	1.02	**6.11**	1.30
12	**3.21**	**6.68**	1.84	0.78	**4.35**	1.62	0.29	0.00	0.88	1.81	0.32	0.78	1.54	0.76	1.33	1.65
13	2.30	1.19	1.25	0.54	1.88	1.62	0.21	0.00	1.38	1.08	0.28	0.88	1.29	0.49	0.72	**7.17**
14	0.75	0.68	0.74	**2.49**	1.87	**5.01**	0.20	0.00	0.71	0.90	0.34	0.64	1.12	0.51	0.70	2.61
15	0.31	0.95	0.56	0.62	0.72	0.79	0.00	0.00	1.14	**2.01**	0.05	**2.24**	0.44	0.38	0.53	1.24

References

Devereux J, Haeberli P, Smithies O (1984) A comprehensive set of sequence analysis programs for the VAX. Nucleic Acids Res 12: 387–395.

Falk K, Rötzschke O, Stevanović S, Jung G, Rammensee H-G (1991) Allele-specific motifs revealed by sequencing of self-peptides eluted from MHC molecules. Nature 351: 290–296.

Falk K, Rötzschke O, Stevanović S, Jung G, Rammensee H-G (1994) Pool sequencing of natural HLA-DR, DQ, and DP ligands reveals detailed peptide motifs, constraints of processing, and general rules. Immunogenetics 39: 230–242.

Friede T, Gnau V, Jung G, et al. (1996) Natural ligand motifs of closely related HLA-DR4 molecules predict features of rheumatoid arthritis-associated peptides. Biochem Biophys Acta, in press.

Metzger JW, Stevanović S, Brünjes J, Wiesmüller K-H, Jung G (1995) Electrospray mass spectrometry and multiple sequence analysis of synthetic peptide libraries. Methods: a companion to Methods in Enzymology 6: 425–431.

Rammensee H-G, Friede T, Stevanović S (1995) MHC ligands and peptide motifs. First listing. Imunogenetics 41: 178–228.

Stevanović S, Jung G (1993) Multiple sequence analysis. Pool sequencing of synthetic and natural peptide libraries. Anal Biochem 212: 212–220.

Use of tandem mass spectrometry for MHC ligand analysis

9.4

Ronald C. Hendrickson[1]
Jonathan C. Skipper[2]
Jeffrey Shabanowitz[1]
Craig L. Slingluff Jr.[3]
Victor H. Engelhard[2]

Donald F. Hunt[4]

[1]Department of Chemistry

[2]Department of Microbiology and Beirne Carter Center for Immunology Research

[3]Department of Surgery

[4]Department of Chemistry and Pathology, University of Virginia, Charlottesville, Virginia, USA

Supported by US Public Health grants AI20963 (to V.H.E.), AI33993 (to D.F.H.), CA57653 (to C.L.S.), and by a Cancer Research Institute postdoctoral fellowship (to J.C.S.).

TABLE OF CONTENTS

Immunology Methods Manual
ISBN 0–12–442712–X

Abstract

Electrospray ionization (ESI) tandem quadrupole mass spectrometry is capable of generating sequence information on individual peptides present in complex biologic mixtures such as peptides extracted from MHC molecules. In addition, mass spectrometry can generate information on the molecular weights of peptides in such mixtures and on the abundance. In this chapter we describe immunoaffinity purification procedures for isolating MHC-associated peptides prior to analysis by mass spectrometry. High-performance liquid chromatography separation conditions useful in purifying peptide mixtures prior to mass spectrometric analysis are also described. Sample loading techniques are discussed along with useful chemical modifications that facilitate peptide sequence interpretation.

Introduction

Cytotoxic T lymphocytes (CTL) recognize peptides which have been presented on the cell surface by major histocompatibility complex (MHC) class I molecules. Generally the antigen processing mechanism involves cytosolic degradation of endogenous proteins including proteins that result from viral infection, cellular transformation, or tissue transplantation. TAP — transporter associated with antigen processing — proteins transport the peptide fragments into the endoplasmic recticulum (ER). In the ER, peptides are loaded into MHC class I molecules and the resulting complex then moves to the cell surface. Characterization and sequencing of MHC class I-associated peptides resulting from particular disease states such as cancer, AIDS, and autoimmune disorders, is an important step towards a more complete understanding of CTL–antigen recognition and may lead to the development of vaccines and other immunotherapeutics.

The methodology for determining amino acid sequence information by tandem quadrupole mass spectrometry was described in the mid-1980s (Hunt et al 1986). Since then, improvements in instrument sensitivity, ionization efficiency, and sample handling techniques have made mass spectrometry suitable for analysis of complex mixtures like MHC-associated peptides. Mixtures of peptides extracted from the class I MHC molecule HLA-A2.1 have been estimated to contain more than 10,000 different species (Cox et al 1994). Peptide extraction and subsequent tandem mass spectrometric analysis have been used to elucidate the residue characteristics for peptides that bind to specific MHC molecules (Hunt et al 1992a, 1992b; Huczko et al 1993), to determine the complexity and length variation of peptides which result from antigen processing or presentation defects (Henderson et al 1992; Sette et al 1992), and to sequence the peptide portions of T cell epitopes (Henderson et al 1993; Cox et al 1994; Castelli et al 1995; den Haan et al 1995). Additionally, mass spectrometry can be used to identify T cell epitopes that result from post-translational modification (Skipper et al 1996). The use of mass spectrometry in biochemical research has been recently reviewed (Siuzdak 1994). The procedures presented here describe extraction and immunoaffinity purification of MHC-associated peptides and their analysis by high-performance liquid chromatography and mass spectrometry.

Preparing samples for analysis by electrospray ionization mass spectrometry: Immunoaffinity purification of MHC-associated peptides

Immunoaffinity purification of specific class I MHC–peptide complexes is achieved using monoclonal antibodies with appropriate specificity coupled to protein A–Sepharose resin. Two methods of immunoaffinity purification have been developed. In the first method, the affinity resin is added directly to a detergent-solubilized cell lysate (immunoprecipitation). The second method, which we estimate to be three times more efficient, involves passing the cell lysate over a column containing the resin. Two protocols for the construction of immunoaffinity columns are described in the next section.

In protocol 1 antibodies that have a high affinity for protein A are used for the construction of immunoaffinity columns (Langone et al 1978). Antibodies that have low or negligible affinity for protein A are covalently linked to a solid matrix (protocol 2). Periodate oxidation of antibody carbohydrate moieties produces aldehyde groups that react with immobilized hydrazides on agarose beads to form stable covalent hydrazone linkages. It is believed that antibody combining sites are oriented away from the matrix since the Fc region carbohydrate is linked to the resin. One might also consider this protocol for antibodies that have a high affinity for protein A if problems associated with blocked combining sites are encountered.

Generally we use approximately 5 mg of murine monoclonal antibody to purify 500–700 µg of class I heavy chain from 1×10^{10} lymphoblastoid cells with an estimated copy number of 0.8–1×10^6 MHC molecules per cell. Some adjustment of quantities might be necessary since the amount of expressed class I differs among cell lines and antibody affinities vary.

This method is based on the selective interaction of an antibody with the peptide–MHC complex.

Construction of immunoaffinity columns

Materials and apparatus

Columns (Econo column 1.5 cm diameter, 10 cm long, BioRad Laboratories, Hercules, CA, USA)

Spectrophotometer

Protocol 1: Construction of an immunoaffinity column with antibodies that bind well to protein A

For purification of up to 700 μg of class I MHC molecules, use a 5 ml column.

Chemicals and solutions

50 mM Tris–HCl, pH 8.0, 150 mM NaCl (Tris-buffered saline) at 4°C

Protein A–Sepharose beads (PAS, Sigma Chemicals), 1:1 slurry in Tris-buffered saline

Antibody specific for MHC molecule (PAS purified)

Procedure

1. Pour 10 ml of the PAS slurry into the Econo column. Allow slurry to settle briefly without flow and then drain buffer until the meniscus is at the top of the resin bed. Wash column with 5 volumes of Tris-buffered saline.
2. Dilute the antibody in 10–15 ml of Tris-buffered saline. Measure the optical density of the antibody solution on the spectrophotometer at 280 nm. Pass 15 ml of the antibody solution over the column twice. Approximately 1 mg of mouse monoclonal antibody of the IgG2a or IgG2b subclasses is bound by 1 ml of PAS. After passage, measure the optical density at 280 nm to determine the extent of antibody binding.

3. Pass 4 column-volumes of Tris-buffered saline (4°C) to wash off unbound antibody and to equilibrate the column.

Protocol 2: Construction of an immunoaffinity column with antibodies that do not bind well to protein A

Chemicals and solutions

100 mM sodium acetate, pH 5.5, 150 mM NaCl

25 mg $NaIO_4$/1.2 ml deionized water. Make immediately before use.

Hydrazide–agarose beads in isopropanol (BioRad Laboratories, Inc. Hercules, CA, USA)

5 ml of 1.5 mg ml^{-1} antibody in 100 mM sodium acetate, pH 5.5, 150 mM NaCl

100 mM sodium acetate, pH 5.5, 0.5 M NaCl

Phosphate-buffered saline (PBS): 4.3 mM $NaHPO_4$ + 1.4 mM KH_2PO_4 + 2.7 mM KCl + 140 mM NaCl

PBS with 0.5 M NaCl

PBS with 0.04% azide

Procedure

1. Add 0.5 ml of freshly prepared $NaIO_4$ solution to 5 ml of the antibody solution in a 17 × 125 mm snap-capped tube. Cover the tube with foil and turn end over end for 1 h at room temperature.
2. Concentrate the solution using an Amicon Ultrafiltration device with a YM100 membrane to remove the unreacted $NaIO_4$ from the antibody. Dilute with 100 mM sodium acetate, pH 5.5, 150 mM NaCl, and reconcentrate to the original volume several times to achieve at least a 1:1000 dilution of the $NaIO_4$.
3. De-gas the antibody solution by placing it under vacuum for 10–15 min. Prolonged storage or dialysis can

cause formation of a precipitate that must be removed before proceeding.

4. To prepare the hydrazide–agarose beads, centrifuge an aliquot of beads at less than 200*g* and remove the isopropanol by pipette. Wash the beads five times to remove the isopropanol by centrifuging with 10 volumes of 100 mM sodium acetate, pH 5.5, 150 mM NaCl.

5. Pour a 1:1 slurry of agarose beads in 100 mM sodium acetate, pH 5.5, 150 mM NaCl into an Econo column (1.5 cm diameter, 10 cm long, BioRad Laboratories) and allow the buffer to drain to the top of column bed.

6. Add the oxidized antibody solution to the top of the column and recirculate for 5–12 h using a peristaltic pump to minimize antibody precipitation from solution.

7. Wash the column with 1 volume of 100 mM sodium acetate, pH 5.5, 0.5 M NaCl.

8. Measure the optical density of the eluent at 280 nm to determine how much antibody has bound.

9. Wash the column, first with 3 volumes of PBS with 0.5 M NaCl, then with 1 volume of PBS with 0.04% azide. Transfer the beads as a 50% slurry to a new Econo column prior to use. Store at 4°C.

10. Gel electrophoresis is performed on 5–15 μl of beads removed from the column and treated with sodium dodecyl sulfate (SDS) sample buffer to confirm the amount of antibody bound based on the amount of light chain released (see step 8 of Protocol 4).

Isolation of peptides bound to MHC molecules

Materials and apparatus

Eppendorf plastic tubes (National Scientific Supply Company, San Rafael, CA, USA)

Cold room, 4°C

Ultracentrifuge

0.2 μm low-protein-binding filter (Gelman Sciences #4192)

SDS-PAGE apparatus

5000-Da cutoff filter units (Millipore Corp., Marlborough, MA, USA, UFC4LCC25)

Chemicals and solutions

Buffered detergent solution: deionized water (0.2 μm filtered and very clean containers) + 150 mM NaCl + 20 mM Tris–HCl pH 8.0, +1% CHAPS (Boehringer Mannheim, Indianapolis, IN, USA)

Phosphate-buffered saline (PBS): 4.3 mM $NaHPO_4$ + 1.4 mM KH_2PO_4 + 2.7 mM KCl + 140 mM NaCl

Protease inhibitors (stock solution concentration; ratio in buffered detergent solution (v/v)):

- 100 μM iodoacetamide (18.5 mg ml^{-1} in water (a); 1:1000)
- 5 μg ml^{-1} aprotinin (5 mg ml^{-1} in water (c); 1:1000)
- 10 μg ml^{-1} leupeptin (10 mg ml^{-1} in water (a); 1:1000)
- 10 μg ml^{-1} pepstatin A (10 mg ml^{-1} in methanol (a); 1:1000)
- 5 mM EDTA (500 mM (b); 1:100)
- 0.04% azide (20% in deionized water (b); 1:500)
- 1 mM PMSF (d); (17.4 mg ml^{-1} in isopropanol (a); 1:100)

(a) Store at −20°C. (b) Store at room

temperature. (c) Store at 4°C. (d) Add PMSF to buffered detergent solution last.

20 mm Tris, pH 8.0, 150 mm NaCl

20 mm Tris, pH 8.0, 1 m NaCl

20 mm Tris, pH 8.0

0.2 n acetic acid

10% acetic acid (1.7 n)

Glacial acetic acid

Immunoaffinity columns from protocol 1

Trouble shooting note

The mass spectrometer is a sensitive detector capable of detecting non-UV-absorbing organic and inorganic compounds. It is therefore important to use high-grade solvents, very clean glassware, and all the washing steps described in the following protocol. In addition, all filters, columns, and containers should be washed with the appropriate buffer or solvent to remove possible residual contaminants introduced in the manufacturing process.

Protocol 3: Solubilization of cellular proteins

Procedure

1. Cultured cells are harvested by centrifugation and pooled. Resuspend the cell pellet and wash twice in ice-cold PBS. Cell pellets may be quick frozen at −80°C at this point. Successive cell harvests can be pooled and extracted at one time.
2. *All subsequent steps should be conducted at 4°C.* Resuspend the fresh or frozen cell pellets in buffered detergent solution containing freshly added protease inhibitors. Do not thaw the frozen cell pellets before adding detergent. Use 1 ml of buffer per 1×10^8 cells and stir or rock gently the resulting suspension for 1 h.
3. Centrifuge at 100,000g for 1 h to remove cellular debris. Decant the supernatant containing soluble proteins. Discard the pellets.
4. To the end of a 60 ml syringe connect a 0.2 μm low-protein-binding filter (Gelman Sciences #4192). Pass the lysed cell supernatant through, changing the filter after every 10–15 ml as it becomes obstructed. One may substitute a Gelman Sciences #4187, which combines a 0.2 μm filter with a 0.8 μm prefilter, if the 0.2 μm filter clogs too quickly.

Protocol 4: Immunoaffinity isolation of class I molecules

Procedure

1. Assemble the following columns in series. The first column, 200–300 μl Sepharose CL-4B, is employed to capture any material remaining in the cell lysate which might clog subsequent antibody columns. Replace this column during the procedure if the flow rate diminishes significantly. The second column in the series should contain an irrelevant antibody that can be used to derive a negative control peptide extract. Next in the series, assemble columns containing relevant class I MHC-specific antibodies for each class I molecule being isolated, equilibrated with buffered detergent solution. Equilibrate the columns by first washing with Tris-buffered saline then with buffered detergent solution. Run the filtered lysate through this series of columns at a rate not to exceed 1 ml min^{-1}, at 4°C.
2. Separate the columns and wash each column with 2 column volumes of buffered detergent solution, 30 column volumes of 20 mm Tris, pH 8.0, 150 mm NaCl, 30 column volumes of 20 mm Tris, pH 8.0, 1.0 m NaCl, and then with 20–30 column volumes of 20 mm Tris, pH 8.0.

3. After the above washing, allow the buffer to drain to the top of the gel bed. Add one column volume of 20 mM Tris, pH 8.0, cover the top of the column with Parafilm, and resuspend the beads by gentle rotation. Remove 20–50 μl of the suspension for quantification of bound class I by sodium dodecyl sulfate–polyacrylamide gel electrophoresis (SDS-PAGE). Allow the beads to settle and the excess buffer to drain.

4. *PAS columns*: Elute with 4 volumes of 0.2 N acetic acid (pH 2.7) and collect 1.5 ml fractions in 1.7 ml Eppendorf tubes. Add 165 μl of glacial acetic acid to each fraction (final concentration 10% acetic acid), and boil for 5 min.

   Hydrazide–agarose columns: Elute with three volumes of 10% acetic acid and collect 1.5 ml fractions in 1.7 ml Eppendorf plastic tubes. Place each fraction in a boiling water bath for 5 min and then chill the tubes on ice.

5. Transfer the contents of each Eppendorf tube into the top reservoir of a prewetted 5000-Da cutoff filter unit. *Important*: Prior to use, the filter unit should be prewetted with 1 ml of 10% acetic acid and spun for 1 h, and all liquid in both reservoirs discarded. Centrifuge the filter units at 3500*g* for ~ 5 h at 4°C. Save both the filtrate and retentate at –80°C.

6. Transfer the filtrate containing the extracted peptides to 1.7 ml Eppendorf tubes and freeze the solution at –80°C.

7. Concentrate the peptide extracts to a final volume of ~100 μl/tube by vacuum centrifugation without heat. Do not allow the samples to thaw completely or to dry. Pool the concentrated peptide in one Eppendorf tube and wash each original tube with 50 μl of 10% acetic acid. Pool the washes with the concentrated peptide and vacuum centrifuge once more. Repeat as necessary until the total extract and washes have been concentrated to a final volume of approximately 250 μl. Refreeze the resulting solution at –80°C.

8. The amount of class I heavy chain present in the beads removed in step 3 is estimated by SDS-PAGE. Microcentrifuge the tube containing the beads, remove the excess liquid, add 40 μl of Laemmli sample buffer and mix the resulting suspension. Boil the suspension for 5 min and then centrifuge lightly to pellet the beads. Analyze 40 μl of the supernatant on a 12% SDS-PAGE gel; stain the bands with Coomassie blue, and compare the intensity of the class I heavy-chain band to the intensities of a series of ovalbumin standards in the range of 0.5–5 μg each.

Reversed-phase HPLC (RP-HPLC) purification of peptide mixture

The mixture of peptides associated with MHC class I molecules contains as many as 10,000 different components (Cox et al 1994). The exact procedure followed will depend on the particular experiment. An individual peptide may be sequenced directly from the mixture by mass spectrometry (Skipper et al 1996) or multiple stages of HPLC purification may be necessary to simplify this mixture prior to analysis by mass spectrometry. Different ion-pairing reagents can be used to change peptide retention behavior in order to achieve the desired separation (Mant and Hodges 1991). The different ion-pairing reagents may be used in series. For example, HFBA may be used in the first separation followed by TFA in the second dimension, or TFA in the first and HFBA in the second, etc. A third ion pairing reagent may be used in the third dimension of chromatography if necessary. The next subsections describe HPLC purification conditions that use three different ion-pairing reagents. T cell epitopes can be reconstituted by adding an aliquot of each HPLC fraction to target cells displaying appropriate MHC class I molecules. Standard ^{51}Cr-release assays are performed to identify HPLC fractions that contain antigenic peptides (Slingluff 1993).

Protocol 5 describes HPLC separation conditions for 0.1–15 nmol of peptide extract using trifluoroacetic acid (TFA). Protocol 6 describes HPLC separation conditions with heptafluorobutyric acid (HFBA) as the ion-pairing agent. HFBA, a hydrophobic counterion, can increase a peptide's affinity for the stationary phase and thus increase the peptide's retention time. This trend has been shown to increase with increasing peptide basicity. Protocol 7 describes the use of hexafluoroacetone (HFA) in HPLC separations. In separations with HFA at basic pH, the peptides show an increase in hydrophilic character and retention times decrease compared with those of TFA or HFBA.

To minimize sample loss, surfaces that come into contact with the peptide mixture should be composed of PEEK or high-pressure Tefzel® rather than stainless steel. These surfaces include the sample loop and column inlet and outlet tubing. HPLC fractions should not be collected in siliconized tubes. Organic solvents leach phthalates from the tube coating. Phthalates are cytotoxic and interfere with ^{51}Cr-release assays.

In this separation method, peptides are partitioned between a mobile liquid phase and a support-bound stationary phase.

Materials and apparatus

Micro bore HPLC (130 A, Applied Biosystems, Foster City, CA, USA)

C18 reversed-phase column (Brownlee narrowbore C18 (2.1 mm × 9 cm, 300 Å 7 μm), Applied Biosystems)

Polypropylene tubes (Sarstedt, Newton, NC, USA)

Protocol 5: RP-HPLC fractionation of 0.1 to 15 nmol of peptide

Chemicals and solutions

NANOpure water (Barnstead, available through VWR Scientific, Bridgeport, NJ, USA)

Trifluoroacetic acid (Applied Biosystems) 0.1% (v/v) in NANOpure water

HPLC-grade acetonitrile (Mallinckrodt, Paris, KY, USA), 60% (v/v) in NANOpure water + 0.085% TFA

Trouble shooting note

Contaminating proteins interfere with the detection of peptides, especially the peptides present at low abundance. Therefore, to ensure that all traces of protein have been removed in the filtration step (see step 5 of protocol 4), analyze an aliquot (1/500) of the immunoaffinity-isolated peptides by ESI mass spectrometry. If contaminating proteins are present, refilter the peptide mixture.

Note: NANOpure water is used throughout these protocols. This water is ultrapure and deionized. HPLC-grade water is usually low in UV absorption but higher in salts which may reduce the mass spectrometer's performance.

Procedure

1. Peptides are injected on to a Brownlee narrow-bore C-18 Aquapore column (2.1 mm × 9 cm, 300Å, 7 μm) and eluted with a 65 min binary gradient increasing from 0 to 100%B at the rate of 3%B min^{-1} for the first 5 min, then 0.9%B min^{-1} for the next 50 min, and finally 4%B min^{-1} for the last 10 min. (Solvent A = 0.1% TFA in NANOpure water; solvent B = 0.085% TFA and 60% acetonitrile in NANOpure water; flow rate 200 μl min^{-1}.)
2. Collect 200 μl fractions at 1 min intervals into 1.5 ml screwcap polypropylene tubes. If fractions are to be used for reconstituting T cell epitopes, vortex the solution, transfer the portion required for biological analysis to another polypropylene tube containing 10–20 μl of water, and

freeze both portions on dry ice. Otherwise, freeze the entire fraction on dry ice.

Protocol 6: RP-HPLC fractionation with heptafluorobutyric acid (HFBA) as the ion-pairing agent

Chemicals and solutions

HFBA (Pierce, Rockford, IL, USA) 0.1% (v/v) in NANOpure water

HFBA 0.1% (v/v) in 60% (v/v) HPLC-grade acetonitrile (Mallinckrodt, Paris, KY, USA) in NANOpure water

Procedure

1. Remove acetonitrile from first dimension fractions by lyophilizing the solution to 1/4 initial volume (from 200 μl to less than 50 μl).
2. Inject the sample onto a Brownlee narrow-bore C-18 Aquapore column (2.1 mm × 9 cm, 300 Å, 7 μm) and elute peptides with a 65 min binary gradient increasing from 0 to 100%B at a flow rate of 200 μl min^{-1}. The gradient used will depend on the elution time of the first dimension fraction. In general, increase solvent B at a rate of 2.5% B min^{-1} until the solvent B concentration is 15% lower than the concentration at which the first dimension fraction eluted. Use a shallower gradient of 1% B min^{-1} increase thereafter.
3. Collect 200 μl fractions at 1 min intervals into 1.5 ml screwcap polypropylene tubes. If fractions are to be used for reconstituting T cell epitopes, vortex the solution, transfer the portion required for biological analysis to another polypropylene tube containing 10–20 μl of HPLC-grade water, and freeze both portions on dry ice. Otherwise, freeze the entire fraction on dry ice.

Protocol 7: RP-HPLC separation using hexafluoroacetone (HFA) as the ion-pairing agent

Chemicals and solutions

0.1% HFA (Pierce, Rockford, IL, USA) in NANOpure water, pH adjusted to 8.1 with 14.8 M ammonium hydroxide

HPLC-grade acetonitrile (Mallinckrodt, Paris, KY, USA)

Procedure

1. Remove acetonitrile from first dimension fractions by lyophilizing to 1/4 initial volume (from 200 µl to less than 50 µl)
2. Inject the sample on to a Brownlee narrow-bore C-18 Aquapore column (2.1 mm × 9 cm, 300 Å, 7 µm) and elute peptides with a 65 min binary gradient increasing from 0 to 100%B at a flow rate of 200 µl min^{-1}. The gradient used will depend on the elution time of the first dimension fraction. In general, increase solvent B at a rate of 2.5%B min^{-1} until the solvent B concentration is 15% lower than the concentration at which the first dimension fraction eluted. Use a shallower gradient of 1%B min^{-1} increase thereafter.
3. Collect fractions at 1 min intervals into screwcapped polypropylene tubes. When fractions are to be used for reconstituting T cell epitopes, vortex the fraction, transfer the portion required for biological analysis to another polypropylene tube containing 10–20 µl of HPLC-grade water, and immediately freeze both portions on dry ice. Otherwise freeze the entire fraction on dry ice.

Peptide sequencing by liquid chromatography– tandem mass spectrometry

Microcapillary liquid chromatography combined with electrospray ionization (ESI) tandem quadrupole mass spectrometry is capable of separating and sequencing individual peptides present in complex mixtures. Samples are introduced into the instrument using a microcapillary HPLC column and converted to protonated gas phase molecules by the process of ESI (Fig. 9.4.1). Mass spectra are recorded on a Finnigan (San Jose, CA, USA) TSQ-7000 triple quadrupole mass spectrometer capable of sequencing peptides at the 5–10 fmol level (peptide loaded on to the HPLC column) and determining the mass-to-charge ratio (m/z) at a factor of 10–100 lower. The following sections describe preparation of microcapillary columns, sample loading techniques to minimize sample loss, and useful chemical modifications that aid in peptide sequence determination.

Protocol 8 describes the procedure adapted from Moseley et al (1991) for packing fused-silica capillary columns with different stationary phases. PRP-1 stationary phase is used for hydrophilic samples that are not retained by silica-based C-18 stationary phase or the polymeric-based stationary phase POROS R/H

II. Polymeric supports, owing to their perfusive nature, are less prone to clogging than silica-based materials, but column resolution drops dramatically when using acetic acid as an ionic modifier. When acetic acid is used for the ionic modifier, the best resolution is achieved with a C-18 stationary phase. Prepared columns are commercially available from LC Packing International, San Francisco, CA, USA.

Protocol 9 describes how we load an aliquot of a fraction directly on to a microcapillary column. Samples are eluted into the mass spectrometer using a gradient of 0–80% acetonitrile in 0.5% acetic acid over 12 min.

The ESI efficiency depends on the source design and the conditions used. With the present instrumentation, ESI efficiency is improved by a factor of three when acetic acid is used in place of TFA as the organic modifier. In the ESI process, protonated peptide molecules are transferred from the solution phase into the gas phase with high efficiency (Fenn et al 1989; Siuzdak 1994). For molecular weight measurements, the first quadrupole mass filter (Q1) is employed to separate ions based on their mass-to-charge ratio (m/z). Quadrupole mass filters two (Q2) and three (Q3) transfer all ions directly to the detector, a conversion

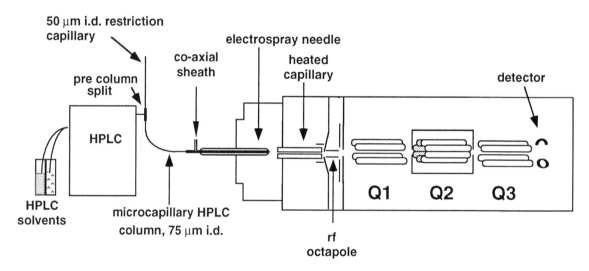

Figure 9.4.1 Schematic diagram of a Finnigen TSQ7000 triple quadrupole mass spectrometer equipped with an electrospray ionization source and a microcapillary HPLC system. An aliquot of a fraction containing peptide is loaded onto a 75 μm i.d. × 190 μm o.d. capillary packed with a 10–15 cm bed of stationary phase. Peptides are eluted with a 12 min linear gradient of 0–80%B (A = 0.1 M acetic acid; B = acetonitrile). We use an Applied Biosystems Model 140A dual syringe pump operating at 180–300 μl of solvent per min and a precolumn split to achieve an operating flow of 0.5 μl min^{-1}. The precolumn split ratio can be varied by adjusting the length of the 50 μl i.d. fused-silica restriction capillary shown. The electrospray needle is operated at a voltage potential of 4.6 kV to the heated capillary with coaxial liquid sheath flow of 1–2 μl min^{-1} of a 70% (v/v) mixture of methanol in NANOpure water plus 0.12% acetic acid. The heated capillary is operated at 150°C. Ions produced are focused into the first quadrupole (Q1) using an octapole and a pair of electrostatic lenses. In MS mode, Q1 filters ions based on their mass-to-charge ratio m/z, while Q2 and Q3 are set to pass all ions. A high-voltage conversion dynode (15 kV) electron multiplier detects the ions. The resulting spectra are recorded. In CAD mode; Q1 is set to pass ions within a 1–2 mass unit window around the peptide ion of interest into Q2, the collision chamber, which is filled with inert gas (3 mTorr argon). The peptide ions collide with the inert gas to cause a single cleavage to occur more or less randomly at the various amide bonds present in the protonated peptide molecules. The resulting fragment ions are passed into Q3, which separates the ions according to mass. To optimize the fragmentation pattern observed from ions of different charges, three different strategies are employed. The range of collision energies for $(M+H)^{+1}$, $(M+2H)^{2+}$ and $(M+3H)^{3+}$ ions are –30 to –38eV, –16 to –26eV, and –15 to –20eV, respectively. Q3 is scanned from 50 mass units to 10 mass units over the molecular mass at a rate of 500 Da s^{-1}. Q3 is operated to pass ions within a 2 amu window to increase sensitivity.

dynode electron multiplier. For the analysis of mixtures containing peptide ligands, spectra are recorded over the mass range (m/z) 300–1400 every 1.5 s. The resulting mass spectra consist of a plot of relative abundance versus m/z. If the total ion current (relative abundance) for each mass spectra is plotted as a function of time (scan number), the result is a reconstructed ion current chromatogram (Fig. 9.4.2A) which looks similar to the elution profile of a conventional HPLC trace. Individual peptides in the mixture can be located by plotting the relative ion abundance at their particular m/z value as a function of time (scan number). The resulting plot is referred to a selected ion current chromatogram (Fig. 9.4.2B). A single mass spectrum containing signals for all the peptides present in the mixture is generated by summing the individual spectrum acquired during the period when the peptides eluted into the mass spectrometer (Fig. 9.4.2C).

Peptide sequence information can be obtained from a CAD (collisional-activated dissociation) mass spectrum. In this experiment, Q1 is set to pass ions at a particular m/z range (called precursor ions) and reject all others. The selected ions enter Q2 which contains a low pressure of an inert gas (argon). The selected protonated peptide molecules (precursor ions) undergo multiple collisions with argon atoms. This process causes a single cleavage to occur more or less randomly at the various amide bonds present in the collection of protonated peptide molecules. The result is a collection of protonated fragments that differ in length by a single amino acid residue. These ions are separated according to their m/z in Q3. The resulting CAD mass spectrum is a plot of relative fragment ion abundance versus m/z. Two major types of fragment ions are observed. Ions of type y contain the C-terminus plus one or more additional residues up to the total number of amino acid residues. Likewise, ions of type b contain the N-terminus plus one or more additional residues (Roepstorff and Fohlman 1984). Subtracting m/z values for any two fragment ions differing by a single amino acid generates a value that specifies the mass (Table 9.4.1) and thus the identity of the amino acid in the larger fragment (Hunt 1986, 1991).

Fig. 9.4.3 is an example CAD spectrum with the deduced amino acid sequence.

The interpretation of data obtained in a CAD mass spectrum is often facilitated by recording additional CAD mass spectra on modified forms of the peptide. Four such modifications are described here. In Protocol 10 (esterification of free carboxylic acids), each carboxylate group is converted into the corresponding methyl or ethyl ester. Methyl esters result in a shift of 14 mass units for each free acid group, including the C-terminus, while ethyl esters result in a 28-mass-unit shift. In a CAD mass spectrum of the modified peptide, ions of type y_n can be identified since they contain the C-terminus and will therefore shift by 14 or 28 mass units. In addition, any ion that contains an acidic residue will also shift by 14 or 28 mass units. An example of a CAD mass spectrum generated on the peptide GXDEXFVQV and the same peptide converted to the corresponding methyl ester is given in Figures 9.4.3A and 9.4.3B respectively. Ions that fail to shift do not contain an acidic residue or the C-terminal residue. Likewise, ions of type b_n will not shift unless they contain an acidic residue. When sequencing peptides with charge states greater than +2 (the mass of the molecular ion plus two protons) ethyl esters are generally more useful because the modifications are easier to detect. For example, consider the case of a +3 ion (the mass of the molecular ion plus three protons). An ethyl ester modification results in an m/z shift of 9.3 mass units, whereas the same methyl ester modification would result in an m/z shift of only 4.7 mass units. The m/z shifts for oxidation of methionine, neutral loss of ammonia, or neutral loss of water would be 5.3, 5.7, and 6, respectively.

Protocol 11 describes how a single cycle of Edman degradation can be performed resulting in the removal of the N-terminal amino acid. Manual Edman degradation is effective in determining the N-terminal amino acid when the y_{n-1} ion and the b_1 ion are absent. The molecular ion and each b ion, since b ions contain the N-terminus, will decrease in mass by the mass of the residue removed from the peptide. Since the y_n ion series does not contain the N-terminus, no mass shift will be observed. An example CAD spectrum generated on the peptide GXDEXFVQV after a single cycle of Edman degradation is given in Fig 9.4.4A). This modification can be particularly helpful when the peptide of interest does not contain an acidic residue and the ion series is predominately b_n.

Protocol 12 describes on-column carboxyamidation, which is useful in verifying the presence and location of a cysteine residue within a peptide. The fourth chemical modification, protocol 13, is N-acetylation. The acetyl group causes a 42-mass-unit increase for the N-terminus as well as for each lysine and cysteine residue present in the peptide. Under mild conditions, without base catalysis, it is possible to acetylate only the N-terminus. This procedure is not commonly used for the analysis of +1 and +2 ions owing to dramatic reduction in signal intensity after acetylation

because of the reduced number of basic sites for ionization. Acetylation is useful when the peptide of interest contains a basic arginine or histidine residue, or if the charge state is greater than +2. Conversely, acetylation can reduce the charge state of multiply charged peptides and simplify sequence interpretation. An example CAD spectrum generated on the peptide GXDEXFVQV after on-column acetylation is shown in Fig. 9.4.4B. Information on up to 10 peptides can be obtained in a single HPLC run, requiring a total of 30 min of instrument time. CAD mass spectra can be generated with as little as 5–10 fmol of material to obtain peptide sequence information, but a higher sample level allows chemical modification and facilitates interpretation.

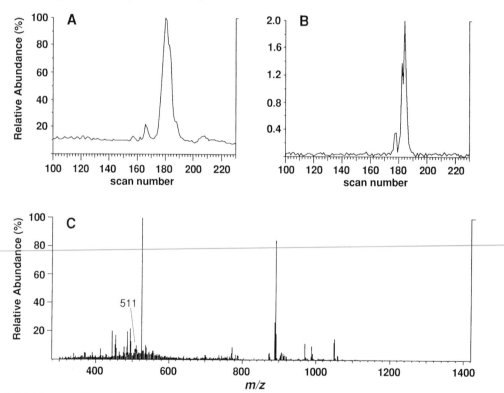

Figure 9.4.2 Microcapillary LC-MS of one HPLC fraction of immunoaffinity-purified MHC-associated peptide extract after two rounds of HPLC purification. (A) Reconstructed ion-current chromatogram (RIC) for the LC-MS analysis of a single HPLC fraction after two rounds of HPLC purification. Shown here is the RIC for MS scans 100–230. (B) RIC from panel (A) replotted to show selected-ion chromatogram for m/z 511. The ordinate is normalized to panel (A). (C) Scans 150 to 200 as shown in panel (A) were summed and the resulting mass spectrum plotted for mass range 300–1400 mass units. Note the relative abundance of ion m/z 511.

Table 9.4.1   Summary of amino acid residue masses.

Amino acid			Residue mass	Amino acid			Residue mass
Glycine	G	Gly	57	Aspartic acid	D	Asp	115
Alanine	A	Ala	71	Glutamine	Q	Gln	128
Serine	S	Ser	87	Lysine	K	Lys	128
Proline	P	Pro	97	Glutamic acid	E	Glu	129
Valine	V	Val	99	Methionine	M	Met	131
Threonine	T	Thr	101	Histidine	H	His	137
Cystine	C	Cys	103	Phenylalanine	F	Phe	147
Leucine	L	Leu	113	Arginine	R	Arg	156
Isoleucine	I	Ile	113	Tyrosine	Y	Tyr	163
Asparagine	N	Asn	114	Tryptophan	W	Trp	186

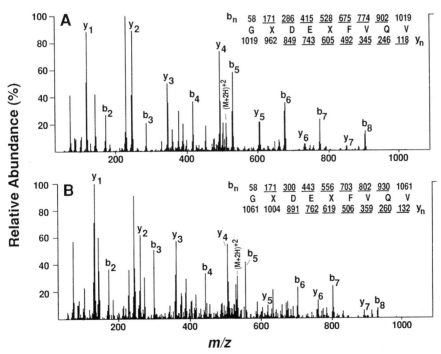

Figure 9.4.3 The CAD mass spectra on the $(M+2H)^{2+}$ ions from the synthetic peptide GXDEXFVQV and its corresponding methyl ester. This derivatization results in a 14-mass-unit shift for each carboxylate group including the C-terminus. Predicted masses for fragment ions of types b_n and y_n are written above and below the peptide sequence, respectively. Ions of type b_n and y_n observed in the spectrum are underlined. X is leucine or isoleucine, which are identical in mass and cannot be distinguished in a CAD spectrum recorded on a triple quadrupole mass spectrometer. (A) CAD spectrum of the $(M+2H)^{2+}$ ions at m/z 511. (B) CAD spectrum of $(M+2H)^{2+}$ ions at m/z 532.

Construction of microcapillary columns

Materials and apparatus

Capillary cleaving tool® (Supelco, Bellefonte, PA, USA)

190 µm o.d. by 75 µm i.d. fused-silica capillary (SGE, Austin, TX, USA)

Stainless steel pressure vessel (see Fig. 9.4.5).

Helium tank and high-pressure regulator

Microtorch (Microflame, Minnetonka, MN, USA)

Stereo microscope

Chemicals and solutions

2-Propanol

C18 stationary phase, 10 µm spherical particles (YMC Corp., Morris Plains, NJ, USA)

Lichrosorb® Si 60 µm spherical silica particles (EM Science, Giggstown, NJ, USA)

POROS II R/H® polymer reversed-phase (PerSeptive BioSystems, Cambridge, MA, USA)

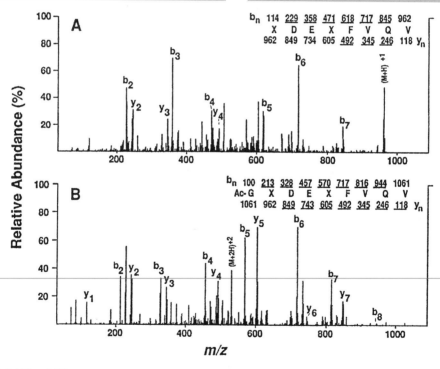

Figure 9.4.4 The CAD mass spectra of protonated peptide molecules generated from GXDEXFVQV after one cycle of Edman degradation or after on-column N-terminal acetylation. Predicted masses for fragment ions of types b_n and y_n are written above and below the peptide sequence, respectively. Ions of type b_n and y_n observed in the spectrum are underlined. X is leucine or isoleucine, which are in identical mass and cannot be distinguished in a CAD spectrum recorded on a triple quadrupole mass spectrometer. (A) CAD spectrum of the $(M+H)^{1+}$ ions at m/z 962. One cycle of Edman degradation results in the removal of the N-terminal residue. (B) CAD spectrum of the $(M+H)^{2+}$ ions at m/z 532. N-Terminal acetylation results in 42-mass-unit shift for the peptide. Ions of type b_n contain the N-terminus and therefore will shift by 42 mass units.

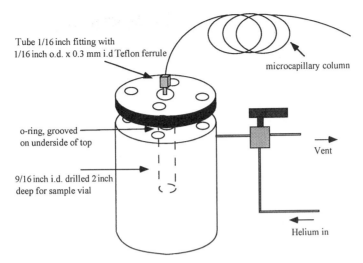

Figure 9.4.5 Stainless steel pressure vessel used for packing columns and for loading samples onto microcapillary columns. Insert the end of the microcapillary column through the lid of the vessel and into an Eppendorf tube containing either a slurry of packing material when making HPLC columns, or a fraction containing sample for analysis. A tube fitting 1/16-inch nut and a 0.3 mm Teflon ferrule forms a seal between the pressure vessel and the microcapillary. The vessel is pressurized with helium to displace contents from the tube into the capillary column.

PRP-1 stationary phase, 10 μm particles (Hamilton Corp., Reno, NV, USA)

Protocol 8: Construction of microcapillary HPLC columns

Procedure

1. Rinse the inside of the capillary with 2-propanol by displacing the solvent from an Eppendorf tube using the pressure vessel. Dry the capillary with a stream of helium gas.
2. Using a capillary cleaving tool, cut a 40–60 cm piece of fused-silica capillary (190 μm o.d., 75 μm i.d.), and inspect for a square end under the microscope.
3. Burn off 3 mm of the polymer coating from one end using a yellow flame from a microtorch.
4. Tap the exposed end of the capillary into Lichrosorb Si 60 silica particles 40–60 times until a small (0.1 mm) plug of particles is formed. Under the microscope, the plug will appear as gray, densely packed material.
5. Quickly pass the plug through a blue flame on the microtorch so that it glows red for a fraction of a second to sinter the particles and form a frit. Repeat 2–3 times. Under the microscope, the plug will appear condensed and adhere to the capillary walls.
6. Test the frit strength by flushing 2-propanol through the capillary at 1000 psi of pressure using the stainless steel high-pressure vessel. The frit will appear translucent. If the frit fails, dry out the capillary, cut off 0.5 cm from the 'fritted' end, and repeat the protocol.
7. Make a slurry of packing material by adding 10–20 mg of C-18, 2–4 mg of POROS II R/H , or 2–4 mg PRP-1 to 0.5 ml of 2-propanol in a 1.5 ml glass vial. Sonicate the slurry to prevent aggregation.
8. Add a small magnetic stirring bar to the slurry and place the vial inside the stainless steel pressure vessel resting upon a magnetic stirring plate. Turn on the stirring plate to continually mix the

slurry.

9. Push the nonfritted end of the fused-silica tubing through the lid of the vessel and into the slurry. Tighten the tube fittings to form a seal.

10. Pressurize the vessel with helium at a pressure of 200–500 psi thus causing a continuous flow of packing material into the tubing. Once the packing material has filled the capillary to the desired length (10–15 cm) depressurize the vessel.

11. Wash the remaining beads to the head of the column and compress the bed by flushing 5% acetic acid in water over the column for 15–25 min at 200–500 psi.

12. Prior to use, condition the column by running two gradients over the column, one blank and one with 10 pmol angiotensin loaded on to the column. The column flow should be at least $0.5\,\mu l\,min^{-1}$ at 500 psi.

13. Store columns with both ends immersed in a test tube half-filled with deionized water.

Sample introduction

Materials and apparatus

Triple quadrupole mass spectrometer equipped with electrospray ionization source (Finnigan MAT TSQ 7000, San Jose, CA, USA)

HPLC 140 A syringe pump solvent delivery system (Applied Biosystems, Foster City, CA, USA)

Stainless steel pressure bomb (Fig. 9.4.5).

Graduated disposal pipettes $1–5\,\mu l$ (cat. no. 21-164-2A, Fisher Scientific, Pittsburgh, PA, USA)

Chemicals and solutions

0.1 M acetic acid (Pierce, Rockford, IL, USA) in NANOpure water

Acetonitrile (Mallinckrodt, Paris, KY, USA)

Protocol 9: Sample introduction into the electrospray source

Procedure

1. Load samples onto the microcapillary columns using the pressure bomb (Fig. 9.4.5).

2. Using a $1–5\,\mu l$ graduated disposable glass pipette, determine the amount of sample loaded by measuring the amount of solvent displaced from the column.

3. Wash the column with 100% solvent A for 5 min prior to beginning the gradient.

4. Elute the sample from the microcapillary column and into the mass spectrometer using a linear gradient of 0–80%B in 12 min (solvent A, 0.1 M acetic acid in NANOpure water; solvent B, acetonitrile). The ideal flow rate is approximately $0.5\,\mu l$ min^{-1} through the microcapillary column. The solvent delivery system operates at $180–300\,\mu l$ of solvent per minute and a precolumn split reduces the solvent flow into the microcapillary column to $0.5\,\mu l\,min^{-1}$. This configuration allows proper mixing with minimal gradient delay. The precolumn split ratio can be varied by adjusting the length of the $50\,\mu m$ i.d. fused-silica restriction capillary (see Fig. 9.4.1).

Chemical modification of peptides

Protocol 10: Chemical modification. Esterification of free carboxylic acids

Materials and apparatus

Vacuum centrifuge

Chemicals and solutions

Acetyl chloride (Aldrich, Milwaukee, WI, USA)

Anhydrous methanol or ethanol

5% acetic acid (v/v) in NANOpure water

Procedure

1. Prepare the esterification solution by adding dropwise with stirring, 160 μl of acetyl chloride to either 1 ml of anhydrous methanol or 1 ml of anhydrous ethanol chilled on ice. Allow reagents to sit at room temperature for 10 min.
2. Add 20–50 μl of the methanolic or ethanolic-HCl solution prepared in step 1 to the peptide solution lyophilized to dryness.
3. Let the reaction proceed at room temperature for 1–2 h for methyl esters and for 3 h for ethyl esters.
4. Remove solvent to dryness by vacuum centrifugation, then resuspend the modified peptides in 15–20 μl of 5% acetic acid for mass spectrometric analysis.

Trouble shooting tips

Under long exposure conditions it is possible to hydrolyze and esterify asparagine and glutamine, resulting in mass shifts of 15 and 29 mass units for the methyl and ethyl ester, respectively.

Protocol 11: Chemical modification. *N*-Terminal acetylation

Chemicals and solutions

NANOpure water

Acetic anhydride

200 mM ammonium acetate solution adjusted to pH 8.5

Procedure

1. Load the peptides on to a microcapillary column as described in protocol 9. Wash with NANOpure water for 5 min.
2. Immediately before use, add 1 μl of acetic anhydride to 99 μl of a 200 mM ammonium acetate solution adjusted to pH 8.5.
    Note: At pH 8.5, both the N-terminus and lysine residues are acetylated (arginine is not acetylated). At pH 6, acetylation occurs mostly at the N-terminus. Partial acetylation can also occur at histidine residues. Ammonium bicarbonate is an alternative to ammonium acetate.
3. Load 3–4 μl of the acetylating reagent through the column, then wash the column with NANOpure water for 5 min at a flow rate of 1 μl min^{-1}.
4. Equilibrate the column with 100% solvent A, then gradient elute the acetylated peptide from the column.

Protocol 12: Chemical modification. Single-cycle Edman degradation

Materials and apparatus

Vacuum centrifuge

Chemicals and solutions

5% phenylisothiocyanate (Pierce, Rockford, IL, USA) in pyridine (Aldrich, Milwaukee, WI, USA)

Trifluoroacetic acid, sequencing grade (Applied Biosystems, Foster City, CA, USA)

n-Butyl acetate (Aldrich)

NANOpure water

Procedure

1. Prepare fresh a solution of 5% phenylisothiocyanate (PITC) in pyridine.
2. Concentrate an aliquot of the HPLC fraction containing the peptide of interest to 1 μl total volume by vacuum centrifugation.
3. Add 2 μl to 10 μl of the PITC solution made in step 1 to the concentrated peptide; overlay the solution with argon, vortex, and incubate the mixture at 45°C for 30–45 min.
4. Lyophilize the sample to dryness, then add 15 μl of concentrated trifluoroacetic acid (sequencing grade) and overlay the solution with argon and incubate for 10 min at 37°C.
5. Vacuum centrifuge the sample to dryness in order to remove the TFA. Redissolve the shortened peptide in 15 μl of NANOpure water and vortex.
6. Extract the sample with 2 × 15 μl n-butyl acetate.
7. Lyophilize the aqueous layer to dryness and resuspend for mass spectrometric analysis by adding 1 μl acetic acid (vortex sample), and 19 μl NANOpure water (vortex sample).
   Note: The reaction of the side-chain amino group of lysine or the thiol group of cysteine with phenylisothiocyanate will result in an increase in the $(M+H)^+$ by 135 mass units.

Protocol 13: Chemical modification. On-column carboxyamidation

Chemicals and solutions

10 mM iodoacetamide in 100 mM Tris–HCl, pH 8.5

Procedure

1. Load the peptide sample on to the microcapillary HPLC column as described in protocol 9. Wash the column with water for 5 min.
2. Flow a solution of 10 mM iodoacetamide in 100 mM Tris–HCl, pH 8.5, through the column at the rate of 0.25 μl min^{-1} (75 μm column) for 15 min.
3. Completely remove Tris buffer by washing with water for 5 min, equilibrate the column with 100% solvent A, and gradient elute peptide into the mass spectrometer.

References

Castelli C, Storkus WJ, Maeurer MJ, et al (1995) Mass spectrometric identification of a naturally processed melanoma peptide recognized by CD8+ cytotoxic T lymphocytes. J Exp Med 181: 363–368.

Cox AL, Skipper J, Chen Y, et al (1994) Identification of a peptide recognized by five melanoma-specific human cytotoxic T cell lines. Science 264: 716–719.

den Haan JM, Sherman NE, Blokland E, et al (1995) Identification of a graft versus host disease-associated human minor histocompatibility antigen. Science 268: 1475–1480.

Fenn JB, Mann M, Meng CK, Wong SF, Whitehouse CM (1989) Electrospray ionization for mass spectrometry of large biomolecules. Science 246: 64–71.

Henderson RA, Michel H, Sakaguchi K, et al (1992) HLA-A2.1-associated peptides from a mutant cell line: a second pathway of antigen presentation. Science 255: 1264–1266.

Henderson RA, Cox AL, Sakaguchi K, et al (1993) Direct identification of an endogenous peptide recognized by multiple HLA-A2.1-specific cytotoxic T cells. Proc Natl Acad Sci USA 90: 10275–10279.

Huczko EL, Bodnar WM, Benjamin D, et al (1993) Characteristics of endogenous peptides eluted from the class I MHC molecule HLA-B7 determined by mass spectrometry and computer modeling. J Immunol 151: 2572–2587.

Hunt DF, Yates JR, Shabanowitz J, Winston S, Hauer CR (1986) Protein sequencing by tandem mass spectrometry. Proc Natl Acad Sci USA 83: 6233–6237.

Hunt DF, Alexander JE, McCormack AL, et al (1991) Mass spectrometric methods for protein and peptide sequence analysis. In Villafranca JJ, ed. Techniques in Protein Chemistry II. Academic Press, London, pp. 441–454.

Hunt DF, Henderson RA, Shabanowitz J (1992a) Characterization of peptides bound to the class I MHC molecule HLA-A2.1 by mass spectrometry. Science 255: 1261–1263.

Hunt DF, Michel H, Dickinson TA, Shabanowitz J, Cox AL, Sakaguchi K, Appella E, Grey HM, Sette A (1992b) Peptides presented to the immune system by the murine Class II major histocompatability complex molecule I-Ad. Science 256: 1817–1820.

Langone JJ, Boyle MD, Borsos T (1978) Studies on the interaction between protein A and immunoglobulin G. I. Effect of protein A on the functional activity of IgG. J Immunol 121: 327–332.

Mant CT, Hodges RS (1991) The effects of anionic ion-pairing reagents on peptide retention in reversed-phase chromatography. In Mant CT, Hodges RS, eds. High-Performance Liquid Chromatography of Peptides and Proteins: Separation, Analysis, and Conformation. CRC Press, Boston, pp. 327–341.

Moseley MA, Deterding LJ, Tomer KB, Jorgenson JW (1991) Nanoscale packed-capillary liquid chromatography coupled with mass spectrometry using coaxial continous-flow fast atom bombardment interface. Anal Chem 63: 1467–1473.

Roepstorff P, Fohlman J (1984) Proposal for a common nomenclature for sequence ions in mass spectra of peptides. Biomed Mass Spectrom 11: 601.

Sette A, Ceman S, Kubo RT, Sakaguchi K. Appella E, Hunt DF, Davis TA, Michel H, Shabanowize J, Rudersdorf R, Grey H, DeMars R (1992) Invariant chain peptides in most HLA-DR molecules of an antigen-processing mutant. Science 258: 1801–1849.

Siuzdak G (1994) The emergence of mass spectrometry in biochemical research. Proc Natl Acad Sci 91: 11290–11297.

Skipper JCA, Hendrickson RC, Golden PH, Brickard V, Van Pel A, Chen Y, Shabanowitz J, Wofel T, Slingluff CL Jr, Boon T, Hunt DF, Engelhard VH (1996) A HLA-H2-restricted tyrosinase antigen on melanoma cells results from posttranslational modification and suggests a novel pathway for processing of membrane proteins. J Exp Med 183: 527–534.

Slingluff CL Jr, Cox AL, Henderson RA, Hunt DF, Engelhard VH (1993) Recognition of human melanoma cells by HLA-A2.1-restricted cytotoxic T lymphocytes is mediated by at least six shared peptide epitopes. J Immunol 150: 2955–2963.

Analysis of MHC class II binding motifs with bacteriophage peptide display libraries

9.5

Juergen Hammer
Francesco Sinigaglia

Roche Milano Ricerche, Milan, Italy

TABLE OF CONTENTS

Immunology Methods Manual
ISBN 0–12–442712–X

Abstract

M13 peptide libraries are large collections of bacteriophages displaying different peptides on their surface. They can be used for the isolation of large pools of MHC class II-binding peptides. The characterization of these peptide pools allows the identification of class II anchor residues and class II allele-specific binding motifs.

Introduction

The definition of MHC class II peptide-binding motifs

The recognition of protein antigens is a major function of the immune system, in which major histocompatibility complex (MHC)-encoded class II molecules play a central role. MHC molecules are highly polymorphic membrane glycoproteins that bind peptide fragments, derived from protein antigens, and display them on the cell surface, evoking effector responses upon recognition by the antigen-specific receptors of CD4$^+$ T-lymphocytes (Germain 1994). To ensure T cell-mediated immunity to the antigenic universe, each allelic form of class II molecules has to bind many different peptides. x-Ray crystal structures indicate that this is due to hydrogen bonding between conserved MHC residues and the peptide main chain that forces different peptides into similar conformations (Stern et al 1994). Peptide main-chain interactions, however, are not the only mode of class II peptide binding. Some of the peptide side-chains contact pockets in the MHC cleft and increase the overall binding affinity and specificity of the associated peptides. These pockets are usually shaped by clusters of polymorphic MHC residues, resulting in strong, allele-specific preferences for interacting with particular amino acid side-chains. The sum of these preferences is defined as the binding motif of an MHC class II molecule (Sinigaglia and Hammer 1994). A breakthrough for the analysis of MHC binding motifs has been the characterization of large, MHC-selected peptide pools, enabling the definition of general rules for peptide binding to MHC molecules (Falk et al 1991; Hammer et al 1993). For class II, purified MHC molecules are used to select large pools of peptides displayed on the surface of M13 bacteriophage. The alignment of these peptides leads to the identification of position-specific preferences for particular amino acid side-chains (class II anchors), and to class II allele-specific binding motifs (Hammer et al 1992, 1993, 1994a).

The principle of screening M13 bacteriophage display libraries with MHC class II molecules

The filamentous bacteriophage M13 is a single-stranded, male-specific DNA phage. It consists of a stretched-out loop of single-stranded DNA, sheathed in a tube composed of thousands of monomers of the major coat protein (gene VIII product). Minor coat proteins are found at the tips of the virion. One of the minor coat proteins, protein III (product of gene III), attaches to the receptor at the tip of the F pilus of the host *Escherichia coli*. Protein pIII folds into two domains. The carboxy-terminal domain interacts with viral coat proteins and is required for viral assembly, while the amino-terminal two-thirds of the protein forms a knob-like domain which projects away from the virion and is responsible for attaching to the F pilus.

In the late 1980s, the ability of filamentous bacteriophages to display foreign peptides on their outer surfaces was demonstrated (Parmeley and Smith 1988). Peptides fused to the NH$_2$ termini of the five copies of the pIII protein are displayed at one tip of each bacteriophage particle and have little or no effect on viral infectivity. Bacteriophage display libraries are large collections of bacteriophages displaying different peptides on their surfaces (Cwirla et al 1990; Devlin et al 1990; Scott and Smith 1990). Screening of an M13 bacteriophage display library with MHC class II molecules can be divided into seven steps (Fig. 9.5.1): (1) Purifica-

tion and biotinylation of MHC class II molecules; (2) incubation of class II molecules with an M13 peptide display library; (3) attachment of the bacteriophage, with peptides able to bind to class II molecules, to a streptavidin–solid phase via the strong biotin–streptavidin reaction, and removal of unbound bacteriophage; (4) elution of bound bacteriophage with acid; (5) amplification of the eluted bacteriophage; (6) monitoring of the enrichment of class II binding bacteriophage during several rounds of screening; (7) sequencing and alignment of the bacteriophage inserts encoding the random peptide region for the identification of anchor residues and class II binding motifs.

The following protocol was developed for the identification of human class II HLA-DR motifs using an M13mp19-based display library with random inserts that are 9 amino acid residues in length and flanked by 4 glycine spacers on each side (Hammer et al 1992). Adjustments might be necessary for other class II isotypes and bacteriophage display libraries.

Materials

- Anti-DR resin: the purified mouse monoclonal anti-HLA-DR antibody L243 is cross-linked to protein A–Sepharose (Pharmacia) with dimethyl pimelimidate according to the manufacturers' instructions to give approximately 3 mg of antibody per ml of settled gel.
- BDR buffer: 50 mM Tris–HCl, pH 7.5, 150 mM NaCl, 2 mM EDTA, and 0.2% Nonidet P-40
- Biotin-XX-NHS solution: 25 mM biotin-XX-NHS (Calbiochem) in dimethylformamide
- Biotinylation buffer: 0.25 M NaHCO$_3$, pH nonadjusted, 0.2% Nonidet P-40
- BSA-blocked streptavidin on 4% beaded agarose (Sigma); to reduce nonspecific binding, the streptavidin on 4% beaded agarose is washed twice with 10 volumes BDR buffer (1 volume = volume of settled streptavidin agarose), incubated for 1 h with 10 volumes BDR buffer containing 5 mg ml^{-1} BSA and washed subsequently three times with 10 volumes BDR buffer
- Cell-lysis buffer: 1% (v/v) Nonidet P-40; 25 mM iodoacetamide; 1 mM phenylmethylsulfonyl fluoride (PMSF); 5 mM ε-amino-n-caproic acid; 10 μg ml^{-1} of each of soybean trypsin inhibitor, antipain, pepstatin, leupeptin, and chymostatin; in 0.05 M sodium phosphate buffer, pH 7.5, containing 0.15 M NaCl
- DOC solution: solution of 5% sodium deoxycholate
- DR-elution buffer: 50 mM

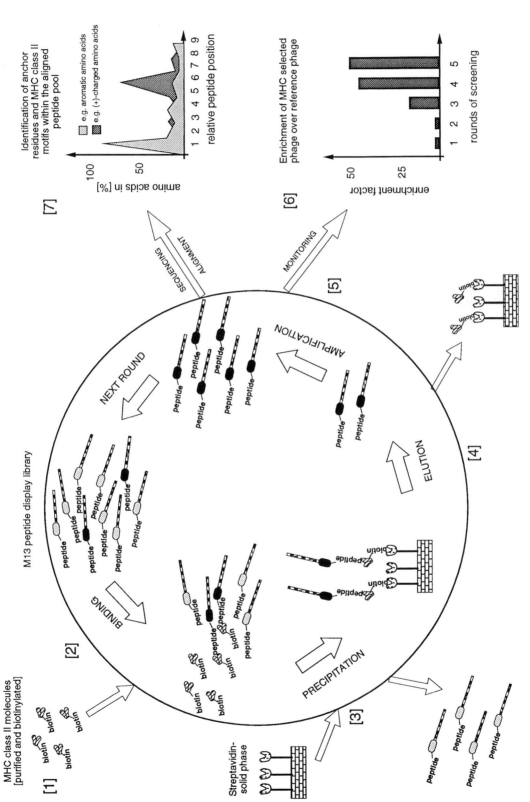

Figure 9.5.1 Schematic representation of the M13 peptide library screening procedure (see text).

diethylamine-Cl, pH 11.5; 1%
octyl-β-D-glucopyranoside; 0.15 M
NaCl; 1 mM EDTA; 10% glycerol;
0.03% NaN$_3$
- *E. coli* XL1Blue plating bacteria:
optical density = 0.5; XL1Blue can be
obtained from Stratagene. Plating
bacteria are prepared as described by
Sambrook et al (1989)
- M13 bacteriophage peptide display
library (amplified) (Hammer et al 1992);
we use amplified M13 display libraries
consisting of 20 million independent
clones, with 9-amino-acids-long
random inserts flanked by two Gly
spacers; the M13mp19-based library
vector contains a β-lactamase gene at
its polylinker site that confers
ampicillin resistance to the host, thus
allowing its propagation, like a
plasmid, independently of
bacteriophage function. In addition,
the disruption of the β-galactosidase
gene fragment by the β-lactamase
gene creates white plaques on X-Gal
indicator plates, while M13mp19, used
as a reference phage in
bacteriophage/MHC binding assays,
produces blue plaques.
- M13mp19 reference bacteriophage
(Stratagene)

- Microsep microconcentrators; 30 k
cutoff (Scan AG)
- PD-10 columns; Sephadex G-25M
(Pharmacia)
- PEG solution; 20% poly(ethylene
glycol) (PEG 8000) in 2.5 M NaCl
- Phage-binding buffer; 150 mM NaCl,
2 mM EDTA, 0.2% Nonidet P-40, 1 mM
PMSF, citrate–phosphate buffer; the
optimal pH is HLA-DR allele-specific
and needs to be determined
empirically (see 'Commentary');
alternatively, screening could be
performed using a range of pH, e.g.,
pH 5, 6 and 7
- Phage-elution buffer; 0.1 N glycine-HCl,
pH 2.2, 1 mg ml^{-1} BSA fraction V
- Phage-storage buffer; 50 mM Tris–HCl,
pH 7.5, 150 mM NaCl
- Phosphate-buffered saline (PBS)
- Wash buffer 1; 50 mM Tris–HCl, pH 8;
0.15 M NaCl; 0.5% Nonidet P-40; 0.5%
sodium deoxycholate; 10% glycerol;
0.03% NaN$_3$
- Wash buffer 2; 50 mM Tris–HCl, pH 9;
0.5 M NaCl; 0.5% Nonidet P-40; 0.5%
sodium deoxycholate; 10% glycerol;
0.03% NaN$_3$
- Wash buffer 3; 2 mM Tris–HCl, pH 8;
1% octyl- β-D-glucopyranoside; 10%
glycerol; 0.03% NaN$_3$

Method

Immunoaffinity purification of DR molecules

1. Harvest 5×10^9 human HLA-DR
homozygous Epstein–Barr
virus-transformed B cells by
centrifugation and wash twice with
PBS.
2. Resuspend cell pellet to a cell density
of 10^8 ml^{-1} in cell-lysis buffer and
incubate on ice for at least 60 min.
3. Clear the lysate of nuclei and debris
by centrifugation at 27,000g for
30 min. If not used immediately, cell
lysates can be stored at –70°C.
4. Add 0.2 volume of DOC solution to the

post-nuclear supernatant and mix for 10 min.

5. Centrifuge at 100,000g for 2 h and filter the supernatant through a 0.45 µm pore-size membrane.
6. Mix the lysate with 5 ml (settled volume) anti-DR resin and rotate at 4°C overnight.
7. Wash the gel mixture twice with 10 volumes of wash buffer 1 and transfer it into a column.
8. Wash the column using a flow rate of 0.5 to 1 ml min^{-1} at 4°C with at least 20 column volumes of wash buffer 1, 5 column volumes of wash buffer 2, and 5 column volumes of wash buffer 3.
9. Elute the DR antigens with two column volumes of DR elution buffer, collect 1 ml fractions, and neutralize fractions with 100 µl of 1 M Tris-Cl, pH 6.8.
10. Analyze aliquots of each fraction by SDS-PAGE. Pool fractions containing most of the DR-antigens, aliquot and keep them frozen at −70°C until use.

Biotinylation of MHC class II molecules

1. Exchange the buffer of 1 mg affinity-purified HLA-DR antigen solution into biotinylation buffer by gel filtration using two PD-10 columns sequentially, according to the manufacturer's instructions.
2. Reduce the volume of the HLA-DR solution to 1 ml by ultradialysis with Microsep Microconcentrators.
3. Add 8 µl of the biotin-XX-NHS solution to the concentrate and rotate for 1 h at room temperature.
4. Remove the excess biotin-XX-NHS by gel filtration using a PD-10 column equilibrated with BDR buffer.

5. Concentrate the biotinylated DR antigens by ultradialysis (see above), aliquot and keep them frozen at −70°C until use.

Screening of bacteriophage display libraries with DR molecules

1. Incubate approximately 1×10^{10} bacteriophages of an amplified M13 peptide display library with 5 µg biotinylated HLA-DR molecules in phage-binding buffer at room temperature for 48 h (total volume 60 µl).
2. Add 500 µl BSA-blocked streptavidin on 4% beaded agarose (total 50 µl settled volume) to the library/HLA-DR incubation mix and rotate for 10 min at room temperature.
3. Purify the M13 bacteriophage/HLA-DR complexes by washing the streptavidin solid phase eight times with BDR buffer using 1 ml per wash step.
4. Elute the bacteriophage with 1 ml phage-elution buffer for 10 min and neutralize with 60 µl 2 M Tris-base.
5. For amplification, infect 1 ml of *E. coli* XL1Blue plating bacteria with 300 µl eluted bacteriophage and transfer cells to 7 ml LB medium (Sambrook et al 1989) in a 25 cm^2 tissue culture flask. After 1 h incubation (37°C, 1200 rpm), add ampicillin to a final concentration of 20 µg ml^{-1} and incubate for further 20 h at 37°C and 200 rpm.
6. For harvesting and purification of the bacteriophage, transfer the cell/bacteriophage mixture to 15 ml tubes (Falcon, polypropylene) and centrifuge at 2500g for 25 min at room

temperature. Transfer 5 ml of supernatant into a new tube, add 1 ml PEG solution and mix by inverting the tube several times. Recover the precipitated bacteriophage particles after 10 min incubation at room temperature by centrifugation using the same conditions as above. Aspirate the supernatant, centrifuge for 30 s, and remove any residual supernatant. Redissolve the bacteriophage pellet in 1.2 ml phage-storage buffer and transfer it into a 1.5 ml tube (Eppendorf). Add 200 μl PEG solution and recover the bacteriophage particles by centrifugation. Redissolve the pellet in 500 μl phage-storage buffer and store at 4°C until used.

7. Use an aliquot of this bacteriophage solution (approximately 1×10^{10} bacteriophage particles) for the next round of screening and amplification (steps 16 to 21).

Monitoring the screening and sequencing the peptide inserts

1. After each round of screening, mix an aliquot of the class-II selected, amplified M13 peptide display library (1×10^9) with a similar number of M13mp19 bacteriophage as a reference and incubate with 10–50 pmol biotinylated DR molecules in phage-binding buffer.

2. After at least 24 h incubation at room temperature, add 250 μl BSA-blocked streptavidin on 4% beaded agarose (25 μl settled volume) and incubate for 10 min.

3. Purify the M13 bacteriophage/DR complexes by washing the solid phase several times with BDR buffer.

4. Elute the bacteriophage with 1 ml phage-elution buffer for 10 min and neutralize with 30 μl 2 M Tris-base.

5. Determine the ratio of the bacteriophage displaying peptides to M13mp19 reference bacteriophage in both the initial mixture and the eluates by plating aliquots on X-Gal indicator plates (Sambrook et al 1989). An enrichment of white (bacteriophage displaying peptides) versus blue plaques (M13mp19) indicates peptide-based binding.

6. Isolate bacteriophage from the first DR-selected peptide pool which exhibits a definite enrichment over M13mp19 bacteriophage and sequence the inserts encoding the displayed peptide region (Sambrook et al 1989).

7. Align the peptide sequences using, for example, the 'pileup' program of the GCG (Genetic Computer Group Inc.) package to resolve the class II motif.

Commentary

Optimization of MHC class II–bacteriophage binding

Since different MHC class II alleles require different peptide-binding conditions, optimization experiments are recommended for each class II allele prior to screening. To optimize binding between bacteriophage-displayed peptides and class II molecules, peptides binding with high affinity (K_d within 10 to 100 nM) to the corresponding class II allele can be placed at the NH_2 terminus of protein III and used in an MHC class II–bacteriophage binding assay (Hammer et al 1992). As for the monitoring of the screening procedure, binding of M13 bacteriophage displaying peptides to DR molecules is determined by their enrichment as compared to the reference bacteriophage M13mp19 (protocol steps 23 to 27, and Fig. 9.5.2). An important parameter to be considered in bacteriophage/MHC binding studies is the pH used for the binding reaction and for washing the streptavidin solid phase (Hammer et al 1993, 1994a). In comparison with other receptors used in screening bacteriophage display libraries, we generally find a rather low enrichment for bacteriophage binding to DR molecules (Fig. 9.5.2). This finding could be due to a specific interaction of DR molecules with sites of the bacteriophage surfaces other than the peptide-insert region. This is best illustrated by the fact that the influenza hemagglutinin peptide HA 307–319 peptide, known to bind to the MHC binding cleft, is able to specifically inhibit some of the background binding of M13mp19 reference bacteriophage.

Library screening

We have observed that the more rounds of screening that are performed, the higher the chance of isolating bacteriophage displaying identical peptides. Because the identification of DR peptide binding motifs requires the alignment of large numbers of peptides derived from independent DR binding bacteriophage, we usually sequence bacteriophage from the round of selection that first exhibits a clear enrichment over the M13mp19 reference bacteriophage.

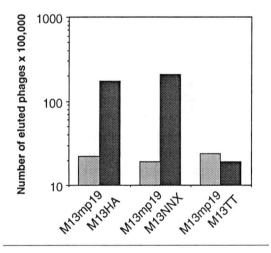

Phage	N-terminal region of pIII	Peptide
M13HA	AEL GGG PKYVKQNTLKLAT GGGG VP - pIII	HA 307-319
M13NNX	AEL GGG NALYKMNAVAAAA GGGG VP - pIII	NNX 83-91
M13TT	AEL GGG SGPDKEQIADEIN GGGG VP - pIII	TT 763-775
M13mp19	- pIII	---------

Figure 9.5.2 Example of a DRB1*0101-bacteriophage binding assay (Hammer et al 1992). Binding of M13 constructs is determined by their enrichment as compared to the reference bacteriophage M13mp19. For M13 constructs the bacteriophage input is 1×10^9, and for M13mp19 it is 2×10^9.

Another problem frequently encountered by screening M13 peptide libraries with class II molecules is the enrichment for non-MHC-specific bacteriophage. This background binding could be due to bacteriophage binding to streptavidin, to biotin, or to agarose. Indeed, whenever binding conditions are not optimized, we observe enrichment for nonspecific bacteriophage displaying peptides carrying several Trp residues. When tested, these bacteriophages bind to any biotinylated protein or to the streptavidin-solid phase.

If the alignment of bacteriophage-derived sequences does not reveal any peptide binding motifs, it is necessary to determine the percentage of DR-binding peptides within the selected peptide pool. This should be done by testing synthetic peptides, based on the bacteriophage sequences, for their capacity to bind the corresponding DR molecules. This test is performed with MHC peptide competition assays (Hammer et al 1994b). Only if a high percentage of the synthetic peptides bind to the corresponding class II allele is it justified to sequence a larger DR-selected peptide pool.

If alignment of bacteriophage-derived sequences reveals anchor positions, anchor-addition and anchor-shifting experiments may subsequently be performed to confirm their role in peptide–MHC interactions. Anchor-substitution binding experiments should also be carried out to evaluate the effect of MHC allele-specific anchors (Hammer et al 1993).

Evaluation of MHC class II motifs derived from screening of bacteriophage display libraries

The screening of bacteriophage peptide display libraries has been used for the identification of several allele-specific class II binding motifs. These motifs generally consist of 3 to 4 anchor positions that are at fixed distances from one another. As these motifs are derived from the characterization of large pools of peptide, they provide better than any other approach general information about major class II anchor positions and their exact spacing. Several lines of evidence indicate, however, that motifs derived from screening of bacteriophage libraries are only the tip of the iceberg, and that the rules for peptide MHC binding are actually much more complex. For example, the presence or absence of peptide side-chains which can interfere with peptide binding (inhibitory residues) are just as important for class II MHC peptide binding as the presence of anchor residues (Hammer et al 1994a, b). Thus, simple motifs derived from bacteriophage library screening provide an important ground for more quantitative studies of anchor and inhibitory residues by synthetic peptide chemistry, which ultimately lead to expanded and quantitative motifs with high predictive values (Hammer, 1995).

References

Cwirla SE, Peters EA, Barret RW, Dower WJ (1990) Peptides on phage: a vast library of peptides for identifying ligands. Biochemistry 87: 6378–6382.

Devlin JJ, Paniganiban LC, Devlin PE (1990) Random peptide libraries: a source of specific protein binding molecules. Science 249: 404–406.

Falk K, Rötzschke O, Stevanovic S, Jung G, Rammensee H-G (1991) Allele specific motifs revealed by sequencing of self peptides eluted from MHC molecules. Nature 351: 290–296.

Germain RN (1994) MHC-dependent antigen processing and peptide presentation: providing ligands for T lymphocyte activation. Cell 7: 287–299.

Hammer J (1995) New methods to predict MHC binding sequences within protein antigens. Curr Opin Immunol 7: 263–269.

Hammer J, Takacs B, Sinigaglia F (1992) Identification of a motif for HLA-DR1 binding peptides using M13 display libraries. J Exp Med 176: 1007–1013.

Hammer J, Valsasnini P, Tolba K, et al (1993) Promiscuous and allele-specific anchors in HLA-DR-binding peptides. Cell 74: 197–203.

Hammer J, Belunis C, Bolin D, et al (1994a) High affinity binding of short peptides to MHC class II molecules by anchor combinations. Proc Natl Acad Sci USA 91: 4456–4460.

Hammer J, Bono E, Gallazzi F, Belunis C, Nagy Z, Sinigaglia F (1994b) Precise prediction of MHC class II-peptide interaction based on peptide side chain scanning. J Exp Med 180: 2353–2358.

Parmeley SF, Smith GP (1988) Antibody-selectable filamentous fd phage vectors: affinity purification of target genes. Gene 73: 305–314.

Sambrook J, Fritsch EF, Maniatis T (1989) Molecular Cloning: A Laboratory Manual. Cold Spring Harbor Press, Cold Spring Harbor, NY.

Scott JK, Smith GP (1990) Searching for peptide ligands with an epitope library. Science 248: 386–390.

Sinigaglia F, Hammer J (1994) Defining rules for the peptide–MHC class II interaction. Curr Opin Immunol 6: 52–56.

Stern LJ, Brown JH, Jardetzky TS, (1994) Crystal structure of the human class II MHC protein HLA-DR1 complexed with an influenza virus peptide. Nature 368: 215–221.

Biochemical analysis of peptide binding to MHC class I

9.6

Anette Stryhn
Lars Østergaard Pedersen
Søren Buus

Institute for Medical Microbiology and Immunology, Panum Institute, University of Copenhagen, Denmark

TABLE OF CONTENTS

Abstract

The T cell receptor has a unique specificity for antigenic peptides presented in the context of MHC molecules on the surface of antigen-presenting cells. To understand and manipulate T cell immunity, one needs to understand and measure MHC function. Here we describe the biochemical analysis of peptide binding to MHC class I (and in principle the methods described can also be used to measure peptide binding to MHC class II). We also describe a modification which will allow the binding of β_2-microglobulin to MHC class I heavy chain to be measured.

MHC restriction, polymorphism, and specificity

T cells are said to be MHC restricted, i.e., T cells recognize foreign antigens presented in the context of MHC. The nature of MHC restriction has been clarified considerably over the past decade (for review see Germain and Margulies 1993). It now appears that the function of MHC molecules is to sample peptides from the intra- or extracellular environment of presenting cells (APC). These peptides have been generated through limited proteolysis of protein antigens in a process known as antigen processing. Subsequent to binding, MHC molecules protect the bound peptides against further degradation and transport them to the APC surface where they are presented to T cells. MHC class I molecules are devoted to the presentation of intracellularly derived peptides to cytotoxic T cells. This preference focuses the attention of cytotoxic T cells on target cells harboring foreign DNA (viral DNA or mutated self-DNA).

A very large number of alleles per MHC locus exists in the population; however, each individual possess only one or two of these alleles at each MHC locus (i.e., corresponding to being homozygous or heterozygous at the particular locus). This is known as MHC polymorphism. The polymorphic residues are concentrated in the outer peptide-binding domains of the MHC molecule, shaping the topography and functionality of this important region of the MHC (Bjorkman et al 1987). Each MHC allele encodes a molecule with a unique, albeit broadly defined, peptide binding specificity based on 'motif' recognition. A motif is defined as a peptide sequence containing one, or a few related, amino acids in a few dominant, or subdominant, 'anchor positions' and no deleterious amino acids. This generates a combinatorial specificity which can recognize a sizable fraction of the universe of peptides. The rules for peptide binding to MHC have been unraveled rapidly in recent years, followed by an improved ability to predict binding (Sette et al 1989; Falk et al 1991). The specificity of a given MHC molecule can be approached from two different angles; one can examine what the MHC has already bound (Falk et al 1991); alternatively, one can examine what it will bind (Sette et al 1989). The former approach has the advantage of being based on *in vivo* peptide binding, but the disadvantage of representing the sum of a number of *in vivo* events (protein repertoire, processing specificities, MHC specificity, etc.). The latter approach has the advantage of dealing with MHC specificity exclusively, but the disadvantage of being based on *in vitro* peptide binding.

Measuring peptide binding to MHC

Defining MHC specificity is just one reason for measuring peptide–MHC interaction — one could have many other scientific, clinical, or commercial reasons for determining peptide–MHC binding. We will describe a strictly biochemical analysis which allows careful quantification of binding. Many other techniques have been suggested and for a more comprehensive discussion of the choice of method the reader is referred to Buus et al (1995). The affinity for peptide binding can be quite high (K_d in the nanomolar range) and typically peptide–MHC complexes are very stable once formed. This stability allows complex formation to be determined by a nonequilibrium technique such as gel filtration. We shall describe two versions of gel filtration: (1) a classical gel filtration method, that is fast, easy, and robust, and (2) an improved spun column gel filtration, which is even faster and can process many samples simultaneously. The classical gel filtration method serves as a reference method for the verification of the spun column assay.

How to choose and generate the reagents

A binding assay, once a working peptide–MHC pair has been obtained, can conveniently be converted into an inhibition assay in which any substance can be tested. Such an inhibition assay is preferable since, once established, it is versatile and easy to use. For each MHC molecule to be examined, only one peptide needs to be labeled, and any unknown preparation can then be tested for inhibition. The less inhibitor needed to block the reaction, the better the binding. The inhibition assay is very sensitive (the presence of 100 fmole of a good peptide binder can be detected). Thus, the first objective is to select the MHC molecule and an agonist peptide that can be labeled, and to generate active preparations of both reactants.

MHC

One should select the proper cell line to produce the MHC molecule of interest and a suitable antibody to purify the MHC. Apart from searching the literature for suitable cell lines, a valuable source of human MHC class I-producing cell lines are the International Histocompatibility Workshop cell lines (which can be obtained, for example, through European Collection of Animal Cell Cultures, ECACC, fax (44) 980 611 315). Preferably one should select homozygous cell lines, but heterozygous cell lines are acceptable as long as the antibody used for purification can yield homogeneous MHC preparations (alternatively, specificity may be resurrected if the agonist peptide to be labeled can discriminate between the MHC molecule of interest and any copurified contaminants). It is worth spending some effort on optimizing yields (screening and/or FACS sorting for high expression), since pure homogeneous preparations are much easier to produce using a good starting material. You should also select a cell line which will grow reasonably well at large scale under *in vitro* cell culture conditions (cell factories, roller bottles, spinner culture, fermenters) — preferably in cheap serum. *In vitro* grown cells are harvested, lysed in detergent, and stored at −70°C until use (Buus et al 1986).

The light chain of MHC class I, the β_2-microglobulin (β_2m), can either be purified from urine by chromatofocusing (Nissen et al 1987) or generated as a recombinant molecule in bacteria (Silver et al 1991) or insect cell lines (Godeau et al 1991). It is also available commercially (e.g., Sigma M4890).

Monoclonal anti-MHC antibodies are obtained through the ATCC (fax (1)-301 231 5826), the ECACC (see above), or, via literature searches, from the originator. The antibodies used for purification should preferably be monoclonal and should be selected for their performance in immunopurification. Most of the antibodies that we have used will elute the MHC under reasonably mild conditions such as a brief exposure to pH 10–11 at 4°C.

One recent alternative involves molecular cloning of the MHC molecules of interest. The engineered molecules can be secreted as soluble molecules from eukaryotic cell lines (Fahnestock et al 1992; Godeau et al 1992; Matsumura et al 1992). Alternatively, they can be expressed in bacteria, extracted as denatured molecules, and purified (Garboczi et al 1992; Silver et al 1991). Subsequently, removal of denaturant and refolding in the presence of β_2m and the proper peptide can be initiated. We shall not deal with the recombinant production of MHC class I, but describe only the purification of natural MHC class I molecules.

MHC purification

1. Frozen lysate ($2–10 \times 10^{10}$ cell equivalents) is thawed slowly (e.g., overnight in the refrigerator) and cleared by high-speed centrifugation ($15,000g$, 20 min, 4°C) followed by 0.45 µm filtration. *All* steps are conducted in the cold (e.g., on ice).

2. The cleared and filtered lysate is applied to an affinity column (5–20 ml) that has been coupled with the relevant antibody at 2–10 mg antibody ml^{-1}.

3. The column is washed extensively (at least 15 column volumes PBS containing 0.5% NP-40 and 0.1% SDS followed by 15 column volumes PBS, and finally equilibrated with 3 column volumes PBS containing 1% β-octyl glucoside (OG)).

4. The MHC molecules are eluted under basic conditions (e.g., 0.15 M NaCl containing 50 mM diethylamine (DEA) and 1% OG, pH 10.5–11.0) and neutralized in 1/20 volume of 2 M Tris, pH 6.3. The eluate is collected in fractions.

5. The fractions obtained are analyzed, e.g., by SDS-PAGE.

6. The MHC-containing fractions are pooled and concentrated by vacuum ultrafiltration to 250–500 µl (Schleicher & Schuell or Sartorius). Add 10 ml PBS containing 0.1% azide and 1% OG and bring the volume down to 250–500 µl again.

7. The concentrated preparation (typically 0.5–5 mg ml^{-1}) is quantified and stored at 4°C.

A typical yield from a mouse lymphoma cell is 50 µg/10^9 cells, whereas the yield from human EBV-transformed cell lines is about 10 times higher. The quality of the MHC varies somewhat from MHC to MHC, from preparation to preparation, but 50% maximal peptide binding can be expected at around 100–1000 nM. Binding analyses are usually compatible with a single equilibrium reaction with a K_d from 2 nM and up. Only a minor fraction, usually between 2% and 5%, of the MHC is available for binding.

The indicator peptide and its ^{125}I labeling

To establish an inhibition assay, an agonist indicator peptide must be identified. The indicator peptide should fulfill the following criteria: It must bind to the MHC in question; it must have an amino acid compatible with the labeling procedure; and it must be compatible with the separation method selected. The search for indicator peptides is based on known immunogenic epitopes reported in the literature. Alternatively, they may be deduced from sequencing of peptides eluted off the MHC (Falk et al 1991). In the latter case the most prevalent amino acid in each position is chosen except for the amino acid to be labeled. The position of this amino acid is determined by trial and error, but the C-1 and C-2 positions are most likely to accept the label. It should be noted that class I epitopes are very restricted in size, and that critical anchor residues must be positioned accurately within a given peptide epitope to match the class I in question.

The best label is currently ^{125}I since it can label peptides to a high specific activity, the detection is sensitive, and it is easy to quantify. This requires the presence of tyrosine for iodination. If it is not known where the tyrosine can be accepted, one may have to make several analogues, scanning the tyrosine through the peptide. Other labeling methods such as biotin followed by chemiluminescence or time-resolved fluorescence (Delphia (E), Wallac) are in use, but unfortunately the amount of free ligand is usually not measured in these protocols, leading to a lack of internal controls and a loss in precision.

Be prepared to test a number of different peptides and aim for a peptide which binds at least 40–50% of the offered label (frequently, close to 90% can be bound) with a background less than 1%. It should be noted that some peptides appear less than 100% bindable, i.e., although you add more and more MHC, binding levels out and reaches a plateau below 100%. An MHC dose–response study should be made for each peptide preparation to determine the maximal bindable fraction of labeled peptide, and all results with this preparation should be corrected accordingly.

Procedure

1. Purify the peptide to homogeneity by HPLC.
2. Take up about 2 nmol (2 µg) in 25 µl 0.3 M phosphate, pH 7.5.
3. Add 10 µl ^{125}I (equaling 1 mCi).
4. Add 5 µl 1 mg ml^{-1} chloramine T, and incubate for 45 s at room temperature.
5. Quench the reaction by adding 5 µl 1 mg ml^{-1} metabisulfite and incubate for 45 s at room temperature.
6. The labeled peptide is gel filtered over a disposable minicolumn (Bio-Spin, Biorad 732-6008) with 2 ml Sephadex G10 (Pharmacia) equilibrated in PBS containing 0.1% azide and 2% ethanol. To reduce loss due to absorption, prime the column with 100 µl 0.05% Tween 20 just before use.
7. Collect 0.2 ml fractions, pool the void volume containing the labeled peptide.
8. Determine specific activity (usually 250 Ci/nmol peptide and 0.5–1 µCi µl^{-1}).
9. Verify purity by HPLC followed by γ-spectrometry. In our experience there is less than 1% free radioactivity after one G10 separation. The peptide appears in 1–2 peaks (the appearance of two peaks probably reflects a mixture of mono- and diiodolabeled tyrosine).
10. Store at 4°C in a lead container. The azide maintains sterility and ethanol acts as a scavenger of radiolysis. Note that a labeled peptide preparation has a rather short shelf-life (about 3–6 weeks). It is time to relabel whenever the MHC binding drops, or when free radioactivity/degradation products start to appear on HPLC analysis.

> 11. Do a dose–response study of MHC to determine the bindability of the labeled peptide.

Biochemical assays of peptide binding

Peptide–MHC complexes are stable enough that bound and free peptide can be separated even by nonequilibrium methods such as simple gel filtration. The loss of complexes due to dissociation during the separation is insignificant and gel filtration can, once equilibrium has been reached, be used to determine accurately the amount of bound and free peptide at equilibrium. Most indicator peptides for MHC class I binding assays will be around 1000 Da in molecular weight, whereas once bound to an MHC molecule (typically in an NP-40 detergent micelle of about 90,000 Da) will be up to 150,000 Da. Peptide associated with MHC will thus appear in the void volume and free peptide will appear in the included volume of a Sephadex G25 to G50 column. Gel filtration has proved to be a fast and reliable separation method for determining peptide–MHC interaction for class I, as well as for class II (Buus et al 1986; Olsen et al 1994). Usually baseline separation can be obtained, allowing accurate quantification. Very little material is required because small volumes can be used. All in all, gel filtration is sensitive, economical, simple, and robust. Two versions will be described below. Classical gel filtration is used as a reference method against which the gradient centrifugation method is checked. The gradient centrifugation method has a much higher throughput and is used routinely.

A dose–response study should be made for each MHC preparation to determine the concentration of receptor to be used (and, conveniently, the bindability of the label can be measured at the same time). As a general rule, a receptor concentration leading to 10–20% binding (after correction for bindability) should be used. Any sample can subsequently be tested for the presence of inhibitory material and the inhibition is expressed as the concentration of inhibitor leading to 50% inhibition (the IC_{50}). Under these conditions the IC_{50} reflects the really interesting parameter, the K_d (in the literature, one frequently, finds the 'ratio of inhibitor versus label' reported as indicating the strength of binding; this is a misconception and the 'ratio of inhibitor versus label' is at best irrelevant).

The optimal binding condition varies depending on the MHC and must be determined empirically. In particular, MHC class I is unstable at 37°C, leading to a biphasic association curve where binding peaks rapidly followed by a slow decline. This makes the assay difficult to handle and interpret. In general, we perform binding assays at 18°C where MHC and peptide–MHC complexes are much more stable. Usually we incubate the reactants for 24–48 h. The buffer is usually PBS (pH 7; add a little phenol red to visually check the pH throughout the experiment) containing a final concentration of 0.1% azide and 0.05% NP-40.

Setting up peptide–MHC class I interaction

Mix the following on ice:

$x\,\mu l$ MHC (enough to lead to 10–20% binding)

$1\,\mu l\ \beta_2 m$ ($1\,mg\,ml^{-1}$ stock in PBS)

$y\,\mu l$ available (e.g., for inhibitor peptide in PBS; $x + y = 11\,\mu l$)

$2\,\mu l$ freshly made protease inhibitor–detergent cocktail ('PI-De-mix'; see below)

$1\,\mu l$ labeled indicator peptide (a total of 50,000 cpm for the classical gel filtration assay and 10,000 cpm for the gradient centrifugation gel filtration assay)

The total volume should be $15\,\mu l$. The reaction is conveniently conducted in V-bottom 96-well microtiter plates. Seal the plate to prevent evaporation. Incubate at 18°C.

The 'PI-De-mix' is a concentrate of various protease inhibitors and detergent. It is mixed from the stock solutions (see below) in the ratio EDTA–NEM–pepstatin A–1,10-phenanthroline–TLCK–TPCK–PMSF–NP-40 = 12:4:4:4:1:1:2:28 by vol. The final reaction mixture contains 1 mM PMSF, 8 mM ethylenediaminetetraacetic acid (EDTA), 1.2 mM 1,10-phenanthroline, 69 μM pepstatin A, 128 μM $N\alpha\text{-}p$-tosyl-L-lysine chloromethyl ketone (TLCK), 135 μM $N\alpha\text{-}p$-tosyl-L-phenylalanine chloromethyl ketone (TPCK), 1 mM N-ethylmaleimide (NEM), and 0.05% NP-40. It is made up freshly each time from stock solutions of the individual components. Some of the reagents are toxic and should be treated with due care. The stock solutions are:

- Serine protease inhibitors: PMSF (e.g., Sigma P-7626, $40.0\,mg\,ml^{-1}$ in isopropanol (extremely toxic); add just before use as it is unstable in aqueous solutions)
- Cysteine protease inhibitor: NEM (e.g., Sigma E-3876, $12.5\,mg\,ml^{-1}$ in isopropanol)
- Metalloprotease inhibitors: EDTA (e.g., Sigma E-4884, $100.0\,mg\,ml^{-1}$ in PBS (pH 7.2)) and 1,10-phenanthroline (e.g., Sigma P-9375, $26.0\,mg\,ml^{-1}$ in isopropanol)
- Aspartic acid protease inhibitor: pepstatin A (e.g., Sigma P-4265, $5.0\,mg\,ml^{-1}$ in isopropanol)
- Serine and cysteine protease inhibitors: TLCK (e.g., Sigma T-7254, $20.0\,mg\,ml^{-1}$ in water: smelly, use in hood) and TPCK (e.g., Sigma T-4376, $20.0\,mg\,ml^{-1}$ in isopropanol)
- Detergent: 0.8% NP-40 in PBS, 0.1% azide

Measuring interaction by Sephadex gel separation

1. Incubate the reaction at 18°C for the time needed to obtain steady state (typically 48 h).
2. Prepare a $1 \times 23\,cm$ Sephadex G50 column and equilibrate it in 0.5% NP-40, 0.1% azide, PBS.
3. Apply the $15\,\mu l$ reaction mixture to the column and elute it in 0.5% NP-40, 0.1% azide, PBS.
4. Collect 1 ml effluent fractions and count them by gamma spectrometry.

Measuring interaction by gradient centrifugation spun column chromatography

Spun columns are made in commercially available 1 ml automatic pipette tips (Greiner, Frickenhausen, Germany, cat. no. 740290) which are equipped with small cotton plugs as frits. When inserted into a 4 ml 70 × 11 mm centrifuge tube (Nunc, Roskilde, Denmark, cat. no. 3 40399), these tips, owing to their high collar, are suspended well above the bottom of the tube, allowing efficient flow during centrifugation. Sephadex G25 SF (Pharmacia) is kept in PBS containing 0.5% NP-40 as a thick slurry (100 g Sephadex suspended to a 600 ml in running buffer) which can be aliquoted into the tips by pipetting. PBS containing 0.5% NP-40 is used as running buffer throughout. The tips are spun in a programmable centrifuge (Sigma 4-1OP; rotor 11140; where 800 and 2400 rpm equals 112g and 1011g, respectively). The best results are obtained in a programmable centrifuge; however, increasing the centrifugation stepwise manually will also work.

1. Make tips with cotton frits. Resuspend the Sephadex and distribute 500 µl slurry into each. We always assay all samples in duplicate (or more).
2. Spin the columns 'semi-dry' at 800 rpm (112g) for 2 min. Visually inspect the columns. The columns should be white throughout at this point (if they are gray (i.e., wet) they should be discarded; if too many columns must be discarded, you are probably using too much cotton frit).
3. Add 25 µl running buffer to the well with the sample, mix and apply 15 µl to the top of each of two 'semi-dry' columns.
4. Spin at 800 rpm for 2 min.
5. Apply 15 µl running buffer to the top of each column.
6. Spin from 0 to 2400 rpm (1011g) over 4 min, and keep spinning at 2400 rpm for another 2 min.

7. The tube now contains the excluded volume with the MHC-bound peptides.
8. The gel in the tip contains the retained free ligand. Cut the tips below the collar with a hot blade, invert the part of the tip containing the gel into a new 4 ml Nunc tube, and spin the gel material into the bottom of the new tube.
9. For each determination you now have two tubes; one containing the bound peptide and one containing the free peptide. If each sample has been assayed in duplicate, you should have four tubes representing bound and free peptide in duplicate pairs.
10. The distribution of bound and free radiolabeled ligand is measured by gamma spectrometry (we have found that the counting efficiencies of the bound and free labeled ligand are identical). Comparison of the gradient centrifugation gel filtration assay with the classical gel filtration assay usually confirms that more than 95% of the bound peptide has been excluded and more than 99% of the free peptide has been retained.

Interpreting the data

For the classical gel filtration assay the ratio of bound peptide to total peptide is calculated as: α (in %) = 100 × (total counts recovered in the void volume peak)/(total counts recovered in the void volume peak + total counts recovered in the included volume). For the spun column assay the ratio of bound peptide to total peptide is calculated as: α (in %) = 100 × (counts recovered in tube containing the excluded volume)/(total counts recovered in the excluded volume tube + the retained volume tube). The spun column assay is usually done in duplicate. Calculate the mean and the standard deviation. The latter is usually less than 1%. Plot the concentration of inhibitor peptide versus bound measured indicator peptide. Interpolate the data to determine the concentration of inhibitor needed to inhibit binding by 50% (IC_{50}). This value approximates the K_d of the inhibitory peptide as long as the receptor concentration is kept below the K_d.

β_2-Microglobulin is quite susceptible to radioiodination. Do not label it to a specific activity of more than $15 \,\mu Ci \,\mu g^{-1}$ (reduce labeling time to 25 s) and be prepared to relabel frequently.

For classical gel filtration: use Sephadex G75. For spun column chromatography: use Sephadex G50. Increase the volume to $830 \,\mu l$ packed slurry. Increase the pre-sample spin to 1500 rpm (395g) for 5 min. Load $12 \,\mu l$ sample. Spin 1500 rpm for 5 min and apply $12 \,\mu l$ running buffer. Spin 0–3500 rpm (2150g) over 4 min and keep spinning at 3500 rpm for another 2 min. Process as described above.

Modifying the gel filtration to measure β_2-microglobulin binding to MHC class I

This assay has previously been described by Pedersen et al (1994).

References

Bjorkman PJ, Saper MA, Samraoui B, Bennett WS, Strominger JL, Wiley DC (1987) The foreign antigen binding site and T cell recognition regions of class I histocompatibility antigens. Nature 329: 512–518.

Buus S, Sette A, Colon SM, Jenis DM, Grey HM (1986) Isolation and characterization of antigen-Ia complexes involved in T cell recognition. Cell 47: 1071–1077.

Buus S, Pedersen LØ, Stryhn A (1995) The analysis of peptide binding to MHC and of MHC specificity. In Zeegers N, Boersma WJA, Claasen E, eds. Immunological Recognition of Peptides in Medicine and Biology. CRC Press, Boca Raton, pp.61–77.

Fahnestock ML, Tamir I, Narhi L, Bjorkman P (1992) Thermal stability comparisons of purified and peptide-filled forms of a class I MHC molecule. Science 258: 1658–1662.

Falk K, Rötzschke O, Stevanović S, Jung G, Rammensee H-G (1991) Allele-specific motifs revealed by sequencing of self-peptides eluted from MHC molecules. Nature 351: 290–296.

Garboczi DN, Hung DT, Wiley DC (1992) HLA-A2-peptide complexes: refolding and crystallization of molecules expressed in Escherichia coli and complexed with single antigenic peptides. Proc Natl Acad Sci USA 89: 3429–3433.

Germain RN, Margulies DH (1993) The biochemistry and cell biology of antigen processing and presentation. Annu Rev Immunol 11: 403–450.

Godeau F, Casanova J-L, Fairchild KD, (1991) Expression and characterization of recombinant mouse β_2-microglobulin type a in insect cells infected with recombinant baculoviruses. Res Immunol 142: 409–416.

Godeau F, Luescher IF, Ojcius DM, et al (1992) Purification and ligand binding of a soluble class I major histocompatibility complex molecule consisting of the first three domains of H-2K^d fused to β_2-microglobulin expressed in the Baculovirus–insect cell system. J Biol Chem 267: 24223–24229.

Matsumura M, Saito Y, Jackson MJ, Song ES, Peterson PA (1992) In vitro peptide binding to soluble empty class I major histocompatibility complex molecules isolated from transfected *Drosophila melanogaster* cells. J Biol Chem 257: 23589–23596.

Nissen MH, Thim L, Christensen M (1987) Purification and biochemical characterization of the complete structure of a proteolytically modified β_2-microglobulin with biological activity. Eur J Biochem 163: 21–28.

Olsen AC, Pedersen LØ, Hansen AS, et al (1994) A quantitative assay to measure the interaction between immunogenic peptides and MHC class I molecules. Eur J Immunol 24: 385–392.

Pedersen LØ, Hansen AS, Ølsen AC, Gerwien J, Nissen MH, Buus S (1994) The interaction between beta 2-microglobulin (β_2m) and purified class I major histocompatibility (MHC) molecules. Scand J Immunol 39: 64–72.

Sette A, Buus S, Apella E, et al (1989) Prediction of major histocompatibility complex binding regions of protein antigens by sequence pattern analysis. Proc Natl Acad Sci USA 86: 3296–3300.

Silver ML, Parker KC, Wiley DC (1991) Reconstitution by MHC-restricted peptides of HLA-A2 heavy-chain with beta 2-microglobulin, *in vitro*. Nature 350: 619–622.

Measuring the capacity of peptides to bind class II MHC molecules and act as TCR antagonists

9.7

Alessandro Sette[1]
John Sidney[1]
Jeff Alexander[1]
Howard M. Grey[2]

[1]Cytel Corporation, San Diego, California, USA
[2]La Jolla Institute for Allergy and Immunology, San Diego, California, USA

TABLE OF CONTENTS

Immunology Methods Manual
ISBN 0–12–442712–X

Abstract

A series of experimental protocols for dissecting the various molecular steps involved in the formation of the trimolecular peptide–MHC–T cell receptor (TCR) complexes are described. Peptide–class II binding is measured by the use of purified MHC and radiolabeled peptide probes in a classical receptor–ligand competition assay. The capacity of test peptides to act as specific antagonists of a given TCR is measured in a prepulse assay, which allows the discrimination of TCR antagonism by peptide competition for binding to MHC molecules.

Introduction

One of the most crucial events during the generation of immune responses is the formation of a trimolecular complex between antigenic peptides, class I or class II MHC molecules, and the antigen-specific T cell (TCR) of $CD4^+$ or $CD8^+$ T cells. Both the induction and effector functions of helper and cytotoxic T cells are dependent on this event. In this context, it can be said that the normal functions of the immune response, such as induction of delayed type hypersensitivity (DTH) responses against parasites, cytotoxic T lymphocyte (CTL) responses against tumors and viruses, as well as induction of specific antibodies against bacteria, are all ultimately dependent on the formation of such trimolecular complexes. This molecular event also underlies such pathological reactions as autoimmunity and allergy. In this light, to gain a molecular insight into the functioning of the immune system, it is essential to quantify the capacity of peptides to bind MHC, as well as the capacity of peptide–MHC complexes to engage the TCR. Since the mid 1980s our group has devoted much of its efforts to the development of systems for studying the various steps involved in this trimolecular reaction. Here we describe assays currently in use in our laboratories to measure peptide–MHC binding, antigenicity, and TCR antagonism.

Alternative methods, and the binding of peptides to class I and class II molecules in general, have been the subject of a number of excellent reviews in the last few years (Barber and Parham 1993; Engelhard 1994; Germain 1993; Germain and Margulies 1993; Joyce and Nathenson 1994; Rothbard 1994; Rötzschke and Falk 1994; Sette and Grey 1992; Sinigaglia and Hammer 1994; Stern and Wiley 1994). Similarly, the molecular mechanisms involved in antigen processing, T cell recognition, and TCR antagonisms have also been reviewed recently (Germain 1993, 1994; Germain and Margulies 1993; Jameson and Bevan 1995; Jorgenson et al 1992; Sette et al 1994), and will not be discussed in this review.

Binding of peptides to class II molecules

The principle of the assay

To study MHC–peptide interactions quantitatively an easily performed, specific, and reproducible technique for the direct measurement of the capacity of peptide ligands to bind MHC molecules is essential. The procedure detailed below does this using purified and detergent-solubilized MHC molecules and radiolabeled peptide ligands.

In essence, purified MHC molecules are incubated with an excess of a radiolabeled probe peptide in the presence of a cocktail of protease inhibitors, and incubated for 2 days at room temperature (Fig. 9.7.1). At the end of the incubation period, MHC–peptide complexes are separated from unbound radiolabeled peptide by size exclusion gel chromatography, and the percentage of bound radioactivity is determined. To determine the binding affinity of a particular peptide for an MHC molecule, various doses of unlabeled competitor peptide are incubated with MHC molecules and then labeled probe peptide. The concentration of unlabeled peptide required to inhibit the binding of the labeled peptide by 50% (IC_{50}) can then be determined by plotting a peptide concentration versus percentage inhibition curve. Because of the conditions utilized, where [label]<[MHC], and $IC_{50} \geq$[MHC], the measured IC_{50} values are reasonable approximations of true K_d values. Class II–peptide interactions are typically in the 1 nM to 50 μM range of affinities at 23–37°C. Association rate constants are relatively low (10^2–10^3 M s^{-1}). Once formed, the complexes tend to be extremely stable, with half-lives at 37°C between a few hours and several days. To date, the methods described below have been successful for establishing over 13 human and 7 mouse class II binding assays (see Table 9.7.1.)

Alternative methods which have been used to determine the affinity of peptide ligands for class II MHC molecules include the use of phage display libraries (Hammer et al 1993) and the use of biotynilated peptides in studies utilizing live antigen-presenting cells (APCs) (Busch et al 1990; Marshall et al 1994). Although the focus of this chapter is on class II–peptide interactions, it may be noted that the same procedure is used for the study of class I–peptide interactions with only minor variations. Alternative methods for the analysis of class I–peptide interactions are described elsewhere in this volume, and have also been discussed previously (Joyce and Nathenson 1994; Sette et al 1992).

Assay validation

As an integral part of developing peptide–MHC binding assays we strongly recommend the validation of assays at the biochemical and biological levels. Validation efforts are essential since class II molecules in some situations can yield low-affinity and artifactual binding.

At the biochemical level it is essential to demonstrate specificity in the sense of the interaction being inhibitable by an excess of unlabeled peptide ligand. It should also be demonstrated that the class II molecule of interest binds some, but not all, peptide ligands.

The biological relevance of an assay may be established by various approaches. A first approach correlates the binding specificity of peptides with known MHC restriction. In most cases, peptides capable of eliciting a response restricted to a given class II molecule will bind

Table 9.7.1 Summary of human and murine class II–peptide binding assays

			Radiolabeled peptide		
Antigen	Allele	Cell line	Source	Sequence	Notes
Human					
DR1	DRB1*0101	LG2	HA Y307–319 (added Y)	YPKYVKQNTLKLAT	Optimal assay pH is 4.5
DR3	DRB1*0301	MAT	MT 65kD Y3–13	YKTIAFDEEARR	
DR4w4	DRB1*0401	Preiss	Nonnatural	YARPQSQTTLKQKT	
DR4w10	DRB1*0402		DR4 nonnatural binder	YARFQRQTTLKAAA	
DR4w14	DRB1*0404	BIN 40	Nonnatural	YARFQSQTTLKQKT	
DR7	DRB1*0701	Pitout	Tetanus toxoid 830–843	QYIKANSKFIGITE	
DR5	DRB1*1101	Sweig	Tetanus toxoid 830–843	QYIKANSKFIGITE	
DR2w2β1	DRB1*1501	GM3107 or L466.1	MBP Y78–101	YGRTQDENPVVHFFKNIVTPRTPPP	
DR2w21β1	DRB1*1601	L242.5	Nonnatural	YAAFAAAKTAAAFA	
DR52a	DRB3*0101	MAT	Tetanus toxoid 1272–1284	NGQIGNDPNRDIL	No NEM in PI mix
DRw53	DRB4*0101	L257.6	Nonnatural	YARFQSQTTLKQKT	
DR2w2β2	DRB5*0101	GM3107 or L416.3	Tetanus toxoid 830–843	QYIRANSKFIGITE	
DR2w21β2	DRB5*0201	L255.1	HA 307–319	PKYVKQNTLKLAT	
DQ3.1	DQA1*0301/ DQB1*0301	PF	Nonnatural	YAHAAHAAHAAHAAHAA	
Murine					
IA^{b}		DB27.4	Nonnatural (ROIV)	YAHAAHAAHAAHAAHAA	Optimal assay pH is 5.5
IA^{d}		A20	Nonnatural (ROIV)	YAHAAHAAHAAHAAHAA	
IA^{k}		CH–12	HEL 46–61	YNTDGSTDYGILQINSR	Optimal assay pH is 5.0
IA^{o}		LS102.9	Nonnatural (ROIV)	YAHAAHAAHAAHAAHAA	
IA^{u}		91.7	Nonnatural (ROIV)	YAHAAHAAHAAHAAHAA	
IE^{d}		A20	Lamda repressor 12–26	YLEDARRKKAIYEKKK	Optimal assay pH is 5.0
IE^{k}		CH–12	Lamda repressor 12–26	YLEDARRKKAIYEKKK	Optimal assay pH is 5.0

with relatively high affinity ($IC_{50} \leq 500$ nM) (Alexander et al 1994a,b; Sidney et al 1992, 1994). Conversely, if a peptide fails to bind to a given MHC molecule, no T cell response to that peptide should be detected in the context of the given MHC molecule (Schaeffer et al 1989).

According to an alternative approach, the correlation between the capacity of unrelated peptides to compete with antigen for class II-restricted antigen presentation and their class II direct binding capacity as measured in the binding assay can also be examined. A good correlation suggests that the binding site involved in the interaction measured by the binding assay is the same site involved in the presentation of antigenic fragments to the T cell receptor (Buus et al 1987, 1988; Lamont et al 1990).

MHC purification

Establishing a binding assay requires appropriate reagents and protocols for production and isolation of purified MHC molecules. We have used a number of different cell lines and cell types successfully. Table 9.7.1 lists some of the cell lines we have used or are routinely using as sources of class II molecules. For historical reasons two different protocols have been used to purify class II molecules of various origins (Alexander et al 1994a,b; Sette et al 1992). The procedure presented here may be used for the isolation of class II from EBV-transformed homozygous cell lines and murine B cell lymphomas, and fibroblasts transfected with either human or murine class II.

Cell lines and antibodies

Cell lines are maintained *in vitro* by culture in RPMI 1640 medium (Flow Laboratories, McLean, VA, USA), supplemented with 2 mM L-glutamine (Gibco, Grand Island, NY, USA), 100 U (100 µg ml^{-1}) penicillin–streptomycin solution (Gibco), and 10% heat-inactivated

Table 9.7.2 Monoclonal antibodies used in class II purification

Monoclonal antibody	Specificity
LB3.1	DR
W6/32	HLA-class I
IVD12	DQ
MKD6	IA^d
10.3.6	IA^k
14.4.4	IE^d, IE^k
Y3JP	IA^b, IA^s, IA^u
M1/42	H-2 class I

FCS (Hazleton Biologics) and grown in T-500 flasks or, for large scale cultures, in roller bottle apparatus.

The various monoclonal antibodies we have utilized in the past or are currently utilizing, and their specificities, are shown in Table 9.7.2.

Purification of MHC molecules

Cells should be lysed at a concentration of 10^8 cells ml^{-1} in 50 mM Tris–HCl, pH 8.5, containing 1% NP-40 (Fluka Biochemika, Buchs, Switzerland), 150 mM NaCl, 5 mM EDTA, and 2 mM PMSF. The lysates are filtered through 0.45 µM filters and cleared of nuclei and debris by centrifugation at 10,000g for 20 min.

MHC molecules are then purified by affinity chromatography as described previously (Gorga 1987; Sette et al 1992). Briefly, DR molecules are purified using the mAb LB3.1 covalently coupled to protein A–sepharose CL-4B. Aliquots of cell lysates equivalent to approximately 10 g of cells are passed sequentially through the following columns: Sepharose CL-4B (10 ml), protein A–Sepharose (5 ml), W6/32–protein A–Sepharose (10 ml), and LB3.1–protein A–Sepharose (10 ml), using a flow rate of 30 ml h^{-1}. The columns are washed with 10 column volumes of 10 mM Tris–HCl, pH 8.0, 1% NP-40 (5 ml h^{-1}); 2 column volumes of PBS; and 2 column volumes of PBS–0.4% n-octylglucoside. The DR from the LB3.1 column is eluted with 0.05 M diethylamine in 0.15 M NaCl containing 0.4% n-octylglucoside and 0.02% NaN$_3$ (pH 11.5), immediately neutralized with 2 M glycine, pH 2.5, and concentrated with a Centriprep 30 microconcentrator (Amicon,

Danvers, MA, USA). Protein content may be evaluated by a BCA protein assay (Pierce Chemical Co., Rockford, IL, USA) and confirmed by SDS-PAGE.

Murine class II molecules are purified using the specific mAb indicated in Table 9.7.2 coupled to Sepharose 4B beads. Lysates are filtered through 0.8 and 0.4 μm filters and then passed over the appropriate anti-Ia columns. Columns are washed with 15 column volumes of 0.5% NP-40–0.1% SDS and 2 column volumes of PBS containing 0.1% n-octylglucoside. Finally, the class II is eluted with 0.05 M diethylamine in 0.15 M NaCl containing 0.4% n-octylglucoside, pH 10.5. A 1/20 volume of 1.0 M Tris, 1.5 M NaCl, pH 6.8, is added to the eluate to reduce the pH to ~7.5, and the class II is then concentrated with a Centriprep 30 microconcentrator.

Identification of a high-affinity ligand

The single most challenging problem in establishing a class II assay is the identification of a high-affinity ligand which can be radiolabeled and will give a clearly detectable signal. Indeed, experimental conditions cannot be optimized and a final protocol established until such a ligand has been identified and some binding detected.

Use of peptides restricted to the class II molecule of interest, with an added N- (or C-) terminal tyrosine, is probably the most expedient way to identify high-affinity ligands. If internal tyrosine residues are present, analogues in which they have been replaced with phenylalanine should also be synthesized, thus eliminating the possibility of labeling potential anchor residues which might be deleterious to the binding of MHC. Also, as long peptides (i.e., those greater than 18 residues in length) are often difficult to work with from a separation standpoint, peptides to be used as potential radiolabeled probes should ideally be between 13 and 18 amino acids in length. Thus, it may

be necessary to synthesize truncated analogues of known epitopes.

Alternatively, good binding peptides can also be identified by random screening of natural or engineered peptides. Such screens should include peptides known to be promiscuous with regard to MHC binding or T cell recognition (Alexander et al 1994a, b; Busch et al 1990; Ceppellini et al 1989; Hill et al 1991; Panina-Bordignon et al 1989; Roche and Cresswell 1991; O'Sullivan et al 1991; Sette et al 1990; Sinigaglia et al 1988). If at least one peptide from such a panel yields a detectable signal, a larger panel of unlabeled peptides may be screened without having to label large numbers of peptides, thus simplifying the task of identifying a higher-affinity peptide that can be used as the radiolabeled probe.

Chloramine T labeling

Peptides to be used as radiolabeled probes are iodinated using the chloramine T method. Details are available elsewhere (Bolton and Hunter 1986; Greenwood et al 1963; Samelson 1994), and only a brief description is given here.

1. Tyrosinated peptides are diluted with 0.3 M phosphate buffer to a concentration of 1 μg ml^{-1}.

2. Peptides are incubated for 40 s sequentially with 1 mCi Na^{125}I (~10 μl with Amersham IMS.30), 10 μl of 0.1 mg ml^{-1} chloramine T in 0.3 M phosphate buffer, and 10 μl of 0.1 mg ml^{-1} of sodium metabisulfate in 0.3 M phosphate buffer. The reaction is quenched by diluting the peptide with 1 ml of 0.05% Tween 20 in PBS.

3. Labeled peptide is separated from free iodine with an 8 cm bed of Sephadex G-10 on a 1.5 × 10 cm disposable column previously flushed with PBS. 1 ml fractions are collected and scanned for activity. The first peak,

typically present around fractions 4–6, represents the labeled peptide. These fractions are pooled in small polyethylene bottles (e.g., 15 ml conicals) with 50 μl of 10% azide in H_2O and 100 μl of ethanol (as a free-radical scavenger).

4. Labeled peptides are stored at 4°C. The length of time that a peptide remains usable as a label is dependent on the particular peptide. Typically, a label remains very active for 2–3 weeks. The integrity of a label may be checked periodically by HPLC. If any sign of degradation is detected, the peptide should be discarded, even if binding is observed. Otherwise, the quantitative aspects of the peptide–MHC assay could be seriously affected.

Performing the binding assay

Materials

Materials used at various stages in the preparation or performance of a binding assay are listed below. In many instances, specific brands have been identified. This is done solely for reference and comparison purposes and, unless specifically stated, other brands (assuming equivalent or improved purity in the case of chemicals) may be substituted. Also, depending on the situation, and potential modifications to the protocol described herein, not all of the items below will be necessary.

Chemicals

Dimethyl sulfoxide (DMSO) (Sigma D-2650)

Nonidet P-40 (NP-40) (Fluka 74385)

Sodium azide (Fisher S227I-500)

Sodium phosphate monobasic (J.T. Baker 3818-01)

Sodium chloride (J.T. Baker 3624-05)

Sodium phosphate dibasic (E.M. SX0720-1)

EDTA (Sigma E5134)

Pepstatin A (Sigma P4265)

NEM (Sigma E3876)

Phenanthroline (Sigma P9375)

PMSF (Sigma P7626)

TLCK (Sigma T7254)

Digitonin (Fluka 37006, or Wako 005–76000)

Chloramine T (Sigma C-9887)

Sodium metabisulfite (Fisher S244-500)

Tween 20 (Sigma P-1379); $Na^{125}I$ (Amersham, IMS.30, or NEN, NEZ-033A).

Stock buffers

PBS pH 6.5 and pH 7.0

PBS pH 7.0 0.1% NaN_3 with 0.82% NP-40

PBS pH 7.0 0.1% NaN_3 with 0.05% NP-40

PBS pH 7.0 0.1% NaN_3 with 0.05% NP-40 and 5% DMSO

PBS pH 7.0 0.1% NaN_3 with 0.05% NP-40 and 15% DMSO

PBS pH 6.5 with 0.5% NP-40 and 0.1% sodium azide

PBS pH 7.0 with 0.5% NP-40 and 0.1% NaN_3

0.3 M phosphate–citrate buffer at various pH values.

Plates

96-well polypropylene round-bottom plates (Costar 3794)

Reagent reservoir (Costar 4870)

Mylar plate sealer with adhesive backing (ICN 76-402–05).

Columns

1.5 × 10 cm (Kontes 'Flex' or Bio-rad 'Econo')

20 mm × 25 cm glass columns

Toso-Haas QC-PAK TSK GFC200 7.8 mm × 15 cm (16215)

Toso-Haas QC-PAK TSK GFC2000 SWXL 7.8 mm × 30 cm (08540)

Pharmacia Sephadex G-10

Pharmacia Sephadex G-50 (medium)

Toso-Haas TSK-gel SWXL top-off.

HPLC hardware

Hewlett-Packard 3396A integrator

Hewlett-Packard 9114B disk drive

Gilson 401 dilutor

Gilson 231 sample injector

Beckman 170 radioisotope detector

Beckman 110B solvent delivery module (pump)

Assay

Peptides are tested in a final assay volume of 15 μl which contains inhibitor peptide, MHC, labeled peptide, and a protease inhibitor cocktail. If separations are to be performed in an automated system (see below), the reaction is performed in 96-well polypropylene plates. For manual separation, the reaction may be performed in 12 × 75 mm culture tubes, snap-cap vials, or other similar vessels, depending on the configuration of the separation scheme and radiodetection method utilized. Figure 9.7.1 shows schematically an assay setup.

1. Test peptide stocks may be made in water, PBS, saline, or DMSO. While both solubility and stability are optimal in DMSO, it should be noted that class II assays do not appear to tolerate more than 5% DMSO concentration in the final assay volume. Therefore, if peptide stocks are prepared in DMSO, they should be sufficiently concentrated that the amount of DMSO is diluted over the range of doses tested. The scheme that we utilize is to dissolve peptides at 4 mg ml^{-1} in 100% DMSO. The first 1:10 dilution is done in 5% DMSO in 0.05% NP-40 PBS. Subsequent 1:10 dilutions are done with 15% DMSO in 0.05% NP-40 PBS. This configuration results in a DMSO concentration of 5% in the final assay volume. Less DMSO may be used if solubility is not a concern.

2. 5 μl of each peptide dose is loaded into the reaction vessel. For positive (i.e., no inhibitor peptide) and negative (no MHC) controls, 5 μl of 15% DMSO in 0.05% NP-40 PBS is loaded. It is recommended to also load a full titration of a standard reference peptide, typically unlabeled probe peptide. The standard peptide we use for each type of assay may be found from Table 9.7.1.

3. A 10 μl/well mix of the remaining ingredients is loaded into each well of the experiment (except the cold control). Mixes should be proportioned such that a small (10%) excess is prepared to comfortably supply all of the wells of the experiment. The mix contains per 10 μl:

   - *MHC*: enough as determined by previous titration to yield approximately 15% binding of labeled peptide
   - *radiolabeled peptide* (see Table 9.7.1): ~40,000 counts per well

- *protease inhibitor cocktail*: 2 µl of cocktail as described below
- *PBS azide pH 7.0*: to bring the final volume to 10 µl. For experiments to be performed at a pH other than 7.0, 6 µl of a citrate buffer at a pH 0.5 units below the desired final pH is included in the mix.

4. The mix is put together in a culture tube in the following order (see Fig. 9.7.1): (a) PBS; (b) protease inhibitor cocktail is made (but not yet added); (c) MHC; (d) protease inhibitor cocktail; (e) labeled peptide.

5. The mix is immediately added to each sample. No mix is added to the cold (negative) control well(s); 2 µl of the protease inhibitor cocktail and the appropriate amount of radiolabel are added to the cold control well.

6. 96-well plates are sealed with Mylar Film, which is trimmed along the edges of the plate. To avoid problems with evaporation of the outer wells of the plate owing to incomplete sealing of the Mylar, the edges of the Mylar should be taped down with general-purpose laboratory tape. If the assay is set up in tubes, the tubes

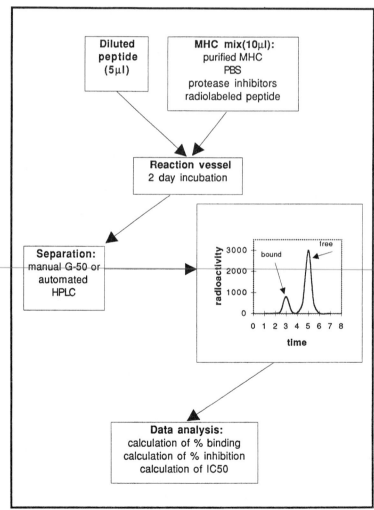

Figure 9.7.1 Summary diagram of an *in vitro* assay system utilizing purified molecules to measure the binding of peptides to HLA class II molecules.

should be tightly stoppered or sealed.
7. The assay is incubated for 2 days at room temperature in the dark.
8. Following incubation, the peptide–MHC complexes are separated from unbound peptide. Two procedures for the separation are described below and data analysis is discussed in the section following. Class II assay plates may be frozen (–20°C) after the 2-day incubation and analyzed at a later time.

The protease inhibitor cocktail

Some of the protease inhibitors utilized are light-sensitive and extremely labile in aqueous solutions. Therefore, it cannot be emphasized enough that the protease inhibitor cocktail must be made fresh at the point indicated in the protocol, and then used immediately.

1. The protease inhibitor stocks are put into solution as described below. These stocks should be stored at –20°C.

   - *EDTA*: 86 mg ml^{-1} in PBS pH 7.0 (if the EDTA does not easily go into solution, it is warmed gently in a water bath; pH should be adjusted to pH 7 with 10 N NaOH)
   - *Pepstatin A*: 5 mg ml^{-1} in methanol
   - *NEM*: 0.8 M in isopropanol
   - *1,10-Phenanthroline*: 26 mg ml^{-1} in ethanol
   - *PMSF*: 40 mg ml^{-1} in isopropanol
   - *TLCK*: 20 mg ml^{-1} in PBS pH 7.0.

2. One 'batch' of protease inhibitor cocktail is mixed in a culture tube as follows: 61 μl 0.82% NP-40 0.1% azide in PBS pH 7.0, or, for murine Ia assays only, 61 μl 20% digitonin (Wako) in water; 12 μl EDTA; 12 μl

pepstatin A; 12 μl NEM; 12 μl 1,10-phenanthroline; 6 μl PMSF; 3 μl TLCK.
3. Because of water–alcohol mixing and alcohol evaporation, each batch yields approximately 100 μl of retrievable cocktail.

Separation of samples

The use of HPLC systems which allow for the automated loading, separation, radioactivity-detection, and data analysis of samples following incubation offers the potential to analyze binding events around the clock. As a result, hundreds of data points may be generated daily. The components listed in the Materials section are specific for the setup we utilize. This arrangement, using a Toso-Haas TSK 200 column (7.8 mm × 15 cm) with a particle size of 5 μm, run at 1.2 ml min^{-1} in PBS pH 6.5 with 0.5% NP-40 and 0.1% azide, affords us the ability to analyze a sample every 7 to 9 min, depending on the condition of the column.

As an alternative, samples may be separated manually using Sephadex G-50 and gravity flow. Samples are eluted with PBS pH 7.0 with 0.5% NP-40 and 0.1% azide. Although this method does not afford the high throughput of automated setups, its low cost is an advantage. Samples are collected in 1 ml fractions and the amount of radioactivity can be determined with a standard gamma scintillation counter.

Data analysis

After chromatographic separation, two peaks are typically present (Fig. 9.7.1). The first peak represents the peptide–MHC complex and the second peak is free peptide. In the automated system that we employ, the first peak appears between 2.5 and 3.5 min after injection. In manual separation systems, the first peak will appear between the 10 and 18 ml fractions when using a 20 mm × 25 cm column. The percentage area of the first peak relative to the total area integrated is recorded as percentage binding. IC_{50} values may be calculated for each test peptide by plotting a dose versus percentage inhibition curve. The percentage inhibition is calculated as follows:

$$\% \text{ Inhibition} = \frac{1 - (\% \text{ binding with test dose})}{(\% \text{ binding with no inhibitor})}$$

Because assays may have sensitivity varying somewhat from day to day, or depending on the batch of purified MHC molecules, to allow the comparison of data obtained in different experiments a relative binding figure (ratio) may be determined for each peptide by calculating the ratio of the IC_{50} of the positive control for inhibition (i.e., the assay standard peptide) to the IC_{50} for each tested peptide:

$$\text{Ratio} = \frac{50\% \text{ dose of the standard peptide}}{50\% \text{ dose of the test peptide}}$$

For database purposes and interexperiment comparisons, relative binding values may be compiled. These values can subsequently be converted back into IC_{50} nM values by dividing the IC_{50} nM of the positive controls for inhibition by the relative binding of the peptide of interest. This method of data compilation has proved to be the most accurate and consistent for comparing peptides that have been tested on different days or with different lots of purified MHC.

Discussion and trouble shooting

The significance of the formation of the trimolecular complex between MHC molecules, peptide antigens, and the TCR was discussed in the general introduction. The assay system described offers a simple, reliable, and quantitative method for studying the initial step in the formation of the trimolecular complex: the binding of peptides to class II MHC molecules.

In general, when high-quality reagents are used, this assay system is easy to perform, and virtually trouble-free. However, we comment on some areas where problems may be encountered.

MHC
Typically, there are many different cell lines available which will express a given MHC of interest. However, we have found that EBV-transformed homozygous cell lines are usually the best first choice when selecting a cell line to use for purification. Because this is not always foolproof, and because some lines express higher or lower levels of MHC than others, it is prudent to have more than one cell line available. It is also best to monitor from time to time the purity and concentration of the MHC preparations used. Once isolated, MHC preparations can be stored at 4°C. We have observed that they maintain good activity for months and even years after isolation.

Detergent
Purified MHC molecules require detergent concentrations above the critical micellar concentration in order to remain in solution. The detergents and detergent concentrations used in this assay system have been worked out over a number of years, and appear to be optimal. If problems are encountered with solubility of MHC molecules it is best to first check the purity of the detergent stocks. This is especially pertinent with regards to digitonin, which is used in the protease inhibitor cocktail for murine Ia assays. As with all aspects of this assay system, the higher the grade of the reagents the fewer problems will be encountered.

Temperature
This binding assay gives the best and most reliable results when incubation is performed at room temperature. If there are temperature fluctuations in the surrounding environment, the assay should be placed in a temperature-controlled setting. The effects of temperature on the assay have been discussed elsewhere (Sette et al 1992).

Protease inhibitors

Even though an MHC preparation may be highly purified, small amounts of proteases may be copurified with the MHC. It is therefore vital that the protease inhibitors are active; proper storage and handling of these reagents is crucial. Most critically, *the protease inhibitor cocktail should be prepared immediately prior to use*, following the schedule outline in the protocol.

Evaporation

Because the final assay volume is so small (15 μl), it must be protected from evaporation. The smaller the vial for incubation, the better. Sealing plates with Mylar film works very well if the edges of the film are taped down.

Radiolabel

As with all other reagents used, fresh material presents the fewest problems. Old and degraded labels often show wide and multi-peaked HPLC profiles. Generally labels last about 2 weeks. Also, the hydrophobicity of some peptides makes them unsuitable for use as radiolabels. These hydrophobic peptides often form high-molecular-weight aggregates that either elute in the area of MHC–peptide complexes, resulting in an uninhibitable front peak, or elute just after peptide–MHC complexes, resulting in poor resolution between the first and second peaks. Similarly, if the peptide used is too large, there may be poor separation of the peaks. In general, peptides between 13 and 18 residues in length are the most problem-free for use as class II radiolabeled probes. Finally, because the chloramine T procedure attaches the ^{125}I to tyrosine, it is essential that each peptide contain at least one, and most preferably only one, tyrosine. In instances where no tyrosines are present, tyrosines may be added to either the N- or C-terminus of the peptide for class II probes. For peptides bearing multiple tyrosines, we have found that in most instances the tyrosines not involved in peptide binding may be substituted with phenylalanine.

Assay pH

For most alleles, we have found that class II assays are optimal at neutral pH. However, for some alleles, the optimal conditions are acidic (see Table 9.7.1). The final assay volume is adjusted to the optimum pH by using in the MHC mix 6 μl of a citrate buffer that is 0.5 pH points below the desired final pH. When developing assays it is prudent to perform direct binding titrations at varying pH values to find the optimal conditions. The influence of pH on binding assays has also been described in detail elsewhere (Sette et al 1992).

Measuring antigenicity and TCR antagonism

Introduction

In the previous subsection we examined experimental protocols to measure the formation of peptide–MHC complexes. Here we shall review protocols which measure the interaction of such complexes with specific TCRs. While we will briefly mention antigenicity assays, particular emphasis will be placed on the TCR antagonism assay (prepulse assay). No mention will be made of direct measurement of TCR–(MHC–peptide complex) interactions by antibody competition or plasma resonance, which has been described elsewhere (Corr et al 1994; Matsui et al 1991).

Antigenicity assays

Antigenicity assays are the most commonly used assays to obtain indirect information regarding whether a particular peptide, once bound by a given MHC molecule, can be recognized by a specific TCR. These assays can be performed in a variety of different ways, but all rely on three crucial components. The T cell population (which can be represented by a T cell clone or hybridoma, or a T cell line or bulk T cell population), the test peptide, and an antigen-presenting cell (APC), which can, in turn, be represented by a purified or clonal cell type (such as spleen adherent cells, activated macrophages, B cell lymphomas, or class II-transfected fibroblasts) or by bulk spleen, lymph node, or PBMC populations.

The test peptide can be used to pulse the APC population before adding the T cells, or simply 'dumped' in the T cell–APC cultures. Typically, a dose range of the 'wild type' antigenic peptides is also assayed, and the doses of wild type antigen and test peptide required to elicit 50% of the maximal response are measured (ED_{50}). The relative antigenicity of any test peptide can then be expressed as the ratio of the wild type peptide/test peptide ED 50%. A relative antigenicity of 0.1 will thus indicate that 10-fold more test peptide than wild type antigen is required to achieve the half-maximal response.

It should be noted that this type of quantification assumes that both the test peptide and the wild type peptide do indeed reach the same plateau maximal response. If this assumption is not met (as in the case of partial agonist peptides) alternative approaches have to be utilized. However, besides this and a few other exceptional causes, it should be noted that this basic antigenicity assay can be utilized in most instances, and irrespective of the particular readout assayed for T cell activity, be it pro-

liferation, lymphokine release, Ca^{2+} fluxes, IP turnover, APC–T cell conjugate formation, and so on.

It should be emphasized that antigenicity assays provide only indirect information regarding the interaction of the various components of the trimolecular complex. In particular, if a peptide is weakly antigenic or nonantigenic, this may be indicative of poor MHC binding capacity, altered capacity of the MHC–peptide complexes to interact with the TCR, or both. It is thus recommended to assay in parallel the same test peptide for MHC binding capacity and antigenicity.

The phenomenon of TCR antagonism

We and others (Allen 1994; Evavold et al 1993a,b; Jameson and Bevan 1995; Sette et al 1994) have reported that antigen analogues can act as powerful and specific inhibitors of T cell activation. This phenomenon was initially observed because of the anomalously high activity of nonstimulatory antigen analogues in inhibition of antigen presentation assays (De Magistris et al 1992). This phenomenon was later demonstrated to cause engagement of the TCR by 'altered peptide ligands' (Alexander et al 1993), formed by MHC and antigen analogues. In this respect, the TCR does behave like many other receptors, whose engagement by altered or suboptimal ligands can have inhibitory effects.

Various studies have revealed that the engagement by TCR antagonists, partial agonists, and agonists of the TCR may result in partial activation, inactivation, or even cell death. Early work (Evavold and Allen 1991) demonstrated that an altered hemoglobin antigen induced IL-4 secretion but no T cell proliferation from a Th2 clone. Evavold and colleagues further observed (1993a,b) that murine Th1 clone cytolysis could be uncoupled from proliferative and lymphokine responses. Our laboratory has focused on inhibition of Th cell proliferation (Alexander et al 1993; De

Magistris et al 1992; Snoke et al 1993). This inhibition was shown to be probably due to inhibition of early T cell signaling events (Ruppert et al 1993). Jameson and colleagues (1993) have also observed TCR antagonism of CTL function, and Evavold and colleagues (1994) have also observed TCR antagonism of superantigen-stimulated Th clones and hybridomas.

The use of antigen analogues has also been a useful tool in the study of the role of affinity of interaction between TCR + MHC–peptide in thymic development. TCR antagonists have been used to induce (Hogquist et al 1994) or inhibit (Spain et al 1994) positive selection. In addition, TCR antagonists have also caused negative selection (Page et al 1994). TCR antagonists and partial agonist peptides have thus become useful tools in research in the areas of thymic education and T cell activation and may ultimately help unravel the intricacies of the various signaling pathways involved in T cell effector functions.

There is also hope that antigen analogues may be useful as therapeutic agents for control of undesired immune responses such as allergies and autoimmunity. Detailed discussion of the molecular mechanisms involved in this effect are beyond the scope of this article, and are discussed in several recent reviews (Allen 1994; Jameson and Bevan 1995; Sette et al 1994;).

Assaying TCR antagonism: the principle

In measuring TCR antagonism, special care must be exercised to exclude the possibility that the inhibitory effects observed might be related to competition of the test peptide and the antigenic peptide for binding to MHC. For this very reason, a specific assay, which we have termed 'prepulse' assay, was developed.

This assay takes advantage of two key features of molecular events involved in T cell activation (Fig. 9.7.2). First, since typically only a very small fraction (0.1% or less) of the MHC molecules on the APC need to be occupied by wild type antigen in order to elicit a measurable T cell response, the APC population is pulsed with a suboptimal antigen dose (Christinck et al 1991; Demotz et al 1990; Harding and Unanue et al 1990). Use of a suboptimal antigen dose (usually corresponding to 30–50% of the maximal response) ensures that the majority of the 'empty' MHC molecules on the APC are available to bind the test analogue peptide, which will be introduced next in the assay system.

Second, since, once formed, MHC–peptide complexes are remarkably stable (Lanzavecchia et al 1992; Sette et al 1992), the peptide-

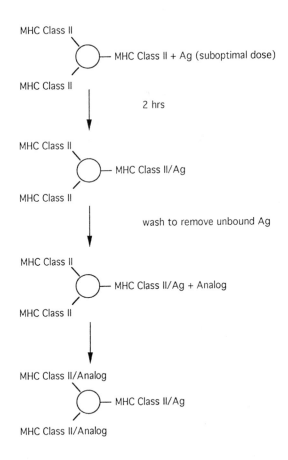

Figure 9.7.2 Prepulse of APCs with antigen excludes Ag-analog competition for MHC class II.

pulsed APC population can be washed to remove unbound antigen. Only after this step is a dose range of the test peptide offered to the APC, with the effect of generating APCs which carry on their surface a constant number of antigen–MHC complexes and increasing numbers of test peptide–MHC complexes.

T cells are added next, the functional readout of choice is measured, and the dose of test peptide yielding 50% inhibition is recorded. Since the antigenic peptide is added in suboptimal amounts and at a different time from the test peptide, competition for MHC molecules can be excluded and TCR antagonism is measured.

Assay variations

Although our group has mostly used the prepulse assay to measure T cell proliferation as a readout, many alternative readouts may be used to study the effects of antigen analogues on T cell function. These include: (1) CTL lysis (Evavold et al 1993a,b; Jameson et al 1993); (2) lymphokine production (Evavold et al 1993a,b; Racioppi et al 1993); (3) early signaling events, i.e., IP turnover, Ca^{2+} influx (Ruppert et al 1993), or ζ-chain phosphorylation (Madrenas et al 1995); (4) apoptosis (Evavold et al 1994); (5) anergy (Alexander et al 1994; Sloan-Lancaster et al 1994); and (6) T cell development (Hogquist et al 1994; Page et al 1994; Sebzda et al 1994; Spain et al 1994). It should be noted that a given peptide may act as an inhibitor (antagonist) when a given T cell readout is utilized, and be stimulatory (agonist) in another readout. This phenomenon has been termed partial agonism/ antagonism and has many parallels in classical pharmacology.

Various types of APCs have also been used in the TCR antagonism assay. In the human Th system, we have used transformed EBV lines or class II-transfected fibroblasts, either mitomycin c-treated or fixed with paraformaldehyde (Alexander et al 1993). In the murine system, we have routinely used irradiated lipopolysaccharide/dextran sulfate-treated spleen blast cells

(Page et al 1994). Evavold and colleagues have routinely used irradiated or fixed spleen cells (Evavold et al 1993a,b; Sloan-Lancaster et al 1994) or mitomycin c-treated B cell lymphomas (Evavold et al 1994) as APCs to demonstrate effects of antigen analogues.

We have demonstrated TCR antagonism, for the most part, in $CD4^+$ T cell clones in both the human and murine systems. However, antigen-induced T cell lines have also been used with equal success (Snoke et al 1993). Others, including Evavold and colleagues (1991, 1993a,b) and Racioppi and colleagues (1993), for the murine system, have studied the effects of antigen analogues on Th1 and Th2 clones. In addition, Evavold has also studied antagonism of superantigen-stimulated T cell hybridomas (1994). $CD8^+$ T cell clones have also been shown by Jameson and colleagues (1993) to be antagonized. As mentioned above, peptide analogues have been used in the study of T cell development. These assays require fetal thymi or adult thymi as a source of T cells (Hogquist et al 1994; Page et al 1994; Sebzda et al 1994; Spain et al 1994).

Materials and apparatus

Materials used at various stages in the preparation or performance of a binding assay are listed below. In many instances, specific brands have been identified. This is done solely for reference and comparison purposes and, unless specifically stated, other brands (assuming equivalent or improved purity in the case of chemicals) may be substituted.

Media

RPMI 1640 (BioWhittaker, cat. no. 12-115B (with 25 mM Hepes)

Fetal calf serum (FCS) (Irvine Scientific, cat. no. 3000)

Human serum AB (Gemini Bioproducts, cat. no. 100-112)

Glutamine (Irvine Science, cat. no. 9317)

Gentamicin (Irvine Science, cat. no. 9355)

MEM nonessential amino acids solution (Gibco BRL, cat. no. 11140-019 MEM sodium pyruvate (Gibco BRL, cat. no. 320-1360AG)

2-Mercaptoethanol (Sigma)

Hanks' balanced salt solution (HBSS) (BioWhittaker, cat. no. 10-508B).

Complete medium

RPMI 1640 with Hepes is supplemented to final concentrations of 10% fetal bovine serum (FCS) for the murine or 5% human serum (HS) for the human system, 2 mM glutamine, 50 μg ml^{-1} gentamicin, 0.05 mM 2-mercaptoethanol, 0.1 mM nonessential amino acids, and 1 mM sodium pyruvate.

Plates

Sterile 96-well U-bottom plate (Costar, cat. no. 3799)

Sterile 96-well flat-bottom plate (Costar, cat. no. 3599).

Chemicals

[Methyl-^{3}H]thymidine (ICN cat. no. 2406601)

Lipopolysaccharide (Difco, cat. no. 3120-25-0)

Dextran sulfate (Pharmacia, Code 17-0340-01)

Paraformaldehyde (Fisher Scientific cat. no. T-353)

Mitomycin c (MMC) (Sigma, cat. no. M-0503).

Equipment

Multichannel pipetter (Oxford Benchmate, cat. no. 21-234 and 21-235)

LKB Wallac cell harvester 1295-001

LKB Wallac betaplate counter 120

LKB Wallac Heat Sealer 1295-012

LKB Wallac Sample Bag 1205-411.

Red blood cell (RBC) lysis

Stock solutions (0.16 M NH$_4$Cl and 0.17 M Tris, pH 7.65). Mix 90 ml of 0.16 M NH$_4$Cl and 10 ml of 0.17 M Tris, pH 7.65;

adjust to pH 7.2 with HCl;

pellet splenocytes and remove supernatant;

resuspend pellet (1 ml lysis buffer/spleen);

incubate 2 to 3 min at room temperature;

add 3 volumes of complete medium and spin;

wash twice with complete medium.

Methods for *in vitro* TCR antagonism assay

We describe herein typical protocols used in the murine and human systems, respectively.

Murine system

- Splenocytes (RBC depleted and 3×10^6 cells ml^{-1}) are activated with lipopolysaccharide (LPS) at 100 μg ml^{-1} and dextran sulfate (DS) at 40 μg ml^{-1} for 3 days and irradiated at 800 rad.
- These APCs (5×10^6 ml^{-1}) are subsequently pulsed for 2 h at 37°C; in complete medium with a suboptimal dose of antigen (giving 30–50% of maximum response), washed twice with complete medium to remove unbound antigen, and plated at

5×10^5 blasts/well (100 µl) in 96-well, flat-bottom microtiter tissue culture plates. An antigen dose–response curve is also run at the same time and under the same conditions. This is to ensure that the antigen dose used for the assay is between 30% and 50% of the maximum proliferative T cell response.

- Tenfold serial dilutions (100 µg ml^{-1} to 10 ng ml^{-1} final concentration) of peptides in complete medium containing 10% FCS are prepared in 96-well U-bottom microtiter plates. These test peptides to be evaluated for TCR antagonism are added (50 µl) to the APCs, and the incubation is continued for 2 h at 37°C, 5% CO_2.

- The T cells, in complete medium containing 10% FCS, are subsequently added (50 µl), resulting in 1×10^5 T cells/well (final volume 200 µl) and the plates are incubated at 37°C, 5% CO_2 for 3 days. The last 16 h of incubation are in the presence of 0.1 µCi of [methyl-^{3}H]thymidine/well.

- [Methyl-^{3}H]thymidine incorporation is determined for duplicate wells by standard liquid scintillation counting by harvesting onto glass fiber filters (LKB Wallac cell harvester) and thymidine incorporation is measured using an LKB Wallac betaplate counter. Percentage inhibition is calculated relative to the no-inhibitor control.

Human system

- Epstein–Barr virus (EBV)-transformed B cell line, homozygous for the appropriate MHC-restricting element or MHC-transfected L cells are either mitomycin c (MMC)-treated (150 µg ml^{-1}, 1 h, 37°C, RPMI–1% FCS, and 5×10^6 cells ml^{-1}) or fixed (0.25% paraformaldehyde, 20 min room temperature) in Hanks' balanced salt solution (HBBS).

- After treatment, the APCs are washed twice in complete medium containing 5% HS and once with complete medium containing 1% FCS.

- The APCs are then pulsed with a suboptimal dose of antigen (2 h, 37°C, 2.7×10^5 APC ml^{-1}, in complete medium containing 1% FCS). The cells are subsequently washed twice in complete medium containing 5% HS and plated (100 µl) in the same medium at 4×10^4 cells/well in a 96-well U-bottom microtiter plate.

- Serial 10-fold dilutions of the test peptides (100 µg ml^{-1} to 10 ng ml^{-1} final concentration) are added (50 µl) in complete medium containing 5% HS and incubated for 2 h at 37°C, 5% CO_2. T cells (50 µl) (2×10^4 well in the case of Cl-1 T cells) (De Magistris et al 1992) are then added.

- A 3-day proliferation assay is then performed as described above for the murine system.

Trouble shooting tips

Both the antigenicity and TCR antagonism assays are relatively straightforward, as long as adequate attention is devoted to certain key elements. One important element is the use, within the confines of each experimental situation, of a suitable source of T cells. Ideally, the T cell source should be such that large numbers of T cells may be reliably and reproducibly obtained, with minimal phenotypic variation amongst different batches, especially with regard to antigen sensitivity.

Another important element of the assay is the type of APC utilized. The choice of the right APC type might be crucial, because of the important influence that APC number and type have on the overall assay antigen sensitivity. We strongly recommend trying different APC types and optimizing for APC number during the course of preliminary assays. It should also be noted that altogether different results may be obtained if different APC are utilized. A peptide which is antigenic with spleen cells might act as an antagonist if B cells are utilized as APC, and so on.

Despite careful assay optimization in certain systems, a considerable 'drift' in antigen sensitivity from experiment to experiment might be difficult or impossible to eliminate. In such a situation, we routinely set up duplicate TCR antagonism experiments which use two different antigen doses 3- to 10-fold apart from each other, to maximize the chance that one of the antigen doses will be in the desired range of 30–50% of the maximal response.

Finally, we should emphasize that, if at all possible, complete parallel dose titrations should be performed for each peptide in both antigenicity and TCR antagonism assays. In murine systems, in particular, antigenic peptides can be poorly antigenic at high doses (high zone inhibition). If such peptides were to be tested for TCR antagonism alone, they could be erroneously identified as TCR antagonists.

References

Alexander J, Snoke K, Ruppert J, et al (1993) Functional consequences of engagement of the T cell receptor by low affinity ligands. J Immunol 150: 1–7.

Alexander J, Ruppert J, Snoke K, Sette A (1994a) TCR antagonism and T cell tolerance can be independently induced in a DR-restricted, hemagglutinin-specific T cell clone. Int Immunol 6: 363–367.

Alexander J, Sidney J, Southwood S, et al (1994b) Development of high potency universal DR-restricted helper epitopes by modification of high affinity DR blocking peptides. Immunity 1: 751–761.

Allen PM (1994) Peptides in positive and negative selection: A delicate balance. Cell 76: 593–596.

Barber LD, Parham P (1993) Peptide binding to major histocompatibility complex molecules. Annu Rev Cell Biol 9: 163–206.

Bolton AE, Hunter WM (1986) Radioimmunoassay and related methods. In Weir DM, ed. Handbook of Experimental Immunology, Vol. 1: Immunochemistry. Blackwell Scientific, Oxford, pp. 26.1–26.56.

Busch R, Strang G, Howland K, Rothbard JB (1990) Degenerate binding of immunogenic peptides to HLA-DR proteins on B cell surfaces. Int Immunol 2: 443–451.

Buus S, Sette A, Colón SM, Miles C, Grey HM (1987) The relation between major histocompatibility complex (MHC) restriction and the capacity of Ia to bind immunogenic peptides. Science 235: 1353–1358.

Buus S, Sette A, Colón SM, Grey HM (1988) Autologous peptides constitutively occupy the antigen binding site on Ia. Science 242: 1045–1047.

Ceppellini R, Frumento G, Ferrara Gt B, Tosi R, Chersi A, Pernis B (1989) Binding of labeled influenza matrix peptide to HLA DR in living B lymphoid cells. Nature 339: 392–394.

Christinck ER, Luscher MA, Barber BH, Williams DB (1991) Peptide binding to class I MHC on living cells and quantification of complexes required for CTL lysis. Nature 352: 67–70.

Corr M, Slanetz AE, Boyd LF, et al (1994) T cell receptor-MHC class I peptide interactions: affinity, kinetics, and specificity. Science 265: 946–949.

De Magistris MT, Alexander J, Coggeshall M, et al (1992) Antigen analog–major histocompatibility complexes act as antagonists of the T cell receptor. Cell 68: 625–634.

Demotz S, Grey HM, Sette A (1990) The minimal number of class II MHC–antigen complexes needed for T cell activation. Science 249: 1028–1030.

Engelhard VH (1994) Structure of peptides associated with class I and class II MHC molecules. Annu Rev Immunol 12: 181–207.

Evavold BD, Allen PM (1991) Separaton of IL-4 production from Th cell proliferation by an altered T cell receptor ligand. Science 252: 1308–1310.

Evavold BD, Williams SG, Chen JS, Allen PM (1991) T cell inducing determinants contain a hierarchy of residues contacting the T cell receptor. Semin Immunol 3: 225–229.

Evavold BD, Sloan-Lancaster J, Allen PM (1993a) Tickling the TCR: Selective T-cell functions stimulated by altered peptide ligands. Immunol Today 14: 602–609.

Evavold BD, Sloan-Lancaster J, Hsu BL, Allen PM (1993b) Separation of T helper 1 cytolysis from proliferation and lymphokine production using analog peptides. J Immunol 150: 3131–3140.

Evavold BD, Sloan-Lancaster J, Allen PM (1994) Antagonism of superantigen-stimulated helper T cell clones and hybridomas by altered peptide ligand. Proc Natl Acad USA 91: 2300–2304.

Germain RN (1993) Antigen processing and presentation. In Paul WE, ed. Fundamental Immunology, 3rd edn. Raven Press, New York, NY, pp. 629–676.

Germain RN (1994) MHC-dependent antigen processing and peptide presentation: providing ligands for T lymphocyte activation. Cell 76: 287–299.

Germain RN, Margulies DH (1993) The biochemistry and cell biology of antigen processing and presentation. Annu Rev Immunol 11: 403–450.

Gorga JC, Horejsi V, Johnson DR, Raghupathy R, Strominger JL (1987) Purification and characterization of class II histocompatibility antigens from a homozygous human B cell line. J Biol Chem 262: 16087–16094.

Greenwood F, Hunter W, Glover J (1963) The preparation of ^{131}I-labeled human growth hormone of high specific radioactivity. Biochem J 89: 114–119.

Hammer J, Valsasnini P, Tolba K, et al (1993) Promiscuous and allele-specific anchors in HLA-Dr-binding peptides. Cell 74: 197–203.

Harding CV, Unanue ER (1990) Quantification of antigen-presenting cell MHC class II–peptide complexes necessary for T-cell stimulation. Nature 346: 574–576.

Hill CM, Hayball JD, Allison AA, Rothbard JB (1991) Conformational and structural characteristics of peptide binding to HLA-DR molecules. J Immunol 147: 189–197.

Hogquist KA, Jameson SC, Heath WR, Howard JL, Bevan MJ, Carbone FR (1994) T cell receptor antagonist peptides induce positive selection. Cell 76: 17–27.

Jameson SC, Bevan MJ (1995) T cell receptor antagonists and partial agonists. Immunity 2: 1–11.

Jameson SC, Carbone FR, Bevan MJ (1993) Clone-specific T cell receptor antagonists of major histocompatibility class I-restricted cytotoxic T cells. J Exp Med 177: 1541–1550.

Jorgensen JL, Reay PA, Ehrich EW, Davis MM (1992) Molecular components of T-cell recognition. Annu Rev Immunol 10: 835–873.

Joyce S, Nathenson SG (1994) Methods to study peptides associated with MHC class I molecules. Curr Opin Immunol 6: 24–31.

Lamont AG, Powell MF, Colón SM, Miles C, Grey HM, Sette A (1990) The use of peptide analogs with improved stability and MHC binding capacity to inhibit antigen presentation in vitro and in vivo. J Immunol 144, 2493–2498.

Lanzavecchia A, Reid PA, Watts C (1992) Irreversible association of peptides with class II MHC molecules in living cells. Nature 357: 249–252.

Madrenas J, Wange RL, Wang JL, Isakost N, Samelson LE, Germain RN (1995) ζ Phosphorylation without Zap-70 activation induced by TCR antagonists or partial agonists. Science 267: 515–517.

Marshall KW, Liu AF, Canales J, et al (1994) Role of the polymorphic residues in HLA-DR molecules in allele-specific binding of peptide ligands. J Immunol 152: 4946–4957.

Matsui K, Boniface JJ, Reay PA (1991) Low affinity interaction of peptide–MHC complexes with T cell receptors. Science 254: 1788–1791.

O'Sullivan D, Arrhenius T, Sidney J, et al (1991) On the interaction of promiscuous antigenic peptides with different DR alleles. Identification of common structural motifs. J Immunol 147: 2663–2669.

Page DM, Alexander J, Snoke K, et al (1994) Negative selection of CD4$^+$ CD8$^+$ thymocytes by T-cell receptor peptide antagonists. Proc Natl Acad Sci USA 91: 4057–4061.

Panina-Bordignon P, Tan A, Termijtelen A, Demotz S, Corradin G, Lanzavecchia A (1989) Universally immunogenic T cell epitopes: promiscuous binding to human MHC class II and promiscuous recognition by T cells. Eur J Immunol 19: 2237–2242.

Racioppi L, Ronchese F, Matis LA, Germain RN (1993) Peptide–major histocompatibility complex class II complexes with mixed agonist/antagonist properties provide evidence for ligand-related differences in T cell receptor dependent intercellular signaling. J Exp Med 177: 1047–1060.

Roche PA, Cresswell P (1991) High-affinity binding of an influenza hemagglutinin-derived peptide to purified HLA DR. J Immunol 144: 1849–1856.

Rothbard JB (1994) One size fits all. Curr Biol 4: 653–655.

Rötzschke O, Falk K (1994) Origin, structure and motifs of naturally processed MHC class II ligands. Curr Opin Immunol 6: 45–51.

Ruppert J, Alexander J, Snoke K, et al (1993) Effect of T-cell receptor antagonism on interaction between T cells and antigen-presenting cells and on T-cell signaling events. Proc Natl Acad Sci USA 90: 2671–2675.

Samelson LE (1994) Iodination of soluble and membrane-bound proteins. In Coligan Current Protocols in Immunology, Vol. 1. Wiley, New York, Section 8.11.

Schaeffer EB, Sette A, Johnson DL, et al (1989) Relative contribution of 'determinant selection' and 'holes in the T cell repertoire' to T cell responses. Proc Natl Acad Sci USA 86: 4649–4653.

Sebzda E, Wallace VA, Mayer J, Yeung RSM, Mak TW, Ohashi PS (1994) Positive and negative thymocyte selection induced by different concentrations of a single peptide. Science 263: 1615–1618.

Sette A, Grey HM (1992) Chemistry of peptide interactions with MHC proteins. Curr Opin Immunol 4: 79–86.

Sette A, Sidney J, Albertson M, et al (1990) A novel approach to the generation of high affinity class II-binding peptides. J Immunol 145: 1809–1813.

Sette A, Southwood S, O'Sullivan D, Gaeta FCA, Sidney J, Grey HM (1992) Effect of pH on MHC class II–peptide interactions. J Immunol 148: 844–851.

Sette A, Alexander J, Ruppert J, et al (1994) Antigen analogs/MHC complexes as specific T cell receptor antagonists. Annu Rev Immunol 12: 413–431.

Sidney J, Oseroff C, Southwood S, et al (1992) DRB1*0301 molecules recognize a structural motif distinct from the one recognized by most DRβ_1 alleles. J Immunol 149: 2634–2640.

Sidney J, Oseroff C, del Guercio M-F, et al (1994) Definition of a DQ3.1-specific binding motif. J Immunol 152: 4516–4525.

Sinigaglia F, Hammer J (1994) Defining rules for the peptide-MHC class II interaction. Curr Opin Immunol 6: 52–56.

Sinigaglia F, Guttinger M, Kilgus J, et al (1988) A malaria T-cell epitope recognized in association with most mouse and human MHC class II molecules. Nature 336: 778–780.

Sloan-Lancaster J, Evavold BD, Allen PM (1994) Th2 cell clonal anergy as a consequence of partial activation. J Exp Med 180: 1195–1205.

Snoke K, Alexander J, Franco A, et al (1993) The inhibition of different T cell lines specific for the same antigen with TCR antagonist peptides. J Immunol 151: 1–7.

Spain LM, Joregensen JL, Davis MM, Berg LJ (1994) A peptide antigen antagonist prevents the differentiation of T cell receptor transgenic thymocytes. J Immunol 152: 1709–1717.

Stern LJ, Wiley DC (1994) Antigenic peptide binding by class I and class II histocompatibility proteins. Structure 2: 245–251.

Human CD8+ T lymphocytes

9.8

Dolores J. Schendel
Barbara Maget
Christine S. Falk
Rudolf Wank

Institute of Immunology, University of Munich, Munich, Germany

TABLE OF CONTENTS

Immunology Methods Manual
ISBN 0–12–442712–X

Abstract

This chapter provides protocols for isolating human lymphocytes from blood or tissues, sensitizing them *in vitro* with antigen-presenting cells, and positively selecting CD8$^+$ cells. Methods for identifying, cloning, and maintaining CD8$^+$ cells in long-term cultures are also provided. Supplementary information regarding preparation of human serum and T cell-conditioned medium is included.

Introduction

CD8$^+$ lymphocytes are one of the two major subsets of T cells that mediate cellular immune responses; they play an essential role in the defense against intracellular pathogens such as viruses and they can also recognize and destroy some types of tumor cells (reviewed in Kupfer and Singer 1989). The T cell receptors (TCR) of CD8$^+$ T cells recognize antigenic fragments of intracellular proteins that are displayed by major histocompatibility complex (MHC) class I molecules at the cell surface. CD8 molecules themselves bind to class I molecules on antigen-presenting cells (APC) and function as coreceptors in T cell activation. Most CD8$^+$ T cells possess cytolytic mechanisms that allow them to specifically destroy cells with which they make direct membrane contact. Some CD8$^+$ T cells also secrete cytokines or interferon-γ that can influence the successful outcome of a cell-mediated immune response.

The nature of the ligands seen by the TCR of CD8$^+$ T cells poses several problems for *in vitro* studies because antigenic peptides are normally derived from proteins synthesized within a cell (reviewed in Yewdell and Resnick 1992). Peptides are processed from proteins in the cytosol and transferred to the endoplasmatic reticulum, where they bind to newly synthesized MHC class I molecules, and this complex is transported to the cell surface. To isolate CD8$^+$ T cells for a specified antigen, a way must be found to achieve adequate presentation of relevant peptides by class I molecules. For example, *in vitro* activation of CD8$^+$ T lymphocytes specific for a particular virus might entail stimulating lymphocytes with APCs directly infected with living virus, in which case ligands will be created from those viral proteins that are efficiently expressed and presented through the natural antigen processing mechanisms of the stimulating cells (Dave 1994; Gotch et al 1987; Lee et al 1993; Malkovsky et al 1988). Alternatively, APC can be infected with recombinant viruses that introduce specific genes whose protein products are expressed in the cytosol; this method allows selected proteins to be studied (Bu et al 1993; Gould et al 1989; Nixon et al 1988; Roberston et al 1993; Townsend et al 1984). Finally, specific peptides that bind to given MHC class I molecules can be used to exogenously pulse appropriate APC (Celis et al 1994; Disis et al 1994; Houbiers et al 1993; Kast et al 1993). Although the major pathway for generating MHC–peptide ligands for CD8$^+$ T cells utilizes proteins that are synthesized in the cytosol, some exogenous proteins taken up by cells via phagocytosis escape to the cytosol in sufficient amounts to be processed and presented in a manner similar to that of endogenously synthesized proteins (Ulmer et al 1994; Kovacsovics-Bankowski and Rock 1995). Various physical means can also be used to force cytosolic uptake of proteins (Chen et al 1993; Donnelly et al 1993; Lopes and Chain 1992; Moore et al 1988).

Standardized protocols for obtaining responding lymphocytes from peripheral blood or lymphoid organs are generally applicable to many experiments, but no single method can suffice for studying the large variety of antigens that come into experimental consideration for stimulation of CD8+ lymphocytes. Selection of the appropriate APC and the form of antigen presentation requires empirical testing. Helpful starting guidelines can be obtained from the extensive literature on human CD8+ T cells specific for many different types of antigens, including proteins of retroviruses and RNA and DNA viruses of many types, as well as proteins of phagocytosed bacteria or soluble proteins presented by macrophages. A number of publications also detail the isolation of CD8+ T cells specific for tumor cells of different histologies. When specific information is not available for human cells, starting ideas can be gleaned from experiments in mice dealing with similar antigen types, with the caveat that it is often not possible to perform directed priming in vivo or to obtain lymph nodes or spleen cells, which form the major sources of lymphocytes for murine experiments.

This chapter details basic procedures for isolating human CD8+ T cells that are applicable to most types of experiments. It also offers some suggestions for preparing APC of several types that can be used for in vitro studies, but of necessity it leaves open the details for obtaining optimal presentation of a given protein or peptide, since these are highly specific for any given experimental situation. Other sections of this volume provide specialized information relating to both of these aspects.

It is assumed that an investigator using these protocols is working in a laboratory suitable for cell culture which is equipped with standard instruments such as a sterile laminar air flow hood, cell culture incubator, centrifuge, light microscope, inverted microscope, water bath, and vortex. It is also assumed that the investigator has basic knowledge regarding the preparation of sterile materials and solutions for use in cell culture. If this is not the case, the reader is encouraged to seek out general information regarding these principles in a handbook on cell culture.

Isolation of human lymphocytes from peripheral blood or tissues

Principle of the method

The most common starting source of material for isolation of human lymphocytes is freshly drawn peripheral blood. Lymphocyte populations can also be isolated from solid tissues such as thymus, spleen, or lymph nodes. Intact tissue is disrupted using a combination of mechanical and enzymatic procedures to obtain a single-cell suspension. Density gradient centrifugation using Ficoll–Hypaque is used to separate the lymphocyte fraction from the remaining cellular components.

Materials and apparatus

Butterfly needle adaptors (for example, Becton Dickinson 19G #38161014) (optional)

Cell dissociation sieve (for example, Sigma CD1 Tissue Grinder Kit)

Centrifuge

Glass beads (2 mm); washed 3 times in distilled water and sterilized

Hemocytometer with coverslip

Incubator: humidified, 6% CO_2, 37°C

Plastic 50 ml conical tubes (for example, Falcon #2070; Costar #6751)

Sterile glass Erlenmeyer flasks (150 ml)

Sterile glass or plastic pipettes (1, 5, 10, 20 ml volumes)

Sterile glass Petri dishes (8–10 cm)

Sterile latex surgical gloves, powder-free (optional)

Sterile scalpels and forceps

Sterile syringes (5, 10, 50 ml)

Solutions and reagents

- Collagenase type IV (for example, Sigma C5138): a 100× stock solution with a concentration of 20 μg ml^{-1} prepared in sterile 0.15 M NaCl
- Deoxyribonuclease 1 (for example, Sigma D4527): a 100× stock solution with a concentration of 2 mg ml^{-1} (~5000 U ml^{-1}) prepared in sterile 0.15 M NaCl
- Fetal bovine serum (FBS) (for example, Life Technologies #10106078): heat-inactivated at 56°C for 40 min; prescreened to support growth of human lymphocytes
- Ficoll–Hypaque solution (for example, Pharmacia Biotech #17084003)
- Human serum (HS): prepared as described in the next subsection
- Hyaluronidase type V (for example, Sigma H6254): a 100× stock solution with a concentration of 10 mg ml^{-1} prepared in sterile 0.15 M NaCl
- L-Glutamine (for example, Life Technologies #25030024)
- Penicillin–streptomycin (for example, Life Technologies #15140114)
- Preservative-free heparin (for example, Novo Nordisk Pharma #I7406797 (25000 I.E.))
- RPMI 1640 (for example, Life Technologies #21875034)
- Serum-free culture medium: RPMI 1640 containing penicillin–streptomycin (100 U ml^{-1}), and L-glutamine (2 mM). Iscove's medium (for example, Life Technologies #21980032) and DMEM medium (for example, Life Technologies #41965039) can also be used.
- Serum-supplemented culture medium: as above and in addition 15% FBS or a mixture of 10% FBS and 5% HS
- Trypan blue solution (for example Sigma T8154)

Basic procedure using heparized blood

The following example is based on preparation of a 50 ml blood sample.

- Draw peripheral blood into a sterile syringe containing 25 units of heparin (5 units/10 ml blood); mix the heparin thoroughly with the blood. Dilute the blood sample 1:1 with serum-free culture medium.
- Place 13–15 ml of room temperature (RT) Ficoll–Hypaque solution into four transparent 50 ml conical tubes and slowly layer 25 ml of diluted blood per tube over the Ficoll–Hypaque solution. Pipetting blood into or under the Ficoll–Hypaque solution can be avoided by holding the tube at a 45° angle and allowing it to flow slowly down the side of the tube.
- Cap the centrifuge tubes tightly to assure sterility and carefully balance

them so that no shaking occurs during centrifugation.

- Centrifuge at 800g for 25–30 min at RT. Rapid acceleration and braking should be avoided since both will disturb the gradient. Carefully place and remove the tubes from the centrifuge.
- The lymphocyte fraction is located at the interface formed between the medium and serum components in the upper fraction and the Ficoll–Hypaque solution of the lower fraction. Red cells and other leukocytes are found in the pellet. Carefully remove and discard most of the upper fraction, leaving only a small amount of medium covering the lymphocyte interface. Using a 10 or 20 ml pipette, completely remove the lymphocytes along with a few ml of the Ficoll–Hypaque solution to assure that all cells are harvested; keep the amount of Ficoll solution to a minimum.
- Transfer the lymphocyte fraction from each gradient to a separate 50 ml conical tube and add serum-free culture medium to a volume of 45 ml. Mix well. Centrifuge for 10 min at 450g. Remove all but 1 ml of the supernatant fluid, disperse the cell pellet in the remaining medium, combine the contents of the four individual tubes in one tube and add serum-free culture medium to a final volume of 45 ml. Repeat the centrifugation step as above. This washing procedure should be repeated once again.
- Disperse the cell pellet by light finger-flicking of the tube and fully resuspend the cells in 5 ml of serum supplemented culture medium. Remove an aliquot and dilute 50 μl of cell suspension with 50 μl of trypan blue solution.
- Determine the number of viable lymphocytes using a hemocytometer and adjust the lymphocyte

concentration to the desired value using serum-supplemented culture medium.

Supplementary procedure using defibrinated blood

- Draw the blood sample into a sterile 50 ml syringe without anticoagulant. Immediately transfer the blood to a 50 ml conical tube together with 5–7 sterile glass beads. The use of more glass beads will increase the size of the fibrin clot, with increased trapping of lymphocytes and lower cell recoveries.
- Cap the tube tightly and rotate it end-over-end for 15 min; perform this procedure without pause and with a very regular timing of about 1 s for each rotation.
- Remove the fibrin-containing clot with sterile forceps; use a separate instrument for each donor sample. Dilute the blood 1:1 with serum-free culture medium and proceed with the Ficoll–Hypaque gradient centrifugation procedure as described in the basic protocol.

Supplementary procedure using lymphatic tissues

- Place the tissue sample in a sterile Petri dish in 5 ml of serum-free culture medium and cut it into small pieces of 1–5 mm^3 size using a sterile scalpel. Transfer the tissue pieces to a 150 ml sterile glass Erlenmeyer flask and add 30 ml of serum-free culture medium containing the digestion enzyme mixture of collagenase type IV (0.2 mg ml^{-1}), hyaluronidase type V (0.1 mg ml^{-1}), and deoxyribonuclease 1 (20 ng ml^{-1}; ~50 U ml^{-1}). Cover the flask and place it in a 37°C CO$_2$ incubator for 4–24 h.
- Transfer the entire contents of the flask to a sterile sieve placed in a

sterile Petri dish. Rinse the flask once with 5 ml of serum-free culture medium and add this to the sieve contents. Remove the plunger from a 5 ml syringe and use its padded end to gently disrupt the tissue pieces. Single cells and small clumps will pass through the sieve.

- Transfer the sieved contents to a 50 ml tube and let the tube stand 5 min. During this time the clumps will settle to the bottom of the tube. Transfer the supernatant medium to a new 50 ml tube leaving the settled debris behind. Centrifuge at 450g for 8 min and resuspend in 40 ml serum-free culture medium. Repeat this procedure once to remove remaining enzyme.
- Resuspend the cell pellet in 20–50 ml of serum-free culture medium. Use the Ficoll–Hypaque density gradient procedure to obtain a lymphocytic fraction free of other cellular components and dead cells.

Discussion and trouble shooting

Following excision, tissue samples should be placed in serum-free culture medium to retain cell viability; this medium can also be supplemented with heparin (0.5 units ml^{-1}). It is important that the starting materials for lymphocyte separation are processed quickly. For solid tissues the best results are obtained if this is done immediately or within a few hours. Peripheral blood lymphocytes can be held for up to 24 h, but it is advisable to dilute the blood 1:1 in serum-free culture medium and hold the samples at room temperature. Ficoll–Hypaque is toxic, so cells should be removed from the gradients and washed immediately. The presence of many platelets can be detrimental in the priming of lymphocytes *in vitro*; when using heparinized blood or tissue lymphocytes, the final centrifugation step should begin with a slower speed (85g) for 5 min which is then increased without interruption to 400g for 15 min. This leads to a substantial depletion of platelets in the cell pellet since many remain in the supernatant fluid. It is also helpful to add heparin (0.5 units ml^{-1}) to the serum-free culture medium to prevent clumping. Heparin should not be added to the serum-supplemented culture medium. While performing the mechanical disruption steps, it is helpful to wear powder-free, sterile surgical gloves to minimize the risk of contamination.

Following Ficoll–Hypaque separation one should obtain a population of lymphocytes with greater than 90% viability. The numbers of lymphocytes per ml of blood and their viability vary among individuals and according to health status and age. To roughly estimate the amount of blood required for an experiment, the figure of 1×10^6 lymphocytes per ml of blood can be used generally for healthy donors in an age range of 20–60 years.

Preparation of a human serum pool

Principle of the method

In many instances human lymphocytes proliferate and differentiate better *in vitro* in medium supplemented with human serum. If serum is prepared from a single donor, AB serum should be used. A serum pool can be prepared using blood of multiple donors; here healthy individuals of mixed ABO types can be used. Serum should be prepared from freshly drawn blood if possible and processed immediately. Heat inactivation is used to inhibit complement activity. Human serum can be stored for periods of months at −80°C.

Materials and apparatus

Butterfly needle adaptors (for example, Becton Dickinson 19G #38161014) (optional)

Centrifuge

Electric rotation device (for example, HETO, Birkerod #R20SP) (optional)

Glass beads (2 mm); washed 3 times in distilled water and sterilized

Mechanical filter device (for example, Millipore Corporation #XX6700P05) (optional)

Plastic 50 ml conical tubes (for example, Falcon #2070; Costar #6751)

Sterile 0.1 μm filter units (for example, Nalgene #1500010)

Sterile 0.45 μm filter units (for example, Nalgene #2450045)

Sterile forceps

Sterile glass or plastic pipettes (1, 5, 10, 20 ml volumes)

Sterile syringes (5,10, 50 ml)

Waterbath, 56°C

Basic procedure

- Draw 200–300 ml of blood from each individual donor, alternating with two 50 ml syringes so that the contents of a single syringe can be distributed immediately to a 50 ml conical tube together with 5–7 glass beads. Syringe exchange is easy to perform if butterfly adaptor needles are used.
- Start the defibrination process immediately for each individual tube. Cap each tube tightly and rotate it end-over-end for 15 min. Perform this procedure without pause and with a very regular timing of about 1 s for each rotation. Electric rotation machines are available for processing larger numbers of tubes.
- Remove the fibrin-containing clots with sterile forceps and cap the tubes tightly. Centrifuge at 1450g for 20 min. Carefully remove the serum from the upper layer without disturbing the cellular sediment. Filter the serum first at 0.45 μm and then at 0.1 μm, distribute in 40 ml aliquots to 50 ml conical tubes, and cap tightly.
- Place the tubes in a 56°C waterbath and incubate for 40 min. Allow the tubes to cool to RT and store at −80°C.

Discussion and trouble shooting

The best growth supplementation for human lymphocytes is seen with serum that is prepared from freshly drawn blood, processed quickly and frozen in working volumes at −80°C. Unused serum can be refrozen or stored for a few days at 4°C. The serum should be free of high concentrations of lipids: these cause major problems in sterile filtration and can be inhibitory for lymphocytes.

It is helpful to obtain blood samples from donors early in the morning and to request that they eat only a light breakfast, avoiding fats and dairy products. Samples that are visibly opaque should not be added to the pool. Healthy male donors can normally donate larger volumes and do not have HLA antibodies that may be present in females who have been pregnant. If large pools of 15–20 donors are made, a mixture of ABO types can be used without detrimental effects on lymphocyte cultures.

Use of prefilters in mechanical filter devices improves the 0.1 μm filtration efficiency. When high-pressure mechanical filtration is not possible, one-way sterile filters can be employed for smaller volumes. It is advisable to filter individual samples first through 0.45 μm filters to improve the 0.1 μm filtration efficiency. Samples are pooled prior to 0.1 μm filtration and they should be filtered prior to heat inactivation.

The 50 ml storage tubes should only be filled to 45 ml and should be able to withstand −80°C.

Preparation of T cell conditioned medium

Principle of the method

Lymphocytes are isolated from defibrinated blood samples, irradiated to block self-usage of cytokines and cultured in medium containing a low percentage of human serum and phytohemagglutinin which stimulates a majority of the lymphocytes to produce cytokines. After 24 h of culture at 37°C in a humidified, CO_2 incubator, the culture medium is harvested and centrifuged to eliminate cell debris. Samples are sterile-filtered at 0.1 μm, aliquoted and stored at −20°C. Highly supportive conditioned medium can be obtained by establishing individual cultures from several donors, pretesting their IL-2 activities as a measure of lymphocyte activation and pooling samples only from those cultures that show high activities.

Materials and apparatus

Cesium-137 irradiation source

Centrifuge

Culture flasks 75 cm^2 (for example, Falcon #3084 or Costar #3376)

Hemocytometer with coverslip

Incubator: humidified, 6% CO_2, 37°C

Light microscope

Mechanical filter device (for example, Millipore Corporation #XX6700P05) (optional)

Plastic 50 ml conical tubes (for example, Falcon #2070; Costar #6751)

Sterile 0.1 μm filter units (for example, Nalgene #1500010)

Sterile glass or plastic pipettes (1, 5, 10, 20 ml volumes)

Solutions and reagents

- Ficoll–Hypaque solution (for example, Pharmacia Biotech #17084003)
- Human serum prepared as described earlier
- L-Glutamine (for example, Life Technologies #25030024)
- Penicillin–streptomycin (for example, Life Technologies #15140114)
- Phytohemagglutinin M (for example, Difco #052857)
- RPMI 1640 (for example, Life Technologies #21875034)
- Serum-free culture medium: RPMI 1640 containing penicillin–streptomycin ($100 \, U \, ml^{-1}$) and L-glutamine (2 mM). Iscove's medium (for example, Life Technologies #21980032) and DMEM medium (for example, Life Technologies #41965039) can also be used.
- Trypan blue solution (for example, Sigma T8154)

Basic procedure for preparing T cell-conditioned medium

- Prepare lymphocytes using the protocol for defibrinated blood given above.
- Determine the number of viable cells and suspend them at a concentration of $10–50 \times 10^6$ cells ml^{-1} medium. Irradiate with 9 Gy to inhibit self-usage of cytokines during the 24 h culture period. Dilute the lymphocytes to a concentration of 1×10^6 cells ml^{-1} in serum-free culture medium and distribute 100–150 ml to 75 cm² culture flasks.
- Add human serum to an end concentration of 1% and phytohemagglutinin to an end concentration of 1%. Lightly cap the flasks to allow gas exchange and incubate them standing upright at 37°C in a humidified incubator with 6% CO_2.
- Transfer the entire contents of individual flasks to 50 ml conical tubes and centrifuge at 450g for 15 min. Remove the supernatant medium, leaving behind the cellular debris, pool, and filter the conditioned medium through an 0.1 μm filter and store in aliquots of 50–100 ml at –20°C.

Discussion and trouble shooting

A major growth factor required for the long-term expansion and cloning of human T cells is interleukin-2, but several other factors provide supplementary support for CD8⁺ T cell growth and differentiation. Many interleukins can be purchased commercially but they are expensive and not all factors are commercially available. An alternative source for several of these factors is conditioned medium derived from phytohemagglutinin-activated lymphocytes. Such conditioned media contain complex mixtures of known and unknown factors that can only be partially defined; nevertheless, conditioned media may support maintenance of some CD8⁺ T cells that is otherwise not possible. Commercial sources of conditioned medium are also available (see 'Maintenance of CD8⁺ T Cell Lines and Clones *In Vitro*'). The residual phytohemagglutinin present in the conditioned medium also provides direct stimulation of the lymphocytes.

An economical and time-saving procedure for making conditioned medium uses lymphocytes from blood processed for preparation of a human serum pool, as described earlier. These cells are more responsive to stimulation and produce better-conditioned medium (Schendel and Wank 1981). After removal of the serum, the cell sediment is resuspended in serum-free culture medium to replace the lost serum volume. Mix well. Dilute the cell samples once again 1:1 with serum-free culture medium and perform Ficoll–Hypaque density separation as described (p. 671) and then proceed with the basic protocol for preparation of conditioned medium.

Highly active sources of conditioned medium will not be produced from cells from every individual donor. Therefore, separate cultures using cells from different individuals should be established and the samples tested individually for their ability to support the growth of IL-2-dependent

cells (Schendel and Wank 1981). This test identifies cultures where good lymphocyte activation has occurred, but it provides no information about the release of cytokines other than IL-2. Those media having the highest levels of IL-2 are pooled, filtered at 0.1 μm and aliquoted for storage at –20°C. When IL-2 testing is performed, the original culture flasks can be stored for up to 2 weeks at 4°C. They should be capped tightly to assure sterility. Such storage often yields richer conditioned medium since cytosolic reserves of cytokines are released from the disintegrating cells.

Activation of human CD8$^+$ T lymphocytes *in vitro*

Principle of the method

Lymphocytes are incubated with appropriate APC *in vitro*. The priming phase can be modified to include multiple rounds of antigen restimulation prior to establishment of T cell lines or clones using supplementary cytokines. Multiple rounds of antigen restimulation allow substantial enrichment for specific responding cells. At the conclusion of the priming phase, unselected responding cells can be cloned directly to isolate cells with unique ligand specificities. Phenotypic surface analysis is used later to identify CD8$^+$ clones. Alternatively CD8$^+$ cells are positively selected using anti-CD8-coupled immunomagnetic beads. Selected populations can be maintained as long-term polyclonal CD8$^+$ T cell lines or cloned to isolate CD8$^+$ cells with unique ligand specificities. CD8$^+$ T cell lines and

clones are maintained in long-term culture through regular restimulation with specific antigen and provision of supplementary interleukins and conditioned medium.

Materials and apparatus

24-well cluster plates (for example, Falcon #3047 or Costar #3424)

96-well flat-bottom microtiter plates (for example, Falcon #3072 or Costar #3596)

96-well round-bottom microtiter plates (for example, Falcon #3077 or Costar 3799)

Cesium-137 irradiation source

Centrifuge

Culture flasks, 25 cm^2 (for example, Falcon #3082 or Costar #3055)

Culture flasks, 75 cm^2 (for example, Falcon #3084 or Costar #3376)

Hemocytometer with coverslip

Incubator: humidified, 6% CO_2, 37°C

Inverted microscope

Laboratory film (for example, Parafilm M)

Light microscope

Plastic 50 ml conical tubes (for example, Falcon #2070; Costar #6751)

Stainless-steel box with loose-fitting lid

Sterile glass or plastic pipettes (1, 5, 10, 20 ml volumes)

Sterile paper or cotton drapes

Solutions and reagents

- Fetal bovine serum (FBS): heat-inactivated; prescreened to support growth of human lymphocytes (for example, Life Technologies #10106078)
- Human serum (HS) as prepared above
- L-Glutamine (for example, Life Technologies #25030024)
- Penicillin–streptomycin (for example, Life Technologies #15140114)
- Serum-free culture medium: RPMI 1640 containing penicillin–streptomycin (100 U ml^{-1}) and L-glutamine (2 mM). Iscove's medium (for example, Life Technologies #21980032) and DMEM medium (for example, Life Technologies #41965039) can also be used
- Serum-supplemented culture medium: as above and in addition 15% FBS or a mixture of 10% FBS and 5% HS.
- Trypan blue solution (for example, Sigma T8154)

Basic method for establishing primary cultures *in vitro*

- Prepare responding lymphocytes according to one of the protocols given above for lymphocyte separation (see p. 671). Dilute the cells to the designated concentration in serum-supplemented culture medium.

Often it is preferable to perform the initial *in vitro* antigen stimulation using culture medium containing only 15% human serum.

- Prepare APC according to the individual experimental specifications. Normally stimulating cells are irradiated to prevent proliferation and utilization of cytokines: the dosage of irradiation varies according to the cell type. Supplementary information regarding several types of stimulating cells is provided below or in other sections of this book.
- Coculture responding lymphocytes and the appropriate APC in tissue culture plates or flasks using the following guidelines:

    For 150 μl volume cultures in round-bottom microtiter plates use 10,000 to 50,000 responding cells per well.

    For 200 μl volume cultures in flat-bottom microtiter plates use 100,000 to 200,000 responding cells per well.

    For 2 ml volume cultures in 24-well cluster plates use 0.5 to 1×10^6 cells/well.

    For 5–10 ml volume cultures in 25 cm^2 culture flasks use 0.5×10^6 cells ml^{-1}. These flasks are incubated standing upright and multiple flasks of this size should be set up if larger numbers of responding cells are used.

- Incubate the culture plates or flasks at 37°C in a well-humidified incubator with 6% CO_2 for approximately 4–7 days, depending on the cell numbers and rate of cellular proliferation.
- Select the next stage according to the individual experimental specifications: For example, restimulate the responding cells with newly prepared APC; clone the responding lymphocytes and select CD8$^+$ clones

of interest; positively select CD8$^+$ lymphocytes and establish lines or clones. Basic protocols are provided for each of these options.

Basic method for restimulation of lymphocytes *in vitro*

- Harvest the lymphocytes from individual wells, pool and determine the number of viable cells. Restimulate these lymphocytes with APC using the basic protocol for the initial priming cultures; retain the same cell numbers and ratios of responding cells to stimulating cells for the various types of culture vessels.
- Visually monitor the cultures starting at day 4 since the rate of cellular proliferation in secondary responses can be faster. If the culture wells are full, split each well into two and feed both wells with serum-supplemented culture medium.
- Restimulation with APC is necessary every 7–10 days; the preferred timing is determined by the rate of cellular proliferation. If cultures show adequate proliferation in between the times of antigen restimulation, they can be split and fed with serum-supplemented culture medium.
- Restimulation with APC can be repeated multiple times; each subsequent restimulation will generally lead to further enrichment of specific responding cells, but this is a highly selective procedure that may result in loss of lymphocytes during each round of culture and may cause loss of cells with specificities of particular interest.

Supplementary information for using lymphoid cells for antigen presentation

Lymphocytes prepared according to one of the protocols given on p. 671 can also be used as stimulating cells for the induction of allogeneic responses or as APC

following infection with certain living viruses or with recombinant viral vectors. Wide ranges of stimulating lymphocytes can be used, varying from 5 stimulating cells per responding lymphocyte to 10 responding lymphocytes per stimulating cell. Irradiation with 20 Gy is normally adequate to inhibit proliferation of normal lymphocytes. Epstein–Barr virus-transformed human lymphoblastoid cell lines (LCL) can also be used as stimulating cells. Since these cells are much larger than normal lymphocytes they should be used at a ratio of 10 responding lymphocytes to one stimulating cell. Irradiation with 60–80 Gy is necessary to inhibit their proliferation.

Supplementary information for using adherent monocytes for antigen presentation

To prepare plates with monolayers of adherent cells that can be pulsed with antigen for use as stimulating cells, the Ficoll–Hypaque interface cells of heparin blood are suspended at concentrations ranging from 5×10^5 to 5×10^6 cells ml^{-1} and seeded into flat-bottomed culture vessels (microtiter plates or 24-well cluster plates) in serum-supplemented culture medium. Volumes of 100–1000 μl are distributed to the various sizes of culture plates. The plates are incubated for 2–4 h at 37°C in a humidified, 6% CO_2 incubator. Visual inspection can be used to monitor the cell attachment process. Wash the plates twice with serum-free culture medium to remove nonadherent cells. Pulse the adherent cells with antigen according to individual experimental specifications. Irradiation with 20 Gy is normally adequate to inhibit proliferation of these cells. The plates should be sealed with Parafilm and wrapped in sterile paper or cloth drapes and carefully transported to the radiation source to avoid contamination.

Supplementary information for using dendritic cells as antigen-presenting cells

See Section 20.

Supplementary information using peptide-pulsed cells as antigen-presenting cells

See elsewhere in this manual.

Supplementary information for using tumor cells as antigen-presenting cells

See chapter 9.9.

Discussion and trouble shooting

The most effective priming of CD8$^+$ T lymphocytes occurs when these cells form close physical contacts with APC and with CD4$^+$ lymphocytes that provide essential factors supporting proliferation and differentiation of the CD8$^+$ T cells. In most instances, supplementary cytokines are not provided during this culture phase since they foster proliferation of T cells with other specificities. However, in some circumstances this may be important to achieve proper activation when using some types of APC. While close cell contact is essential, it is also important that overcrowding does not occur. Responding lymphocytes grow optimally to a concentration of about 1×10^6 cells ml^{-1}. If the APC can also proliferate, their utilization of space and culture medium will influence the optimal development of the responding lymphocytes; therefore it is often advisable to inhibit their proliferation through irradiation. The irradiation dosage necessary to inhibit cellular proliferation but not disturb optimal antigen presentation varies among cell types and should be determined empirically.

The numbers, sizes, and types of stimulating cells determine the type of culture vessels that should be used. Careful attention should be paid to finding the optimal numbers of responding cells and APC for each type of priming; several variations should be tested to establish a standard experimental protocol.

The progress of the priming phase can be estimated through several visual parameters. The pH of the medium will become more acidic in active cultures, so that bicarbonate-buffered medium will acquire a more yellow hue. Inspection of culture plates through an inverted microscope will reveal large clusters of lymphocytes if active proliferation is occuring. Activation and differentiation of CD8$^+$ T cells is usually accompanied by the appearance of comma-shaped lymphocytes. If the stimulating cells can be differentiated morphologically from the responding lymphocytes, as is the case when monocytes or tumor cells are used as APC, the presence of CD8$^+$ cytotoxic lymphocytes usually leads to rapid disappearance of the stimulating cells after several days. Culture wells plated with APC cells but not responding lymphocytes or wells plated with both responding lymphocytes and APC but without antigen serve as useful controls. When making visual inspection of cultures, caution should be taken not to expose the plates to sources of contamination: the microscope stage should be disinfected and the plates should never be placed on nonsterile surfaces. It is also advisable to wear sterile gloves.

Under no conditions should plate observation be done too often: frequent changes in temperature and CO$_2$ exposure are extremely detrimental to lymphocyte development. Cluster formation is also very important in the priming phase, thus agitation of the cultures, particularly shaking of culture flasks, can disturb important cellular interactions. A general rule of thumb is that culture plates and flasks should be placed at the back of the incubator shelf and left totally undisturbed for 4 days. Thereafter, quick visual observation can be made to assess the progress of the priming, but under no circumstances should any plate or flask be kept out of the incubator for more than 2 minutes.

Lymphocyte cultures are very susceptible to evaporation, which results in changes of salt concentrations in the culture medium. Such changes are highly toxic for lymphocytes. Evaporation is less extreme with larger culture vessels and generally presents no problem in culture flasks as long as the humidity in the incubator is adequate. Evaporation is first apparent in culture plates if cells in the central wells thrive while those in the periphery do not grow or die unexpectedly. It may be advisable only to plate lymphocytes in the central 60 wells of a microtiter plate or in the central 8 wells of a 24-well plate and to fill the outer wells with sterile water or serum-free culture medium. Evaporation can also be minimized by placing the plates in lightly covered sterile stainless-steel boxes to which sterile water is added. Plates should be placed on metal or plastic supports so they do not make contact with the water. The boxes should be sterilized and exchanged weekly to prevent fungal growth.

Lymphocyte cultures are also extremely susceptible to alterations in the pH of the culture medium. Old medium that is too basic should not be used. When lymphocytes proliferate rapidly, the culture medium pH will begin to change; this is a positive sign of lymphocyte activation. At this stage careful observation is essential. The lymphocytes should be allowed to achieve maximal activation in the rich cytokine milieu that is present in the culture medium, but if it becomes too acidic or its nutritional value has been exhausted lymphocytes will stop development or die. The correct judgment whether to continue or stop cultures at this stage is very difficult and is a matter of experience. For beginners in cell culture, the following strategy can be helpful for learning how to proceed. When cultures start to change color to a decided yellow, make a careful visual observation of the lymphocytes and then carry out the following variations: (1) leave some wells undisturbed; (2) remove 50% of the culture medium from several wells and resupplement them with fresh serum-containing medium; and (3) split the contents of several wells into two wells and add fresh serum-containing medium to replace the original volume. Observe the lymphocyte status after 24 h to determine under which conditions they continue to grow optimally. Cells that have been split too early or too late will take on a dark brown, granular appearance and individual cells will be difficult to distinguish; signs of apoptosis may also be apparent.

Isolation of CD8$^+$ T cells

Principle of the method

Stimulation of human lymphocytes with APC *in vitro* leads to activation of both CD4$^+$ and CD8$^+$ T lymphocytes and some non-CD3 cells can also be expanded in such cultures. Indirect or direct immunofluoresence staining of the activated lymphocyte populations with antibodies specific for CD8 molecules is used to estimate the numbers of CD8$^+$ cells obtained during the priming phase. Individual T cell clones are isolated by limiting dilution and CD8$^+$ clones identified by immunofluorescence staining with CD8-specific monoclonal antibodies. Alternatively, CD8$^+$ T cells are selected by immunomagnetic bead separation using

CD8 monoclonal antibodies. These highly purified cells are cloned by limiting dilution or maintained in long-term cultures *in vitro* as polyclonal populations through regular antigen restimulation and growth in the presence of supplementary cytokines and conditioned medium.

Materials and apparatus

96-well round-bottom microtiter plates (for example, Falcon #3077 or Costar #3799)

Absorbent paper towels

Aluminum foil

Automated flow cytometer (optional)

Centrifuge

Eppendorf tubes (for example, Eppendorf #3810)

Hemocytometer with coverslip

Ice and ice container

Immunofluorescence microscope

Incubator: humidified, 6% CO_2, 37°C

Light microscope

Magnetic particle concentrator (Dynal MPC1 #12001)

Plastic 10–15 ml round-bottom tubes (for example, Greiner #163160)

Plastic 50 ml conical tubes (for example, Falcon #2070; Costar #6751)

Rock'n'roller (for example, Denley, Mixer 10)

Sealing foil (for example, Nunc #25082)

Sterile glass or plastic pipettes (1, 5, 10, 20 ml volumes)

Vortexer (for example, Scientific Industries, Vortex-Genie)

Solutions and reagents

- FBS-supplemented PBS: PBS with 10% FBS and 0.1% sodium azide
- FITC-labeled mouse anti-human CD8 monoclonal antibody (for example, Dianova #0451)
- FITC-labeled rabbit anti-mouse immunoglobulin (for example, Dakopatts F261)
- Human recombinant interleukin-2 (for example, Proleukin, Cetus)
- L-Glutamine (for example, Life Technologies #25030024)
- Mouse anti-human CD8 monoclonal antibody (for example, antiCD8(DK25) Dako #M707 or T8, Coulter Electronics)
- Mouse anti-human CD8-coated magnetic beads (Dynal #11107)
- Paraformaldehyde (for example, Merck #4005): 1% solution in PBS
- Penicillin–streptomycin (for example, Life Technologies #15140114)
- Phosphate-buffered saline (PBS) (for example, Life Technologies #14040091)
- RPMI 1640 (for example, Life Technologies #21875034)
- Serum-free culture medium: RPMI 1640 containing penicillin–streptomycin (100 U ml^{-1}), and L-glutamine (2 mM). Iscove's medium (for example, Life Technologies #21980032) and DMEM medium (for example, Life Technologies #41965039) can also be used.
- Serum-supplemented culture medium: as above and in addition 15% FBS or a mixture of 10% FBS and 5% human serum (see p. 675)
- Sodium azide (for example, Merck #6688): 10% stock solution prepared in PBS
- T cell-conditioned medium (prepared as on p. 676) or purchased commercially (for example, Lymphocult T Biotest #811011) (optional)

Basic procedure for immunofluorescence staining of CD8⁺ T cells

If directly labeled antibodies specific for CD8 are available then a simple one-step antibody binding procedure can be used for detecting CD8$^+$ cells. When unlabeled mouse anti-human CD8 antibodies are used in the first step, detection of positive cells is made using a second FITC-labeled antibody that recognizes mouse immunoglobulin. The direct method requires fewer cells and less time. Moreover, studies of two surface molecules can be made simultaneously using reagents labeled with two different fluorescent markers. The procedure described here uses 96-well round-bottom microtiter plates. Samples can also be prepared in Eppendorf tubes, but more cells (100,000–200,000 cells/tube) should be used. Isotype control antibodies as well as other markers for T cells such as CD3 or CD4 can be tested in parallel.

Direct method using FITC-labeled anti-CD8 monoclonal antibody

- Count the responding lymphocytes harvested from the priming cultures and determine the number of viable cells required for antibody labeling: 50,000–100,000 cells/well are needed per test antibody. Centrifuge the cells at 450g for 5 min, discard the supernatant medium, and resuspend the cells in serum-free PBS at a concentration of 50,000–100,000 cells per 150 μl.
- Distribute 150 μl of cell suspension per well. Cover the plates with sealing foil and centrifuge at 450g for 5 min. Remove the sealing foil and discard the supernatant medium by inverting the plate and tapping it once on a flat surface covered with absorbent paper towels. Replace the sealing foil and vortex the plate lightly, moving the plate over the vortex head to be

certain that the cell pellets in each well are vortexed. Resuspend the cells in 40 μl of serum-free PBS.
- Place the plate on a bed of ice; make certain that the base of the plate makes direct contact with the ice. Distribute 4 μl of directly labeled antibody to each well, cover again with sealing foil, and incubate for 45 min on ice. Protect the entire plate from light by covering it with aluminum foil.
- At the conclusion of the incubation period pipette 100 μl of cold FBS-supplemented PBS to each well. Seal and centrifuge as above. Discard the supernatant medium by inverting and tapping the plate on absorbent paper towels. Repeat this washing procedure a second time. After the second wash, pipette 150 μl of cold PBS with 1% paraformaldehyde to each well.
- Evaluate the positive cells with the aid of an immunofluorescence microscope or by automated flow cytometry.
- Paraformaldehyde-fixed cells are stable for evaluation for 3–5 days, but the plates should be held in the dark.

Indirect method using unlabeled anti-CD8 monoclonal antibody and FITC-labeled rabbit anti-mouse immunoglobulin

- Count the responding lymphocytes harvested from the priming cultures and determine the number of viable cells required for antibody labeling: 100,000–200,000 cells are needed per test antibody.
- Centrifuge the cells at 450g for 5 min, discard the supernatant medium and resuspend the cells in serum-free PBS at a concentration of 100,000–200,000 cells per 100 μl. Distribute 100 μl of cell suspension per well.
- Cover the plates with sealing foil and centrifuge at 450g for 5 min. Remove

the sealing foil and discard the supernatant medium by inverting the plate and tapping it once on a flat surface covered with absorbent paper towels. Replace the sealing foil and vortex the plate lightly, moving the plate over the vortex head to be certain that the cell pellets in each well are vortexed.

- Place the plate on a bed of ice; make certain that the base of the plate makes direct contact with the ice. Distribute 50 µl of unlabeled anti-CD8 antibody diluted 1:50 in FBS-supplemented PBS to each well, cover again with sealing foil, and incubate for 90 min on ice.
- Pipette 100 µl of cold FBS-supplemented PBS to each well. Seal and centrifuge as above. Discard the supernatant medium by inverting and tapping the plate on absorbent paper towels. Repeat this washing procedure a second time.
- After the second wash, pipette 50 µl of FITC-labeled rabbit anti-mouse antibody, diluted 1:50 in FBS-supplemented PBS, to each well, cover the plate with sealing foil, place the plate back on the bed of ice, cover with aluminum foil, and incubate for 30 min on ice. Wash the cells twice with 100 µl of cold FBS supplemented PBS. Add 150 µl of cold PBS with 1% paraformaldehyde to each well.
- Evaluate the positive cells with the aid of an immunofluorescence microscope or by automated flow cytometry.

Basic procedure for positive selection of CD8$^+$ lymphocytes

This protocol utilizes commercial immunomagnetic beads that are directly coupled with anti-CD8 antibody; thus separation is achieved with a minimum number of steps and in a minimum amount of time. These conditions are better for activated lymphocytes that are to be cultured or cloned.

- Transfer a specified number of primed cells to a sterile, 10 ml round-bottomed tube that fits to the magnet base.
- Wash the cells once in cold 2% FBS supplemented PBS and suspend in 2% FBS-supplemented PBS at a concentration of 10×10^6 cells/ml (minimal volume is 200 µl).
- Determine the number of beads required for the rosetting, calculating 2–4 beads per estimated CD8$^+$ lymphocyte. Wash the designated number of immunomagnetic beads twice in cold 2% FBS-supplemented PBS with the aid of a commercial magnet.
- Add the washed beads to the lymphocyte suspension at the selected ratio. Roll the mixture at 4°C for 20–30 min.
- Control rosette formation by removing a 10 µl sample and examining it under a light microscope. Add a 5-fold excess of cold 2% FBS-supplemented PBS to the tube and place it in the appropriate commercial magnet. Let the tube stand for 2 min. The cells binding the magnetic beads will cling to the side of the tube. Carefully pipette away the unrosetted cells. Remove the tube from the magnet and add 5 ml of cold 2% FBS-supplemented PBS, mix gently, and place the tube once again on the magnet to remove unrosetted cells.
- Perform this procedure a total of five times. If only small numbers of responding cells (i.e. fewer than 10×10^6) were used, one can reduce the total number of washing steps to three to reduce the cell loss.
- At this stage the cells can be resuspended in serum-supplemented culture medium containing 20–50 units of rIL-2 and 10–20% conditioned medium, prepared as described on

p. 676 or purchased commercially.
- Distribute the cells to culture plates plating 400,000 cells per well of a 24-well plate. The magnetic beads will detach by themselves over the next 2–4 days. To remove the beads after this time, harvest the cells and dilute them to at least 5 ml in culture medium; place the tube on the magnet and carefully transfer the magnetic-particle-free cell suspension to a new tube, leaving behind the detached magnetic beads on the side of the tube. Repeat this procedure once again to remove all remaining beads.
- Pool the magnetic-particle-free suspensions containing the CD8+ selected cells. Their purity can be determined by immunofluoresence staining.
- At this stage the positively selected cells can be cloned by limiting dilution or expanded as polyclonal lines in culture, as described below.

Discussion and trouble shooting

Immunofluorescence staining of cells using CD8-specific monoclonal antibodies is a simple but powerful procedure for gauging the development of CD8+ T cells *in vitro*. It can be applied at each stage of an experiment, thereby helping the investigator to determine more precisely how to proceed. For example, it can be used to monitor the increases of CD8 cells during each round of antigen restimulation, or it can be used to assess the purity of positively selected cells. The protocol described here uses FITC-labeled reagents; antibodies coupled with other fluorescent markers can be substituted. If access to a fluorescent microscope or flow cytometer is not available, other detection methods can be used (Holzmann and Johnson 1983).

The immunomagnetic bead separation procedure involves only one round of separation. If enough cells are available, this can be repeated a second time; it can also be applied to cells that have been cultured after one round of separation.

Maintenance of CD8+ T cell lines and clones *in vitro*

Principle of the method

CD8+ lymphocytes are normally dependent upon CD4+ cells for provision of important growth and differentiation factors. CD4+ cells are prevalent in the lymphocyte populations isolated from heparinized or defibrinated blood or lymphoid tissue samples that are used in the initial priming cultures and cytokine supplementation is not necessary. If multiple rounds of restimulation are done to select highly enriched populations of antigen-specific CD8+ cells then the number of CD4+ cells becomes limiting and addition of exogenous cytokines is needed to maintain further proliferation and function *in vitro*. Following positive selection with immunomagnetic beads, maintainence of CD8⁺ growth is dependent upon provision of supplementary cytokines. Some factors can be provided in recombinant form, whereas conditioned medium is a useful and cost-effective supplement if recombinant cytokines are not available. In general, addition of rIL-2 to the culture medium is recommended so that provision of sufficient amounts can be controlled; clones encountering a deficit of rIL-2 will quickly die. CD8⁺ T cells cloned under limiting dilution conditions also require supplementation with exogenous cytokines and conditioned medium can foster optimal development.

Materials and apparatus

24-well cluster plates (for example, Falcon #3047 or Costar #3424)

96-well flat-bottom microtiter plates (for example, Falcon #3072 or Costar #3596)

96-well round-bottom microtiter plates (for example, Falcon #3077 or Costar #3799)

Cesium-137 irradiation source

Centrifuge

Culture flasks, 25 cm² (for example, Falcon #3082 or Costar #3055)

Hemocytometer with coverslip

Incubator: humidified 6% CO_2, 37°C

Plastic 50 ml conical tubes (for example, Falcon #2070; Costar #6751)

Sterile glass or plastic pipettes (1, 5, 10, 20 ml volumes)

Sterile Pasteur pipettes (cotton plugged, wide mouth)

Sterile stainless-steel box with loose-fitting lid

Solutions and reagents

- Fetal bovine serum (FBS): heat-inactivated; prescreened to support growth of human lymphocytes (for example, Life Technologies #10106078)
- Human recombinant interleukin-2 (for example, Proleukin, Cetus)
- L-Glutamine (for example, Life Technologies #25030024)
- Penicillin–streptomycin (for example, Life Technologies #15140114)
- RPMI 1640 (for example, Life Technologies #21875034)
- Serum-free culture medium: RPMI 1640 containing penicillin–streptomycin (100 U ml⁻¹) and L-glutamine (2 mM). Iscove's medium (for example, Life Technologies #21980032) and DMEM medium (for example, Life Technologies #41965039) can also be used.
- Serum-supplemented culture medium: as above and in addition 15% FBS or a mixture of 10% FBS and 5% human serum (see p. 675)
- T cell-conditioned medium, prepared as on p. 676 or purchased commercially (for example, Lymphocult T Biotest #811011) (optional)
- Trypan blue solution (for example, Sigma T8154)

Basic procedure for isolating CD8⁺ T cell clones

The following procedure describes a general cloning approach that utilizes serum-supplemented culture medium containing rIL-2 and conditioned medium peripheral blood lymphocytes or lymphoblastoid cell lines as feeder or APC, and positively selected CD8 cells as responding cells. Further adaptations should be made according to individual experimental specifications.

- Prepare APC according to individual experimental specifications and suspend them at a concentration of $1 \times 10^5\,ml^{-1}$ when using normal lymphocytes or $2 \times 10^4\,ml^{-1}$ for lymphoblastoid cell lines (LCL) in serum-supplemented culture medium without conditioned medium or rIL-2.
- Irradiate the normal lymphocytes with 40 Gy and the LCL with 80 Gy.
- Distribute 50 µl of this suspension to the central 60 wells of several round-bottom microtiter plates. Fill the peripheral wells with 150 µl of serum-free culture medium and place the plates in the incubator to maintain a 37°C temperature and optimal pH in the culture medium.
- Isolate CD8$^+$ responding cells by the immunomagnetic bead separation method described on p. 682. Cells must be free of immunomagnetic beads.
- Determine the number of viable CD8$^+$ cells and prepare a suspension containing 300 cells ml^{-1} in serum-supplemented culture medium. Perform serial dilutions to achieve a suspension containing 0.3 cells/100 µl in serum-supplemented culture medium containing 30–40 units ml^{-1} rIL-2 and 20% conditioned medium.
- Distribute 100 µl of this suspension to the central 60 wells of the microtiter plates preseeded with APC. Place the plates in a sterile stainless-steel box with a loose fitting cover that allows gas exchange. Add 10–20 ml sterile water to the bottom of the box. The plates should be stacked on metal or plastic supports to avoid direct contact with the water.
- Incubate the plates in a humidified, 6% CO_2 incubator at 37°C for 7–10 days. Visually control the plates to determine the status of the cultures. Good proliferating cultures will show large clusters of cells at the bottom of the culture well and single cells can appear around the perimeter moving up the sides of the wells. When the cells cover the entire bowl of the culture well they are ready for transfer.
- Prepare flat-bottom microtiter plates with irradiated feeder/APC as described above, but increase the numbers of stimulating cells 5-fold. Use sterile, cotton-plugged, wide-mouth Pasteur pipettes to mix and transfer clones from individual wells to the new plates. Do not induce air bubbles during mixing. Feed each well with an additional 50 µl of serum-supplemented culture medium with rIL-2 (20–50 units ml^{-1}) and conditioned medium (20%).
- Incubate the plates as before and perform visual inspection after 4–5 days. If major proliferation has occurred during this period so that the cells cover more than 90% of the well, they should be split into two wells and each fed with 100 µl of cytokine-containing serum medium.
- Depending on the rate of proliferation of individual clones it may be necessary to add fresh irradiated stimulating cells to the flat-bottom plates and make an exchange of 50% of the used culture medium for fresh medium before the cells can be transferred to 24-well cluster plates. For rapidly growing clones, pool the cells from 2–4 wells and transfer to one well of a 24-well cluster plate. Be certain not to mix cells of different clones; only pool cells that have been split from one original microtiter well.

Basic procedure for maintaining CD8$^+$ T cell lines *in vitro*

Identical procedures for preparation of APC, splitting and feeding of cultures, and transfer of T cell lines from plates to flasks are used for polyclonal T cell lines as described above for CD8$^+$ T cell clones. The kinetics of restimulation need to be established for each line, but generally fall in the time span of 7–21 days. Adequate doses of irradiation should be used to prevent outgrowth of APC cultured in the presence of high amounts of exogenous cytokine.

Discussion and trouble shooting

All general considerations for judging the progress of cultures described under 'Activation of human CD8$^+$ T lymphocytes *in vitro*' apply to cloning and maintaining CD8$^+$ T cells in long-term cultures. Clones generally need restimulation with specific antigen every 7–10 days, although some will continue to proliferate for as long as 3 weeks without restimulation. To determine the numbers of cells to be used, follow the guidelines given for activation of CD8$^+$ cells. The irradiation doses should be maintained at the higher levels recommended above to prevent growth of stimulating cells in culture medium supplemented with cytokines and conditioned medium. In between the stimulation period cultures should be split and fed with cytokine-supplemented culture medium according to their individual growth rates. It may be easiest to expand clones in multiple wells of 24-well cluster plates since the rate of proliferation can be visually controlled. Some clones can be transferred to 25 cm^2 culture flasks

by pooling the cells of 3–5 cluster wells and transferring them to an upright culture flask. Add 2–4 ml of cytokine-supplemented culture medium. If the clones continue to grow very quickly, the flask can eventually be cultured lying flat. Not all CD8$^+$ clones adapt easily to growth in flasks.

Since generation of antigen specific CD8$^+$ T cell clones can be regulated by accessory cells, such as CD4$^+$ T cells, a dilution of 0.3 cells/well may lead to a very low cloning efficiency. In this case use of a series of dilutions (for example, 3 cells/well, 2 cells/well and 1 cell/well) can be helpful in establishing oligoclonal sublines. Once these grow well, subsequent cloning at 0.3 cells/well is more feasible to isolate defined CD8$^+$ T cell clones.

Immunofluorescence staining with anti-CD8 monoclonal antibodies as described in 'Isolation of CD8$^+$ T cells' should be done to certify phenotype. Molecular analyses of TCR, described elsewhere in this volume, can be used to substantiate clonality. If mixtures of TCR are detected, then recloning according to the same procedure can be performed. It is possible, however, for the progeny of a single T cell to express two TCR (Padovan et al 1993).

References

Bu D, Domenech N, Lewis J, Taylor-Papadimitriou J, Finn OJ (1993) Recombinant vaccinia mucin vector: *in vitro* analysis of expression of tumor-associated epitopes for antibody and human cytotoxic T-cell recognition. J Immunother 14: 127–135.

Celis E, Tsai V, Crimi C, et al (1994) Induction of anti-tumor cytotoxic T lymphocytes in normal humans using primary cultures and synthetic peptide epitopes. Proc Natl Acad Sci USA 91(6): 2105–2109.

Chen W, Carbone FR, McCluskey J (1993) Electro-poration and commercial liposomes efficiently deliver soluble protein into the MHC class I presentation pathway. Priming *in vitro* and *in vivo* for class I-restricted recognition of soluble antigen. J Immunol Methods 160: 49–57.

Dave VP, Allan JE, Slobod KS, et al (1994) Viral cross-reactivity and antigenic determinants recognized by human parainfluenza virus type 1 specific cytotoxic T-cells. Virology 199(2): 376–383.

Disis ML, Smith JW, Murphy AE, Chen W, Cheever MA (1994) *In vitro* generation of human cytolytic

T-cells specific for peptides derived from the HER-2/neu protooncogene protein. Cancer Res 15 54(4): 1071–1076.

Donnelly JJ, Ulmer JB, Hawe LA, et al (1993) Targeted delivery of peptide epitopes to class I major histocompatibility molecules by a modified *Pseudomonas* exotoxin. Proc Natl Acad Sci USA 90: 3530–3534.

Gotch F, McMichael A, Smith G, Moss B (1987) Identification of viral molecules recognized by influenza-specific human cytotoxic T lymphocytes. J Exp Med 165: 408–416.

Gould K, Cossins J, Bastin J, Brownlee GG, Townsend A (1989) A 15 amino acid fragment of influenza nucleoprotein synthesized in the cytoplasm is presented to class I-restricted cytotoxic T lymphocytes. J Exp Med 170: 1051–1056.

Holzmann B, Johnson JP (1983) A beta-galactosidase-linked immunoassay for the cell analysis of antigen on individual cells. J Immunol 60: 359–367.

Houbiers JGA, Nijman HW, van der Burg SH, et al (1993) *In vitro* induction of human cytotoxic T lymphocyte responses against peptides of mutant and wild-type p53. Eur J Immunol 23: 2072–2077.

Kast WM, Brandt RM, Drijfhout JW, Melief CJ (1993) Human leukocyte antigen-A2.1 restricted candidate cytotoxic T lymphocyte epitopes of human papillomavirus type 16 E6 and E7 proteins identified by using the processing-defective human cell line T2. J Immunother 14(2): 115–120.

Kovacsovics-Bankowski M, Rock KL (1995) A phagosome-to-cytosol pathway for exogenous antigens presented on MHC class I molecules. Science 267: 243–246.

Kupfer A, Singer SJ (1989) Cell biology of cytotoxic and helper T-cell functions. Annu Rev Immunol 7: 309–337.

Lee SP, Thomas WA, Murray RJ, et al (1993) HLA A2.1-restricted cytotoxic T cells recognizing a range of Epstein–Barr virus isolates through a defined epitope in latent membrane protein LMP2. J Virol 67(12): 7428–7435.

Lopes LM, Chain BM (1992) Liposome-mediated delivery stimulates a class I-restricted cytotoxic T cell response to soluble antigen. Eur J Immunol 22: 287–290.

Malkovsky M, Philpott KA, Dalgleish G, et al (1988) Infection of B lymphocytes by the human immunodeficiency virus and their susceptibility to cytotoxic cells. Eur J Immunol 18: 1315–1321.

Mehrotra PT, Wu D, Crim JA, Mostowski HS, Siegel JP (1993) Effects of IL-12 on the generation of cytotoxic activity in human CD8$^+$ T lymphocytes. J Immunol 151: 2444–2452.

Moore MW, Carbone FR, Bevan MJ (1988) Introduction of soluble protein into the class I pathway of antigen processing and presentation. Cell 54: 777–785.

Nixon DF, Townsend ARM, Elvin JG, Rizza CR, Gallwey J, McMichael AJ (1988) HIV-1 gag-specific cytotoxic T lymphocytes defined with recombinant vaccinia virus and synthetic peptides. Nature 336: 484–487.

Padovan E, Casorati G, Dellabona P, Meyer S, Brockhaus M, Lanzavecchia A (1993) Expression of two T cell receptor a chains: dual receptor T cells. Science 262: 422–424.

Robertson MN, Buseyne F, Schwartz O, Riviere Y (1993) Efficient antigen presentation to cytotoxic T lymphocytes by cells transduced with a retroviral vector expressing the HIV-1 Nef protein. AIDS Res Hum Retroviruses 9(12): 1217–1223.

Schendel DJS, Wank R (1981) Production of human T cell growth factor. Hum Immunol 2: 325–332.

Townsend ARM, McMichael AJ, Carter NP, Huddleston JA, Brownlee GG (1984) Cytotoxic T cell recognition of the influenza nucleoprotein and hemagglutinin expressed in transfected mouse L cells. Cell 39: 13–25.

Ulmer JB, Donnelly JJ, Liu MA (1994) Presentation of an exogenous antigen by major histocompatibility complex class I molecules. Eur J Immunol 24(7): 1590–1596.

Yewdell JW, Resnick JR (1992) Cell biology of antigen presenting to major histocompatibility complex class I molecule-restricted T lymphocytes. Adv Immunol 52: 1–23.

Cloning of genes coding for antigens recognized by cytolytic T lymphocytes

9.9

Etienne De Plaen[1]
Christophe Lurquin[1]
Vincent Brichard[1]*
Pierre van der Bruggen[1]
Jean-Christophe Renauld[1]*
Pierre Coulie[1]
Jean-Pierre Szikora[1]
Thomas Wölfel[2]
Aline Van Pel[1]
Thierry Boon[1]

[1]Ludwig Institute for Cancer Research, Brussels Branch, Brussels, and Cellular Genetics Unit, Université Catholique de Louvain, Brussels, Belgium
[2]Medizinische Klinik der Johannes Gutenberg-Universitat, Mainz, Germany

This work was partially supported by the Belgian program on Interuniversity Poles of Attraction initiated by the Belgian State, Prime Minister's Office, Office for Science, Technology and Culture (OSTC). The scientific responsibility is assumed by the authors. This research was partially subsidized by the Fonds Maisin, by the Association contre le Cancer (Belgium), and by an EEC grant (Biomed I program).

*V. Brichard is Senior Research Assistant and J.-C. Renauld is Research Associate with the Fonds National de la Recherche Scientifique, Belgium

TABLE OF CONTENTS

Introduction

Most murine tumors express antigens that are targets for a T cell-mediated rejection response in the syngeneic animal. Such antigens were found not only on tumors obtained after viral infection or after treatment with chemical carcinogens or ultraviolet radiation, but also on some spontaneous tumors (Klein et al 1960; Kripke 1974; Old et al 1962; Prehn and Main 1957; Van Pel et al 1983). In some tumor systems, highly specific cytolytic T lymphocyte (CTL) clones directed against mouse tumor antigens have been isolated (Brunner et al 1980; Levy and Leclerc 1977; Maryanski et al 1982). One of these tumor systems is mastocytoma P815, a tumor that appeared in a DBA/2 mouse after methylcholanthrene treatment. A clonal cell line named P1 was derived from the tumor and a panel of anti-P1 CTL clones was obtained. They were used to select antigen-loss variants that were resistant to lysis (Uyttenhove et al 1983). These variants proved resistant to subsets of the panel of anti-P1 CTL, leading to the definition of four distinct antigens on P1.

We resorted to a transfection approach to clone the gene encoding one of these tumor antigens. Because P1 was a poor DNA recipient cell line, we selected from it a highly transfectable cell line, named P1.HTR (Van Pel et al 1985). This cell line, however, was not suitable, because it expressed the four antigens of P1. Additional selection with anti-P1 CTL yielded clone P0.HTR which had lost three of the four antigens (Van den Eynde et al 1991). We used a cosmid library and a CTL proliferation assay to obtain transfectants expressing tumor antigen P815A. From these cosmid transfectants, we were able to retrieve gene P1A, which transferred the expression of both P815A and P815B. This gene is identical to that present in normal tissues, but appears to be expressed only by the tumor cells. Activation of gene P1A in the mastocytoma cell line is therefore responsible for the production of

antigens P815A and P815B (Van den Eynde et al 1991).

For human tumors, mixed cultures of tumor cells and lymphocytes of the same patient (autologous) often generate CTL that lyse the autologous tumor cells (Anichini et al 1987; Knuth et al 1984; Mukherji and MacAlister 1983). Using lymphocytes of patient MZ2, we obtained a panel of CTL clones that lysed the autologous melanoma cell line MZ2-MEL (Hérin et al 1987). These CTL clones did not lyse targets of natural killer cells or autologous control cells such as fibroblasts or EBV-transformed B lymphocytes. Antigen-loss variants of MZ2-MEL cells were obtained by selecting tumor cells that were resistant to some of the CTL clones. This demonstrated that six different antigens were recognized on the tumor cells by the autologous CTL (Van den Eynde et al 1989). We set out to clone the gene encoding antigen MZ2-E and used as recipient cell line antigen-loss variant MZ2-MEL.2.2, which was found to be transfectable with good efficiency (Traversari et al 1992a).

A cosmid library was prepared with the DNA of an MZ2-MEL subclone and transfected in MZ2-MEL2.2. Cosmid transfectants expressing antigen MZ2-E were identified by their ability to stimulate the production of tumor necrosis factor (TNF) by the appropriate CTL clone (Traversari et al 1992a). From a cosmid transfectant, we rescued gene MAGE-1 that directs the expression of antigen MZ2-E (van der Bruggen et al 1991). The sequence of the gene found in lymphocytes of patient MZ2 was identical to that found in MZ2-MEL cells. MAGE-1 is not expressed in normal tissues, except testis, but it is expressed in tumors of various histological types. The activation of MAGE-1 gene is responsible for the appearance of the tumor antigen.

The recovery of the gene encoding an antigen from the genome of cosmid transfectants is a tricky and time-consuming process. We therefore adopted a faster procedure that involves the transfection of cDNA expression libraries into simian COS-7 cells (Seed and Aruffo 1987). In these cells, the episomal replication of the transfected plasmids ensures that large amounts of the proteins and therefore of the antigens encoded by the cDNA inserts are produced within 2 days. Using this method, we were able to characterize two antigens present on melanoma cell line SK29-MEL. One is encoded by the tyrosinase gene (Brichard et al 1993), the other by a new gene which was named *Melan-A* or *MART-1* (Coulie et al 1994). These two genes are expressed in melanocytes as well as melanoma. The antigens they encode are melanocytic differentiation antigens.

Antigens recognized by CTL are small peptides derived from cellular proteins that are presented on the cell surface by class I major histocompatibility complex (MHC) molecules (Bjorkman et al 1987; Rötzschke et al 1990; Townsend et al 1986). In our case, the identification of the region coding for the antigenic peptide was greatly facilitated by the fact that transfection of small gene fragments containing this region regularly led to the expression of the antigen. When the region encoding the antigenic peptide was narrowed down, candidate synthetic peptides were synthesized to identify those that sensitize appropriate targets to lysis by the antitumor CTL (de Bergeyck et al 1994; Lurquin et al 1989; Traversari et al 1992b).

We describe here in detail the techniques that were used to clone genes producing antigens recognized by CTL clones and to subsequently identify the antigenic peptide encoded by these genes.

Cloning of genomic sequences

Selection of antigen-loss variants

CTL clones are used to select *in vitro* antigen-loss tumor cell variants. The analysis of the pattern of resistance of these variants to different CTL clones allows for an evaluation of the number of different antigens present on a tumor cell line. Antigen-loss variants are also useful as recipient cell lines for cloning strategies based on DNA-mediated gene transfer.

Tumor cells are mixed with the appropriate CTL clone in conditions of CTL-to-target ratio and incubation time that ensure lysis of at least 99% of the tumor cells. Owing to the variable efficacy of the different CTL clones, these conditions vary from a CTL-to-target ratio of 0.5 during 3 h to a ratio of 3 during 24 h. Approximately one-half of the selected cell suspension is put into culture to serve as a source of cells for a subsequent selection, if necessary. The remaining selected cells are plated in limiting dilution in 96-well flat-bottom microplates. Using this procedure, we obtained stable resistant variants at a frequency that varied between 10^{-3} and 10^{-7}.

Transfection of genomic DNA

The initial approach followed to isolate genes coding for antigens recognized by CTL was based on DNA transfer by the calcium phosphate precipitation method. It involves the transfection of a highly transfectable recipient cell line that does not express the relevant antigen with DNA of a cell expressing this antigen. This DNA is mixed with DNA of a dominant selectable marker, such as a geneticin- or hygromycin-resistance gene, to select the rare cells that have integrated transfected DNA. Considering that the size of the haploid genome is 3×10^6 kb and that mouse cell transfectants have been reported to integrate 1000 kb of exogenous DNA (Perucho et al 1980), the frequency of transfectants expressing a gene present as single copy per haploid genome should be of the order of 1 in 3000 (Fig. 9.9.1). This frequency should be 1 in 6000 for a gene present as single copy per diploid genome. This was observed with *tum⁻* genes, which are mutated alleles producing antigens recognized by CTL. The frequency of genomic P1.HTR transfectants expressing such antigens was found to be about 1/13,000 (Sibille et al 1990; Szikora et al 1990; Wölfel et al 1987).

Because the amount of DNA incorporated per human cell transfectant had been reported to be lower than in mouse cell lines (Hoeijmakers et al 1987; Perucho et al 1980), we tried to estimate the average amount of DNA integrated in human MZ2-MEL.2.2 transfectants.

Cotransfection experiments suggested that these transfectants integrate 5 times less DNA than the mouse P1.HTR transfectants. This led to an estimation of about 200 kb of integrated DNA per MEL.2.2 transfectant. The transfection efficiency should therefore vary between 1 in 15,000 and 1 in 30,000. One genomic MZ2-MEL.2.2 transfectant expressing antigen MZ2-E was found in a population of 70,000 geneticin-resistant transfectants (Traversari et al 1992a).

Transfection of genomic DNA and selection of transfected cells

An outline of our transfection and detection procedure is presented in Fig. 9.9.2. Genomic DNA of the cell expressing the relevant antigen is purified according to Gross-Bellard (1973). The method is based on the use of proteinase K in the presence of sodium dodecyl sulfate (SDS) and ethylenediaminetetraacetate (EDTA). The genomic DNA is then cotransfected with

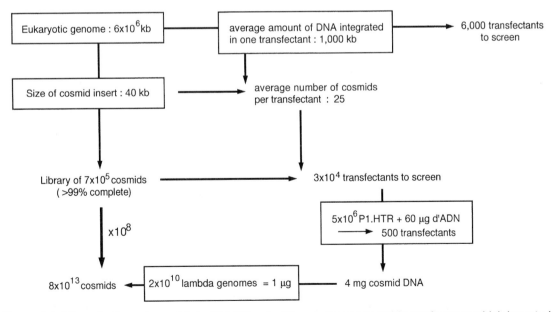

Figure 9.9.1 Transfection of cosmids in P1.HTR cells. The number of cosmid transfectants which have to be screened for expression of an antigen is related to the average amount of DNA integrated in each transfectant and to the size of the mammalian genome. This number is multiplied by a factor of 5 for MZ2-MEL.2.2 transfectants which incorporate only 200 kb of exogenous DNA.

plasmid pSVtkneoβ (Nicolas and Berg 1983) or with cosmid pHMR272 (Bernard et al 1985) which confer resistance to G418 (geneticin) and to hygromycin B (HmB), respectively. Plasmid DNA is isolated from bacterial cultures by the clear lysis procedure followed by a CsCl–ethidium bromide equilibrium density gradient centrifugation. Recipient cells that have integrated DNA are selected with antibiotics G418 or hygromycin B and can then be screened for the expression of the relevant antigen.

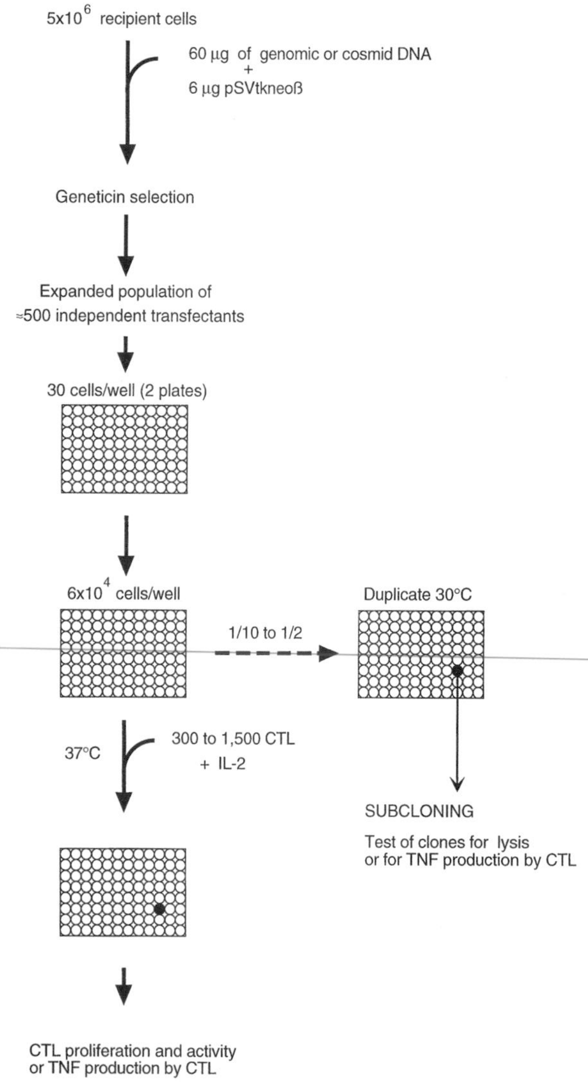

5x10^6 recipient cells

60 µg of genomic or cosmid DNA
+
6 µg pSVtkneoβ

Geneticin selection

Expanded population of
≈500 independent transfectants

30 cells/well (2 plates)

6x10^4 cells/well   Duplicate 30°C

1/10 to 1/2

37°C   300 to 1,500 CTL
+ IL-2

SUBCLONING

Test of clones for lysis
or for TNF production by CTL

CTL proliferation and activity
or TNF production by CTL

Figure 9.9.2 Outline of the procedure for stable transfection and detection of transfectants expressing an antigen recognized by CTL.

We use a transfection procedure described by Corsaro and Pearson (1981).

• Genomic DNA (60 µg) and pSVtkneoβ (6 µg) or pHMR272 (2 µg) are mixed in 940 µl of 1 mM Tris–HCl, pH 7.9, 0.1 mM EDTA; 312.5 µl of CaCl$_2$ 1 M is then added. This DNA solution is slowly added under constant bubbling to 1.25 ml of 2× Hanks' buffered solution (HBS) (50 mM Hepes, 280 mM NaCl, 1.5 mM Na$_2$HPO$_4$, adjusted to pH 7.1 with NaOH). The calcium phosphate–DNA precipitates are allowed to form for 30–45 min at room temperature.

• The precipitates are incubated with the recipient cells that do not express the relevant antigen. The DNA recipients have to be highly transfectable, sensitive CTL targets and to express the appropriate class I molecule. We currently use as recipient cells, P1.HTR, a highly transfectable murine P815 cell line (Van Pel et al 1985), and human melanoma line MZ2-MEL.2.2 (Traversari et al 1992a). Geneticin-resistant P1.HTR transfectants are usually obtained at a frequency of 4×10^{-4}, MZ2-MEL.2.2 transfectants at a frequency of 10^{-5} to 10^{-4}. P1.HTR cells grow in suspension, whereas MZ2-MEL.2.2 cells grow attached to the dish: the protocols to transfect them are therefore slightly different.

• Groups of 5×10^6 P1.HTR cells are centrifuged for 10 min at 400g, and the pellets are resuspended directly in the medium containing the DNA precipitates (2.5 ml). The mixture is incubated for 20 min at 37°C and then added to an 80 cm^2 tissue culture flask containing 22.5 ml of culture medium supplemented with 10% fetal calf serum (FCS). For MZ2-MEL.2.2 cells, the calcium phosphate–DNA precipitates are applied to 80 cm^2 tissue culture flasks seeded one day before with 3×10^6 cells in 22.5 ml of culture medium supplemented with 10% FCS.

- After 24 h, the medium is replaced or diluted 2-fold with fresh medium. Forty-eight hours after transfection, the cells are harvested and counted. For each group, 4×10^6 to 8×10^6 cells are seeded per 80 cm^2 flask. Selection of transfected cells in mass cultures is carried out in medium containing G418 (P1.HTR, 1.5 mg ml^{-1}; MZ2-MEL.2.2, 2 mg ml^{-1}) or HmB (P1.HTR, 350 μg ml^{-1}). The optimal dose of antibiotic has to be established for each recipient cell line.
- The frequency of drug-resistant transfectants is estimated as follows. For P1.HTR transfectants, 1×10^6 cells from each group are plated 2 days after transfection in 5 ml DMEM with 10% FCS, 0.4% bactoagar (Difco), and either 1.2 mg ml^{-1} G418 or 300 μg ml^{-1} of HmB. Colonies are counted 7 days later. A correction is made for the cloning efficiency of G418-resistant P1.HTR cells in agar, of approximately 0.3. For MZ2-MEL.2.2 transfectants, drug-resistant colonies are counted 15 days after transfection.

Screening of transfected cells for the expression of the antigen

- Eight to fifteen days after transfection, the antibiotic-resistant transfectants are recovered and maintained in nonselective medium for 1 or 2 days.
- P1.HTR transfectants are usually separated from dead cells by density centrifugation on Ficoll-Paque (Pharmacia). At that stage, the transfectants are screened for the presence of an antigen-expressing cell. Screening with murine CTL clones

is based on a proliferation assay or on a TNF release assay. Screening with human CTL is performed only by the TNF release assay.

Stimulation of cytolytic T cell proliferation

This assay is based on the observation that some CTL clones require the presence of the relevant antigen for optimal proliferation, even in the presence of IL-2. This effect is particularly marked when the CTL clones have been deprived of their antigen for a few days. CTL proliferation observed visually is confirmed by testing the lytic activity of the culture. We have shown that using this assay, populations of P1.HTR transfectants containing 3% of cells expressing the relevant antigen could be clearly distinguished from populations containing only P1.HTR (Wölfel et al 1987).

- Transfected cells are plated in 96-well round-bottom microplates at 30 living cells per well in nonselective medium (Fig. 9.9.2). The number of microcultures ensures that every transfectant is present in several wells.
- After 5 days, aliquots (1/10) from all wells are transferred to duplicate plates which are incubated at 30°C. The master plates are centrifuged and the medium is removed. 1×10^2 to 5×10^2 CTL are added together with 10^6 irradiated (2000 rad from a ^{137}Cs source) syngeneic feeder spleen cells and incubated with the P1.HTR transfected cells (6×10^4 to 1×10^5) in 200 μl cultures. The medium is supplemented with recombinant IL-2 (10–20 units ml^{-1}). Hypoxanthine, aminopterin, thymidine (HAT) is added to kill the stimulator cells that are tk-mutants.
- Five to seven days later, the wells are examined visually for CTL proliferation. Aliquots of 100 μl from all wells are transferred to a separate plate

containing ^{51}Cr-labeled target cells (2000–4000/well) expressing the relevant antigen. Chromium release is measured after 4 h.

- It is our experience that the microcultures where considerable proliferation has been observed visually invariably display high lytic activity. The replica microcultures corresponding to the microcultures showing high CTL activity are expanded and cloned by limiting dilution in culture medium with 10% FCS. After 5 days, between 150 and 300 P1.HTR transfectant clones are screened in a visual lysis assay with CTL (Maryanski and Boon 1982) as follows. Aliquots (10^3 cells) are incubated either alone or with 2×10^4 CTL in culture medium with 10% FCS in 96-well round-bottom microplates. Before use, CTL are cleared from dead feeder and stimulator cells by density centrifugation with Ficoll-Paque. One to two days later, the plates are examined microscopically for the presence of surviving tumor cells.

Stimulation of tumor necrosis factor (TNF) release by the cytolytic T cells

Unfortunately, in contrast to mouse CTL, human CTL often fail to show a significant and reproducible proliferation when they are stimulated with antigen-expressing cells in a short-term culture (Traversari et al 1992a). On the basis of observations made by N. Ohta and L. Old, we developed a test system that evaluates the stimulation of CTL by their release of tumor necrosis factor β (TNF-β). The presence of TNF in the supernatant is indicated by its cytotoxicity on murine cell line WEHI-164 clone 13 (W13), as described by Espevik and Nissen-Meyer (1986). A significant production of TNF is observed when a CTL clone is stimulated with transfected cell populations containing only 3% of antigen-expressing cells.

- Transfected MZ2-MEL cells are plated in 96-well microtiter plates at 30 living cells/well in 200 µl of culture medium with 20% FCS. After 10 days, the wells contain approximately 6×10^4 cells.
- The cells are detached and 2/5 of each microculture is transferred to a duplicate plate which is incubated at 30°C. After 6 h, the cells adhere again to the original microwells and the medium is removed.
- 1500 CTL are added to each well in 100 µl of medium containing 25 units ml^{-1} of IL-2. After 24 h, 50 µl of supernatant is withdrawn and transferred to a new microplate containing 3×10^4 W13 cells in 50 µl of culture medium supplemented with 2 µg ml^{-1} of actinomycin D and 40 mM LiCl (Beyaert et al 1989).
- After 20 h of incubation at 37°C, the percentage of dead cells is measured by a colorimetric assay (Hansen et al 1989). This test relies on the ability of living cells to produce a colored tetrazolium product. 50 µl of MTT (3-(4,5-dimethylthiazol-2-yl)-2,5-diphenyltetrazolium bromide; Merck) at 2.5 mg ml^{-1} in phosphate-buffered saline is added and the plates are incubated for 2 h at 37°C. The dark blue formazan crystals are dissolved by adding 10 µl of a lysis solution. This solution is prepared by mixing 1 volume of N,N-dimethylformamide with 2 volumes of 30% SDS; the pH is adjusted to 4.7 by adding 2.5% of 80% acetic acid and 2.5% 0.1 M HCl. The plates are incubated overnight at 37°C. The optical densities (OD) at 570 nm are measured with a 96-well multiscanner (Molecular Devices, Menlo Park, CA, USA) using 650 nm as control wavelength (Espevik and Nissen-Meyer 1986). The concentration (pg ml^{-1}) of TNF in the supernatant is estimated by comparing

the measured OD to the values obtained for known concentrations of human rTNF-β (R and D Systems, Minneapolis, MN, USA).

Transfection of cosmids and recovery of transfected cosmids

Because we had observed that DNA fragments of 30 kb were capable of transferring the expression of an antigen recognized by a specific CTL (Wölfel et al 1987), we decided to transfect cosmid libraries obtained from a cell line expressing the antigen. We set up a procedure to obtain cosmid transfectants and to rescue the integrated sequences with the help of the cosmid sequences.

To obtain transfectants of a complete cosmid library (700,000 cosmids), it is necessary to generate approximately 3×10^4 to 1.5×10^5 independent transfectants, assuming an average integration of 5 to 25 cosmids (Fig. 9.9.1). Since P1.HTR and MZ2-MEL.2.2 cells have a transfection efficiency of about 10^{-4}, this

implies that 3×10^8 to 1.5×10^9 cells must be submitted to transfection. We have found that groups of 5×10^6 cells are optimally transfected with 60 μg of DNA. Thus, it is necessary to produce 4 to 20 mg of cosmid DNA, and this requires at least a 10^8-fold amplification of the library. Fortunately, we chose cosmid vector c2RB (Bates and Swift 1983) and found that cosmid libraries constructed with this vector could be amplified 10^8- to 10^9-fold without important loss of diversity.

Construction of cosmid libraries

Preparation of cosmid arms

- Cosmid c2RB (6.8 kb) is linearized at a unique SmaI site located between the two cos sites (Fig. 9.9.3). To inactivate the enzyme, the DNA solution is extracted once with phenol and once with chloroform–isoamyl alcohol (24:1).
- After ethanol precipitation, the DNA is resuspended at a concentration of about 100 μg ml^{-1} in 50 mM Tris–HCl, 0.1 mM EDTA, pH 8.

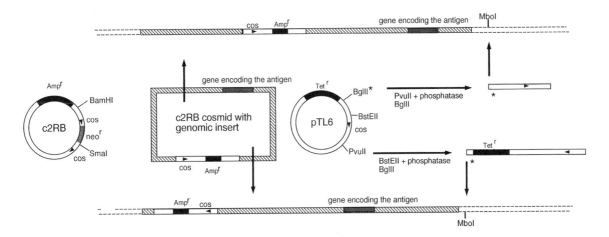

Figure 9.9.3 Relative orientation of the gene coding for a CTL-defined antigen, of the cohesive-end (cos) site and of the ampicillin-resistance (Amp^r) marker in the transfectant according to where the circular cosmid is opened during transfection (large arrows). Each configuration can be rescued only by one pTL6 arm as indicated. Tet^r is the tetracycline-resistance gene of pTL6.

- To dephosphorylate the blunt ends, 0.05 U of calf intestinal phosphatase is added per μg of DNA and the solution is incubated 30 min at 37°C. To inactivate the enzyme, one volume of SDS 1%, STE 2× (10× STE is 100 mM Tris–HCl pH 8, 1 M NaCl, 10 mM EDTA) is added and the solution is heated at 68°C for 30 min. The solution is extracted 3 times with phenol and 3 times with chloroform–isoamyl alcohol.
- After ethanol precipitation, the cosmid DNA is digested with BamHI restriction enzyme, producing the two cosmid arms of 1.7 and 5.1 kb. The preparation is extracted with phenol–chloroform–isoamyl alcohol (25:24:1 by vol) and precipitated with ethanol.

Partial digestion and size fractionation of genomic DNA

The genomic DNA is digested according to Grosveld et al (1981, 1982, 1983) with some modifications.

- 120 μg of high-molecular-weight DNA (>100 kb) is preheated at 37°C and partially digested by MboI that recognizes a 4 bp sequence (0.05 units μg^{-1}) in a volume of 1.2 ml.
- During 75 min, aliquots of 6 μg are withdrawn every 5 min, put on ice, and adjusted to 25 mM EDTA to stop the reaction. 3 μg of each of those fractions is analyzed in 0.3% agarose gels at 4°C. The run lasts 24 h at 2.5 V cm^{-1} in 1× TAE buffer (10× TAE is 0.4 M Tris (48.46 g l^{-1}), 50 mM sodium acetate (6.8 g l^{-1}), 10 mM EDTA (3.72 g l^{-1}) adjusted to pH 7.8 with acetic acid).
- We compare the fragments of the different fractions with a DNA size ladder and select the products of three digestion times, a short time (DNA mostly ≥45 kb), an intermediate time (DNA mostly 35–45 kb), and a long time (DNA mostly <35 kb).
- We usually check that in the selected fractions most fragments have intact MboI cohesive ends. For this purpose, 3 μg of each of these fractions is precipitated with ethanol and resuspended at a concentration of 300 μg ml^{-1} in 2× ligation buffer (10× concentrated buffer is 660 mM Tris–HCl pH 7.5, 50 mM MgCl$_2$, 50 mM dithiothreitol, 10 mM ATP). 4 units of T4 DNA ligase are added to each of the samples. After 4 h at 14.5°C, they are analyzed on a 0.3% agarose gel. The ligation products should show a size comparable to that of the nondigested genomic DNA.
- The same conditions (enzyme concentration, time, temperature and DNA concentration) are then used to digest 200–400 μg of genomic DNA with the same batch of MboI enzyme. 120 μg is removed at the three chosen digestion time points. An aliquot of 6 μg is taken from each of these three fractions: 3 μg is analyzed on agarose gel to check the average size of the fragments and 3 μg is ligated to check the MboI ends.
- The remainder of the three fractions is pooled. This pool is extracted once with phenol, once with chloroform–isoamyl alcohol and precipitated with ethanol. After centrifugation, the DNA is resuspended in 500 μl of TE (Tris 10 mM pH 8, EDTA 1 mM) and fractionated on a sodium chloride gradient.
- The digested DNA fragments are fractionated on a 13.5 ml NaCl gradient (1.25–5 M NaCl in Tris 20 mM pH 7.5, EDTA 1 mM). The DNA in a volume of 500 μl is loaded on the gradient. The tube is centrifuged at 18°C for 4.5 h at 195,000g (SW40Ti Beckman rotor, 33,000 rpm).
- When the run is finished, 18 fractions of about 750 μl are collected from the

bottom of the tube. Three volumes of Tris 20 mM pH 7.5, EDTA 5 mM are added to each fraction. The solutions are precipitated with ethanol. Each DNA pellet is resuspended in 500 μl of TE buffer.

- An aliquot of each fraction (40 μl) is analyzed in a 0.3% agarose gel: fractions containing fragments of 35–50 kb are chosen and their concentration is measured by UV spectrophotometry.

Ligation and packaging of the cosmids

- 1.5 μg of cosmid arms and 1.5 μg of inserts (the vector-to-insert molar ratio is 6) are coprecipitated in a microtube treated with dimethyldichlorosilane.
- After centrifugation, the pellet is resuspended in 10 μl of 2× ligation buffer and 1.5 μl (4 units) of T4 DNA ligase. The mixture is incubated at 14.5°C for 16 h.
- At the end of the reaction, 1 μl (0.3 μg) of the ligation product is removed and analyzed in a 0.4% agarose gel.

We usually observe that the ligase has converted some cosmid arms (1.7 and 5.1 kb) to dimers of 10.2 kb, 6.8 kb, and 3.4 kb and the eukaryotic DNA to high-molecular-weight concatamers.

In vitro packaging of the DNA into lambda phage heads is performed with commercial extracts (Stratagene, Gigapack II Gold).

- To 10 μl of the freeze-thaw lysate (FTL), a maximum of 4 μl of the ligation mixture and 15 μl of the sonic extract (SE) are added. We usually add 2 or 3 μl of the ligation mixture, which corresponds to 0.3–0.45 μg of insert.
- The solution is gently mixed with a

micropipette. The tube is centrifuged briefly. After 2 h of incubation at 22°C, the reaction is diluted with 500 μl of SM (0.1 M NaCl (5.84 g l^{-1}), 10 mM MgSO$_4$·7H$_2$O (2.46 g l^{-1}), 50 mM Tris (6.06 g l^{-1}), 2% gelatin (20 g l^{-1}) adjusted to pH 7.5). A drop of chloroform is added.

- The packaging reaction can be stored for months at 4°C.

Titration and amplification of cosmid libraries

- In a 100 ml flask, 20 ml of LB medium (Luria-Bertani; 10 g Bacto Tryptone, 5 g Bacto Yeast Extract, and 10 g NaCl per liter adjusted to pH 7.5 with NaOH 1 M) containing 0.4% maltose is inoculated with a single colony of *E. coli* ED8767(*recA56, SupE, SupF, hsdS$^-$*, met$^-$, *lacY*).
- The culture is shaken vigorously at 37°C for 18 h and reaches an OD at 600 nm of 2 to 3. It is then centrifuged for 20 min at 4000*g*. The bacterial pellet is resuspended in 8 ml of 10 mM MgCl$_2$.
- Although these bacteria are stable for 5 days, it is preferable to use fresh overnight cultures.
- 5 μl (1%) of the packaging reaction is diluted with 15 μl of SM buffer and 100 μl bacteria in 10 mM MgCl$_2$. The mixture is incubated for 20 min at 37°C without shaking and then diluted with 1 ml of LB medium.
- The suspension is incubated for 1 h at 37°C to allow expression of the ampicillin resistance gene present in cosmid c2RB.
- Infected bacteria are plated on a semi-solid LB medium containing 1.5% Bacto Agar and 25 μg ml^{-1} ampicillin. These plates are incubated at 37°C until colonies are large enough to be counted (usually 18 h). This

enables us to estimate the number of cosmids that can be obtained per unit volume of the packaging reaction.

We frequently obtain 2×10^5 Ampr colonies per µg of chromosomal DNA.

The cosmid library is usually prepared in the form of 20 groups of 35,000 cosmids.

- An aliquot of the packaging reaction corresponding to 35,000 cosmids is diluted to 500 µl with SM and used to infect 2 ml of ED8767 bacteria in 10 mM $MgCl_2$.
- The mixture is incubated for 20 min at 37°C, diluted with 20 ml of LB medium, and then incubated for 1 h at 37°C.
- To the 22.5 ml of the infected mixture, we add 40 ml of fresh LB medium and ampicillin to a final concentration of 25 µg ml^{-1}. The culture is incubated with shaking for about 16 h.
- An aliquot of the culture is then withdrawn and titrated on agar plates to evaluate the amplification of the library.
- Several aliquots of the amplified culture are mixed with 1 volume of LB medium containing 20% glycerol, immediately frozen in liquid nitrogen and stored at −70°C.
- The preparation of the cosmid DNA is preceded by a second amplification. One-third of the amplified culture is used to inoculate 2 liters of LB medium in the presence of ampicillin. The culture is shaken at 37°C for 18 h. Total cosmid DNA is isolated by the alkaline lysis procedure and purified on CsCl gradients.

Transfection of cosmid DNA

- For the transfection of a c2RB cosmid library, groups of 5×10^6 P1.HTR cells are cotransfected with 60 µg of cosmid DNA of the amplified library and 2 µg of cosmid pHMR272, which carries a hygromycin-resistance gene. The procedures are similar to those described for the transfection of genomic DNA.
- For human MZ2-MEL.2.2 cells, 4.5×10^6 cells attached to 600 cm^2 tissue culture flasks (Singletray Unit, Nunc) containing 180 ml of medium are treated with 20 ml of the calcium phosphate–DNA precipitate of 240 µg of cosmid DNA and 24 µg of pSVtkneoβ.
- Cosmid transfectants are obtained by hygromycin or G418 selection. Transfectants expressing the relevant antigen are detected by the CTL proliferation or TNF release assays as described above (Fig. 9.9.2).

Recovery of cosmids

Direct packaging

Lau and Kan reported that a cosmid containing a globin gene, which had been used to obtain transfectants, could be recovered by directly packaging the DNA of these transfectants into λ phage components (Lau and Kan 1983). Presumably, this was made possible by occasional tandem integration of two cosmids producing a pair of cos sites that are separated by 39–52 kb of DNA, the size that can be packaged in λ phage heads. For direct packaging, high-molecular-weight DNA of the cosmid transfectants is purified. 1 µg is packaged in Gigapack extracts (Stratagene). The product is titrated on ED8767 with ampicillin selection. The number of colonies produced by the direct packaging has been found to vary between 20,000 and 30,000 per µg of DNA. Cosmid DNA is prepared from single colonies or from groups of colonies and the DNA is cotransfected with pSVtkneoβ in the recipient cell line. The transfectants are then screened for the expression of the antigen. Cosmids from some, but not all, transfectants transfer the expression of the antigen. This is expected since obviously not every transfected gene is integrated between two adequately spaced *cos* sites. We therefore developed another gene recovery strategy whereby we are able to recover sequences integrated near a *cos* site irrespective of the presence of a second *cos* site and of their position relative to the first *cos* site and the drug-resistance marker of the cosmid. This recovery procedure involves cosmid pTL6.

pTL6 rescue

Lund et al set up a method which involves the ligation of cosmid arms of pTL6 to partially digested DNA of the transfectant (Lund et al 1982). Two types of cosmid arms are prepared by digestion with either *Pvu*II or *Bst*EII, treatment with alkaline phosphatase, and digestion with *Bgl*II (Fig. 9.9.3). DNA of the transfectants is partially digested with *Mbo*I and fragments of 50–100 kb are prepared by centrifugation on a NaCl gradient. Cosmid arms (1.5 μg) are mixed with 1.5 μg of insert DNA. After ligation, packaging, and infection of ED8767, Tet^r or Amp^r colonies are selected to recover the cosmids obtained with *Bst*EII arms and *Pvu*II arms, respectively. As shown in Fig. 9.9.3, each of the two types of cosmid arms can rescue a sequence integrated in one of two main possible configurations. We showed that, from each transfectant, either one or the other arm of pTL6 was able to rescue an integrated tum^- sequence (De Plaen et al 1988).

Cloning of cDNA sequences

The cosmid transfection approach led to the identification of six genes coding for tumor antigens, but failed for others (Szikora et al 1990). We therefore initiated another approach involving the transfection of cDNA libraries. Reconstruction experiments were performed to recover the mouse P1A gene after stable transfection of a P1 cDNA expression library into P0.HTR cells. The presence of gene P1A was ascertained by hybridization of the library, but no stable transfectant that expressed antigen P815A could be detected. Transient expression of cDNA in COS cells appeared to be an attractive alternative.

Simian cells, which are permissive for SV40 virus, replicate the SV40 genome to a high copy number after virus infection. The SV40 large T antigen (SV40-LT), an early gene product, binds to the SV40 origin of replication (SV40 ori) and allows the host-cell DNA polymerases to start a bidirectional DNA synthesis (Danna and Nathans 1972; Li and Kelly 1985). This results in a 1000-fold amplification of the viral genome 48 h after the infection (Eckhart 1990). The origin of replication of SV40 consists of a 100 bp region with binding sites for the SV40 large T protein (Deb et al 1986; Fareed et al 1972; Fiers et al 1978). When cloned into a plasmid, the SV40 ori fragment promotes plasmid amplification in monkey cells infected with an SV40 helper virus. SV40-LT is the only viral protein required for viral replication. In 1981, Gluzman transfected a cell line, named CV-1, derived from the kidney of an African green monkey, with a variant of SV40 that had been rendered replication defective by a 3 bp deletion in the SV40 ori region (Gluzman 1981). Three cell lines were derived: COS-1, COS-3 and COS-7 (COS for **C**V-1, **o**rigin, **S**V40) that constitutively express the SV40 large T antigens, and therefore support the replication of plasmids that contain the SV40 ori.

For a number of reasons, COS cells have been used extensively as a mammalian expression system. These cells are easy to maintain in culture and to transfect. A large number of mammalian expression vectors are available that contain the SV40 ori and are therefore amplified to a high copy number in transfected COS cells, allowing the production of high levels of mRNA and therefore of proteins. This transient expression system has been a powerful tool for various purposes: the study of the regulation of mammalian gene expression (Mellon et al 1981) and the production of recombinant proteins for structural and functional studies (Rose and Bergmann 1982). It has also been used for quick evaluation of expression constructs in a transient assay before their stable incorporation into recipient cell lines (Fouser et al 1992). The COS cell expression system has been developed extensively by Aruffo and Seed for the rapid cloning of cDNA encoding cell-surface molecules recognized by monoclonal antibodies (Allen and Seed 1989; Aruffo and Seed 1987; Seed and Aruffo 1987; Simmons et al 1992). Furthermore, it has been proposed for cloning of cDNA producing CTL-defined antigens (Karttunen et al 1992).

A similar approach involving mouse cells is based on another papovavirus, the murine polyoma virus. Plasmids containing the polyoma origin of replication are amplified in murine cells such as WOP or COP that constitutively express the polyoma large T antigen. Because of our observation that a tumor antigen encoded by the human tyrosinase gene was not recognized by the appropriate CTL when the tyrosinase cDNA was expressed in murine P1.HTR cells or when these cells were pulsed with the tyrosinase nonapeptide, we felt that this amplification system in murine cells might be less suitable for the recognition of human tumor antigens. We have therefore used the COS system extensively.

The isolation of a cDNA encoding a defined tumor antigen by transfection into COS cells requires the prior identification of the HLA molecule presenting the antigenic peptide to the CTL clone. A cDNA library must be prepared with the RNA of the tumor cells. The cDNA library is then transfected into COS cells and the transfectants are screened for the expression of the antigen. The cDNA clone encoding the tumor antigen is then recovered from the library.

Identification of the class I molecule presenting the tumor antigen

Among the six HLA alleles (two HLA-A, two HLA-B, and two HLA-C) expressed by a tumor cell line and identified by HLA typing of a blood sample, only one of them is able to present the antigenic peptide recognized by the appropriate CTL clone. Since COS cells do not express human class I molecules, the relevant HLA gene must be cotransfected in the COS cells with the cDNA library. The relevant HLA molecule must therefore be identified.

A primary tool to identify the HLA-presenting molecule is to inhibit, with appropriate anti-HLA monoclonal antibodies, the lysis of the target cells or TNF release by the CTL (Darrow et al 1989; Wölfel et al 1989). In chromium-release or TNF assays, target cells are preincubated with the antibody at 37°C for 15–60 min before adding the cytolytic effectors. An inhibition curve is drawn with values obtained for decreasing concentrations of the antibody available as an ascitic fluid or a culture supernatant of hybridoma cells. Controls always include incubation of the antibody at the highest concentration with target cells alone to exclude any antibody-dependent mechanism of cell lysis in the absence of CTL. Monoclonal antibody W6/32, a mouse IGg2a that reacts with a monomorphic determinant on HLA-A, -B and -C (Barnstable et al 1978) is used to confirm that the antigenic peptide is presented by a HLA class I molecule. If the tumor cell line expresses HLA-A2, we try to inhibit the lysis or the TNF release with monoclonal antibodies BB7.2, MA2.1, and PA2.1 that react with determinants on HLA-A2, -A2/-B17 and -A2/-A28, respectively (McMichael et al 1980; Parham and Bodmer 1978; Parham and

Brodsky 1981). Inhibition with monoclonal antibody B1.23.2 (Viret et al 1993) directed against HLA-B, -C, and -A24,-A31,-A32, and with monoclonal antibody 4E (Yang et al 1984), which recognizes a common HLA-B epitope, can show that the antigenic peptide is presented by the HLA-B or -C molecules.

A second way to identify the HLA-presenting molecule is to correlate recognition by the CTL clone and expression of a definite class I allele. Surmising that the gene encoding the antigenic peptide is frequently expressed in tumor cell lines, we (and others) tested the production of TNF by the CTL clone in the presence of several allogeneic tumor cell lines that share one class I allele with the autologous tumor (Darrow et al 1989; Kawakami et al 1992; Wölfel et al 1993). Allogeneic tumor cell lines can also be stably or transiently transfected with candidate HLA genes and tested for their ability to stimulate TNF release by the CTL. For transient transfections, most tumor cell lines can be transfected in 96-well plates using 20–50 ng of DNA of the HLA construct and 1–2 μl of lipofectAMINE (Gibco BRL) per well. The CTL are added 24–48 h later. Considering the polymorphism affecting class I products within serologically homogeneous groups, it is always preferable to work with the HLA genes of the autologous tumor cell line. These HLA genes are obtained by screening an oligo(dT)-primed cDNA library or by cloning PCR products obtained with primers common to several class I genes. For PCR, we use thermostable DNA polymerases which have proofreading activity in order to decrease the rate of error during amplification (Ennis et al 1990). The PCR products are then sequenced and used in transient transfection assays.

If the presenting HLA cannot be definitely identified, transfection of the cDNA library into COS cells together with two or three putative class I candidates is still conceivable.

Transfection of cDNA libraries

Construction of cDNA libraries

Preparation of poly(A)$^+$ RNA

The quality of the mRNA used for the cDNA library is important for the efficiency of the reverse transcription. Kits designed for mRNA extraction from cells are easy to use but the presence of residual ribosomal RNA reduces the yield of mRNA available for reverse transcription. We prefer a two-step method involving first the extraction of total RNA followed by the purification of polyadenylated RNA.

At least 2–5 μg of mRNA is required for a cDNA library. Since 10^6 cells contain about 10 μg of total RNA, 0.5–2% of which is poly(A)$^+$ RNA, we start from 50–100×10^6 cells. These cells are homogenized in guanidine thiocyanate and the lysate is loaded on top of a CsCl cushion (Sambrook et al 1989). Homogenized lysates of cells in guanidine thiocyanate can be preserved for several months at –20°C until use. Total RNA is recovered and processed through two successive oligo(dT)-cellulose columns (mRNA purification kit, Pharmacia). The eluted RNA contains essentially mRNA, separated from ribosomal RNA, transfer RNA, and degraded mRNA, thus enhancing the synthesis of first-strand cDNA and minimizing reverse transcription of non-mRNA.

cDNA libraries

cDNA can be cloned in one or either orientation with respect to the promoter of the expression vector (Fig. 9.9.4). Directional libraries are obtained with an oligo(dT) primer containing a NotI restriction site. BstXI adaptors are added to both ends of the double-stranded cDNA molecules. After digestion with NotI, the cDNA is ligated to an expression vector cut by BstXI and NotI. For nondirectional libraries, cDNA obtained with oligo(dT) primers or random hexamers is ligated to BstXI adaptors and inserted into the BstXI site of the expression vector. The BstXI nonpalindromic adaptors are

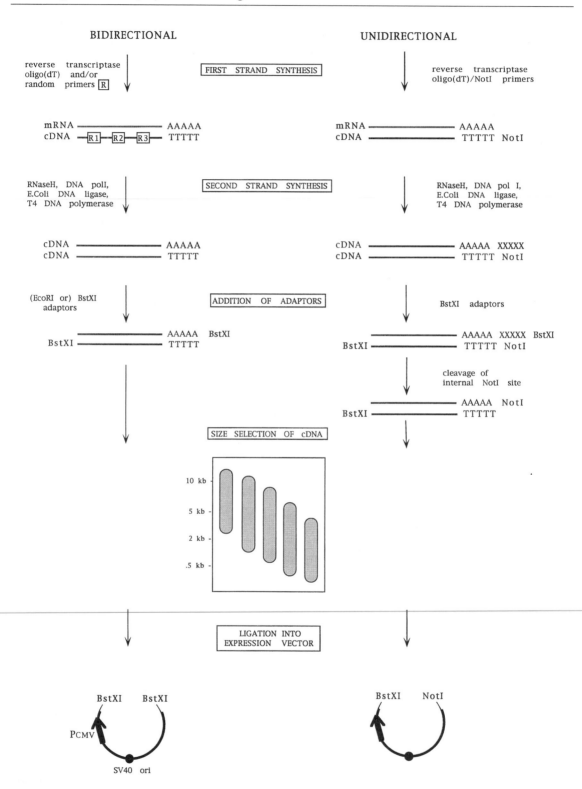

Figure 9.9.4 Schematic diagram of the procedure for constructing bidirectional and unidirectional cDNA libraries.

used because they reduce the concatemerization of adaptors or cDNA molecules by having noncomplementary overhangs.

The oligo(dT) priming method offers two major advantages: only poly(A)$^+$ RNA is reverse-transcribed, and the 3' ends of the cDNAs are complete up to the poly(A) tail. However, frequently reverse transcription of long RNA or of RNA containing strong secondary structures does not generate full-length cDNA molecules. Random hexamers, which may be used either alone or in combination with oligo(dT), increase the proportion of cDNA molecules containing the 5' ends of the mRNAs (Ozkaynak et al 1990).

A ready-to-use kit (SuperScript Choice System, BRL) has been successful in our hands. A crucial step is the synthesis of first-strand cDNA. When the yield is inferior to 10%, it is definitely useful to check the quality and the amount of mRNA used. The first strand is synthesized with a reverse transcriptase that is devoid of RNAase H activity (Superscript II RNAse H$^-$ Reverse Transcriptase, BRL), therefore eliminating degradation of the RNA molecule during first-strand cDNA synthesis and improving the yield of full-length cDNA molecules. After second-strand cDNA synthesis and ligation to adaptors, cDNA molecules are separated on a column prepacked with Sephacryl S-500 HR. This eliminates the excess of adaptors. It also separates the cDNA according to size and five fractions are collected (Fig. 9.9.4).

Titration and amplification of cDNA libraries

cDNAs of every size fraction are ligated to vectors derived from pCDM8 (Seed 1987). Plasmid pcDNAI/Amp (Invitrogen Corporation, Oxon, UK) carries the immediate-early cytomegalovirus (CMV) promoter, SV40 transcription termination and RNA processing signals, and the SV40 origin for episomal replication in COS-7 cells. Another derivative, pcDNA3, contains in addition the neomycin resistance gene for the selection of G418 resistant stable transfectants. The product of the ligation is entirely electroporated into bacteria such as DH5αF'IQ

(BRL) or TOP10F' (Invitrogen). Optimal efficiencies are obtained by using less than 20 ng of the ligation product to transform 40 μl of electrocompetent bacteria (~10^9 cells). After electroporation, the bacteria are diluted with 1 ml of LB medium. For every fraction, 4 ml of electroporated bacteria are plated on two agar plates (22 × 22 cm); 20 μl and 100 μl are plated on small agar plates (8 cm diameter). Bacteria are incubated overnight at 37°C. The colonies present on the plates are counted to evaluate the number of clones obtained with each fraction. We frequently obtain 20,000 colonies ng^{-1} of cDNA. Plasmid DNA is extracted from 24 colonies corresponding to every fraction and analyzed by restriction enzyme digestion for the presence and size of inserts. In directional libraries, 90% of the clones contain inserts with sizes of 0.5 to 4 kb. The colonies grown on the two large plates are scraped in 30 ml of LB medium. Bacteria representing every fraction of the library are then frozen in 5 master tubes (1 ml) and 20 small tubes (30 μl).

Preparation of pools of bacteria

One tube of 30 μl is titrated (Fig. 9.9.5). The usual titer of the frozen sample is 5×10^8 bacteria. Another aliquot is then plated on an agar plate so as to obtain a set of 10,000 colonies which is divided into smaller pools. DH5αF'IQ and TOP10F' colonies exhibit a homogeneous size and we therefore believe that the diversity of the cDNA library is conserved when this library is amplified in liquid culture. To evaluate the maximal number of colonies that can be collected in a pool, we diluted DNA of a vector carrying a *MAGE-1* cDNA into DNA of unrelated recombinant vectors and transfected these mixtures into COS cells. We screened the COS transfectants by the TNF release assay. We found transfectants expressing the antigen derived from *MAGE-1* only in groups transfected with a dilution of the *MAGE-1* cDNA of less than 1/250 to 1/500. We usually divide the cDNA library into pools of 100–200 colonies.

The pools are amplified at 37°C for 3–5 h (or overnight) in 5 ml LB medium supplemented with ampicillin (50 μg ml^{-1}). An aliquot (0.5 ml) is frozen and kept at –80°C for eventual recovery of the antigen-coding cDNA clone from a pool of bacteria identified as positive (as described below). Plasmid DNA is extracted from the culture (yield 2–4 µg) and resuspended in 50 µl of TE containing 20 µg ml^{-1} RNAase. A minimum of 10^5 cDNA clones is tested in transfection.

Transfection of COS-7 cells

- One day before transfection, COS cells are seeded in a 96-well flat-bottom microplate in 100 µl of DMEM supplemented with 10% FCS (Fig. 9.9.5). Typically, 15,000–20,000 cells per well are seeded and a

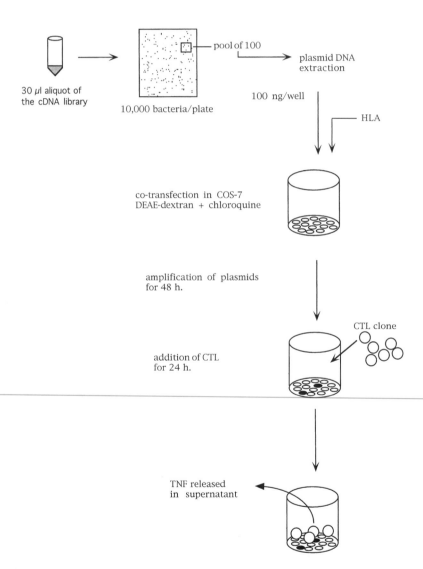

30 µl aliquot of the cDNA library

10,000 bacteria/plate

pool of 100

plasmid DNA extraction

100 ng/well

HLA

co-transfection in COS-7 DEAE-dextran + chloroquine

amplification of plasmids for 48 h.

addition of CTL for 24 h.

CTL clone

TNF released in supernatant

Figure 9.9.5 Transfection of COS-7 cells. COS cells seeded in 96 microwells are cotransfected with the relevant HLA and 100 ng of plasmid DNA extracted from a pool of 100 bacteria. After 2 days of plasmid replication, transfectants are incubated with the appropriate CTL and then tested for the expression of the antigen upon measurement of the amount of TNF released in the supernatant.

confluent monolayer is obtained on the day of transfection.

- Cells are transfected with 100 ng of DNA of the cDNA pool and 100 ng of vector containing the HLA coding sequence in a final volume of 35 μl/well. The DNA is dissolved in DMEM supplemented with DEAE-dextran (400 μg ml^{-1}), chloroquine (200 μM) and heat-inactivated NuSerum (10%, Collaborative Biomedical Products, Two Oak Park, Bedford, MA, USA). Plasmid DNA degradation is limited by the use of the DNAase-free NuSerum and of chloroquine, which blocks the lysosomal degradation within transfected cells. The DEAE-dextran concentration may be reduced to 200 μg ml^{-1} without affecting the efficiency of expression of the antigen. Chloroquine, however, which is toxic to the cells, cannot be used at a lower concentration without decreasing the efficiency of the transfection.
- After 4 h at 37°C in an 8% CO_2 atmosphere, microwells are gently flicked and 50 μl of PBS containing 10% DMSO is added. After 2 min at room temperature, PBS–DMSO is removed and 200 μl of fresh DMEM supplemented with 10% FCS is added to the microwells.
- After 48 h at 37°C, the medium is discarded and 1000–2000 CTL are added in 100 μl of Iscove medium containing 10% human serum and 25 U ml^{-1} IL-2. After 24 h, the amount of TNF released into the supernatant is measured as described in under 'Cloning of genomic sequences'.

Recovery of a positive cDNA clone

If transfectants expressing the relevant antigen are obtained with a pool of bacteria, this pool is recovered from the frozen tube and subcloned. The positive bacteria are isolated by testing smaller pools of bacteria or single colonies as follows.

- Individual colonies are seeded in U-bottom microwells in 0.3 ml of TYGPN medium (Ausubel et al 1993). Bacteria are cultured for 48 h at 37°C. Plasmid DNA is then prepared in the microplates by the alkaline lysis method (Ausubel et al 1993).
- The DNA is recovered by isopropanol precipitation and resuspended in 50 μl of Tris 10 mM, EDTA 1 mM, pH 7.4, containing 20 μg ml^{-1} RNAase. Strains DH5αF'IQ and TOP10F' are devoid of endonuclease A activity (endA1), thus improving the yield of plasmid DNA recovery from cultures in microplates in the absence of phenol–chloroform extraction.
- Half of the plasmid DNA is transfected into COS cells as above. For these experiments involving small pools of bacteria, the time of transfection can be reduced to 24 h. Transfectants are tested for their ability to stimulate TNF release by the appropriate CTL.

Advantages and limitations of the COS system

Transient expression offers several advantages over long-term expression. It obviates the time-consuming procedure of selecting cells that have integrated exogenous DNA. It also obviates the variability introduced by the integration of the transfected DNA, since the level of expression of each stable transfectant is highly dependent on the location and number of integrated copies. Transient expression allows the investigator to look at gene products within hours of DNA uptake. The COS approach does not require the selection of antigen-loss variants that are used as recipients for cosmid transfection. Besides the difficulty of the selection procedure, a variant that is resistant to lysis by the CTL may also have lost the expression of the presenting HLA molecule. These HLA-loss variants are sometimes difficult to distinguish from genuine antigen-loss variants and their use in cosmid transfection can lead to the cloning of the HLA gene (van der Bruggen et al 1994). On the other hand, the use of COS cells requires the cloning of the relevant HLA. Taken together, the COS approach is definitely faster and more convenient than the cosmid transfection approach.

The COS-cDNA approach has led to the identification of several genes coding for tumor antigens recognized by autologous CTL. We believe that this amplification system can be used widely for the identification of other genes. However, the episomal multiplication of plamids could lead to the identification of inappropriate cDNAs that are recognized by a CTL because of the overexpression of a peptide that is normally ignored. It is therefore mandatory to verify that a positive cDNA also confers the expression of the antigen when incorporated into stable transfectants which express a lower amount of antigen.

But sometimes the COS approach is impossible because some CTL clones recognize COS cells either untransfected or transfected with the relevant HLA gene. Interestingly, the latter cross-reaction does not occur with COS-7 cells transfected with the nonrelevant class I genes. COS cells presumably produce a peptide that resembles the original antigenic peptide recognized by the CTL. The high level of transient expression of the HLA does not account for these cross-reactions since COS cells stably transfected with the HLA are recognized by the CTL as well. This type of cross-reaction presents a challenge that will have to be overcome in order to identify some of the antigens recognized on tumors by autologous CTL.

Identification of antigenic peptides encoded by cloned genes

When the gene that transfers the expression of the antigen is isolated, the next step is to identify the antigenic peptide recognized by the CTL. One approach is based on the observation that peptides binding to a given HLA molecule share certain anchor residues which interact with amino acids present in pockets of the HLA groove (Falk et al 1991; Madden et al 1991). Such consensus anchor residues have now been identified for several HLA molecules (reviewed in Engelhard 1994). Peptides carrying these potential anchor residues can be deduced from the nucleotide sequence, synthesized, and tested directly for their ability to sensitize to lysis by the CTL target cells expressing the relevant HLA molecule. The predominant length of peptides associated with most class I molecules analyzed to date is 9 residues, but several peptides of 10 or 11 residues have now been identified (reviewed in Engelhard 1994).

Another approach has to be used when no information is available for the binding of peptides to the HLA-presenting molecule. We have used a method involving the transfection of subgenic fragments to localize the region encoding the antigenic peptide.

Transfection of subgenic fragments

Subgenic fragments are easily generated by PCR amplification and cloned into vectors carrying the origin of replication of SV40, such as pcDNAI/Amp (Invitrogen) or pCR™3 (Invitrogen). The constructs are then tested for their ability to code for the antigen by transfection into COS-7 cells together with the gene coding for the HLA-presenting molecule.

A start codon with a consensus sequence for translation initiation is incorporated in the sense oligonucleotide primer and a STOP codon in the antisense primer (Fig. 9.9.6). Restriction sites are also incorporated into the primers to facilitate subsequent directional cloning into the expression vector. PCR products are purified and inserted into the corresponding sites of the expression vectors, using standard techniques.

Alternatively, PCR fragments can be generated and cloned directly into pCR™3 using the Eukaryotic TA cloning kit from Invitrogen. *Taq* polymerase has a non template-dependent terminal transferase activity which adds a single deoxyadenosine to the 3' ends of duplex molecules. The linearized vector has single 3' deoxythymidine residues. This enables the ligation of PCR products at efficiencies 50 times greater than blunt-end ligation. Since the PCR fragment is subsequently inserted into the vector in either orientation, individual recombinant clones need to be analyzed by restriction mapping or sequencing to determine the orientation of the insert.

Although these two techniques were very efficient in our hands, other procedures for cloning PCR products or fragments generated by digestion with restriction enzymes can also be used.

COS cells are then transfected by the DEAE–dextran–chloroquine method with the various gene fragments together with the gene coding for the HLA presenting molecule and tested for their ability to stimulate TNF release by the CTL (Fig. 9.9.7).

Another approach to identify the region of the gene that codes for the antigenic peptide is to generate a large number of deleted clones by exonuclease III digestion. The ability of these clones to code for the antigen is then tested by transfecting them into recipient cells. Exonuclease III is used to progressively digest insert DNA from a 5' protruding or blunt end. The adjacent vector sequence is protected from digestion by a 3' protruding end or by an α-phosphorothioate filled end. For example, if a cDNA is inserted into the *Eco*RI site of pcDNAI/Amp, the recombinant plasmid is cleaved in the polylinker with *Not*I (5' protruding end) and *Sph*I (3' protruding end). Exonuclease III treatment is then performed with the Erase-a-base System (Promega, Madison, WI, USA) to generate progressive deletions starting from the *Not*I end. Samples are removed at timed intervals. After treatment with S1 nuclease and Klenow DNA polymerase, T4 DNA ligase is added to circularize the deleted vectors. The ligation mixtures are electroporated in bacteria and the plasmids are selected with an antibiotic. Approximately 10 clones are isolated for each time point. DNA is extracted from each clone and cotransfected into COS-7 cells with the HLA construct. Transfectants expressing the relevant antigen are detected by their ability to stimulate TNF release by the CTL. At least 5 positive and 5 negative deleted cDNAs are sequenced. By comparing the sequences of the longest negative and the shortest positive cDNA, it is usually possible to delineate a short region that encodes the C-terminal part of the antigenic peptide (Fig. 9.9.8).

Before transfection, it is advisable to check for the presence of a translation initiation codon in the sequence of the vector, between the eukaryotic promoter and the cloning site and in-frame with the desired product. In that case, the translation product is longer than the expected one. By transfecting subgenic fragments, it is often easier to determine the C-terminal part of the antigenic peptide than its N-terminal part.

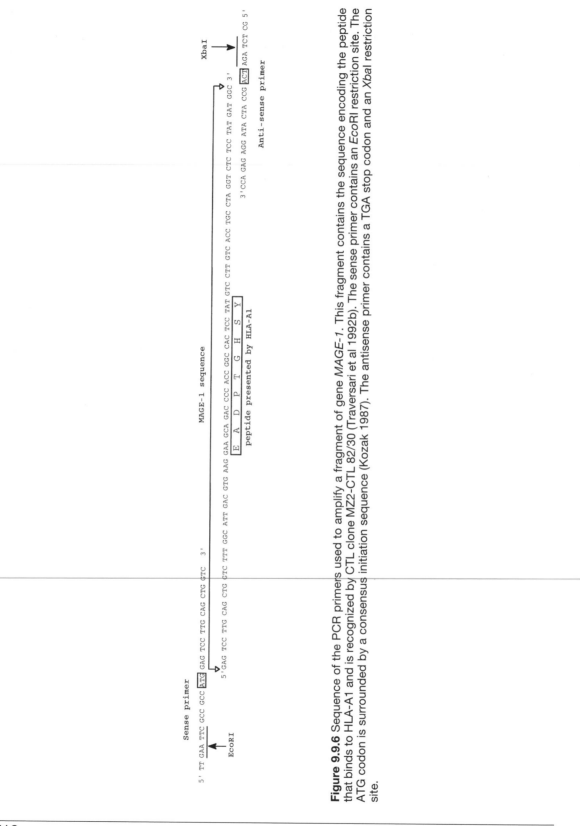

Figure 9.9.6 Sequence of the PCR primers used to amplify a fragment of gene *MAGE-1*. This fragment contains the sequence encoding the peptide that binds to HLA-A1 and is recognized by CTL clone MZ2-CTL 82/30 (Traversari et al 1992b). The sense primer contains an *Eco*RI restriction site. The ATG codon is surrounded by a consensus initiation sequence (Kozak 1987). The antisense primer contains a TGA stop codon and an *Xba*I restriction site.

Recognition assay of target cells incubated with synthetic peptides

When a small sequence that codes for the antigenic peptide is identified, several overlapping peptides encoded by this sequence are synthesized and tested for their ability to sensitize target cells to lysis by the relevant CTL. These peptides are usually 14–15 residues long; although such peptides bind poorly to class I molecules, the preparations are frequently contaminated with shorter peptides, including the one that is recognized by the CTL. By testing in a first step several long peptides, we reduce the number of peptides that have to be synthesized and tested. When such a long peptide is found to sensitize target cells to lysis, shorter peptides are then tested to identify the minimal peptide recognized by the CTL (Fig. 9.9.7). If the HLA-presenting molecule is known, short peptides carrying potential anchor residues for binding to this particular HLA molecule can be deduced from the sequence of the small region producing the antigenic peptide.

The different synthetic peptides are tested for their ability to sensitize target cells to lysis by the CTL in ^{51}Cr-release assays. Any target cell expressing the appropriate HLA molecule can be used, but EBV-transformed lymphoblastoid cell lines are convenient because they usually express HLA molecules at high level. If the peptide is presented by HLA-A2.1, the human CEM $\times$ 721.174.T2 (T2) mutant cell line (Salter et al 1985) is very useful. T2 cells exhibit an antigen-processing defect resulting in an increased capacity to present exogenous peptides. It is important to mention that pretreating target cells with monoclonal antibodies MA2.1 or W6/32 frequently increases their capacity to

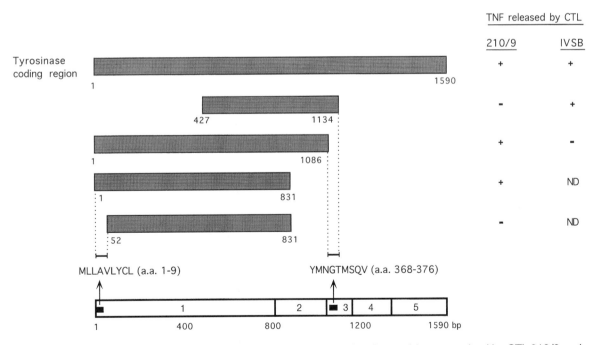

Figure 9.9.7 Location of tyrosinase sequences coding for the antigenic peptides recognized by CTL 210/9 and IVSB. The coding part of the cDNA, which comprises 1590 bp, is shown with the inter-exon boundaries. Nucleotides are numbered from the start codon. The limits of the cDNA fragments that were cloned into pcDNAI/Amp and transfected into COS cells together with HLA-A2 are indicated. (+) or (–): presence or absence of TNF release by the CTL stimulated with the transfected COS cells (Wölfel et al 1994).

present exogenous peptides. Pretreatment with MA2.1 has been reported to reduce by a factor of 100 the concentration of an influenza A matrix peptide needed to induce lysis by the appropriate CTL clone (Bodmer et al 1989). We observed a similar effect for the tyrosinase (Fig. 9.9.9) and the MAGE-3 peptides (Gaugler et al 1994). In the peptide sensitization assay, target cells are ^{51}Cr-labeled for 1 h at 37°C and washed extensively. Target cells (1000) are then incubated in 96-well microplates in the presence of various concentrations of peptide for 30 min at 37°C. Peptide concentrations vary between 30 pM and 30 mM in a first assay. CTL are then added in an equal volume. The lymphocyte-to-target ratio is chosen so as to

obtain a high percentage of specific lysis. Chromium release is measured after 4 h at 37°C.

Another assay for testing synthetic peptides is to transfect COS-7 cells with the appropriate HLA construct. Twenty-four hours later, transfected COS-7 cells are pulsed 1 h at 37°C with a synthetic peptide, washed twice, and tested for their ability to stimulate TNF release by the CTL.

These procedures can help to define the minimal peptide recognized by a CTL clone. This does not imply that such a peptide is fully identical to the corresponding naturally processed peptide. Naturally processed peptides can be detached from the class I MHC

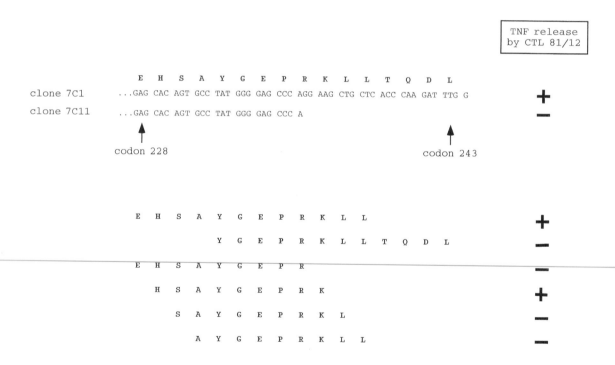

Figure 9.9.8 Amino acid and nucleotide sequence of the region of *MAGE-1* coding for the antigenic peptide that binds to HLA-Cw*1601 and is recognized by CTL clone MZ2-CTL81/12 (van der Bruggen et al 1994). The 3' end of the shortest truncated cDNA that confers recognition by CTL 81/12 (clone 7C1) and of the longest cDNA that does not (clone 7C11) are indicated. These truncated cDNA clones were tested by transfection into COS-7 cells together with an HLA-Cw*1601 construct for their ability to stimulate TNF release by MZ2-CTL 81/12. Different synthetic peptides corresponding to this short region were tested: COS-7 cells transfected with HLA-CW*1601 were pulsed 1 day later with a 1 mM concentration of these peptides. CTL were added and the TNF content of the supernatant was estimated 1 day later by testing its toxicity on WEHI-164 clone 13 cells. + means production by the CTL of more than 20 pg ml^{-1} TNF and − means production of less than 4 pg ml^{-1} TNF. These concentrations were estimated according to a standard curve of human rTNFβ.

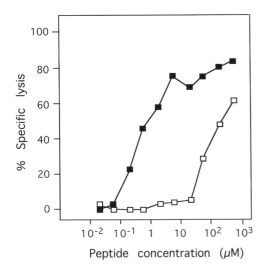

Figure 9.9.9 Enhancement of recognition of tyrosinase peptide 368–376 by anti-HLA-A2 monoclonal antibody MA2.1. T2 target cells were chromium-labeled in the presence (■) or absence (□) of MA2.1. They were then incubated with the peptide for 90 min. CTL IVSB were then added at an effector-to-target ratio of 40 and chromium release was measured after 6 h.

molecules by a mild acid treatment (Falk et al 1990; Rötzschke et al 1990). Peptides are then separated by high-performance liquid chromatography (HPLC), and individual peptides are analyzed by sophisticated mass spectrometry (Hunt et al 1992). In a few cases, this technique has led to the identification of the CTL-defined tumor antigens (Cox et al 1994).

References

Allen JM, Seed B (1989) Isolation and expression of functional high affinity Fc receptor complementary DNAs. Science 243: 378–381.

Anichini A, Fossati G, Parmiani G (1987) Clonal analysis of the cytolytic T-cell response to human tumors. Immunol Today 8: 385–389.

Aruffo A, Seed B (1987) Molecular cloning of a CD28 cDNA by a high-efficiency COS cell expression system. Proc Natl Acad Sci USA 84: 8573–8577.

Ausubel FM, Brent R, Kingston RE, et al (eds) (1993) Current Protocols in Molecular Biology. Greene Publishing Associates and Wiley). New.

Barnstable CJ, Bodmer WF, Brown G, et al (1978) Production of monoclonal antibodies to group A erythrocytes, HLA and other human cell-surface antigens — new tools for genetic analysis. Cell 14: 9–20.

Bates PF, Swift RA (1983) Double *cos* site vectors: simplified cosmid cloning. Gene 26: 137–146.

Bernard H-U, Krammer G, Rowekamp WG (1985) Construction of a fusion gene that confers resistance against hygromycin B to mammalian cells in culture. Exp Cell Biol 158: 237–243.

Beyaert R, Vanhaesebroeck B, Suffys P, Van Roy F, Fiers W (1989) Lithium chloride potentiates tumor necrosis factor-mediated cytotoxicity *in vitro* and *in vivo*. Proc Natl Acad Sci USA 86: 9494–9498.

Bjorkman PJ, Saper MA, Samraoui B, Bennett WS, Strominger JL, Wiley DC (1987) The foreign antigen binding site and T cell recognition regions of class I histocompatibility antigens. Nature 329: 512–518.

Bodmer H, Ogg G, Gotch F, McMichael A (1989) Anti-HLA-A2 antibody-enhancement of peptide association with HLA-A2 as detected by cytotoxic T lymphocytes. Nature 342: 443–446.

Brichard V, Van Pel A, Wölfel T, et al (1993) The tyrosinase gene codes for an antigen recognized by autologous cytolytic T lymphocytes on HLA-A2 melanomas. J Exp Med 178: 489–495.

Brunner KT, MacDonald HR, Cerottini J-C (1980) Antigenic specificity of the cytolytic T lymphocyte (CTL) response to murine sarcoma virus-induced tumors. II. Analysis of the clonal progeny of CTL precursors stimulated *in vitro* with syngeneic tumor cells. J Immunol 124: 1627–1634.

Corsaro CM, Pearson ML (1981) Enhancing the efficiency of DNA mediated gene transfer in mammalian cells. Somat Cell Mol Genet 7: 603–616.

Coulie PG, Brichard V, Van Pel A, et al (1994) A new gene coding for a differentiation antigen recognized by autologous cytolytic T lymphocytes on HLA-A2 melanomas. J Exp Med 180: 35–42.

Cox AL, Skipper J, Chen Y, et al (1994) Identification of a peptide recognized by five melanoma-specific human cytotoxic T cell lines. Science 264: 716–719.

Danna KJ, Nathans D (1972) Bidirectional replication of simian virus 40 DNA. Proc Natl Acad Sci USA 69: 3097–3100.

Darrow TL, Slingluff CLJ, Seigler HF (1989) The role of HLA class I antigens in recognition of melanoma cells by tumor-specific cytotoxic T lymphocytes. Evidence for shared tumor antigens. J Immunol 142: 3329–3335.

de Bergeyck V, De Plaen E, Chomez P, Boon T, Van Pel A (1994) An intracisternal A particle sequence codes for an antigen recognized by syngeneic cytolytic T lymphocytes on a mouse spontaneous leukemia. Eur J Immunol 24: 2203–2212.

De Plaen E, Lurquin C, Van Pel A, et al (1988) Immunogenic (tum⁻) variants of mouse tumor P815: cloning of the gene of tum⁻ antigen P91A and identification of the tum⁻ mutation. Proc Natl Acad Sci USA 85: 2274–2278.

Deb S, DeLucia AL, Baur C-P, Koff A, Tegtmeyer P (1986) Domain structure of the simian virus 40 core origin of replication. Mol Cell Biol 6: 1663–1670.

Eckhart W (1990) Polyomavirinae and their replication. In: D.M.K.e.a. B.N. Fields, ed. Virology. Raven Press, New York, pp. 1593–1607.

Engelhard VH (1994) Structure of peptides associated with class I and class II MHC molecules. Annu Rev Immunol 12: 181–207.

Ennis PD, Zemmour J, Salter RD, Parham P (1990) Rapid cloning of HLA-A,B cDNA by using the polymerase chain reaction: frequency and nature of errors produced in amplification. Proc Natl Acad Sci USA 87: 2833–2837.

Espevik T, Nissen-Meyer J (1986) A highly sensitive cell line, WEHI 164 clone 13, for measuring cytotoxic factor/tumor necrosis factor from human monocytes. J Immunol Methods 95: 99–105.

Falk K, Rötzschke O, Rammensee H-G (1990) Cellular peptide composition governed by major histocompatibility complex class I molecules. Nature 348: 248–251.

Falk K, Rötzschke O, Stevanovic S, Jung G, Rammensee H-G (1991) Allele-specific motifs revealed by sequencing of self-peptides eluted from MHC molecules. Nature 351: 290–296.

Fareed GC, Garon CF, Salzman NP (1972) Origin and direction of simian virus 40 deoxyribonucleic acid replication. J Virol 10: 484.

Fiers W, Contreras R, Haegeman G, et al (1978) The complete nucleotide sequence of SV40 DNA. Nature (London) 273: 113–120.

Fouser LA, Swanberg SL, Lin B-Y, et al (1992) High level expression of a chimeric anti-ganglioside GD2 antibody: genomic kappa sequences improve in COS and CHO cells. Biotechnology 10: 1121–1127.

Gaugler B, Van den Eynde B, van der Bruggen P, et al (1994) Human gene MAGE-3 codes for an antigen recognized on a melanoma by autologous cytolytic T lymphocytes. J Exp Med 179: 921–930.

Gluzman Y (1981) SV40-transformed simian cells support the replication of early SV40 mutants. Cell 23: 175–182.

Gross-Bellard M, Oudet P, Chambon P (1973) Isolation of high-molecular-weight DNA from mammalian cells. Eur J Biochem 36: 32–38.

Grosveld FG, Dahl H-HM (1983) Cosmid libraries. In Flavell RA, Techniques in the Life Sciences, Techniques in Nucleic Acid Biochemistry, Vol. B5. Elsevier, Ireland, pp. 1–18.

Grosveld FG, Dahl H-HM, de Boer E, Flavell RA (1981) Isolation of β-globin-related genes from a human cosmid library. Gene 13: 227–237.

Grosveld FG, Lund T, Murray EJ, Mellor, AL, Dahl H-HM, Flavell RA (1982) The construction of cosmid libraries which can be used to transform eukaryotic cells. Nucleic Acids Res 10: 6715–6732.

Hansen MB, Nielsen SE, Berg K (1989) Re-examination and further development of a precise and rapid dye method for measuring cell growth/cell kill. J Immunol Methods 119: 203–210.

Hérin M, Lemoine C, Weynants P, et al (1987) Production of stable cytolytic T-cell clones directed against autologous human melanoma. Int J Cancer 39: 390–396.

Hoeijmakers JHJ, Odijk H, Westerveld A (1987) Differences between rodent and human cell lines in the amount of integrated DNA after transfection. Exp Cell Res 169: 111–119.

Hunt DF, Henderson RA, Shabanowitz J, et al (1992) Characterization of peptides bound to the class I MHC molecule HLA-A2.1 by mass spectrometry. Science 255: 1261–1263.

Karttunen J, Sanderson S, Shastri N (1992) Detection of rare antigen-presenting cells by the *lacZ* T-cell activation assay suggests an expression cloning strategy for T-cell antigens. Proc Natl Acad Sci USA 89: 6020–6024.

Kawakami Y, Zakut R, Topalian SL, Stotter H, Rosenberg SA (1992) Shared human melanoma antigens. Recognition by tumor-infiltrating lymphocytes in HLA-A2.1 transfected melanomas. J Immunol 148: 638–643.

Klein G, Sjogren H, Klein E, Hellstrom KE (1960) Demonstration of resistance against methylcholanthrene-induced sarcomas in the primary autochthonous host. Cancer Res 20: 1561–1572.

Knuth A, Danowski B, Oettgen HF, Old L (1984) T-cell mediated cytotoxicity against autologous malignant melanoma: analysis with interleukin-2-dependent T-cell cultures. Proc Natl Acad Sci USA 81: 3511–3515.

Kozak M (1987) An analysis of 5'-noncoding sequences from 699 vertebrate messenger RNAs. Nucleic Acids Res 15: 8125–8132.

Kripke ML (1974) Antigenicity of murine skin tumors induced by ultraviolet light. J Natl Cancer Inst 53: 1333–1336.

Lau YF, Kan YW (1983) Versatile cosmid vectors for the isolation, expression and rescue of gene sequences: studies with the human alpha-globin gene cluster. Proc Natl Acad Sci USA 80: 5225–5229.

Levy JP, Leclerc JC (1977) The murine-sarcoma-virus-induced tumor: exceptions or general model in tumor immunology. Adv Cancer Res 24: 1–59.

Li JJ, Kelly TJ (1985) Simian virus 40 DNA replication *in vitro*: specificity of initiation and evidence for bidirectional replication. Mol Cell Biol 5: 1238–1246.

Lund T, Grosveld FG, Flavell RA (1982) Isolation of transforming DNA by cosmid rescue. Proc Natl Acad Sci USA 79: 520–524.

Lurquin C, Van Pel A, Mariame B, et al (1989) Structure of the gene coding for tum⁻ transplantation antigen P91A. A peptide encoded by the mutated exon is recognized with Ld by cytolytic T cells. Cell 58: 293–303.

Madden DR, Gorga JC, Strominger JL, S, Wiley DC (1991) The structure of HLA-B27 reveals nonamer self peptides bound in an extended conformation. Nature 353: 321–325.

Maryanski J, Boon T (1982) Immunogenic variants obtained by mutagenesis of mouse mastocytoma P815. IV. Analysis of variant-specific antigens by selection of antigen-loss variants with cytolytic T cell clones. Eur J Immunol 12: 406–412.

Maryanski JL, Van Snick J, Cerottini JC, Boon T (1982) Immunogenic variants obtained by mutagenesis of mouse mastocytoma P815. III. Clonal analysis of the syngeneic cytolytic T lymphocyte response. Eur J Immunol 12: 401–406.

McMichael AJ, Parham P, Rust N, Brodsky F (1980) A monoclonal antibody that recognizes an antigenic determinant shared by HLA-A2 and B17. Hum Immunol 1: 121–129.

Mellon P, Parker V, Gluzman Y, Maniatis T (1981) Identification of DNA sequences required for transcription of the human α1-globin gene in a new SV40 host–vector system. Cell 27: 279–288.

Mukherji B, MacAlister TJ (1983) Clonal analysis of cytotoxic T cell response against human melanoma. J Exp Med 158: 240–245.

Nicolas JF, Berg P (1983) Regulation of expression of genes transduced into embryonal carcinoma cells. CSH Conf Cell Prolif 10: 469–485.

Old LJ, Boyse EA, Clarke DA, Carswell EA (1962) Antigenic properties of chemically-induced tumors. Ann NY Acad Sci 10: 80–106.

Ozkaynak E, Rueger DC, Drier EA, et al (1990) OP-1 cDNA encodes an osteogenic protein in the TGF-β family. EMBO J 9: 2085–2093.

Parham P, Bodmer WF (1978) Monoclonal antibody to a human histocompatibility alloantigen, HLA-A2. Nature (London) 276: 397–399.

Parham P, Brodsky FM (1981) Partial purification and some properties of BB7.2. A cytotoxic monoclonal antibody with specificity for HLA-A2 and a variant of HLA-A28. Hum Immunol 3: 277–299.

Perucho M, Hanahan D, Wigler M (1980) Genetic and physical linkage of exogenous sequences in transformed cells. Cell 22: 309–317.

Prehn RT, Main JM (1957) Immunity to methylcholanthrene-induced sarcomas. J Natl Cancer Inst 18: 769–778.

Rose J, Bergmann J (1982) Expression from cloned cDNA of cell-surface secreted forms of the glycoprotein of vesicular stomatitis virus in eukaryotic cells. Cell 30: 753–762.

Rötzschke O, Falk K, Deres K, et al (1990) Isolation and analysis of naturally processed viral peptides as recognized by cytotoxic T cells. Nature 348: 252–254.

Salter RD, Howell DN, Cresswell P (1985) Genes regulating HLA class I antigen expression in T-B lymphoblast hybrids. Immunogenetics 21: 235–246.

Sambrook J, Fritsch EF, Maniatis T (1989) Molecular Cloning. A Laboratory Manual. Cold Spring Harbor Laboratory Press, Cold Spring Harbor, NY.

Seed B (1987) An LFA-3 cDNA encodes a phospholipid-linked membrane protein homologous to its receptor CD2. Nature 329: 840–842.

Seed B, Aruffo A (1987) Molecular cloning of the CD2 antigen, the T-cell erythrocyte receptor, by a rapid immunoselection procedure. Proc Natl Acad Sci USA 84: 3365–3369.

Sibille C, Chomez P, Wildmann C, et al (1990) Structure of the gene of tum⁻ transplantation antigen P198: a point mutation generates a new antigenic peptide. J Exp Med 172: 35–45.

Simmons DL, Satterthwaite AB, Tenen DG, Seed B (1992) Molecular cloning of a cDNA encoding CD34, a sialomucin of human hematopoietic stem cells. J Immunol 148: 267–271.

Szikora J-P, Van Pel A, Brichard V, et al (1990) Structure of the gene of tum⁻ transplantation antigen P35B: presence of a point mutation in the antigenic allele. EMBO J 9: 1041–1050.

Townsend A, Rothbard J, Gotch F, Bahadur G, Wraith D, McMichael A (1986) The epitopes of influenza nucleoprotein recognized by cytotoxic T lymphocytes can be defined with short synthetic peptides. Cell 44: 959–968.

Traversari C, van der Bruggen P, Van den Eynde B, et al (1992a) Transfection and expression of a gene coding for a human melanoma antigen recognized by autologous cytolytic T lymphocytes. Immunogenetics 35: 145–152.

Traversari C, van der Bruggen P, Luescher IF, et al (1992b) A nonapeptide encoded by human gene MAGE-1 is recognized on HLA-A1 by cytolytic T lymphocytes directed against tumor antigen MZ2-E. J Exp Med 176: 1453–1457.

Uyttenhove C, Maryanski J, Boon T (1983) Escape of mouse mastocytoma P815 after nearly complete rejection is due to antigen-loss variants rather than immunosuppression. J Exp Med 157: 1040–1052.

Van den Eynde B, Hainaut P, Hérin M, et al (1989) Presence on a human melanoma of multiple antigens recognized by autologous CTL. Int J Cancer 44: 634–640.

Van den Eynde B, Lethe B, Van Pel A, De Plaen E, Boon T (1991) The gene coding for a major tumor rejection antigen of tumor P815 is identical to the normal gene of syngeneic DBA/2 mice. J Exp Med 173: 1373–1384.

van der Bruggen P, Traversari C, Chomez P, et al (1991) A gene encoding an antigen recognized by cytolytic T lymphocytes on a human melanoma. Science 254: 1643–1647.

van der Bruggen P, Szikora J-P, Boel P, et al (1994) Autologous cytolytic T lymphocytes recognize a MAGE-1 nonapeptide on melanomas expressing HLA-Cw*1601. Eur J Immunol 24: 2134–2140.

Van Pel A, Vessière F, Boon T (1983) Protection against two spontaneous mouse leukemias conferred by immunogenic variants obtained by mutagenesis. J Exp Med 157: 1992–2001.

Van Pel A, De Plaen E, Boon T (1985) Selection of a highly transfectable variant from mouse mastocytoma P815. Somat Cell Genet 11: 467–475.

Viret C, Davodeau F, Guilloux Y, et al (1993) Recognition of shared melanoma antigen by HLA-A2-restricted cytolytic T cell clones derived from human tumor-infiltrating lymphocytes. Eur J Immunol 23: 141–146.

Wölfel T, Van Pel A, De Plaen E, Lurquin C, Maryanski JL, Boon T (1987) Immunogenic (tum⁻) variants obtained by mutagenesis of mouse mastocytoma P815. VIII. Detection of stable transfectants expressing a tum⁻ antigen with a cytolytic T cell stimulation assay. Immunogenetics 26: 178–187.

Wölfel T, Klehmann E, Muller C, Schutt K-H, Meyer zum Buschenfelde K-H, Knuth A (1989) Lysis of human melanoma cells by autologous cytolytic T cell clones. Identification of human histocompatibility leukocyte antigen A2 as a restriction element for three different antigens. J Exp Med 170: 797–810.

Wölfel T, Hauer M, Klehmann E, et al (1993) Analysis of antigens recognized on human melanoma cells by A2-restricted cytolytic T lymphocytes (CTL). Int J Cancer 55: 237–244.

Wolfel T, Van Pel A, Brichard V, et al (1994) Two tyrosinase nonapeptides recognized on HLA-A2 melanomas by autologous cytolytic T lymphocytes. Eur J Immunol 24: 759–764.

Yang SY, Morishima Y, Collins NH, et al (1984) Comparison of one-dimensional IEF patterns for serologically detectable HLA-A and B allotypes. Immunogenetics 19: 217–231.

Peptide translocation by the transporters associated with antigen processing (TAP)

9.10

Anne Neisig
Jacques J. Neefjes
Monique Grommé

Division of Cellular Biochemistry, The Netherlands Cancer Institute, Amsterdam, The Netherlands

TABLE OF CONTENTS

Immunology Methods Manual
ISBN 0–12–442712–X

Abstract

Presentation of peptides by MHC class I molecules is preceded by translocation of the peptides from the cytosol into the lumen of the endoplasmic reticulum (ER) by the peptide transporter TAP. This is an essential step in class I-restricted antigen presentation and confers additional selectivity. Here, we describe assays for following TAP-dependent peptide translocation into the ER, allowing the determination of those peptides arriving at the site of contact with MHC class I molecules.

Introduction

Antigen presentation by MHC class I molecules

Major histocompatibility complex (MHC) class I molecules display a variety of antigenic peptides on the surface of cells for presentation to cytotoxic $CD8^+$ T lymphocytes. MHC class I molecules are composed of a polymorphic heavy (H) chain noncovalently associated with $\beta2$-microglobulin (β_2m). These subunits assemble into a heterodimer in the endoplasmic reticulum (ER). Binding of a peptide is essential for stability and transport of MHC class I molecules from the ER to the cell surface. Generally, the antigenic peptides have a length of 8–11 amino acids and are derived from endogenous antigens (Falk et al 1990, 1991; Jardetzky et al 1991; Van Bleek and Nathenson 1990, 1991). The breakdown of intracellular protein antigens to peptides destined for presentation by MHC class I molecules occurs in the cytosol, probably by proteasomes (Goldberg and Rock 1992). Until 1990, it was not clear how these antigenic peptides were translocated from the cytosol into the ER, where binding to newly assembled MHC class I molecules takes place. The study of mutant human and murine cell lines and intra-MHC recombinant rat strains, all defective in antigen presentation, has led to the characterization of two genes encoding multimembrane-spanning molecules. They are encoded in the MHC class II region and are required for proper peptide loading of MHC class I molecules.

Identification of TAP

The murine RMA-S and the human LCL 721.174 and T2 cell lines all have low MHC class I surface expression owing to retention of unstable class I H chain-β_2m heterodimers in the ER (DeMars et al 1984, 1985; Salter et al 1985). MHC class I molecules escaping ER retention could be stabilized at the cell surface by exogenously added peptide (Townsend et al 1989, 1990; Hosken and Bevan 1990; Schumacher et al 1990), suggesting that these mutant cell lines were defective in the transport of peptides from the cytosol into the ER (Townsend et al 1989; Cerundolo et al 1990). The genetic elements responsible for the defect in these mutant cell lines were mapped to the MHC class II region (Monaco et al 1990; Deverson et al 1990; Trowsdale et al 1990; Spies et al 1990). This region contains two genes previously designated as HAM1/2, PSF1/2, or RING4/11 and MTP1/2 in mouse, human, and rat, respectively, but now known as TAP1 and TAP2 for transporter associated with antigen processing (WHO Nomenclature

Committee for Factors of the HLA system 1992). The amino acid sequence comparison of the TAP proteins revealed high homology to the ATP-binding cassette (ABC) family of membrane translocators. Members of this family all have a hydrophobic domain predicted to span the membrane six to eight times, and a cytosolic ATP-binding domain. Both domains are repeated twice, either on a single polypeptide or by dimerization of separate polypeptides. The different ABC translocators all actively transport a wide variety of substrates, including oligopeptides, across cell membranes (reviewed by Higgins 1992).

It is proposed that TAP1 and TAP2 form a heterodimer that transports peptides into the lumen of the ER (Kelly et al 1992; Spies et al 1992; Yewdell et al 1993). Peptide translocation by TAP requires hydrolysis of ATP (Neefjes et al 1993; Shepherd et al 1993; Androlewicz et al 1993), which has been shown to bind to the C-terminal domain of TAP1 (Muller et al 1994). TAP is mainly localized to the ER (Kleijmeer et al 1992) and can physically associate with newly synthesized MHC class I H chain/β_2m heterodimers (Ortmann et al 1994; Suh et al 1994; Neisig et al 1996). This interaction of TAP and 'empty' MHC class I complexes may serve to facilitate peptide binding to class I molecules by exposing them to high local concentrations of translocated peptides in the vicinity of the transporter.

TAPs are involved in peptide transport *in vivo*

Transfection of the cDNAs of TAP1 and/or TAP2 could restore stable surface MHC class I expression in the mutant cell lines that were deficient in functional TAP1 and/or TAP2 (Bahram et al 1991; Spies and DeMars 1991; Powis et al 1991; Attaya et al 1992; Arnold et al 1992; Momburg et al 1992). Conversely, mice with a targeted deletion in the TAP1 gene showed a drastically reduced expression of MHC I molecules (Van Kaer et al 1992) and consequently failed to develop a normal CD8$^+$ T cell reper-

toire (Ashton-Rickardt et al 1993; Aldrich et al 1994). In addition, an inherited mutation in the human TAP2 gene has been described, also resulting in low MHC class I surface expression and a low number of CD8$^+$ T cells (de la Salle et al 1994). In some human tumors and transformed cell lines, downregulated expression of the peptide transporter genes correlates with reduced levels of cell surface MHC class I, reducing the immunogenicity of these tumor cells (Restifo et al 1993; Cromme et al 1994; Rotem-Yehudar et al 1994).

In rat strains that express the *cim*^{a} allele, the MHC class I molecule RT1A^a is transported normally to the cell surface, whereas on a *cim*^{b} background, intracellular transport of RT1A^a is impaired, presumably owing to a lack of suitable peptides in the ER. This so-called class I modifier, or *cim* phenomenon was attributed to TAP2. The rat TAP2 alleles can differ by up to 29 amino acids and this polymorphic variation causes distinct sets of peptides to be loaded into the MHC class I molecule RT1A^a (Powis et al 1992). Murine and human TAP genes are not highly polymorphic (Colonna et al 1992; Carrington et al 1993; Pearce et al 1993; Powis et al 1993; Szafer et al 1994).

TAP-mediated peptide transport *in vitro*

When radiolabeled peptides containing a consensus sequence for *N*-linked glycosylation (Neefjes et al 1993) or anchor residues for binding to MHC class I molecules (Androlewicz et al 1993) are incubated with permeabilized cells or microsomes, a TAP- and ATP-dependent accumulation of glycosylated or MHC class I-bound peptides can be detected. Since TAP supplies MHC class I molecules with peptides in the ER and is encoded in the polymorphic MHC class II region, it was of interest to determine its substrate specificity. TAP efficiently translocates peptides of 8–16 amino acids; thus, the size selectivity of TAP is broader than that of MHC class I molecules (8–11 amino acids). Longer (up to 40 amino acids) and shorter peptides are less efficiently translocated (Androlewicz et al 1993; Momburg et al 1994b; Heemels and Ploegh, 1994).

The C-terminal residue of a peptide is a key determinant for efficient translocation by TAP. Murine TAP and the rat TAP2[u] allele select for hydrophobic and aromatic C-terminal residues, whereas human TAP and the rat TAP2[a] allele show a broader selectivity in that they efficiently translocate peptides with hydrophobic or basic C-termini (Heemels et al 1993; Schumacher et al 1994; Momburg et al 1994a). Interestingly, proteasomes that contain the LMP2 and LMP7 subunits, which map close to the TAP1 and TAP2 genes in the MHC class II region, generate peptides with predominantly basic or hydrophobic amino acids at the C-terminus corresponding to the TAP selectivity (Driscoll et al 1993; Gaczynska et al 1993). Whether any link exists between the expression of LMP2 or LMP7 and the selectivity of TAP is unclear, but expression of LMP2 and LMP7 is not an absolute requirement for antigen presentation (Arnold et al 1992; Momburg et al 1992; Yewdell et al 1994).

The influence on translocation efficiency of amino acid substitutions at other positions in a peptide is only minor. However, proline at position 2 or 3 negatively influences translocation (Heemels and Ploegh 1994; Neisig et al

1995; Neefjes et al 1995). Addition of flanking residues to a peptide with a proline at position 3 significantly increases its translocation rate (Heemels and Ploegh 1994; Neisig et al 1995).

After translocation, peptides are rapidly removed from the ER unless they bind to MHC class I molecules or become glycosylated. Peptide release from the ER requires ATP but does not seem to involve TAP (Schumacher et al 1994; Momburg et al 1994b; Roelse et al 1994). A fraction of the peptides released from the ER escapes complete degradation in the cytosol and recycles back to the ER in a TAP-dependent fashion (Roelse et al 1994).

Cell lines used for peptide translocation studies

A series of mutants of human and murine cell lines were produced by γ-irradiation and/or mutagen exposure and sequential treatments with complement and monoclonal antibodies against MHC class I and II molecules. All the mutants were selected for low expression of MHC class I molecules at the cell surface (DeMars et al 1984, 1985; Salter et al 1985;

Table 9.10.1 Human and murine mutant cell lines and their expression of TAP genes.

	TAP expression
Human cell lines	
LCL 721	TAP1[+]/TAP2[+] (Spies et al 1992)
LCL 721.45	TAP1[+]/TAP2[+] (Spies et al 1992)
LCL 721.174	1 Mb homozygous genomic deletion including TAP1 and TAP2 (Spies et al 1990)
LCL 721.134	TAP1[-]/TAP2[+] (Spies et al 1990; Spies and DeMars 1991)
BJAB-B95.8.6	TAP1[+]/TAP2[+]
BM28.7	TAP1[+]/TAP2[+]
BM36.1	TAP1[+]/2 bp deletion in ATP-binding domain of TAP2 (Kelly et al 1992)
T1 (721.174×CEM[R].3)	TAP1[+]/TAP2[+]
T2	TAP1[-]/TAP2[-]
T2/TAP1+2[a]	T2 transfected with rat TAP1 and TAP2[a] (Momburg et al 1992)
T2/TAP1+2[u]	T2 transfected with rat TAP1 and TAP2[u] (Momburg et al 1994a)
Murine cell lines	
RBL-5	TAP1[+]/TAP2[+]
RMA	TAP1[+]/TAP2[+]
RMA-S	TAP1[+]/point mutation in TAP2 resulting in a premature stop codon (Yang et al 1992)

Ljunggren and Karre 1985; Karre et al 1986). Table 9.10.1 gives an overview of these cell lines and their expression of TAP genes.

TAP-independent peptide translocation

Peptides containing a signal sequence enter the ER lumen in a TAP-independent fashion (Walter and Lingappa 1986). It has been shown that MHC class I molecules of TAP-deficient cells present fragments derived from peptides preceeded by a signal sequence or from the signal sequence itself (Anderson et al 1991; Henderson et al 1992; Bacik et al 1994). These TAP-deficient cells can also present viral peptides (Zhou et al 1993a,b) suggesting the existence of another mechanism of peptide supply to MHC class I molecules. However, the contribution of this pathway to antigen presentation appears to be minor since TAP-deficient cell lines or mice have low surface expression of MHC class I molecules.

TAP-dependent peptide translocation

TAP-dependent translocation of peptides from the cytosol to the ER lumen has been followed *in vitro* in permeabilized cells (Neefjes et al 1993; Androlewicz et al 1993) or in purified microsomes (Shepherd et al 1993). The protocols for both assays are described in this subsection.

by including in the assay permeabilized cells lacking TAP. Therefore, cell lines with counterparts deficient in TAP, such as the murine cell lines RMA and RMA-S (Ljunggren and Karre 1985; Karre et al 1986) or the human cell lines LCL 721 (DeMars et al 1984) and T2 (Salter et al 1985), are preferentially used for translocation studies (see earlier). Alternatively, assays performed in the absence of ATP can serve as control.

Peptide translocation in streptolysin O-permeabilized cells

To follow TAP-dependent translocation of peptides from the cytosol into the ER lumen, introduction of peptides into the cytosol is a prerequisite. This can easily be achieved by permeabilization of the cell membrane with the bacterial toxin streptolysin O (Neefjes et al 1993; Androlewicz et al 1993). When used in appropriate concentrations, the streptolysin O affects only the plasma membrane, leaving the ER membrane intact. However, integrity of the ER membrane should always be controlled for

Permeabilization of cells

Permeabilization of cell membranes can be achieved using detergents like digitonin or bacterial toxins such as α-toxin from *Staphylococcus aureus*, or streptolysin O or S from *Streptococcus* strains (Buckingham and Duncan 1983; Bhakdi et al 1985). We and others (Neefjes et al 1993; Androlewicz et al 1993) find streptolysin O the most suitable reagent for membrane permeabilization since it is easy to use and, importantly, can be applied to a wide variety of cell types.

Streptolysin O binds to cholesterol-containing target membranes and assembles into supramolecular ring structures generating transmembrane channels exceeding 15 nm in diameter (Buckingham and Duncan 1983; Bhakdi et al 1985). Pore formation is irreversible and allows large molecules like proteins, but also peptides, ions, nucleotides, and carbohydrates, to enter the cytosol. The binding of streptolysin O to the cell membrane occurs at temperatures between 0°C and 37°C, whereas pore formation requires temperatures higher than 20°C. The efficiency of permeabilization is dependent on both the cell type used and the concentration of streptolysin O. High levels of streptolysin O result in total cell lysis (Buckingham and Duncan 1983) and low concentrations lead to inefficient membrane permeabilization. Streptolysin O is therefore titrated for each cell type to determine the optimal concentration required for permeabilization of 60–80% of the cells.

The quality and activity of streptolysin O available from manufacturers varies, and we recommend the products of Wellcome or Biomerieux, which work well in our hands. Most streptolysin O preparations are impure, and contaminants may affect the outcome of the experiment. Contaminants can be removed from the streptolysin O preparation by incubating cells with streptolysin O at 0°C (to allow binding to the cell membrane), removing the non-bound fraction (including most contaminants) by washing, followed by permeabilization at 37°C (see protocol 1b below).

Streptolysin O is reversibly inactivated by atmospheric oxygen (e.g., due to storage). Since reduced cysteine is required for activation, the toxin can be (re)activated by incubation with the reducing agent dithiothreitol (DTT) at low concentration prior to use.

Entry of peptides into the ER

The entry of peptides into the ER lumen of permeabilized cells has been followed using different protocols: (1) accumulation of glycosylated peptide in the ER, (2) peptide binding to MHC class I molecules in the ER.

A convenient and reproducible method for studying peptide translocation in permeabilized cells is to use radioiodinated peptides containing an N-linked glycosylation consensus sequence (Neefjes et al 1993). Upon arrival in the ER, the peptide will become glycosylated (Kornfeld and Kornfeld 1985) and can easily be recovered with Con A-Sepharose and quantified. Peptides entering the ER lumen have been shown to be degraded or even released from the ER by an as yet unknown mechanism (Schumacher et al 1994; Momburg et al 1994b; Roelse et al 1994). However, when peptides are glycosylated, they become stable in the ER and are not exported to the cytosol (Momburg et al 1994b; Roelse et al 1994). The assay requires peptides containing a tyrosine (Y) for iodination and an N-linked glycosylation site Asn-X-Thr/Ser (where X is any amino acid except proline) for glycosylation. Most naturally processed peptides do not contain these specific amino acids, but TAP-dependent translocation of such peptides can be studied by competing for translocation of a glycosylatable model peptide, as described under 'Peptide Substrates'.

Alternatively, peptide translocation has been studied by following the binding of a radiolabeled peptide to MHC class I molecules in the ER of permeabilized cells (Androlewicz et al 1993). Solubilized MHC class I molecules with bound, radiolabeled peptides were subsequently immunoprecipitated and the amount of bound peptide quantified. However, using this technique two different events are being studied at the same time; namely peptide translocation by TAP and peptide binding by MHC class I molecules. It is impossible to distinguish the two processes and therefore impossible to determine at what level peptide competition occurs. Furthermore, radiolabeled peptides can bind to MHC class I molecules expressed at the cell surface and thereby seriously influence the result.

ATP

Peptides bind to TAP in the absence of ATP (Van Endert et al 1994; Androlewicz and Cresswell 1994), whereas the hydrolysis of ATP is required for the subsequent TAP-dependent peptide translocation across the ER membrane

(Neefjes et al 1993; Shepherd et al 1993). It is therefore essential to use ATP or an ATP-regenerating system in the translocation assays. We recommend the use of ATP since it is more convenient (and cheaper) and because it does not introduce further contaminants that may be present in the creatine kinase preparation of the ATP-regenerating system. ATP is acidic and it is important to neutralize the ATP solution before use, since peptide translocation is inhibited at acidic pH (Roelse et al 1994).

Peptidase activity

In translocation assays using permeabilized cells, peptidase activity was observed (Momburg et al 1994b; Roelse et al 1994). Peptides of various lengths added to permeabilized cells were degraded within minutes into a mixture of shorter peptides. The input 16-mer, 20-mer or 40-mer peptides had not been transported as such, but smaller fragments (degradation intermediates) had (Momburg et al 1994b). The rate of degradation decreases with increasing length of the peptide. Partial reduction in peptidase activity can be achieved by washing the permeabilized cells to remove the cytosol.

Protocol 1a. Peptide translocation in streptolysin O-permeabilized cells using peptides with an *N*-linked glycosylation consensus sequence (Neefjes et al 1993)

Materials

- Iodinated peptides
- Incubation buffer:

130 mM	KCl
10 mM	NaCl
1 mM	CaCl$_2$
2 mM	EGTA
2 mM	MgCl$_2$
5 mM	Hepes (pH 7.3)

- Lysis mix:

1% (v/v)	Triton X-100
150 mM	NaCl
10 mM	MgCl$_2$
50 mM	Tris–HCl (pH 7.5)

- Streptolysin O
- ATP
- Concanavalin A-Sepharose (Con A-Sepharose) beads, or any lectin binding high-mannose carbohydrate coupled to beads.

Method

1. Dissolve 10 μg of peptide in 10–50 μl PBS and radiolabel it by addition of Na^{125}I and 10 μl chloramine T (10 mg ml^{-1}) freshly dissolved in PBS. We routinely iodinate peptides with 0.2 to 1 mCi iodine. After 5 min incubation at room temperature, free iodine is removed from peptide-bound iodine by adding DOWEX OH^{-}, which binds free iodine. The iodinated peptides remain in solution and are stored at –20°C.

   Note: Some peptides also bind to DOWEX. These peptides can be separated from iodine by gel filtration using Sephadex G10.

2. Dissolve streptolysin O in incubation buffer or PBS to an optimal concentration for cell permeabilization. The optimal concentration depends on the cell type used and the activity of the streptolysin O preparation. Streptolysin O solution must be freshly made and can be (re)activated with 0.2 mM DTT (15 min on ice) prior to use.

3. Harvest cells and wash once with PBS in a 1.5 ml Eppendorf tube. Pellet the cells by centrifugation for 2 min at 2000 rpm and remove the supernatant.

4. Permeabilize cells (3 × 10^6 cells/translocation) with 50 μl activated streptolysin O in incubation buffer during 10 min incubation at 37°C; 60–80% of the cells should be

permeabilized as measured by trypan blue uptake. After permeabilization, cells are kept on ice to prevent lysis. *Optional*: To decrease peptidase activity, the cells can be washed carefully in incubation buffer after permeabilization.

5. Make a 100 mM ATP solution in incubation buffer and adjust to pH 7 with 15% (v/v) 1 M Tris-base. ATP solution is freshly made and stored on ice.

6. Mix 50 μl permeabilized cells with 30 μl incubation buffer and 10 μl (~1 μM) of radiolabeled peptide in an Eppendorf tube and add 10 μl of 100 mM ATP to a final reaction volume of 100 μl.

7. Incubate for 5 min at 37°C; lower or higher temperatures will decrease translocation efficiency.

8. Stop translocation by addition of 1 ml lysis mix. Vortex and keep lysate on ice for 30 min.

9. Centrifuge lysate for 5 min at 14,000 rpm (4°C) to remove nuclei, and transfer supernatant to a new tube.

10. Incubate supernatant with 50–100 μl Con A-Sepharose for 1–2 h at 4°C (rotating) to recover glycosylated peptides.

11. Wash Con A-Sepharose beads 4 times with 1 ml cold lysis buffer. Pellet the beads by centrifugation for 2 min at 6000 rpm between the washes.

12. Quantify the Con A-Sepharose-bound peptides by gamma-counting.

Protocol 1b. Peptide translocation in streptolysin O-permeabilized cells using peptides binding to MHC class I molecules (Androlewicz et al 1993)

Materials

- Iodinated peptide
- Incubation buffer:

78 mM	KCl
4 mM	$MgCl_2$
8.37 mM	$CaCl_2$
10 mM	EGTA
50 mM	Hepes (pH 7.0)
1 mM	DTT, bovine serum albumin (4 mg ml^{-1})

- Lysis mix:

1%	Triton X-100
150 mM	NaCl
10 mM	Tris (pH 7.4)
5 mM	iodoacetamide
0.5 mM	phenylmethylsulfonyl fluoride

- Streptolysin O
- ATP-regenerating system: 2.2 mM ATP, 0.44 mM GTP, 22.2 mM creatine phosphate, creatine phosphate kinase (31.8 units ml^{-1})
- Antibody against MHC class I molecule
- Protein A-Sepharose beads
- Wash buffer:

20 mM	Tris (pH 8.0)
150 mM	NaCl
0.5%	Triton X-100
0.5%	deoxycholate
0.05%	SDS
0.02%	NaN_3

Methods

1. Iodinate peptide (see protocol 1a).

2. Harvest cells ($1–2 \times 10^7$) and resuspend in 1 ml of serum-free medium containing 4 mM DTT.

3. A modification of the permeabilization method described in protocol 1a was used: Incubate cells on ice for 10 min with streptolysin O dissolved in incubation buffer to an appropriate concentration.

4. Wash cells three times in serum-free medium to remove unbound streptolysin O and resuspend in 1 ml incubation buffer containing ATP-regenerating system.

5. Incubate for 20 min at 37°C to allow cell permeabilization and peptide translocation.

6. Wash cells once in incubation buffer.

7. Lyse cells in 1 ml lysis mix containing excess (10 μM) of unlabeled peptide to prevent post-lysis association of labeled peptide with its corresponding MHC class I molecules.

8. Centrifuge lysate at 10,000g for 5 min.

9. Immunoprecipitate solubilized MHC class I molecules from 0.5–1.0 ml of the lysate with specific antibody (5 μg). Incubate with 25 μl protein A-Sepharose beads for 1 h at 4°C.

10. Wash beads three times with 1 ml of wash buffer and quantify MHC class I-bound peptides by gamma-counting.

Peptide translocation in microsomes

Generation of microsomes

Peptide translocation by TAP can be studied in purified microsomes (Shepherd et al 1993; Heemels et al 1993; Schumacher et al 1994; Roelse et al 1994; Heemels and Ploegh 1994). Microsomes have been made from various tissues, cell lines, and recently from insect cells overexpressing TAPs. The microsomes from Sf9 insect cells infected with hTAP1 and hTAP2 recombinant baculovirus DNA contain 10–30 times more TAP1 and TAP2 proteins than do human B cells (Meyer et al 1994; Van Endert et al 1994). Large quantities of microsomes can easily be made and stored at −70°C. However, they should not be frozen/thawed more than once, since this will reduce their quality. Microsomes may be leaky, resulting in ATP-independent translocation of peptides, and therefore it is very important to control for ATP-dependent entry of peptide into the ER via TAP. Furthermore, a proper negative control (e.g., microsomes deficient in TAP) should be included.

Entry of peptides into microsomes

Different methods have been applied to study the entry of peptides into microsomes: (1) accumulation of glycosylated peptides in the ER, (2) sedimentation of translocated peptides in microsomes.

The first method was described for permeabilized cells under 'Entry of Peptides into the ER' and also applies for microsomes. In the second method used to follow peptide translocation in microsomes, radiolabeled (nonglycosylatable) peptide is added to purified microsomes (Shepherd et al 1993). Nontranslocated peptides are removed by washing, the microsomes are pelleted, and retained peptides are quantified. During the extensive washing to remove nontranslocated peptide, loss of microsomes and translocated peptides may occur. Furthermore, peptides entering the microsomes are rapidly exported again at physiological temperatures (Shepherd et al 1993). The translocation assay should therefore be performed during a very short period (5 min) or at lower temperature (23°C) to minimize the effect of peptide release.

ATP

ATP is included in the translocation assay performed with microsomes, as described earlier for permeabilized cells.

Peptidase activity

As for the permeabilized cell system, the microsomes do contain some peptidase activity. Translocation studies of a 16-mer peptide demonstrated that mainly proteolytic fragments of the peptide, rather than the 16-mer itself, were transported. The proteolysis was suggested to take place prior to translocation into the microsomes (Heemels and Ploegh 1994). The effect of proteolysis of peptides on the outcome of translocation experiments can be minimized by reducing the time of the translocation assay.

Protocol 2a. Preparation of microsomes from cell cultures

Materials

- STKMM:

  250 mM   sucrose
  50 mM   TEA-HCl (pH 7.5)
  50 mM   potassium acetate (KAc)
  5 mM   magnesium acetate (MgAc)
  0.1%   β-mercaptoethanol

- RM:

  250 mM   sucrose
  50 mM   TEA-HCl (pH 7.5)
  50 mM   KAc
  2 mM   MgAc
  1 mM   DTT

Method

1. Harvest 3×10^9 cells by centrifugation for 10 min at 1500 rpm (4°C) and wash once in PBS.
2. Resuspend cells in 20 ml STKMM and centrifuge 10 min at 1500 rpm (4°C). *Keep cells on ice throughout the preparation.*
3. Resuspend cells in 10 ml H_2O and homogenize in a douncer (15 strokes). *Avoid air bubbles.*
4. Add 30 ml STKMM and centrifuge 10 min at 7500 rpm (4°C) in an ultracentrifuge (Beckmann rotor GA/18). Continue with supernatant.
5. Spin the supernatant for 40 min at 18,000 rpm (4°C) to pellet the microsomes.
6. Resuspend pellet in 20 ml STKMM and dounce again (15 strokes).
7. Centrifuge 40 min at 18,000 rpm to collect microsomes.
8. Resuspend microsomes in 1 ml RM in a small douncer (15 strokes).
9. Freeze microsomes in small aliquots in liquid nitrogen.
10. Store microsomes at −70°C.

Protocol 2b. Preparation of microsomes from insect cells (Meyer et al 1994; Van Endert et al 1994)

Materials

- Cavitation buffer:

  250 mM   sucrose
  50 mM   Tris (pH 7.4)
  25 mM   KAc
  5 mM   MgAc

0.5 mM CaAc with a mix of protease inhibitors (aprotinin, phenylmethylsulfonyl fluoride, EDTA, benzamidine, and pepstatin)

- Gradient buffer:

150 mM KAc
50 mM Tris (pH 7.4)
5 mM MgAc with a mix of protease inhibitors (aprotinin, phenylmethyl-sulfonylfluoride, EDTA, benzamidine and pepstatin)

Method

1. Harvest $5–8 \times 10^7$ cells by centrifugation and wash once in cold PBS.
2. Resuspend cells in 800 µl cavitation buffer and fractionate by repeated drawing through a 26-gauge needle.
3. Remove nuclei and large cell fragments by centrifugation for 5 min at 1000 rpm (4°C).
4. Dilute supernatant to a final volume of 5.6 ml with a 2.5 M sucrose solution and overlay with 2.9 ml of 2.0 M and 1.3 M sucrose (all sucrose solutions in gradient buffer).
5. Centrifuge for 5–16 h at 85,000g (4°C) and collect the microsomes accumulated at the interface of the 2.0 and 1.3 M sucrose layers.
6. Wash the pooled microsomes once in cold PBS including 1 mM DTT (1–2 h centrifugation at 100,000–190,000g, 4°C).
7. Resuspend the microsomes in PBS including 1 mM DTT by homogenization by drawing through a 25-gauge needle and snap-freeze microsomes in liquid nitrogen.
8. Store microsomes at –70°C.

Protocol 3a. Peptide translocation in microsomes using peptides with an N-linked glycosylation consensus sequence

Materials

See protocol 1a

Method

1. Thaw microsomes on ice.
2. Mix 30 µl incubation buffer, 5 µl microsomes and 10 µl iodinated peptide and add 5 µl 100 mM ATP (pH 7.0) to a final reaction volume of 50 µl.
3. Incubate for 5 min at 37°C.
4. Stop translocation by adding 1 ml of lysis mix. Vortex and keep lysate on ice for 30 min.
5. Remove debris by centrifugation for 5 min at 14,000 rpm (4°C) and transfer the supernatant to a new tube with 100 µl Con A-Sepharose.
6. Incubate for 1–2 h at 4°C while rotating.
7. Wash and quantify the Con A-Sepharose-bound radioactivity as in protocol 1a.

Protocol 3b. Peptide translocation in microsomes using sedimentation of translocated peptides (Shepherd et al 1993)

Materials

- Iodinated peptides
- ATP regenerating system: 50 µM ATP, 250 µM UTP, 2.5 mM creatine

phosphate, 8 U of rabbit muscle creatine kinase
- Incubation buffer: 50 mM Hepes (pH 7.3), 150 mM KAc (pH 7.5), 5 mM MgAc, 250 mM sucrose, 1 mM DTT
- Sucrose cushions: 1 M KAc, 500 mM sucrose, 50 mM Hepes (pH 7.0), 10–25 µM unlabeled peptide (to block free peptide-binding sites during pelleting)
- Lysis mix: 0.5% NP-40, 150 mM Tris–HCl (pH 7.4), 150 mM NaCl, 5 mM MgCl$_2$

Method

1. Prepare for each translocation reaction a 50 µl assay mixture containing 5 µl microsomes, ATP regenerating system, incubation buffer, and 0.5 µM iodinated peptide. The transport assay mixtures are assembled on ice.
2. Incubate for 10 min at 23°C.
3. Transfer samples to ice.
4. Load samples on to the surface of ice-cold 1-ml sucrose cushions in thick-walled polypropylene centrifuge tubes (Beckman #349622).
5. Spin samples for 15 min at 67,000 rpm (4°C) in a TL100 ultracentrifuge (Beckman TLA 100.2 rotor).
6. Aspirate the supernatant and resuspend the pellet in 150 µl lysis mix by repeated pipetting.
7. Incubate 30 min on ice.
8. Remove debris by centrifugation at 14,000g for 4 min.
9. Quantify soluble radioactivity by scintillation counting.

Peptide substrates

Peptide length

TAP translocates peptides of a similar size to those binding to MHC class I molecules (8–11 amino acids) (Momburg et al 1994b). However, the size selectivity of TAP is broader, preferring peptides of 8–16 amino acids in length. Peptides shorter than 8 amino acids and longer than 16 amino acids are also translocated, albeit less efficiently. The translocation efficiency decreases with increasing peptide size (Momburg et al 1994b), but a clear upper limit is not defined.

Peptide sequence

In addition to the size of the peptide, its sequence also affects the efficiency of peptide translocation by TAP. The C-terminal amino acid, in particular, influences translocation in a species-specific fashion. The rat TAP2[u] allele and murine TAP prefer peptides with a hydrophobic C-terminus, whereas the rat TAP2[a] allele and human TAP preferentially translocate peptides with hydrophobic or basic C-termini (Heemels et al 1993; Momburg et al 1994a; Schumacher et al 1994). In addition, it has recently been demonstrated that proline at positions 2 and 3 greatly decreases the efficiency of translocation (Heemels and Ploegh 1994; Neisig et al 1995; Neefjes et al 1995).

It is very difficult to study translocation of hydrophobic peptides because they do not dissolve well in water or PBS. The use of DMSO, ethanol, or NaOH to dissolve these peptides will influence the activity of TAP. Furthermore, it is difficult to determine the exact concentration of hydrophobic peptides in solution because the peptides may form aggregates or bind to membranes or the reaction tubes. In addition, hydrophobic peptides are very inefficiently iodinated, and are not recommended for use in translocation studies.

Peptide modifications

Chemical modifications of peptides have been demonstrated to influence the efficiency of TAP-dependent peptide translocation (Momburg et al 1994a; Schumacher et al 1994). Peptides with unmodified N- or C-termini are most efficiently transported by TAP, whereas an acetylated N- or amidated C-terminus decreases translocation efficiency considerably. In addition, inclusion of D-amino acids in model peptides negatively affects peptide translocation by TAP and circular peptides, like cyclosporin A, are not substrates for TAP (unpublished observations). Thus, unmodified peptides, as presented by MHC class I molecules, are most efficiently translocated. Any chemical modification of the peptide may affect its translocation by TAP.

Design of glycosylatable peptides

Peptides containing the N-linked glycosylation consensus sequence Asn-X-Thr/Ser (where X is any amino acid except Pro) will become glycosylated upon arrival in the ER lumen (Kornfeld and Kornfeld 1985). Such peptides are protected against degradation in and release from the ER (Momburg et al 1994b; Roelse et al 1994). Glycosylated peptides can easily be recovered through lectins coupled to Sepharose beads, and are useful for studying TAP-dependent translocation of nonglycosylatable peptides by competition (Neefjes et al 1993; Heemels et al 1993; Schumacher et al 1994; Heemels and Ploegh 1994, Neisig et al 1995). However, in designing the model peptides, it is important to note that proline interferes with the rate of glycosylation when placed between Asn and Thr/Ser in the glycosylation site or on either side of the glycosylation consensus sequence (our own observation). Furthermore, the glycosylation site should not be placed at the very N-terminal position of the peptides, but should be preceded by at least one amino acid for efficient glycosylation. When peptides are designed with the glycosylation site and the tyrosine at opposite ends, then recovery of glycosylated, radiolabeled peptide indicates that the actual translocated peptide is identical to the input peptide, or shortened by at most a single residue.

Competition for peptide translocation

Since most peptides do not contain a tyrosine for radioiodination and a glycosylation consensus sequence which allows recovery of the translocated peptide by Con A-Sepharose, TAP-dependent translocation of these peptides can only be analyzed by competition for translocation of a model peptide (Shepherd et al 1993; Neefjes et al 1993; Heemels et al 1993; Momburg et al 1994b; Androlewicz and Cresswell 1994; Schumacher et al 1994; Heemels and Ploegh 1994).

The translocation of a glycosylatable, radiolabeled model peptide can be inhibited by adding increasing amounts of competitor peptide. The concentration of competitor peptide resulting in 50% inhibition of translocation of the model peptide (IC_{50} value) can then be determined and compared.

Remarkably, when peptide translocation is performed in the presence of high concentrations of competing peptides, the efficiency of translocation of the model peptide may increase because the competitor peptide also competes for peptidase activity. This effect can be overcome by decreasing the time of the competition experiment, thereby reducing the contribution of peptide degradation.

Inhibition of TAP-dependent translocation of the model peptide by the competitor peptide does not necessarily imply that the competitor peptide is translocated into the ER, since peptide binding and peptide transport by TAP are two different processes. However, the ability of a peptide to compete for translocation of the reporter peptide usually correlates with its transport rates as measured directly in the translocation assay (Neefjes et al 1995).

Protocol 4. Competition assays

Materials

- Iodinated model peptide containing an *N*-linked glycosylation site (see protocol 1a)
- Incubation buffer (see protocol 1a)
- Lysis mix (see protocol 1a)
- Streptolysin O-permeabilized cells (see protocol 1a or 1b) or microsomes (see protocol 2a or 2b)
- ATP (see protocol 1a or 1b)
- Con A-Sepharose

Method

Competition assays can be performed in both permeabilized cells and microsomes.

1. Prepare peptide mixtures containing different concentrations of competing peptides in 30 µl incubation buffer. We routinely use none or 0.1–100 µM competitor peptides.
2.
   - *Cells*: Permeabilize cells in 50 µl streptolysin O in incubation buffer.
   - *Microsomes*: Thaw microsomes.

3.
   - *Cells*: Add 10 µl iodinated model peptide (0.05–0.5 µM) and 50 µl permeabilized cells to the different peptide mixtures.
   - *Microsomes*: Add 10 µl iodinated model peptide (0.05–0.5 µM) and 5 µl microsomes to the different peptide mixtures.

4.
   - *Cells*: Add 10 µl 100 mM ATP (pH 7) to a final reaction volume of 100 µl.
   - *Microsomes*: Add 5 µl 100 mM ATP (pH 7) to a final reaction volume of 50 µl.

5. Incubate for 5 min at 37°C.
6. Stop translocation by addition of 1 ml lysis mix, isolate the glycosylated model peptide with Con A-Sepharose and quantify by gamma-counting as described in protocol 1a.
7. Plot the concentration of the competitor peptide against the percentage of translocation of the model peptide (the cpm value of the translocation assay performed without competitor peptide is taken as 100% translocation of the model peptide) and determine the concentration of competitor peptide at which the translocation of the model peptide is 50% (IC_{50}).

References

Aldrich CJ, Ljunggren H-G, Van Kaer L, Ashton-Rickardt PG, Tonegawa S, Forman J (1994) Positive selection of self- and alloreactive CD8+ T cells in *Tap*-1 mutant mice. Proc Natl Acad Sci USA 91: 6525–6528.

Anderson KS, Cresswell P, Gammon M, Hermes J, Williams A, Zweerink H (1991) Endogenously synthesized peptide with an endoplasmic reticulum signal sequence sensitizes antigen processing mutant cells to class I-restricted cell-mediated lysis. J Exp Med 174: 489–492.

Androlewicz MJ, Cresswell P (1994) Human transporters associated with antigen processing possess a promiscuous peptide-binding site. Immunity 1: 7–14.

Androlewicz MJ, Anderson KS, Cresswell P (1993) Evidence that transporters associated with antigen processing translocate a major histocompatibility complex class I-binding peptide into the endoplasmic reticulum in an ATP-dependent manner. Proc Natl Acad Sci USA 90: 9130–9134.

Arnold D, Driscoll J, Androlewicz MJ, Hughes E, Cresswell P, Spies T (1992) Proteasome subunits encoded in the MHC are not generally required for the processing of peptides bound by MHC class I molecules. Nature 360: 171–174.

Ashton-Rickardt PG, Van Kaer L, Schumacher TNM, Plough HL, Tonegawa S (1993) Peptide contributes to the specificity of positive selection of CD8+ T cells in the thymus. Cell 73: 1041–1049.

Attaya M, Jameson S, Martinez CK, et al (1992) *Ham*-2 corrects the class I antigen-processing defect in RMA-S cells. Nature 355: 647–649.

Bacik I, Cox JH, Anderson R, Yewdell JW, Bennink JR (1994) TAP (transporter associated with antigen processing)-independent presentation of endogenously synthesized peptides is enhanced by endoplasmic reticulum insertion sequences located at the amino- but not carboxyl-terminus of the peptide. J Immunol 152: 381–387.

Bahram S, Arnold D, Bresnahan M, Strominger JL, Spies T (1991) Two putative subunits of a peptide pump encoded in the human major histocompatibility complex class II region. Proc Natl Acad Sci USA 88: 10094–10098.

Bhakdi S, Tranum-Jensen J, Sziegoleit A (1985) Mechanism of membrane damage by streptolysin-O. Infect Immun 47: 52–60.

Buckingham L, Duncan JL (1983) Approximate dimensions of membrane lesions produced by streptolysin S and streptolysin O. Biochem Biophys Res Commun 729: 115–122.

Carrington M, Colonna M, Spies T, Stephens JC, Mann DL (1993) Haplotypic variation of the transporter associated with antigen processing (TAP) genes and their extension of HLA class II region haplotypes. Immunogenetics 37: 266–273.

Cerundolo V, Alexander J, Anderson K, et al (1990) Presentation of viral antigen controlled by a gene in the major histocompatibility complex. Nature 345: 449–452.

Colonna M, Bresnahan M, Bahram S, Strominger JL (1992) Allelic variants of the human putative peptide transporter involved in antigen processing. Proc Natl Acad Sci USA 89: 3932–3936.

Cromme FV, Airvey J, Heemels M-T, et al (1994) Loss of transporter protein, encoded by the TAP-1 gene, is highly correlated with loss of HLA expression in cervical carcinomas. J Exp Med 179: 335–340.

De La Salle H, Hanau D, Fricker D, et al (1994) Homozygous human TAP peptide transporter mutation in HLA class I deficiency. Science 265: 237–241.

DeMars R, Chang CC, Shaw S, et al (1984) Homozygous deletions that simultaneously eliminate expressions of class I and class II antigens of EBV-transformed B-lymphoblastoid cells. I. Reduced proliferative responses of autologous and allogeneic T cells to mutant cells that have decreased expression of class II antigens. Hum Immunol 11: 77–97.

DeMars R, Rudersdorf R, Chang CC, et al (1985) Mutations that impair a posttranscriptional step in expression of HLA-A and -B antigens. Proc Natl Acad Sci USA 82: 8183–8187.

Deverson EV, Gow IR, Coadwell WJ, et al (1990) MHC class II region encoding proteins related to the multidrug resistance family of transmembrane transporters. Nature 348: 738–741.

Driscoll J, Brown MG, Finley D, Monaco JJ (1993) MHC-linked *LMP* gene products specifically alter peptidase activities of the proteasome. Nature 365: 262–264.

Falk K, Rötzschke O, Rammensee H-G (1990) Cellular peptide composition governed by major histocompatibility complex class I molecules. Nature 348: 248–251.

Falk K, Rötzschke O, Stevanovic S, Jung G, Rammensee H-G (1991) Allele-specific motifs revealed by sequencing of self-peptides eluted from MHC molecules. Nature 351: 290–296.

Gaczynska M, Rock KL, Goldberg AL (1993). Gamma interferon and expression of MHC genes regulate peptide hydrolysis by proteasomes. Nature 365: 264–267.

Goldberg AL, Rock KL (1992) Proteolysis, proteasomes and antigen presentation. Nature 357: 375–379.

Heemels M-T, Ploegh HL (1994) Substrate specificity of allelic variants of the TAP peptide transporter. Immunity 1: 775–784.

Heemels M-T, Schumacher TNM, Wonigeit K, Ploegh HL (1993) Peptide translocation by variants of the transporter associated with antigen processing. Science 262: 2059–2063.

Henderson RA, Michel H, Sakaguchi K, et al (1992) HLA-A2.1-associated peptides from a mutant cell line: a second pathway of antigen presentation. Science 255: 1264–1266.

Higgins CF (1992) ABC transporters: from microorganisms to man. Annu Rev Cell Biol 8: 67–113.

Hosken NA, Bevan MJ (1990) Defective presentation of endogenous antigen by a cell line expressing class I molecules. Science 248: 367–370.

Jardetzky TS, Lane WS, Robinson RA, Madden DR, Wiley DC (1991) Identification of self peptides bound to purified HLA-B27. Nature 353: 326–329.

Karre K, Ljunggren HG, Pointek G, Kiesling R (1986) Selective rejection of H-2-deficient lymphoma variants suggests alternative immune defence strategy. Nature 319: 675–678.

Kelly A, Powis SH, Kerr L-A, et al (1992) Assembly and function of the two ABC transporter proteins encoded in the human major histocompatibility complex. Nature 355: 641–644.

Kleijmeer MJ, Kelly A, Geuze HJ, Slot JW, Townsend A, Trowsdale J (1992) Location of MHC-encoded transporters in the endoplasmic reticulum and cis-Golgi. Nature 357: 342–344.

Kornfeld R, Kornfeld S (1985) Assembly of asparagine-linked oligosaccharides. Annu Rev Biochem 54: 631–664.

Ljunggren HG, Karre K (1985) Host resistance directed selectivity against H-2-deficient lymphoma variants. Analysis of the mechanism. J Exp Med 162: 1745–1759.

Meyer TH, van Endert PM, Uebel S, Ehring B, Tampe R (1994) Functional expression and purification of the ABC transporter complex associated with antigen processing (TAP) in insect cells. FEBS Lett 351: 443–447.

Momburg F, Ortiz-Navarrete V, Neefjes JJ, et al (1992) Proteasome subunits encoded by the major histocompatibility complex are not essential for antigen presentation. Nature 360: 174–177.

Momburg F, Roelse J, Howard JC, Butcher GW, Hammerling GJ, Neefjes JJ (1994a) Selectivity of MHC-encoded peptide transporters from human, mouse and rat. Nature 367: 648–651.

Momburg F, Roelse J, Hammerling GJ, Neefjes JJ (1994b) Peptide size selection by the major histocompatibility complex-encoded peptide transporter. J Exp Med 179: 1613–1623.

Monaco JJ, Cho S, Attaya M (1990) Transport protein genes in the murine MHC: possible implications for antigen processing. Science 250: 1723–1726.

Muller KM, Ebensperger C, Tampe R (1994) Nucleotide binding to the hydrophilic C-terminal domain of the transporter associated with antigen processing (TAP). J Biol Chem 269: 14032–14037.

Neefjes JJ, Momburg F, Hammerling GJ (1993) Selective and ATP-dependent translocation of peptides by the MHC-encoded transporters. Science 261: 769–771.

Neefjes J, Gottfried E, Roelse J, Grommé M, Obst R, Hämmertling GJ, Momburg F (1995) Analysis of the fine specificity of rat, mouse and human TAP peptide transporters. Eur J Immunol. 25: 1133–1136.

Neisig A, Roelse J, Sijts AJAM, et al (1995) Major differences in transporter associated with antigen presentation (TAP) dependent translocation of MHC class I-presentable peptides and the effect of flanking sequences. J Immunol 154: 1273–1279.

Neisig A, Wubbolts R, Zang X, Melief C, Neefjes J (1996) Allele-specific differences in the interaction of MHC Class I molecules with transporters associated with antigen processing. J Immunol 156: 3196–3206.

Ortmann B, Androlewicz MJ, Cresswell P (1994) MHC class I/β_2-microglobulin complexes associate with TAP transporters before peptide binding. Nature 368: 864–867.

Pearce RB, Trigler L, Svaasand EK, Peterson CM (1993) Polymorphism in the mouse TAP-1 gene. Association with abnormal CD8[+] T cell development in the nonobese nondiabetic mouse. J Immunol 151: 5338–5347.

Powis SJ, Townsend ARM, Deverson EV, Bastin J, Butcher GW, Howard JC (1991) Restoration of antigen presentation to the mutant cell line RMA-S by an MHC-linked transporter. Nature 354: 528–531.

Powis SJ, Deverson EV, Coadwell WJ, et al (1992) Effect of polymorphism of an MHC-linked transporter on the peptides assembled in a class I molecule. Nature 357: 211–215.

Powis SH, Tonks S, Mockridge I, Kelly A, Bodmer JG, Trowsdale J (1993) Alleles and haplotypes of the MHC-encoded ABC transporters TAP1 and TAP2. Immunogenetics 37: 373–380.

Restifo NP, Esquivel F, Kawakami Y, et al (1993) Identification of human cancers deficient in antigen processing. J Exp Med 177: 265–272.

Roelse J, Gromme M, Momburg F, Hammerling G, Neefjes J (1994) Trimming of TAP-translocated peptides in the ER and in the cytosol during recycling. J Exp Med 180: 1591–1597.

Rotem-Yehudar R, Winograd S, Sela S, Coligan JE, Ehrlich R (1994) Downregulation of peptide transporter genes in cell lines transformed with the highly oncogenic adenovirus 12. J Exp Med 180: 477–488.

Salter RD, Howell DN, Cresswell P (1985). Genes regulating HLA class I antigen expression in T-B lymphoblast hybrids. Immunogenetics 21: 235–246.

Schumacher TNM, Heemels M-T, Neefjes JJ, Kast WM, Melief CJM, Ploegh H (1990) Direct binding of peptide to empty MHC class I molecules on intact cells and *in vitro*. Cell 62: 563–567.

Schumacher TNM, Kantesaria DV, Heemels M-T, et al (1994) Peptide length and sequence specificity of the mouse *TAP1/TAP2* translocator. J Exp Med 179: 533–540.

Shepherd JC, Schumacher TNM, Ashton-Rickardt PG, et al (1993) TAP1-dependent peptide translocation *in vitro* is ATP dependent and peptide selective. Cell 74: 577–584.

Spies T, DeMars R (1991) Restored expression of major histocompatibility class I molecules by gene transfer of a putative peptide transporter. Nature 351: 323–324.

Spies T, Bresnahan M, Bahram S, et al (1990) A gene in the human major histocompatibility complex class II region controlling the class I antigen presentation pathway. Nature 348: 744–747.

Spies T, Cerundolo V, Colonna M, Cresswell P, Townsend A, DeMars R (1992) Presentation of viral antigen by MHC class I molecules is dependent on a putative peptide transporter heterodimer. Nature 355: 644–646.

Suh W-K, Cohen-Doyle MF, Fruh K, Wang K, Peterson PA, Williams DB (1994) Interaction of MHC class I molecules with the transporter associated with antigen processing. Science 264: 1322–1326.

Szafer F, Oksenberg JR, Steinman L (1994) New allelic polymorphisms in TAP genes. Immunogenetics 39: 374.

Townsend A, Ohlen C, Bastin J, Ljunggren H-G, Foster L, Karre K (1989) Association of class I major histocompatibility heavy and light chains induced by viral peptides. Nature 340: 443–448.

Townsend A, Elliott T, Cerundolo V, Foster L, Barber B, Tse A (1990) Assembly of MHC class I molecules analyzed *in vitro*. Cell 62: 285–295.

Trowsdale J, Hanson I, Mockridge I, Beck S, Townsend A, Kelly A (1990) Sequences encoded in the class II region of the MHC related to the 'ABC' superfamily of transporters. Nature 348: 741–743.

Van Bleek GM, Nathenson SG (1990) Isolation of an endogenously processed immunodominant viral peptide from the class I H-2K^b molecule. Nature 348: 213–216.

Van Bleek GM, Nathenson SG (1991) The structure of the antigen-binding groove of major histocompatibility complex class I molecules determines specific selection of self-peptides. Proc Natl Acad Sci USA 88: 11032–11036.

Van Endert PM, Tampe R, Meyer TH, Tisch R, Bach J-F, McDevitt HO (1994) A sequential model for peptide binding and transport by the transporters associated with antigen processing. Immunity 1: 491–500.

Van Kaer L, Ashton-Rickardt PG, Ploegh HL, Tonegawa S (1992) TAP1 mutant mice are deficient in antigen presentation, surface class I molecules, and CD4$^-$8+ T cells. Cell 71: 1205–1214.

Walter P, Lingappa VR (1986) Mechanism of protein translocation across the endoplasmic reticulum membrane. Annu Rev Cell Biol 2: 499–516.

WHO Nomenclature Committee for Factors of the HLA System (1992) Nomenclature for factors of the HLA system, 1991. Immunogenetics 36: 135–148.

Yang Y, Fruh K, Chambers J, et al (1992) Major histocompatibility complex (MHC)-encoded HAM2 is necessary for antigenic peptide loading onto class I MHC molecules. J Biol Chem 267: 11669–11672.

Yewdell JW, Esquivel F, Arnold D, Spies T, Eisenlohr LC, Bennink JR (1993) Presentation of numerous viral peptides to mouse major histocompatibility complex (MHC) class I-restricted T lymphocytes is mediated by the human MHC-encoded transporter or by a hybrid mouse–human transporter. J Exp Med 177: 1785–1790.

Yewdell JW, Lapham C, Bacik I, Spies T, Bennink JR (1994) MHC-encoded proteasome subunits LMP2 and LMP7 are not required for efficient antigen presentation. J Immunol 152: 1163–1170.

Zhou S, Glas R, Liu T, Ljunggren H-G, Jondal M (1993a) Antigen processing mutant T2 cells present viral antigen restricted through H-2K^b. Eur J Immunol 23: 1802–1808.

Zhou S, Glas R, Momburg F, Hämmerling GJ, Jondal M, Ljunggren H-G (1993b) TAP2-defective RMA-S cells present Sendai virus antigen to cytotoxic T lymphocytes. Eur J Immunol 23: 1796–1801.

Purification of heat shock protein–peptide complexes for use in vaccination against cancers and intracellular pathogens

9.11

Pramod K. Srivastava

Department of Biological Sciences, Fordham University, New York, USA

TABLE OF CONTENTS

Immunology Methods Manual
ISBN 0–12–442712–X

Abstract

Methods for purification of heat shock proteins gp96, hsp90, and hsp70 from murine and human tissues and cell lines are described. In the preparations obtained by these methods, the heat shock proteins are associated non-covalently with an array of cellular peptides. The use of such heat shock protein–peptide complexes for eliciting prophylactic and therapeutic immunity to cancers and infectious diseases is described.

Introduction

Heat shock proteins (HSPs) are present in all cells of all life forms (Lindquist 1988). They constitute a superfamily of several distinct families of proteins (such as hsp110, hsp90, hsp70, hsp65, hsp27, etc.), some of which are known to facilitate folding and assembly of other cellular proteins. Selected HSPs, such as members of the hsp90 and hsp70 family, also associate with a broad array of cellular peptides which are generated during protein degradation (Li and Srivastava 1993; Udono and Srivastava 1993; Srivastava 1994a,b). The HSP–peptide complexes are quite stable and withstand conventional biochemical fractionation; thus, a purified hsp90 or hsp70 preparation carries a range of noncovalently associated peptides. As the HSP-associated peptides are derived from cellular proteins and reflect the cellular protein composition (H. Udono, E. Nakayama, and P.K. Srivastava, unpublished), the HSP preparations derived from any given cell type contain the peptide repertoire of that particular cell type. This becomes a matter of immunological interest as vaccination of mice and rats with HSP–peptide complexes leads to a powerful antigen-specific, $CD8^+$ cellular response against the peptides chaperoned by the HSPs, but not against the HSP itself (Blachere et al 1993; Udono and Srivastava 1993, 1994; Udono et al 1994).

The immunogenicity of HSP–peptide complexes has obvious applications in vaccination against cancers and infectious diseases (Srivastava 1993). With respect to cancers, HSP–peptide complexes isolated from a patient's cancer can serve as customized, patient-specific therapeutic vaccines. In spite of the current enthusiasm for shared antigens of human tumors (see Houghton 1994; Pardoll 1994), the need for patient-specific vaccines is highlighted by the classical observation that tumors of individual mice and rats are individually distinct with respect to their antigenicity (see Srivastava and Old (1989) for a fuller discussion of the merits of shared versus unique antigen vaccines of human cancers). With respect to infectious diseases, the immunogenicity of HSP–peptide vaccines provides an opportunity to complex known antigenic peptides with HSPs and use such complexes to elicit peptide-specific cytolytic T lymphocyte (CTL) response, in spite of their exogenous presentation. The unique and profound advantages of such vaccines have been discussed in detail in published research.

This chapter describes the methods used in purification of a number of HSPs, along with their peptide cargo, the methods of dissociation of peptides from the HSPs, and the use of HSP–peptide complexes in prophylactic and therapeutic vaccination against cancer.

Materials

For purification of gp96

- 30 mM sodium bicarbonate, pH 7, 0.1 mM phenylmethylsulfonyl fluoride (homogenization buffer)
- Concanavalin A–Sepharose (Pharmacia)
- Phosphate-buffered saline (Sigma) with and without 2 mM Ca^{2+} and 2 mM Mg^{2+} (for Con A affinity chromatography)
- 10% (w/v) α-methyl pyrannoside (Sigma) in PBS containing calcium and magnesium (elution buffer for Con A affinity chromatography)
- Ammonium sulfate (Sigma)
- 5 mM sodium phosphate buffer, pH 7, without and with 1 M sodium chloride (for ion exchange chromatography)
- Anti-gp96/grp94 monoclonal antibody (rat anti-mouse gp96); recognizes mouse, rat, and human gp96 (cat. no. RT-102-P; NeoMarkers, 4880 Creekwood Drive, Fremont, CA 94555, USA; Fax 510 792 8932)

For purification of hsp90

- 20 mM sodium phosphate, pH 7.4, 1 mM EDTA, with 200 mM or 250 mM sodium chloride (for dialysis)
- 20 mM sodium phosphate, pH 7.4, 1 mM EDTA, with 200 mM or 600 mM sodium chloride (for ion exchange chromatography)
- Anti-hsp84/86 (90) polyclonal antibody; recognizes mouse and human hsp90 (cat. no. RB-118-P and RB-119-P; NeoMarkers)

For purification of hsp70

- 10 mM Tris-acetate buffer, pH 7.5, 0.1 mM EDTA with 10 mM sodium chloride (for dialysis)
- 20 mM Tris-acetate buffer, pH 7.5, 0.1 mM EDTA, 15 mM 2-mercaptoethanol, with 20 mM or 500 mM sodium chloride (for ion exchange chromatography)
- Ammonium sulfate
- ATP–agarose (Sigma, cat. no. A-5394)
- 20 mM Tris-acetate buffer, pH 7.5, 0.1 mM EDTA, 15 mM 2-mercaptoethanol, 20 mM sodium chloride, 3 mM magnesium chloride (for ATP–agarose chromatography).
- Anti-hsp72/73 monoclonal antibody (rat anti-mouse hsp70); recognizes mouse and human hsp70 (cat. no. MS-124-P; NeoMarkers)

General reagents for tissue culture, SDS-PAGE, immunoblotting, etc. are not listed here.

Description of methods

Purification of gp96

The procedure described here is a modification of our earlier method (Srivastava et al 1986). The same method may be used for murine, rat or human cell lines or tissues (Maki 1991).

- A cell pellet is suspended in 4 volumes of 30 mM sodium bicarbonate, pH 7, 1 mM PMSF, and cells are allowed to swell for 20 min on ice.
- The cell suspension is homogenized in a Dounce homogenizer (the appropriate clearance of the homogenizer will vary for each cell type) and homogenized until >95% cells are lysed.
- For isolation of gp96 from cancerous or normal tissues, the same buffer may be used; however, a swelling period is unnecessary and the homogenization is best carried out by a Polytron or other mechanical device. The use of sonication should be avoided.
- The lysate is centrifuged to remove unbroken cells, nuclei, and other debris. The pellet from this centrifugation is discarded and the supernatant is further centrifuged at $100,000g$ for 90 min at 4°C. Gp96 may be purified from the supernatant as well as the pellet derived from this step.
- For purification from the supernatant, it is brought to 50% ammonium sulfate saturation by adding the appropriate amount of ammonium sulfate very slowly, while gently stirring the solution in a beaker placed in a shallow tray containing ice-water. The solution is stirred for anywhere

between 2 and 12 h at 4°C and is centrifuged at 6000 rpm in an SS34 rotor. The supernatant from this step is removed carefully and brought to 70% ammonium sulfate saturation, and centrifuged as described before.

- The pellet from this step is saved and suspended in a solution of PBS brought to 70% ammonium sulfate saturation, in order to rinse the pellet free from contaminating proteins. This mixture is centrifuged as before and the pellet is dissolved carefully in PBS containing 2 mM each of Ca^{2+} and Mg^{2+}, at a final protein concentration of approximately 2–5 mg ml^{-1}. Any undissolved material is removed by a brief centrifugation at 15,000 rpm in an SS34 rotor.
- The solution is now mixed gently with Con A–Sepharose beads, previously equilibrated with PBS containing 2 mM each of Ca^{2+} and Mg^{2+}. (As a rule of thumb, we use 0.5 ml bed volume of Con A–Sepharose for each ml wet weight of starting cells or tissues.)
- The mixture is shaken gently in a rotary shaker at 4°C for 2–6 h. The slurry is packed into a column and washed with PBS containing Ca^{2+} and Mg^{2+} until the absorbance at 280 nm drops to near zero and is stable.
- The bound proteins are eluted with 10% α-methylmannoside dissolved in PBS containing Ca^{2+} and Mg^{2+}: one-third of the column bed volume of this eluant is applied gently to the column (e.g., for a column with a bed volume of 1 ml, we initially apply 0.3 ml α-MM)
- This volume is allowed to run through the column, after which the column is closed at the bottom with a piece of parafilm and incubated at room temperature for 15 min. The parafilm at

the bottom of the column is then removed and 5 column volumes of eluant is applied to the column. Fractions (1/2 column volume each) are collected and tested by SDS-PAGE. This material can be between 40% and 99% pure gp96, depending upon the cell type, the tissue-to-lysis buffer ratio used, and the experimenter.

- The gp96-containing fractions are pooled and transferred to 5 mM sodium phosphate buffer, pH 7, 300 mM sodium chloride, by dialysis, or preferably by buffer exchange on a Sephadex G25 or equivalent column (such as the PD10 column of Pharmacia).

- After equilibration, the solution is mixed with DEAE–Sepharose beads previously equilibrated with 5 mM sodium phosphate buffer, pH 7, 300 mM sodium chloride. As a rule of thumb, we use the same volume of DEAE–Sepharose as Con A–Sepharose used in a previous step. The protein solution and the beads are mixed gently for 1 h and poured into a column, which is washed with 5 mM sodium phosphate buffer, pH 7, 300 mM sodium chloride, until the absorbance at 280 nm drops to near zero and is stable. The column is eluted with 5 volumes of 5 mM sodium phosphate buffer, pH 7, 700 mM sodium chloride and fractions (1/3 the column volume) are collected, until the absorbance of the eluate at 280 nm drops to near zero and is stable.

- Protein-containing fractions are pooled and diluted 4-fold with 5 mM sodium phosphate buffer, pH 7, so as to bring the salt concentration to 175 mM. This material is now applied to a Mono Q column equilibrated with 5 mM sodium phosphate buffer, pH 7, and eluted first with a step elution of 400 mM sodium chloride and then by a continuous gradient from 400 mM to 600 mM sodium chloride.

- The fractions eluting between 400 mM and 600 mM salt are analyzed by SDS-PAGE and silver staining. Apparently homogeneous preparations of gp96 are observed to be present in fractions eluting between approximately 450 mM and 550 mM salt.

- For purification of gp96 from the 100,000g pellet, the pellet is suspended in 5 volumes of PBS containing 0.1% octyl glucopyrannoside and allowed to stand on ice for 1 h. The suspension is centrifuged at 20,000g for 30 min and the supernatant is dialyzed against several changes of PBS extensively to remove any detergent. The dialysate is centrifuged at 100,000g for 90 min and the supernatant is used for further purification. Calcium and magnesium are added to the supernatant to a final concentration of 2 mM each. Further purification is carried out exactly as described above for the 100,000g supernatant.

Approximately 15–30 µg of apparently homogeneous preparations of gp96 are obtained from the 100,000g supernatant of each gram wet weight of cells/tissues by this method. The yield from the 100,000g pellet is more variable and may be anywhere between 5 and 20 µg, per gram wet weight of cells/tissues, depending on the cellular volume and the extent of efficacy of the detergent under the conditions used. We estimate that the total recovery of gp96 by this method is between 10% and 20%. The identity of the purified preparation as gp96 may be confirmed by immunoblotting with the anti-gp96 monoclonal antibody.

It may be instructive to discuss here some additional suggestions and comments on the phenomena most commonly encountered during purification of gp96. It is always advisable to use autoclaved buffers, or buffers made in autoclaved water when autoclaving of the buffer may be inappropriate (such as in the case of the hypotonic homogenization buffer).

Similarly, the use of protease inhibitors such as PMSF, TLCK, and leupeptin is highly recommended, especially during and after the Con A–Sepharose chromatography. This is so because of the presence in the DEAE-eluted material, of a powerful protease activity. Further, if the yield of gp96 is substantially lower than the values reported here, the Con A–Sepharose chromatography is likely to be the step where the loss occurred. This is generally because of insufficient levels of Ca^{2+} in the buffers during this chromatography and can be rectified easily. One must make sure that Ca^{2+} is not precipitating out of the buffers in form of calcium phosphate. Also, the time periods mentioned for each step in the entire procedure can be considerably shortened, especially for smaller quantities of starting materials.

One may encounter the problem of instability in purified gp96 preparations. A preparation which appeared homogeneous immediately after purification may develop a number of lower, and occasionally even higher, molecular weight bands after a week of storage. (The higher molecular weight bands appear to arise from an SDS-resistant association of the intact gp96 with some of the degradation products.) All these bands will still be detected by an antibody to gp96. The degradation is seen more reproducibly in more concentrated preparations and therefore we do not recommend storing gp96 at concentrations higher than 0.1 mg ml^{-1}. The basis of this instability is presently unclear. Preliminary evidence suggests that gp96 molecules may contain a protease activity and we have so far been unable to prove that this activity is a contaminant (Li and Srivastava 1994; A. Menoret and P.K. Srivastava, unpublished). Further, the purified preparation may occasionally appear to be a set of 2, 3, 4, or even 5 very closely spaced bands, between 94 and 110 kDa size, depending on the type of starting material (M. Daou and P.K. Srivastava, unpublished). This heterogeneity, whose structural basis is presently unclear, is not due to differences in glycosylation and does not appear to be a consequence of degradation (Feldweg and Srivastava, 1995).

Purification of hsp90

The procedure described here is a modification of that developed by Marc (1988). The Concanavalin A-unbound fraction from the purification of gp96 may be used for purification of hsp90.

- The unbound fraction is dialyzed against 100 volumes, three times (12 h each), of 20 mM sodium phosphate, pH 7.4, 1 mM EDTA, 250 mM NaCl, 1 mM PMSF.
- The dialyzed sample is centrifuged at 17,000 rpm (Sorvall SS34 rotor) for 20 min.
- The supernatant is applied to a Mono Q (Pharmacia) column equilibrated with 20 mM sodium phosphate, pH 7.4, 1 mM EDTA, containing 200 mM NaCl. The bound proteins are eluted with the same buffer by a linear salt gradient up to 600 mM sodium chloride. Fractions are collected and analyzed by SDS-PAGE followed by silver staining. Several fractions containing apparently homogeneous preparations of hsp90 are detected.

Approximately 150–300 µg of apparently homogeneous hsp90 are obtained from the 100,000g supernatant of each gram wet weight of cells/tissues by this method. We estimate that the total recovery of hsp90 by this method is approximately 70%. The identity of the purified preparation as hsp90 may be confirmed by immunoblotting with an anti-hsp90 antibody.

Purification of hsp70 species

Purification of hsp70 is more complex than that of gp96 or hsp90. Although several publications have described purification of hsp70 (Welch and Feramisco 1985; Nandan et al 1994), they utilize ATP-agarose chromatography or some other form of exposure of proteins to ATP. While this results in homogeneous preparations of hsp70, these preparations are also devoid of the peptides chaperoned by hsp70 because of the ATP-sensitivity of the hsp70–peptide association (Flynn et al 1989). For this reason, such procedures for purification of hsp70 are ineffective for immunological purposes, where the peptides chaperoned by the hsp70 have a role.

We have attempted to develop a protocol for purification of hsp70 *without* using ATP–agarose chromatography and have met with limited success. The difficulties arise from the fact that hsp70 appears to elute on an ion exchange column across the entire salt range used for elution; similarly, in ammonium sulfate precipitations, hsp70 appears to precipitate at all saturation levels (P. Peng and P.K. Srivastava, unpublished). Interestingly, it appears that the hsp70 molecules eluting at different salt concentrations from an ion exchange column, or precipitating at different levels of ammonium sulfate saturation, are distinct species, as each species appears to elute or precipitate 'true'. The biochemical basis of this vast heterogeneity is being examined, and until it is resolved satisfactorily the prospects for purification of hsp70–peptide complexes with a reasonable recovery are rather remote. In the interim, however, we use the following protocol, through which we have isolated immunogenic, peptide-associated hsp70 species from a number of murine cancers (Udono and Srivastava 1993).

The Concanavalin A-unbound fraction from the purification of gp96 may be used for purification of hsp70.

- The unbound fraction is dialyzed against 100 volumes, three times (12 h each), of 10 mM Tris-acetate, pH 7.5, 0.1 mM EDTA, 10 mM NaCl, 1 mM PMSF.
- The dialyzed sample is centrifuged at 17,000 rpm (Sorvall SS34 rotor) for 20 min.
- The supernatant is applied to a Mono Q (Pharmacia) column equilibrated with 20 mM Tris-acetate, pH 7.5, 0.1 mM EDTA, 15 mM 2-mercaptoethanol, containing 20 mM NaCl. The bound proteins are eluted with the same buffer in a linear salt gradient up to 500 mM sodium chloride.
- Fractions are collected and analyzed by SDS-PAGE followed by silver staining. All fractions contain several protein bands. Immunoblotting of the fractions with an antibody against the constitutive form of hsp70 will result in identification of hsp70-positive fractions. Only fractions which are very strongly positive for hsp70 are pooled and brought to 50% and then to 70% ammonium sulfate saturation, as described in the purification of gp96.
- The 50–70% protein pellet is collected by centrifugation, washed with 70% saturated ammonium sulfate and solubilized in the low-salt buffer of the Mono Q column described earlier. This preparation is desalted on a G25 or equivalent column and may be apparently homogeneous or close to it. If further purification is desired, it can be reapplied to the Mono Q column as described earlier.

Approximately 300 μg of hsp70 is obtained from the 100,000*g* supernatant of each gram wet weight of cells/tissues by this method. It is our belief that this represents a poor recovery and we are exploring other methods for a better recovery. The identity of the purified preparation as hsp90 may be confirmed by immunoblotting with the anti-hsp70 monoclonal antibody described earlier.

If the peptides associated with hsp70 are not required, the hsp70 preparation obtained as above can be repurified by ATP–agarose chromatography.

- The ATP agarose column is equilibrated with:

    20 mM Tris-acetate (pH 7.5)

    20 mM NaCl

    0.1 mM EDTA

    15 mM 2-mercaptoethanol

    3 mM $MgCl_2$.

- $MgCl_2$ is also added to the sample to a final concentration of 3 mM and the sample is applied repeatedly to the column.
- The column is washed extensively with the equilibration buffer containing 0.5 M NaCl, followed by an extensive wash with the buffer alone.
- Hsp70 is eluted in 3 mM ATP in the equilibration buffer.

Detection of immunogenicity of HSP preparations

Tumor rejection

The gp96, hsp90, and hsp70 preparations obtained from tumor cells or tissues can be used to elicit tumor immunity (Srivastava and Das 1984; Srivastava et al 1986; Palladino et al 1987; Udono and Srivastava 1993, 1994; Udono et al 1994). Typically, HSPs are injected subcutaneously under the nape of the neck in 100–500 µl volume of PBS, twice at weekly intervals. The animals are challenged with live cancer cells injected intradermally one week after the second immunization and the kinetics of tumor growth is monitored. While the preceding account seems relatively straightforward, a number of parameters should be observed carefully for a satisfactory tumor rejection assay. The following precautions are recommended.

1. Mice or rats should be immunized with graded doses of HSPs. As an example, groups of 5 mice each should be immunized with 1, 5, 10, or 20 µg gp96 per mouse per injection. Immunogenicity of HSPs can be significantly dose-dependent and inclusion of a number of doses insures that one is operating within a suitably broad range (Srivastava et al 1986; Udono and Srivastava 1994).

2. The number of live cancer cells used for challenge should be determined carefully, by prior titration of different doses of cancer cells. Unless one is very comfortable with a given challenge dose, animals vaccinated with HSPs should be challenged with three different doses bracketing the estimated appropriate dose. Thus, for each of the four doses of HSPs described earlier, one may

consider three groups of three different challenge levels (see Udono and Srivastava 1994).

3. A control, unimmunized group, is included, naturally. However to establish a higher degree of confidence in the control group, two groups of control animals may be used, one group to be challenged before and the other after the immunized groups have been challenged. This is recommended because, occasionally, challenging a large number of mice may take up to 2 hours, during which period the cells used for challenge may lose viability. In this case, mice immunized later will show an artificially high tumor rejection response.

4. In the same context, the viability of cells used for challenge should be between 98% and 100%. The nonviable cells serve as effective vaccines and, thus, a given number of viable cells will be more tumorigenic by itself than if mixed with some nonviable cells. Cells should be placed at 4°C in an appropriate medium without calf serum and not in PBS while waiting to be used in challenge.

5. The tumor challenges should be given intradermally, rather than subcutaneously, unless specifically intended to be subcutaneous. The kinetics of tumor growth can be monitored with greater accuracy for intradermal tumors.

6. The biochemical integrity of the HSP preparation must be determined by SDS-PAGE the day it is used for immunization or challenge. This is significant because some HSPs are prone to autodegradation (Srivastava et al 1986; Li and Srivastava 1994; A. Menoret and P.K. Srivastava, unpublished) and others to loss due to aggregation.

7. Animals should also be immunized with HSPs purified from normal tissues and/or other tumors as negative controls. Similarly, animals should be immunized with the intact irradiated cancer cells from which the HSPs were derived and with other cancer cells.

8. The tumors should be measured every other day along two perpendicular axes. If feasible, tumor thickness should also be measured. The data should be plotted in terms of average tumor diameter and also tumor volume, in order to get a complete picture of tumor growth. Tumors should also be examined attentively for texture, signs of inflammation, and necrosis.

9. While it may be obvious that the health of animals used in vaccination is of seminal significance, it is worth stating this explicitly. Animals carrying bacterial or viral infections can be artificially permissive or resistant to tumor growth and may easily mask a vaccination effect. Similarly, the genotype of animals used for vaccination must be identical (and not merely close) to that of the tumors used in the study. Observation of this sacred dictum may sometimes require an extensive survey of previous literature and several unsatisfactory telephone calls to determine the source of animals used for induction of tumors. However, attention to this detail during planning of the experiments will 'vaccinate' the investigator against future shock (see Parham 1989).

The observances recommended above lead to large and perhaps cumbersome tumor rejection assays easily involving over a hundred mice in several groups. However, precisely because these assays are so expensive and time-consuming, it is essential that they be done well, if they are to be done at all. Although only the results from prophylactic vaccination have been reported so far, therapeutic administration of HSP preparations has also resulted in significant life prolongation, tumor remissions, and complete regressions in tumor-bearing mice (S. Janetzki, N.E. Blachere, M. Daou and P.K. Srivastava, unpublished; Y. Tamura and P.K. Srivastava, unpublished).

Generation of antigen-specific, MHC class I-restricted, cytotoxic T lymphocytes

The gp96 and hsp70 preparations obtained from tumor cells or virus-infected or transformed cells can be used to elicit cognate antigen-specific, MHC class I-restricted, cytotoxic T lymphocytes (Blachere et al 1993; Srivastava et al 1994; N.E. Blachere and P.K. Srivastava, unpublished; S. Janetzki, N.E. Blachere and P.K. Srivastava, unpublished; Suto and Srivastava 1995; Arnold et al 1995). Such experiments have not been attempted so far with hsp90 preparations. Typically, mice are immunized twice by weekly subcutaneous injection of HSP preparations derived from cognate cells. The HSP preparation is delivered in PBS and without any adjuvants. Indeed, the use of a number of adjuvants, such as Freund's complete and incomplete adjuvants, the Gerbu adjuvant and colchicine, has led not only to lack of an immuno-enhancing effect but to an abrogation of the response observed with HSP in PBS alone (S. Janetzki, M. Daou and P.K. Srivastava, unpublished). Typically between 5 and 20 µg HSP preparations are used per injection and the precise optimal quantity varies between cell lines. Spleens are harvested between 4 and 7 days after the final immunization and splenocytes are cocultured with the cognate stimulator cells for 5–7 days, as appropriate.

At the end of this period, the T cells are tested for cytotoxicity against a panel of target cells by a chromium-release assay. The cytolytic T lymphocyte (CTL) activity may also be tested by cytokine release by activated T cells. One may observe better detection of CTL response with one assay than another, depending upon the particular system used. The CTL responses generated in this manner are antigen-specific and MHC class I- and CD8-restricted. Our results also suggest that the CTLs generated by vaccination with HSP–peptide complexes are qualitatively different from the CTLs elicited by vaccination with intact cells bearing the appropriate antigen (N.E. Blachere and P.K. Srivastava, unpublished). This indicates a hitherto unsuspected diversity among CTLs.

An interesting aspect of vaccination with HSP–peptide complexes is their ability to cross-prime, i.e., to elicit an MHC-restricted antigen-specific CTL response in mice of any haplotype, as long as the HSPs are derived from cells which express the cognate antigen (Srivastava et al 1994; Suto and Srivastava 1995; Arnold et al 1995). This is possible presumably because the association of peptides with HSPs occurs before their association with the MHC molecules and thus, in contrast to the MHC-associated peptides, the HSP-associated peptides are not preselected with respect to any particular haplotype (Srivastava et al 1994). This phenomenon has obvious applications in generation of vaccines against diseases where the protective antigens have been identified.

References

Arnold D, Faath S, Rammensee H-G, Schild H (1995) Cross-priming of minor histocompatibility cytotoxic T cells upon immunization with the heat shock protein gp96. J Exp Med 182: 885–889.

Blachere NE, Udono H, Janetzki S, Li Z, Heike M, Srivastava PK (1993) Heat shock protein vaccines against cancer. J Immunother 14: 352–356.

Feldweg AF, Srivastava PK (1995) Molecular heterogeneity of the tumor rejection antigen/heat shock protein gp96. International J Cancer 63: 310–314.

Flynn GC, Chappell TG, Rothman JE (1989) Peptide binding and release by proteins implicated as catalysts of protein assembly. Science 245: 385–390.

Houghton AN (1994) Cancer antigens: immune recognition of self and altered self. J Exp Med 180: 1–4.

Li Z, Srivastava PK (1993) Tumor rejection antigen Gp96/ Grp94 is an ATPase: implications for protein folding and antigen presentation. EMBO J 12: 3143–3151.

Li Z, Srivastava PK (1994) A critical contemplation on the role of heat shock proteins in transfer of antigenic peptides during antigen presentation. In: Rammensee H-G, Hammerling G, eds. MHC Molecules: Structural and Functional Aspects. BIM Press.

Lindquist S, Craig EA (1988) The heat shock proteins. Annu Rev Genet 22: 631–77.

Maki RG (1991) The human homologue of the mouse tumor rejection antigen gp96. PhD thesis, Cornell University Medical College, New York.

Marc D (1988) Two-step purification and N-terminal amino acid sequence analysis of the rat M_r 90,000 heat shock protein. Anal Biochem 173: 405.

Nandan D, Daubenberger C, Mpimbaza G, Pearson TW (1994) A rapid, single step purification method for immunogenic members of the hsp70 family: validation and application. J Immunol Methods 176: 255–263.

Palladino MA, Srivastava PK, Oettgen HF, DeLeo AB (1987) Expression of a shared tumor-specific antigen by two chemically induced BALB/c sarcomas. I. Detection by a cloned cytotoxic T cell line. Cancer Res 47: 5074–5079.

Pardoll DM (1994) The new look of tumor antigens in the 90's. Nature 369: 357–358.

Parham P (1989) Alien antigens return to fold. Immunol Today 10: 206–212.

Srivastava PK (1993) Peptide-binding heat shock proteins in the endoplasmic reticulum: role in immune response to cancer and in antigen presentation. Adv Cancer Res 62: 153–177.

Srivastava PK (1994a) Stress-induced proteins: basis for a new generation of anti-cancer vaccines. In: van Eden W, Young DB, eds. Stress Proteins in Medicine. Marcel Dekker, Amsterdam.

Srivastava PK (1994b) Heat shock proteins as priming agents: Implications for cancer immunotherapy. In: Browning M, Dalgleish AG, eds. Tumor Immunology. Cambridge University Press, Cambridge.

Srivastava PK, Das MR (1984) Serologically unique surface antigen of a rat hepatoma is also its tumor-associated transplantation antigen. Int J Cancer 33: 417–422.

Srivastava PK, Old LJ (1989) Identification of the human homologue of the murine tumor rejection antigen gp96. Cancer Res 49: 1341–1343.

Srivastava PK, DeLeo AB, Old LJ (1986) Tumor rejection antigens of chemically induced sarcomas of inbred mice. Proc Natl Acad Sci USA 83: 3407–3411.

Srivastava PK, Udono H, Blachere NE, Li Z (1994) Heat shock proteins transfer peptides during antigen processing and CTL priming. Immunogenetics 39: 93–98.

Suto R, Srivastava PK (1995) A mechanism for the specific immunogenicity of heat shock protein-chaperoned peptides. Science 269: 1585–1588.

Udono H, Srivastava PK (1993) Heat shock protein 70-associated peptides elicit specific cancer immunity. J Exp Med 178: 1391–1396.

Udono H, Srivastava PK (1994) Relative immunogenicities of heat shock proteins gp96, hsp90 and hsp70 against chemically induced tumors. J Immunol 152: 5398–5403.

Udono H, Levey DL, Srivastava PK (1994) Definition of T cell sub-sets mediating tumor-specific immunogenicity of cognate heat shock protein gp96. Proc Natl Acad Sci USA 91: 3077–3081.

Welch WJ, Feramisco JR (1985) Rapid purification of mammalian 70,000 d stress proteins. Affinity of the proteins for nucleotides. Mol Cell Biol 3: 1–13.

Outlook

9.12

Hans-Georg Rammensee

Abteilung Immunologie, Institut für Zellbiologie, University of Tübingen, Germany

How will our field, the physiology of MHC molecules in particular, and immunology in general, look 10 years from now?

In the mid-eighties, we knew protein sequences of MHC molecules but had no notion of their further structure. There was only a faint idea that there must be some way of interaction with peptides. The identification of T cell receptor genes and proteins was a 'hot new thing'. In another ten years, we will know the fine structure of TCR–peptide–MHC complexes, and we will know exactly how the MHC ligands are produced in the cell: which accessory molecules are involved, where these are located in the cellular compartments, and where the genes are encoded. Quite a number of the genes sequenced by HUGO will have something to do with antigen processing in particular, or intracellular protein degradation and translocation in general. We will also know which do what and when in antigen processing, and we will know exactly which peptides are transported by TAP, and what the relative contribution of endopeptidases and exopeptidases to antigen processing is. We will know the function of DM/DP molecules in class II-restricted processing, and we will know the structure of CLIP/class II complexes.

Will all this mean the end of the road for our field? Of course not, because there is always another layer of complexity. For example, who would have thought five years ago that pocketology, the science of the interaction of peptide side-chains with corresponding structures in the MHC molecule, would become such a thriving, fascinating, and fashionable pastime? And who would have thought that natural killer cells exploit such a complex system of receptors for their antigen nonrecognition? And apart from all the progress into new spheres, some of the old problems will still not be completely solved. For example, we will still not understand all aspects of self-tolerance, and we will still not know the real function of $\gamma\delta$ T cells. Then there is the encyclopedic aspect. The growth rate of the lists of CD molecules, of cytokines, and of MHC alleles will slow down but not come to an end. We will still not understand in detail the immune systems of more than a handful of species, although each species might have added its own special feature to solving the problem of cleansing itself from parasites. For example, neither mouse nor human have functionally polymorphic TAPs — only the rat.

This was a view of the theoretical side. On the applied side, I expect even more spectacular progress. Based on our knowledge of the workings of the system, I expect the first forms of rationally designed intervention into immune

Immunology Methods Manual
ISBN 0–12–442712–X

responses. Especially in tumor immunology, I expect to see the successful application of antigen-specific, T cell-mediated immunotherapy for at least a few types of cancer, such as the melanoma. The immunization of patients with a well-defined and immunogenic preparation of tumor associated antigen — e.g., a mixture of MAGE 1–3, tyrosinase, Mart-1/gp100, and so on — either with a nucleic acid vector, or as a protein, or as peptides, could be successful; or perhaps the adoptive immunotherapy with antigen-specific CTL expanded *in vitro* with the help of cytokines, dendritic cells, and peptides. On the other hand, as the history of tumor immunology has taught us, disappointments and drawbacks are notorious in this field, making it advisable to retain a healthy portion of skepticism. Apart from tumor immunology, I also expect useful applications of rational immune intervention for autoimmune and infectious diseases.

The collection of methods in this Section should contribute to progress — be it on the applied or basic side. If this is indeed the case, the purpose of these chapters will have been accomplished.

Recommended reading

9.13

Falk K, Rötzschke O, Stevanović S, Jung G, Rammensee H-G (1991) Allele-specific motifs revealed by sequencing of self-peptides eluted from MHC molecules. Nature 351: 290–296.

Klein J (1986) Natural History of the Major Histocompatibility Complex. Wiley, New York.

Rammensee H-G (1995) Chemistry of peptides associated with MHC class I and class II molecules. Curr Opin Immunol, 7: 85–96.

Rammensee H-G, Falk K, Rötzschke O (1993) Peptides naturally presented by MHC class I molecules. Annu Rev Immunol 11: 213–244.

Rammensee H-G, Friede T, Stevanović S (1995) MHC ligands and peptide motifs. First listing. Immunogenetics 41: 178–228.

Stern LJ, Wiley DC (1994) Antigenic peptide binding by class I and class II histocompatibility proteins. Structure 2: 245–251.

Section

10

Screening of T and B Cell Epitopes

Section Editor
Siegfried Weiss

LIST OF CHAPTERS

A brief review on epitope screening

10.1

Siegfried Weiss

Department of Cell Biology and Immunology, GBF National Research Center for Biotechnology, Braunschweig, Germany

The search for epitopes, the antigenic determinants which are recognized by antibodies or the T cell receptor, has a long tradition. In the early days of immunochemistry, the antibody responses characterized were mainly directed against blood-group substances and bacterial cell wall components. Therefore, the epitopes defined at that time mainly consisted of carbohydrates. The carbohydrate chemistry known at that time allowed the synthesis of simple carbohydrates to characterize minimal antigenic determinants (Kabat 1976). Only with the introduction of modern peptide chemistry and molecular biology was it possible to investigate the more complex epitopes found on various proteins; this field finally started to blossom when it was recognized that the receptor of T cells recognized short peptides bound to molecules of the major histocompatibility complex (Townsend et al 1986; Babbitt et al 1985).

From the structural studies of antibody–antigen complexes it is clear that antibodies have several contact regions on the surface of the antigenic molecule (Fischmann et al 1991). Therefore, epitopes can only be defined for antibodies which recognize linear stretches of the protein sufficient in length to allow binding.

Epitopes that depend on the native confirmation of the protein cannot be defined. Thus it is not possible at present to assign an epitope to every antibody. For T cell receptors the situation is different, since all conventional T cell receptors recognize short proteolytic fragments which are presented by (bound to and displayed on the cell surface by) molecules of the major histocompatibility complex (MHC). It should therefore be possible to define the peptide epitope for all conventional T cell receptors.

For the study of epitopes of antibodies, proteolytic fragments or fragments derived from chemical degradation have usually been used for epitope localization. In the meantime, however, several expression systems in *E. coli* and other organisms have been established which allow the derivation of fragments of various lengths of the proteins in question (Sassenfeld 1990). This now provides the possibility of correlating unknown open reading frames of viral genomes with antiviral antibodies found in patients or defining binding sites of neutralizing antibodies or of antibodies interfering with protein–protein interaction, thus allowing the definition of functional domains or sites.

The production of these protein fragments often strongly interferes with the physiology of the bacterium. This could result in the loss of the desired protein fragment. Therefore, the expression systems usually use inducible promoters like the *lac* promoter of *E. coli* or temperature-sensitive promoters of bacteriophages (Sassenfeld 1990). Small filamentous bacteriophages (Scott and Smith 1990) have also been used successfully as well as λgt10 and λgt11, which are normally used for expression cloning (Huynh et al 1985). The latter systems cirumvent the above problems, since the protein is exported or the bacterium is lysed; however, they introduce problems for the large-scale isolation of the protein fragment.

The desired protein fragments are usually expressed as fusion proteins. The partners used for fusion provide several important features to the fragments, e.g. they provide stability to the small fragments they are fused with. More important, they provide a means to easily isolate the fusion protein by affinity chromatography. Peptide tags have been developed for which monoclonal antibodies exist for single-step isolations. Similarly, oligohistidine tags have been used in combination with chelating chromatography for single-step purification (Sassenfeld 1990) or streptags, consisting of the peptide recognized by streptavidin in combination with streptavidin columns (Schmidt and Skerra 1993). Many commercial suppliers provide kits for the production and purification of fusion proteins with particular properties and it can be expected that in future more and easier to use systems will become available. Here we will restrict ourselves to two well-established systems: (a) the maltose binding protein of *E. coli* under the control of the *lac* promoter and (b) of the β-galactosidase of *E. coli* under the control of the temperature-inducible derivative of the left promoter of bacteriophage λ. The former can be isolated over amylose resins by using maltose for elution, i.e. under nondenaturing conditions. The latter, owing to its large size, can often be purified by molecular sieve chromatography, thus also avoiding denaturing conditions for purification.

Several systems include a protease-sensitive site constructed between the protein fragment and the fusion partner. However, since the conformation of the fusion protein is unpredictable, often the protease does not cleave the two polypeptides or, worse, the protease degrades the protein fragment of interest. However, for epitope screening usually the fusion partner does not interfere with the immunological assays. The possibility of removing the fusion partner from the fragment is therefore not considered in the following chapter. Except for dialysis against a physiological buffer, the fusion proteins described can be used 'straight' in most immunological assays such as immunoblot, ELISA, or dot blots for antibodies and proliferation or cytokine release for CD4$^+$ T cells. For cytotoxic CD8$^+$ T cells these proteins cannot be used without further manipulations, because CD8$^+$ cells usually recognize proteins which are derived from antigens produced by the presenting cell itself, like viral proteins (Yewdell and Bennink 1992). Therefore, the antigen has to be expressed by transfection or the fusion proteins have to be introduced into the MHC class I pathway by osmotic loading (Moore et al 1988) or by electroporation (Harding 1992) or cotransferred with listeriolysin (Darji et al 1995).

After the binding region has been narrowed down by fusion proteins or proteolytic fragments, the epitope can be mapped precisely with sets of overlapping synthetic peptides. As mentioned above for fusion proteins, this assumes that a linear determinant of sufficient length is recognized by the antibody. For T cells this generates no problem since most T cell receptors recognize short linear determinants bound to a groove on the MHC class I and class II molecules (Bjorkman et al 1987; Brown et al 1993).

To perform a complete scan of a protein or protein fragment with overlapping peptides on a routine basis, efficient methods for the synthesis of low amounts of many peptides need to be used. Several ways of simultaneous synthesis of multiple peptides have been established to date and some will be presented in this section. The possibility of automation of some of these methods and of miniaturization reduces the amount of work and chemicals involved in the actual peptide synthesis to an extent which allows the simultaneous synthesis of even hundreds of peptides. Thus, it should be possible to screen small linear antibody epitopes of large proteins even without the prior characterization by fusion protein.

Screening of T cell epitopes requires the elution of the peptides from the matrices they were synthesized on. This introduces many additional handling steps. First of all, linkers may have to be introduced which can remain on the peptide after elution and might interfere with binding to at least the MHC class I molecules. In addition, the cell cultures have to be handled. However, here also semi-automation is possible and should allow the screening of even large proteins with reasonable effort.

Systematic investigation of peptides bound to MHC molecules either eluted from these molecules or defined by screening has revealed that characteristic amino acids are found at certain positions, the so-called anchor residues, which are responsible for the high-affinity binding of the peptide towards the particular MHC molecules (Falk and Rötzschke 1993). This allows the prediction of possible epitopes and thus reduces the number of peptides needed for definition of a T cell epitope. However, until all binding motifs for MHC

molecules are known and precise predictions become available, the procedure described here will be necessary for uncharacterized MHC molecules. In addition, immunogenicity cannot be predicted only on the basis of an acceptable binding affinity of the peptides to the MHC molecules. The induction of an immune response includes many complex biological processes such as protease sensitivity of the determinant, selection of T cell receptor repertoires, and many other aspects. Therefore, even with the possibility of prediction of T cell epitopes, several peptides will have to be synthesized for precise mapping.

Recently, it has been described that peptide variants at non-anchor residues might antagonize the stimulation of T cells by the nominal antigen or peptide (De Magistris et al 1992). This is due to induction of signals insufficient to stimulate proliferation or cytokine release. Since antagonists act despite the fact that the agonistic peptide is bound to the MHC molecule, this phenomenon has raised great hopes for the ability to interfere with ongoing auto-immune diseases. In addition, antagonists are supposed to be involved in positive selection of T cells in the thymus (Hogquist et al 1994). At the moment the variations which need to be introduced to change a peptide from an agonist to a true antagonist are not predictable. Therefore, screening for potential antagonists requires multiple peptides, best performed by the simultaneous synthesis described here.

References

Babbitt BP, Allen PM, Matsueda G, Haber E, Unanue ER (1985) Binding of immunogenic peptides to Ia histocompatibility molecules. Nature 317: 359–361.

Bjorkman PJ, Saper MA, Samraoui B, Bennett WS, Strominger JL, Wiley DC (1987) Structure of the human class I histocompatibility antigen, HLA-A2. Nature 329: 506–512.

Brown JH, Jardetzky TS, Gorga JC, Stern LJ, Urban RG, Strominger JL, Wiley DC (1993) Three-dimensional structure of the human class II histocompatibility antigen HLADR1. Nature 364: 33–39.

Darji A, Chakraborty T, Wehland J, Weiss S (1995) Listeriolysin provides access to the MHC class I presentation pathway for soluble proteins in vitro and in vivo. Eur J Immunol 25: 2967–2971

De Magistris MT, Alexander J, Coggeshall M, Altman A, Gaeta FCA, Grey HM, Sette A (1992) Antigen analog-major histocompatibility complexes act as antagonists of the T-cell receptor. Cell 68: 625–634.

Falk K, Rötzschke O (1993) Consensus motifs and peptide ligands of MHC class I molecules. Semin Immunol 5: 8194.

Fischmann TO, Bentley GA, Bhat TN, Boulot G, Mariuzza RA, Phillips SE, Tello D, Poljak RJ (1991) Crystallographic refinement of the three-dimensional structure of the FabD1.3-lysozyme complex at 2.5-Å resolution. J Biol Chem 20: 12915–12920.

Harding III CV (1992) Electroporation of exogenous antigen into the cytosol for antigen processing and class I major histocompatibility complex (MHC) presentation: weak base amines and hypothermia (18°C) inhibit the class I MHC processing pathway. Eur J Immunol 22: 1865–1869.

Hogquist KA, Jameson SC, Heath WR, Howard JL, Bevan MJ, Carbone FR (1994) T cell receptor antagonist peptides induce positive selection. Cell 76: 17–27.

Huynh TV, Young RA, Davis RW (1985) Constructing and screening cDNA libraries in λgt10 and λgt11. In Glover DM, ed. DNA Cloning. IRL Press, Oxford, pp. 49–78.

Kabat EA (1976) Antigenic determinants and the size of the antibody combining site; determinants of cell-mediated immunity. In Structural Concepts in Immunology and Immunochemistry. Holt, Rinehart and Winston, New York, pp. 119–166.

Moore MW, Carbone FR, Bevan MJ (1988) Introduction of soluble protein into the class I pathway of antigen processing and presentation. Cell 54: 777–785.

Sassenfeld HM (1990) Engineering proteins for purification. Trends Biotechnol 8: 88–93.

Schmidt TGM, Skerra A (1993) Protein Engineering 6: 109–122.

Scott JK, Smith GP (1990) Searching for peptide ligands with an epitope library. Science 249: 386–390.

Townsend ARM, Rothbard J, Gotch FM, Bahadur G, Wraith D, McMichael AJ (1986) The epitopes of influenza nucleoprotein recognized by cytotoxic T lymphocytes can be defined with short synthetic peptides. Cell 44: 959–968.

Yewdell JW, Bennink JR (1992) Cell biology of antigen processing and presentation to major histocompatibility complex class I molecule-restricted T lymphocytes. Adv Immunol 52: 1–123.

Expression of protein fragments as fusion proteins in E. coli

10.2

Siegfried Weiss[1]
Werner Lindenmaier[2]

[1]Department of Cell Biology and Immunology, [2]Department of Molecular Biotechnology, GBF National Research Center, Braunschweig, Germany

TABLE OF CONTENTS

Immunology Methods Manual
ISBN 0–12–442712–X

Abstract

For screening epitopes on proteins or testing open reading frames for immunological reactivity, *E. coli* is a very suitable expression system. The polypeptide of interest is produced as a fusion protein with either β-galactosidase or maltose binding protein. This provides stability and a simple way to purify the polypeptide.

The constructs are driven by inducible promoters which allow the separation of the growth phase of the bacteria and the expression phase. Large amounts of proteins can be produced by this method in a short time and with minimal effort.

General considerations

To characterize the function, immunological features, or other properties of proteins in most cases it is more convenient to use the indirect approach of protein expression by recombinant DNA technology rather than the traditional biochemical approach. Besides the production of large amounts, expression allows the addition of particular modifications, construction of specific fragments, targeting to specific cellular compartments, etc. More insights will be gained this way than by handling the protein itself. The choice of the expression system obviously depends on the particular question to be asked but also on the properties of the protein. Here we describe two expression systems in *E. coli* in which expression of polypeptides is facilitated by a fusion to β-galactosidase or to the maltose binding protein.

The expression vectors used for construction provide all the necessary sites for expression such as translation start sites or termination sites and they include inducible promoters (pλR and *TAC*) since hyperexpression of heterologous proteins usually interferes strongly with the growth of the bacteria. Owing to the inducible promotors, the expressor strain can be grown to a certain density before expression is allowed.

The vector-derived polypeptide of the fusion protein introduces important features to the expressed protein, e.g. labile proteins might be stabilized, inclusion body formation can be favored or avoided, properties for single-step purification are supplied, and the immunogenicity is improved when used for immunization owing to a carrier effect, to name a few. Especially when correct folding of the protein is not an essential prerequisite of the experiment, these systems are the method of choice, since they are easy to handle, large numbers of clones can be screened using the established cloning procedures in bacteria, and amounts of proteins in the range of 10 to 100 mg can be produced with a minimal effort in time and resources.

The test system where these proteins are used needs careful consideration. As already mentioned, the correct folding of the protein cannot be predicted, especially if proteins derived from inclusion bodies have to be solubilized by denaturing agents. Similarly, the polypeptide used as fusion partner might interfere with the assay. Although most of the fusion systems provide a protease-sensitive site between the two polypeptides to allow the removal of the fusion partner, very often it is not possible to obtain a stable product because either the particular conformation of the fusion proteins makes the site inaccessible to the protease or the polypeptide in question is also digested by the protease during the procedure.

Another problem often disregarded is the presence of bacterial products which might interfere with the assay, e.g. in immunological systems which involve cells reactive to lipopolysaccharide or lipoprotein from *E. coli*. These compounds have to be removed or the experiments have to be carefully controlled.

Nevertheless, many proteins expressed as fusion proteins have helped to solve biological questions. With the introduction of more versatile expression systems and better understanding of protein folding and refolding, the use of fusion proteins will become a standard procedure in molecular biology.

ORF expression as β-galactosidase fusion protein

Introductory remarks

Expression of protein fragments as β-galactosidase fusion proteins has proved an efficient tool for analyzing antigenic determinants recognized by the immune system. In addition to screening on the basis of antibody–fusion protein interaction, T cell epitopes can also be defined.

Vectors and inserts

A large number of *E. coli* expression vectors have been established which use β-galactosidase fusion protein expression for the detection and stabilization of foreign antigens. λ-Phage vectors, like λgt11 (Young and Davies 1983), have been used extensively for cloning of cDNAs and immunological screening of plaques (Jahn et al 1987; Mehra et al 1986; Plachter et al 1990; Young and Davies 1983). Plasmid vectors, which improve gene manipulation and production of substantial amounts of fusion protein, have predominantly been used for epitope mapping and production of recombinant antigen (Koenen et al 1982; Stanley and Luzio 1984; Weinstock et al 1983).

Fusions can be located at the N-terminus of β-galactosidase, in which case a tripartite gene is expressed and two in-frame cloning steps are required (Amann 1985; Ellis et al 1985; Gray et al 1982; Koenen et al 1982; Rüther and Müller-Hill 1983; Watson et al 1983; Weis et al 1983). When random fragments are inserted in such an N-terminal cloning site, only 1 in 18 clones will be in frame with the fusion partner at both sides in addition to the correct orientation and will express the fusion protein. In most cases these fusion proteins will retain β-galactosidase activity, which simplifies detection and purification by affinity chromatography (Ullmann 1984; Koenen et al 1982). Insertion of DNA fragments at the C-terminus of β-galactosidase in the most commonly used vectors leads to the expression of a β-galactosidase fusion protein, with a probability of 1/6 (Stanley and Luzio 1984; Young and Davies 1983). The fusion protein is usually enzymatically inactive.

For ORF-cloning from genomic DNA, random fragments of a few hundred base pairs are suitable for a library covering the genomic region of interest, e.g. a viral genome. Randomness of fragmentation is approached by digestion with DNAase I in the presence of Mn^{2+}-ions or by mechanical shearing (Ellis et al 1985; Lindenmaier et al 1990; Mehra et al 1986).

Promotors

Since compatibility and stability of the foreign gene product cannot be predicted with certainty, it is a general strategy to separate growth phase and production by using tightly regulated promoters like *lacZ* (Koenen et al 1982), *TAC* (Amann et al 1983) or λ_{PR} and λ_{PL} (Botterman and Zabeau 1987; Stanley and Luzio 1984) in combination with the corresponding repressors (*lacI* and λ_{CI}). Strains which overexpress these repressors or vectors that carry their own repressors have been developed for strict regulation of expression. In our hands *lacZ*-P/O derived systems frequently caused problems due to leakiness of repression leading to instability and loss of clones most probably due to toxic effects of the fusion proteins (Ellis et al 1985; Lindenmaier et al 1990; Stanley 1983). The λ_{PR} and λ_{PL} promoters in combination with the temperature-sensitive λ_{CI}-repressor which is inactivated by incubation at 42°C seem to be more tightly regulated. Under these conditions predominantly insoluble fusion protein is formed. This leads to an improvement of fusion protein yields due to stabilization of proteins in inclusion bodies (Stanley 1983).

Detection and purification of β-galactosidase fusion proteins

Isolation of reactive clones from ORF libraries can be achieved by high-density screening. Plaques can be identified by direct immunoscreening (Young and Davies 1983). When plasmid vectors are used, replica plating and lysis of *E. coli* cells is necessary to make the antigens accessible (Stanley 1983; Ellis et al 1985; Verweij et al 1985). Immunoblotting of proteins from individual clones allows the analysis of the interaction of antibodies and antigen. To avoid interference with cross-reactivity

due to contaminating *E. coli* proteins, the fusion proteins should be purified. Enzymatically active fusion proteins can be isolated by affinity chromatography on TEPG–Sepharose (Sassenfeld 1990; Ullmann 1984). Inactive proteins and proteins from inclusion bodies can be enriched by fractionated extraction with urea and size exclusion chromatography (Stanley, 1983; Ellis et al 1985; Lindenmaier et al 1990). When the β-galactosidase moiety of the fusion proteins interferes with the assays (Ellinger et al 1989; Knobloch et al 1987), e.g. owing to reactivity of patient sera with the β-galactosidase, this reactivity has to be removed either by absorption of β-galactosidase-reactive antibodies or by removal of the β-galactosidase part from the fusion protein. A number of expression systems include protease-sensitive sites between the β-galactosidase moiety and the antigen, which allows the use of sequence-specific proteases (for example, factor X_a (Ellinger et al 1989; Maina et al 1988; Nagai and Thøgerson 1984); IgA protease (Pohlner et al 1993); renin (Haffey et al 1987); or collagenase (Scholtissek and Grosse 1988)) to release the antigen from the fusion protein. These enzymes, however, are expensive and not terribly efficient. The digestion conditions have to be optimized and are dependent on the individual protein. As an alternative method for the isolation of β-galactosidase free antigen expression vectors that are easily modified to allow expression of inserts in an unfused form can be used (Banting et al 1991; Watson et al 1983). However, for most screening purposes, diagnostic use or production of antigen-specific antisera purified fusion proteins can be applied (Landini et al 1990; Lindenmaier et al 1990; Peterhans et al 1987; Ripalti et al 1994; Scholl et al 1988; Scott et al 1984).

In the following protocol we describe the use of pEX-vectors for ORF-cloning of antigenic determinants, immunological screening, and the isolation of β-galactosidase fusion proteins for immunization and serological diagnosis. In an analogous way other expression vectors in combination with genomic DNA or cDNA fragments can be used. For additional information on basic protocols see Sambrook et al (1989).

Materials

- Standard laboratory chemicals (ethanol, phenol, etc.) of highest quality available
- Vector DNA pEX1
- Genomic DNA or cloned genomic DNA (e.g. cosmid)
- Restriction enzyme SmaI
- SmaI-restriction enzyme buffer (10×):

  100 mM Tris–HCl
  70 mM MgCl$_2$
  500 mM KCl
  10 mM DTT pH 7.5 at 37°C

- DNAaseI
- DNAaseI-Mn^{2+}-buffer (1×):

  10 mM MnCl$_2$
  33 mM Tris–HCl pH 7.5

- TE:

  10 mM Tris–HCl
  0.1 mM EDTA pH 7.5

- Klenow DNA polymerase
- DNA polymerase buffer (10×):

  100 mM MgSO$_4$
  1 mM   DTT
  500 mM Tris–HCl pH 7.2

- T4 DNA ligase
- Blunt end DNA ligase buffer (10×):

  5 mM   ATP
  50 mM   MgCl$_2$
  100 mM   DTT
  500 μg ml^{-1} bovine serum albumin
                  500 mM Tris–HCl pH 7.6

- Calf intestine phosphatase (CIP)
- CIP buffer (10×):

  10 mM ZnCl$_2$
  10 mM MgCl$_2$
  100 mM Tris–HCl pH 8.3

- E. coli DH1 [pRK248clts]
  Other E. coli strains expressing the temperature-sensitive phage λ repressor from integrated prophage (E. coli NS4830) or plasmids (e.g. pRK248clts (Bernard and Helinski 1980)) can be used accordingly.
- TFB (transformation buffer)

  10 mM MES (potassium morpholinoethansulfonate)
  100 mM RbCl
  45 mM MnCl$_2$·H$_2$O
  3 mM   HA-CoCl$_3$ (hexaminocobalt(III) chloride)

- SOB medium:

  20 g l^{-1} bactotryptone
  5 g l^{-1} yeast extract
  10 mM   NaCl
  2.5 mM   KCl
             After autoclaving adjust to
  10 mM   MgCl$_2$
  10 mM   MgSO$_4$

- SOC medium = SOB, 20 mM glucose
- Nitrocellulose filters (e.g. Millipore, Schleicher & Schüll)
- Lysis buffer:

  50 mM   Tris–HCl pH 7.5
  150 mM   NaCl
  5 mM   MgCl$_2$
  0.1%   Tween-20
  1 μg ml^{-1} DNAaseI
  40 μg ml^{-1} lysozyme

- TBS:

  20 mM Tris–HCl pH 7.5
  500 mM NaCl

- T-TBS:

  0.05% Tween-20 in TBS

- Blocking buffer:

  3%   gelatine in T-TBS

- Antibody incubation buffer:

  1%   gelatine in T-TBS

- HRPO staining solution:

  1 volume   3 mg ml^{-1} chloronaphthol dissolved in methanol

5 volumes	TBS
1/2000 volume	30% H_2O_2

- Urea gel sample buffer:

20 mM	Tris–HCl pH 7.5
9 M	urea
4%	SDS
4%	β-mercaptoethanol
20%	glycerol
250 μg ml⁻¹	bromophenol blue

10 min. Cool to room temperature, extract with phenol–chloroform, adjust to 0.3 M sodium acetate by addition of 1:10 volume of a 3 M stock solution, and precipitate DNA with 2 volumes of ethanol (15 min, 0°C).

6. Redissolve linearized, dephosphorylated vector DNA in 100 μl 10 mM Tris–HCl pH 7.5, 0.1 mM EDTA.

Preparation of genomic fragments

Protocols

Subcloning of random fragments of genomic DNA into ORF expression vector pEX1

Preparation of vector DNA

1. Digest supercoiled pEX1 DNA (10–20 μg) in 100 μl of 1× SmaI buffer with 2- to 3-fold excess of SmaI for 1 h at 25°C.
2. Control completeness of digestion by running 0.2 μg supercoiled DNA and 0.2 μg digested DNA in parallel on a 0.5% agarose gel.
3. When digestion is complete, remove restriction enzyme by extraction with phenol–chloroform (1:1). Precipitate DNA with 2 volumes of ethanol for 15 min at 0°C. After centrifugation (12,000g, 10 min, 4°C) dissolve pellet in 100 μl of CIP buffer (1×).
4. Dephosphorylate blunt ended vector DNA by addition of CIP (1 unit per pmol DNA). Incubate 15 min at 37°C. Add the same amount of CIP once more and incubate for 45 min at 55°C.
5. Add EDTA (pH 8.0) to 5 mM and inactivate CIP by heating to 75°C for

Establishment of conditions for DNAaseI digestion

- Dilute DNAaseI 0.5 μg ml⁻¹ in 500 μg ml⁻¹ BSA, 1 mM $MgCl_2$. Set up three reactions with different amounts of DNAaseI. 5 μg of genomic DNA in 30 μl of DNAaseI-Mn^{2+} buffer is mixed with 5, 2, and 0.5 μl of DNAaseI dilution. 10 μl aliquots are removed after 5, 10, and 20 min at room temperature and the digestion is stopped by addition of 2 μl of 0.1 M EDTA. The degree of digestion is determined by running the aliquots on a 1.5% agarose gel in the presence of suitable size markers.

Preparative isolation of genomic DNA fragments

- 10 to 100 μg of genomic DNA is digested with DNAaseI using the conditions for optimal yield of DNA fragments in the desired size range established by the pilot experiment described above. After phenol extraction and ethanol precipitation, the DNA is dissolved and subjected to preparative agarose gel electrophoresis. Fragments of the size range desired (~400 to 800 bp) are isolated using the Jetsorb kit (Genomed) according to the instructions of the manufacturer. A large number of other commercially available kits for DNA fragment isolation can be used. Alternatively, fragments can be isolated by electroelution into dialysis bags (Sambrook et al 1989). Finally, resuspend DNA fragments in TE at ~100 µg ml^{-1}.

- To increase the number of blunt ended fragments, 5' protruding ends are filled in by Klenow DNA polymerase. Add 1/10 volume of 10× DNA polymerase buffer and 1/25 volume of a mixture of dATP, dCTP, dGTP, and dTTP (2 mM each). Add Klenow DNA polymerase (10 U ml^{-1}) and incubate for 30 min at room temperature. 1/25 volume of EDTA (0.5 M, pH 8.0) is used to stop the reaction. DNA fragments are dissolved in TE at a concentration of 100 µg ml^{-1} after phenol–chloroform extraction and ethanol precipitation.

Ligation of blunt ended genomic fragments into dephosphorylated vector

Ligation of blunt ended DNA by T4 DNA ligase is improved by low concentration of ATP (0.5 mM), high concentrations of ligase (50 Weiss units ml^{-1}) and high concentration of DNA termini. The probability of ligating unrelated genomic fragments increases with fragment concentration.

- Mix 5 µg of vector fragment and equimolar amounts of genomic fragments (see above). Add H$_2$O to 90 µl, 10 µl of blunt end DNA ligase buffer and 5 U of T4 DNA ligase. Incubate 3 h at room temperature. Use 0.2 µl, 1 µl and 5 µl of ligation reaction to transform competent E. coli DH1 [pRK248clts] cells at 30°C.

Transformation

1. Grow E. coli DH1 [pRK248clts] at 30°C on LB plates supplemented with 20 µg ml^{-1} tetracycline.
2. Inoculate 50 ml SOB with 2–5 fresh single colonies. Grow at 30°C with good aeration to OD$_{550}$ 0.45–0.55.
3. Transfer to centrifuge tube and cool to 0°C for 10 min. Collect cells by centrifugation at 4°C, 1000g for 10 min. Carefully resuspend pellet in 1/3 volume TFB, incubate at 0°C for 10 min. Centrifuge as above and resuspend in 1 ml TFB and incubate for 5 min at 0°C.
4. Add 7 µl 2.25 M DMSO per 200 µl aliquot. After 5 min add 7 µl of 2.25 M DTT and after further incubation for 5 min add another 7 µl of 2.25 M DTT and continue incubation for another 10 min. Cells are now competent for transformation.
5. Use 0.2 µl, 1 µl and 5 µl of ligation reaction to transform competent E. coli DH1 [pRK248clts] cells at 30°C. Mix cells with DNA and incubate for 30 min at 0°C. Heat to 30°C for 90 s. Cool on ice for 2 min and add 800 µl of SOC medium.
6. Shake for 1 h at 30°C to allow expression of resistance to antibiotics and plate aliquots on LB agar plates containing 50 µg ml^{-1} ampicillin and 20 µg ml^{-1} tetracycline to determine transformation efficiency.

Immunological screening

1. Plate about 5000 transformants per 85 mm diameter LB agar plate containing 50 µg ml^{-1} ampicillin and 20 µg ml^{-1} tetracycline. Grow at 30°C until small colonies are visible.

2. Prepare marked replica filters, place on fresh LB plates and induce production of fusion proteins by incubation at 42°C for 3 h.

3. Expose filters for 20 min to an atmosphere saturated with vapor of chloroform. Incubate each filter separately for 16 h at room temperature in 20 ml of lysis buffer under gentle agitation. Remove cellular debris by washing intensively with 50 mM Tris–HCl, pH 7.5, 150 mM NaCl.

4. Block unspecific binding of antibodies by incubation with blocking buffer for 30 min at room temperature. Wash filters twice with T-TBS.

5. Dilute sera or monoclonal antibodies in antibody incubation buffer and preabsorb antibodies reactive to *E. coli* proteins by incubation for 30 min at room temperature with protein lysates of the *E. coli* host strains, e.g. DH1 [pRK248clts, pEX1], at a concentration of 200 μg *E. coli* protein ml^{-1}.

6. Incubate filters with preabsorbed antibody solution (5–10 ml/filter) for 16 h.

7. Wash twice with T-TBS.

8. Incubate with horse radish peroxidase (HRPO)-labeled secondary antibody in incubation buffer (preabsorbed as above if necessary).

9. Wash twice with T-TBS and twice with TBS to remove unbound antibodies.

10. Incubate filters in HRPO staining solution (5–10 ml/filter) to visualize bound HRPO-labeled antibody.

11. Pick positive clones from master plates using sterile tooth picks.

Production and purification of β-galactosidase fusion proteins

Small scale

1. Inoculate 1 ml L-broth containing 50 μg ml^{-1} ampicillin and grow overnight at 30°C.

2. Dilute 1:5 into fresh medium and incubate another hour at 30°C.

3. Divide into two 2.5 ml aliquots and incubate one aliquot for 1 h at 42°C and 2 h at 37°C, the other one for a further 3 h at 30°C.

4. Collect cells from induced and uninduced cultures by centrifugation.

5. Resuspend pellets in 100 μl of 40 mM Tris–HCl, pH 8, 5 mM EDTA, 0.3 mg ml^{-1} lysozyme. Add an equal volume of 20 mM Tris–HCl, pH 7.5, 20 mM MgCl$_2$, 10 μg ml^{-1} DNAaseI. Incubate 10 min, 0°C. Add 10 μl of 10% sodium deoxycholate.

6. For analysis by SDS-PAGE add 100 μl of urea gel buffer and incubate for 1 h at 37°C. Apply 10 to 20 μl of protein lysate per gel lane.

Larger scale

1. Grow *E. coli* clones to OD$_{600}$ = 0.3 in L-broth containing 50 μg ml^{-1} ampicillin at 30°C. Shift temperature to 42°C for 30 min. Incubate for a further 2 h at 37°C.

2. Collect cells by centrifugation.

3. Resuspend in 1/20 volume of 40 mM Tris–HCl, pH 8, 5 mM EDTA, 0.3 mg ml^{-1} lysozyme. Incubate 5 min at 0°C. Add an equal volume of 20 mM Tris–HCl, pH 7.5, 20 mM MgCl$_2$, 10 μg ml^{-1} DNAaseI. Incubate 5 min at 0°C. Lyse cells by sonication.

4. Pellet inclusion bodies by centrifugation. Redissolve fusion

protein from inclusion bodies in 8 M urea, 50 mM Tris–HCl, pH 8.0 and remove bacterial debris by centrifugation.

5. Dialyze against 50 mM Tris–HCl, 200 mM NaCl, pH 8.3. Clear dialyzate by centrifugation (8000 rpm, 5 min). Soluble fusion proteins can then be further purified by Sepharose 6B-CL chromatography in 100 mM Tris–HCl, pH 7.0, 100 mM NaCl, 0.1% SDS. Fractions containing fusion protein are pooled and concentrated and can be used for the preparation of antisera, for immunoblots, for ELISAs and for stimulation of T cells.

Discussion and troubleshooting

In most applications the protocols described above should enable establishment of a β-galactosidase ORF expression library, which allows immunological screening for B and T cell epitopes. The probability of expression of partial ORFs as fusion protein is of course also dependent on the kind of insert fragments used. cDNAs with 5' untranslated sequences most likely will not produce fusion protein and the mode of fragmentation and the size distribution will determine the probability of ORF expression from genomic DNA. Ligation conditions sometimes have to be optimized to avoid artificial joining of fragments.

In some cases, e.g. for the production of soluble, correctly folded antigens, heat induction may not be ideal. This problem can be partially overcome by reducing induction and production temperatures to 37°C and 30°C, respectively. Alternatively, use of the tightly regulated tetracycline resistance gene promoter, which is inducible by tetracycline at low temperature, may help to solve this problem (Skerra 1994). For purification and direct detection of fusion protein, enzymatically active fusion protein may be interesting. This can be achieved by reconstituting a complete β-galactosidase moiety. After attachment of the fusion partner at the C-terminal residue of an untruncated β-galactosidase, enzymatically active fusion proteins can be produced. However, only the soluble fraction seems to be active.

Expression of overlapping polypeptide fragments fused with maltose binding protein

Introductory remarks

The system for producing fusion proteins with maltose binding protein is one of several systems with similar properties. Most use an inducible promotor which is controlled by a repressor gene also encoded on the expression vector, by which expansion of the bacteria and expression of the fusion protein can be separated.

Expression of fusion protein in *E. coli*   767

The desired polypeptide is usually cloned into a small polylinker. In addition to the polylinker these vectors often contain an in-frame *lacZ* gene fragment which allows alpha complementation of β-galactosidase if the sequence is not interrupted by an insert. Thus this provides in principle the possibility of screening for successful cloning by testing for blue or white colonies when IPTG and X-Gal are included in the plates. However, most of the time this screening system is not useful, since induced bacteria often do not survive to be propagated further.

The desired polypeptide is cloned into the 3' part of the fusion partner since this requires only a single cloning step. Under this set-up the fusion partner provides all the necessary sites for expression. In addition, the polypeptide used as fusion partner provides stability and usually retains its property of binding to a particular ligand, which can be used for affinity purification.

We have concentrated on the pMAL-c originally developed by Guan et al (1987) and commercially distributed by BioLabs. The vector contains the *malE* gene of *E. coli* encoding the maltose binding protein containing a deletion of the signal peptide. This results in exclusive intracellular expression of the fusion protein. The construct is driven by the TAC promotor (Aman et al 1985) and repressed by the product of the *lacI* gene also encoded by the vector. Addition of IPTG to the medium results in expression of the fusion protein. The gene or gene fragment of interest is cloned into a small polylinker 3' of the maltose binding protein (Fig. 10.2.1). The sites indicated are in frame with the maltose binding protein, i.e. the first nucleotide of the restriction site also represents the first nucleotide of a codon. A factor Xa-sensitive site exists between the upstream *Bam*HI and *Stu*I sites which is removed if *Bam*HI is used for cloning. If *Stu*I is used with another blunt end enzyme, the fusion partner is precisely removed when cut with factor Xa. However, this was often found to be problematic and therefore it is not included in the described procedure. Similarly, in the unmanipulated vector a *lacZ* gene fragment is fused in frame with the maltose binding protein, which in principle allows α-complementation of β-galactosidase in appropriate strains, but might cause problems when used for screening (see above). Therefore, this is also not considered in this protocol. The fusion with the β-galactosidase fragment is abolished when the fragment of interest which is cloned into the polylinker provides its own stop codon.

Efficient translation start sites and termination sites are provided by the vector. This, however, generates the most serious problem of this expression system. Unfortunately, the TAC promoter is not completely inert in the noninduced state. A fusion protein might interfere with the growth of the bacteria containing this particular plasmid already at a very low expression level, often resulting in a deletion of the DNA fragment of interest. Several experimental conditions can be changed to diminish this problem; however, some fragments might be expressed with extreme difficulty. Nevertheless, good results are usually achieved using this vector.

To obtain the most information for epitope screening by fusion proteins, the polypeptide of interest is expressed as fragments that overlap, e.g. a protein of 300 amino acids would be expressed in two fragments of 200 amino acids overlapping by 100 amino acids. Thus the epitope of an antibody or a T cell which is positioned in the middle of the protein would reside on both, while an N-terminal or C-terminal epitope would be found on only one of the fusion proteins. The particular fragments are amplified by PCR, which allows the introduction of suitable restriction sites for forced

	BamHI		StuI	EcoRI		BamHI	XbaI	SalI	PstI		HindIII*		
malE -----			GGATCC	AGGCCTGAATTC		GGATCCTCTAGAGTCGACCTGCAG					AAGCTT	-----	lacZα

↑
factor Xa cleavage site

Figure 10.2.1 Polylinker of pMAL-c (not all sites are shown). *HindIII is not in frame.

cloning. Since the *Taq* polymerase sometimes introduces mutations into the amplified fragment, the construct should be completely sequenced, or at least two independently derived fusion proteins of the same type should be tested.

For basic information on molecular methods, see also Sambrook et al (1989) or Harlow and Lane (1988).

Materials

Bacteria

JM 109, BL 21

Plasmid

pMAL-c

Enzymes

Taq polymerase (Perkin Elmer)

T4 ligase (Gibco BRL)

Chemicals

Low-melting-point agarose (Gibco BRL)

Mineral oil ultrapure (Perkin Elmer)

dNTPs 100 mM (Pharmacia)

Diethylimine (Sigma)

Amylose resin (New England BioLabs)

IPTG (Sigma)

Media and plates

LB medium:

10 g	Bacto-Trypton
5 g	Bacto-Yeast extract
10 g	NaCl

Add H_2O to 1 liter.

Adjust pH to 7.0 with NaOH and autoclave.

Ampicillin $50 \mu g \, ml^{-1}$ final concentration is added just before use if required.

SOB medium:

20 g	Bacto-Trypton
5 g	Bacto-Yeast extract
0.5 g	NaCl

Dissolve with 950 ml H_2O.

Add 10 ml of 250 mM KCl and make up to 1 liter with H_2O

Adjust pH to 7.0; autoclave.

Add 10 ml of 1 M $MgCl_2$ and 10 ml of 1 M $MgSO_4$ before use.

SOC medium: (SOB medium containing 20 mM glucose)

Buffers and solutions

$10\times$ PCR buffer:

100 mM	Tris pH 8.3
500 mM	KCl
15 mM	$MgCl_2$
0.01%	gelatine

Sterilize using a $0.02 \mu m$ Millipore filter and store in aliquots at $+4°C$

$5\times$ Ligase buffer (supplied by the vendor of T4 ligase)

Lysis buffer:

10 mM	Na_2HPO_4
30 mM	NaCl
10 mM	β-mercaptoethanol
10 mM	EDTA
10 mM	EGTA

PMSF: 100 mM PMSF (Sigma) is dissolved in EtOH and stored at $-20°C$ not longer than 1 week.

Column buffer:

200 mM	NaCl
1 mM	EDTA
20 mM	Tris-Cl pH 7.4
1 mM	β-Mercaptoethanol

Elution buffer: column buffer containing 10 mM maltose.

Apparatus

PCR thermocycler
FPLC
French press sonicator

Protocols

PCR primer designing

Primers of at least 18 nucleotides are used. In addition, they contain one of the restriction sites shown in Fig. 10.2.1. These sites have been found to be easily cut when PCR products have to be cloned. Usually a few extra Gs or Cs are included 5' of the restriction site of the primer to insure that the site is found complete on the PCR product.

PCR

Take 1–100 ng of plasmid containing the gene of interest, add 100 μmol of each primer, 200 μmol of ATP, CTP, CiTP, TTP, 4.5 μl of 10× PCR buffer and make up to 45 μl with H$_2$O. Overlay with mineral oil. Heat mixture to 84°C before adding 5μl containing 2.5 U *Taq* polymerase in 1× PCR buffer and run 25–30 cycles of 30 s 94°C, 1 min 52°C, and 3 min 72°C. The hybridization conditions might have to be optimized by increasing or lowering the temperature. Check 5 μl of reaction mix on 0.8% agarose gel for successful amplification.

Propagation of PCR products

Remove mineral oil by addition of chloroform or freezing. Digest PCR product with the appropriate enyzmes. Separate digested PCR product from primer and small digestion products over 0.6% low-melting point (LMP) agarose. Isolate fragment from gel by heating for 10 min at 65°C and extraction two times with phenol and once chloroform. Concentrate by ethanol precipitation. Alternatively, use normal agarose and the Jet sorb kit (Genomed) or similar kits.

Preparation of pMAL-c

Digest pMAL-c with the appropriate restriction enzymes. Check successful digestion on 0.8% agarose gel. Separate on LMP gel (see above).

Ligation

50–100 ng of digested pMAL-c and 50–100 ng of PCR product (ratio of vector to PCR product ~1:10) are ligated in 20 μl with 1 U T4 ligase in 1× ligase buffer at room temperature for 30–60 min. This time is usually sufficient to obtain a good ligation. The ligation buffer might interfere with the transformation of the bacteria, therefore the ligation mix is dialyzed before use. Float a VS membrane (Millipore, 0.025 μm) on 2 ml H$_2$O. Pipette ligation mix onto the filter. Leave for 20–30 min. Remove ligation mix from the filter and use for transformation.

Preparation of transformation-competent bacteria

In principle most strains can be used. However, some differences in stability of the constructs have been observed and the strain commonly used in the laboratory might not be the best choice. We have good experience with JM 109 (Yanisch-Perron et al 1985) and BL 21 (Grodberg and Dunn 1988).

Preparation of electrocompetent cells

Inoculate 1000 ml LB medium with 10 ml of an overnight culture, shake at 180 rpm at 37°C until 0.75 OD_{600} is reached. Chill on ice for 15 min; centrifuge at 5000 rpm for 15 min in a GS3 rotor (Sorvall) at 0°C. Resuspend pellet in 1000 ml cold deionized or distilled H_2O. Centrifuge again and resuspend in 500 ml cold H_2O. Centrifuge again and resuspend pellet in cold 10% glycerol. Centrifuge for 15 min at 5000 rpm in SS34 (Sorvall); resuspend pellet in 3 ml cold 10% glycerol; distribute 50 μl aliquots in precooled tubes and snap-freeze in dry ice–ethanol or liquid nitrogen. Store at –70°C. Test efficiency by using 1 ng of pUC or Bluescript. 10^8–10^9 colonies should be obtained per μg of control plasmid.

Preparation of CaCl₂ competent cells

Alternatively to electrotransformation, cells can be treated with $CaCl_2$ to make them transformation-competent. Especially with BL 21 this method gave more reproducible results. Inoculate 100 ml of LB with 1 ml of an overnight culture and shake at 180 rpm at 37°C until 0.5 OD_{600} (for JM 109 0.7–0.8 OD_{600}) is reached. Centrifuge for 10 min at 3000 rpm in GS3 rotor; wash pellet twice in cold 100 mM $CaCl_2$ solution. Resuspend pellet in 7.5 ml cold 100 mM $CaCl_2$ plus 2.5 ml 50% glycerol. Snap-freeze in dry ice–ethanol or liquid nitrogen and store at –70°C.

Transformation

Electrotransformation

Thaw electrocompetent cells on ice. Add 1–2μl of ligation mix. Pipette immediately into a precooled electrotransformation cuvette (0.2 cm). Pulse at 25 μF 2.25 kV and 200 Ω. The resulting time constant should be between 4 and 5 ms. Resuspend electrotransformed cells immediately in 1 ml prewarmed SOC medium. Shake for 1 h at 37°C and plate on LB plates containing ampicillin.

Transformation with CaCl₂-competent cells

Thaw cells on ice. Take 300 μl of cells and mix with 10–20 μl ligation mix. (Do not use polystyrene tubes for transformation!) Incubate on ice 30–40 min. Incubate mixture for 2 min at 42°C (heat shock) and chill for 5 min on ice. Add 500 μl of LB medium and shake for 30–40 min at 37°C. Plate on LB plates containing ampicillin.

Minipreparation of plasmids for insert screening

Pick several colonies and grow them overnight at 37°C in 3–5 ml cultures in LB medium containing 50 µg ml^{-1} ampicillin. Use 1.5 ml of this culture for a plasmid minipreparation (Birnboim and Doly 1979; Ish-Horowicz and Burke 1981). Test for insert with appropriate restriction enzymes and save positive clones using part of the remaining culture by adding glycerol and storing at −70°C. This is very important because owing to instability of some clones, retransformation with isolated plasmids often results in clones with deletions.

Induction test

Cloning is usually very efficient therefore it is not always necessary to test for inserts in plasmid minipreparation, rather expression can be tested directly. In addition, minipreps of BL 21 often do not result in good plasmid preparation.

Inoculate 3 ml of LB containing ampicillin with 100 µl of a positive culture from the preceding protocol. Grow at 37°C to 0.5 OD$_{600}$. Take 1 ml, centrifuge for 30 s at 15,000 rpm and resuspend the pellet in SDS sample buffer (uninduced culture). To the remaining 2 ml add 0.1 M IPTG to obtain 0.3 mM final concentration and incubate for a further 2 h at 37°C. Take 0.5 ml, centrifuge for 20 s at 15,000 rpm and resuspend in 100 µl SDS sample buffer (induced culture). Heat both samples 5 min at 95°C and run 15 µl on a 10% SDS-PAGE including untransformed bacteria treated the same way. Stain gel with Coomassie Blue for 20 min. Destain by heating the gel in H$_2$O in a microwave oven for 2.5 min twice with an exchange of H$_2$O between. (The maltose binding protein–β-galactosidase fusion protein as found in the vector if no insert is present is of molecular weight 52 kDa. The molecular weight of the maltose binding protein is 42 kDa. An expressed fusion protein should therefore be correspondingly larger.) A

clear difference between the uninduced and the induced sample should be visible. Several clones should be tested to obtain a stable high-expressing clone. Prepare a maxiprep of the positive clones for sequencing and storage (but see preceding protocol).

Purification of fusion proteins

Inoculate 500 ml of LB containing ampillicin with 5 ml of an overnight culture. Shake with good aeration for ~2 h at 37°C until ~0.5 OD$_{600}$ is reached. Take a 1 ml sample and treat it as in the induction test. To the remaining culture add 1.5 ml of 0.1 M IPTG. Shake for another 2–4 h at 37°C. Take a 0.5 ml sample and treat it as in the induction test above. Centrifuge remaining culture for 10 min at 5000 rpm in GSA rotor (Sorvall). Resuspend pellet in 20–30 ml of lysis buffer. Freeze at 20°C. Thaw the bacteria rapidly in a 37°C water bath under constant motion. Add 30 µl of 0.1 M PMSF. Lyse bacteria by running them twice through a French Press at 16,000 psi. Sonicate suspension for 5 min on ice. Centrifuge at least twice until supernatant is clear. Keep pellet for analysis. To the supernatant add again 30 µl 0.1 M PMSF.

To precipitate DNA and RNA, add diethylimine to a final concentration of 0.1%. Incubate 10 min on ice. Centrifuge for 10 min at 10,000 rpm (SS34). Resuspend the pellet in 10 ml lysis buffer and keep for analysis. To the supernatant add, slowly on ice, solid ammonium sulfate to reach 50% saturation. Incubate 10 min on ice and centrifuge for 10 min at 10,000 rpm in SS34. Resuspend the pellet in 20 ml column buffer. Keep an aliquot for analysis.

The supernatant of the 50% AMS precipitation is further propagated. Keep an aliquot of the supernatant for analysis and add ammonium sulfate to the remaining solution to reach 60% saturation. Continue as for the 50% AMS precipitation.

The resuspended pellet of the 50% AMS precipitation is run over the affinity column. For reproducible results, purification is best achieved on a FPLC system with a 1.5×20 cm column containing amylose resin equilibrated in column buffer. The sample is applied via the superloop using 25 ml of column buffer. The column is then washed with 100 ml of column buffer and the fusion protein is eluted with 100 ml of elution buffer. Flow rate is kept constant at 1 ml min^{-1} through the whole run and 5 ml fractions are collected. 10 µl of each fraction is mixed with $2\times$ SDS sample buffer and analyzed on 10% SDS-PAGE. Fractions containing the fusion protein are pooled and dialysed versus PBS.

Regeneration of the amylose column

It is important to regenerate the column rather quickly because the resin can be destroyed by contaminating microorganisms.

The column is washed with 100 ml H$_2$O, 60 ml 0.1% SDS, 200 ml H$_2$O and 50 ml column buffer. For long-term storage the resin should be equilibrated in 20% ethanol.

Discussion and trouble shooting

When protein fragments are expressed as fusion proteins using the method described, in our experience about 50% can be expressed in JM 109 without any problem. Plating and the culture can be performed at 37°C and the column runs at room temperature without any obvious degradation. For the remaining 50%, conditions have to be optimized. In the following some suggestions are given.

No PCR product: Optimize hybridization conditions, design new primers hybridizing at a slightly different position.

No insert in plasmid minipreparation, no fusion protein in induction assay: Switch to BL 21; grow minicultures at 30°C or 28°C; design new primers at a slightly different position; use primers that result in a smaller fragment.

No fusion protein eluted from the column: Samples have been taken at all steps — analyze by SDS-PAGE where the fusion protein was lost. Up to now we have mainly lost the fusion protein during the upscaling, i.e. fusion protein was present in the induction assay but not in the 500 ml culture. To avoid loss during upscaling, grow bacteria at 30°C or 28°C. Under our conditions we have never found the fusion protein in inclusion bodies. However, if inclusion bodies are formed, the protein will remain in the insoluble fraction after lysis of the bacteria. Switch strain, or grow at lower temperature, or use lower IPTG concentrations for induction. Alternatively, isolate the fusion protein from the inclusion bodies (see earlier).

References

Amann E (1985) Plasmid vectors for the regulated, high level expression of eukaryotic genes in *Escherichia coli*. Dev Biol Stand 59: 11–22.

Amann E, Brosius J, Ptashne M (1983) Vectors bearing a hybrid trp-lac promoter useful for regulated expression of cloned genes in *Escherichia coli*. Gene 25: 167–178.

Banting G, Luzio JP, Braghetta P, Brake B, Stanley KK (1991) pUBEX/pUBSEX: A versatile expression vector system for production of fusion and non-fusion proteins in *Escherichia coli*. Gene 107: 127–132.

Bernard HU, Helinski DR (1980) Bacterial plasmid cloning vehicles. In Setlow JK, Hollaender A, eds. Genetic Engineering — Principles and Methods. Plenum Press, New York, pp 133–167.

Birkelund S, Larsen B, Holm A, Lundemose AG, Christianson G (1994) Characterization of a linear epitope on *Chlamydia trachomatis* Serovar L2 DnaK-like protein. Infect Immun 62: 2051–2057.

Birnboim HC, Doly J (1979) A rapid alkaline extraction procedure for screening recombinant plasmid DNA. Nucleic Acids Res 7: 1513–1523.

Botterman J, Zabeau M (1987) A standardized vector system for manipulation and enhanced expression of genes in *Escherichia coli*. DNA 6: 583–591.

Ellinger S, Glockshuber R, Jahn G, Pluekthun A (1989) Cleavage of procaryotically expressed human immunodeficiency virus fusion proteins by factor X_a and application in Western blot (immunoblot) assays. J Clin Microbiol 27: 971–976.

Ellis RW, Keller PM, Lowe RS, Zivin RA (1985) Use of a bacterial expression vector to map the Varicella-Zoster virus major glycoprotein gene, gC. J Virol 53: 81–88.

Gray MR, Colt HV, Guarente L, Rosbash M (1982) Open reading frame cloning: identification, cloning, and expression of open reading frame DNA. Proc Natl Acad Sci USA 79: 6598–6602.

Grodberg J, Dunn JJ (1988) *ompT* encodes the *Escherichia coli* outer membrane protease that cleaves T7 RNA polymerase during purification. J Bacteriol 170: 1245.

Guan C, Li P, Riggs PD, Inouye H (1987) Vectors that facilitate the expression and purification of foreign peptides in *Escherichia coli* by fusion to maltose-binding protein. Gene 67: 21–30.

Haffey ML, Lehman D, Boger J (1987) Site specific cleavage of a fusion protein by renin. DNA 6: 565–571.

Harlow E, Lane D (1988) Antibodies — A Laboratory Manual. Cold Spring Harbor Laboratory Press, Cold Spring Harbor, NY.

Ish-Horowicz D, Burke JF (1981) Rapid and efficient cosmid cloning. Nucleic Acids Res 9: 2989–2998.

Jahn G, Kouzarides T, Mach M, et al (1987) Map position and nucleotide sequence of the gene for the large structural phosphoprotein of human cytomegalovirus. J Virol 61: 1358–1367.

Knobloch J, Schreiber M, Grokllovsky S, Scherf A (1987) Specific and nonspecific immunodiagnostic properties of recombinant and synthetic *Plasmodium falciparum* antigens. Eur J Clin Microbiol 6: 547–550.

Koenen M, Rüther U, Müller-Hill B (1982) Immunoenzymatic detection of expressed gene fragments cloned in the *lacZ* gene of *E. coli*. EMBO J 4: 509–512.

Landini MP, Guan MX, Jahn G, et al (1990) Large-scale screening of human sera with cytomegalovirus recombinant antigens. J Clin Microbiol 28: 1375.

Lindenmaier W, Necker A, Krause S, Bonewald R, Collins J (1990) Cloning and characterization of major antigenic determinants of human cytomegalovirus Ad169 seen by the human immune system. Arch Virol 113: 1–16.

Maina CV, Riggs PD, Grandea AG III, et al (1988) An *Escherichia coli* vector to express and purify foreign proteins by fusion to and separation from maltose-binding protein. Gene 74: 365–373.

Mehra V, Sweetser D, Young RA (1986) Efficient mapping of protein antigenic determinants. Proc Natl Acad Sci USA 83: 7013–7017.

Nagai K, Thøgerson C (1984) Generation of β-globin by sequence specific proteolysis of a hybrid protein produced in *E. coli*. Nature 309: 310.

Peterhans A, Mecklenburg M, Meussdoerfer F, Mosbach K (1987) A simple competitive enzyme linked immunosorbent assay using antigen-β-galactosidase fusions. Anal Biochem 163: 470–475.

Plachter B, Klages S, Hagelmann S, Britt W, Landini MP, Jahn G (1990) Procaryotic expression of phosphorylated tegument protein pp65 of human cytomegalovirus and application of recombinant peptides for immunoblot analyses. J Clin Microbiol 28: 1229–1235.

Pohlner J, Krämer J, Meyer TF (1993) A plasmid system for high-level expression and *in vitro* processing of recombinant proteins. Gene 130: 121–126.

Ripalti A, Ruan Q, Boccuni MC, Campanini F, Bergamini G, Landini MP (1994) Construction of polyepitope fusion antigens of human cytomegalovirus ppUL32: reactivity with human antibodies. J Clin Microbiol 32: 358–363.

Rüther U, Müller-Hill B (1983) Easy identification of cDNA clones. EMBO J 2: 1791.

Sambrook J, Fritsch EF, Maniatis T (1989) Molecular Cloning. A Laboratory Manual. Cold Spring Harbor Laboratory Press, Cold Spring Harbor, NY.

Sassenfeld HM (1990) Engineering proteins for purification. Trends Biotechnol 6: 88.

Scholl B-C, von Hintzenstern J, Borisch B, Traupe B, Broker M, Jahn G (1988) Prokaryotic expression of immunogenic polypeptides of the large phosphoprotein (pp150) of human cytomegalovirus. J Gen Virol 69: 1195–1204.

Scholtissek S, Grosse F (1988) A plasmid vector system for the expression of a triprotein consisting of beta-galactosidase, a collagenase recognition site and a foreign gene product. Gene 62: 55–64.

Scott MO, Kimelmann D, Norris D, Picciaci RP (1984) Production of a monospecific antiserum against the early region 1 A proteins of adenovirus 12 and adenovirus 5 by an adenovirus 12 early region 1A-β-galactosidase fusion protein antigen expressed in bacteria. J Virol 50: 895.

Skerra A (1994) Use of the tetracycline promoter for the tightly regulated production of a murine antibody fragment. Gene 151: 131–135.

Stanley KK (1983) Solubilization and immune detection of β-galactosidase hybrid proteins carrying foreign antigenic determinants. Nucleic Acids Res 11: 4077–4092.

Stanley KK, Luzio JP (1984) Construction of a new family of high efficiency bacterial expression vec-

tors: identification of cDNA clones coding for human liver proteins. EMBO J 3(6): 1429–1434.

Ullmann A (1984) One-step purification of hybrid proteins which have β-galactosidase activity. Gene 29: 27–31.

Ullmann A, Danchin A (1994) Role of cyclic AMP in regulatory mechanisms in bacteria. Trends Biochem Sci 1980(4): 95–96.

Verweij CL, de Vries CJM, Distel B, et al (1985) Construction of cDNA coding for human van Willebrand factor using antibody probes for colony screening and mapping of the chromosomal gene. Nucleic Acids Res 13: 4699.

Watson RJ, Weis JH, Salstrom JS, Enquist LW (1983) Bacterial Synthesis of herpes simplex virus type 1 and 2 glycoprotein D antigen. J Invest Dermatol 83: 102s–111s.

Weinstock GM, ap Rhys C, Berman ML, et al (1983) Open reading frame expression vectors: a general method for antigen production in *Escherichia coli* using protein fusions to beta-galactosidase. Proc Natl Acad Sci USA 80: 4432–4436.

Weis JH, Enquist LW, Salstrom JS, Watson RJ (1983) An immunologically active chimaeric protein containing herpes simplex virus type 1 glycoprotein D. Nature 302: 72–74.

Yanisch-Perron C, Vieira J, Messing J (1985) Improved M13 phage cloning vectors and host strains: nucleotide sequence of the M13mp18 and pUC18 vectors. Gene 33: 103–119.

Young RA, Davies RW (1983) Efficient isolation of genes by using antibody probes. Proc Natl Acad Sci USA 80: 1194–1198.

Expression of fusion protein in *E. coli*

Principles of simultaneous synthesis of multiple peptides

10.3

Ronald Frank

Department of Cell Biology and Immunology, GBF National Research Center for Biotechnology, Braunschweig, Germany

TABLE OF CONTENTS

Introduction

Substantial improvements in peptide chemistry, solid phase synthesis techniques and process automation during the past two decades have made access to synthetic peptides for immunologists relatively easy. Several companies offer custom synthesis at reasonable prices and extend their services even to the production of anti-peptide antisera and monoclonal antibodies. In particular, for the systematic analysis of immunogenic determinants in protein antigens with series of peptide fragments and substitution analogues, large numbers of different peptide sequences are required. For their preparation, adequate simultaneous multiple synthesis techniques have been developed. Moreover, the perfection and simplification of peptide synthesis chemistry led to the establishment of routine synthesis protocols which do not require sophisticated expensive instrumentation and can be followed with quite good success even by researchers not particularly trained in chemistry. A selection of such techniques and their application to epitope analysis is given in this section.

Chemical assembly of peptide sequences

Modern methods of peptide synthesis exclusively follow the solid phase synthesis (SPS) principle developed by Merrifield (1963). Essential steps in the chemical assembly of a peptide chain by successive addition of suitably protected amino acid building units (B_n) are outlined in Fig. 10.3.1. The general composition of ▶-B'- here is X-AA(Y)-OH where X is the temporary protecting group for the amino function, AA is one of the α-L-amino acids, Y is a permanent protecting group for the side-chain function of a trifunctional AA, and -OH is the free carboxyl function. Via this function the first unit (B_1) is coupled to a reactive anchor group on the surface of an insoluble polymeric support material. The anchor ⎌ may be a spacer arm (noncleavable) or a linker arm (cleavable). The terminal amino protecting group is then removed (step 1, deblocking) and the liberated amino function is coupled with the preactivated free carboxyl function of the next building unit B_2 (step 2, coupling). Nonreacted functions of B_1 may be blocked in a 'capping reaction', e.g. acetylation, to avoid formation of deletion peptides during the following elongation cycle. Repeating these two steps, the peptide chain is assembled from C- to N-terminus. The final treatments remove all permanent protecting groups and leave the peptide immobilized via the spacer arm on the support or additionally cleave the linker arm to release the peptide into solution.

The classical 'Merrifield' synthesis utilizes the Boc (*tert*-butyloxylcarbonyl) group for temporary amino protection, which is removed with 20–50% trifluoroacetic acid (TFA) in dichloromethane and benzyl-type permanent side-chain protecting groups which are sufficiently stable to TFA but have to be removed after peptide assemblies with very strong acids such as liquid hydrogen fluoride (HF). The latter is an extremely aggressive reagent and requires special equipment for its handling. A much milder and safer chemistry utilizes the Fmoc (9-fluorenylmethoxycarbonyl) group for temporary amino protection, which is removed with

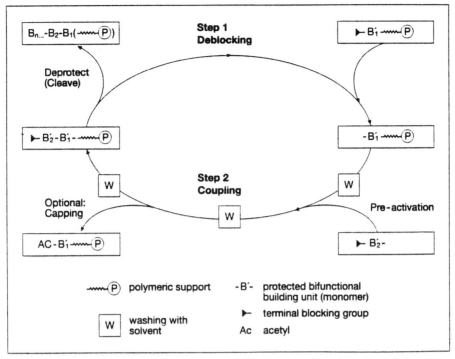

Figure 10.3.1 Essential steps in the chemical assembly of a peptide chain by successive addition of suitably protected amino acid building units (B_n).

organic bases such as piperidine in dimethylformamide. Consequently, the side-chain-protecting groups can be chosen for easy cleavage by mild acids such as TFA and are of a tert-butyl type (reviewed by Fields and Noble 1990).

The classical support material consists of beads of cross-linked polystyrene which is chemically modified to carry reactive groups for the anchoring of the growing peptide chain (Merrifield 1963). The support beads are placed into a suitable reaction chamber (reactor), solvents and reagents are fed in either manually or automatically and are removed through a bottom sinter plate. A variety of manual and automated devices are available that can perform one to almost 200 syntheses in parallel (reviewed by Jung and Beck-Sickinger 1992). Several alternative reactor designs, support materials and formats are also employed, particularly to devise special multiple synthesis techniques (Table 10.3.1). Of these, experimental details for Spots, Multipin, T-Bag, and Multiple Column Synthesis are given in this section.

The chemical nature of the bond that links the peptide chain to the support, and hence its stability towards different reagents, can be selected by insertion of so-called linker compounds or handles. By this means a variety of different C-terminal peptide modifications such as free acid or amide as well as special cleavage strategies can be realized (Atherton et al 1979). In particular, the synthesis and handling of large numbers of different peptide preparations is greatly facilitated by the use of 'safety-catch linkers' as proposed by Bray et al (1990). Such linkers allow the final cleavage of protecting groups and removal of chemicals while the peptides remain attached to the solid support followed by the direct release into an aqueous buffer solution for use in bioassays.

Synthetic peptides can be applied in different ways for presentation in immunological assay systems. B cell epitope analysis, in principle, is the identification of peptide ligands to antibodies. Complex formation may be assayed by competition with a labeled antigen or by affinity trapping of the antibody with the peptides immobilized on a solid carrier. This carrier may

Table 10.3.1   Features of simultaneous multiple peptide synthesis techniques

Method	Support material	Technique[a]	Reference
Multipin	Grafted polyethylene pin	Parallel; 8×12 arrays of pins; pin heads fitting into the wells of a Teflon microtiter plate	Geysen et al (1984)
T-bag	Conventional resin beads sealed in porous plastic bags	Multiple; bags are grouped into reaction flasks	Houghten (1985)
Filter	Cellulose paper or cotton fabric	Multiple; disks or clippings are grouped into reaction flasks or columns	Frank and Döring (1988) Eichler et al (1991)
ABACUS	Conventional resin beads sealed in porous Teflon wafers	Multiple; wafers are grouped into reaction columns	Beattie and Frost (1987)
VLSIPS	Planar glass surface	Photolithographic addressable reaction sites	Fodor et al (1991)
Spots	Cellulose paper	Parallel; in situ formed arrays of spot reactors	Frank (1992)
Pilot	Coated polyethylene sinter disks (winks)	Parallel; 10×10 array of winks in a holder plate	Cass et al (1994)

[a]Parallel: coupling reactions are carried out by distributing activated amino acid derivates to an array of positionally fixed, individual reactors. Multiple: segmented support materials are sorted into groups that are reacted with the same activated amino acid derivatives in common reaction vessels.

be the synthesis support itself, such as beads (Smith et al 1977), pins (Geysen et al 1984), or spots on paper (Frank 1992). Alternatively, soluble peptides can be presented on the surface of the wells of a microtiter plate by different means: direct coating, covalent attachment to a chemically activated surface, coupling to a carrier protein such as BSA followed by coating the conjugate, and last but not least binding to a coated or covalently attached tag-binding protein such as streptavidin if the peptide is N- or C-terminally modified with a tag such as biotin. A variety of immunoassay configurations are described by Harlow and Lane (1988). The use of peptides in screening for T cell epitopes requires intact effector cell (T cells, antigen-presenting cells, CTLs) and, thus, soluble peptides are a prerequisite.

Systematic peptide screening approaches in epitope analysis include the following.

- Localization in a protein sequence with series of overlapping peptide fragments (Kazim and Atassi 1980; Geysen et al 1984; Van der Zee et al 1989). Only linear but not conformational epitopes are accessible by this strategy.
- Boundary analysis to determine the minimal epitope structure with series of stepwise truncated peptides.
- Substitution analysis to determine the contribution of individual amino acid residues to epitope function (Geysen et al 1984; Houghten 1985; Evavold et al 1993).
- *A priori* delineation utilizing modern peptide library techniques (Geysen et al 1986); without knowing the protein antigen itself, epitope peptides can be selected from complete random peptide pools; according to the mimotope approach, peptide ligands to antibodies may be identified that can mimick a conformational (discontinuous) epitope.

The following chapters will present some alternative synthetic methods for the convenient preparation of large numbers of different peptide sequences and their application in epitope screening.

References

Atherton E, Gait MJ, Sheppard RC, Williams BJ (1979) The polyamide method of solid phase peptide and oligonucleotide synthesis. Bioorg Chem 8: 351–370.

Beattie KL, Frost JD (1987) Porous wafer for segmented synthesis of biopolymers. US Patent 5,175,209.

Bray AM, Maeji NJ, Geysen HM (1990) The simultaneous multiple production of solution phase peptides; assessment of the Geysen method of simultaneous peptide synthesis. Tetrahedron Lett 31: 5811–5814.

Cass R, Dreyer ML, Giebel LB, et al (1994) Pilot, a new peptide lead optimization technique and its application as a general library method. In Hodges RS, Smith JA, eds. Peptides: Chemistry Structure and Biology. ESCOM Science Publishers, Leiden, pp. 975–977.

Eichler J, Bienert M, Stierandova A, Lebl M (1991) Evaluation of cotton as a carrier for solid-phase peptide synthesis. Peptide Res 4: 296–307

Evavold BD, Sloan-Lancaster J, Allen PM (1993) Tickling the TCR: selective T-cell functions stimulated by altered peptide ligands. Immunol Today 14: 602–609.

Fields GB, Noble RL (1990) Solid phase peptide synthesis utilizing 9-fluorenylmethoxycarbonyl-amino-acids. Int J Peptide Protein Res 35: 161–214.

Fodor SPA, Read JL, Pirrung MC, Stryer L, Lu AS, Solas D (1991) Light-directed, spatially addressable parallel chemical synthesis. Science 251: 767–773.

Frank R (1992) Spot-synthesis: an easy technique for the positionally addressable, parallel chemical synthesis on a membrane support. Tetrahedron 48: 9217–9232.

Frank R, Doring R (1988) Simultaneous multiple peptide synthesis under continuous flow conditions on cellulose paper disks as segmental solid supports. Tetrahedron 44: 6031–6040.

Geysen HM, Meloen RH, Barteling SJ (1984) Use of peptide synthesis to probe viral antigens for epitopes to a resolution of a single amino acid. Proc Natl Acad Sci USA 81: 3998–4002.

Geysen HM, Rodda SJ, Mason TJ (1986) A priori delineation of a peptide which mimics a discontinuous antigenic determinant. Mol Immunol 23: 709–715.

Harlow E, Lane D (1988) Antibodies: A Laboratory Manual. Cold Spring Harbor Laboratory Press, Cold Spring Harbor, NY.

Houghten RA (1985) General method for the rapid solid-phase synthesis of large numbers of peptides: specificity of antigen–antibody interaction at the level of individual amino-acids. Proc Natl Acad Sci USA 82: 5131–5135

Jung G, Beck-Sickinger A (1992) Methoden der multiplen Peptidsynthese und ihre Anwendungen. Angew Chem 104: 375–391.

Kazim AL, Atassi MZ (1980) A novel and comprehensive synthetic approach for the elucidation of protein antigenic structures. Biochem J 191: 261–264.

Merrifield RB (1963) Solid phase peptide synthesis. I. The synthesis of a tetrapeptide. J Am Chem Soc 85: 2149–2154.

Smith JA, Hurrell JGR, Leach SJ (1977) A novel method for the delineating antigenic determinants: peptide synthesis and radioimmunoassay using the same solid support. Immunochemistry 14: 565–568.

Van der Zee R, Van Eden W, Meloen RH, Noordzij A, Van Embden JDA (1989) Efficient mapping and characterization of a T-cell epitope by the simultaneous synthesis of multiple peptides. Eur J Immunol 19: 43–47.

Multiple peptide synthesis with SPOT technique

10.4

Ronald Frank

Department of Cell Biology and Immunology, GBF National Research Center for Biotechnology, Braunschweig, Germany

TABLE OF CONTENTS

Immunology Methods Manual
ISBN 0–12–442712–X

Abstract

SPOT synthesis is a facile and very flexible technique for the simultaneous parallel chemical peptide synthesis on membrane supports. The method gives immunologists rapid and low cost access to large numbers of peptides both as solid phase bound and solution phase products for systematic epitope analysis.

Principles of the method

Upon dispensing small droplets of liquid onto a predefined array of positions on a porous membrane, the droplets are absorbed and form individual separate spots (Fig. 10.4.1). Using a solvent of low volatility containing appropriate reagents, such spots can form open reactors for chemical conversions involving reactive functions anchored to the membrane matrix, e.g. conventional solid phase synthesis. A great number of distinct spots can be arranged on a large membrane sheet and each of these is individually addressable by manual or automated delivery of the corresponding reagent solutions. The volume dispensed and the absorptive capacity of the membrane determine the spot size. According to the specific functionality of the matrix, the spot size correlates with the particular scale of the synthesis. The spot size also controls the minimal distance between spot positions and thereby the maximum density of the array. Synthetic steps common to all spot reactors are carried out by treating the whole membrane with respective reagents and solvents.

Chemical and technical performance of this type of simultaneous parallel solid phase synthesis has so far been optimized for the assembly of arrays of peptide sequences up to a length of 20 residues utilizing conventional Fmoc/tBu chemistry (Frank et al 1991; Frank 1992). The membrane supports are of specially selected, pure cellulose chromatography paper and are chemically derivatized to carry spots of dipeptide anchors for the preparation of either immobilized (βAla-βAla-anchor, entry 1 in Table 10.4.1) or solution phase peptides (Lys-Pro-anchor, entry 2 in Table 10.4.1). Novel types of safety-catch linkers have recently been developed by us to cleave also C-terminally unmodified peptide acids or amides directly into neutral aqueous buffers. Arrays of spots providing suitable anchor functions for peptide assembly on cellulose membranes are most easily generated by a two-step procedure including first the preparation of 'amino'-paper through esterification of an αN-Fmoc protected amino acid to available hydroxyl functions on the cellulose fibers of the whole sheet followed by Fmoc-cleavage and, second, spot-wise

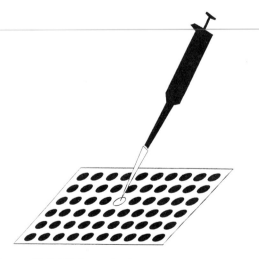

Figure 10.4.1 Schematic diagram showing the principle of SPOT synthesis.

Table 10.4.1   Orthogonal safety-catch linkages and anchors used with SPOT synthesis

Entry	Anchor/linker[a]	Peptide product	Reference
1	peptide-βAla-βAla—COO-resin	Immobilized via C-terminus	Frank (1992)
2	(pyrrolidine)N–COO-resin / CO / peptide-NH-(CH$_2$)$_4$–NH-Boc	Soluble, C-terminally modified with diketopiperazine	Bray et al (1990)
3	H, COOC(CH$_3$)$_3$ / N / N / R / (CH$_2$)$_n$ / peptide-O–CO-βAla-resin	Soluble, C-terminal acid	Hoffmann and Frank (1994)
4	O–R^1 / peptide-NH–CO-βAla-resin	Soluble, C-terminal amide	Hoffmann and Frank (1996), submitted
5	S–R^2 / peptide-NH–CO-βAla-resin	Soluble, C-terminal amide (optional release after biological assay)	Hoffmann and Frank (1996), submitted
6	COO-TBDMS / (CH$_2$)$_m$ / peptide-NH–CO-βAla-resin	Immobilized, cyclic[b]	Kramer et al (1994)

[a]$n = 0$ (R = H, CH$_3$),1 (R = H); $m = 1,2$; R^1 = methyl, isopropyl, *tert*-butyl, *tert*-butyldimethylsilyl (TBDMS), trimethylsilylethoxymethyl (SEM), methoxymethyl (MOM), methoxyethoxymethyl (MEM); R^2 = ethyl, isopropyl, *tert*-butyl, trityl.
[b]After selective removal of TBDMS and cyclization.

coupling of another Fmoc-amino acid or suitable linker compound (Table 10.4.1). During the second derivatization step, the array of spot reactors is generated and all residual amino functions between spots are blocked by acetylation. This array formation step in particular requires very accurate pipetting. During peptide assembly, slightly larger volumes are dispensed and the spots then formed exceed those initially formed, which helps to avoid incomplete couplings at the edges.

SPOT synthesis is particularly flexible with respect to numbers and scales that can be accomplished. The arrays are freely selectable to fit the individual needs of the experiment by variation of paper quality, thickness, specific anchor loading, and spot size (Table 10.4.2). The standard format used in manual SPOT synthesis was adapted to the 8 × 12 array of a microtiter plate with 96 spots. Recently, an automated SPOT synthesizer has been developed at ABIMED Analysen-Technik GmbH (Langenfeld, Germany) based on a Gilson pipetting workstation. This instrument can handle simultaneously up to six standard membrane sheets. Moreover, automated spotting

Table 10.4.2 Standard SPOT-array configurations (each fitted to a sheet of the size of a microtiter plate, 8 × 12 cm).

Format	Membrane type[a]	Anchor[b] (µmol/cm²)	Spotted Volume[c] (µl)	Spot size (mm)	Positional distance (mm)	Synthesis scale[d] (nmol)
8 × 12=96	540	0.2–0.4	0.5/0.7	7	9	25
7 × 10=70	Chrl	0.4–0.6	1.0/1.5	8	10	50
17 × 25=425	540	0.2–0.4	0.1/0.15	3	4	6
40 × 50=2000	50	0.2–0.4	0.03/0.05	1	2	1

[a]Chromatography paper products from Whatman, Maidstone, UK.
[b]Typical derivatization with first β-alanine.
[c]Volumes are given for array generation step/peptide assembly step.
[d]Mean values.

can be exploited to reduce the size of spots and, thus, increase the number per area considerably (Table 10.4.2), which otherwise is extremely tedious to achieve by hand. Presently, up to 2500 spots can be generated on a microtiter-plate-sized sheet. The instrument so far performs only the pipetting work; all washing steps are carried out manually.

Free amino functions on the spots can be stained with bromophenol blue (Krchnák et al 1988) prior to the coupling reactions. This allows visual monitoring of proper performance of all synthesis steps such as correct dispensing, quantitative coupling and capping, and effective removal of piperidine from the Fmoc-deblocking steps. A standard membrane for SPOT synthesis displays an array of light blue spots on a white background. Each spot can be marked by writing a number with pencil next to it (Fig. 10.4.2). These numbers refer to the corresponding peptide sequences that are assembled on them and are a guide for rapid manual distribution of the solutions of activated amino acid derivatives at each elongation cycle. An example of a corresponding, convenient pipetting protocol generated with a computer is shown in Fig. 10.4.3. For automated pipetting no pencil marking is necessary. Exact fixing of the membranes is assured by the perforation for the holder pins. The dry membranes are placed in a flat, chemically inert trough or fixed on the platform of the synthesizer. As soon as the droplets of activated amino acid solutions are added to the spots,

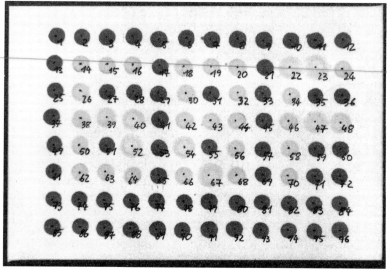

Figure 10.4.2 A microtiter plate adapted format of a manually prepared 96-spot membrane. Dark spots are blue, light spots have turned to yellow after coupling with an Fmoc-amino acid HOBt ester.

coupling proceeds with a conversion of free amino groups to amide bonds. After all amino groups have been consumed, the blue color of the spots changes to yellow, thus indicating a quantitative reaction. Because the solvent within the spots is slowly evaporating over the reaction time, additional drops may be added to the same position without enlarging the spots and risking overlap with their neighbors. In this way difficult coupling reactions may also be brought to completion by double or triple couplings.

```
SPOTscan - Wed May 06 18:29:42 1992

Sequence name: cmv26.seq
Sequence length: 58
Peptide length: 6
Offset: 1
53 spots on 1 membrane

Amino acid usage:
   A=12 D=24 E=8 G=137 H=7 I=1 K=11
   L=42 M=12 N=18 P=12 R=10 S=24

Spot  Mol.Wt.  Peptide sequence
----  -------  ----------------
  1     658    IEGRGK
  2     632    EGRGKS
  3     659    GRGKSR
  4     659    RGKSRG
  5     560    GKSRGG
  6     560    KSRGGG
  7     489    SRGGGG
  8     459    RGGGGG
  9     360    GGGGGG
 10     360    GGGGGG
 11     390    GGGGGS
 12     446    GGGGSL
 13     476    GGGSLS
 14     506    GGSLSS
 15     562    GSLSSL
 16     576    SLSSLA
 17     603    LSSLAN
 18     561    SSLANA
 19     531    SLANAG
 20     501    LANAGG
 21     501    ANAGGL
 22     567    NAGGLH
 23     568    AGGLHD
 24     612    GGLHDD
 25     612    GLHDDG
 26     652    LHDDGP
 27     596    HDDGPG
 28     572    DDGPGL
 29     572    DGPGLD
 30     571    GPGLDN
 31     629    PGLDND
 32     645    GLDNDL
 33     719    LDNDLM
 34     720    DNDLMN
 35     734    NDLMNE
 36     717    DLMNEP
 37     733    LMNEPM
 38     677    MNEPMG
 39     659    NEPMGL
 40     602    EPMGLG
 41     530    PMGLGG
 42     546    MGLGGL
 43     472    GLGGLG
 44     472    LGGLGG
 45     416    GGLGGG
 46     416    GLGGGG
 47     416    LGGGGG
 48     360    GGGGGG
 49     360    GGGGGG
 50     360    GGGGGG
 51     431    GGGGGK
 52     502    GGGGKK
 53     582    GGGKKH
```

```
cmv26.seq - Pipetting Schedule for Cycle 1
A: 16 18
D: 23 24 29 31
E: 35
G: 4 5 6 7 8 9 10 19 20 25 27 38 40 41 43 44 45 46 47 48 49 50
H: 22 53
K: 1 51 52
L: 12 15 21 28 32 39 42
M: 33 37
N: 17 30 34
P: 26 36
R: 3
S: 2 11 13 14


cmv26.seq - Pipetting Schedule for Cycle 2
A: 17 19
D: 24 25 30 32
E: 36
G: 1 5 6 7 8 9 10 11 20 21 26 28 39 41 42 44 45 46 47 48 49 50 51
H: 23
K: 2 52 53
L: 13 16 22 29 33 40 43
M: 34 38
N: 18 31 35
P: 27 37
R: 4
S: 3 12 14 15
```

⬇

```
cmv26.seq - Pipetting Schedule for Cycle 5
A: 20 22
D: 27 28 33 35
E: 1 39
G: 2 4 8 9 10 11 12 13 14 23 24 29 31 42 44 45 47 48 49 50 51 52 53
H: 26
K: 5
L: 16 19 25 32 36 43 46
M: 37 41
N: 21 34 38
P: 30 40
R: 3 7
S: 6 15 17 18


cmv26.seq - Pipetting Schedule for Cycle 6
A: 21* 23*
D: 28* 29* 34* 36*
E: 2* 40*
G: 3* 5* 9* 10* 11* 12* 13* 14* 15* 24* 25* 30* 32* 43* 45* 46* 48* 49* 50*
   51* 52* 53*
H: 27*
I: 1*
K: 6*
L: 17* 20* 26* 33* 37* 44* 47*
M: 38* 42*
N: 22* 35* 39*
P: 31* 41*
R: 4* 8*
S: 7* 16* 18* 19*
```

Figure 10.4.3 Some features of a SPOT-synthesis protocol form for rapid manual performance. Each peptide in the list is coded by an arabic number. In this example, 53 overlapping hexapeptide sequences with a shift (offset) of one amino acid residue were generated to scan the 58 amino acid residue CMV26 protein sequence. Only peptide code numbers have to be followed for the distribution of amino acid derivatives during the assembly cycles. Numbers marked with an asterisk indicate peptide chains that are finished after the respective cycle.

Multiple peptide synthesis with spots

Table 10.4.3 Strategies for the deconvolution of individual sequences by activity screening of random pools[a]

1. Iterative search starting with one or more defined positions[b]

first screen	X-X-1-2-X-X	400 pools (each 160,000 components)
second screen	X-1-0_3-0_4-2-X	400 pools (each 400 components)
third screen	1-0_2-0_3-0_4-0_5-2	400 pools (each 1 components)

2. Positional scanning with single fixed positions[c], one single screen

1-X-X-X-X-X	20 pools (each 3,2 Mio. components)
X-1-X-X-X-X	20 pools (each 3,2 Mio. components)
X-X-1-X-X-X	20 pools (each 3,2 Mio. components)
X-X-X-1-X-X	20 pools (each 3,2 Mio. components)
X-X-X-X-1-X	20 pools (each 3,2 Mio. components)
X-X-X-X-X-1	20 pools (each 3,2 Mio. components)

3. Dual-positional scanning[d], one single screen

1-2-X-X-X-X	400 pools (each 160,000 components)
X-1-2-X-X-X	400 pools (each 160,000 components)
X-X-1-2-X-X	400 pools (each 160,000 components)
X-X-X-1-2-X	400 pools (each 160,000 components)
X-X-X-X-1-2	400 pools (each 160,000 components)

[a]Special codes to describe the pool compositions are: 0_n=unvaried position in a particular screen occupied by single amino acid residues; 1,2,3,…= positions systemically varied by single amino acid residues in a particular screen; X = position occupied by a set of (e.g. all 20 L-) amino acid residues.

[b]'Mimotope' approach (Geysen et al 1986).

[c]With only 120 pools of hexapeptides, acceptor preferences for certain amino acid residues at all positions are obtained (consensus sequence); sequences of individual active sequences have to be identified by synthesizing and testing all possible combinations of hits from this 'positional scanning library' screen (Dooley and Houghten 1993).

[d]With 2,000 pools of hexapeptides, acceptor preference for certain dipeptide combinations at all positions are obtained; from matching overlaps of these, sequence connectivities can be directly delineated (Frank et al 1995).

Any series of individual peptide sequences can be freely arranged as a two-dimensional array on a SPOT membrane for systematic epitope analysis with, for example, overlapping fragments derived from a protein sequence (SPOTscan), stepwise N- or C-terminally truncated fragments (SPOTsize), substitution analogues (SPOTsalogue), etc. Some examples are described in the following chapters. Moreover, modern peptide library screening approaches allowing the a priori delineation of peptide epitopes have been incorporated into SPOT synthesis (Frank 1994; Kramer et al 1993). These approaches exploit the preparation of arrays of defined peptide mixtures (or pools) and the presentation of entire peptide libraries (e.g., all 64 Mio hexapeptides) as strategic sets of sublibraries. Some current strategies for the deconvolution of individual sequences by activity screening of random pools are given in Table 10.4.3.

The introduction of randomized positions (X) within a peptide sequence assembled on a spot is quite reliably achieved by coupling with equimolar amino acid mixtures, applying these at a submolar ratio with respect to available amino functions on the spots (this is to allow all activated derivatives to react quantitatively during a first round of spotting), then completing all peptide elongations by two to three successive repeats of spotting. Using this coupling procedure, any position in a peptide sequence can easily be randomized without special considerations or increase in technical effort. Figure 10.4.4 gives an example of a peptide library array probed with a monoclonal antibody.

Because of their hydrophilic nature, the cellulose membranes are particularly well suited for the presentation of immobilized peptides to a biological assay system. The whole membrane (or parts of it, see 'Hints and

Trouble-shooting Tips' later) is incubated with, for example, an antiserum and detection of antibody molecules bound to peptide spots can be achieved by conventional solid phase ELISA or 'Western blot' procedures. Other labeling techniques such as radioisotopes or fluorescent dyes are also fully compatible. The choice of a detection system, however, should assure that peptide spots will not become chemically or otherwise irreversibly modified, because peptide arrays on cellulose membranes are reusable many times (>20) when treated properly. Signal patterns obtained from peptide arrays on spots can be documented and quantitatively evaluated utilizing modern image analysis systems (Fig. 10.4.4a) as used with other 2D analysis media such as electrophoresis gels and blotting membranes.

If solution phase peptides are to be prepared, then synthesis is carried out on correspondingly derivatized membranes, preferably utilizing a safety-catch linkage (Table 10.4.1). These linkages are stable during peptide assembly, cleavage of protecting groups, and acidic washings to remove chemicals. After this procedure, the membrane is dried and cut into pieces, each carrying one spot, or the spots are punched out with the help of a suitable device. For the standard 8 × 12 formats a puncher is commercially available that places the cut spots directly into the wells of a microtiter plate. Then an appropriate buffer of pH 7–8 is added to individual membrane pieces and the peptides are released from them into solution.

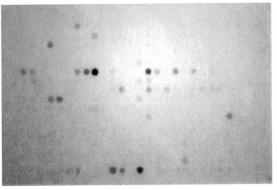

(A)

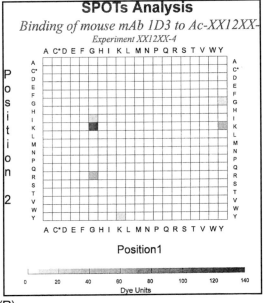

(B)

Figure 10.4.4 Example of a first screening round for the *a priori* delineation of epitope peptides from peptide libraries on spots (see Table 10.4.3, entry 1). (A) Electronic image of a membrane with 400 spots of a hexapeptide library in the format Ac-XX12XX-plus 25 reference spots (upper lane) after probing with the monoclonal antibody 1D3. Bound antibody was visualized with an alkaline phosphatase-conjugated secondary antibody and a color reaction (Frank 1992). (B) Spectral diagram display of (A) showing the correlation of quantified signals to the amino acid residues at positions 1 and 2; C* = Cys(Acm). The natural epitope of mAb 1D3 is -NYG-KYE-.

Materials

A kit that includes all necessary items for manual SPOT synthesis is currently available from Genosys Biotechnologies, Inc. (The Woodlands, TX, USA).

- A DotPunch from Bibby Dunn Labortechnik GmbH, Asbach, Germany

Equipment

- Flat reaction trough with a good closing lid made of chemically inert material (glass, Teflon, polyethylene) whose dimensions are slightly larger than those of the membranes used
- A micropipette adjustable from 0.5 to 10 µl (Eppendorf or Gilson) with corresponding plastic tips
- Small (1 ml) plastic tubes (e.g., Eppendorf, safe-lock) as reservoirs for amino acid solutions
- A rocker table
- Appropriate bench space *in a hood*!
- –70°C freezer
- Software: Special DOS-PC computer programs for the generation of peptide lists and pipetting protocols are included in the synthesis kit and the operation software of the spotting robot.

Optionally

- A spotting robot, model ASP 222, available from Abimed AnalysenTechnik GmbH, Langenfeld, Germany
- Two dispensers adjustable from 5 to 50 ml for DMF and alcohol containers
- A Teflon tubing line connected to a 5 l container which is connected to a vacuum line for collection of solvents and solutions aspirated from the reaction trough

Chemicals

- Chromatography paper type Chr1, 50, 540 and 3MM from Whatman, Maidstone, UK
- Bromophenol blue indicator (BPB) from, e.g., E. Merck, Darmstadt, Germany. A stock solution of 10 mg per 1 ml DMF is used throughout.
- *N,N*-Dimethylformamide (DMF) should be free of contaminating amines and thus of highest affordable purity, at least p.a. grade. Amine contamination is checked by addition of 10 µl of BPB stock solution to 1 ml of DMF. If the resulting color is yellow, this batch can be used without further purification. *Check each new batch*!
- 1-Methyl-2-pyrrolidinone (NMP) should be of highest available purity. Amine contamination is checked by addition of 10 µl of BPB stock solution to 1 ml of NMP. If the resulting color is yellow, the NMP can be used without further purification. Most commercial products, however, are not acceptable. Then deionize 1 liter of NMP with mixed bed ion-exchange resin AG 501-X8 (Bio-Rad Laboratories, Richmond, CA, USA) until a 1 ml aliquot gives a yellow BPB test. Filter off the resin and dry by addition of 100 g of molecular sieve MS 4 Å (e.g., E. Merck or Fluka). Gently agitate overnight. Decant from the MS beads and filter through a

- 20 μm polyethylene sinter or cellulose membrane. Divide the clear liquid into 100 ml portions and store tightly closed at −20°C.
- N-Hydroxybenzotriazole (HOBt) can be obtained from several suppliers only as hydrate. Dehydrate in a desiccator over phosphorus pentoxide at 50°C and 10^{-3} bar for 3 days. Store in a tightly closed container at room temperature.
- N-Methylimidazole (NMI) is distilled from solid sodium hydroxide and stored in small aliquots (5 ml) over molecular sieve MS 4 Å at −70°C.
- N,N'-diisopropylcarbodiimide (DIC), purum, from Fluka, Heidelberg, Germany, or other suppliers.
- Fmoc-amino acid derivatives are available from several suppliers in sufficient quality. In situ prepared HOBt-esters of these in NMP are used throughout for spotting reactions. Side-chain protection is Cys(Acm) or Cys(Trt), Asp(OtBu), Glu(OtBu), His(Boc) or His(Trt), Lys(Boc), Asn(Trt), Gln(Trt), Arg(PBf), Ser(tBu), Thr(tBu), Trp(Boc), and Tyr(tBu). Dissolve 1 mmol of each in each 5 ml NMP containing 0.25 M HOBt to give 0.2 M solutions. Divide into 100 μl aliquots in correctly labeled Eppentorf tubes. Close tightly, freeze in liquid nitrogen and store at −70°C.
- Acetylation mix: a 2% (v/v) solution of acetic anhydride (p.a.) in DMF

- Piperidine, p.a.: a 20% (v/v) solution in DMF is used to cleave the Fmoc protecting group; piperidine is toxic and should be handled with gloves under a hood!
- Alcohol (methanol or ethanol) of technical grade
- Deprotection mix containing by volume 50% trifluoroacetic acid (TFA, synthesis grade from, e.g., Merck, Darmstadt, Germany), 3% triisobutylsilane (TIBS, from, e.g., Aldrich Chemicals), 2% water, and 45% dichloromethane (DCM, p.a., from, e.g., Merck, Darmstadt, Germany). Trifluoroacetic acid is very agressive and should be handled with gloves under a hood!
- Derivatized SPOT membranes: the preparation of these requires additional chemical expertise and is described by Frank (1992). Ready-to-use membranes in an 8 × 12 spot format with 96 spots of βAla-βAla anchors are available from Genosys Biotechnologies, Inc. (The Woodlands, TX, USA). Blank membranes derivatized with βAla only are available from Abimed Analysen-Technik and for the generation of spot arrays with the ASP robot follow the instructions of the supplier. Commercial availability of membranes with other types of anchors and linkers (Table 10.4.1) is in preparation. A brief outline of the derivatization procedure is included below.

Methods

All volumes given below hold for a standard Whatman 50 or 540 paper sheet of 8 × 12 cm and have to be adjusted for more sheets or other paper qualities or sizes. Solvents or solutions used in washing/incubation steps are gently agitated on a rocker plate and aspirated or decanted after the time indicated.

Derivatization of membranes

In this step the esterification of the first anchor component to the hydroxyl functions of the cellulose is carried out.

1. Cut out a sheet of paper plus two small pieces of 1 cm^2 and dry overnight in a desiccator at 10^{-3} bar. Use only thin and acid-hardened paper qualities such as Whatman 50 or 540 for multiple use of peptide arrays in binding assays.
2. Soak the sheet and pieces in 2 ml of a solution containing 0.2 M Fmoc-β-alanine or Fmoc-proline, 0.24 M DIC and 0.3 M NMI in dry NMP. Keep in a closed container for 1 to 4 h.
3. Take out one piece and place in a small beaker. Wash with DMF, 20% (v/v) piperidine in DMF for 5 min, then DMF again and then stain with 1% (v/v) BPB stock in DMF until the supernatant remains yellow. Wash with alcohol and dry with a fan. Then place in a glass vessel containing 5 ml 20% (v/v) piperidine. Determine the optical density of the deep blue solution and calculate the loading with amino functions per cm^2, which is roughly the amount of BPB eluted ($\varepsilon_{605} = 95,000$). If OK proceed with step 4. Typical values are given in Table 10.4.2.
4. Wash the sheet in a flat trough twice with 20 ml of acetylation mix and then leave overnight in 20 ml of acetylation mix.
5. Wash three times each with 20 ml DMF.
6. Float in 20 ml of 20% (v/v) piperidine–DMF for 15 min.
7. Wash with DMF (four times with 20 ml) for 2 min, alcohol (3 times with 20 ml)

for 2 min, and then dry in a desiccator at 10^{-3} bar overnight. Store at −20°C.

Generation of the spots array

Generate the list of peptides to be prepared and define the array(s) required for the particular experiment according to number, spot size, and scale (Table 10.4.2). Mark the spot positions on the membranes with pencil dots for manual synthesis and place in the reaction trough or fix membranes on the platform of the spot robot.

1. Prepare a solution containing 0.3 M of second anchor compound (e.g. Fmoc-βAla-OH or Boc-Lys(Fmoc)-OH), 0.45 M HOBt and 0.35 M DIC in NMP. Leave for 30 min and then spot aliquots of this solution to all positions according to the chosen array configuration (e.g., Table 10.4.2). Leave for 15 min. Repeat spotting once and let react for 30 min.
2. Wash with 20 ml acetylation mix for 30 s, once again for 2 min, and finally leave overnight in acetylation mix.
3. Wash three times with 20 ml DMF.
4. Incubate for 10 min with 20 ml 20% (v/v) piperidine in DMF.
5. Wash four times with 20 ml DMF.
6. Incubate with 20 ml of 1% (v/v) BPB stock in DMF (must be a yellow solution!). Repeat if traces of remaining piperidine turn it into a dark blue solution. Spots should be stained only light blue!
7. Wash twice with 20 ml alcohol.
8. Dry between a folder of 3MM paper using cold air from a hairdrier and store sealed in a plastic bag at −20°C.

Assembly of peptides on spots

Number the spot positions on the membranes with a pencil according to the peptide list for manual synthesis and place in separate reaction troughs. Alternatively, fix membranes on the platform of the synthesizer. Follow the pipetting protocol or start spotter for the respective elongation cycle.

1. Take one set of Fmoc-amino acid stock solutions from the freezer and activate by addition of DIC (4 µl per 100 µl vial; ~0.25 M). Leave for 30 min and then spot aliquots of these solutions or start spotter. Leave for 15 min. Repeat spotting once and let react for 30 min.
2. Wash with 20 ml acetylation mix for 30 s, once again for 2 min, and then incubate for about 10 min until all remaining blue color has disappeared.
3. Wash three times with 20 ml DMF.
4. Incubate for 5 min with 20 ml 20% (v/v) piperidine in DMF.
5. Wash four times with 20 ml DMF.
6. Incubate with 20 ml of 1% (v/v) BPB stock in DMF (must be a yellow solution!). Repeat Step 6 if traces of remaining piperidine turn it into a dark blue solution. Spots should be stained only light blue!
7. Wash twice with 20 ml alcohol.
8. Dry between a folder of 3MM using cold air from a hairdrier. Then go back to step 1 for the next elongation cycle.

After the final cycle, all peptides can be N-terminally acetylated by additionally carrying out steps 2, 3, 7 and 8. See also 'Hints' below.

Assembly of peptide pools on spots

This procedure follows essentially the preceding steps except that the amino acid stock solutions are diluted to 50 mM with NMP and activated with only one-quarter of DIC. For X-couplings, an equimolar mixture of the amino acid derivatives to be represented at random positions is prepared by combining equal aliquots of the diluted stock solutions (e.g. all 20). This mixture is activated and applied in the same way as the other individual amino acid solutions. Spotting per elongation cycle is repeated four times.

Side-chain deprotection

Must be performed under a hood! Trifluoro-acetic acid is very agressive.

1. Prepare 40 ml of deprotection mix.
2. Place the dried paper in the reaction trough, add 20 ml deprotection mix, close the trough tightly and leave for 1 h. Replace the deprotection mix with the remaining 20 ml and leave again for 1 h.
3. Wash four times for 10 min each with 20 ml DCM.
4. Wash three times for 10 min each with 20 ml DMF.
5. Wash three times for 10 min each with 20 ml 1 M acetic acid in water.
6. Wash three times for 2 min each with 20 ml alcohol.

The sheet may now be dried with cold air and stored at –20°C or directly further processed as described in the next chapter.

Peptide elution

This procedure is applicable for safety-catch linkers of types 2, 3, and 4 of Table 10.4.1.

> Proceed as in steps 1, 2, and 3 of 'Side-chain Deprotection'.
>
> 4. Wash three times for 15 min each with 20 ml of methanol–water (1:1) containing 0.1% HCl.
> 5. Wash three times for 10 min each with 20 ml of 1 M acetic acid in water.
> 6. Dry under vacuum in a desiccator overnight.
> 7. Cut the membrane into pieces containing individual spots or punch out with a suitable device. Place the pieces into separate plastic tubes or wells of a microtiter plate.
> 8. Add a buffer of pH 7–8 and sufficient capacity, e.g. 0.1 M phosphate. You may add an organic solvent such as ethanol or dimethyl sulfoxide to dissolve more hydrophobic peptides. Final concentrations of peptide and other buffer constituents must be compatible with the bioassay to be performed!
> 9. Check pH with a reference piece of membrane.
> 10. Close or seal the tubes and shake overnight at 30–37°C such that the membrane pieces are gently agitated.
> 11. Remove aliquots and dilute as required for the assay. Store remainder at –20°C.

Hints and trouble-shooting tips

1. *Special chemical derivatives*

   - Free thiol functions of cysteine may be problematic because of postsynthetic uncontrolled oxidation. To avoid this, you may replace Cys by serine (Ser), alanine (Ala) or α-aminobutyric acid (Abu). Alternatively, choose the hydrophilic Cys(Acm) and leave protected.
   - For the simultaneous preparation of peptides of different sizes with free amino terminus, couple the terminal amino acid as αN-Boc derivatives such that they will not become acetylated during the normal elongation cycle. Boc is removed during the final side-chain deprotection procedure.
   - Special labels can be attached to the N-termini by spotting respective derivatives. We have successfully added biotin via its *in situ*-formed HOBt-ester (normal activation procedure) or fluorescein via its isothiocyanate (FITC) dissolved in DMF.

2. Membranes can easily be cut into parts or pieces with scissors prior to or after synthesis as required for the particular experiment. Simply mark the cutting lines with a pencil. Spot positions should be also marked with a pencil dot if treated individually after the synthesis. The pencil marking is sufficiently stable during the synthesis procedure.

3. Bromophenol blue staining is a very

helpful visual aid for monitoring the efficiency of the steps during peptide assembly. It is, however, not a quantitative technique. Some peptide sequences and terminal amino acid residues such as C, N, or D give a particularly weak staining. The indication of free amino functions is quite sensitive and a color change to green is sufficient for a good coupling.

4. Remember that the peptides are linked to the cellulose by ester bonds which are labile to alkaline hydrolysis in aqueous media of pH >7.

References

Bray AM, Maeji NJ, Geysen HM (1990) The simultaneous multiple production of solution phase peptides; assessment of the Geysen method of simultaneous peptide synthesis. Tetrahedron Lett 31: 5811–5814.

Dooley CT, Houghten RA (1993) The use of positional scanning synthetic peptide combinatorial libraries for the rapid determination of opioid receptor ligands. Life Sci 52: 1509–1517.

Frank R (1992) Spot-synthesis: an easy technique for the positionally addressable, parallel chemical synthesis on a membrane support. Tetrahedron 48: 9217–9232.

Frank R (1994) Spot-Synthesis: An easy and flexible tool to study molecular recognition. In Epton R, ed. Innovations and Perspectives in Solid Phase Synthesis 1994. Mayflower Worldwide, Birmingham, pp. 509–512.

Frank R, Guler S, Krause S, Lindenmaier W (1991) Facile and rapid 'spot-synthesis' of large numbers of peptides on membrane sheets. In Giralt E, Andreu D, eds. Peptides 1990. ESCOM Leiden, pp. 151–152.

Frank R, Kieß M, Lahmann H, Behn Ch, Gausepohl H (1995) Combinatorial synthesis on membrane supports by the SPOT technique. In Maia HLS, ed. Peptides 1994. ESCOM, Leiden, pp. 479–480.

Geysen HM, Rodda SJ, Mason TJ (1986) *A priori* delineation of a peptide which mimics a discontinuous antigenic determinant. Mol Immunol 23: 709–715.

Hoffmann S, Frank R (1994) A new safety-catch peptide–resin linkage for the direct release of peptides into aqueous buffers. Tetrahedron Lett 35: 7763–7766.

Hoffmann S, Frank R (1996) submitted.

Kramer A, Volkmer-Engert R, Malin R, Reineke U, Schneider-Mergener J (1993) Simultaneous synthesis of peptide libraries on single resin and continuous cellulose membrane supports: examples for the identification of protein, metal and DNA binding peptide mixtures. Peptides Res 6: 314–319.

Kramer A, Schuster A, Reineke U, Malin R, Volkmer-Engert R, Landgraf C, Schneider-Mergener J (1994) Combinatorial cellulose-bound peptide libraries: screening tools for the identification of peptides that bind ligands with predefined specificity. In: METHODS (Comp Methods Enzymol) 6: 912–921.

Krchnák V, Vágner J, Safár P, Lebl M (1988) Noninvasive continuous monitoring of solid phase peptide synthesis by acid–base indicator. Collect Czech Chem Commun 53: 2542–2548.

Screening of antibody epitopes and regions of protein–protein interaction sites using SPOT peptides

10.5

Kirsten Niebuhr
Jürgen Wehland

Department of Cell Biology and Immunology, GBF National Research Center for Biotechnology, Braunschweig, Germany

TABLE OF CONTENTS

Immunology Methods Manual
ISBN 0–12–442712–X

Abstract

SPOT synthesis of overlapping peptides covering the amino acid sequence of polypeptides is an extremely potent method for identifying the binding sites of antibodies or sites of protein–protein interaction. The peptides remain immobilized to the cellulose paper matrix, which can be handled like an immunoblot membrane. By applying stringent stripping conditions, the sheets containing the peptides can be reused for several mappings.

Introduction

The combination of monoclonal antibodies which inhibit the function of a protein and the mapping of the epitopes recognized by these antibodies is a powerful way to obtain insights into functional domains of proteins (Darji et al 1996). This approach is not limited to antibodies but can be extended to other proteins as well given that short peptides are sufficient for binding (Niebuhr et al unpublished).

For this assay the primary sequence of the polypeptide under investigation is subdivided into overlapping peptides with an average length of 10 to 15 amino acid residues, which are synthesized according to the SPOT method (Frank 1992). The actual binding assay (overlay) itself is relatively fast and easy: the membrane can be probed like a dot or immunoblot. Moreover, the peptide sheet can be reused several times with the application of stringent washing or stripping procedures (Darji et al 1996).

However, a prerequisite for this method is that the respective interacting site is linear since binding domains that depend on the three-dimensional structure of the protein (i.e. conformational epitopes) can hardly be displayed by the relatively short synthetic peptides. Nevertheless, the chances of mapping epitopes of antibodies generated against proteins are usually quite good.

Binding assay: materials

- Immobilized peptides prepared by SPOT synthesis (see chapter 10.4)
- Protein (antibody) under investigation, either directly labeled (for example with ^{125}I, biotin, etc.) or detectable with (secondary) antibodies
- Ethanol
- *Blocking solution*: 'Sue's reagent' (10% (w/v) skim milk powder, 0.1 M maleic acid, 0.1 M NaCl), commercially available as 'CBR' Concentrated Blocking Reagent from ICI/Cambridge Research Chemicals diluted 1:5 in TBS-T (see below) and 5% (w/v) sucrose
- *Washing solutions*: TBS-T, Tris-buffered saline (20 mM Tris–HCl, 140 mM NaCl, pH 7.6) with 0.1% (v/v) Tween-20; TBS-T with 0.5 M NaCl and TBS-T with 0.5% (v/v) Triton X-100
- *Detection reagents for bound protein*:

not necessary when protein is radioactively labeled; biotinylated probes are detected by enzyme-coupled streptavidin; otherwise detection via primary and secondary antibodies

- *Enzyme reaction*: preferably developed with a chemoluminescence kit, which is usually very sensitive and avoids colored precipitates on the cellulose sheet; for example, ECL (Enhanced Chemoluminescence Kit, Amersham)

for horseradish peroxidase-coupled detection reagents

- '*Stripping buffers*':
  A: 8 M urea, 1% (w/v) SDS, 0.5% (v/v) β-mercaptoethanol
  B: 10% (v/v) acetic acid, 50% (v/v) ethanol
  DMF (*N,N*-dimethylformamide), needed to strip filters where enzyme reactions are visualized by a colored precipitate.

Protocol

The procedure basically resembles that of developing a dot blot or immunoblot.

Important: Check the secondary antibodies first to make sure that they do not give an unspecific signal on the immobilized peptides.

1. If the peptide sheet is dry, moisten it with ethanol.
2. Wash three times (5 min each) with TBS-T and saturate filter sheet with blocking buffer for 2 h at room temperature or preferably overnight at 4°C.
3. Wash sheet with TBS-T and incubate with antibody or protein of interest in blocking reagent for 1 h (or longer) at room temperature (RT) (use antibodies to be tested at appropriate dilutions).
4. Wash at least 5 min each with TBS-T, TBS-T + NaCl and TBS-T + Triton and TBS-T.

   In the case of radioactive labeling dry the peptide sheet and expose it to x-ray film, otherwise continue as follows.

5. For epitope mapping of antibodies, dilute an adequate secondary antibody according to the instructions of the vendor, incubate peptide sheet for 1 h at RT. For other proteins an additional step of incubation with a primary antibody followed by the same washing procedure as described in step 4 is necessary.
6. Wash peptide sheet (see step 4) and perform enzyme reaction to detect bound antibody.

The respective binding site can be determined from the sequence that the detected synthetic peptides have in common. Generally the smaller the offset of the overlapping peptides, the more precisely the binding site can be delineated.

Regeneration of the peptide sheet

After a stringent washing procedure the membrane can be regenerated and reprobed several times. Each washing step is performed three times for at least 10 min at room temperature.

1. Wash with water.
2. Wash with 'stripping buffer' A.

3. Wash with 'stripping buffer' B.
4. Wash with ethanol.

Afterwards the peptide sheet is dried and stored at −20°C or is saturated and probed with another antibody or protein.

Trouble shooting

The stringent stripping procedure usually removes all proteins and/or antibodies quantitatively. In some cases, however, the affinity of the antibody/protein to be tested is rather high and traces are still left on the respective spots. These might be recognized by the antibodies of a subsequent assay. Freezing and thawing of the peptide sheet usually reduces this background.

If the enzyme reaction is developed with substrates that give a colored precipitate on the membrane, start the stripping procedure as follows:

1. Wash with water twice.
2. Wash with DMF (N,N-dimethylformamide) three times and sonicate.
3. Continue as described above.

References

Darji A, Niebuhr K, Hense M, Wehland J, Chakraborty T, Weiss S (1996) Neutralizing monoclonal antibodies against listeriolysin: mapping of epitopes involved in pore formation. Infect Human 64: 2356–2358.

Frank R (1992) Spot-synthesis: an easy technique for the positionally addressable, parallel chemical synthesis on a membrane support. Tetrahedron 48: 9217–9232.

Screening of T cell epitopes and antagonists using SPOT peptides

10.6

Ayub Darji[1]
Brigitta Stockinger[2]
Siegfried Weiss[1]

[1]Department of Cell Biology and Immunology, GBF National Research Center for Biotechnology, Braunschweig, Germany
[2]Department of Molecular Immunology, National Institute for Medical Research, London, UK

TABLE OF CONTENTS

Immunology Methods Manual
ISBN 0–12–442712–X

Abstract

Soluble peptides can be synthesized by the SPOT method in large numbers, which facilitates screening for T cell epitopes in proliferation or cytokine release assays even for proteins with high molecular weight. Similarly, peptides agonizing a specific T cell response can be screened for using peptide variants synthesized by this method.

Introductory remarks

In contrast to B cells which recognize native antigens with their immunoglobulin receptor, T cells only recognize short peptides, derived by processing of proteins, bound to molecules of the major histocompatibility complex (MHC). The peptides bound to MHC class I, which are recognized by cytotoxic CD8$^+$ cells, are well-defined in length (8–10 amino acids), since the binding site of MHC I is closed at both sides and the N- as well as the C-terminus of the peptide contributes to binding (Engelhard 1994; Stern and Wiley 1994). Peptides bound to MHC class II molecules, which are recognized by CD4$^+$ T cells are on the average 14 amino acids in length but, since the class II binding site is open at both ends, peptides containing an epitope might be heterogeneous (Engelhard 1994; Stern and Wiley 1994). In both types of peptides amino acid residues can be defined which are the major contact residues for the MHC molecules and which are responsible for the specificity of the interaction with the highly polymorphic MHC molecules (see Section 9). In the meantime, for many of the MHC molecules these so-called anchor residues are known (Rammensee et al 1995). If the protein of interest is presented by such an MHC molecule, it is possible to narrow epitopes down to a few peptides (Pamer et al 1991) which contain the right amino acids at the appropriate positions (see Section 9). Testing is nevertheless required because some of these peptides will not bind and some will not be immunogenic (Wipke et al 1993). For uncharacterized MHC

molecules screening of the whole antigen with overlapping peptides will be necessary.

Recently it has been discovered that variants of peptides which are changed at T cell receptor contact residues can block a T cell response even in the presence of the correct peptide (for review see Sette et al 1994). This antagonism is not due to competition for the peptide binding site of the MHC molecule but rather interferes with signaling of the T cell receptor (Sloan-Lancaster et al 1994; Madrenas et al 1995). In the future, T cell antagonists will play an important role in medicine since it might be possible to treat ongoing autoimmune diseases or allergies with them. In addition, natural antagonists have been found encoded by variants of viruses (Bertoletti et al 1994; Klenerman et al 1994). They also might be important in the selection of the T cell repertoire (Hogquist et al 1994).

Strong antagonists often contain conservative exchanges of T cell receptor contact residues. However, which exchange will give the best results is not predictable; in particular, a good antagonist of a T cell might be stimulatory for another. Therefore, to obtain effective antagonists, many peptide variants should be screened.

Here we describe epitope and antagonist screening with soluble peptides produced with the SPOT synthesis. Since the uncoupling of the peptide from the cellulose matrix will result in a Lys-Pro-diketopiperazin at the C-terminus, SPOT peptides are currently only suitable for screening MHC class II bound epitopes. The peptides used for scanning the whole protein sequence are usually 15 amino acids in length with an offset of 3 amino acids, but longer peptides or smaller offsets are also feasible. To subsequently map the epitope precisely, peptides containing the sequence of the positive peptide are synthesized starting with a 15–20mer the sequence of which is truncated from both ends. Only the mouse system is described here but SPOT peptides have been used successfully in the human system as well (Adler et al 1994).

Materials

Animals and cell lines

Mice of appropriate strain:

CTL-L (Gillis et al 1978)

X63/IL-2 (Karasuyama and Melchers 1988)

Chemicals

Mitomycin C (Sigma)

Lympholyte-M (Camon)

[^{3}H]Thymidine TRA 310 (Amersham)

Recombinant IL-2 or supernatant of the IL-2 transfectant X63/IL-2

Medium and buffer

Erythrocyte lysis buffer:

17 mM	Tris
180 mM	NH$_4$Cl

Adjust pH to 7.2

Medium: RPMI 1640 (Gibco)

Add to 1 liter:

100 ml	selected FCS (heat-inactivated: 65°C 2 h
10 ml	L-glutamine 200 mM (Gibco)
5 ml	gentamycin 10 mg ml^{-1} (Gibco) or 10 ml penicillin/streptomycin 10 mg ml^{-1} (Gibco)
1 ml	β-mercaptoethanol 0.25 mM

Material and apparatus

Syringes and needles (Becton Dickinson)

Low-speed centrifuge (Beckman or Heraeus)

x-Ray or γ-source

Glass filter mats (Skatron)

Cell harvester

CO$_2$ incubator

Protocols

Epitope screening by T cell proliferation

Preparation of peptides

A 1 cm^2 spot size for peptide synthesis is selected, resulting in ~25 nm of peptide. Peptides should be eluted from the cellulose sheet as described (see chapter 10.4) and dissolved in 500 μl PBS. Some of the peptides might be hydrophobic and difficult to dissolve. Therefore, 10% (v/v) DMSO or 20% (v/v) EtOH should be included in the buffer. This does not interfere with the T cell assays. Do not use polystyrene tubes, since they might be damaged by the solvents. Using a sonication bath for 1–5 min might also accelerate dissolution of the peptides. Some firms offer holders with tubes in the microtiter format which allow the use of multiple pipettes to test large numbers of peptides quickly. Since all media contain antibiotics and the assays are short-term, the peptides do not need to be sterile filtered.

Preparation of antigen presenting cells (APCs)

Remove the spleen of a mouse of the appropriate MHC type and place in a sterile Petri dish containing 10 ml erythrocyte lysis buffer. Take a 5 ml syringe with a bent 26 G needle and flush cells out of the spleen. Remove cells with a pipette, avoiding debris, and centrifuge 1000 rpm 5 min in a low-speed centrifuge. Wash twice with medium. Resuspend in 1 ml medium and count the cells. Irradiate with 3500–4000 Rad in a x-ray or γ-source if available.

Alternatively, treat with mitomycin C by incubating cells at 37°C with 50 μg ml^{-1} mitomycin C for 30 min in medium. Wash cells very well (three times) with cold medium.

Mitomycin C is very labile when dissolved. Aliquots should be kept at –70°C and not stored too long. Do not refreeze once thawed.

Preparation of T cells

T cells should not be taken earlier than 5 days after the last restimulation (the optimal time might depend on the particular clone). Cells are removed from the culture flask and washed once in medium. If too many dead cells are present you will need to enrich the living cells over a density gradient. Take a 15 ml transparent tube and add 5 ml Lympholyte-M. Overlay carefully with cell suspension and centrifuge 2300 rpm 20 min at room temperature. Do not use the brake to stop the centrifuge. Living cells will collect at the interface. Remove cells carefully with a sterile Pasteur pipette and wash twice with medium. Resuspend in 1 ml of medium and count the cells.

Proliferation assay

If possible this assay should be always performed in triplicate, but for a first screening of a large number of peptides or if T cells are limiting, fewer cultures might be sufficient. Add 5–10 µl peptide to a microtiter well, add a mixture of $1–5 \times 10^5$ irradiated APCs and 5×10^4 T cells in medium to obtain a culture volume of 200 µl. Incubate at 37°C for 48 h. Add 1 µCi [^{3}H]thymidine per well and incubate for another 18 h. Harvest cells on glass filters using a cell harvester and count the filters in a β-scintillation counter depending on the particular system available. Include as negative control T cells alone, APCs alone, T cells plus APC, and as positive control T cells plus APCs plus protein under investigation.

to destroy accidentally cotransfered living cells. Add 10^4 CTL-L (Gillis et al 1978) cells in 100 µl medium. (The growth of the T cell line CTL-L is dependent on IL-2, it should be grown with 100 U IL-2 ml^{-1} and washed carefully before use in the biological assays. Some CTL-L clones may proliferate with IL-4 to some extent. Anti-IL-4 antibodies can be included in the assay, if exact identification of the cytokine is required.) Include CTL-L alone as negative control and CTL-L plus recombinant IL-2 as positive control. Incubate for 24 h at 37°C. Add 1 µCi [^{3}H]thymidine. Incubate for another 18 h, harvest on glass filters, and count (see earlier for screening by T cell proliferation).

Epitope screening by cytokine release (IL-2)

Since T cell hybridomas proliferate continuously, they have to be tested in assays which are not based on cell proliferation, e.g. IL-2 secretion. These assays are also suitable for T cell clones. IL-2 can be quantified after 24 h of incubation by sandwich-ELISA, which is also commercially available. On the other hand, IL-2 can be measured more easily and more sensitively in a biological assay.

Peptides and APCs should be prepared as described above. APCs do not need to be irradiated or treated with mitomycin C.

Cytokine release

Add 5–10 µl of peptides to a microtiter well, add a mixture of $1–5 \times 10^5$ APCs and 5×10^4 T cells or T cell hybridomas in medium to the well to obtain a volume of 200 µl. Include controls as described for proliferation assay. Incubate for 24 h at 37°C. Gently remove 100 µl of supernatant from each well and transfer into another microtiter plate. Freeze at –20°C and thaw

Screening for T cell antagonists

It is claimed that conservative exchanges at T cell receptor contact residues of antigenic peptides might result in strong antagonists. However, we have found strong antagonists also among peptides with drastic changes. Therefore, at present, as many variants as possible should be screened to find good candidates.

Titration of the antigenic peptide

The antigenic peptide should be synthesized in the conventional way. Alternatively, SPOT synthesis can be scaled up and several spots can be pooled. The concentration should be suboptimal for the assay used. Therefore, the antigenic peptide needs to be carefully titrated.

Pulse appropriate APCs (see 'Preparation of Antigen-presenting Cells') for 1–2 h with various concentrations of the peptide. Wash twice with medium. Determine proliferation or IL-2 release using 2×10^5 APCs as described above. The concentration of peptide resulting in ~50% of the maximal stimulation should be used.

Testing the stipulatory capacity of peptide variants

The variant peptides could be fully stimulatory, fully antagonistic, partially antagonistic, or inert. Some variants might no longer bind to the MHC molecule. Their stimulatory capacity should be tested before use in the antagonist assay. If you wish to screen for binding, see Section 9.

Use 5–10 μl of SPOT peptides and test for proliferation and IL-2 release as described in the preceding two protocols.

Testing antagonistic capacity of peptide variants

Pulse appropriate APCs with the predetermined concentration of the antigenic peptide for 1–2 h. Wash twice with medium. Add various amounts of variant peptides to microtiter well and add pulsed APCs plus T cells in medium to obtain a volume of 200 μl. Test for proliferation or cytokine release as described above.

Conclusion

The synthesis of SPOT peptides can be performed by pipetting robots (see chapter 10.4) and large numbers of peptides can easily be synthesized. Therefore, the handling of the functional assays for screening epitopes and antagonists is the limiting step. By using microtiter format during synthesis and later, it is possible to use multiple pipettes for the proliferation and cytokine assays. Thus it should be possible to scan even large proteins or to screen large numbers of variants.

References

Adler S, Frank R, Lanzavecchia A, Weiss S (1994) T cell epitope analysis with peptides simultaneously synthesized on cellulose membranes: fine mapping of two DQ dependent epitopes. FEBS Lett 352: 167–170.

Bertoletti A, Sette A, Chisari FV, et al (1994) Natural variants of cytotoxic epitopes are T-cell receptor antagonists for antiviral cytotoxic T cells. Nature 369: 407–410.

Engelhard VH (1994) Structure of peptides associated with class I and class II MHC molecules. Annu Rev Immunol 12: 181–207.

Gillis S, Ferm MM, Ou W, Smith KA (1978) T cell growth factor: parameters of production and a quantitative microassay for activity. J Immunol 120: 2027–2032.

Hogquist KA, Jameson SC, Heath WR, Howard JL, Bevan MJ (1994) T cell receptor antagonist peptides induce positive selection. Cell 76: 17–27.

Karasuyama H, Melchers F (1988) Establishment of mouse cell lines which constitutively secrete large quantities of interleukin 2, 3, 4 or 5, using modified cDNA expression vectors. Eur J Immunol 18: 97–104.

Klenerman P, Rowland-Jones S, McAdam S, et al (1994) Cytotoxic T-cell activity antagonized by naturally occurring HIV-1 Gag variants. Nature 369: 403–406.

Madrenas J, Wange RL, Wang JL, Isakov N, Samelson LE, Germain RN (1995) Phosphorylation without ZAP-70 activation induced by TCR antagonists or partial agonists. Science 267: 515–518.

Pamer EG, Harty JT, Bevan MJ (1991) Precise prediction of a dominant class I MHC-restricted epitope of *Listeria monocytogenes*. Nature 353: 852–855.

Rammensee H-G, Friede T, Stevanovi S (1995) MHC ligands and peptide motifs: first listing. Immunogenetics 41: 178–228.

Sette A, Alexander J, Ruppert J, et al (1994) Antigen analogs/MHC complexes as specific T cell receptor antagonists. Annu Rev Immunol 12: 413–431.

Sloan-Lancaster J, Shaw AS, Rothbard JB, Allen PM (1994) Partial T cell signaling: Altered phospho- and lack of Zap70 recruitment in APL-induced T cell anergy. Cell 79: 913–922.

Stern LJ, Wiley DC (1994) Antigenic peptide binding by class I and class II histocompatibility proteins. Structure 2: 245–251.

Wipke BT, Jameson SC, Bevan MJ, Pamer EG (1993) Variable binding affinities of listeriolysin O peptides for the H-2Kd class I molecule. Eur J Immunol 23: 2005–2010.

Multiple peptide synthesis with commercially available Multipin™ kits

10.7

Andrew M. Bray

Chiron Mimotopes Pty. Ltd., Clayton, Victoria, Australia

TABLE OF CONTENTS

Abstract

The Multipin method is an effective, low-cost, simultaneous multiple peptide synthesis technology which gives researchers ready access to large numbers of peptides. This chapter gives an overview of a selection of commercially available Multipin kits, with an emphasis on the multiple cleavage methodologies used with these kits.

Commercially available Multipin kits for peptide synthesis

Chiron Mimotopes currently markets several Multipin kits for multiple peptide synthesis. Four of these are outlined in Table 10.7.1 together with suggested applications. Each kit allows a researcher to prepare up to 96 discrete peptides simultaneously, using relatively low-cost equipment. In all cases, synthesis is performed with Fmoc-protected amino acids. Where required, side-chain functionality is protected with trifluoroacetic acid (TFA)-labile protecting groups, hence avoiding the need to use hydrogen fluoride. Synthesis is performed on the heads of polyethylene pins (crowns). The detachable crowns are radiation-grafted with an appropriate polymer and subsequently derivatized so that they are ready for synthesis. The pins are mounted on a plastic holder in an 8×12 matrix, which matches the common

Table 10.7.1 Four commercially available Multipin kits for peptide synthesis

Kit type		Pin loading (µmol)	Cleavage method	Suggested applications and comments
NCP	Noncleavable multipin peptide synthesis kit	0.1	–	Antibody epitope analysis with pin-bound peptides
DKP	Cleavable peptide kit – diketopiperazine, C-termini	1	pH 7–8 buffer/40% MeCN[a]	Helper T cell epitope analysis; biotinylated peptides for antibody epitope analysis
GAP	Cleavable peptide kit – glycine acid or amide C-termini	1	(1) 0.1M NaOH/ 40% MeCN(aq)[a] (2) Neutralize	Cytotoxic T cell epitope analysis
MPS	Multiple peptide synthesis kit	7–8	TFA/scavengers	General peptide synthesis kit

[a]Cleaved peptide solution can be used directly in assay, or dried down and reconstituted.

Chapter 10.7

96-well microtiter tray format. Coupling reactions are performed in the wells of polypropylene microtiter trays, whereas washes or common reactions such as Fmoc deprotection, are performed in baths.

Chemistry used in peptide assembly

Activating agents

Owing to their long-term stability, we recommend the use of free acids (as opposed to active esters) together with an activating agent and 1-hydroxybenzotriazole (HOBt). For simplicity of use we recommend diisopropylcarbodiimide (DIC) as activating agent: it is a liquid, it reacts very selectively with carboxylic acids in the presence of amines, and it is economical.

Typical coupling conditions are Fmoc-amino acid/DIC/HOBt (100 mM, 100 mM, 120 mM) in DMF. Recommended coupling times are 2–16 h. When preparing coupling solutions, the activating agent is always added last, and the solution should be allowed to stand for 10 min prior to coupling to the solid support. For peptides of moderate length (>20mer), better results may be obtained using Castro's reagent ((benzotrazolyloxy)tris(dimethylamino)-phosphonium hexafluorophosphate) (BOP) (Fournier et al 1988) as follows: Fmoc-amino acid/BOP/N-methylmorpholine (NMM)/HOBt (140 mM, 140 mM, 210 mM, 140 mM) in DMF. Other activating agents such as PyBOP® (Coste et al 1990) and HATU (O-(7-azabenzotriazol-1-yl)-1,1 : 3,3-tetramethyluronium hexafluorophosphate) Carpino and El-Faham 1994) have also been applied to pins with good results. In the case of BOP, PyBOP, and HATU, a base such as NMM must be included for effective coupling to occur. Furthermore, these reagents should not be used in excess (with respect to the amino acid) as they can react with support-bound peptide amines. Care must be taken when handling any activating agent. In particular, BOP releases the known carcinogen hexamethylphosphoramide during the activation process. To our knowledge, the safety aspects of the many proposed BOP alternatives have not been fully studied.

Amino acid protection

Where Fmoc-protected amino acids require acid-labile side-chain protection, we recommend Arg(Pmc), Asp(OtBu), Cys(Trt), Glu(OtBu), His(Boc), Lys(Boc), Ser(tBu), Thr(tBu), Trp(Boc), Tyr(tBu). Although Fmoc-His(Boc)-OH is thermally labile and must be stored at –20°C, we have found that it couples more efficiently than Fmoc-His(Trt)-OH. If BOP, PyBOP, or HATU are used, Fmoc-Asn(Trt)-OH and Fmoc-Gln(Trt)-OH are recommended; the non-side-chain protected derivatives of Asn and Gln can be used with DIC.

Peptide cleavage

Following assembly, peptides must be side-chain deprotected and cleaved for use in cellular assays. Three side-chain deprotection/cleavage methodologies are available. The DKP and GAP kits are designed for easy multiple cleavage into physiologically compatible, aqueous-based solutions, hence allowing researchers to perform assays directly with the cleaved peptide solutions if desired. The MPS kit is a general peptide synthesis kit, which employs TFA as the cleavage reagent.

DKP kit

Background

The diketopiperazine C-termini kit is based on a linker which cleaves via a cyclization mechanism at neutral pH, following side-chain deprotection (Maeji et al 1990; Bray et al 1991; Valerio et al 1993). Of the three multiple cleavage methods presented here, the DKP method is the simplest to perform, with cleavage being triggered by incubation of the pins in buffer. The kit was designed for helper T cell epitope analysis. Peptides produced by this method carry a C-terminal diketopiperazine moiety, which has been found not to interfere in the assay (see chapter 10.9). When biotinylated, peptides produced by this method have been used in antibody epitope mapping (see chapter 10.8).

Peptide cleavage into neutral buffer

Prior to cleavage, side-chain protection is removed by acidolysis with a solution of TFA and cation scavengers in an open polypropylene bath. Typically, deprotection is performed with TFA–ethanedithiol–anisole (38:1:1, v/v/v, 50 ml/96 pins) for 2.5 h at 25°C. Immediately following side-chain deprotection, potentially toxic reagents are removed from the pin surface by a rigorous washing protocol, which constitutes a 10 min methanol soak (full immersion), followed by a 1 h soak in 0.5% AcOH in MeOH/H_2O (1:1, v/v) and then a 5 min soak in sterile water.

The block-bound pins (96) are then immersed into cleavage buffer (0.8 ml/pin) contained in racked 1 ml BioRad tubes. To ensure efficient cleavage, all buffers should contain 40% (v/v) acetonitrile (MeCN). Examples of cleavage solutions include 0.05 M Hepes (pH 7.6), and 0.1 M phosphate buffer (pH 7.6) in 40% MeCN(aq). Ideally, cleavage is performed for 1 h in an ultrasonic bath (Bray et al 1994a). If an ultrasonic bath is not available, the pins can be left to soak in the cleavage buffer for 16 h. To avoid the presence of carried-over MeCN in subsequent assays, cleavage can be performed in the volatile buffer 0.1 M ammonium bicarbonate (pH 8.4) in 40% MeCN(aq). As the buffer is volatile, the peptide solutions can be dried in a vacuum centrifuge attached to a freeze drier. The dry peptides can then be reconstituted in a solution appropriate for the subsequent bioassay.

GAP kit

Background

The GAP kit is designed to yield peptides with glycine acid or amide C-termini (Valerio et al

1991; Bray et al 1994b). As in the case of the DKP cleavage, side-chain protection is first removed by acidolysis and the pins are subjected to a washing protocol. Peptides are then cleaved with dilute NaOH, to give C-terminal acids. Peptides generated by this procedure have been used in cytotoxic T cell epitope analysis (see chapter 10.9). Alternatively, peptides can be cleaved with ammonia vapor to give a glycine amide C-terminus, and the cleaved peptides then eluted from the pins with a solvent of choice.

Cleavage to give C-terminal glycine acid

Cleavage to give peptide acids is affected by treatment of the support-bound peptides with 0.1 M NaOH in 40% MeCN (0.73 ml/pin) in racked 1 ml BioRad tubes. As NaOH reacts with MeCN on standing, this solution must be prepared immediately prior to use. Ideally, cleavage is performed for 30 min in an ultrasonic bath (Bray et al 1994a). If a sonication bath is not available, the cleavage time can increased to up to 2 h. It should be noted that some sequences are base-sensitive. For example, Asn-Gly and N-terminal Gln cyclize to aspartimide and pGlu, respectively. Met and Cys will be oxidized. Immediately after cleavage, the pins are removed and the basic solution is neutralized to pH 7 with 2 M NaH$_2$PO$_4$ (0.07 ml/pin), or to pH 5 with 0.166 M acetic acid in 1 M HCl. (*Do not use commercial standard 1 M HCl!* Commercial standard HCl solutions contain HgCl$_2$.) The neutralized peptide solutions can be freeze-dried if desired or used directly.

Cleavage to give C-terminal glycine amide

Peptides can also be cleaved from the support with ammonia vapor to yield C-terminal glycine amides (Bray et al 1994b). This is a more difficult procedure than the base cleavage method as it involves the condensation of ammonia gas into tetrahydrofuran at −78°C to give a 30% solution (by volume). The side-chain-deprotected and washed pins are stored in a desiccator together with the ammonia solution for 24 h. The peptides are then eluted from the pins with a solvent of choice. We refer prospective users to the manual for a detailed procedure.

MPS kit

Background

The multiple peptide synthesis (MPS) kit allows the user to concurrently prepare 96 peptides with native C-termini. In this case, peptide cleavage is performed by acidolysis, which allows for simultaneous cleavage and side-chain deprotection (Valerio et al 1994). The kit is based on large crowns (Macro Crowns) which have loadings of ~8 μmol/crown. It is intended as a general-purpose kit.

Linkers used in MPS kit

Peptide acids are synthesized on crowns which are prederivatized with the desired C-terminal residue. In most cases, the amino acid is coupled to the crown via the 4-(hydroxymethyl)phenoxyacetyl (HMPA) handle (Atherton et al 1981). Proline, however, is coupled via the hindered 4-(1',1'-dimethyl-1'-hydroxypropyl)-phenoxyacetyl handle (Akaji et al 1990) in order to prevent diketopiperazine formation during peptide assembly. To avoid racemization during their manufactures Asp and Glu are side-chain-linked to the HMPA handle. Similarly, Asn and Gln are prepared by coupling Fmoc-Asp-OtBu and Fmoc-Glu-OtBu, respectively, to Rink amide handle (Rink 1987) via their side-chains. Peptide amides are prepared on the Rink amide handle, the desired C-terminal residue being incorporated in the first coupling step.

Cleavage by acidolysis

Cleavage is performed in 10 ml polypropylene centrifuge tubes. The pins are removed from the block and placed into their respective labeled tubes. Each pin is cleaved with TFA–EDT–anisole (38:1:1, by volume) (1.5 ml/tube) for 2.5 h at 25°C. The pins are removed from the tubes and placed back on to the original block. Note: the pins can be washed with a small amount of TFA to minimize peptide loss. Note that peptides which contained Trt protection will give yellow solutions. Cleavage solutions are then concentrated. This can be performed by blowing a gentle nitrogen stream on to the solutions; multiple samples can be dried using a Pierce Reacti-vap™. Alternatively, peptide solutions can be evaporated with a TFA-resistant vacuum centrifuge and vacuum source. The centrifuge manufacturer should be consulted in this regard. Vacuum alone is not suitable as the solutions will bump, resulting in loss and cross-contamination.

Residual TFA, EDT, anisole, and side-chain-deprotection byproducts are then removed by extraction with petroleum ether–ethyl ether (2:1, v/v) (8 ml/tube). Both solvents should be of AR grade and either the ether should be anhydrous grade or it should be dried before use. The solution is added and the tubes are capped, vigorously shaken, or vortexed and then stored at –20°C for 30 min. The tubes are then centrifuged (2000–6000 rpm) for 6 min. *Caution*: ensure that the centrifuge is spark-proof — ether vapor is explosive. If a precipitate has formed, decant the solution and repeat the process with 4 ml of ether solution. If little or no precipitate is observed, the solution should be evaporated to dryness under a stream of nitrogen. The extraction can be reattempted with 1.5 ml of ether solution. If an unsatisfactory result is once again observed, the solution should be evaporated to dryness and used as is. This is only likely to occur with either very short sequences or very hydrophobic sequences. To prevent Cys oxidation, 0.1% 2-mercaptoethanol can be added to the first wash. It is recommended that the peptides are then reconstituted or suspended in 1% AcOH in 40% MeCN(aq) and dried down or freeze-dried. The procedure yields peptides which are relatively free of EDT, anisole, TFA, and deprotection byproducts.

Safety aspects

Many of the reagents used in peptide synthesis are potentially hazardous. All synthetic steps must be performed in a fully functioning fume cupboard and the use of gloves and safety glasses is strongly advised. All coupling reagents should be considered toxic. DIC is a known sensitizing agent, and BOP should be treated as a potential carcinogen. The toxicological properties of reagents such as protected amino acids are not fully described. Many of the solvents are toxic and flammable. Ether is extremely flammable and must be used in a spark-free area. TFA is a toxic, corrosive acid. TFA and DMF waste should not be mixed as an exothermic reaction can result. EDT is a noxious smelling reagent. Equipment contaminated with EDT can be decontaminated by soaking in dilute aqueous hydrogen peroxide. *Caution*: EDT and TFA/EDT solutions react violently with hydrogen peroxide — do not mix! Before using any reagent, the manufacturers' chemical safety data sheets should be consulted. The researcher should also be aware of local waste disposal requirements.

Paul Ave., Milwaukee, WI 53233, USA (Fax +1 414 273 4979).

Ultrasonic Baths (suggested model, Branson B2200 (60 W)): Branson Ultrasonics Corporation, Eagle Rd, Daburg, CT 06810 1961, USA.

Vacuum centrifuge: Savant Instruments Inc., 103–110 Bi-County Blvd, Farmington, NY, USA (Fax +1 516 249 4639).

Suppliers

Multipin™ kits, peptides, protected amino acids:* Chiron Mimotopes Pty. Ltd., PO Box 1415, Clayton South, Victoria 3169, Australia (Phone: +61 3 9565 1111, Fax +61 3 9565 1199); or San Diego, CA, USA (Phone +1 619 558 5800, Fax +1 619 558 5810); or France, (Phone +33 1 41 38 9400, Fax +33 1 41 38 9409).

Protected amino acids: Bachem Feinchemikalien AG, Hauptstrasse 144, CH-4416 Bubendorf, Switzerland (Fax +41 61 931 2549).

Protected amino acids, BOP, PyBOP: Carlbiochem-Novabiochem AG, Weidenmattweg 4, CH-4448 Laufelfingen, Switzerland (Fax +41 62 69 1992).

Racked BioRad Tubes: Bio-Rad Laboratories, 3300 Regatta Blvd., Richmond, CA 94804, USA.

Reacti-vap™: Pierce, 3747 North Meridian Rd, PO Box 117, Rockford, IL 61105, USA.

Solvents, DIC and general reagents: Aldrich Chemical Company, Inc., 1001 West Saint

*Since this article was written, Chiron Mimotopes has release a number of new Multipin™ kits for solid phase organic synthesis. There is a choice of either polystyrene or polyacrylamide-based graft polymers, and there are currently seven linker types. Two loadings are available: 1 μmol/crown and 5 μmol/crown. Contact Chiron Mimotopes for further details.

References

Akaji K, Kiso Y, Carpino LA (1990) Fmoc-based solid- phase peptide synthesis using a new *t*-alcohol type 4-(1',1'-dimethyl-1'-hydroxypropyl)phenoxyacetyl handle (DHPP)-resin. J Chem Soc, Chem Commun 584–586.

Atherton E, Logan CJ, Sheppard RC (1981) Peptide synthesis. Part 2. Procedures for solid-phase synthesis using N^α-fluorenylmethoxycarbonyl-amino-acids on polyamide supports. Synthesis of substance P and of acyl carrier protein 65–74 decapeptide. J Chem Soc, Perkin Trans I 538–546.

Bray AM, Maeji NJ, Valerio RM, Campbell RA, Geysen HM (1991) Direct cleavage of peptides from a solid support into aqueous buffer. Application in simultaneous multiple peptide synthesis. J Org Chem 56: 6656–6666.

Bray AM, Jhingran AG, Valerio RM, Maeji NJ (1994a) Simultaneous multiple synthesis of peptide amides by the Multipin method. Application of vapor-phase ammonolysis. J Org Chem 59: 2197–2203.

Bray AM, Lagniton LM, Valerio RM, Maeji NJ (1994b) Sonication-assisted cleavage of hydrophobic peptides. Application in multiple peptide synthesis. Tetrahedron Lett 35: 9079–9082.

Carpino LA, El-Faham A (1994) Effect of tertiary bases on *O*-benzotriazolyluronium salt-induced peptide segment coupling. J Org Chem 59: 695–698.

Coste J, Le-Nguyen D, Castro B (1990) PyBOP®: a new peptide coupling reagent devoid of toxic byproduct. Tetrahedron Lett 31: 205–208.

Fournier A, Wang C-T, Felix AM (1988) Applications of BOP reagent in solid phase synthesis. Int J Peptide Protein Res 31: 86–97.

Maeji NJ, Bray AM, Geysen HM (1990) Multi-pin peptide synthesis strategy for T cell determinant analysis. J Immunol Methods 134: 23–33.

Rink H (1987) A new acid-labile anchor group for the solid phase synthesis of C-terminal peptide amides by the Fmoc method. Tetrahedron Lett 28: 3787–3790.

Valerio RM, Benstead M, Bray AM, Campbell RA, Maeji NJ (1991) Synthesis of peptide analogues using the Multipin peptide synthesis method. Anal Biochem 197: 168–177.

Valerio RM, Bray AM, Campbell RA, et al (1993) Multipin peptide synthesis at the micromole scale using 2-hydroxyethyl methacrylate grafted polyethylene supports. Int J Peptide Protein Res 42: 1–9.

Valerio RM, Bray AM, Maeji NJ (1994) Multiple peptide synthesis on acid-labile handle derivatized polyethylene supports. Int J Peptide Protein Res 44: 158–165.

Antibody epitope mapping with Multipin™ peptides

10.8

Gordon Tribbick

Chiron Mimotopes Pty. Ltd., Clayton, Victoria, Australia

TABLE OF CONTENTS

Immunology Methods Manual
ISBN 0–12–442712–X

Abstract

Multipin peptides facilitate the efficient mapping of sequential epitopes on a protein. Sets of peptides encompassing all the linear peptide sequences are synthesized on pins. The ELISA (enzyme-linked immunosorbent assay) has become the method of choice for detecting the interaction between an antibody and its specific epitope. ELISA has been used successfully with the Multipin system to scan sets of peptides and define specific epitopes. These protocols are intended for horseradish peroxidase-labeled-anti-species immunoglobulins using ABTS as the chromogenic substrate; however, other substrates and variations in the test procedure are equally satisfactory.

Materials

Peptides required

Appropriate sets of peptides may be obtained from Chiron Mimotopes Peptide Systems (CMPS), ready for immediate testing; alternatively, Multipin kits are available for the synthesis of peptides (see chapter 10.7). For B cell epitope mapping we recommend that all sequential 8-mer sequences (overlap of seven residues) be represented in the set of peptides tested (Geysen et al 1987, 1988). To reduce the number of peptides required to cover long protein sequences, longer peptides (up to 15-mers) may be synthesized with a greater offset between successive peptides, but maintaining the overlap of seven residues. For the pin bound peptides we recommend that the N-terminus of the peptide be acetylated. The peptides remain covalently coupled to the pins and can be tested many times (≥50).

For epitope mapping using biotinylated peptides, each peptide incorporates biotin at either the N- or C-terminus and a spacer sequence to separate the biotin from the peptide sequence homologous with the antigen being investigated. The peptides are cleaved from the pins and immobilized for solid-phase immunoassays using avidin or streptavidin. This method of immobilization is independent of the sequence of the peptide being tested.

Each peptide has the general format Biotin-SGSG-PEPTIDE-Cleavable linker. For the C-terminal ending a C-terminal amide or diketopiperazine (Maeji et al 1990) can be chosen. One set of peptides is sufficient to perform ≥1000 scans.

Reagents required

Phosphate-buffered saline (10× concentration)

(PBS 10×) stock solution is used to prepare working strength PBS for use in ELISA tests.

53.7 g $Na_2HPO_4 \cdot 2H_2O$
15.6 g $NaH_2PO_4 \cdot 2H_2O$
340 g NaCl

Dissolve salts in hot distilled water to give a final volume of 4 liters. Mix thoroughly and allow the solution to cool to room temperature. Adjust the pH to 7.2 with either 50% (w/v) NaOH or concentrated

(37% w/w) HCl (take appropriate precautions when using these reagents).

Phosphate-buffered saline pH 7.2 (PBS)

Dilute the PBS 10× stock solution with distilled water when required for use.

Precoat diluent

This buffer is used to precoat the pins. This reduces nonspecific binding and gives a better signal-to-background ratio in ELISA assays. This buffer is also used as the diluent for the test antibodies.

PBS with 0.1% (v/v) Tween 20 (PBS/Tween) and 0.1% (w/v) sodium azide, pH 7.2

Conjugate diluent

This is the diluent for the goat anti-species conjugate in the pin ELISA. It is PBS with 1% (v/v) sheep serum, 0.1% (v/v) Tween 20, 0.1% (w/v) sodium caseinate, pH 7.2. Store the solution in a refrigerator for up to 24 h. For longer-term storage, freeze the solution at –20°C. Thaw as required. Do not add sodium azide as a preservative since the activity of the horseradish peroxidase will be destroyed.

Note: This diluent is totally unsuitable for use in assays where the primary antibody to be tested is of sheep or goat origin. A modified diluent substituting normal rabbit serum for sheep serum may be used in these cases. We chose sheep serum because of its ready availability and because the conjugates used are made from goat antibodies (sheep and goat sera are similar).

Substrate diluent

This buffer is used as the solvent for the chromogenic substrate in ELISA.

35.6 g $Na_2HPO_4 \cdot 2H_2O$

33.6 g citric acid monohydrate.

Dissolve the salts in a small amount of hot distilled water. Make up to 2 liters with distilled water at room temperature. Adjust the pH to 4.0 if necessary with 1 M Na_2HPO_4 solution or 0.8 M citric acid.

Substrate diluent should be stored in the refrigerator, and the amount required should be brought to room temperature before use each day. Substrate diluent should be used within 2 weeks of preparation and should be checked for signs of contamination immediately before use. If contamination is detected, discard the solution.

ABTS substrate solution

This is the chromogenic substrate for the horseradish peroxidase conjugate used to detect antibodies. Prepare the substrate solution immediately before use. Prepare 20 ml of substrate solution per block of pins or 12 ml for each microtiter plate being tested. For accuracy, it is recommended that a minimum of 100 ml of substrate solution be prepared. Equilibrate the substrate diluent to room temperature. Dissolve 0.5 mg ml^{-1} of diammonium 2,2'-azino-bis[3-ethylbenzthiazoline-6-sulfonate] (ABTS) in substrate diluent. When the ABTS has completely dissolved, add sufficient hydrogen peroxide solution (35% (w/w), 120 vol) to give a final concentration of 0.01% (w/v). Because the concentration of H_2O_2 in the concentrated stock bottle can change over time, we recommend that it be checked regularly. This can be done volumetrically by titration against permanganate (1 ml of 0.1 N potassium permanganate is equivalent to 1.701 mg of acidified H_2O_2; see the European Pharmacopeia for details). Alternatively, the concentration can be assessed spectrophotometrically at 240 nm, taking the molar extinction coefficient as 43.6 M^{-1}cm^{-1}.

Disruption buffer

0.1 M phosphate buffer with 1% (w/v) sodium dodecyl sulphate (SDS), 0.1% (v/v) 2-mercaptoethanol, pH 7.2 at 60°C (adjust with 50% (w/v) NaOH or orthophosphoric acid).

Note: Mercaptoethanol is a toxic chemical that can be absorbed by the skin and lungs: handle with care.

Special equipment

Sonication bath, located in a fume cupboard.

Suppliers

Diammonium 2,2'-azino-bis[3-ethylbenzthiazo-line-6-sulfonate] (ABTS) Boehringer Mannheim GmbH. Catalog No 122 661.

Horseradish peroxidase-labeled goat anti-species conjugates. Kirkegaard and Perry Laboratories, Inc., 2 Cessna Court, Gaithersburg, MD 20879, USA (Phone +1 301 948 7755. Fax +1 301 948 0169).

Microtiter plates suitable for streptavidin coating, Nunc-Immuno MaxiSorb plates (Nunc catalog No. 442404).

ELISA protocol for testing pin-bound peptides

Conjugate test

The conjugate used in an ELISA test is usually an antibody. Therefore it is possible that it will react directly and specifically with pin-bound peptides. This reaction could mask or be mistaken for the peptide binding of the primary antibody being tested. We therefore recommend that the reactivity of the conjugate with the peptides be checked before any testing with a primary antibody is carried out. If it is intended that antibodies from more than one species are to be tested, each of the conjugates should be checked. The actual concentration of the conjugate should be optimized. The ideal concentration is found by assaying with different concentrations of the conjugate. The sensitivity of the assay will tend to plateau with increasing concentration. The optimum concentration is the lowest concentration (highest dilution) that gives the maximum sensitivity. This concentration will also minimize interference to the assay by specific reaction of peptides with the conjugate. We have found that the Kirkegaard & Perry conjugates are satisfactory when used at a concentration of 0.25 μg ml^{-1}.

Precoating of pins (blocking step)

Precoating of pins reduces nonspecific binding, so giving a better signal-to-background ratio.

Dispense 200 μl of the precoat buffer into each well of a microtiter plate. Place the pins in the wells and incubate for 60 min at room temperature, preferably on a shaker table (at 100 rpm).

Wash

Remove the pins from the precoat buffer and flick any excess buffer from them. Wash the pins in a bath of PBS for 10 min at room temperature with agitation.

Note: The bath should hold at least 1 liter of wash solution for each block being tested and the reactive pin's surface should be well submerged.

Conjugate reaction

Dilute the appropriate conjugate preparation to its working concentration with conjugate diluent that has equilibrated to room temperature. Dispense 200 µl/well into microtiter plates. Flick excess PBS from the pins. Place the washed pins into the wells filled with conjugate and incubate at room temperature for 60 min with agitation.

Wash

Remove pins from the conjugate solution and wash the pins in a bath of PBS for 10 min at room temperature with agitation.

Substrate reaction

Dispense 200 µl/well of ABTS substrate solution into ELISA reading plates (e.g., flat-bottom polystyrene plates). Flick any excess PBS wash solution from the pins and place them in the substrate solution. Incubate the pins at room temperature on a shake table (100 rpm) for 45 min. Stop the enzyme action by removing the pins from the substrate solution.

Reading results

The optical density of the substrate in each well should be read as soon as possible after incubation. The reacted substrate solutions should be evenly mixed in the wells before reading.

The optical density is read at 405 nm in a suitable plate reader. If the plate reader is capable of dual wavelength mode of operation, the optical density at 492 nm can be subtracted to correct for background. Background in this case includes color present in the substrate solution before use, and optical defects of individual wells of the ELISA plate.

Primary antibody test

The assay with the primary antibody is a typical, indirect ELISA, with extra steps added to the conjugate reactivity test described above.

After the precoat blocking step, flick off excess precoat buffer then dispense 200 µl/well of suitably diluted primary antibody into microtiter plates (antiserum is normally diluted in precoat buffer containing 0.1% (w/v) sodium azide).

Note: The precoat incubation and the primary antibody incubation test may be performed by dispensing the solutions into suitable baths. The bath dimension should be ~11.5 × 8 × 2.5 cm. Lids from pipette tip boxes are suitable for use as baths. The volume of solution used in the bath should be sufficient to cover the reactive pin tips (gears).

Incubate the pins in the microtiter plate or bath overnight in a refrigerator (or cold room) on a shaker table (set at 100 rpm if the microtiter plate is used). If a bath is used, set the shaker table to a speed sufficient to allow the solution to wash over the gears.

We have found the following dilutions of antibodies to be suitable for the first test (note that the test should be repeated with

monoclonal antibody as ascites fluid, and human antisera	1:1000
hyperimmunized animal sera	1:5000

supernatant of hybridoma cell cultures	1:20
naturally infected (as opposed to hyperimmunized) animal sera	1:1000

If using the bath incubation method you can expect a 10-fold increase in the sensitivity of detection of antibodies: i.e., the dilutions for primary antibodies recommended should be increased 10-fold; e.g., test ascites at a dilution of 1:10,000.

Remove pins from the primary antibody preparation and wash once in a bath of PBS. The wash should be for 10 min at room temperature with agitation. Complete the ELISA following steps from 'Conjugate reaction' on the previous page.

Note: Substrate reaction. The reaction can be stopped by removing the pins from the plate. This can be done after 10 min if it appears that the reaction will give an optical density of 2 or greater (this is off the scale of many plate readers) if allowed to continue. However, this is not recommended and it is better to adjust the concentration of the primary antiserum so that on-scale readings are given with a 45 min incubation. Ensure that the contents of the wells are well mixed before reading the plate. Precipitation of the chromophore may occur when very strong color has developed during the substrate reaction.

Removal of antibody

A major benefit of the Multipin system is that peptides can be reused after testing. The peptides are covalently coupled to pins and can be treated thoroughly to ensure that all antibody is removed.

Disruption procedure

- Preheat the disruption buffer to 60°C and add to the sonication bath. Prepare sufficient disruption buffer to allow the blocks of pins to float freely in the solution.
- Turn on the bath and allow the temperature to equilibrate to between 55 and 65°C. Higher temperatures may damage the peptides, and lower temperatures may lead to ineffective removal of antibody.
- Place the blocks in the bath with the pins downward. Sonicate for 10 min then rinse the blocks twice in distilled water, preheated to 60°C, for 30 s.
- Wash the blocks on a shaker table in a bath of distilled water at an initial temperature of 60°C for at least 30 min. Shake off any excess water and totally immerse the blocks in hot methanol (about 60°C) for at least 15 s. Extreme care should be taken at this step to avoid exposure to methanol fumes and to avoid sources of ignition. Methanol is toxic and highly flammable..
- Allow the blocks to air dry for at least 15 min. They are ready for a further test. Store the blocks sealed in a plastic bag with desiccant in a refrigerator if they are not to be used immediately.

Before discarding the used disruption buffer, add 2 ml of hydrogen peroxide (120 vol) per liter and allow to stand for 5 min. This will destroy any remaining 2-mercaptoethanol.

Confirmation of antibody removal

The effectiveness of antibody removal can easily be checked by repeating the test for conjugate reactivity as described above. If antibodies have not been completely removed, the scan may be similar before and after disruption. Any peaks in this test for carry-over of bound antibody should be lower than the first scan, indicating partial, but not complete, removal of antibody.

If antibody is shown to be present, repeat the sonication step. Do not increase the temperature above 65°C as this may adversely affect the peptides. It is better to repeat the sonication step rather than make the disruption conditions harsher.

Discussion and trouble shooting

The reactivity by ELISA of positive (PLAQ) and negative (GLAQ) control peptides may be used as a measure of the quality of the peptide synthesis carried out. As control peptides are prepared on each block of peptides synthesized, they provide some measure of the variation in quality of synthesis from block to block and also from synthesis to synthesis.

Two sets of control peptide pins are supplied with the kit; one set has the positive control sequence PLAQGGGG, the other set has the negative control sequence GLAQGGGG. These control pins are distinguished from other pins by their stem colors. The positive control has a red stem and the negative control pin has a green stem. These control peptides have been tested at our laboratories and gave satisfactory reactivity by ELISA with the control antibody provided. The reactivity of these peptides with the control antibody may be used as an indication of the sensitivity being achieved in your ELISA assay. Comparison with the block controls made by you gives an indication of the quality of the synthesis.

The monoclonal antibody provided (in the bottle marked Control Antibody) reacts with the PLAQ but not the GLAQ peptide. It should be reconstituted with the indicated volume of purified water and requires no further dilution prior to use. Ensure that the antibody is completely dissolved before dispensing. After reconstitution, the monoclonal antibody preparation should be stored at 4°C. Control peptides may be tested independently of the rest of the peptide set, using the recommended ELISA procedure.

Where possible, test the preimmune serum before testing the hyperimmune serum from the same animal. Preexisting cross-reactive antibodies may be present that are able to bind to peptides.

Where possible, peptides that have been identified as binding antibodies should be confirmed to be epitopes by showing that the same antibodies are able to bind to the target antigenic protein. This can be done by using a competition assay, or by using the peptides as affinity supports to fractionate peptide binding antibodies (Tribbick et al 1991). Alternatively, antibodies can be affinity purified before they are tested on the peptides.

ELISA protocol for testing Multipin synthesized biotinylated peptides

Additional reagent — streptavidin stock solution

Make a stock solution of streptavidin in purified water at a concentration of $2\,mg\,ml^{-1}$. This stock solution can be stored indefinitely if frozen at $-20°C$.

Coat plates with streptavidin

The microtiter plates for the assay must be coated with streptavidin. Select plates that are designed to absorb proteins to their surface. Prepare a solution of streptavidin by diluting the streptavidin stock solution 1:400 in purified water. The final concentration of streptavidin is $5\,\mu g\,ml^{-1}$. Dispense $100\,\mu l$ of the solution into each well of the plates. Leave the plates exposed to the air at 37°C overnight to allow the solution to evaporate to dryness. The plates can then be stored in a refrigerator at 4°C. Pack the coated plates in a plastic bag with silica gel as a desiccant before storage. The coated plates will retain their activity for at least 4 weeks. Each well is coated with enough streptavidin to bind approximately 30 pmol of biotinylated peptide.

Block non-specific absorption

Dispense $200\,\mu l$ of PBS/Tween 20 blocking diluent into each well of the dry, streptavidin-coated plate. We find that the Tween 20 is satisfactory for blocking nonspecific interactions. However, if you wish, you can dispense $200\,\mu l$ of a protein-supplemented blocking buffer (which contains BSA or sodium caseinate as a blocking protein) as an alternative. Allow to incubate for 1 h at 20°C. After incubation, flick out the solution from the wells.

Wash

Wash the plates with PBS/Tween 20 solution four times. This is done by flooding the plate with PBS/Tween 20 solution, ensuring that all wells are completely full of solution. Then vigorously flick the solution from the wells. Remove excess solution by vigorously slapping the plates, well side down, on a benchtop that has been covered with an absorbent material (e.g. paper towels).

Reconstitute peptides

If the peptides are dry powders, dissolve them. We recommend that the peptides be reconstituted in $200\,\mu l$ of either a pure solvent (e.g. dimethyl sulfoxide or dimethylformamide) or solvent–water mixture (e.g. 40% acetonitrile in water).

The redissolved peptide is at $\sim 5\,mM$. Store peptides in a freezer between uses (i.e. at $-20°C$ or lower). This reconstitution step is not needed if the peptide was cleaved directly into aqueous solution; for example, self-synthesized DKP peptides. In this case, the cleavage solution can be used as is for dilution to the working concentration.

For use, dilute to a working strength of $1/5000$ ($1\,\mu M$) with precoat diluent. The peptide stock solution can be diluted further (down to 1/25,000); however, there may be some loss in ELISA sensitivity at too great a dilution. These solutions can be stored for one day at 4°C. For longer storage, the diluted peptide solutions should be frozen.

Peptide immobilization

Transfer 100 µl of each of the diluted peptide solutions into the corresponding well positions of the streptavidin-coated plate. If you are going to use software supplied by CMPS, it is important that the solutions are placed in wells in the same plate positions as synthesized. Failure to do this will mean that reactions read using the software will be identified with the wrong peptides. Place the plate on a shaker table and allow the reaction to proceed for 1 h at room temperature. For convenience, several sets of immobilized peptides may be prepared simultaneously and assayed at a later date.

Wash

After incubation, flick out the solution and repeat the washing procedure. If the plates are not going to be used immediately, they should be dried at 37°C before storing in the dry state at 4°C.

Antibody binding

Dilute the serum to be tested, using precoat diluent. The optimum dilution will depend on the source and amount of antibodies present. The recommended dilutions are 1:5000 for hyperimmune serum from experimental animals and ascites fluid from hybridoma-bearing mice, and 1:500 for human serum.

Add 100 µl of the diluted serum to each of the wells of the plates containing captured peptides. Place the plate on a shaker table and incubate with agitation for 1 h at 20°C. Alternatively, the serum can be incubated overnight at 4°C. Remove the incubation mixture by flicking the plate and repeat the washes as described above.

Conjugate reaction

Dispense 100 µl of the dilute conjugate into each well and incubate at 20°C for 1 h (see 'ELISA Protocol for Testing Pin-bound Peptides'). Remove the incubation mixture by flicking the plate and repeat the washes as described. Finally, wash the plate twice with PBS only (i.e. containing no Tween 20) to remove traces of Tween.

Substrate reaction

Detect the presence of peroxidase by adding 100 µl of freshly prepared ABTS substrate solution to each well. This can be incubated for up to 45 min at 20°C. The plates are then read at a wavelength of 405 nm. A suitable reference wavelength is 492 nm if your plate reader has a dual wavelength mode.

Discussion and trouble shooting

A test should be done using preimmune, negative, or normal serum to verify that any binding observed was due to specific antibodies. We recommend that this test be done in parallel with the test antibody. Similarly, a negative control test can be performed by omitting the incubation with specific antibody. Any positives in this test would be due to conjugate binding directly to the peptide. See under 'Conjugate test', p. 820 for additional comments.

Suppliers

Nunc, Postbox 280, DK 4000, Roskilde, Denmark (Phone +4542359065 Fax +4542350105).

Multipin kits, peptide sets: Chiron Mimotopes Pty. Ltd., PO Box 1415, Clayton South, Victoria 3169, Australia (Phone +61 3 9565 1111, Fax +61 3 9565 1199); or San Diego, CA, USA (Phone +1 619 558 5800, Fax +1 619 558 5810); or France, (+33 41 38 9400, Fax +33 41 38 9409).

Sodium caseinate: United States Biochemical Corporation, Cleveland, OH 44128. Catalog no. 12865.

Sonication bath: Suggested model, Branson B2200 (60 W) Branson Ultrasonics Corporation, Eagle Rd, Daburg, CT 06810 1961, USA.

Streptavidin: Sigma catalog no. S-4762 (affinity purified, salt-free streptavidin) is suitable.

Tween 20: Sigma catalog no. P 1379 is suitable.

References

Geysen HM, Rodda SJ, Mason TJ, Tribbick G, Schoofs PG (1987) Strategies for epitope analysis using peptide synthesis. J Immunol Methods 102: 259–274.

Geysen HM, Mason TJ, Rodda SJ (1988) Cognitive features of continuous antigenic determinants. J Mol Recognition 1: 32–41.

Maeji NJ, Bray AM, Geysen HM (1990) Multipin peptide synthesis strategy for T cell determinant analysis. J Immunol Methods 134: 23–33.

Tribbick G, Triantafyllou B, Lauricella R, Rodda SJ, Mason TJ, Geysen HM (1991) Systematic fractionation of serum antibodies using multiple antigen homologous peptides as affinity ligands. J Immunol Methods 139: 155–166.

T cell epitope mapping with Multipin™ peptides

10.9

Stuart J. Rodda

Chiron Mimotopes Pty. Ltd., Clayton, Victoria, Australia

TABLE OF CONTENTS

Abstract

Screening of protein sequences for T cell epitopes is easily accomplished with sets of overlapping peptides made by the Multipin system. Antigen-presenting cells are exposed to peptide, T cells are added, and the response of the T cells is measured. T cell proliferation is one of the simpler parameters of T cell response to measure and is the one described in detail here. Differences between test conditions for T cell clones and polyclonal T cells are highlighted.

Introduction

Multipin peptides allow the systematic scanning of protein sequences for the presence of cytotoxic (Tc) or helper (Th) epitopes, and also allow detailed analoging studies which can identify antagonistic peptides or cross-reacting peptides for a particular defined epitope. The synthesis of the peptides is dealt with in Chapter 10.7. Peptides are added to antigen-presenting cells (APC) before or at the time of adding T cells and the presence or absence of recognition is determined by measuring a parameter of T cell response such as proliferation, cytotoxicity, or cytokine release.

Decisions have to be made as to the length of the peptides, the extent of overlap, and the type of end-capping groups. Peptides for Th mapping should be in the 13 to 20 residues length range, whereas peptides for Tc mapping may be in this range or can be as short as 9 residues. The choice depends on the purpose: for initial screening, longer peptides are usually used, while the shorter peptides are used for final detailed identification of single epitopes. Overlap of succeeding peptides is necessary to avoid gaps in the coverage of a protein sequence, because it is not possible to rule out the existence of an epitope in a given sequence using predictive methods. For Th epitopes, it is preferable to overlap so that no continuous sequence of fewer than 12 residues is missed. For Tc epitopes, it is important that no continuous sequence of fewer than 10 residues is missed.

Peptide capping groups can affect the efficiency of peptides in epitope mapping procedures. A choice of ending is required when the synthesis and cleavage method is chosen. Capping of the N-terminus by acetylation can increase the effective length of the peptide and may help it to bind to class II MHC, as well as to survive longer in cell culture medium. The C-terminus can be capped with a diketopiperazine (DKP) group, as occurs when the mild aqueous cleavage method is used (Maeji et al 1990). Peptides with a C-terminal DKP group function efficiently in Th assays (Mutch et al 1991; Reece et al 1994). Alternatively, the C-terminus can be amidated, or may have a constant C-terminal glycine residue, or be the 'native' free acid form. Native or C-terminal glycine are preferred for Tc epitope mapping, as it is (usually) necessary for the peptide to be presented to the Tc cell as a nonamer on the surface of class I MHC (Guo et al 1992).

Materials

- Cloned T cells (10,000 cells per assay point) or polyclonal T cells such as spleen or peripheral blood mononuclear cells (PBMC) from immune animals (100,000–200,000 cells per assay point).
- Gamma-irradiated APC (e.g. 100,000 histocompatible spleen cells per assay point). Spleen cells or PBMC as APC can be irradiated with 3000–5000 rad; EBV-transformed B cells usually require a higher dose, 5000–10,000 rad. Mitomycin C treatment of APC to prevent replication is a satisfactory alternative to irradiation for those not having access to a gamma-irradiation apparatus.
- Peptide set (as dry powder), e.g. from Chiron Mimotopes.
- Solvent for peptides, e.g. AR dimethyl sulfoxide (DMSO) or dimethylformamide (DMF) (Merck)
- Gassed (5% CO_2 in air), humidified incubator (37°C) or equivalent.

- Cell culture vessels, e.g. 96-well sterile flat- or round-bottom sterile tissue culture grade plate with lid (Nunc cat. no. 167008 or 163320).
- Culture medium, e.g. RPMI 1640 (ICN cat. no.12-602) supplemented with 10% heat-inactivated fetal bovine serum, 10 mM Hepes, 2 mM glutamine, and antibiotics (e.g. 20 μg ml^{-1} gentamicin). For mouse T cells, 20 μM of 2-mercaptoethanol must be added. For human T cells, it may be preferable to use autologous serum or pooled, screened human AB serum.
- [$methyl$-^{3}H]Thymidine, specific activity ~40–85Ci mmol^{-1}, 0.25–1 μCi per assay point (e.g. Amersham cat. no. TRK758 or TRK637)
- Cell harvester
- Beta-particle radiation counter, e.g. multiple harvester (Skatron), multiple scintillation counter (LKB Betaplate 1205 counter) or equivalent from Packard, etc.

Choice of cells

Th clones are usually propagated using cycles of antigen-specific stimulation (by antigen on histocompatible APC), alternating with non-specific stimulation using IL-2 or an uncharacterized mixture of cytokines (Taylor et al 1987). The specificity of the clone for the antigen in question should naturally have been established beforehand. For epitope mapping, T cells must be in a state where they will respond to specific antigen stimulation, i.e. they should be 'rested' or not have had any stimulation for at least 3 days before the epitope mapping test.

Polyclonal T cells taken directly from an animal immunized with the antigen in question can usually be tested immediately in $vitro$. In both cases, controls must be included to ensure that the recognition of the peptides is specific and relevant to the antigen in question.

Th or CD4$^+$ cells recognize peptide in the context of self class II MHC molecules. As these are only expressed on certain 'professional' APC, a source of appropriate APC should be chosen. This can include mixed spleen cells (suitable for many experimental animal systems), blood monocytes and B cells (as found mixed in PBMC), or pure B cells (suitable for human systems, especially when immortalized Epstein–Barr virus-transformed B cells are available). The usual methods of detection of a specific response by Th cells are cell proliferation or cytokine release. Detection of proliferation can be by incorporation of tritiated thymidine or by a metabolic test such as the MTT assay. Detection of cytokines can be by ELISA or bioassay.

Procedure for Th cells

1. Dissolve the peptides in water to make a 1 mM stock solution. Note that peptides as supplied may contain salts or impurities and weighing the solid will not necessarily give the target amount of peptide. Amino acid analysis is the best method for absolute quantification of a peptide which has salts present. Sonication of the peptide suspension may be necessary to assist the dissolution process. If the peptides fail to dissolve freely in water, the suspension can be freeze-dried and a solvent such as DMSO or DMF can then be tried, provided the stock solution can be diluted 1 : 300 or more into the final assay medium. Alternatively, all peptides can be dissolved in DMSO initially to avoid the possibility of having to deal with water solubility problems on an individual basis. Use the peptides immediately and store any unused peptides frozen at −20°C, or colder, to prevent degradation.

2. Dilute the peptides 1:10 (to give 10 μM final) or 1:100 (to give 1 μM final) in phosphate-buffered saline (PBS, 0.01 M sodium phosphate in 0.15 M sodium chloride, pH 7.2) and add 20 μl to each of three or more replicate wells in the microtiter plate. A target concentration of 1 μM is usually sufficient to stimulate clones. If resources permit, it is desirable to test more than one concentration and indeed to perform a dose-range titration. Diluent-only negative controls and whole antigen positive controls should be set up in parallel, as well as controls on the APC and T cells (see below).

3. Add the irradiated APC in 80 μl of culture medium. A typical dose would be 10,000–100,000 spleen cells (mouse) or 10,000 EBV-transformed B cells (human) per well. These can then be incubated for an hour, or the T cells can be added immediately (see below). EBV-transformed human B cells are more radiation-resistant than spleen cells and the irradiation conditions sufficient for each line must be established before they are used as APC in an epitope mapping experiment.

4. Add the T cells in 100 μl medium. A typical number of cells for a T cell clone would be 10,000 per well. In the case of proliferation tests with polyclonal T cells, such as whole spleen, it is necessary to use larger numbers (20,000 to 200,000) of cells per well and to omit the addition of irradiated APC, since there are plenty of APC in the fresh spleen cell preparation. The low frequency of specific precursor T cells in such

polyclonal cell preparations makes it necessary to establish many more replicates (we suggest a minimum of 8) and to analyze the results differently (see below).

5. Set up controls in parallel with the test groups. As mentioned above, negative controls (no antigen) and positive controls (whole antigen, no peptide) should be set up. In addition, it is useful to include 'APC only' and 'T cells only' as a check on the source of problems if the 'no antigen' negative control is found to exhibit proliferation.

6. Incubate the peptide/APC/T cells at 37°C in the gassed, humidified incubator. It is worthwhile to place the plates in a plastic lunchbox with a loose-fitting lid prior to placing them in the incubator. The lunchbox can also be prepared with a moistened paper towel in the bottom. The use of the lunchbox minimizes fluctuation of temperature and carbon dioxide concentration resulting from each opening of the incubator, and also minimizes the possibility of evaporation of the medium. If uncontrolled, these factors can produce an 'edge effect' in 96-well plates.

7. The cells are incubated undisturbed for an appropriate length of time before addition of a pulse of tritiated thymidine (see below). Incubation for 2–3 days (48–72 h) is sufficient for T cell clones but polyclonal cells should be incubated for 3–6 days, with the best signal/noise ratio usually found around 4 days.

8. Add to each well an accurately measured aliquot (10 or 20 µl) of medium containing 0.25 µCi tritiated thymidine. Variation in amount of thymidine added per well will be reflected in the measure of incorporated thymidine. As this is a terminal labeling step, high specific activity thymidine (40–85 Ci mmol^{-1}) (trace labeling) can be used rather

than flood labeling with low specific activity thymidine. The radiotoxicity of high specific activity thymidine is unimportant in this system, and it is much cheaper to use as only 0.25 µCi need be added per well compared with the 1 µCi commonly chosen when low specific activity thymidine is used.

9. Incubate at 37°C in the gassed, humidified incubator for 6 h (for same day harvest) or 18 h (for overnight labeling).

10. Harvest the tritiated DNA by collecting the cells on a filter and washing with water. This lyses the cells, releases the DNA which remains trapped on the filter, and washes away unincorporated thymidine. Any of the common cell harvesters can be used, such as a Skatron model 11019 with glass fiber filtermats, which can then be processed for counting in an LKB Betaplate 1205 scintillation counter. This allows all 96 wells to be harvested and counted without the cost and time required for individual handling of vials. A similar system from Packard allows simultaneous counting of the tritiated DNA from all 96 wells without the use of scintillant (gas ionization detector).

11. Count the tritium to a predetermined precision, e.g. SD ≤ 5% (accumulate at least 400 counts per sample for this precision level). As there is little variation in counting efficiency from sample to sample, either the corrected (dpm) or the raw counts (cpm) can be used for analysis of the data.

12. Analyze the data. Convert all counts to cpm. For tests on clones, take means and SD of replicates. Compare positive and negative controls. If the difference between positive and negative controls is not significant ($p < 0.05$), the entire experiment is invalid. Apply a statistical test to the differences between the means of the negative controls and each of the test groups, and only score as positive

data which is significantly different from the negative control group. Where multiple antigen doses have been used, plot the dose–response curve. Look for positive data which could be due to recognition of the same peptide sequence (where overlapping peptides have been used). Deduce the likely 'minimal' epitope. For polyclonal T cells, cpm in individual wells are not a guide to the overall strength of the response because of the low frequency of Th cells able to respond to individual epitopes. Each replicate culture should be scored as 'positive' (responding) or 'negative' (no response) and a precursor frequency can then be estimated (Reece et al 1993).

13. *Further experiments*. It may be of value to identify more precisely the peptide sequence being recognized. This may be accomplished by synthesizing and testing a set of shorter peptides spanning the deduced epitope. It is frequently also of interest to determine the allotype of class II MHC which is being used for recognition by the Th. This may be accomplished with sets of APC which are homozygous for particular alleles and which present each of the donor's class II allotypes against a class II allotypes of 'irrelevant' (nonhomologous with the Th phenotype) class II allotypes.

Discussion and troubleshooting

Responses of T cells to externally supplied peptide may or may not depend on the ability of the APC to process the peptide. Helper epitopes in general can probably be presented passively by uptake onto class II MHC molecules on APC, whereas for cytotoxic T cell epitopes, unless a peptide is already the minimal epitope (usually a 9-mer for Tc recognition) it will have to be broken down in some way in the test system in order to be recognized. Proteases in serum or on cell surfaces may aid this process. Failure to get a T cell response to a 'known' epitope could be due to: too low a peptide concentration (below the threshold of responsiveness) or even too high a peptide concentration (toxicity, or altered kinetics of the response); insufficient or histoincompatible APC; T being cells in an unresponsive state as a result of stimulation being too recent; incorrect peptide format (too short, too long, or with incorrect sequence); inadequate medium or

incubation conditions; degraded thymidine; or presence of mycoplasma in the T cells (competition for thymidine). Inclusion of a range of controls is the first line of defence against such difficulties.

In rare cases, the peptide recognized by the T cells may be a sequence generated by post-translational modification of the genetically coded sequence.

Tests for antagonistic peptides require the simultaneous addition of the test peptide and a control peptide known to be stimulatory (De Magistris et al 1992).

Peptides are not always sufficiently soluble to allow the concentrated stock solution to be prepared, even when DMSO or DMF are used as the solvent. In such cases, it is often possible to dissolve the peptide in a larger volume of 40% acetonitrile (HPLC grade) in water and use the peptide at a 1:20 to 1:50 dilution (final) in the T cell assay. The resulting

2–0.8% acetonitrile appears to have no effect on T cell responses.

Method for Tc cells (general remarks)

The methods for testing recognition of peptides by Tc are quite different from those for Th and will not be presented in detail here. The major differences are as follow.

- Peptides are presented to Tc on class I MHC, which is present on most cell types in the body. Thus, normal cultured cells such as fibroblasts or epithelial cells can be used as targets.
- The cytotoxicity test, which is a rapid (4–6 h) test is most commonly used. The parameter measured may be direct (cell viability), or indirect (chromium release, cytokine release).
- The peptide being presented by class I MHC must be a short peptide with 'native' ends, i.e. free amino and free carboxyl groups. When longer peptides are added, it is therefore essential for them to be broken down or processed before they are presented. Detection of the presence of a Tc epitope within a longer peptide would logically be followed by synthesis and testing of a set of minimal sequences (usually 9-mers) to identify the epitope precisely.
- It is frequently of interest to determine the allotype of class I MHC which is being used for recognition by the Tc. This may be accomplished with sets of APC (target cells) which are homozygous for particular alleles and which present each of the donor's class I allotypes against a background of 'irrelevant' (nonhomologous with the Tc phenotype) class I allotypes.

Suppliers

Radioisotopes: Amersham International Plc, Amersham Place, Little Chalfont, Bucks, UK. (Phone +44 1494 54 4000, Fax +44 1494 54 4350).

AR reagents: E. Merck, PO Box 4119, D-6100 Darmstadt, Frankfurterstrasse 250, Germany. (Phone +49 61 51 72 0, Fax +49 61 51 72 20 00).

Cell culture media: ICN Pharmaceuticals, 3300 Hyland Ave., Costa Mesa, CA, USA (Phone +1 714 545 0113, Fax 800 334 6999).

Cell harvester and beta counter: Pharmacia LKB Biotechnology AB, Bjorkgatan 3Q, S-75182 Uppsala, Sweden (Phone +46 18 16 3000, Fax +46 18 14 38 20).

Cell harvester and beta counter: Packard Instrument Company, One State St, Meriden, CT 06450, USA (Phone +1 203 238 2351, Fax +1 203 235 1347).

Plasticware: A/S Nunc, Postbox 280, Kamstrup, DK4000, Roskilde, Denmark (Phone +45 2 35 90 65, Fax +45 2 35 01 05).

Peptides and peptide sets: Chiron Mimotopes Pty. Ltd., PO Box 1415, Clayton South, Victoria 3169, Australia (Phone +61 3 9565 1111, Fax +61 3 9565 1199); or San Diego, CA, USA (Phone +1 619 558 5800, Fax +1 619 558 5810); or France (Phone +33 1 41 38 94 00, Fax +33 1 41 38 94 09).

References

De Magistris MT, Alexander J, Coggeshall M, et al (1992) Antigen analog–major histocompatibility complexes act as antagonists of the T cell receptor. Cell 68: 625–634.

Guo H-C, Jardetzky TS, Garrett TPJ, Lane WS, Strominger JL, Wiley DC (1992) Different length peptides bind to HLA-Aw68 similarly at their ends but bulge out in the middle. Nature 360: 364–366.

Maeji NJ, Bray AM, Geysen HM (1990) Multi-pin peptide synthesis strategy for T cell determinant analysis. J Immunol Methods 134: 23–33.

Mutch DA, Rodda SJ, Benstead M, Valerio RM, Geysen HM (1991) Effects of end groups on the stimulatory capacity of minimal length T cell determinant peptides. Peptide Res 4: 132–137.

Reece JC, Geysen HM, Rodda SJ (1993) Mapping the major human T helper epitopes of tetanus toxin. The emerging picture. J Immunol 151: 1–10.

Reece JR, McGregor DL, Geysen HM, Rodda SJ (1994) Scanning for T helper epitopes with human PBMC using pools of short synthetic peptides. J Immunol Methods 172: 241–254.

Taylor PM, Thomas DB, Mills KHG (1987) In vitro culture of T cell lines and clones. In Klaus GGB, ed. Lymphocytes. IRL Press, Oxford, pp. 133–147.

The generation of peptide combinatorial libraries using tea-bag synthesis: Identification of B-cell epitopes

10.10

Clemencia Pinilla
Jon R. Appel
Richard A. Houghten

Torrey Pines Institute for Molecular Studies, 3550 General Atomics Ct., San Diego, California, USA

TABLE OF CONTENTS

Immunology Methods Manual
ISBN 0–12–442712–X

Abstract

This chapter describes the synthesis of two peptide libraries in different formats using the simultaneous multiple peptide synthesis (SMPS) method, commonly known as the 'tea-bag' approach. Each of these libraries can be used for the identification of antigenic determinants recognized by monoclonal antibodies.

Included in this chapter is a list of the materials and reagents required to carry out the screening of each peptide library using a competitive ELISA. Protocols for optimizing the ELISA conditions, which are critical for the successful screening of each peptide library, are also given.

Introduction

Synthetic peptide libraries composed of millions of different sequences offer a fundamental, practical advance in the study of antigen–antibody interactions, as well as for the potential development of therapeutic drugs (Geysen et al 1986; Pinilla et al 1992; Houghten et al 1991, 1992a; Lam et al 1991). Of the various peptide library approaches described, only soluble synthetic peptide combinatorial libraries (SPCLs), prepared using the tea-bag approach, offer the advantage of working with free peptides in solution, enabling their concentrations to be adjusted to accommodate particular assay systems. SPCLs have been successfully used for the identification of a number of different antigenic determinants (or B cell epitopes) recognized by antibodies (Appel et al 1992; Pinilla et al 1992, 1993, 1994a and b). SPCLs have also proved useful for the identification of novel peptide sequences, such as enzyme inhibitors (Eichler and Houghten 1993), receptor antagonists and agonists (Houghten and Dooley 1993; Dooley et al 1993, 1994), and potent antimicrobial agents (Blondelle et al 1994, 1995a and b), as well as for the potential development of therapeutic drugs.

Epitope mapping approaches using peptide libraries have yielded detailed information on the positional importance of each residue in the epitope as well as the degree of polyspecificity of the antibody due to the enormous diversity of peptide sequences present in each library. This chapter describes how two differently formated hexapeptide libraries and competitive ELISA can be used for the identification of individual peptide sequences that inhibit the binding of a target antibody to its respective antigenic peptide or protein. Owing to the enormous number of peptides in each peptide mixture of the SPCLs, the specific ELISA conditions used for each antigen–antibody interaction study must be optimized to achieve the highest level of sensitivity.

Synthesis of peptide libraries

The dual fixed-position acetylated SPCL is shown in Table 10.10.1. The first two positions of each hexapeptide mixture of this SPCL are individually defined with a single L-amino acid (represented as O), while the remaining four positions are present as mixtures (represented as X), incorporating 19 of the 20 natural L-amino acids (cysteine excluded). This SPCL is represented as $Ac-O_1O_2XXXX-NH_2$ which yields 400 different peptide mixtures. Since each peptide mixture consists of 130,321 (19^4) different peptides, the peptide library is made up of 52,128,400 ($400 \times 130,321$) different hexapeptides. As an alternative, a similarly formated, nonacetylated SPCL can be used. Also, the two defined positions can be located at any of the six positions of the sequence.

The second peptide library, termed a positional scanning SPCL (PS-SPCL), is shown in Table 10.10.2. This PS-SPCL is also acetylated and is made up of six individual peptide libraries, each with one position defined. Each positional peptide library consists of hexapeptides in which one position is defined with one of the 20 L-amino acids, while the remaining five positions are mixtures of 19 amino acids (cysteine excluded). This library is represented as $Ac-O_1XXXXX-NH_2$, $Ac-XO_2XXXX-NH_2$, ... $Ac-XXXXXO_6-NH_2$. Each peptide mixture represents more than 2 million (19^5) individual sequences; the entire peptide library thus contains more than 52 million different hexamers.

Two different approaches to the synthesis of peptide mixtures making up each peptide library are used: 'physical' and 'chemical'. Both approaches can be carried out easily using the tea-bag synthesis method. The first method can generally be termed the divide, couple, and recombine (DCR, Houghten et al 1991, 1992a) physical method because it involves physically dividing and recombining the peptide resins after each coupling step. Each resin bead contains only a single peptide. It is, however, limited to peptides of four to five residues in length if the 20 natural L-amino acids are used, since increasing the number of mixture positions requires 20 times more starting resin of each position added to ensure the presence of a statistical number of beads for every peptide. Also, for practical reasons, the defined positions can only be fixed to the amino end of the mixture positions.

The second approach, termed the chemical ratio method, prepares peptide mixture resins using a specific ratio of amino acids, empirically defined to give an equimolar incorporation of each amino acid at each coupling step (Pinilla et al 1992; Ostresh et al 1994). The degree of equimolarity of the resulting peptide mixtures is not as exact as that of the physical method owing to the differences in amino acid coupling rates. Each resin bead contains a mixture of peptides. Approximate equimolar representation can then be confirmed by amino acid analysis.

Table 10.10.1 Dual fixed-position SPCL

$Ac-O_1O_2XXXX-NH_2$

1.	$Ac-AAXXXX-NH_2$
2.	$Ac-ACXXXX-NH_2$
3.	$Ac-ADXXXX-NH_2$
:	
398.	$Ac-YVXXXX-NH_2$
399.	$Ac-YWXXXX-NH_2$
400.	$Ac-YYXXXX-NH_2$

Table 10.10.2 Positional scanning SPCL (PS-SPCL)

$Ac-O_1XXXXX-NH_2$
$Ac-XO_2XXXX-NH_2$
$Ac-XXO_3XXX-NH_2$
$Ac-XXXO_4XX-NH_2$
$Ac-XXXXO_5X-NH_2$
$Ac-XXXXXO_6-NH_2$

Synthesis of dual-fixed position SPCL (Ac-O_1O_2XXXX-NH$_2$)

Peptide mixture resins are prepared using the process termed divide, couple, and recombine (DCR, Houghten et al 1991) in combination with simultaneous multiple peptide synthesis (SMPS, Houghten 1985). Methylbenzhydryl-amine (MBHA) polystyrene resin is used in conjunction with t-BOC chemistry, although other peptide chemistry strategies (i.e., FMOC and different resins) can also be used.

- *Divide* starting MBHA resin into 19 aliquots containing 20 grams. Place each aliquot into individual, labeled porous polypropylene packets (referred to as 'tea bags').
- *Couple* each of the 19 protected N-α-t-BOC natural L-amino acids (cysteine is excluded) to one of 19 porous polypropylene packets. Monitor coupling reaction for completion (>99.5%) using ninhydrin (Kaiser et al 1970), picric acid (Gisin 1972), or bromophenol blue (Krchnak et al 1989). Dry each packet of resin completely.
- *Recombine* the resins from each packet and mix thoroughly. This one-position resin mixture is represented as X-resin.
- *Divide* the resin mixture again into 19 portions of equal weight and place these back into porous polypropylene packets. This is followed by N-α-t-BOC protecting group removal and neutralization on all of the packets.
- Separately *couple* the 19 individual activated amino acids to the resin packets to yield the 361 dipeptide combinations (OX-resin; $19^2 = 361$ sequences).
- *Recombine* resins and repeat the DCR steps twice to obtain a final mixture of 130,321 (19^4) protected tetrapeptide mixture resins (XXXX-resin).
- *Divide* the peptide resin (XXXX-resin) into 400 aliquots and place each aliquot into numbered (1–400) porous, polypropylene packets.

Synthesize the next two defined positions using SMPS. If desired, acetylate the N-terminal. Deprotect and cleave the peptide mixtures from the resin using the low–high hydrogen fluoride method in a multiple cleavage apparatus (Houghten et al 1986). Extract peptide mixtures with water or dilute acetic acid, lyophilize peptide solutions twice, and reconstitute peptide mixtures (1–5 mg ml^{-1}) in water. Sonication is often helpful to assist in the solubilization of peptide mixtures containing hydrophobic amino acids (F,I,L,W) in the defined positions.

Synthesis of positional scanning SPCL (PS-SPCL)

The peptide mixture resins making up either the nonacetylated or the acetylated PS-SPCL are prepared using the chemical mixture approach (i.e., a specific ratio of a mixture of amino acids) (Houghten and Dooley 1993; Ostresh et al 1994) in conjunction with SMPS (Houghten 1985) and MBHA resin and t-BOC chemistries.

Amino acid analysis, using the physically divided, coupled, and recombined mixture resins as a control, is used to identify each amino acid in approximately equimolar concentration. The cleavage and extraction of peptide mixtures from the resin was carried out as described earlier for other SPCLs. Individual peptide purity and identity were characterized by RP-HPLC and MALDI-TOF mass spectrometry, respectively.

Materials used to screen SPCLs for the identification of antigenic determinants

- 96-Well microtiter plates (high binding polystyrene, 1/2 area, Costar #3690)
- Polypropylene tubes: 1-ml (for peptide mixture dilutions) and 50-ml (for antigen and antibody dilutions)
- 8- or 12-channel pipetter and repetitive pipetter
- Peptide or protein antigen of interest
- Monoclonal antibody against antigen of interest
- PBS; 0.3 M bicarbonate pH 9.3 buffer; and 1% (w/v) BSA/PBS as blocking buffer
- Peptide libraries (sold by Houghten Pharmaceuticals, Inc.; also available

from Torrey Pines Institute for Molecular Studies on a collaborative basis, or prepared following the protocols described above)
- Horseradish peroxidase (HRPO)-conjugated anti-mouse antibody specific for isotype of monoclonal antibody
- Enzyme substrate and developing reagents. For peroxidase use 3% H_2O_2 and o-phenylenediamine (Sigma #P-8287)
- 4 N Sulfuric acid
- 96-Well microtiter plate spectrophotometer with 492-nm filter

Library screening

ELISA standardization and optimization of each antigen–antibody interaction is essential for the successful identification of antigenic determinants using SPCLs and PS-SPCLs, owing to the number of peptide sequences in each peptide mixture. Thus, for each of the 400 different peptide mixtures of Ac-O_1O_2XXXX-NH_2 at 5 mg ml^{-1}, each individual peptide is present at a concentration of 50 nM. When used at 5 mg ml^{-1}, each peptide in the mixtures of the PS-SPCL is present at a concentration of 3 nM. Thus, the binding threshold of the PS-SPCL is lower than the dual fixed-position SPCL owing to the numbers of individual peptides making up each peptide mixture (2×10^6 vs 130,321, respectively).

Direct ELISA

A direct ELISA is used to determine the optimum starting concentrations of antigen and antibody.

Coat microtiter plates with 50 µl/well antigen in PBS or bicarbonate buffer (try both buffers separately to see which might give better results). Perform 2-fold serial dilutions of the antigen across the plate, starting with a concentration of 10 µg ml^{-1}. Incubate for either 2 h at 37°C or 18 h (overnight) at room temperature in a moistened box. Always incubate plates in a moistened box to avoid evaporation of reagents. Typically, 100 pmol/well is used for synthetic peptides that are 15 residues in length. Less peptide is used for those that are longer than 20 residues (25 pmol/well or lower). Shake out liquid from wells and wash plates ten times with deionized water. Remove residual water from wells by rapping plates upside down over paper towels. Avoid complete drying of wells. Repeat washing step after each incubation. Block plates for nonspecific binding by adding 100 μl/well of 1% (w/v) BSA/PBS to microtiter plates and incubate plates for 1 h at 37°C. Add 50 μl/well of antibody, performing 2-fold serial dilutions in 1% BSA/PBS down the plate. Incubate plates overnight at 4°C. For most of the antigen–antibody systems we have examined, better assay sensitivity is obtained when this step is carried out at 4°C. However, incubation at 37°C for 1 h may give reasonable results in some cases. Add 50 μl/well of secondary antibody–enzyme conjugate (goat–anti-mouse peroxidase) at manufacturer's specified dilution in 1% BSA/PBS. Incubate plates for 1 h at 37°C. Develop plates. For each plate, dissolve 1 tablet of OPD in 6 ml deionized water, and add 25 μl of 3% hydrogen peroxide. Add 50 μl/well of developing solution to plates and develop in the dark for 10 to 15 min. Terminate developing reaction with 25 μl/well of 4 N sulfuric acid. Read plates on microplate spectrophotometer at 492 nm. Choose the concentrations for antibody and antigen that give the optimum results, i.e., lowest antigen and antibody concentrations that still give high OD values (1.5–2.0).

Use these conditions for competitive ELISA.

Competitive ELISA

Coat microtiter plates with the control antigen at the predetermined concentration (see above protocol) and incubate plates in a moist box for 18 h at 25°C (or 2 h at 37°C). Wash plates 10 times with deionized water and after each subsequent incubation. Block for nonspecific binding as in direct ELISA. Add 25 μl/well of blocking buffer to each plate. Add 25 μl/well of control antigen (10, 100, and 1000 times the amount of control antigen on the plate) to the top row and perform 2-fold serial dilutions down the plate. A fixed dilution of monoclonal antibody (25 μl/well) is added to each well. Do the same for antibody concentrations 2–5 times higher and lower than the selected concentration. Incubate plates for 18 h at 4°C. Add goat–anti-mouse peroxidase conjugate, develop, and read plates as in direct ELISA. Determine the IC_{50} (inhibiting concentration of antigen in solution that yields 50% of antibody binding to antigen on the plate) of the control antigen for the three antibody concentrations. Choose the antibody concentration that gives the lowest IC_{50} while maintaining an acceptable signal-to-noise ratio (10:1). We have found that the lowest antibody concentration taken from the top of the linear portion of the saturation binding curve gives the best sensitivity for the competitive ELISA.

Use these conditions to screen the peptide library.

Screening dual fixed-position SPCL

Determination of the appropriate conditions using the direct and competitive ELISA described above is essential for successful screening results. The screening of this library will yield information about the most active amino acids in the first and second positions of a hexapeptide.

Test each of the 400 peptide mixtures of the SPCL (Ac-O_1O_2XXXX-NH_2 or O_1O_2XXXX-NH_2) at a final peptide concentration of 2.5 mg ml^{-1} against target monoclonal antibody using the conditions established by competitive ELISA. We routinely use 10 microtiter plates to assay the 400 peptide mixtures of the SPCL. The first column of each microtiter plate is used for 100% antibody binding to the antigen on the plate (no inhibitor). The second column is used for the antigen in solution serially diluted as a competitive control to ensure the assay is working in a sensitive manner. The remaining 80 wells on each plate are used to test 40 peptide mixtures with copies. We use 1-ml polypropylene tubes to aliquot the peptide library. The peptide mixtures are then arranged in vertical strips of 8, which facilitates pipetting when using a multichannel pipetter. Thus, columns 3 and 4 (as a copy) will contain peptide mixtures #1–8, columns 5 and 6 for peptide mixtures #9–16, and so on. Once the SPCL has been added to the plates, the antibody is added using a repetitive multichannel pipetter at a fixed dilution (25 µl/well)

previously determined to effectively compete for binding of the control antigen in solution with the control antigen on the plate. Express inhibitory activity of peptide mixtures as optical density (OD) values (inhibition = low OD) or convert peptide mixture activity to % inhibition relative to the binding of antibody to the control antigen from column 1 of each plate. Retest peptide mixtures at lower peptide concentrations that were found to have good inhibitory activities (>50% inhibition). Determine inhibitory concentrations at 50% of antibody binding (IC_{50}) for each peptide mixture in order to select the most active peptide mixture(s) from the peptide library.

Screening PS-SPCL

The screening of a hexapeptide PS-SPCL will yield the most active amino acids at each position of the hexapeptide. The screening of the PS-SPCL is carried out in a similar manner as for the dual-fixed position SPCL by competitive ELISA. Only 3 microtiter plates are required to screen the 120 peptide mixtures of the PS-SPCL using the same plate layout and controls as the previous SPCL screening. Determine the IC_{50} values of the most active peptide mixtures at each position.

From screening results to individual sequences

Upon screening a dual-fixed position SPCL against an antibody and selecting the most effective peptide mixture, an iterative process is then carried out in which the subsequent X positions are individually defined with each of

the 20 natural L-amino acids. It is recommended that the previous peptide mixture that was followed is also synthesized as a control in order to confirm the initial screening results. Each iteration is composed of 20 new peptide mixtures, which are then assayed by competitive ELISA. This iterative process involves ranking, selecting, and reducing the number of peptide sequences while synthetically defining one more position at each step. An example of the data obtained from screening a library and identification of an individual peptide through an iterative synthesis and selection process is given in Table 10.10.3. Depending on the relative importance of the position defined for each iteration, one should see an increase in activity ranging from 2- to 10-fold over the peptide mixture from the previous iteration. The iterative synthesis and selection process is repeated until individual peptide sequences are identified.

It is sometimes necessary to move forward with several peptide mixtures that are found to have similar activities upon screening the peptide library, or in any iterative step. Often, the difference between these peptide mixtures is due to a conservative substitution at the same position (i.e., a valine residue substituted for an isoleucine residue). Since each iteration requires the synthesis of 20 new peptide mixtures, it is important to select only the most effective peptide mixtures that are significantly different in chemical character to keep the total number synthesized to a minimum. A typical synthesis using the SMPS approach (Houghten 1985) consists of 100–150 peptide mixtures, enough for five to seven iterations. Those cases not pursued initially, however, can always be moved forward at a later date.

The PS-SPCL method is much more rapid for the identification of high-affinity peptide ligands than the dual fixed-position SPCL approach. Although the screening of either library against an antibody takes only a single day, the PS-SPCL method yields information for each of the six positions. If the results indicate good inhibition for one peptide mixture for each of the six positions, then the peptide sequence can be determined directly from the ELISA data. If the sequence of the immunogen is known, the screening results can be used to quickly locate the antigenic determinant recognized by the antibody. When more than two or three peptide mixtures are found to have good activity at any of the six positions of the PS-SPCL, then individual peptides, representing the combinations of the most effective amino acids in the defined positions, need to be synthesized. These individual peptides are then tested by competitive ELISA in order to identify the most effective peptide(s) resulting from the screening of the PS-SPCL.

The number of amino acids selected from each position that will be used to synthesize the individual peptides should be minimized. For example, if two amino acids are selected from each position, one would need to synthesize 64 peptides. In some cases, not every position of the library will yield a clear profile. It should be noted that since the PS-SPCL is composed of six separate positional SPCLs, each one can be considered independent of the others. Therefore, each positional SPCL can be independently screened and pursued using the iterative synthesis and selection process described in the screening of the dual fixed-position SPCL.

Table 10.10.3 SPCL screening and iterative process against mAb 17D09

Step	Peptide	Defined sequences × mixture combinations	Total number of sequences	IC_{50} (nM)
Screening	Ac-OOXXXX-NH$_2$	400 × 130,321	52,128,400	
Selection	**Ac-DVXXXX-NH$_2$**	1 × 130,321	130,321	58,000
Synthesis/screening	Ac-DVOXXX-NH$_2$	20 × 6,859	137,180	
Selection	**Ac-DVPXXX-NH$_2$**	1 × 6,859	6,859	15,000
Synthesis/screening	Ac-DVPOXX-NH$_2$	20 × 361	7,220	
Selection	**Ac-DVPDXX-NH$_2$**	1 × 361	361	1,100
Synthesis/screening	Ac-DVPDOX-NH$_2$	20 × 19	380	40
Selection	**Ac-DVPDYX-NH$_2$**	1 × 19	19	40
Synthesis/screening	Ac-DVPDYO-NH$_2$	20 × 1	20	
Peptide	**Ac-DVPDYA-NH$_2$**	1	1	2

Conclusions

SPCLs and PS-SPCLs screened against monoclonal antibodies raised against peptides should result in the clear identification of specific peptide sequences. As a general observation, the more specific a position is, the smaller will be the number of amino acids in the peptide mixtures found to be effective. Through the iterative process of a peptide mixture, one can see the relative importance of each position of the peptide as it is defined. At a particular iterative step, if the position being defined is highly specific, then only one peptide mixture will show an approximately 10-fold increase in activity over the previous peptide mixture. On the other hand, if the position being defined is redundant or replaceable, then little or no increase in activity is found for a majority of the peptide mixtures relative to the previous peptide mixture. In other instances, if the position being defined is moderately replaceable, then a number of peptide mixtures having similar chemical characteristics are found to show an increase in activity of 3- to 5-fold.

It is important to optimize ELISA conditions for low-affinity antibodies ($K_a < 10^6$ M) to ensure successful identification. A nonacetylated SPCL may be a better choice for screening, if no significant inhibition is found or for those antigen–antibody interactions that require a free amine on the N-terminal of the peptide for antibody recognition. For example, a non-acetylated SPCL was successfully used for the identification of the antigenic determinant of β-endorphin as recognized by mAb 3E7 since the first residue of the antigenic determinant is the first residue of β-endorphin, namely tyrosine, which has a free amine group (Pinilla et al 1993).

Antibodies that have been raised against proteins often recognize discontinuous antigenic determinants. Since peptide libraries are composed of linear hexapeptide sequences, one may assume that such libraries are not applicable for identifying discontinuous determinants. However, we have examined a number of such antigen–antibody systems in which specific sequences were identified (Pinilla et al 1995). It should be noted that such sequences, while effective antigens, are likely to be only a portion of a much larger antigenic determinant. New peptide libraries of different formats and lengths (Pinilla et al 1994a and b), or constrained libraries using disulfide bridges (Blondelle et al 1995a and b), have been designed and are being prepared for future use in the identification of antigenic determinants.

References

Appel JR, Pinilla C, Houghten RA (1992) Identification of related peptides recognized by a monoclonal antibody using a synthetic peptide combinatorial library. Immunomethods 1: 17–23.

Blondelle SE, Takahashi E, Weber PA, Houghten RA (1994) Identification of antimicrobial peptides using combinatorial libraries made up of unnatural amino acids. Antimicrob Agents Chemother 38: 2280–2286.

Blondelle SE, Perez-Paya E, Dooley CT, Pinilla C, Houghten RA (1995a) Chemical combinatorial libraries, peptidomimetics and peptide diversity. Trends Anal Chem, 14: 83–92.

Blondelle SE, Takahashi E, Dinh KT, Houghten RA (1995b) The antimicrobial activity of hexapeptides derived from synthetic combinatorial libraries. J Appl Bacteriol, 78: 39–46.

Dooley CT, Chung NN, Schiller PW, Houghten RA (1993) Acetalins: Opioid receptor antagonists determined through the use of synthetic peptide combinatorial libraries. Proc Natl Acad Sci USA 90: 10811–10815.

Dooley CT, Chung, NN, Wilkes BC, et al (1994) An all D-amino acid opioid peptide with central analgesic activity from a combinatorial library. Science 266: 2019–2022.

Eichler J, Houghten RA (1993) Preparation of synthetic peptide combinatorial libraries on cotton carriers and their application to the identification of trypsin inhibitors. In Schneider CH, Eberle AN, eds. Peptides 1992. ESCOM, Leiden, pp. 320–321.

Geysen HM, Rodda SJ, Mason TJ (1986) A priori delineation of a peptide which mimics a discontinuous antigenic determinant. Mol Immunol 23: 709–715.

Gisin BF (1972) The monitoring of reactions in solid-phase peptide synthesis with picric acid. Anal Chim Acta 58: 248–249.

Houghten RA (1985) General method for the rapid solid-phase synthesis of large numbers of peptides: specificity of antigen–antibody interaction at the level of individual amino acids. Proc Natl Acad Sci USA 82: 5131–5135.

Houghten RA, Dooley CT (1993) The use of synthetic peptide combinatorial libraries for the determination of peptide ligands in radio-receptor assays: opioid peptides. BioMed Chem Lett 3: 405–412.

Houghten RA, Brays MK, De Graw ST, Kirby CJ (1986) Simplified procedure for carrying out simultaneous multiple hydrogen fluoride cleavages of protected peptide resins. Int J Pept Protein Res 27: 673–678.

Houghten RA, Pinilla C, Blondelle SE, Appel JR, Dooley CT, Cuervo JH (1991) Generation and use of synthetic peptide combinatorial libraries for basic research and drug discovery. Nature 354: 84–86.

Houghten RA, Appel JR, Blondelle SE, Cuervo JH, Dooley CT, Pinilla C (1992a) The use of synthetic peptide combinatorial libraries for the identification of bioactive peptides. Biotechniques 13: 412–421.

Houghten RA, Blondelle SE, Cuervo JH (1992b) Development of new antimicrobial agents using a synthetic peptide combinatorial library involving more than 34 million hexamers. In Epton R, ed. Innovation and Perspectives in Solid Phase Synthesis. Solid Phase Conference Coordination, Ltd. Andover, UK, pp. 237–239.

Kaiser ET, Colescott RL, Blossinger CD, Cook PI (1970) Color test for detection of free terminal amino groups in the solid-phase synthesis of peptides. Anal Biochem 34: 595–598.

Krchnak V, Vagner J, Eichler J, Lebl M (1989) Color monitored solid phase peptide synthesis. In Jung G, Bayer E, eds. Peptides 1988. Walter de Gruyter, Berlin, pp. 232–234.

Lam KS, Salmon SE, Hersh EM, Hruby VJ, Kazmierski WM, Knapp RJ (1991) A new type of synthetic peptide library for identifying ligand-binding activity. Nature 354: 82–84.

Ostresh JM, Winkle JH, Hamashin VT, Houghten RA (1994) Peptide libraries: Determination of relative reaction rates of protected amino acids in competitive couplings. Biopolymers 34: 1681–1689.

Pinilla C, Appel JR, Blanc P, Houghten RA (1992) Rapid identification of high affinity peptide ligands using positional scanning synthetic peptide combinatorial libraries. Biotechniques 13: 901–905.

Pinilla C, Appel JR, Houghten RA (1993) Synthetic peptide combinatorial libraries (SPCLs): identification of the antigenic determinant of β-endorphin recognized by monoclonal antibody 3E7. Gene 128: 71–76.

Pinilla C, Appel JR, Houghten RA (1994a) Investigation of antigen–antibody interactions using a soluble, nonsupport-bound synthetic decapeptide library composed of four trillion sequences. Biochem J 301: 847–853.

Pinilla C, Appel JR, Blondelle SE, et al (1994b) Versatility of positional scanning synthetic combinatorial libraries for the identification of individual compounds. Drug Dev Res 33: 133–145.

Pinilla C, Appel JR, Houghten RA (1995) Detailed studies of antibody specificity using synthetic peptide combinatorial libraries. In: Brown F, Chanock R, Ginsberg H, Norrby E, eds. Vaccines 1995: molecular approaches to the control of infectious diseases. Cold Spring Harbor Laboratory Press, Cold Spring Harbor, USA. pp. 13–17.

Multiple peptide synthesis on resin

10.11

Arne Holm[1]
Søren Østergaard[1]
Robert S. Hodges[2]
Devon Husband[2]
Søren Buus[3]

[1]*Research Center for Medical Biotechnology, Chemistry Department, The Royal Veterinary and Agricultural University, Copenhagen, Denmark*

[2]*Protein Engineering Network of Centers of Excellence, Biochemistry Department, University of Alberta, Edmonton, Canada*

[3]*Institute of Medical Microbiology and Immunology, Panum Institute, Copenhagen, Denmark*

TABLE OF CONTENTS

Immunology Methods Manual
ISBN 0–12–442712–X

Objective and principle of method

In recent years a number of methods for synchronous chemical solid phase synthesis of a multitude of peptides have been published (reviewed in this manual and in Gallop et al 1994). The synthesis of sequence-overlapping peptides and of analogues derived from the primary structure of a given protein, or of representative peptide libraries can be undertaken and used to identify the specificities of monoclonal antibodies, T cell receptors, as well as other receptors. Among these methods, the tea-bag and the MCPS method offer the possibility of producing milligram-scale quantities sufficient for various biological investigations and spectroscopic measurements, such as nuclear magnetic resonance (NMR).

In this chapter the MCPS (multiple column peptide synthesis) method will be discussed (Holm et al 1988; Meldal et al 1993). This method is based on a parallel array of small packed columns (or wells) in a reaction block. Solvents and deprotecting reagents are dispensed from two washers in a parallel fashion. Activated and protected amino acids are transferred, e.g. 8 at a time, from a dispenser tray as solutions.

Experimental procedure

Schematic representation of the synthesis apparatus

A schematic representation of the apparatus is shown in Fig. 10.11.1 depicting a part of the synthesis unit (2) with columns and drainage system (3) and part of the liquid introduction system (1). The synthesis unit (2) is made of an inert material such as Teflon, comprising a number of columns, e.g. 4×4 or 10×10. Each column is open at the top and has a liquid outlet at the bottom. Essentially it consists of two parts, an upper part containing the columns and a lower part containing a drainage chamber (3). The individual column consist of two bores, an upper bore of relatively large diameter in which the resin for peptide synthesis is placed, and a narrow lower bore into which a tightly fitting tube made of stainless steel or glass is inserted. To retain the resin a Teflon filter is placed at the bottom of the upper bore. Liquid chemical substances such as solutions of the protected and activated amino acids in DMF* can be introduced into the synthesis chambers in any suitable manner, but transfer with a multipipette is preferred. Liquid within the columns may be controlled by connecting the drainage chamber to either a pressure source (nitrogen) to keep liquid in the synthesis unit or a vacuum source to empty the synthesis unit.

Solvents (e.g. DMF) and deprotecting reagents (e.g. piperidine) are introduced into the columns by one or two liquid introduction units (optimally one unit for washing and one unit for deprotection). The special design of the liquid introduction unit allows liquid to be delivered into all synthesis chambers simultaneously and evenly.

After completion of the synthesis, the synthesis unit is disassembled and cleavage of the

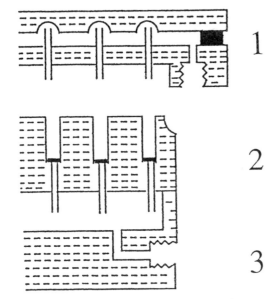

1

2

3

Figure 10.11.1 Schematic representation of apparatus for multiple peptide synthesis on resin showing (1) the liquid introduction system, (2) the synthesis unit and (3) the columns and drainage system.

peptide from the resin is carried out directly in the columns, whereafter the products are collected into glass tubes placed one under each column.

Information about the commercial aspect of the machine or about requisition of a complete set of drawings may be obtained from Dr RS Hodges.

Peptide synthesis

Peptide resins like NovaSyn K (Calbiochem-Novabiochem), a composite of polydimethyl-acrylamide and kieselguhr, and TentaGel (Rapp polymere), a graft polymer of poly(ethylene glycol) onto polystyrene have been used with

*DMF = dimethylformamide; Opfp = pentafluorophenyl; DhbtOH = 3,4-dihydro-3-hydroxy-4-oxo-1,2,3-benzotri-azine; DIC = diisopropylcarbodiimide; BOP = benzo-triazole-1-yl-oxy-tris-(dimethylamino)-phosphonium; HPTU = 2-(1H-benzotriazole-1-yl)-1,1,3,3-tetramethyluronium hexafluorophosphate; HOBt = 1-hydroxybenzotriazole; Fmoc = 9-fluorenylmethyloxycarbonyl; DIEA = N,N-diiso-propylethylamine; TFA = trifluoroacetic acid; EDT = ethane-dithiol.

success in the MCPS apparatus. Both resins are commercially available preloaded with a handle and an Fmoc-amino acid.

Two different coupling strategies can be employed. The amino acids may either be coupled as their Opfp* esters together with DhbtOH* as catalyst, or activated *in situ* by the commonly used reagents DIC*, BOP* or HBTU* in combination with HOBt*. The former method allows the detection of the light yellow DhbtOH anion as a probe for monitoring the completion of the simultaneous coupling reactions (Cameron et al 1988). The latter method allows the preparation of stock solutions of Fmoc-amino acids together with HOBt and these solutions can be stored for several weeks at –20°C without any side reactions.

In a typical peptide synthesis experiment a block containing 10×10 wells is used. In each well 125 mg TentaGel resin with Rink amide linker (substitution = 0.24 mmol g^{-1} or 0.03 mmol/column) is added.

Synthesis cycles consist of the following operations:

- Removal of the Fmoc-group with 2×25 ml 20% (v/v) piperidine in DMF (2×5 min).
- After deprotection the wells are washed with 5×45 ml DMF (5×1 min).
- Stock solutions of Fmoc-amino acids/HOBt (0.2 M) are activated with a solution of HBTU (0.2 M) and DIEA (1.5 eq. relative to Fmoc-amino acid).
- After activation, 3 eq. relative to resin substitution of Fmoc-amino acids are added to the wells and allowed to couple for 2 h or more followed by a double coupling.
- Excess Fmoc-amino acid is removed by washing with 5×45 ml DMF (5×1 min).
- After washing with ethanol and drying overnight, cleavage of the peptides from the support is accomplished by TFA–H$_2$O–EDT–thioanisole (90:5:3:2 by volume), or TFA/H$_2$O (95:5, v:v),

---

Multiple peptide synthesis on resin

depending on side-chain-protecting groups. The TFA solutions are concentrated in a Speed Vac to 1/5

volume and precipitated and washed in diethylether and lyophilized.

Applications

MCPS for the preparation of defined peptides

As described elsewhere in this chapter, synthetic peptides can be used to induce and characterize immunological reactions. To raise B and/or T cell responses, crude peptides in conjunction with suitable adjuvants (e.g. complete Freund's adjuvant) can be used to immunize experimental animals. MCPS yields sufficient amounts of peptides to immunize cohorts of animals. It also yields enough material for purification (e.g. by high-pressure liquid chromatography (HPLC)) and verification (e.g. by mass spectrometry) prior to analytical work. B cells mainly recognize the three-dimensional structure of antigens and consequently peptides derived from the original protein sequence predominantly represent linear B cell epitopes (Birkelund et al 1994; Ibsen et al 1993). In contrast, T cells always recognize short peptides presented in association with MHC molecules and their specificities are well addressed using synthetic peptides. Overlapping peptides derived from the sequence of interest, and sets of analogue peptides, can be used for a detailed analysis of T cell epitopes (Petersen et al 1992, 1993). MCPS can easily handle the large number of different peptides needed for these approaches. Finally, major histocompatibility complex (MHC) class I molecules tend to recognize short peptides with free amino and carboxy termini, i.e. they prefer soluble peptides. MCPS yields soluble peptides and is therefore suitable for the identification of MHC class I restricted cytotoxic T cell responses. MCPS has also been used for glycopeptide synthesis (Peters et al 1992) and

in connection with NMR studies (Wishart et al, 1995).

MCPS for the preparation of soluble peptide libraries

Peptide libraries are valuable tools for epitope mapping. Ideally, a library should be unbiased and it should represent all possible peptides of its sort, e.g. a library of nonamer peptides containing the 20 naturally occurring L-amino acid should include $20^9 = 512 \times 10^9$ peptides. Such diversity cannot be achieved using any of the solid phase peptide synthesis techniques (e.g. the 'portioning, mix and recombine procedure'; Furka et al 1991). Monoclonal antibodies can be screened using such libraries whether the epitope is known or not (for review see Houghten 1993; Scott and Craig 1994). However, the specificities identified may be completely unrelated to the original inducing antigen (the term mimeotopes has been used for this phenomenon).

The MCPS described here generates soluble peptides and is suitable for the synthesis of synthetic peptide combinatorial libraries (SPCLs) of much larger diversities than can be achieved by the portioning–mix procedure. These SPCLs can be based on the iterative strategy of Hougthen and colleagues (1991) represented as OOXXXX, termed dual defined SPCL, or based on the scanning strategy, O_1XXXXX, XO_2XXXX, ... $XXXXXO_6$, termed positional scanning SCL (PS-SPCL) (Pinilla et al 1992). It is important that every amino acid is incorporated equally. If an equimolar mixture of amino acids are coupled in excess to the resin, then the incorporation of amino acids will be reflected by the difference in coupling rates of the individual amino acids. This can be avoided if the mixture is coupled with an amount that is 1–1.1 equivalents relative to the resin substitution and followed by a double coupling (Kramer et al 1993; Wong et al 1994a).

We have used MCPS for the preparation of dual defined positional scanning SPCLs represented as O_1O_2XXXX, XXO_3O_4XX and $XXXXO_5O_6$ (Wong et al 1994a, b). For that purpose a block with 100 wells has been constructed in which an amount of 0.03 mmol peptide can be synthesized in each well. We have not used all 20 naturally occurring amino acids, but rather only 10 representative amino acids selected for ease of synthesis. The resulting peptides are of a better quality, the libraries are still reasonably representative, and their complexity has been reduced (e.g. in the example above only a total of 300 peptide sublibraries has to be synthesized). We have used this approach successfully to identify B cell epitopes (Wong et al 1994a, b) and MHC binding epitopes (Stryhn et al, in press).

References

Birkelund S, Larsen B, Holm A, Lundemose AG, Christiansen G (1994) Infection and Immunity 62: 2051–2057.

Cameron LR, Holder JL, Meldal M, Sheppard RC (1988) J Chem Soc, Perkin Trans 1 2895–2901.

Furka A, Sebestyen F, Asgedom M, Dibo G (1991) Int J Peptide Protein Res 37: 487–493.

Gallop MA, Barrett RW, Dower WJ, Fodor SPA, Gordon EM (1994) J Med Chem 37: 1233–1251.

Holm A, Meldal M (1989) In Bayer E, Jung G, eds. Peptides 1988. Walter de Gruyter, Berlin, p. 208.

Houghten RA (1993) Trends in Genetics 9: 235–239.

Houghten RA, Pinilla C, Blondelle SE, Appel JR, Dooley CT, Cuervo JH (1991) Nature 354: 84–86.

Ibsen PH, Holm A, Petersen JW, Olsen CE, Heron I (1993) Infection and Immunity 61: 2408–2418.

Kramer A, Volkmer-Engert R, Malin R, Reineke U, Schneider-Mergener J (1993) Peptide Res 6: 314–319.

Meldal M, Bisgaard Holm C, Boejesen G, Havsteen Jakobsen M, Holm A (1993) Int J Peptide Protein Res 41: 250–260.

Peters S, Bielfeldt T, Meldal M, Bock K, Paulsen H (1992) J Chem Soc, Perkin Trans 1 1163–1171.

Petersen JW, Holm A, Ibsen PH, Hasløv K, Capiau C, Heron I (1992) Infection and Immunity 60: 3962–3970.

Petersen JW, Holm A, Ibsen PH, Hasløv K, Heron I (1993) Infection and Immunity 61: 56–63.

Pinilla C, Appel JR, Blanc P, Houghten RA (1992) BioTechniques 13: 901–905.

Scott JK, Craig L (1994) Current Opinion Biotechnol 5: 40–48.

Stryhn A, Pedersen LØ, Romme T, et al (1996) Eur J Immunology, in press.

Wishart DS, Bigam CG, Holm A, Hodges RS, Sykes BD (1995) J Biomol NMR 5: 67–81.

Wong WY, Sheth HB, Holm A, Irvin RT, Hodges RS (1994a) In Hodges RS, Smith JA, eds. Peptides: Chemistry, Structure and Biology. ESCOM, Leiden, pp. 175–177.

Wong WY, Sheth HB, Holm A, Irvin RT, Hodges RS (1994b) In Methods: A Companion to Methods in Enzymology. Academic Press, Orlando, 6: 404–410.

Outlook

10.12

Siegfried Weiss

Department of Cell Biology and Immunology, GBF National Research Center for Biotechnology, Braunschweig, Germany

TABLE OF CONTENTS

Immunology Methods Manual
ISBN 0–12–442712–X

Outlook

The knowledge of epitopes recognized by T cells and antibodies is an essential prerequisite in understanding immune responses. Besides the immunological importance, the knowledge of epitopes recognized by antibodies which interfere with functional aspects of protein antigens (e.g. substrate binding by enzymes or the binding of receptor to ligands) contributes much to the definition of functional domains or other features of proteins under investigation.

The screening for T cell epitopes will be facilitated by the possibility of narrowing the epitope to a few possible peptides because they have to be presented by molecules of the major histocompatibility complex (MHC) to be recognized by the T cell receptor. Since binding to the MHC molecule needs particular amino acids at certain positions, possible epitopes can be predicted. This situation will be improved further as more data about MHC–peptide interaction are accumulated (see Section 9). The recent finding that peptide analogues of T cell epitopes might antagonize a stimulatory signal provided by the nominal epitope has raised great hopes for the treatment of ongoing autoimmune diseases. Interaction of the T cell receptor and the MHC–peptide complex is too little understood to allow prediction of an antagonistic peptide. Systematic screening will therefore be necessary, and in this section several ways in which this can be achieved with minimal effort have been outlined.

Any reasonable prediction of B cell or antibody epitopes will not be possible within the foreseeable future. Thus, the methods described in this section will provide valuable help. It is to be expected that some of the problems indicated regarding construction and expression of fusion proteins, such as instability, will be solved in the near future with the use of inducible promoters which are very tightly controlled, thus avoiding the counterselection against the fusion protein that is often observed. The use of fusion proteins might be overtaken by the automation of simultaneous peptide synthesis, which might make obsolete the use of fusion proteins for epitope screening, even for large proteins.

Thus, continuous technological improvements and automation rather than new concepts are to be expected. This is best demonstrated by the use of mass spectrometry in epitope definition. This technology used to be limited to experts and has been successfully applied for the study of peptide binding to MHC molecules (see chapter 9.4). Newly developed table top machines for resolving peptides are available in the meantime which can be handled by a reasonably skilled laboratory technician. Antibody epitopes can now be narrowed down by analyzing protein fragments with and without preincubation with the particular antibody (Zhao and Chait 1994).

In a similar way, the availability of peptide libraries, some of them described in this section, will improve epitope screening and will make the epitopes of antibodies accessible even where the antigen is not known, e.g. immunoglobulins derived from B cell lymphoma cells and so on. These libraries were originally designed to define confirmation-dependent epitopes which could not be defined by the conventional approach of overlapping peptides containing the entire sequence of the protein. These so-called mimotopes should make even some of the confirmational epitopes available to analysis by extrapolation from the peptide sequence recognized in the library to the amino acids recognized on the native antigen (Geysen et al 1986).

Another approach to screening for epitopes is the use of phage display libraries. They represent a collection of filamentous bacteriophages expressing in one of their coat proteins one of a set of random peptides which were engineered into the corresponding gene (Hoess 1993). Some of these libraries are commercially available and, since high-titered phage stocks can be obtained, a large number of phages can easily be handled. By several rounds of selection with antibodies and expansion of the bound phages, the epitope can be determined by sequencing the DNA of the enriched bacteriophage. This method has also been used successfully for determining binding motifs for MHC molecules (see Section 9).

In conclusion, the future of epitope screening will primarily be improvement of the automation of the chemical synthesis and screening procedures. Many of these improvements are already foreseeable.

References

Geysen HM, Rodda SJ, Mason TJ (1986) *A priori* delineation of a peptide which mimics a discontinuous antigenic determinant. Mol Immunol 23: 709–715.

Hoess RH (1993) Phage display of peptides and protein domains. Current Opinion Struct Biol 3: 572–579.

Zhao Y, Chait BT (1994) Protein epitope mapping by mass spectrometry. Anal Chem 66: 3723–3726.

Recommended reading

10.13

Jung G, Beck-Sickinger AG (1992) Methods and applications of multiple peptide synthesis. Angew Chem Int Ed Engl 31: 367–383

Sassenfeld HM (1990) Engineering proteins for purification. Trends Biotechnol 8: 88–93.

Section 11

Development of Cells of the B Lineage

Section Editor
Christopher J. Paige

LIST OF CHAPTERS

Introduction

11.1

Christopher J. Paige

The Wellesley Hospital Research Institute and Department of Immunology, University of Toronto, Toronto, Ontario, Canada

Studying the process of hematopoietic cell growth and differentiation has engaged the energies of many investigators for decades. This developmental system comprises multiple cell types, complex pathways of differentiation and an interactive network of soluble and membrane-bound proteins. Although many of the cellular components have been well characterized, the underlying genetic basis for commitment and progression along a particular developmental pathway remains to be established. Invariably, increased understanding has been accompanied by the introduction of novel or improved techniques. Functional assays which allow the *in vitro* growth of single cells revealed precursor product relationships and also led to the identification of cytokines which regulate growth and maturation. Analytical tools for examining both intracellular and cell surface properties of developing progenitors, many which are sensitive to the single cell level, have permitted the characterization of cells which initiate and emerge from these functional assays. This section brings together a set of methods which can be used to study the growth and development of B lymphocytes generated from uncommitted progenitors. Cellular cloning and analysis strategies for both murine and human cells are included.

Detection of multipotent hematopoietic cells in pre-liver embryos

11.2

Sylvie Delassus
Philippe Kourilsky
Ana Cumano

Unité de Biologie Moléculaire du gène, Institut Pasteur, Paris, France

TABLE OF CONTENTS

Introduction

Lymphocytes are continuously generated throughout life. Like all other members of the hematopoietic system B and T cells are derived from multipotent hematopoietic stem cells (Wu et al 1967). The developmental pathway which leads from multipotent stem cells to committed lymphocytes is characterized by a series of differentiation steps. *In vivo* reconstitution experiments have been used to detect the presence of cells that can reconstitute the host hematopoietic system but they cannot be used to identify intermediate stages of the developmental process. This method is also not useful for identifying growth factors and cellular interactions involved in hematopoiesis. In the last few years, *in vitro* culture conditions have been defined which support lymphocyte differentiation from uncommitted multipotent hematopoietic precursors (Cumano et al 1992; Godin et al 1995; Hirayama et al 1992; Kee et al 1994). We describe here recent advances in the development of *in vitro* culture systems which allow the differentiation of immunoglobulin-secreting plasma cells, T cells and myeloid-erythroid cells, from multipotent hematopoietic cells, isolated from pre-liver embryos. We will also describe a PCR-based analysis of the CDR3 region of the immunoglobulin heavy chain that allows one to follow the diversification of the repertoire of B lineage cells that develop *in vitro* from one single hematopoietic precursor.

An *in vitro* assay system that provides conditions for myeloid and lymphoid cell development from single precursors

Multipotent precursors capable of differentiating into all hematopoietic cells are present in the hematopoietic organs throughout life in the mouse. They constitute a minor fraction of bone marrow and fetal liver cells. In pre-liver embryos (fetal liver contains hematopoietic precursors starting at day 10 of gestation) hematopoietic precursors can be detected in the yolk sac and in the para-aortic splanchnopleura (Godin et al 1993).

Para-aortic splanchnopleura was isolated by dissection under microscope from C56BL/6 embryos at the stages between 8 and 25 somites, corresponding to day 8 to day 10 of gestation, and single-cell suspensions were made by passing splanchnopleuras 8 to 10 times through a 26-gauge needle attached to a syringe. Cells were then cloned under limiting dilution conditions, at 5 or 10 cells per well, in 96-well plates (this cell density gives around 10 growing clones per 100 wells seeded). Alternatively, whole embryos were disrupted using the same procedure. Embryos older than 20 somites were enriched for the presence of AA4.1-positive cells (a marker for multipotent hematopoietic precursors) by panning procedures (Cumano et al 1993). Micromanipulation was done by cloning AA4.1$^+$ cells at 1 cell/well in 10 μl in a Terasaki plate. Cells were allowed to settle for 30 min and the plates were inspected under microscope for the presence of a single cell. Individual cells were then transferred into single wells of 96-well plates in the presence of S17 stromal cells and IL-7; in some experiments kit-ligand and IL-3 were added. After 10 to 12 days, wells containing multiple cell types were identified by microscope inspection and clones were divided into three different culture conditions: (1) S17 cells and IL-7, allowing the development of B lineage cells; (2) irradiated fetal thymic lobes from C57BL/Ka congenic Ly5.1 embryos providing the conditions for T cell differentiation; (3) S17 cells and a mixture containing c-kit ligand, GM-CSF, M-CSF and IL-3 allowing the detection of multilineage myeloid precursors. Erythropoietin was added in some experiments for mature erythrocyte differentiation.

Using this procedure we could detect multipotent hematopoietic precursors in the para-aortic splanchnopleura starting at the stage of 8–10 somites and their absolute numbers increased until the stage of 25 somites (Godin et al 1995). Every clone analyzed containing B cell precursors also contained multiple myeloid and T cell precursors showing that most B cell precursors in pre-liver embryos are multipotent hematopoietic precursors.

Reagents

Medium

The culture medium used in the assay is Opti-MEM medium (Gibco BRL), made from powder and filtered through a 0.22 μm 500 ml filter, supplemented with 10% (v/v) fetal calf serum (FCS), 100 IU ml^{-1} penicillin, 100 μg ml^{-1} streptomycin and 5×10^{-5} M 2-mercaptoethanol.

S17 cells are expanded in Opti-MEM medium containing 2% (v/v) FCS, 100 IU ml^{-1} penicillin and 100 μg ml^{-1} streptomycin.

EBSS (Gibco BRL) is a balanced salt solution used for all cell suspensions, for isolation and irradiation of thymic lobes, and in all panning procedures. We make it from powder and it is buffered to pH 7.4 with Tris–HCl.

Panning procedure

Panning is performed as described (Kincade et al 1981) in Optilux 100-mm Falcon 1001 plastic Petri dishes (Becton-Dickinson, CA, USA). For these experiments, a two-step procedure is used to coat plates, as previously described (Cumano and Paige 1992). The first coating antibody is an affinity-purified polyclonal mouse anti-rat IgG (50 μg/plate) (Jackson Immunoresearch Laboratories. Jackson, ME, USA). After 3 washings with cold EBSS, the plates are subsequently incubated with saturating amounts of supernatant from the hybridoma cell line AA4.1 (McKearn et al 1985). A maximum of 2×10^7 cells are seeded per plate. Nonadherent cells are recovered by two gentle washes with ice-cold EBSS–2% (v/v) FCS. Adherent cells are recovered using a rubber policeman, either directly or after additional washes (between 6 and 8, depending on the enrichment required and the number of cells applied initially).

Cytokines and growth factors

Recombinant IL7 and IL-3 are obtained from culture supernatant of cell lines, transfected with the BMG vector containing the corresponding cDNAs (from Fritz Melchers, Basle, Switzerland) (Karasuyama and Melchers 1988) and are titrated on the 2E8 IL-7-dependent cell line, a kind gift from P. Kincade, and on IL-3-dependent bone marrow mast cells. The c-kit ligand, also called stem cell factor, is obtained from stably transfected CHO cells with the cDNA for c-kit ligand (Genetics Institute, Boston, MA, USA); the supernatant was titrated on c-kit ligand-dependent mast cells from bone marrow.

We use WEHI-3 conditioned medium as a source of granulocyte-macrophage and macrophage colony-stimulating factor (GM-CSF and M-CSF) (a kind gift from C. Roth, in our laboratory). Human recombinant erythropoietin was obtained from I. Godin (Institut d'embryologie, Nogent-sur-Marne, France).

Stromal cells

Stromal cells are necessary to support the differentiation of multilineage progenitors to IL-7-responsive pre-B cells, as well as the transition of these IL-7-responsive cells to the LPS-reactive stage. In our assay, S17 stromal cells (from K. Dorshkind, University of California at Riverside, CA, USA) (Collins and Dorshkind 1987) are used. For the development of precursors, about 2×10^3 cells are seeded in each well of a 96-well plate (Costar, Cambridge, MA, USA) in a final volume of 50 μl of complete medium and cultured overnight at 37°C, 5%

CO_2. The plates are then irradiated with 2,000 rad using a cesium source.

Limiting dilution

Micromanipulated single cells or 5–10 cells per well are seeded in 96-well plates with S17 cells and IL-7. Although S17 cells are sufficient to support multilineage hematopoietic cell differentiation (they express mRNA for GM and M-CSF as well as for c-kit ligand), we observed that the addition of exogenous c-kit ligand and IL-3 generates larger clones in shorter periods of time (5–6 days as compared to 10–12 days). Overgrowth of basophils/mast cells is, however, a potential problem under these conditions. Presently, we supplement cultures with exogenous c-kit ligand and follow carefully the development of the clones. The plates are fed every 5 days by substituting 100 μl of medium supplemented with growth factors. After 10–12 days clones detected by microscope inspection are divided into three different culture conditions that support B cell, T cell and myeloid cell differentiation.

Detection of B lymphocytes

The detection of B cell precursors is done in medium supplemented with IL-7 in the presence of S17 cells. These conditions allow proliferation of committed B cell precursors and sIg+ cells can be detected after 5–10 days of further expansion. Lipopolyssacharide (LPS) is used as mitogen inducing the proliferation and the differentiation of mature B cells into immunoglobulin-secreting plasma cells. The first LPS-reactive cells are detected 13–15 days after the initiation of culture. For LPS stimulation, cells obtained from a single well of a 96-well plate (typically $1–10 \times 10^4$ cells) are resuspended in 100 μl and seeded into one well containing irradiated S17 cells and LPS (*Salmonella typhosa* WO901; Difco), at the final concentration of $25 \mu g \, ml^{-1}$. The final volume is $200 \mu l$/well and supernatants are collected after 12 days and tested for the presence of immunoglobulin by ELISA, performed on the supernatants as previously described (Cumano and Paige 1992).

Detection of myeloid-erythroid cells

For detection of myeloid-erythroid cells, medium supplemented with WEHI-3 conditioned medium, IL-3, c-kit ligand and eventually erythropoietin is added to one-third of the clone. Under these conditions basophils/mast cells can rapidly overgrow. Cultures are isolated 5–10 days later and spotted on to a glass slide, in a cytospin. They are then stained using the May–Grunwald–Giemsa kit (Biolyon, France).

Detection of T cells

Thymic lobes from C57BL/6 day 14–15 embryos are dissected and the two lobes are separated in a Petri dish containing EBSS. C57BL/Ka congenic Ly5.1 embryos are used as a source of fetal thymuses allowing us to distinguish between donor- and recipient-derived cells, as previously described (Uchida and Weissman 1992). The thymic lobes are irradiated with 3,000 rad. Cell suspensions (around 10^4 cells) are seeded in a Terasaki plate (max. $28 \mu l$/well) and one lobe is added to each well. Three to four lobes are used for one clone analyzed. The plate is then inverted to form a 'hanging drop' (Jenkinson et al 1982) with the lobe at the bottom of the drop, and incubated at 37°C, 5% CO_2, as previously described. After 24 to 48 h, the lobes are transferred to $0.8 \mu m$ ATTP filters (Millipore) floating in Petri dishes containing 3 ml of complete Opti-MEM. After 12 to 20 days, lobes are disrupted between two needles and cells are analyzed by flow cytometry for $TcR\alpha\beta$, $\gamma 6$, CD4, and CD8 expression in the Ly5.2$^+$ cell population (mAb A20.1).

PCR analysis of the diversity of the immunoglobulin heavy-chain repertoire

The complexity of a mature B lymphocytes repertoire (which has been calculated to be in the order of 10^8 for an adult mouse) is achieved by the association of variable (V_H), diversity (D) and joining (J_H) genes of the heavy chain and the variable and joining genes of the light chains (Tonegawa 1983), increased by the insertion of N and P nucleotides at the different junctions (Lafaille et al 1989, Tonegawa 1983). The resulting hypervariable region, or complementary determining region 3 (CDR3) is diverse in size and sequence. We followed the diversification of the immunoglobulin heavy-chain rearrangements in clones derived from a single multipotent hematopoietic precursor, by the analysis of the diversity of CDR3 lengths generated after V_H-D-J_H rearrangement.

Principle of the method

The first step in the analysis of diversity of rearrangements is a PCR amplification using a V_H-specific primer and a primer specific for the constant region of IgM. This amplification product contains a heterogeneous population corresponding to all the rearranged genes using one member of a given V_H family. An aliquot of this product is then submitted to an elongation reaction (run-off reaction) primed with an antisense fluorescent primer specific for one of the J_H genes. As this elongation product encompasses the V_H-D-J_H junction of the rearranged immunoglobulin gene, it is composed of several fragments whose sizes vary according to the length variation of the rearrangements. These fragments are then separated on an automated sequencer and size determination of the run-off products is performed using software written for this purpose (Pannetier et al 1993).

Procedure

RNA and cDNA

Cells are pelleted and resuspended in 4 M guanidine thiocyanate (GuT) containing 1% (v/v) 2-mercaptoethanol. RNA is purified in cesium chloride gradients (Chirgwin et al 1979) and resuspended in water at 1 μg μl^{-1}. To obtain single-strand cDNA, 10 g of RNA are reverse-transcribed using AMV-reverse transcriptase (Boehringer) as described (Maniatis et al 1982) and the final samples are resuspended in water.

Oligonucleotides

The specific primers used for the PCR amplifications are the following:

J558 (sense): AAGGCCACACTGACTGTAGAC
Q52 (sense): AGACTGAGCATCAGCAAAGAC
7183 (sense): GCGAATTCGATTCATCATCTCCAGAGAC
IgM.3' (antisense): CTGGATCCGGCACATGCAGATCTC

For the detection of the specific amplified products, fluorescent labeled oligonucleotides specific for the four J_H sequences are used:

J_H1 (antisense): XGACGGTGACCGTGGTCCCTGT
J_H2 (antisense): XGACTGTGAGAGTGGTGCCTTG
J_H3 (antisense): XGACAGTGACCAGAGTCCCTTG
J_H4 (antisense): XGACGGTGACTGAGGTTCCTTG

where X stands for the fluorescent dye (Fam) which was linked as recommended by the supplier (Applied Biosystems).

PCR and run-off reactions

Classical PCR reactions with a sense primer specific for one of the V_H families and antisense primer specific for the constant region (IgM) are performed on cDNA in a final volume of 25 μl, as described (Pannetier et al 1993). Amplified product (2 μl) is then submitted to a run-off elongation with a fluorescent primer specific for one of the J_H genes: in a final volume of 10 μl, 2 μl of the PCR product is added to 0.2 mM dNTPs, 0.1 μM of the fluorescent primer, Taq buffer, 3 mM MgCl$_2$ and 0.1 u of Taq polymerase. A single cycle of elongation, consisting of 2 min at 94°C, 1 min at 60°C, and 15 min at 72°C is performed in a Thermal Cycler (Perkin Elmer).

Electrophoresis

Elongation product (2 μl) mixed with the same amount of 95% (v/v) formamide/10 mM EDTA is loaded on an automated DNA sequencer (Applied Biosystems). As this elongation product encompasses the V_H-D-J_H junction of the rearranged immunoglobulin, it is composed of several fragments whose sizes vary according to the length variation of the rearrangements. Size determination of the run-off products is performed using software written for this purpose and introducing in each gel a set of size standards. This software provides an image of the gel by analyzing each detected band as a peak, the area of which is proportional to the intensity of the fluorescence (Pannetier et al 1993). Figure 11.2.1 provides an example of the CDR3 profiles obtained from cDNA from a B cell population developing *in vitro* from a single multipotent hematopoietic cell, with the J558 V_H primer and the J_H4 fluorescent primer (upper panel), compared with the profile obtained in cDNA from total spleen cells from an adult mouse (lower panel). The intensity of the bands detected on the gel (in arbitrary units) is given as a function of the length of the amplified fragment. The diversity of rearrangements detected by this method is indistinguishable in

J558/JH4

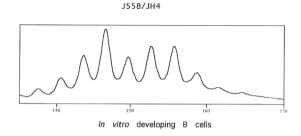

In vitro developing B cells

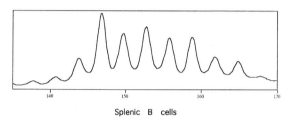

Splenic B cells

Figure 11.2.1 RNA was extracted from spleen cells and from equivalent numbers of B220⁺ cells that developed from a single multipotent hematopoietic cell isolated from a para-aortic splanchnopleura from a 22-somite embryo (day 9–9.5 of gestation). cDNA and PCR assays were done as described. Shown are the run-off profiles obtained for the J558/J_H4 primer combination.

B cells obtained from *in vitro* differentiation of a single hematopoietic precursor and in B cells in the spleen of an adult mouse.

General comments

The PCR analysis described here has previously been used to characterize the T cell repertoire in the mouse (Cochet et al 1992; Pannetier et al 1993), the repertoire of intestinal intraepithelial T lymphocytes (Regnault et al 1994), and the T cell response to hen-egg lysozyme (Cibotti et al 1994). In B cell repertoire analysis this technique can be used to analyze rearrangements in genomic DNA, circumventing differences in Ig mRNA content present in different stages of B cell differentiation (Delassus et al, 1995).

Trouble shooting

One major problem in the detection of multiple hematopoietic cells from multipotent precursors has been the differential growth rate of different cell types. IL-3 and c-kit ligand are two growth factors known to be required by multiple myeloid precursors for development. However, they induce strong proliferation of granular basophils/mast cells that can rapidly invade the cultures and possibly inhibit other cell types from developing. We suspended the addition of IL-3 but have continued using exogenous c-kit ligand. Daily observation of cultures and rapid recognition of this cell type are important. When mast cells are identified in the culture, we transfer the content of the well into a fresh S17 feeder layer without c-kit ligand but with the remaining growth factors required.

For fetal thymic organ cultures, we prefer thymic lobe irradiation to the dioxyguanosine treatment of thymic lobes for endogenous thymocyte depletion prior to reconstitution. In our experience, 2000 rad is insufficient to deplete endogenous thymocytes. With an irradiation of 3000–3500 rad no endogenous lymphocytes are detected after 9 days of culture, whether or not the lobes are reconstituted by hematopoietic cells.

The fetal bovine serum used in all these experiments has been obtained from different sources. We usually test it for the expected efficiency when cloning B cell precursors from a day 12 fetal liver with stromal cells and IL-7. Other members of our laboratory test it in other conditions, e.g. transfections and CTL assays, and typically the best serum in their assays is also the best for us. Many tested batches give satisfactory results.

Acknowledgments

This work was done in collaboration with I. Godin and F. Dieterlen-Livre from the Institut d'Embryologie du CNRS et du College de France, Nogent-sur-Marne.

References

Chirgwin J, Przybyla A, MacDonald R, Rutter W (1979) Isolation of biologically active ribonucleic acid from sources enriched in ribonuclease. Biochemistry 18: 5294–5299.

Cibotti R, Cabaniols J-P, Pannetier C, et al (1994) Public and private Vb T cell receptor repertoires against hen egg white lysozyme (HEL) in non-transgenic versus HEL transgenic mice. J Exp Med 180: 861–872.

Cochet M, Pannetier C, Regnault A, Darche S, Leclerc C, Kourilsky P (1992) Molecular detection and *in vivo* analysis of the specific T-cell response to a protein antigen. Eur J Immunol 22: 2639–2647.

Collins L, Dorshkind K (1987) A stromal cell line from myeloid long-term bone marrow cultures can support myelopoiesis and B lymphopoiesis. J Immunol 138: 1082–1087.

Cumano A, Paige C (1992) Enrichment and characterization of uncommitted B-cell precursors from fetal liver at day 12 of gestation. EMBO J 11: 593–601.

Cumano A, Paige CJ, Iscove NN, Brady G (1992) Bipotential precursors of B cells and macrophages in murine fetal liver. Nature 356: 612–615.

Cumano A, Furlonger C, Paige C (1993) Differentiation and characterization of B-cell precursors detected in the yolk sac and embryo body of embryos beginning at the 10- to 12-somite stage. Proc Natl Acad Sci USA 90: 6429–6433.

Delassus S, Gey A, Darche S, Cumano A, Roth C, Kourilsky P (1995) PCR-based analysis of murine immunoglobulin heavy-chain repertoire. J Immunol Methods 184: 219–229.

Godin I, Garcia-Porrero J, Coutinho A, Dieterlen-Lievre F, Marcos M (1993) Para-aortic splanchnopleura from early mouse embryos contains B1a cell progenitors. Nature 364: 67–70.

Godin I, Dieterlen-Lievre F, Cumano A (1995) Emergence of multipotent hematopoietic cells in the yolk sac and para-aortic splanchnopleura of 8.5 dpc mouse embryos. Proc Natl Acad Sci USA 92: 773–777.

Hirayama F, Shih J, Awgulewitsch A, Warr G, Clark S, Ogawa M (1992) Clonal proliferation of murine lymphohematopoietic progenitors in culture. Proc Natl Acad Sci USA 89: 5907–5911.

Jenkinson EJ, Fanchi LL, Kingston R, Owen JJT (1982) Effect of deoxyguanosine on lymphopoiesis in the developing thymus rudiment in vitro: application in the production of chimeric thymus rudiments. Eur J Immunol 12: 583–592.

Karasuyama H, Melchers F (1988) Establishment of mouse cell lines which constitutively secrete large quantities of interleukin 2, 3, 4 or 5, using modified cDNA expression vectors. Eur J Immunol 18: 97–104.

Kee B, Cumano A, Iscove N, Paige C (1994) Stromal cell independent growth of bipotent B cell-macrophage precursors from murine fetal liver. Int Immunol 6: 401–407.

Kincade P, Lee G, Watanabe T, Sun L, Scheid M (1981) Antigens displayed on murine B lymphocyte precursors. J Immunol 127: 2262–2269.

Lafaille J, DeCloux A, Bonneville M, Takagaki Y, Tonegawa S (1989) Junctional sequences of T cell receptor $\gamma\delta$ genes: implications for $\gamma\delta$ T cell lineages and for a novel intermediate of V-D-J joining. Cell 59: 859–870.

McKearn J, McCubrey J, Fagg B (1985) Enrichment of hematopoietic precursor cells and cloning of multipotential B-lymphocyte precursors. Proc Natl Acad Sci USA 82: 7414–7418.

Maniatis T, Fritsch E, Sambrook J (1982) Molecular Cloning: A Laboratory Manual. Cold Spring Harbor Laboratory Press, Cold Spring Harbor, N.Y.

Pannetier CMC, Darche S, Casrouge A, Zoller M, Kourilsky P (1993) The sizes of the CDR3 hypervariable regions of the murine T-cell receptor β chains vary as a function of the recombined germline segments. Proc Natl Acad Sci USA 90: 4319–4323.

Regnault A, Cumano A, Vassalli P, Guy-Grand D, Kourilsky P (1994) Oligoclonal repertoire of the CD8$\alpha\alpha$ and the CD8$\alpha\beta$ TCR$\alpha\beta$ murine intraepithelial lymphocytes: evidence for the random emergence of T cells. J Exp Med 180: 1345–1350.

Tonegawa S (1983) Somatic generation of antibody diversity. Nature 302: 575–581.

Uchida N, Weissman IL (1992) Searching for hematopoietic stem cells: Evidence that Thy-1.11^{lo} Lin$^-$ Sca$^+$ cells are the only stem cells in C57BL/Ka-Thy1.1 bone marrow. J Exp Med 175: 175–184.

Wu A, Till J, Siminovitch L, McCulloch E (1967) Cytological evidence for a relationship between normal hemopoietic colony-forming cells and cells of the lymphoid system. J Exp Med 127: 455–463.

Clonal culture assay for murine lympho-hematopoietic progenitors

11.3

Fumiya Hirayama
Makio Ogawa

Department of Medicine, Medical University of South Carolina and Ralph H. Johnson Veterans Affairs Medical Center, Charleston, South Carolina, USA

TABLE OF CONTENTS

Immunology Methods Manual
ISBN 0–12–442712–X

Introduction

The existence of pluripotent lymphohemato-poietic stem cells has long been postulated through observations in bone marrow transplantation in man. Also, studies of murine hematopoiesis using radiation-induced chromosomal markers (Abramson et al 1977) and, more recently, retroviral labeling of individual stem cells (Dick et al 1985; Keller et al 1985; Lemischka et al 1986) have indicated a single cell origin for lymphoid and myeloid precursors. However, despite all of the *in vivo* evidence, it was not possible to quantify the lymphohema-topoietic progenitors *in vitro* until recently. In 1992, investigators in three laboratories independently described culture assays for lympho-hematopoietic progenitors. Baum et al (1992) and Cumano et al (1992) developed culture assays for human fetal marrow and murine fetal liver progenitors, respectively, using coculture with murine stromal cells. In our laboratory, we developed a two-step methylcellulose clonal culture system for adult murine progenitors that have the capacity for differentiation along myeloid as well as B cell lineages (Hirayama et al 1992).

Because our culture assay does not require stromal cells, it enabled us to characterize the positive and negative cytokine regulation of early lymphohematopoiesis. In our culture system the combinations of two factors that consisted of steel factor (SF, also called ligand for c-kit, mast cell growth factor and stem cell factor) and one of interleukin (IL)-6, IL-11, granulocyte-colony-stimulating factor (G-CSF) (Hirayama et al 1992), and IL-12 (Hirayama et al 1994a) effectively supported proliferation and differentiation of the lymphohematopoietic pro-

genitors in primary culture. Differentiation into myeloid and B cell lineages was confirmed by reculture of the primary colonies into secondary cultures consisting of IL-3, conditioned medium from pokeweed mitogen-stimulated spleen cells (PWM-SCM), and erythropoietin (Ep) and cultures with SF and IL-7, respectively. Although somewhat weaker than SF, the ligands for flt3/flk2 (Hirayama et al 1995) and IL-4 (Hirayama et al 1992) were also effective in support of B cell potential of the progenitors in the primary culture when combined with one of IL-6, IL-11, and G-CSF. Interestingly, IL-3 could neither replace nor act synergistically with SF to support lymphoid potential of the progenitors in the primary culture, although it supported myeloid proliferation effectively (Hirayama et al 1992). In addition, IL-3 suppressed B-lymphoid potential of the colonies in the primary culture when added to permissive cytokine conditions (Hirayama et al 1994b). After screening available lymphohematopoietic cytokines, we found that IL-1α also possesses inhibitory effects on B-lymphoid potential. In the primary culture there appears to be a single wave of differentiation of the lymphohematopoietic progenitors. Their commitment to the B cell lineage takes place on days 6 and 7 of incubation in the presence of SF and IL-11 and concludes by day 9, followed by a wave of proliferation of the committed B cell progenitors (Ball et al, in press). Therefore, this assay system will be useful for studies of cellular and molecular mechanisms of the commitment of lympho-hematopoietic progenitors and the physiology of early committed B cell progenitors.

Reagents for culture

- Minimum essential medium, α-modification (α-MEM): The medium is prepared according to the manufacturer's instructions. We add 100,000 units penicillin and 100 mg streptomycin to 1 liter of medium.
- Fetal calf serum (FCS): It is necessary to screen lots of FCS for identification of the best myeloid colony and pre-B cell colony formation. FCS is decomplemented at 56°C for 30 min.
- Bovine serum albumin (BSA): BSA (fraction V) is also screened for its ability to support colony formation. 10% stock solution is prepared as follows (Katayama et al 1994).

  1. Add 10 g of BSA powder to 44.2 ml of distilled water in a 100 ml beaker; let stand at 4°C for 3 h to overnight for BSA to become completely dissolved.
  2. To deionize the BSA solution, add 1 g of analytical grade mixed bed resin (AG-501-X8(D), Bio-Rad, Hercules, CA, USA) to the solution and leave at 4°C for 2 h, swirling the solution gently every 30 min. Remove the resin by passing the solution through gauze or coarse filter paper, add 1 g fresh resin, and repeat the above procedure.
  3. Remove the resin.
  4. Mix the BSA solution with an equal volume of 2× α-MEM.
  5. Sterilize the solution by filtration, aliquot in 10 ml quantity and store at −20°C.
  6. Before use, adjust the pH of the solution to 7.4 with 7% (w/v) $NaHCO_3$.

- Methylcellulose, 1500 centipoise: Methylcellulose immobilizes cells in culture medium, thus allowing clonal studies of hematopoiesis. Therefore, methylcellulose culture provides an excellent opportunity for characterization of clonal descendants because it is easy to harvest cells from it. Stock solution is prepared as follows (Katayama et al 1994).

  1. Boil distilled water.
  2. Pour 300 ml of boiling water into a 2–3 liter flask containing a sterile stir bar.
  3. Add 30 g of 1500 centipoise methylcellulose to the flask slowly while stirring continuously. Mix for about 30 min to ensure that all the methylcellulose fibers are wet.
  4. Add 200 ml of 4°C distilled water and then 500 ml of cold 2× α-MEM. The final concentration of methylcellulose is 3.0% (w/v).
  5. Shake the flask vigorously to separate large clumps of gel.
  6. Stir for 24 to 48 h at 4°C until the solution becomes clear.
  7. Aliquot the mixture in 100 ml quantities. Freeze.

Procedure

Preparation of progenitors

Bone marrow cells are harvested from femurs and tibiae of mice 2 days after intravenous injection of 5-fluorouracil (5-FU, Adria Laboratories, Colombus, OH, USA). The marrow cells are then purified by a combination of metrizamide density separation, BSA density separation, negative immunomagnetic selection using lineage-specific monoclonal antibodies, and cell sorting as described previously (Shih et al 1992).

5-FU treatment of mice

Male or female BDF1 mice, 10–20 weeks old, are used. 5-FU is administered through the tail vein of mice at $150\,mg\,kg^{-1}$ body weight.

Preparation of bone marrow cells

Femurs and tibiae are harvested 2 days after 5-FU injection.
Bone marrow cells are flushed from the bones.
Pooled marrow cells are made into single-cell suspension in α-MEM by repeated pipetting and passing through 401 μm nylon mesh.

Metrizamide density separation

Two stock metrizamide (Accurate Chemical and Scientific, Westbury, NY, USA) solutions with density of 1.1000 and $1.0560\,g\,ml^{-1}$ each at 25°C are made in 1% BSA Krebs–Ringer–Tris solution (125 mM NaCl, 5 mM KCl, 1.2 mM $MgSO_4$, 35 mM Tris–HCl, 1 mM sodium phosphate, pH 7.4). Solutions with densities of 1.0631 and 1.0770 are prepared by mixing the two stock solutions.
Three ml of lower density (1.0631) solution containing bone marrow cells ($<1 \times 10^8$ cells) from 5-FU-treated mice are layered over 3 ml of higher density (1.0770) solution in a 15 ml tube.
The tube is centrifuged for 30 min at 1000g at room temperature. Cells in the interface are collected.

BSA density separation

The cells recovered after metrizamide density centrifugation are suspended in Ca^{2+}, Mg^{2+}-free PBS (PBS$^-$) containing 0.1% BSA (PBS$^-$/BSA).
The cells were suspended on top of 3 ml of 10% BSA solution in a tube and then centrifuged for 10 min at 100g at room temperature to remove debris and platelets.
The cells in the bottom of the tube are collected.

Negative immunomagnetic selection

Rat anti-mouse antibodies, anti-Mac-1 (M1/70.15.11.5.HL, American Type Culture Collection (ATCC), Rockville, MD, USA), anti-Gr-1 (RB6-8C5, a gift from Dr R.L. Coffman), anti-B220 (14.8, ATCC), anti-L3T4 (GK1.5, ATCC), anti-Ly-2 (53.6.72, ATCC), and TER119 (erythroid lineage marker, a gift from Dr T. Kina, Kyoto, Japan) are used for eliminating mature cells.

The density-separated cells are incubated with a cocktail of monoclonal antibodies for 20 min on ice.

After washing, the cells are resuspended in 1 ml PBS⁻/BSA, mixed with Dynabeads M-450 sheep anti-rat IgG (Dynal Inc., Great Neck, NY, USA) at a 1:20 cell:beads ratio and incubated at 4°C for 45 min with constant agitation.

After diluting the mixture, the tube is placed in a Dynal MPC-1 magnetic particle concentrator for 10 min and bead-free cell fraction is collected.

FACS cell sorting

Density-separated, lineage-negative (Lin⁻) cells are further enriched for primitive progenitors by sorting for Ly-6A/E positive cells (stained with D7 monoclonal antibody; PharMingen, San Diego, CA, USA) using a FACStar Plus cell sorter (Becton Dickinson, Mountain View, CA, USA); 30–45% of Lin⁻ Ly-6A/E⁺ cells give rise to myeloid lineage colonies (approximately 800-fold enriched), 40% of which also have B lineage potential (Hirayama et al 1992). Other cell surface markers, such as J11d (Shih and Ogawa 1993), c-kit (Katayama et al 1993), and CD43 (Hirayama and Ogawa 1994) are

useful for further enriching the progenitors when combined with Ly-6A/E.

Two-step methylcellulose culture assay

Methylcellulose culture

Methylcellulose cell culture is carried out in 35 mm suspension culture dishes with α-MEM containing 25% (v/v) FCS, 1% (w/v) BSA, 1×10^{-4} M 2-mercaptoethanol (2-ME), and 1.2% (w/v) methylcellulose. The protocol for quadruplicate dishes is as follows.

1. Place 1.25 ml FCS, 0.5 ml 10% (w/v) BSA, 0.05 ml 1×10^{-2} M 2-ME, 2 ml 3.0% (w/v) methylcellulose, growth factors and appropriate number of cells into a 15 ml polystyrene tube. Adjust the volume to 5 ml with α-MEM. Use a 15-gauge aluminum hub hypodermic needle to dispense the 3.0% methylcellulose. Shake the tube well or mix well using a vortex mixer.
2. Aliquot 1 ml of the mixture into each of 4 dishes using a 5 ml syringe with an 18-gauge needle. Gently tilt individual dishes so that the culture medium covers the entire surface.

Primary culture

Routinely, 50 Ly-6A/E⁺ cells or 25 Ly-6A/E⁺ J11d⁺, Ly-6A/E⁺ c-kit⁺, or Ly-6A/E⁺ CD43^bright cells per dish are plated in culture in the presence of combinations of cytokines that support the proliferation of lymphohematopoietic progenitors (Hirayama et al 1992). Dishes are incubated at 37°C in a humidified atmosphere flushed

Table 11.3.1 Analysis of B cell potentials in primary colonies supported by combinations of recombinant cytokines

Exp.	SF	IL-3	IL-4	IL-6	IL-11	G-CSF	IL-12	No. of primary colonies	Cell no. of primary colonies ($\times 10^{-4}$)	No. of pre-B cell colonies[a]
1[b]	+	+						20 ± 4	6.0	0
	+		+					2 ± 2	0.2	0
	+			+				19 ± 3	9.4	38 ± 4
	+				+			19 ± 3	5.6	71 ± 9
	+					+		18 ± 3	6.4	54 ± 1
		+		+				16 ± 2	5.6	0
		+			+			19 ± 3	4.9	0
		+				+		15 ± 6	3.7	0
			+	+				15 ± 6	2.8	5 ± 1
			+		+			18 ± 3	2.4	9 ± 2
			+			+		0		
2[c]	+				+			24 ± 6	5.5	18 ± 4
	+						+	11 ± 4	0.43	9 ± 1
		+			+			23 ± 2	3.3	0
		+					+	14 ± 3	2.8	0

Lin⁻ Ly-6A/E⁺ cells (50 cells/dish) were cultured with designated cytokines. On day 11 or 13, 20 primary colonies were picked and replated in the secondary culture containing SF and IL-7. Single factors failed to support primary colony formation except for IL-3. No pre-B cell colonies were produced from these colonies.
[a] Colony number derived from 5% of 20 pooled primary colonies.
[b] Primary colonies were replated on day 11.
[c] Replated on day 13.

with 5% CO_2/95% air for 8 to 13 days. Fifteen to twenty-five colonies per dish arise as shown in Table 11.3.1. A photomicrograph of a representative colony present on day 11 is shown in Fig. 11.3.1a.

Growth of more than 25 colonies per dish usually results in premature degeneration of the colonies, probably because of exhaustion of nutrients.

Secondary culture

On day 8–13 of primary culture, the resulting primary colonies are individually lifted from the medium using a 10 μl Eppendorf micropipette under direct microscopic visualization, pooled, and washed. Then 2.5% or 5% of the pooled samples are plated in secondary methylcellulose cultures containing SF and IL-7. When appropriate combinations of cytokines are used in the primary culture, large compact unicentric colonies consisting of small round cells are observed in the secondary culture after 10 to 12 days of incubation (Fig. 11.3.1b).

Because this culture system is not very favorable for expression of the myeloid lineages, coexisting macrophage colonies are small. May–Grunwald–Giemsa staining of the lymphocyte colonies reveals lymphoblast-like cells with a nucleus that is sometimes cleft or lobulated. The pre-B cell nature of the cells has been established by their expression of cell surface markers, immunoglobulin μ-chain mRNA, as well as their ability to reconstitute B cell, but not T cell, compartments of scid mice upon adoptive cell transfer.

The timing for the replating of primary colonies into secondary pre-B cell culture is critical. The precursors for pre-B cell colonies start to appear in the primary colonies on day 6 or 7, increase in number until around day 11, and then decrease (Hirayama et al 1994b). Therefore, replating needs to be carried out on several days between day 8 and day 13. The number of pooled cells to be replated is also important. When too many cells are recultured, secondary colonies may fuse and deteriorate prematurely.

Cytokine regulation of early stages of lymphohematopoiesis

Positive regulation

Of known cytokines that regulate the early stages of hematopoiesis, we have tested SF, IL-6, G-CSF, IL-11, IL-12, IL-3, and IL-4. Single

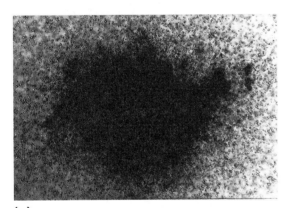

(a)

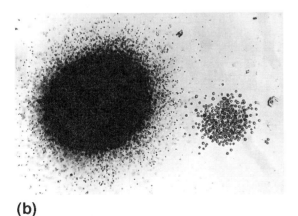

(b)

Figure 11.3.1 Colonies identified in the primary culture and secondary culture: (a) a typical primary colony; (b) a large lymphocyte colony and a small macrophage colony in the secondary culture containing SF and IL-7.

factors failed to support primary colony formation except IL-3. Therefore, we focused on two-factor combinations that had been shown in our laboratory to work synergistically in support of multilineage myeloid colony formation (Ogawa 1993). The results of representative experiments are presented in Table 11.3.1. The combinations containing SF plus one of IL-6, IL-11, G-CSF, or IL-12 are effective in supporting the B cell potential of the primary colonies. In contrast, none of the combinations containing IL-3 nor IL-3 alone supports B cell capability in the primary colonies even though the number and size of the primary colonies are comparable to the colonies detected in the SF-containing cultures. IL4-based factor combinations such as IL-4 plus IL-6 and IL-4 plus IL-11 are also effective, but less efficient than SF-based combinations. The recently cloned ligand (FL) for the flt3/flk2 tyrosine kinase receptor also synergizes with IL-6, IL-11 and G-CSF in support of the B cell potential of the progenitors in primary culture (Hirayama et al 1995).

Myeloid lineage expression may be examined by reculturing the primary colonies into secondary culture containing a combination of growth factors, such as IL-3, PWM-SCM, and Ep.

Negative regulation

Because of the ineffectiveness of IL-3 in supporting the B-lymphoid potential in the primary culture, we examined possible inhibitory effects of IL-3 and other lymphohematopoietic cytokines, including IL-1α, IL-2, IL-4, IL-5, IL-6, IL-9, IL-10, IL-11, IL-12, FL, tumor necrosis factor (TNF)-β, and leukemia inhibitory factor (LIF). Each of these cytokines was added to the primary culture along with permissive cytokine combinations, such as SF plus IL-11, and resulting primary colonies were pooled and tested for pre-B cell colony-forming ability by replating into secondary methylcellulose culture containing SF and IL-7. IL-1α, and IL-3 strongly inhibited the B cell potential of the primary colonies supported by SF and IL-11 and other permissive growth factor combinations (Table 11.3.2). No other cytokines that were tested proved to be inhibitory.

Table 11.3.2 Effects of addition of IL1-α or IL-3 on the size and B cell potential of the primary colonies supported by SF and IL-11

Test cytokine in primary culture	No. of primary colonies	Mean cell no. of primary colonies ($\times 10^{-4}$)	No. of pre-B cell colonies
None	17 ± 2	3.8	122 ± 2
IL-1α	22 ± 6	1.6	0
IL-3	20 ± 3	5.8	0

Discussion

We have established a two-step clonal methylcellulose culture assay for murine lymphohematopoietic progenitors and demonstrated cytokine regulation of early B-lymphopoiesis.

In the presence of two cytokines, one from a cytokine group consisting of SF, IL-4, and FL and the other one from another group of cytokines consisting of IL-6, IL-11, G-CSF and IL-12, lymphohematopoietic progenitors will proliferate and give rise to committed B-lymphoid progenitors, which respond to SF and IL-7 and differentiate to pre-B cells. This culture system provides a method for study of the mechanism of commitment of lymphohematopoietic progenitors and for isolating early committed B cell progenitors for analysis of the cellular and molecular processes of early B-lymphopoiesis.

References

Abramson S, Miller RG, Phillips RA (1977) The identification in adult bone marrow of pluripotent and restricted stem cells of the myeloid and lymphoid systems. J Exp Med 145: 1567–1579.

Ball TC, Hirayama F, Ogawa M (1995) Modulation of early B-lymphopoiesis by interleukin-3. Exp Hematol: in press.

Baum CM, Weissman IL, Tsukamoto AS, Buckle AM, Peault B (1992) Isolation of a candidate human hematopoietic stem-cell population. Proc Natl Acad Sci USA 89: 2804–2808.

Cumano A, Paige CJ, Iscove NN, Brady G (1992) Bipotential precursors of B cells and macrophages in murine fetal liver. Nature 356: 612–615.

Dick JE, Magli MC, Huszar D, Phillips RA, Bernstein A (1985) Introduction of a selectable gene into primitive stem cells capable of long-term reconstitution of the hemopoietic system of W/W^v mice. Cell 42: 71–79.

Hirayama F, Ogawa M (1994) CD43 expression by murine lymphohemopoietic progenitors. Int J Hematol 60: 191–196.

Hirayama F, Shih JP, Awgulewitsch A, Warr GW, Clark SC, Ogawa M (1992) Clonal proliferation of murine lymphohematopoietic progenitors in culture. Proc Natl Acad Sci USA 89: 5907–5911.

Hirayama F, Katayama N, Neben S, et al (1994a) Synergistic interaction between interleukin-12 and steel factor in support of proliferation of murine lymphohematopoietic progenitors in culture. Blood 83: 92–98.

Hirayama F, Clark SC, Ogawa M (1994b) Negative regulation of early B-lymphopoiesis by interleukin-3 and interleukin-1α. Proc Natl Acad Sci USA 91: 469–473.

Hirayama F, Lyman SD, Clark SC, Ogawa M (1995) The FLT3 ligand supports proliferation of lymphohematopoietic progenitors and early B-lymphoid progenitors. Blood 85: 1762–1768.

Katayama N, Shih JP, Nishikawa S, Kina T, Clark SC, Ogawa M (1993) Stage-specific expression of c-kit protein by murine hematopoietic progenitors. Blood 82: 2353–2360.

Katayama N, Ogawa M (1994) Assay for murine blast cell colonies. In: Freshney RI, Pragnell IB, Freshney MG, eds. Culture of Haematopoietic Cells. Wiley-Liss, New York, pp. 41–54.

Keller G, Paige C, Gilboa E, Wagner EF (1985) Expression of a foreign gene in myeloid and lymphoid cells derived from multipotent haematopoietic precursors. Nature 318: 149–154.

Lemischka IR, Raulet DH, Mulligan RC (1986) Developmental potential and dynamic behavior of hematopoietic stem cells. Cell 45: 917–927.

Ogawa M (1993) Differentiation and proliferation of hematopoietic stem cells. Blood 81: 2844–2853.

Shih JP, Zeng HQ, Ogawa M (1992) Enrichment of murine marrow cells for progenitors of multilineage hematopoietic colonies. Leukemia 6: 193–198.

Shih JP, Ogawa M (1993) Monoclonal antibody J11d.2 recognizes cell cycle-dormant, primitive-hematopoietic progenitors of mice. Blood 81: 1155–1160.

Murine lymphohematopoietic progenitors

Development of B lineage cells from uncommitted progenitors *in vitro*

11.4

Barbara L. Kee
Christopher J. Paige

The Wellesley Hospital Research Institute and Department of Immunology, University of Toronto, Toronto, Ontario, Canada

This work was supported through grants from Medical Research Council of Canada, The National Cancer Institute and Canadian Cancer Society, and a Terry Fox Program Project Grant

TABLE OF CONTENTS

Introduction

A number of *in vitro* assays have been developed which allow the identification and quantification of B lineage progenitors. However, few of these assays have been shown to provide the requirements for the development of B lineage cells from lineage-unrestricted progenitors. In 1992, we described an assay which supports the development of B lymphocytes and macrophages from single bipotent progenitors isolated from murine fetal liver at day 12 of gestation (Cumano et al 1992). The bipotent progenitors are purified on the basis of their expression of AA4.1 and Ly6A and the absence of B220 and Mac-1, markers associated with the B and myeloid lineages, respectively. Approximately 1 in 10 of the fetal liver cells with this phenotype give rise to both B lymphocytes and macrophages when cultured in the presence of the stromal cell line S17 and the pre-B cell growth factor IL-7. These conditions are also sufficient to support the development of B lineage progenitors from multipotent progenitors isolated from the embryo body and yolk sac prior to day 12 of gestation (Cumano et al 1993; Godin et al 1995). The B lineage cells which develop under these conditions are able to secrete immunoglobulin when transferred to a second culture containing S17 and the B cell mitogen lipopolysaccharide (Cumano et al 1990). Recently, we described two stromal cell derived factors, IL-11 and mast cell growth factor, which are sufficient to replace the requirement for S17 in supporting the development of IL-7 responsive pre-B lymphocytes and macrophages from single bipotent progenitors (Kee et al 1994). However, these factors do not replace the requirements for S17 in the transition of pre-B lymphocytes to mitogen responsive B cells (Cumano et al 1994; B.L. Kee, unpublished). In this chapter we describe the method for the enrichment of bipotent progenitors from fetal liver of day 12 of gestation and for their growth and differentiation *in vitro*.

Materials and method

Panning materials

- Optilux 100 mm Petri dishes (Falcon no. 1001)
- Coating buffer (0.05 M Tris-Cl pH 9.8, 0.15 M NaCl)
- Mouse anti-rat IgG ($5 \mu g \, ml^{-1}$)
- Hybridoma supernatants: α-B220 (14.8) (Kincade et al 1981), α-Mac-1 (M1/70) (Springer et al 1979), α-Ly-6A (E13 161-7) (Aihara et al 1986) and α-AA4.1 (AA4.1) (McKearn et al 1984)

- 500 ml 5% fetal calf serum (FCS)/phosphate buffered saline (PBS)

Culture materials

- 96-well culture dishes
- OptiMEM (Gibco) containing 5×10^{-5} M 2-mercaptoethanol, 10% FCS, $100 \mu g \, ml^{-1}$ streptomycin, and $100 \mu g \, ml^{-1}$ penicillin
- S17 stromal cell line (Collins and Dorschkind 1987)

- IL-11 (100 ng ml^{-1}) (Paul et al 1990)
- MGF (100 ng ml^{-1}) (Copeland et al 1990; Huang et al 1990; Martin et al 1990; Williams et al 1990; Zsebo et al 1990a,b)
- IL-7 (125 U ml^{-1}) (Namen et al 1988)
- LPS (10 μg ml^{-1}) (Gibco)

Description of method

Selection of AA4.1$^+$B220$^-$Mac-1$^-$Ly6A$^+$ cells from fetal liver of day 12 of gestation.

Fetal livers are isolated by conventional methods from C57BL/6 or (C57BL/6 × BALB/c) F$_1$ embryos 12 days after mating; the appearance of a vaginal plug is considered day 0. All incubations and reagents are at 4°C and the panning plates should be maintained on a level surface. The panning procedure described here is a modification of previously published methods (Mage et al 1977; Wysocky and Sato 1978).

1. Coat Optilux 100 mm Petri dish for a minimum of 2 h at 4°C with 10 ml of mouse anti-rat IgG diluted to a final concentration of 5 μg ml^{-1} in coating buffer.
2. Wash the plates 3 times with 10 ml of 5% FCS/PBS.
3. Apply 3 ml of hybridoma supernatant at a final concentration of 10–50 μg ml^{-1} for a minimum of 1 h.
4. Wash the plate 3 times with 10 ml of 5% FCS/PBS.
5. Apply the suspension of fetal liver cells (<2 × 10^7 cells) to the AA4.1-coated plate in 3 ml of 5% FCS/PBS.
6. Leave the cells for 30 min and then gently swirl the plate to redistribute the nonadherent cells. Leave the cells for a further 30 min.

7. To recover the adherent cells, aspirate the nonadherent layer with a Pasteur pipette attached to a vacuum aparatus. Apply 3 ml of 5% FCS/PBS to the edge of the Petri dish, gently swirl the liquid on the plate and aspirate with the Pasteur pipette, making certain to aspirate the meniscus of each plate. Wash the plates in a similar manner 6 to 8 times, until the nonadherent cells are no longer visible under the light microscope.
8. Apply 2 ml of 5% FCS/PBS to the plate; gently scrape the surface of the Petri dish with a plastic scraper. Wash the plate twice in 2 ml of PBS/5% FCS to collect the majority of AA4.1$^+$ cells.
9. Apply the cells (in a 3 ml volume) to a Petri dish coated with the 14.8 and M1/70 hybridoma supernatant and repeat step 6.
10. To recover the nonadherent cells, gently swirl the plate and remove the medium with a 5 ml pipette. Apply an additional 3 ml of PBS/5% FCS, gently swirl the plate and remove the medium with a 5 ml pipette.
11. Apply the remaining cells (in a 3 ml volume) to a Petri dish coated with the E13 161-7 supernatant and repeat steps 6 and 7.

Limiting dilution analysis

1. Establish an adherent layer of the S17 stromal cell line by plating 500–2000 cells per well (5000–20,000 cells ml^{-1} 100 μl/well) for 4–18 h in a 96-well dish. All stromal-cell containing plates are irradiated (2000 rad) prior to addition of fetal liver progenitors.
2. The selected fetal liver cells are plated on the established stromal cell layer in 100 μl of medium containing 250 U IL-7 ml^{-1} (final concentration of 425 U IL-7 ml^{-1}). A minimum of three cell

concentrations should be used, typically 3.3, 10, and 30 cells per well over 32 or 48 wells.

3. The cultures are maintained at 37°C in a humidified incubator containing 5% CO_2 for 10 days. On days 4 and 7 after the initiation of culture 100 μl of culture supernatant is removed and replaced with fresh medium containing IL-7.

4. On day 10 of culture the contents of each well are resuspended in 200 μl of medium and 25 μl is transferred to a new plate containing 500–2000 irradiated S17 stromal cells and 10 μg ml^{-1} LPS in a 200 μl volume.

5. The LPS-containing cultures are maintained, without replacement of the culture medium, for 15 days.

6. Mitogen-responsive B cell progenitors can be identified under the light microscope as large blast colonies and the production of IgM can be measured in the culture supernatant using a standard ELISA.

7. The frequency of clonable B cell progenitors in the initial culture was calculated based on the Poisson distribution:

$$F_r = \frac{\mu^r}{r!} \cdot e^{-\mu}$$

where F_r = fraction of wells containing r cells, μ = mean number of cells/well, r = actual number of precursors/well. Given the unique case when $\mu = 1$ and $r = 0$, then $F_0 = e^{-1} = 0.3679 = 37\%$. By plotting the fraction of nonresponding wells against the input cell concentration, the frequency of responding B cell progenitors is determined as the number of input cells per well where 37% of the wells are nonresponding.

IL-11 (100 ng ml^{-1}) and MGF (100 ng ml^{-1}) can be used to replace the S17 stromal cell line in steps 1–3. When these factors are used, they should be added on days 4 and 7 in addition to IL-7 when the culture supernatant is replaced in step 3.

Discussion and trouble shooting

The method described in the above has been used routinely in our laboratory for the detection of B lineage progenitors. The bipotent progenitors require a minimum of 8 days in culture before they develop the capacity to secrete immunoglobulin in the presence of S17 and LPS. We routinely transfer the progeny of bipotent progenitors to S17 and LPS on day 10 of culture; at this time essentially 100% of the IL-7-responsive cells are able to give rise to cells which produce IgM in the S17 and LPS cultures. By day 15 of culture, and day 25 of culture, 50% and 10% respectively of the IL- 7-responsive cells give rise to progeny which can secrete IgM in the S17 and LPS cultures. The optimal time for transferring progenitors from different developmental stages should be determined empirically.

On day 10 of culture the B lineage progeny of the bipotent progenitors are readily detectable as small clusters of cells with typical lymphoid morphology. These cells are easily distinguished from the adherent macrophage progeny and the granulocytic cells which develop from other progenitors in the AA4.1$^+$ B220-Mac-1-Ly-6A$^+$ population. However, after 10 days of culture in the presence of IL-11, MGF, and IL-7 the myeloid cells present in these cultures remain as nonadherent, undifferentiated progenitors which closely resemble lymphocytes. Under these conditions, it is best to quantify B cell progenitors on the basis of IgM secretion.

The most frequent source of variation in this assay is in the selection for Ly-6A-expressing cells. This antigen is expressed at very low levels on hematopoietic progenitors at day 12 of gestation and is expressed only on cells of the $Ly6^b$ haplotype (Spangrude and Brooks 1993). Generally we use C57BL/6 embryos; however, when (C57BL/6 $\times$ BALB/c) F_1 embryos are used, the level of Ly-6A expression may be even lower; therefore, less stringent washing conditions may be required. Preliminary experiments should be performed to optimize the washing conditions for the selection of Ly-6A-expressing progenitors.

A second source of variation occurs when the number of input cells per well is altered to detect progenitors which are present at low frequencies in the test population. When the concentration of input cells is low, the expansion of progenitors is such that by day 10 of culture each well is below its maximum cellular capacity. However, at higher cell concentrations there is a danger of overcrowding with myeloid progenitors which may lead to the death of the B lineage cells. As a general rule, if a well reaches one-third confluence (approximately 2×10^4 cells per well), one-quarter of the nonadherent cells should be transferred to a new well containing the identical growth conditions. The problem of overcrowding is more acute in cultures containing IL-11, MGF, and IL-7 as the absolute number of cells present by day 10 of culture is frequently much higher than in cultures containing S17 and IL-7.

References

Aihara Y, Buhring HJ, Aihara M, Klein J (1986) An attempt to produce 'pre-T' cell hybridomas and to identify their antigens. Eur J Immunol 16: 1391.

Collins LS, Dorshkind K (1987) A stromal cell line from myeloid long-term bone marrow cultures can support myelopoiesis and B lymphopoiesis. J Immunol 138: 1082.

Copeland NJ, Gilbert GJ, Cho BC, et al (1990) Mast cell growth factor maps near the *Sl* locus and is structurally altered in a number of Steel alleles. Cell 63: 175.

Cumano A, Dorshkind K, Gillis S, Paige CJ (1990) The influence of S17 stromal cells and interleukin 7 on B cell development. Eur J Immunol 20: 2183.

Cumano A, Iscove NN, Paige CJ, Brady G (1992) Bipotential progenitors of B-cells and macrophages from fetal liver at day 12 of gestation. Nature 356: 612.

Cumano A, Furlonger C, Paige CJ (1993) Differentiation and characterization of B-cell precursors detected in the yolk sac and embryo body of embryos beginning at the 10- to 12-somite stage. Proc Natl Acad Sci USA 90: 6429.

Cumano A, Kee BL, Ramsden DA, Marshall A, Paige CJ, WGE (1994) Development of B lymphocytes from lymphoid committed and uncommitted progenitors. Immunol Rev 137: 5.

Godin I, Dieterlen-Lievre F, Cumano A (1995) Emergence of multipotent hematopoietic cells in the yolk sac and para-aortic splanchnopleura of 8.5 dpc mouse embryos. Proc Natl Acad Sci 92: 773–777.

Huang E, Nocka K, Bier DR, et al (1990) The hematopoietic growth factor KL is encoded by the steel locus and is the ligand for the *c-kit* receptor, the product of the *W* locus. Cell 63: 223.

Kee BL, Cumano A, Iscove NN, Paige CJ (1994) Stromal cell independent growth of bipotent B-cell-macrophage precursors from murine fetal liver. Int Immunol 6: 401.

Kincade PW, Lee G, Watanabe S, Scheid MP (1981) Antigens displayed on murine B lymphocyte precursors. J Immunol 127: 2262.

McKearn JP, Baum C, Davie JM (1984) Cell surface antigens expressed by subsets of pre-B cells and B cells. J Immunol 132: 332.

Mage MG, McHugh LL, Rothstein RL (1977) Mouse lymphocytes with and without surface immunoglobulin: preparative scale separation in polystyrene tissue culture dishes coated with specifically purified anti-immunoglobulin. J Immunol Methods 15: 47.

Martin FH, Suggs S, Langley KE, et al (1990) Primary structure and functional expression of rat and human stem cell factor. Cell 63: 203.

Namen AE, Lupton S, Hjerrild K, et al (1988) Stimulation of B-cell progenitors by cloned murine interleukin 7. Nature 333: 571.

Paul SR, Bennett F, Calvetti JA, et al (1990) Molecular cloning of a cDNA encoding interleukin 11, a stromal cell-derived lymphopoietic and hematopoietic cytokine. Proc Natl Acad Sci USA 87: 7512.

Spangrude GJ, Brooks DM (1993) Mouse strain variability in the expression the hematopoietic stem cell antigen Ly-6A/E by bone marrow cells. Blood 82: 3327.

Springer T, Galfre G, Secher DS, Milstein C (1979) Monoclonal xenogeneic antibodies to murine cell surface antigens: identification of novel leukocyte differentiation antigens. Eur J Immunol 9: 301.

Williams DE, Eisenman J, Baird A, et al (1990) Identification of a ligand for the c-*kit* proto-oncogene. Cell 63: 167.

Wysocky LJ, Sato VL (1978) 'Panning' for lymphocytes: a method for cell selection. Proc Natl Acad Sci USA 75: 2844.

Zsebo KM, Williams DA, Geissler EN, et al (1990a) Stem cell factor is encoded at the *Sl* locus of the mouse and is the ligand for the c-*kit* tyrosine kinase receptor. Cell 63: 213.

Zsebo KM, Wypych J, McNiece IK, et al (1990b) Identification, purification and biological characterization of stem cell factor from Buffalo rat liver conditioned medium. Cell 63: 195.

Use of long-term bone marrow cultures and cloned stromal cell lines to grow B lineage cells

11.5

Kenneth Dorshkind[1]
Kenneth S. Landreth[2]

[1]Division of Biomedical Sciences, University of California, Riverside, California, USA

[2]The Department of Microbiology and Immunology and the Mary Babb Randolph Cancer Center, West Virginia University Health Sciences Center, Morgantown, West Virginia, USA

TABLE OF CONTENTS

Immunology Methods Manual
ISBN 0–12–442712–X

Introduction

In 1977 Dexter and colleagues described culture conditions permissive for long-term, stromal cell-dependent myelopoiesis (Dexter et al 1977), and in 1982 Whitlock and Witte reported the ability to grow B cells and their precursors *in vitro* on an adherent stromal cell layer.

These two culture systems have been invaluable in defining mechanisms by which extracellular signals from the hematopoietic microenvironment regulate primary B cell development (reviewed in Kincade 1987; Kincade et al 1989). In particular, the generation of clonal populations of both stromal cells and B lineage cells from these cultures has made it possible to study developmental interactions that occur between these cells and to identify molecules that regulate B cell development. The aim of this chapter is to describe the various long-term bone marrow cultures and to provide examples of how stromal cell lines that have been cloned from the cultures are used to grow heterogeneous and clonal populations of B lineage cells.

Long-term bone marrow cultures

Hematopoiesis in long-term bone marrow cultures is dependent on an adherent layer composed of cells that is thought to duplicate the *in vivo* hematopoietic microenvironment. The latter compartment includes endothelial cells that line the venous sinusoids, fibroblast-like reticular cells present in the intersinusoidal spaces, and selected hematopoietic cells such as macrophages (Dorshkind and Witte 1987; Dorshkind 1990).

As noted above, there are two principal types of long-term bone marrow cultures. The differentiated cells present in Dexter type long-term bone marrow cultures include neutrophilic granulocytes, macrophages, and myeloid precursors. Mature B cells and their immediate precursors are not present in the cultures, but an immature B cell precursor is maintained (Dexter et al 1977; Dorshkind and Phillips 1982). The precise developmental state of these cells is not known. Whitlock–Witte cultures are optimal for the growth of B lineage cells. While a few mature macrophages are present in the adherent layer, myeloid cells and their precursors are not present in established cultures (Whitlock and Witte 1982).

It is possible to induce the B cell precursor present in Dexter cultures to differentiate into B cells by transferring an established culture to Whitlock–Witte conditions. Over the course of 4 weeks, myelopoiesis subsides and B lymphopoiesis initiates (Dorshkind 1986). These 'switch' cultures can be used to examine events that occur as B cells develop from immature precursors.

Long-term bone marrow cultures are relatively easy to initiate and maintain for any laboratory in which tissue culture is routine.

Dexter myeloid long-term bone marrow cultures

Materials and apparatus

- $25\,cm^2$ tissue culture flasks
- 3 ml syringes and 25-gauge needles
- 5% CO_2, 95% air mixture for gassing cultures
- α-Minimal essential medium (MEM)
- Penicillin–streptomycin
- Horse serum
- Hydrocortisone sodium succinate
- Sterile tissue culture hood
- Automatic CO_2 tissue culture incubator (5% CO_2 in air) set at 33°C

Procedure

- Dexter cultures are initiated and maintained in α-MEM supplemented with 20% horse serum, 10^{-6}–10^{-7} M hydrocortisone sodium succinate, $100\,\mu g\,ml^{-1}$ penicillin, and 100 units ml^{-1} streptomycin. It is possible to initiate cultures in 10^{-7} M steroids, but the use of 10^{-6} M steroids facilitates establishment of cultures.
- Add 8 ml of medium to each $25\,cm^2$ flask. Using a 3 ml syringe fitted with a 25-gauge needle, remove 3 ml of medium from a flask and use it to flush the contents of a femur and tibiae directly back into the flask. Repeat this procedure if additional flushing of bones is required. It is not necessary to resuspend any clumps.
- Gas each flask with a mixture of 5% CO_2/95% air for approximately 10 s. Cap the flask tightly, and place it in a nonhumidified, 5% CO_2 and air incubator set at 33°C. It is important when handling flasks to prevent medium from touching the cap of the flask, as this can increase the risk of contamination. Contamination can be further minimized by placing cultures in a nonhumidified incubator.
- Feed 1 week later by removing half the medium, and any nonadherent cells and debris, and replace with 4 ml of fresh medium.
- By 2 weeks post-initiation, an adherent layer of cells with some associated hematopoietic foci will have established. However, hematopoietic cells often die during this initial period of culture when the stromal cell layer establishes. Therefore, at the 2-week point the cultures are recharged with 10^6 bone marrow cells. All medium is removed from the flasks, including any nonadherent cells, and replaced with 8 ml of medium containing the 10^6 cells in suspension.
- Cultures are fed weekly thereafter by removing half of the supernatant, along with any nonadherent cells, and replacing it with fresh medium. The cells removed can be used for various assays.

Discussion

There is evidence that the most immature progenitors in the cultures are present in the adherent layer (Dorshkind and Phillips 1983). Often, it is possible to harvest these adherent hematopoietic cells by vigorous pipetting. This may also dislodge the adherent cells as well. Alternatively, adherent cells can be harvested by treating the cultures with sterile collagenase dispase (0.15% in PBS) or trypsin (0.05% trypsin + 0.53 mM EDTA in MEM). If trypsinization is being used, be sure to rinse flasks thoroughly with medium, because any remaining serum can interfere with trypsin action. Use of collagenase may be advantageous, because it will not remove cell surface antigens from cells.

The harvested adherent layer will contain both hematopoietic and stromal cells. If necessary, hematopoietic cells in the adherent layer can be separated from stromal cells by adherence on plastic tissue culture plates or passage of the cell suspension through nylon wool or Sephadex G10 columns.

However, these procedures may not yield pure populations of hematopoietic cells and may deplete some populations.

The efficiency of culture initiation can vary depending on the lot of horse serum used. Therefore, it may be necessary to screen several serum lots to find the one most suitable. One general screening method is to initiate cultures in several lots of serum simultaneously. At the end of the 2-week period, harvest adherent cells from nonrecharged cultures and measure the number of myeloid progenitors in a CFU-S or CFU-GM assay. The lot that maintains the most precursors without recharging will usually be suitable for long-term cultures.

The mouse strain employed may also influence culture initiation and longevity (Sakakeeny and Greenberger 1982). However, the use of both a femur and tibiae in initiating cultures usually can overcome strain limitations.

While Dexter cultures can be maintained for several months, it is not clear whether all progenitor populations survive. Therefore, cultures used within 6-weeks post-recharge may provide the best experimental results.

Whitlock–Witte lymphoid long-term bone marrow cultures

Materials and apparatus

$10\,cm^2$ tissue culture plates

3 ml syringe and 25-gauge needle

RPMI-1640

Fetal calf serum (FCS)

2-Mercaptoethanol (2-ME)

Sterile tissue culture hood

Automatic CO_2 incubator (5% CO_2 in air) at 37°C

Procedure

Remove femurs and tibiae from mice and prepare a single-cell suspension of bone marrow cells at 10^6 cells ml^{-1} in RPMI1640, 5% FCS, and 5×10^{-5} M 2-ME. Although these cultures can be established using cells from mice at any age, the best results are achieved when mice are 3–4 weeks old.

Plate 13.5 ml of the above cell suspension per $10\,cm^2$ plate. Place cultures in a humidified 5% CO_2 and air incubator at 37°C.

Cultures are fed twice a week. 5.0 ml of fresh medium is added to cultures during the first feeding. Do not remove any medium at this initial feeding. At the second feeding, 80% of the medium is removed from the cultures, taking care not to remove any nonadherent cells unless needed for assay, and replaced with 10 ml of fresh medium. Because antibiotics are usually not added to these cultures, sterile technique is critical.

A confluent adherent layer should establish by 2 weeks post-initiation of cultures. Adherent foci, which will consist of B lineage cells, can form by this time or may not appear until 1 or 2 weeks later. It is not necessary to recharge these cultures. Many of the lymphoid cells are buried in the adherent layer. These can be harvested by the enzymatic treatments described above.

Discussion

Whitlock–Witte cultures will contain B lineage cells at various stages of differentiation, including CD45R(B220)$^+$/Ig$^-$ pro-B cells, pre-B cells expressing cytoplasmic immunoglobulin (Ig) μ heavy chains, and surface Ig expressing B cells.

Long-term B cell cultures can be maintained for several months. However, as they age, a few or even a single clone of B lineage cells can predominate in the cultures. This pauciclonality usually occurs in cultures after approximately 8 weeks post-initiation (Whitlock and Witte 1982; Dorshkind and Witte 1987).

Dexter–Whitlock/Witte switch cultures

Materials and apparatus

Same as above for Dexter and Whitlock–Witte cultures

Procedure

All nonadherent cells and medium are removed from an established Dexter culture. The flask is then gently rinsed with medium.

Eight ml of medium used to initiate and maintain Whitlock–Witte cultures is added to the flasks and the cultures are transferred to lymphoid conditions at 37°C.

Cultures are fed twice weekly according to the protocol for Whitlock–Witte cultures.

B lineage cells should predominate in cultures by 4 weeks after change in culture conditions.

Discussion

It may be necessary to screen batches of fetal calf serum that will support these switch cultures, although lots that support primary lymphoid cultures usually work.

Generation of stromal cell lines from long-term bone marrow culture adherent layers

The use of stromal cell lines is particularly advantageous when the goal is to measure the B cell developmental potential of a hematopoietic population isolated from fetal or adult tissues or cultured bone marrow. As opposed to heterogeneous feeder layers derived from primary bone marrow cultures, the use of stromal cell lines circumvents the possibility that any ensuing lymphopoiesis is derived from precursors present in the adherent layer. The culture conditions used in such experiments are the same as described above for initiating Whitlock–Witte long-term lymphoid cultures.

Many stromal cell lines that support various stages of B cell development have been identified (Kincade et al 1989), and these can be maintained by continuous passage. The lines that are most useful in seeding experiments are those which form confluent adherent layers in culture and do not interfere with the development of hematopoietic populations owing to uncontrolled growth. The procedure used successfully for generation of the S17 stromal cell line in our laboratory is outlined below. This particular stromal cell line was cloned from the adherent layer of a Dexter myeloid long-term bone marrow culture (Collins and Dorshkind 1987), but the adherent layers of Whitlock-Witte cultures may also be utilized.

Materials and apparatus

Same as above for Dexter cultures, plus:

$10 \, cm^2$ tissue culture plates

Cloning rings

Mycophenolic acid (MPA)

Collagenase

Procedure

Initiate a Dexter myeloid long-term bone marrow as described above, but do not recharge the cultures.

Eliminate any hematopoietic cells present in the cultures by addition of MPA. Remove all culture medium and replace it with α-MEM supplemented with 20% horse serum, 10^{-6} M hydrocortisone, and $4 \, \mu g \, ml^{-1}$ MPA. MPA is prepared by first dissolving 10 mg of the drug in $400 \, \mu l$ of 95% ethanol at room temperature. The dissolved MPA is then diluted in PBS to make a $1 \, mg \, ml^{-1}$ working solution.

Three days after MPA addition to cultures, replace all of the drug-containing medium with fresh medium. After an additional 3-day incubation, treat cultures a second time with MPA.

The MPA-treated stromal cell cultures are used as the starting population for isolation of stromal cell lines. To select for stromal cell populations with long-term growth potential, passage stromal cell cultures

once a week for 6 weeks. Adherent layers to be passaged are harvested by treating cultures with 2 ml of collagenase (0.15% in PBS) solution. Cultures can be maintained in α-MEM plus 5% FCS.

Following the last passage at week 6, the heterogeneous population of stromal cells is harvested and washed twice, and aliquots of 100 cells in α-MEM supplemented with 20% FCS are added to $10 \, cm^2$ tissue culture plates and placed in a humidified 5% CO_2 and air incubator at 37°C.

By one day after plating of cells, individual stromal cells that attach to the plates can be observed under an inverted phase-contrast microscope, and their position on the plate is marked. After 3–5 days of growth, foci containing up to 20 cells develop.

These foci are harvested. All medium is removed from the plate, and a $5 \, mm \times 7 \, mm$ high stainless steel cloning ring with a thin coat of silicone grease around its edges is sealed in place over the stromal cell colony. One hundred microliters of 0.15% collagenase in PBS is then pipetted into the cylinder and the entire plate is incubated at 37°C for 15 min. This treatment detaches the adherent stromal cells from the plate.

The detached cells are transferred to $2 \, cm^2$ tissue culture wells in 2 ml of α-MEM supplemented with 10% FCS. Upon reaching confluency, the cells can be passaged to larger flasks. Cultures are fed weekly.

Discussion

Not all of the stromal cell colonies will ultimately yield stromal cell lines. Therefore, it is important to pick at least 20 colonies.

An alternative to using cloning rings is to plate stromal cells obtained from the MPA-treated heterogeneous stroma at limiting dilution in wells of 96-well plates.

Growth of clonal populations of B lineage cells

In order to address specifics of intracellular events that regulate production of B lymphocytes, to estimate precursor frequencies, or to investigate how growth and differentiation signals directly affect B cell development, it is advantageous to generate clonal populations of non-transformed B lineage cells. Two methods for the growth of clonal populations of pro-B and pre-B cells are outlined below.

Growth of pre-B cell colonies in semisolid medium

The ability of B cell progenitors to proliferate in response to various growth stimulatory molecules such as IL-7 (Namen et al 1988) can be measured in either short-term liquid cultures or in a colony assay in semisolid medium such as agar or methylcellulose (Lee et al 1989; Dorshkind et al 1989). Either approach can be used to quantify the frequency of responding cells in a population. In addition, use of the colony assay makes it possible to pick individual colonies for further phenotypic and molecular analysis on a clone of cells (Henderson et al 1992).

In addition to IL-7, colony growth can also be potentiated in the presence of conditioned media from selected stromal cell lines (Dorshkind et al 1989).

Growth of colonies in methylcellulose

Materials

3 ml syringe and 18-gauge needle

35 × 10 mm Petri dishes

Methylcellulose

Autoclaved double-distilled H_2O

2× α-MEM

Fetal calf serum (FCS)

α-MEM

2-Mercaptoethanol (2-ME)

Stromal cell conditioned medium or recombinant interleukin-7

Procedure

Prepare a 2.1% stock methylcellulose solution. Add 28 g autoclaved methylcellulose powder to 490 ml double-distilled water, bring to a boil, and stir for 3 h. Subsequently, add 490 ml 2× MEM, stir overnight and then aliquot into 100 ml lots that can be stored frozen indefinitely.

Prepare an 0.8 % methylcellulose solution (100 ml = 40 ml of 2.1% methylcellulose stock, 15 ml FCS, 5×10^{-5} M 2-ME; use 1×Iscoue's medium to bring volume to 100 ml). If stromal cell conditioned medium is used, add it at 5–10% of the final volume, and reduce the amount of MEM proportionally. If recombinant IL-7 is used, a range of concentrations should be tested to determine the optimal dose.

Determine the number of target cells per plate. This will be dependent upon the frequency of responding cells in the population; if unknown, a range of cell concentrations ranging from 10^3 to 10^6 cells/plate should be tested. Mix three times the number of target cells/plate in 0.1 ml with 2.9 ml of the 0.8% methylcellulose solution.

Using a 3 ml syringe fitted with an 18-gauge needle, deliver 1 ml to each of two plates. Because of the viscosity of the methylcellulose solution, which results in loss in the tube and syringe, only two Petri dishes can be plated for every 3 ml of methylcellulose medium.

Cultures are placed in a 5% CO_2 and air incubator at 37°C. Methylcellulose cultures dry out very readily. Therefore, if humidity is low in the incubator, place cultures in a larger covered dish containing a smaller, open Petri dish filled with water. On day 5 to 7, lymphoid colonies are enumerated under a dissecting microscope. These colonies usually consist of 300–500 cells that are tightly packed together.

Growth of colonies in agar

Materials and apparatus

Same as above plus:

Bacto-Agar

Heated water bath

Sodium pyruvate

Sodium bicarbonate

Essential and nonessential amino acids

Vitamins

L-Glutamine

L-Serine

L-Asparagine

Procedure

Resuspend three times the number of target cells per plate in 2.7 ml of α-MEM containing 15% FCS, 5×10^{-5} M 2-ME, recombinant IL-7 sodium pyruvate, sodium bicarbonate, essential and nonessential amino acids, vitamins, L-glutamine, L-serine, and L-asparagine as described (Lee et al 1989).

Add 0.3 ml of warm 3% agar to the cell suspension, mix by pipetting, and distribute 1 ml per dish. This step should be performed rapidly, as the agar tends to solidify in the plates.

Incubate plates and count colonies as described above.

Discussion

The number and size of colonies generated in response to IL-7 can be increased in the presence of additional proliferation cofactors such as insulin-like growth factor I (Landreth et al 1992) or c-kit ligand (McNeice et al 1991).

It is very easy to pick colonies from methylcellulose cultures using a finely drawn glass pipette. Cells in these colonies can be assayed using various phenotypic and molecular approaches. For example, a few hundred cells from a single colony have been assayed for status of Ig gene rearrangements by the polymerase chain reaction (Henderson et al 1992).

Growth of cloned pro-B cell lines on bone marrow stromal cells

The growth of normal pro-B cells has been accomplished by several laboratories by passaging lymphoid cell populations on bone marrow stromal cell lines in the presence of IL-7 (Billips et al 1990; Rolink et al 1991; Gibson et al 1993). Cells from fetal liver (gestational day 11 to term), bone marrow, or long-term cultures expand in the presence of selected stromal cell lines and IL-7 and can be cloned by limiting dilution.

The predominant clonal population of cells resulting from selection on stromal cells in the presence of IL-7 have D to J_H Ig gene rearrangements on both heavy-chain chromosomes but retain unrearranged V_H and light-chain gene loci (Rolink et al 1991; Gibson et al 1993). These cells do not express cytoplasmic Ig heavy-chain or Ig light-chain gene products but can be identified by expression of the pre-B specific marker CD45R (B220) and therefore have been classified as pro-B cells. Largely because these pro-B clones mature to express surface Ig *in vitro*, populate the vacant lymphoid tissues of SCID mice *in vivo*, do not form tumors in SCID mice, and die by the process of apoptosis in the absence of stromal cell underlayers, they are considered to retain normal options for differentiation in the B lineage.

Materials and apparatus

$100 \, mm^2$ tissue culture dishes

α-Minimal Essential Medium (MEM)

2-Mercaptoethanol (2-ME)

Fetal calf serum (FCS)

Penicillin–streptomycin

Recombinant murine IL-7

Sterile tissue culture hood

Automatic CO_2 tissue culture incubator (5% CO_2 in air)

Procedure

Fetal liver cells are removed aseptically from gestational day 11–19 mouse fetuses and immediately dissociated into single-cell suspensions by repetitive pipetting with a Pasteur pipette followed by gentle aspiration through successively smaller bore hypodermic needles (18–22 gauge). The resulting cell suspension is then spun through an underlayer of fetal calf serum in a 15 ml conical bottom tube ($400g$ for 7 min) to separate hematopoietic cells from hepatocyte debris. Hematopoietic cell recovery from one fetal liver (gestational day 14–18) is normally $30–40 \times 10^6$ cells.

Stromal cell cultures are grown to approximately 80% confluence in $100 \, mm^2$ tissue culture dishes in α-MEM containing 5% FCS, 5×10^{-5} M 2-ME, $100 \, \mu g \, ml^{-1}$ penicillin, and 100 units ml^{-1} streptomycin. Immediately before adding fetal liver cells, 50 units ml^{-1} rIL-7 is added to each plate (maximum volume 10 ml/plate). Fetal liver cells are seeded on to these plates at $5 \times 10^4 – 5 \times 10^5$ cells ml^{-1} and incubated at 37°C in a humidified incubator.

Nonadherent cells ($10^5 \, ml^{-1}$) are transferred to new stromal cell layers under the same conditions every 4–6 days for 4 weeks. During this initial selection period, cells which are capable of more limited proliferation under these conditions are lost and proliferating nonadherent cells which remain can be cloned at limiting dilution in 96-well plates containing stromal cells and IL-7 as described above.

Resulting cells should be evaluated microscopically for lymphoid morphology and analyzed by flow cytometry for expression of B lineage restricted antigens, e.g. CD45R (B220), BP-1, and Ig (both membrane-bound and cytoplasmic). Ig gene rearrangement status of these cells can be evaluated by methods described elsewhere in this volume.

Approximately 10^7 nonadherent cells can be collected from each 100 mm^2 tissue culture dish every 2–4 days.

Discussion

Pro-B cell lines can be initiated from fetal liver, bone marrow, or long-term bone marrow cultures; however, the frequency of initiating cells differs in these tissues, with the highest frequency present in fetal tissue. It is much more difficult to initiate pro-B cell lines from bone marrow cells at any age tested. It should also be noted that, whereas pro-B cell cultures can be readily initiated from Dexter myeloid long-term bone marrow cultures, this is more difficult with Whitlock–Witte lymphoid long-term bone marrow cultures.

Many of the cells that initially proliferate in these cultures to form lymphopoietic foci do not survive more than a few days. If the frequency of cells capable of long-term proliferation in the presence of stromal cells and IL-7 is to be determined, growth of cells in limiting dilution assays should be scored at greater than one week.

Pro-B cells initially proliferate beneath stromal cell layers in these cultures and subsequently are present as foci of loosely adherent cells on the stromal cell surface or as nonadherent cells. These cells have not been found to differ in expression of cell-associated molecules, Ig gene rearrangement status, proliferative potential, or cloning efficiency in the presence of stromal cells and IL-7.

Stromal cell lines differ in their ability to support pro-B cell growth, and this difference appears to be related to stromal cell production of negative regulatory signals rather than failure of production of positive stimuli for cell proliferation (particularly since IL-7 is supplied exogenously). Several stromal cell lines should be screened for their ability to support long-term growth of pro-B cells before initiating experiments. Of particular note, the bone marrow stromal cell line S17 (Collins and Dorshkind 1987), which has been particularly useful for studies of pro-B cell differentiation, does not support the long-term growth of normal murine pro-B cells in vitro.

Pro-B cells are exquisitely sensitive to cell crowding and should be split at low cell concentration on to new stromal cell layers every 5–7 days to prevent overcrowding. Within hours after a critical cell concentration is reached, pro-B cells die rapidly by apoptosis.

As with cells in long-term bone marrow cultures described above, growth requirements for pro-B cells may change over extended periods of time in culture and pro-B cells should be replaced with fresh cells approximately monthly to insure consistency of experimental results.

Murine cells respond equally well to murine or human IL-7. The optimal concentration of IL-7 for stimulation of pro-B cell proliferation is 20–50 units ml^{-1}, however, concentrations of IL-7 to 500 units ml^{-1} do not result in any diminution of maximal proliferation.

For proliferation assays, pro-B cell clones should be collected on the third day following splitting on to new adherent layers by gentle pipetting to dislodge loosely adherent cells. Adherent cells which contaminate these cell preparations can be easily removed by passage of cells over G10-Sephadex columns. Cell proliferation is best measured in cultures of 5×10^5 cells ml^{-1} at 24 h following cytokine stimulation (Gibson et al 1993). Alternatively, proliferation can be measured in cultures of 10^5 cells/200 µl in 96-well tissue culture plates.

Human bone marrow cells have not been shown to proliferate in a similar manner in the presence of human bone marrow stromal cells or recombinant human IL-7.

References

Billips LG, Petitte D, Landreth KS (1990) Bone marrow stromal cell regulation of B lymphopoiesis: IL-1 and IL-4 regulate stromal cell support of pre-B cell production *in vitro*. Blood 75: 611–619.

Collins LS, Dorshkind K (1987) A stromal cell line from myeloid long-term bone marrow cultures can support myelopoiesis and B lymphopoiesis. J Immunol 138: 1082–1087.

Dexter TM, Allen TD, Lajtha LG (1977) Conditions controlling the proliferation of haematopoietic stem cells *in vitro*. J Cell Physiol 91: 334–344.

Dorshkind K (1986) *In vitro* differentiation of B lymphocytes from immature hemopoietic precursors present in long-term bone marrow cultures. J Immunol 136: 422–429.

Dorshkind K (1990) Regulation of hemopoiesis by bone marrow stromal cells and their products. Annu Rev Immunol 8: 111–137.

Dorshkind K, Phillips RA (1982) Maturational state of lymphoid cells in long-term bone marrow cultures. J Immunol 129: 2444–2450.

Dorshkind K, Phillips RA (1983) Characterisation of early B lymphocyte precursors present in long-term bone marrow cultures. J Immunol 131: 2240–2245.

Dorshkind K, Witte ON (1987) Long-term murine hemopoietic cultures as model systems for analysis of B lymphocyte differentiation. Current Topics Microbiol Immunol 135: 25–41.

Dorshkind K, Johnson A, Harrison Y, Landreth KS (1989) A colony assay system that detects B-cell progenitors in fresh and cultured bone marrow. J Immunol Methods 123: 93–101.

Gibson LF, Piktel D, Landreth KS (1993) Insulin like growth factor-1 potentiates expansion of interleukin-7 dependent pro-B-cells. Blood 82: 3005–3011.

Henderson AJ, Narayanan R, Collins L, Dorshkind K (1992) Status of KL chain gene rearrangements and c-kit and IL-7 receptor expression in stromal cell-dependent pre-B cells. J Immunol 149: 1973–1979.

Kincade PW (1987) Experimental models for understanding B lymphocyte formation. Adv Immunol 41: 181–267.

Kincade PW, Lee G, Pietrangeli CE, Hayashi SI, Gimble JM (1989) Cells and molecules that regulate B lymphopoiesis in bone marrow. Annu Rev Immunol 7: 111–143.

Landreth KS, Narayanan R, Dorshkind K (1992) Insulin-like growth factor-1 regulates pro-B cell differentiation. Blood 80: 1207–1212.

Lee G, Namen AE, Gillis S, Ellingsworth LR, Kincade PW (1989) Normal B cell precursors responsive to recombinant murine IL-7 and inhibition of IL-7 activity by transforming growth factor-β. J Immunol 142: 3875–3883.

McNeice IK, Langley KE, Zsebo KM (1991) The role of recombinant stem cell factor in early B cell development. J Immunol 146: 3785–3790.

Namen AE, Lupton S, Hjerrild K, et al (1988) Stimulation of B-cell progenitors by cloned murine interleukin-7. Nature 333: 571–573.

Rolink A, Kudo A, Karasuyama H, Kikuchi Y, Melchers F (1991) Long-term proliferating early pre-B cell lines and clones with the potential to develop to surface Ig-positive, mitogen reactive B cells *in vitro* and *in vivo*. EMBO J 10: 327–336.

Sakakeeny MA, Greenberger JS (1982) Longevity of granulocytes in corticosteroid-supplemented continuous bone marrow cultures varies significantly between 28 different mouse strains and outbred stocks. J Natl Cancer Inst 68: 305–317.

Whitlock CA, Witte ON (1982) Long-term cultures of B lymphocytes and their precursors from murine bone marrow. Proc Natl Acad Sci USA 79: 3608–3612.

Methods for purification and growth of human B cell precursors in bone marrow stromal cell-dependent cultures

11.6

Julie A.R. Pribyl*
Nisha Shah*
Bonnie N. Dittel
Tucker W. LeBien

Department of Laboratory Medicine and Pathology and the University of Minnesota Cancer Center, University of Minnesota Medical School, Minneapolis, Minnesota, USA

This work was supported by NIH grants R01 CA31685 and P01 CA21737.

*The first two authors contributed equally to the
development of this chapter.

TABLE OF CONTENTS

Abstract

Studies of human B cell ontogeny have been facilitated by the development of *in vitro* bone marrow stromal cell cultures that support the growth and differentiation of B cell precursors. Methodology is described for the isolation and purification of B cell precursors, establishment and preservation of bone marrow stromal cell cultures, and quantitative assessment of B cell precursor growth. These assays can be used to elucidate the patterns of gene expression and growth factor requirements of normal and leukemic human B cell precursors.

Introduction

Studies of human lymphohematopoiesis have relied heavily on the development of *in vitro* culture systems that attempt to mimic the lymphohematopoietic microenvironment *in vivo*, e.g., in bone marrow. The cytokines and colony-stimulating factors essential for survival, growth, and differentiation of pluripotent stem cells and their committed progenitors of the myeloid, monocytoid, megakaryocytoid, and erythroid lineages are well characterized (Quesenberry 1995). Implicit in the development of these culture systems has been the assumption that the more closely the *in vitro* culture conditions mimic the bone marrow microenvironment *in vivo*, the greater the probability that the cultured stem cells will manifest normal developmental programs of survival and death. In this context, it is common knowledge that most supportive culture systems contain a population of bone marrow stromal cells, a generic term commonly used to describe the collective population of adherent cells that support the growth of stem cells. Nonetheless, *in vitro* culture systems have limitations in their capacity to completely recapitulate the bone marrow microenvironment, and it is difficult to be certain that any *in vitro* culture system has the right three-dimensional landscape.

Human bone marrow stromal cell-dependent culture systems that support the growth of cells at various stages of human B cell ontogeny have recently been developed (McGinnes 1991;

Wolf et al 1991; Ryan et al 1992; Moreau et al 1993; Manabe et al 1994). Our original culture system requires the presence of a bone marrow stromal cell microenvironment consisting largely of myofibroblast-like cells, which supports the IL-7-dependent growth of normal B cell precursors for several weeks (Wolf et al 1991). We have subsequently utilized this culture system to address a number of questions about the growth regulation and stromal cell requirements of normal B cell precursors (Dittel et al 1993; Larson and LeBien 1994; Dittel and LeBien 1995a,b). We will resist the temptation to discuss the biological implications of our studies, since readers can draw their own conclusions by reading the original papers. The purpose of this chapter is to provide detailed methodology for the isolation and purification of normal human B cell precursors, the establishment and preservation of normal human bone marrow stromal cells, and the growth of B cell precursors.

If readers have questions about the methodologies described herein, they should contact Julie Pribyl or Nisha Shah (telephone 612-626-4839; fax 612-624-2400; e-mail rehma001@maroon.tc.umn.edu or shahx002@maroon.tc.umn.edu).

Materials, reagents, and safety

Materials

5 ml polystyrene tubes (Falcon #2052, Lincoln Park, NJ, USA)

5 ml polystyrene tubes with cap (Falcon #2054)

15 ml polypropylene tubes (Falcon #2097)

15 ml polystyrene tubes (Falcon #2099)

19–20 G needle

35 ml syringe

37°C, 5% CO_2 humidified incubator (Nuaire, Plymouth, MN, USA)

50 ml polypropylene tubes (Falcon #2098)

$75\,cm^2$ canted neck flask (Falcon #3111)

$75\,cm^2$ straight neck flask (Falcon #3024)

96-well flat bottom plates (Costar #3595, Cambridge, MA, USA)

FACScan (Becton Dickinson, San Jose, CA, USA)

FACStar Plus (Becton Dickinson)

Flame/bunsen burner

Forceps

Latex gloves

Magnet (Dynal #120.02, Lake Success, NY, USA)

Petri dish, 60 and 100 mm (Falcon #3004 and #3003)

Polystyrene 6 μm microspheres (Polysciences #07312, Warrington, PA, USA)

Rotating mixing wheel (Dynal #159.02)

Rubber policeman

Scalpel and blades (22 G)

Spill pad

Reagents

Antibodies (see Table 11.6.1)

Cell dissociation solution (Sigma #C-5789, St. Louis, MO, USA)

Fetal bovine serum (FBS, Hyclone #A-1111, Logan, UT, USA)

Gentamicin 50 mg ml^{-1} (Sigma #G-1397)

Goat F(ab')$_2$ anti-mouse FITC (BioSource International #4350, Camarillo, CA, USA)

Histopaque (Sigma #1077-1)

Isopropyl alcohol

Magnetic Dynabeads M450 goat anti-mouse IgG-coated (Dynal #110.06)

PBS

Recombinant human interleukin-7 (IL-7, PeproTech #200-07, Rocky Hill, NJ, USA)

Trypan blue (Sigma #T-8154)

RPMI medium

RPMI-1640 (Life Technologies #31800-071, Grand Island, NY, USA)		
50 U ml^{-1}	penicillin (Life Technologies #15070-022)	
50 μg ml^{-1}	streptomycin (Life Technologies #15070-022)	

Table 11.6.1 Antibodies used for isolation and analysis of B lineage cells

	Antibody	Name	Isotype	Target	Source[a]
Antibodies for B lineage isolation **Basic cocktail**	Anti-CD11b	OKM-1	IgG2b	Myeloid cells	ATCC #CRL8026
	Anti-CD2	13B3	IgG2a	T cells	LeBien Laboratory
	Anti-CD33	LeuM9	IgG1	Myeloid cells	Becton Dickinson #347780
	Anti-glycophorin A	10F7MN	IgG1	Erythroid precursors	ATCC #HB8162
	Anti-CD7	T3-3A1	IgG1	T/NK precursors	ATCC #HB2
	+				
	Anti-μ and/or	DA4.4	IgG1	Surface Ig$^+$ cells	ATCC #HB57
	Anti-κ/Anti-λ	TB28.2/ HP6054	IgG1/IgG2a	Surface light chain$^+$ cells	ATCC #HB61/ CRL1763
Additional enrichment	Anti-CD34	HPCA-1	IgG1	Pro-B cells	Becton Dickinson #347660
Antibodies for sorting	Anti-CD10-PE	J5	IgG2a	B lineage cells	Coulter #6604120
	Anti-CD34-FITC	HPCA-2	IgG1	Pro-B cells	Becton Dickinson #348057
	Anti-CD19-biotin	25C1	IgG1	B lineage cells	S. Peiper (Univ. of Louisville)
Antibodies for immunophenotyping	Anti-CD10-PE	J5	IgG2a	B lineage cells	Coulter #6604120
	Anti-CD19-biotin	25C1	IgG1	B lineage cells	S. Peiper
	Anti-μ-biotin	DA4.4	IgG1	Surface Ig$^+$ cells	ATCC #HB57
	Anti-κ-biotin/ Anti-λ-FITC	TB28.2/ HP6054	IgG1/IgG2a	Surface light chain$^+$ cells	ATCC #HB61/ CRL1763
Secondary reagents	Goat anti-mouse-FITC (H and L chain)			Unconjugated antibodies	BioSource International #6250
	Streptavidin-PE (SA-PE)			Biotinylated antibodies	Caltag #SA-1004

[a] ATCC, Rockville, MD; Becton Dickinson, San Jose, CA; Coulter, Hialeah, FL; BioSource International, Camarillo, CA; Caltag, South San Francisco, CA, USA.

Store at 4°C.

Stromal cell medium (SCM)

500 ml	Ex-Cell 300 (JRH Bioscience #14-032, Lenexas, KS, USA)
50 U ml^{-1}	penicillin
50 μg ml^{-1}	streptomycin

Store at 4°C.

X-VIVO medium

X-VIVO 10 (BioWhittaker #04-380, Walkersville, MD, USA)

2 mM	L-glutamine (Life Technologies #25030-0160)

Store at 4°C.

Trypsin–EDTA

450 ml	sterile PBS
50 ml	10× trypsin–EDTA (Life Technologies #15400-013)

Store at 4°C.

Purification and growth of human B cell precursors

Fluorescence buffer

2.5%	newborn calf serum (NBCS, Sigma #N-4762)
0.02%	sodium azide (Sigma #S-2002)
PBS	

Filter through a 0.45 µm membrane and store at 4°C.

1% Paraformaldehyde

100 ml	PBS
1 g	paraformaldehyde (Sigma #P-6148)

Filter through a 0.45 µm membrane and store at 4°C.

Propidium iodide (PI) buffer

PBS	
2 µg/ml	PI (Sigma #P-4170)

Sterile technique and safety precautions

All work is performed in a tissue culture laminar flow hood using *sterile* technique.

Occupational Safety and Health Administration (OSHA) and institutional safety standards must be observed when working with human tissue. These include but are not limited to:

- Latex gloves must be worn at all times and should be changed frequently.
- A lab coat or other appropriate garment must be worn.
- The face must be shielded from the tissue.
- The work area should be decontaminated after processing human tissue by cleansing it with bleach.
- All tissue must be disposed of according to proper OSHA and institutional guidelines.
- All other biohazardous materials (chemicals) must be disposed of according to proper OSHA and institutional guidelines.
- Exercise caution when working with alcohol and flames in the laminar flow hood.

Isolation and purification of fetal bone marrow (BM) B lineage cells

This method is used to isolate and purify distinct populations of fetal BM B lineage cells and their progenitors. Table 11.6.2 provides an overview of the entire purification protocol, including individual steps, time involved, and the expected outcome. Figure 11.6.1 shows a minimal antigenic map for stem cell to mature B cell differentiation. The pool of cells or specific

Table 11.6.2 Flow table for purification of B lineage cells from fetal BM

Step	Time	Outcome	Comments
Process fetal BM	60–90 min	Single-cell suspension	18–21 week fetal BM
Ficoll–hypaque separation	60 min	Lymphohematopoietic and stromal cell enrichment	Layer single-cell suspension over ficoll–hypaque
Adherent depletion	90–120 min	Cells ready for Ab/magnetic bead incubation	Incubate cells in canted neck flasks with RPMI medium–10% FBS. Yield of ~10^8 cells post-adherence Flasks of adherent cells can be used to establish fetal BM stromal cells
Antibody incubation	60 min	Antibody coated non-B lineage cells	Incubate cells in antibody cocktail (see Table 11.6.1); e.g., basic antibody cocktail + anti-μ
Magnetic bead incubation	60–90 min	Antibody/bead coated non-B lineage cells	Incubate antibody-coated cells with magnetic beads
Magnet	5–10 min	Purified negatively selected cell population	Yield of ~2–3×10^7 BCP. Check BCP for purity by immunofluorescence and use in an experimental assay or sort.
Fluorescent antibody incubation	30–45 min	Fluorescent antibody-coated cells	Incubate BCP with desired antibodies (see Table 11.6.1); e.g., anti-CD34-FITC and anti-CD10-PE.
FACS	120–240 min	Highly purified stem cells, pro-B, or pre-B cells	Sort antibody-stained cells (see Fig. 11.6.3c) Yield of ~1.5×10^6 pro-B cells Yield of ~1-1.5×10^6 pre-B cells

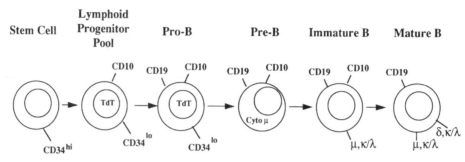

Figure 11.6.1 Minimal antigenic map of human B cell ontogeny in fetal BM. TdT = terminal deoxynucleotidyl transferase.

populations of cells (stem cells, pro-B cells, pre-B cells) shown in this figure can be isolated by the method described below.

Processing fetal long bones

Fetal bones are obtained in accordance with the University of Minnesota Committee on the Use of Human Subjects in Research. The tissue is generally obtained from 19-21-week fetuses.

1. Fetal long bones are transported in a 50 ml conical tube containing RPMI medium which includes RPMI 1640 supplemented with 50 U ml^{-1} penicillin, 50 μg ml^{-1} streptomycin, and 2% FBS (RPMI medium–2% FBS).
2. A sterile work area is set up on a spill pad with the items and reagents necessary to perform isolation of fetal BM cells. Connective tissue and cartilage are removed from the bones in a 100 mm Petri dish. A separate 100 mm Petri dish containing 5 ml of RPMI medium–2% FBS supplemented with 0.1 ml of gentamicin (final concentration 1 mg ml^{-1}) is used to hold cleaned bones. A sterile scalpel, blade, and forceps are used to remove connective tissue and mince the

bones. The scalpel and forceps are placed in isopropyl alcohol and flamed before use, and this step is repeated frequently to maintain sterility.
3. Using the forceps to hold the bone secure, scrape off (longitudinally) the connective tissue with a scalpel. Transfer the bones to the 100 mm Petri dish containing RPMI medium–2% FBS–gentamicin solution.
4. Flush the individual bones with a 19–20-gauge needle attached to a 35 ml syringe filled with RPMI medium–2% FBS. Hold the bone longitudinally with the forceps over a 50 ml cell collection tube and push the needle into the soft marrow of the bone. Gently force (inject) the medium through the bone. Repeat on both ends of the bone. Return the bones to the RPMI medium–2% FBS–gentamicin solution. The BM color will change from red to pink as the cells are flushed out of the marrow cavity.
5. With a new scalpel blade, finely chop/mince the bones into small cross-sectional pieces in the RPMI medium–2% FBS–gentamicin solution. Transfer the medium in the Petri dish into the same 50 ml collection tube used to collect the flushed cells (described in step 4). Continue to mince, rinse and collect medium from the bones until the bones are very faintly pink or white. This mincing process is critical to obtaining a high

yield of nucleated lymphohematopoietic cells from the fetal BM tissue.

6. The single-cell suspension is now ready for ficoll–hypaque (Histopaque) separation and adherent cell depletion.

Isolation of fetal BM lymphohematopoietic precursors

The fetal BM single-cell suspension can readily be depleted of erythrocytes, mature granulocytes, and adherent cells (stromal cells and monocytes). Removal of these cells enriches the preparation for lymphohematopoietic precursors and reduces the amount of antibodies and magnetic beads (Dynabeads) used later in the isolation procedure.

1. Red blood cells and granulocytes are removed by centrifugation over a ficoll–hypaque (Histopaque) gradient. In a 15 ml polystyrene tube gently *layer* 10 ml of the fetal BM single-cell suspension over 5 ml of ficoll–hypaque. Do not disrupt the ficoll–hypaque layer. Centrifuge the preparation for 30 min at 500g, 4°C.

2. After centrifugation collect the low-density cellular interface fraction, which contains all lymphohematopoietic cells and fetal BM stromal cell precursors. Red blood cells and granulocytes (bone spicules may also be present) are found in the pellet following ficoll–hypaque centrifugation and can be discarded. The excess medium above the low-density cellular interface can be aspirated, followed by collection of the interface cells with a Pasteur pipette. Dilute the interface cells 1:3 with RPMI medium–2% FBS and wash twice for 10 min at 700g, 25°C. After the last

wash, resuspend the cell pellet in RPMI medium–10% FBS and quantify the nucleated cells by trypan blue exclusion.

3. Add 10 ml of the cell suspension to a 75 cm^2 canted neck flask, such that each flask has ~5–7.5 $\times$ 10^7 cells. Lay the flask flat in a 37°C, 5% CO$_2$ humidified incubator for 90–120 min. This step will deplete adherent cells (monocytes and BM stromal cell precursors).

4. Collect the nonadherent fraction from each flask with a 10 ml pipette and rinse the flasks with 10 ml of RPMI medium–2% FBS to disassociate remaining nonadherent cells. Pool the nonadherent fractions for subsequent B lineage cell purification (as described in the next subsection).

5. To each flask add 10 ml of stromal cell medium containing Ex-Cell 300, 50 U ml^{-1} penicillin, 50 μg ml^{-1} streptomycin, supplemented with 10% FBS (SCM–10% FBS). Return the flasks to a 37°C, 5% CO$_2$ humidified incubator. The flasks can be used for stromal cell establishment as described under 'Growth and Maintenance of Bone Marrow Stromal Cells'.

Antibody/magnetic bead depletion for purification of B lineage cells

This part details the enrichment of distinct B lineage cell populations (from the adherent cell depleted fetal BM) by antibody/magnetic bead depletion. Approximately 5 $\times$ 10^6 cells are reserved for eventual flow cytometric comparison of nondepleted and enriched cells.

1. Antibody and magnetic bead incubations are performed with

adherent cell depleted fetal BM cells at 1×10^7 cells ml^{-1}.

2. Select the appropriate combination of depletion antibodies, examples of which are shown in Table 11.6.1. The 'basic' cocktail will deplete non-B lineage committed cells (e.g., myeloid, erythroid, and T lineage cells), and the addition of anti-μ will target surface Ig-positive immature B cells (also shown in Fig. 11.6.1).

3. Prepare the antibody cocktail. The depletion antibodies must be titrated on appropriate target cells prior to use to determine the antigen saturating concentration (generally $10 \, \mu g \, ml^{-1}$). Depletion antibodies can be aliquoted at a $10 \times$ concentration in RPMI medium–2% FBS and stored at $-20°C$ until use. For 1×10^8 nonadherent fetal BM cells to be depleted, 1 ml of each of the antibodies at a $10 \times$ concentration is needed.

4. Centrifuge the nonadherent fetal BM cells and resuspend the cells in RPMI medium–2% FBS containing the antibody cocktail. 1×10^8 cells are incubated in a total volume of 10 ml. For volumes >10 ml, use more than one 15 ml polypropylene tube to assure adequate mixing. Incubate the cell/antibody suspension on a rotating mixing wheel for 30–45 min, 4°C.

5. Centrifuge the antibody/cell suspension for 5 min at 700g, 25°C and remove the supernatant. Wash the antibody-coated cells twice by adding 10 ml of RPMI medium–2% FBS and centrifuging for 5 min at 700g, 25°C. Remove any cells which do not disassociate after resuspension of the cell pellet. These are nonviable cells that have 'clumped' together.

6. Prepare the magnetic beads. First, determine the total number (no.) of magnetic beads necessary to adequately target antibody-coated cells (5–10 beads/antibody-coated cell).

Next, based on the total number of cells to be depleted, calculate the volume of beads needed (the magnetic bead package insert lists the bead concentration; e.g., Dynabeads are typically at 4×10^8 beads ml^{-1}). In a basic depletion protocol (basic cocktail + anti-μ or anti-λ) dedicated to enriching for surface Ig$^-$ BCP, ~60% of the cells are non-B lineage or surface Ig$^+$ (antibody-coated cells). A sample calculation based on the strategy outlined in Table 11.6.2 is as follows.

Total no. of beads needed =

$$\text{(Starting cell no.)} \times (\text{\% of antibody-coated cells}) \times \left(\frac{7 \text{ beads}}{\text{antibody-coated cell}} \right)$$

Thus if we had 1×10^8 cells for depletion,

$(1 \times 10^8) \times (0.6) \times (7) = 4.2 \times 10^8$ beads

Total volume of beads needed =

$$\frac{\text{Total no. of beads needed}}{\text{No. of beads/ml}}$$

= ml of beads

$$\frac{4.2 \times 10^8}{4 \times 10^8} = 1.05 \text{ ml of Dynabeads}$$

7. Wash the magnetic beads with RPMI medium–2% FBS to eliminate the sodium azide present in the original magnetic bead reagent. Measure the beads into a 15 ml polypropylene tube with 5 ml of RPMI medium–2% FBS. Place the bead suspension against the magnet. An obvious change in the medium occurs as the iron (brown) colored beads collect against the magnet and the medium returns to its pink color. After the beads have collected against the magnet (~5 min),

remove the supernatant with a Pasteur pipette and repeat. This step, as all other steps, is done using sterile technique and in a laminar flow hood.

8. After the final wash of the cell suspension (step 5 above), combine the cells and magnetic beads in a 15 ml polypropylene tube to the volume used in the antibody incubation. 1×10^8 cells are incubated in a total volume of 10 ml. For volumes >10 ml, use more than one 15 ml tube. Incubate the magnetic bead/cell suspension on a rotating mixing wheel for 60–90 min, 4°C.

9. Place the magnetic bead/cell suspension against the magnet for 5–10 min 25°C. Cells coated with antibody and magnetic beads will adhere to the magnet. Remove the nonadherent cell suspension with a Pasteur pipette. The enriched negatively selected cells contain the pool of hematopoietic stem cells (<5%), pro-B cells (~30%), and pre-B cells (~60%). There are generally <5% 'contaminating' non-B lineage cells representing myeloid/monocytoid cells.

10. Additional antibody/magnetic bead depletions can be performed to further enrich or selectively purify subpopulations using the above method. The percentage of antibody-targeted cells must be determined to calculate accurately the number of magnetic beads to use in subsequent rounds of depletion. Target antigens for subsequent rounds of depletion (e.g., CD34) are listed in Table 11.6.1.

Immunofluorescent analysis of enriched BCP

1. Select the appropriate antigen-specific and isotype-matched control antibodies for the phenotypic analysis of the enriched BCP, and cells removed before antibody/magnetic bead depletion. At a minimum, cells should be stained with anti-CD19 or anti-CD10, anti-μ, anti-κ and anti-λ. Antibodies can be used alone (single-color, FITC or PE) or in combination (two-color, FITC and PE).

2. Incubate 5×10^5 cells/tube with a saturating concentration of antibody (10 μg ml^{-1}) in a total volume of 50 μl fluorescence buffer. (Fluorescence buffer is PBS containing 2.5% newborn calf serum, 0.02% sodium azide, which is filtered through a 0.45 μm membrane and stored at 4°C for up to 2 months.) Incubate the cells for 30 min on ice. Wash the cells twice in fluorescence buffer. (If immunofluorescent staining is performed indirectly, then wash once and incubate with 50 μl of the appropriate secondary staining reagent (FITC-GAM or SA-PE) for 30 min on ice, and wash as above.)

3. Resuspend the antibody-stained cells in 300 μl of PI buffer or 1% paraformaldehyde buffer (1 g paraformaldehyde/100 ml PBS stirred overnight at room temperature, filtered through a 0.45 μm membrane and stored at 4°C for up to 6 months) after the second wash. In general cells are resuspended in the PI buffer and the acquisition and analysis are performed as described in the next two steps. Cells in PI buffer must be stored at 4°C and acquired within 60–120 min. However, if immediate acquisition is not possible, then the cells should be resuspended in paraformaldehyde buffer and stored at 4°C for up to 1 week.

4. Acquire and analyze the cells on a FACScan. Use the isotype-matched negative control to adjust light scatter and fluorescent settings. If the cells are resuspended in the PI solution, then the nonviable cells can be gated

Development of cells of the B lineage

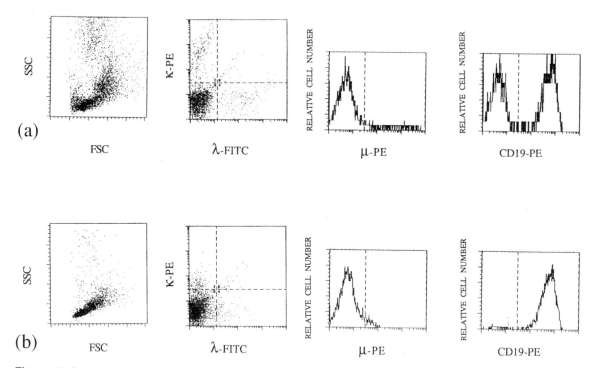

Figure 11.6.2 Flow cytometric analysis of fetal BM before and after antibody/magnetic bead depletion. (a) Ficoll–hypaque, adherent depleted fetal BM, and (b) CD19$^+$/μ^-/κ^-/κ^- cells enriched by antibody/magnetic bead depletion of fetal BM. From left to right: light scattering profile (FSC = forward scatter, SSC = side scatter), κ (HB61-biotin) and κ (HP6054-FITC) light-chain expression, μ (HB57-biotin) heavy-chain expression, and CD19 (25C1-biotin) expression. All of the antibodies were used at 10 μg ml^{-1}; biotinylated antibodies were detected with streptavidin-PE. The acquisition and analysis were performed using FACScan Research software.

out and eliminated from the acquisition and analysis. PI is intercalated into the nucleic acid of nonviable cells and PI$^+$ (i.e., dead) cells can be electronically gated out. Acquire the isotype-matched control, followed by acquisition of the antibody-stained cells.

5. Figure 11.6.2 compares the immunophenotype before and after antibody/magnetic bead depletion. The light scatter profile in Fig. 11.6.2a shows a heterogeneous population of lymphoid, myeloid/monocytoid and other cells, contrasted with the uniform low forward/90° angle light scatter representative of lymphoid cells shown in Fig. 11.6.2b. Greater than 90% enrichment of CD19$^+$ B cells with <5% κ^+, λ^+, or μ^+ B lineage cells

is evident as shown by the phenotypic analysis of enriched cells in Fig. 11.6.2b contrasted with the nonantibody/magnetic bead depleted cells in Fig. 11.6.2a. The enriched cells can be used in experimental assays or they can be further purified by cell sorting. An assay for BCP growth is described later.

6. Enriched BCP with >10% surface μ^+ cells should undergo a second round of antibody/magnetic bead depletion with the anti-μ antibody.

Fluorescence activated cell sorting (FACS)

Cell sorting is used to obtain 95–99% purity of the desired cell population. Specific B lineage subpopulations shown in Fig. 11.6.1 can be isolated by this method. Enriched B lineage cells from the antibody/magnetic bead depletion described above are the starting population for FACS.

1. Define the populations to be sorted based on antigen expression (see Fig. 11.6.1) and choose the appropriate antibody combination.
2. This procedure must be performed using sterile technique and reagents. Resuspend the cells in a saturating concentration of directly conjugated antibodies in PBS, such that the final cell suspension is at 2×10^7 cells/ml. Also stain 1×10^6 cells with the appropriate isotype-matched control antibodies for use as a negative control. Incubate the cells on a rotating mixing wheel for 30 min, 4°C. Wash the cells twice in PBS. (If immunofluorescent staining is performed indirectly, then wash once and incubate with the appropriate secondary staining reagent (FITC-GAM or SA-PE) on a rotating mixing wheel for 30 min, 4°C, and wash as above.) Resuspend the cells in PBS at 1×10^7 cells ml^{-1} in a sterile, capped 5 ml polystyrene tube; aliquot no more than 3 ml/tube. Set aside 1×10^6 cells to compare the presort fluorescent staining profile with the postsort fluorescent staining profile. The cells are now ready to be sorted.
3. The method of cell sorting is conducted using a FACStar Plus or other comparable cell sorter equipped for at least 6-parameter analysis. First, using the isotype-matched negative

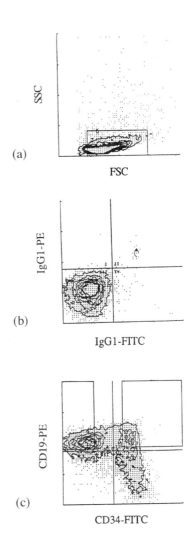

Figure 11.6.3 Sort gates for optimal separation of B cell precursor populations. (a) Light scattering profile of CD19$^+$/μ^-/κ^-/κ^- cells enriched by antibody/magnetic bead depletion of fetal BM; gate is set around lymphoid population based on low to medium forward angle and 90° light scattering. (b) Control (IgG1-FITC and IgG2a-PE) staining of CD19$^+$/μ^-/κ^-/κ^- cells; vertical and horizontal markers are set to define background immunofluorescence. (c) CD19$^+$/μ^-/κ^-/κ^- cells stained with anti-CD34 (HPCA-2-FITC, Becton Dickinson, San Jose, CA, USA) and anti-CD10 (J5-PE, Coulter, Hialeah, FL, USA); stringent gates are set around the pre-B (CD10$^+$/CD34$^-$) and pro-B (CD10$^+$/CD34$^+$) cells. Acquisition, analysis, and sorting were performed using a FACStar Plus.

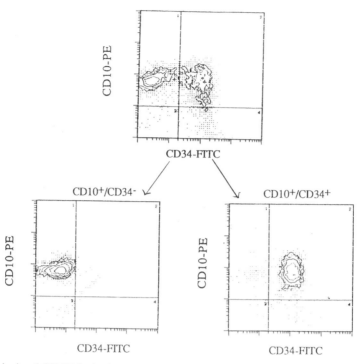

Figure 11.6.4 Analysis of CD19$^+$/μ^-/κ^-/κ^- cells before and after FACS. The top panel shows CD19$^+$/μ^-/κ^-/κ^- cells stained with anti-CD34 (HPCA-2-FITC) and anti-CD10 (J5-PE); markers were set on the basis of negative control staining. The bottom panel shows >95% purified pre-B (CD10$^+$/CD34$^-$) cells (left) and >95% purified pro-B (CD10$^+$/CD34$^+$) cells (right). Acquisition and analysis were performed on a FACScan using FACScan Research software.

control, set a gate around the desired cell population based on light scattering characteristics, as shown in Fig. 11.6.3a. Cells with low to medium forward and low 90° angle light scatter are gated. Second, adjust the log fluorescent gains and set markers for the isotype-matched negative control (Fig. 11.6.3b). Third, using the positively stained cells, set the sort gates around the desired cell populations based on positive fluorescence. Fig. 11.6.3c shows effective sort gates used to purify CD10$^+$/CD34$^+$pro-B cells and CD10$^+$/CD34$^-$ pre-B cells. Only ~30% of the desired populations are gated to enhance the purity of the sorted populations and to eliminate 'bleed-over' of unwanted cells. Sorted populations are collected into sterile

5 ml polystyrene tubes containing 1.5 ml of RPMI medium–10% FBS.
4. Post-sort flow-cytometric analysis is performed to determine the purity of the sort. Remove 5×10^4 or fewer of the sorted cells for acquisition and analysis on the FACScan. The FACScan is used because this instrument is used for subsequent acquisition and analysis of the cultured cells. The FACStar Plus can also be used for post-sort acquisition and analysis. Use the isotype-matched negative control to adjust light scattering and fluorescent settings. Acquire the negative control, the unsorted cells (from step 2, above) and the sorted cells and compare the fluorescent profiles as shown in Fig. 11.6.4. The 95–99% purified pro-B and pre-B cells can be used in

experimental assays. An assay for
BCP growth is described on p. 920.

Troubleshooting

Ideally, purification of B lineage cells from fetal
BM to obtain BCP should be performed within
8–12 h of receiving the tissue. However, cells
can be stored after adherence depletion, or
after antibody/magnetic bead depletion (before
sorting), without damage for an additional
12–24 h. Cells should be stored at 4°C in RPMI
medium–10% FBS. Table 11.6.2 provides a
timeline.

The antibody cocktail for magnetic bead
depletion can be modified as necessary. The
suggested 'basic' cocktail is against populations of non-B lineage cells with high antigen
expression and surface Ig$^+$ B cells. Additional
depletions and sorts are performed on enriched
B cell populations to more effectively and
efficiently target specific populations. In particular, a second round of μ depletion will more
effectively target and eliminate cells with low
levels of surface Ig.

Some BM cells may die during the BCP
enrichment and purification; this is usually
observed as 'clumping' of the cells. Remove
any clumps of cells which do not disassociate
upon resuspension of the cell pellet.

Dynabead magnetic spheres are suggested
because of their excellent performance in
obtaining high yields of highly enriched, viable,
and functional cells. Other magnetic particles
may damage cell integrity owing to the irregu-
larity of the particles.

If there are microscopically detectable magnetic beads in the enriched BCP population,
place the cell suspension against the magnet
for an additional 5–10 min, and then remove the
nonadherent cells.

Growth and maintenance of bone marrow stromal cells

In vitro studies have demonstrated that BM
stromal cells are required for the growth of
BCP, although the exact mechanism by which
the stromal cells and associated extracellular
matrix exert their supportive effects is
unknown. This section will describe the method
for establishing and maintaining BM stromal
cells for use in functional assays.

used to initiate BM stromal cell cultures can be
obtained from the light density fraction of ficoll–
hypaque gradients. Since processing of fetal
BM to obtain BCP includes ficoll–hypaque
(Histopaque) centrifugation (as described earlier), this section will describe processing of
adult BM.

Isolation of human BM stromal cells

Stromal cells are established from fetal or adult
BM specimens. With either source the cells

1. Adult BM obtained as 5–10 ml
   aspirates from normal donors is
   collected in Minimal Essential Medium
   containing heparin sodium.
2. Each ml of the adult BM aspirate is
   diluted in 2 ml SCM–2% FBS. Thirty ml
   of diluted adult BM is then *layered*
   over 15 ml of ficoll–hypaque in 50 ml

polypropylene tubes. Do not disrupt the ficoll–hypaque layer. Centrifuge the preparation for 30 min at 500g, 4°C.

3. The low-density cellular interface fraction, which contains all lymphohematopoietic cells (except erythrocytes and granulocytes) and various adult BM stromal cell precursors, is carefully removed. Aspirate the excess medium above the low-density cellular interface fraction, then remove the interface cells with a Pasteur pipette and dilute them 1:3 into SCM–2% FBS. The interface cells are washed twice in SCM–2% FBS for 10 min at 700g, 25°C.

4. The nucleated cell pellet is quantified by trypan blue exclusion, and the cells are adjusted to a concentration of 1–5×10^6 cells ml^{-1} in SCM–10% FBS. Cells (10–12 ml) are then added to individual 75 cm^2 straight neck culture flasks, and the flasks are placed flat in a 37°C, 5% CO$_2$ humidified incubator.

Establishment and maintenance of adult BM stromal cells

1. Following an initial incubation period of 3–7 days, the nonadherent cells are removed from the culture by gently rinsing the bottom of the flask with the spent medium in the flask. The medium is removed and replaced with 10 ml of fresh SCM–10% FBS. Examination of the culture on an inverted microscope will reveal a small number of fibroblastic cells adherent to the bottom of the flask within the first week of culture. Some residual nonadherent cells remain and will be washed away during subsequent medium changes and stromal cell passage.

2. A complete replacement of the medium in each flask is repeated once or twice a week until adherent adult BM stromal cells reach confluence (i.e. when the stromal cell monolayer covers 80–95% of the flask), generally within 3–4 weeks.

3. When BM stromal cell cultures reach confluence they can be passaged using trypsin–EDTA (described in the next subsection). However, cultures can also be maintained in SCM–0% FBS at this point (refer to Fig. 11.6.5). The absence of FBS will inhibit stromal cell proliferation while still maintaining viability. Continue to monitor these cultures and replace the medium once a week with SCM–0% FBS.

Passage of stromal cells

When BM stromal cell cultures reach confluence, the rate of proliferation slows owing to contact-dependent mechanisms. Cells can be detached by trypsin–EDTA and passaged into new flasks at a lower density, thereby facilitating proliferation and expansion of stromal cells.

1. Remove SCM from a confluent flask of BM stromal cells and replace with 10 ml of 1× trypsin–EDTA solution (0.5%(w/v) trypsin, 5.3 mM EDTA). Trypsin's enzymatic activity works best at 37°C, therefore it is best to warm the 1× trypsin–EDTA solution before adding to the flask. Incubate the flasks in a 37°C, 5% CO$_2$ humidified incubator for no longer than 7 min (excessive exposure to trypsin will damage the cells).

2. To detach the majority of stromal cells, hold the flask upright with the cap tightly sealed and slap the back of the flask 1–2 times with an open palm.

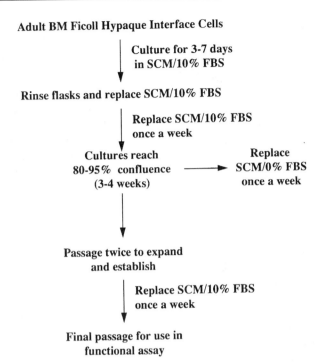

Figure 11.6.5 Flow chart for establishing and maintaining adult BM stromal cells. SCM = stromal cell medium (see text).

The stromal cells will detach, sometimes as large aggregates or sheets of cells, and float in the trypsin–EDTA solution. If the majority of stromal cells do not detach, replace the solution in the flask with fresh 1× trypsin–EDTA, and reincubate in a 37°C, 5% CO_2 humidified incubator for 1–2 min more.

3. Transfer the stromal cells into a 50 ml collection tube. Rinse the flask with 10 ml SCM–10% FBS and add to the 50 ml collection tube. Centrifuge for 10 min at 700g, 25°C. FBS will neutralize the trypsin's enzymatic activity.

4. Resuspend the stromal cell pellet in 30 ml of SCM–10% FBS per flask of trypsinized BM stromal cells. Mix the stromal cells vigorously or vortex to assure a uniform single-cell suspension. One 75 cm^2 straight-neck flask of confluent adult BM stromal cells yields on average of 1–2 × 10^6 stromal cells.

5. Aliquot 10 ml of BM stromal cell solution into individual 75 cm^2 straight-neck flasks (0.3–0.6 × 10^6 stromal cells/flask), and maintain under the conditions described above in the previous protocol until confluence is reached.

6. Stromal cells should be passaged at least twice before use in a functional assay. BM stromal cells passaged more than three times frequently exhibit a decreased ability to reattach to tissue culture flasks and undergo expansion.

Troubleshooting

If BM stromal cells grow to 100% confluence in SCM–10% FBS they may detach or 'peel off'. To prevent stromal cells from reaching 100% confluence, switch the culture to a serum-free medium when the flask is 80–95% confluent.

During maintenance of stromal cells in medium containing no serum, stromal cells may start to detach. If this occurs, trypsinize and passage the stromal cells (procedure described earlier). If stromal cells fail to reattach and grow, the cultures will have to be discarded.

Cryopreservation of BM stromal cells is not recommended. The yield and growth of cryopreserved stromal cells is very poor, but our experience with this is limited. If cryopreservation is attempted, then ficoll–hypaque interface cells should be used (prior to stromal cell establishment).

Stromal cells established from adult BM must be monitored for the outgrowth of Epstein–Barr virus (EBV)-transformed B cells. In our experience up to 20% of adult BM specimens cultured give rise to EBV-transformed cells. EBV-transformed cells can readily be observed on an inverted microscope as small clusters of lymphoblastic cells. Outgrowth of EBV-transformed cells is generally seen after 4–6 weeks of culture. If this occurs, then all stromal cells from this sample must be discarded since it is difficult to 'purge' the stromal cells of the EBV transformants.

Establishment and maintenance of fetal BM stromal cells

The procedure for isolating fetal BM ficoll–hypaque interface cells is described earlier under 'Isolation of Fetal BM Lymphohematopoietic Precursors'. The flasks used to adherent-deplete fetal BM ficoll–hypaque interface cells can be used to establish fetal BM stromal cells.

The growth, maintenance, and passage of fetal BM stromal cells is the same as for adult BM stromal cells with the following differences.

1. Fetal BM stromal cells adherent to plastic are smaller and less fibroblastic than adult BM stromal cells. Therefore, the yield from one confluent 75 cm² straight neck flask is $2–3 \times 10^6$ fetal BM stromal cells, versus $1–2 \times 10^6$ adult BM stromal cells.
2. There is no problem with EBV transformation since cells in fetal BM have not been exposed to EBV.
3. Fetal BM stromal cells become confluent within 1–3 weeks after initiation or passage, compared to adult BM stromal cells which reach confluence in 2–4 weeks.

Harvest of stromal cells for flow cytometric analysis

1. Stromal cells that have been passaged at least twice are trypsinized (as described earlier), passaged into 60–100 mm Petri plates, and allowed to attach for 2–4 days. Approximately 2×10^5 fetal BM stromal cells are passaged into one 60 mm Petri plate.
2. To detach the stromal cells for phenotyping, all medium is removed from the Petri plate and 3–5 ml of cell dissociation solution is added. Do not use trypsin to detach the stromal cells since many cell surface epitopes recognized by monoclonal antibodies are trypsin sensitive.
3. Incubate for 15–30 min in a 37°C, 5% CO_2 humidified incubator and then detach the stromal cells from the surface of the Petri dish using a rubber policeman.
4. Rinse the Petri dish with fluorescence buffer and centrifuge the stromal cells for 2 min at 1000g, 4°C.
5. The supernatant is aspirated off and the stromal cells are stained with

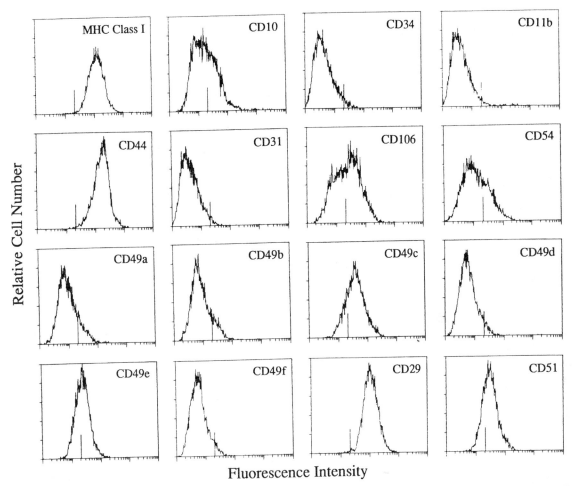

Figure 11.6.6 Expression of cell surface antigens on fetal BM stromal cells after three passages. The vertical bar in each histogram represents background staining in which greater than 95% of the events fall to the left of the vertical bar.

monoclonal antibodies. Since stromal cells have a high number of Fc receptors on their cell surface it is recommended that the secondary antibody used be a F(ab')$_2$ goat anti-mouse Ig conjugated to FITC or PE.

Figure 11.6.6 demonstrates a typical monoclonal antibody staining profile of fetal BM stromal cells.

Growth of BCP

Initiation of BCP cultures

An important logistical aspect in growing BCP is to coordinate BM stromal cell establishment with BCP isolation so that both cell types are in optimal condition for a growth assay at the same time. Since purified BCP should be plated within 24 h after purification, and establishment of BM stromal cells requires 4–6 days, it is best to prepare stromal cells in anticipation of available BCP (see Fig. 11.6.7).

1. BM stromal cells that have been passaged twice are trypsinized (as described above in 'Passage of Stromal Cells') and plated in 96-well flat bottom tissue culture plates in SCM–10% FBS (200 µl/well). Adult BM stromal cells are plated at 4×10^3 cells/well and fetal BM stromal cells are plated at 8×10^3 cells/well.

2. Stromal cells in SCM–10% FBS will reach confluence in 4–5 days. Once this occurs, all the medium is carefully removed, to avoid disruption of the stromal cell matrix, and replaced with 200 µl/well of X-VIVO serum-free medium. FBS is left out to prevent continued stromal cell proliferation. This step also eliminates the need to irradiate the stromal cells. Stromal cells should be incubated in a 37°C, 5% CO_2 humidified incubator for 24 h, and are optimal for use in a growth assay for 1–7 days after switching the culture to serum-free medium.

3. BCP enriched from fetal BM by magnetic bead depletion or subsets purified by FACS are resuspended in X-VIVO medium at $4\times$ the final desired concentration (e.g., $0.5–1.0 \times 10^6$ cells ml^{-1}). Successful growth will consistently occur when the BCP are plated from 5.0×10^3 to 5.0×10^4 cells/well.

4. Cytokines to be added should be diluted into X-VIVO medium at $4\times$ the final desired concentration (e.g., IL-7 at 40 ng ml^{-1}).

5. The growth assay is initiated by removing half the medium (100 µl) from each well containing confluent BM stromal cells. Add 50 µl of BCP suspension to each well (to yield a final concentration of $2.5–5.0 \times 10^4$ cells/well) and then add 50 µl of the cytokine solution to each well (e.g., IL-7 at 40 ng/ml to yield a final concentration of 10 ng ml^{-1}).

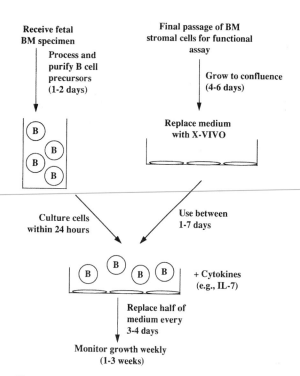

Receive fetal BM specimen

Process and purify B cell precursors (1-2 days)

Final passage of BM stromal cells for functional assay

Grow to confluence (4-6 days)

Replace medium with X-VIVO

B B B B

Culture cells within 24 hours

Use between 1-7 days

B B B B + Cytokines (e.g., IL-7)

Replace half of medium every 3-4 days

Monitor growth weekly (1-3 weeks)

Figure 11.6.7 Flow chart for establishing and maintaining BCP growth assay.

Maintenance and monitoring of growth assay

1. Feed the culture every 3–4 days by replacing half the medium in each well with 100 μl of cytokine at 1× concentration in fresh X-VIVO medium (e.g., IL-7 at 10 ng ml⁻¹). Carefully pipette 100 μl of fresh medium down the side of the well to avoid disrupting BCP/stromal cell interactions at the bottom of the well.

2. Monitor the growth of B lineage cells weekly over a 1–3 week period, either by trypan blue exclusion or by the microsphere/flow cytometry growth quantification assay (Larson 1994). Figure 11.6.8 shows the results of a typical growth assay of BCP on BM stromal cells using this quantification method. In brief, this microsphere/flow cytometry growth quantification assay is conducted as follows.

(a) Vigorously pipette the contents of a single assay well and then transfer contents to a 5 ml polystyrene tube. Triplicate wells are harvested for each time point.

(b) Trypsin–EDTA solution (1×, 200 μl) is added to the assay well for 5 min at 37°C. The well is mixed and pooled with the contents in the 5 ml tube.

(c) The cells are washed with fluorescence buffer and stained with anti-CD19 conjugated to PE or FITC as described for 'Immunofluorescent Analysis of Enriched BCP'.

(d) Cells are resuspended in 200 μl PI buffer.

(e) Polystyrene 6 μm microspheres are diluted in fluorescence buffer to approximately $1\text{--}2 \times 10^6$ spheres ml⁻¹ and stored at 4°C. Microspheres are mixed vigorously and quantified each time they are used, by carefully counting on a hemacytometer.

(f) Accurately add 30 μl of microsphere solution to the cell suspension and acquire and analyze on a FACScan.

(g) Total viable CD19⁺ cells/well are calculated as [(total number of viable CD19⁺ events/total number of microsphere events) × total number of microspheres added per tube].

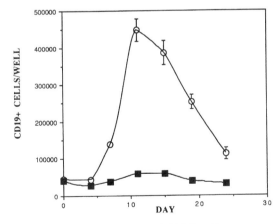

Figure 11.6.8 Growth of fetal BM B cell precursors on adult BM stromal cells. CD10⁺/surface IgM⁻ BCP were cultured on adult BM stromal cells in the presence (○) or absence (■) of 20 ng ml⁻¹ IL-7. Data points represent CD19⁺ cells recovered from each well over a 3-week period using the microsphere/flow cytometry growth quantification assay. Values represent the mean + SD of 3 replicates.

Troubleshooting

Myeloid cell outgrowth may occur if cytokines such as IL-3 are added to the culture. This myeloid outgrowth could lead to a noticeable decrease in BCP growth. Two causes for myeloid cell outgrowth are: (a) contaminating myeloid cells in the BCP enriched from fetal BM, or (b) stem cells differentiating to the

Hematology, 5th edn. McGraw-Hill, New York, pp. 211–228.

Ryan DH, Nuccie BL, Abboud CN (1992) Inhibition of human bone marrow lymphoid progenitor colonies by antibodies to VLA integrins. J Immunol 149: 3759–3764.

Wolf ML, Buckley JA, Goldfarb A, Law CL, LeBien TW (1991) Development of a bone marrow culture for maintenance and growth of normal human B cell cultures. J Immunol 147: 3327–3330.

Wolf ML, Weng W-K, Stieglbauer KT, Shah N, LeBien TW (1993) Functional effect of IL-7-enhanced CD19 expression on human B-cell precursors. J Immunol 151: 138–148.

Utilization of poly(A) PCR to establish gene expression patterns in single cells

11.7

Ian McDermott
Christopher J. Paige

The Wellesley Hospital Research Institute and Department of Immunology, University of Toronto, Toronto, Ontario, Canada

This work was supported through grants from Medical Research Council of Canada, The National Cancer Institute and Canadian Cancer Society, and a Terry Fox Program Project Grant.

TABLE OF CONTENTS

Introduction

The method described here is a modification of poly(A) PCR (Brady and Iscove 1993). This approach allows representative amplification of all available mRNA within a single cell. Single cells are acquired by making use of the Automatic Cell Deposition Unit, ACDU, of the FACStar Plus (Becton Dickenson) and sorted directly into a PCR lysis buffer. Without any further purification steps, a combination of M-MLV and AMV reverse transcriptases are used to synthesize the first strand of cDNA from all available mRNA. Time-limited incubation insures production of cDNAs with an average length of 500 bases. Newly synthesized cDNAs are tailed with dATPs using terminal deoxynucleotidyl transferase, allowing the annealing of a pd(T)-X primer. Keeping the average length of the cDNAs the same allows representative amplification of each cDNA produced so that PCR products appear as smears, as visualized with ethidium bromide on agarose gels, ranging in size from 100 to 700 bp with a concentration around 500 bp. To determine gene expression patterns, these products are transferred to nylon membranes and hybridized with a variety of probes from the genes of interest. These probes must be derived from the 3' regions of the genes. Owing to the limited amount of primary amplified PCR product, reamplification of samples is often required to allow further study of gene expression. Reamplified products will give gene expression patterns that are consistent with any pattern developed with primary PCR products.

Materials

Equipment

ACDU-FACStar Plus (Becton Dickenson)

PCR 9600 (or 2400) System (Perkin Elmer)

Chemicals

Tris

KCl

$MgCl_2$

Nonidet P-40

BSA (molecular grade)

Triton X-100

Items as required for agarose gel electrophoresis and hybridization protocols (Maniatis et al 1989)

Reagents

Inhibit Ace (FivePrime-ThreePrime)

RNAguard (Pharmacia)

dNTPs (Boehringer Mannheim)

dATP (Boehringer Mannheim)

pd(T)$_{19-24}$ (Pharmacia)

M-MLV reverse transcriptase (Canadian Life Technologies, Gibco/BRL)

AMV reverse transcriptase (Canadian Life Technologies, Gibco/BRL)

Terminal deoxynucleotidyl transferase (Canadian Life Technologies, Gibco/BRL)

pd(T)$_{24}$-X

Taq polymerase (Canadian Life Technologies, Gibco/BRL)

Description of method

cDNA production

Single cells are acquired using the Automatic Cell Deposition Unit, ACDU, of the FACStar Plus. Cells are sorted directly into 0.2 ml tubes, held in the Perkin Elmer retaining tray as this tray has the same configuration as a 96-well tissue culture plate. Cells are sorted based on viability and are often stained with monoclonal antibodies conjugated to FITC and/or PE. In general, a single cell is sorted in an approximate volume of 0.005 μl into a lysis mix of 4.5 μl.

Lysis mix

96%	cDNA lysis buffer
1%	Inhibit Ace, 0.5–1 μl, (5'–3')
1%	RNAguard, 24,600 U ml^{-1} (Pharmacia)
2%	cDNA prime mix (diluted appropriately)

cDNA lysis buffer

52 mM	Tris-HCL pH 8.3
78 mM	Tris-HCL pH 8.3
3.1 mM	MgCl$_2$
0.52%	Nonidet P-40 (Sigma)

cDNA prime mix

12.5 mM	dNTPs (Boehringer Mannheim)
6.125 OD$_{260}$ ml^{-1}	pd(T)$_{19-24}$ primer (Pharmacia)

The cDNA lysis buffer can be made and stored at room temperature for several months. The cDNA prime mix is stored at −20°C. The lysis mix is made fresh each day.

The optimal dilution of the cDNA prime mix, used in the lysis mix, must be determined for each new batch of prime mix created. This can be accomplished by using a series of cDNA prime mix dilutions, 1:5 to 1:25, in the lysis mix. Using this range of lysis mixes, global PCR should be performed on a range of 1–50 cells.

When sorting cells with the FACStar Plus, up to 50 cells may be sorted into the original 4.5 μl of lysis mix without appreciably altering the volume of the reaction. Prior to addition of a cell(s), the lysis mix can be aliquoted into the PCR tubes and left on ice for up to 30 min. After the addition of cells, the lysis mix should be mixed by orbital rotation such as that achieved with the Eppendorf mixer. This will ensure lysis of the cells. A combination of M-MLV and AMV reverse transcriptases are used to synthesize the first strand of cDNA from all available mRNA. This reverse transcriptase is added to the cell lysis mixture in a volume of 0.5 μl. Note that the amount of enzyme added at any step in this process will keep the enzyme/glycerol content at 5% of the total reaction volume, 5% being the optimal enzyme/glycerol content recommended by the enzyme manufacturers. The reverse transcriptase mix should be made immediately prior to use and kept on ice for the briefest time possible.

Reverse transcriptase mix

25%   M-MLV reverse transcriptase, 200 U µl⁻¹ (Canadian Life Technologies, Gibco/BRL)
25%   AMV reverse transcriptase, 2.5 U µl⁻¹ (Canadian Life Technologies, Gibco/BRL)
50%   cDNA lysis buffer

Samples should be mixed by orbital rotation to ensure an even mixing of cell lysate and reverse transcriptase. Incubation at 37°C for 15 min allows production of cDNAs with an average length of 500 bases. Further cDNA production is prevented by incubation for 10 min at 65°C, thus inactivating the reverse transcriptases. These reactions have been carried out in the Perkin Elmer PCR 9600 machine, which prevents sample from evaporating and condensing on the tube lids. Water baths can also be used, ensuring that only the bottom half of the tube is emersed in the water. After incubating in a water bath, the tubes must be spun briefly to collect all 5 µl at the bottom of the tube. Samples can be spun in a microcentrifuge, with the proper adaptors, or in a table-top centrifuge with microplate carriers. Samples are then placed on ice for a minimum of 2 min.

TDT mix

5%   terminal deoxynucleotidyl transferase, 15–20 U µl⁻¹ (Canadian Life Technologies, Gibco/BRL)
5%   distilled H_2O
90%   2× tailing buffer

2 × Tailing buffer

44.4%   5× TDT buffer (Gibco/BRL)
1.7%   100 mM dATP (Boehringer Mannheim)
53.9%   distilled H_2O

The 2× tailing buffer can be aliquoted and stored at –20°C.

Add an equal volume of TDT mix, 5 µl, to the cDNA reaction mixture and mix by orbital rotation. Ensure that sample does not end up on the wall of the tube; if this does occur the sample can be briefly centrifuged, as described for the collection of sample after cDNA production. Samples are then incubated for a minimum of 15 min at 37°C, followed by 10 min at 65°C to inactivate the TDT. Samples are placed on ice to allow sufficient cooling prior to amplification. Single-cell cDNA products are amplified directly and not stored for any length of time.

cDNA tailing

Newly synthesized cDNAs are tailed with dATPs using terminal deoxynucleotidyl transferase (TDT), allowing the annealing of a pd(T)-X primer. The TDT mix should be made fresh each day and kept on ice prior to use.

Amplification

Efficient amplification of the cDNA is directly affected by both the primer pd(T)$_{24}$-X and the $MgCl_2$ concentration and, therefore, must be tested accordingly. This should be done by amplifying cDNA, from single cells, using a range of $MgCl_2$ and primer concentrations. The 10× *Taq* polymerase buffers should have a range of 15–50 mM $MgCl_2$ in 5 mM increments.

The primer should be tested at a final concentration of 0.5–4 OD_{260} units ml^{-1} in the PCR reaction mixture.

10× *Taq* polymerase buffer

100 mM	Tris-HCl pH 8.3
500 mM	KCl
15–50 mM	$MgCl_2$ in increments of 5 mM
1 mg ml^{-1}	BSA (molecular grade)
0.5%	Triton X-100

PCR mix

12.5%	10× *Taq* polymerase buffer
x%	primer pd(T)$_{24}$-X
4.7%	25 mM dNTPs
5%	*Taq* polymerase, 2.5 U μl^{-1}
(77.8–x)%	distilled H_2O

To each tailed cDNA reaction mixture add 40 μl of the PCR mix and ensure proper mixing of the sample. This will give a final reaction volume of 50 μl. If required, overlay the sample with mineral oil before amplification.

The primer used for the amplification of the cDNA is a 60-mer: 5'-ATG TCG TCC AGG CCG CTC TGG ACA AAA TAT GAA TTC plus 24 Ts-3', reverse phase purified primer is suitable for PCR.

Amplification protocol

25 cycles of 1 min at 94°C
2 min at 42°C
6 min at 72°C
Followed by 25 cycles of 1 min at 94°C
1 min at 42°C
2 min at 72°C
Followed by a hold at +4°C.

Analysis

A fraction of the PCR product, 10–20% of the reaction volume, is examined by agarose gel electrophoresis utilizing 1.5% agarose, ethidium bromide and 1× TBE buffer. Smears from 100 to 700 bp are expected with a concentration around 500 bp. Samples are transferred to nylon membranes by alkaline capillary transfer (Maniatis et al 1989) and probed with a variety of 3' probes. DNA probes from the 3' regions of the genes of interest are radiolabeled with [α-^{32}P]dCTP using random priming techniques (Maniatis et al 1989). Prehybridization and hybridizations (Maniatis et al 1989) are carried out at 42°C with a final wash of 0.1× SSC/65°C.

Reamplification

Reamplification of samples are often required to allow further study of gene expression patterns. Reamplification uses 1 μl of the original PCR product diluted into 100 μl of the re-amp PCR mix.

Re-amp PCR mix

10%	10× *Taq* polymerase buffer
0.8%	25 mM dNTPs
5%	*Taq* polymerase
x%	primer pd(T)$_{24}$-X
(84.2–x)%	distilled H_2O

Re-amp protocol

1 min at 94°C
1 min at 42°C
2 min at 72°C.

The primer in the reamplification should be used at a final concentration of 0.5–2 OD_{260} ml^{-1} in the re-amp PCR mix. New

PCR products are analyzed in the same way as the primary products.

Discussion and trouble shooting

This method of cDNA production from direct lysis of a cell can be applied to more than one cell. However, if more than 50 cells are to be sorted using the FACStar Plus, then the volume of lysis mix should be reduced to 4 µl. At no time should the lysis mix drop below this volume. Samples in excess of 100 cells do not yield representatively amplified products. cDNA production that is allowed to continue longer than 15 min can lead to nonrepresentative amplification of smaller cDNA products. The time required for tailing newly synthesized cDNA is a minimum of 15 min and can be extended for longer periods. Note that cDNA produced from a single cell is not stable and will degrade if stored overnight, even at –86°C. Once tailed, cDNA from single cells is amplified as quickly as possible. PCR products can then be stored at +4°C overnight, at –20°C for short-term storage, and at –86°C for long-term storage.

The primer pd(T)-X used for amplification of the cDNA is custom synthesized and should undergo a reverse-phase purification. Highly concentrated primer is kept in small aliquots at –86°C for long-term storage, while primer in current use can be kept at –20°C for a few weeks. In general, if a primer does not work properly it is probably not worth trying to purify it any further. Have a new primer synthesized. Note that the primer has an *Eco*RI restriction site, allowing the PCR products to be cloned into the appropriate vectors creating 3' cDNA libraries.

Amplified samples are first hybridized with a 'housekeeping' gene to look at the integrity of the sample. In general, if the 'housekeeping' gene hybridization does not give a positive result, the sample is considered to be non-representative of the mRNA available within the cell and should be discarded.

References

Brady G, Iscove NN (1993) Methods in Enzymology 255: 611–623.

Maniatis T, Fritsch E, Sambrook J (1989) Molecular Cloning: A Laboratory Manual. Cold Spring Harbor Laboratory Press, Cold Spring Harbor, NY.

Outlook

11.8

Christopher J. Paige

The Wellesley Hospital Research Institute and Department of Immunology, University of Toronto, Toronto, Ontario, Canada

Techniques which allow the *in vitro* growth of developing cells are essential for investigating the fundamental questions of differentiation. Some of these are: What are the origins of lymphocytes? What constitutes an appropriate environment for lymphocyte development and growth? What factors are needed as progenitors progress through the stages of commitment and selection? And, what genes underlie the commitment process itself? Although the techniques outlined in this section specifically address developmental issues of importance for B lymphopoiesis, the insights gained may well reach beyond the B cell horizon. For example, methods which provide environments for B cell growth and development may lead to better methods for growth of human hematopoietic progenitors in general. This in turn will find ample clinical applications which require selective cell growth prior to transplantation. Likewise, we anticipate that the discovery of new genes involved in the process of commitment and progression of a lineage is likely to have consequences for understanding abnormalities which lead to immunodeficiency and malignancy.

Section 12

Ex vivo and *in vitro* Methods for Studying B Lymphopoiesis in Mouse and Man

Section Editor
Antonius Rolink

Introduction

12.1

Antonius Rolink

Basel Institute for Immunology, Basel, Switzerland

TABLE OF CONTENTS

B Lymphopoiesis

B cell development in bone marrow of mouse can be dissected into several stages based on (1) the differential expression of a number of cell surface and intracellular markers; (2) the rearrangement status of the IgH and L chain loci; (3) the cell cycle status; and (4) the capacity to grow on stromal cells in the presence of IL-7 or IL-3 *in vitro*. On the basis of these criteria we have grouped the different stages of B cell development into pro-B, pre-B-I, pre-B-II, immature B and mature B cells (Fig. 12.1.1). All stages express the pan B-cell specific marker CD45R (B220). Pro- and pre-B-I cells can only be dissected by the rearrangement status of the IgH chain locus, i.e. pro-B cells have all IgH chain loci in germline configuration while the majority of pre-B-I cells have them $D_H J_H$ rearranged (Rolink et al 1993b). The most characteristic cell surface

marker of pro/pre-B-I cells is the tyrosine kinase c-kit (Fig. 12.1.2A) (Ogawa et al 1991; Rolink et al 1991b). Pro/pre-B-I cells also express TdT and the surrogate light-chain proteins, $\lambda 5$ and V_{pre-B}, but not CD25 (Rolink et al 1994a). About 30% of these cells are in S, G2, and M phase of the cell cycle (Fig. 12.1.2D) (Rolink et al 1994a). Moreover, these cells can be grown long-term as clones or lines *in vitro* on stromal cells plus IL-7 or IL-3 (Fig. 12.1.3) (Rolink et al 1991a; Winkler et al 1995a). In fact, limiting dilution analysis with sorted B220+ c-kit+ pro/pre-B-I cells indicated that at least 1 out of 5 of these cells possesses such an *in vitro* growth capacity (Rolink et al 1991a; Winkler et al 1995a). In the presence of stromal cells and growth factor, the cells keep their pro/pre-B-I phenotype and genotype. Removal of the growth factor from the culture system leads

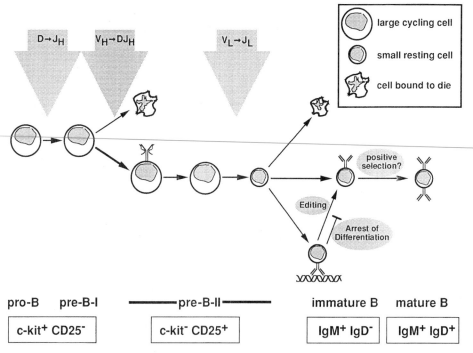

Figure 12.1.1 A model of B cell development in normal mouse bone marrow based on the analysis described in Rolink and Melchers (1991), Rolink et al (1991a, 1994a), ten Boekel et al (1995), Winkler et al (1995a).

to the differentiation of the cells, resulting in the appearance of sIgM$^+$, mitogen-reactive B cells (Rolink et al 1991a). This differentiation process is, however, accompanied by high apoptotic activity (Rolink et al 1991a). This high rate of apoptosis can, however, be inhibited by the expression of a bcl-2 transgene (Rolink et al 1993a). Upon injection into immunodeficient mice these long-term growing pro/pre-B-I cells can, at least partially, reconstitute a functional mature B cell compartment (Rolink et al 1991a; Reininger et al 1992; Rolink et al 1994b).

The most characteristic cell surface marker of pre-B-II cells is CD25 (TAC) (Fig. 12.1.2B and

C) (Rolink et al 1994a).Based on the cell cycle status, pre-B-II cells can be subdivided into large actively cycling and small resting cells. About 70–80% of the large pre-B-II cells and fewer than 5% of the small resting pre-B-II cells are in S, G2, and M phase of the cell cycle (Fig. 12.1.2E,F) (Rolink et al 1994a). From the large pre-B-II cells, 30% express the surrogate L chain encoded by the pre-B cell specific genes $\lambda 5$ and V_{pre-B} together with μH chain protein on the surface (Winkler et al 1995b). All pre-B-II cells are at least on one H chain allele productively V_H-$D_H J_H$ rearranged since they are all $c\mu^+$ (Fig. 12.1.4) (Rolink et al 1994a). Light-

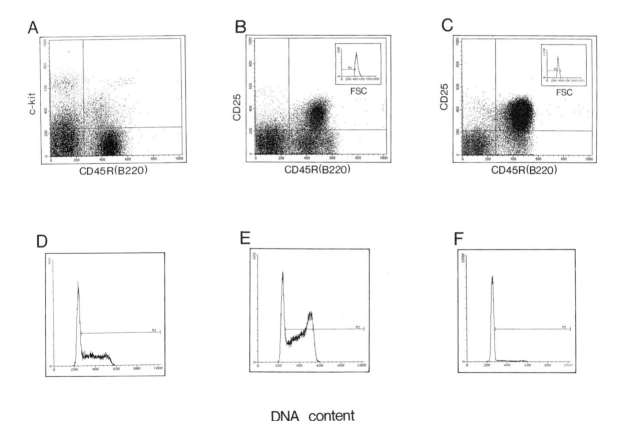

DNA content

Figure 12.1.2 Dual labeling analysis of sIgM depleted mouse bone marrow for CD45R (B220)/c-kit (A) and CD45R (B220)/CD25 (TAC) ((B) and (C)). For the expression of CD45R (B220)/c-kit total nucleated bone marrow cells were analyzed while CD45R (B220)/CD25 (TAC) expression was analyzed by forward and side scattering gating (see inserts) on large (B) and small (C) cells. The percentages of double positive cells in 4–5 week-old normal mouse bone marrow are CD45R (B220)/c-kit, 3.5%; CD45R (B220)/CD25 (TAC) large, 6.2%; CD45R (B220)/CD25 (TAC) small, 18.0%. DNA content analysis in nuclei prepared from sorted CD45R (B220)$^+$/c-kit$^+$ (D), large CD45R (B220)$^+$/CD25 (TAC)$^+$ (E) and small CD45R(B220)$^+$/CD25 (TAC)$^+$ (F) bone marrow-derived cells. The marker indicates the percentage of cells in S and G2/M phases of the cell cycle. These percentages are D 40%, E 75%, F 5%.

chain gene loci rearrangements are still absent in large pre-B-II cells and present in small pre-B-II cells (ten Boekel et al 1995). All pre-B-II cells have lost the capacity to grow on stromal cells plus IL-7 or IL-3 *in vitro* (Rolink et al 1993b). Immature B cells are the first cells to express, although at low levels, sIgM (Fig. 12.1.5a). They do not yet express sIgD. Cell cycle analysis revealed that less than 5% of these cells are in S, G2, and M phase of the cell cycle. Mature B cells are characterized by the expression of IgM and IgD (Fig. 12.1.5a).

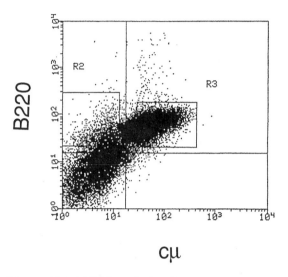

Figure 12.1.4 FACS analysis of intracellular expression of μH chains in sIgM-depleted CD45R (B220)$^+$ normal mouse bone marrow cells. About 80% of the CD45R (B220)$^+$ cells express μH chains cytoplasmically.

Figure 12.1.3 Mouse bone marrow-derived pre-B-I cells growing on stromal cells in the presence of 100–200 U ml^{-1} IL-7 ($\times$ 250).

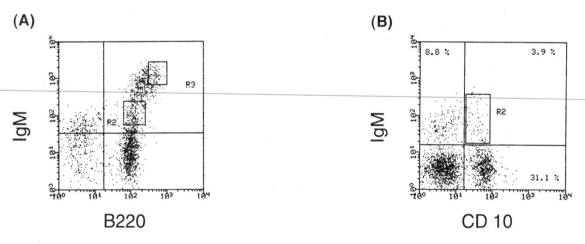

Figure 12.1.5 sIgM staining of mouse (A) and child (B) bone marrow. (A) R2 gate identifies low-expressing B220$^+$ B cells with low amount of immunoglobulin on the surface, corresponding to the immature B cells. On the contrary, R3 gate identifies B cells expressing high levels of immunoglobulin together with high amount of B220 on the cell membrane (mature B cells). In human bone marrow, immature B cells are characterized by the expression of CD10 together with sIgM (gate R2 in (B)), while the majority of sIgM$^+$ sIgD$^+$ mature B cells are CD10 negative. It is not possible to distinguish, as in the mouse, immature (CD10$^+$) cells of human bone marrow which express different levels of sIgM.

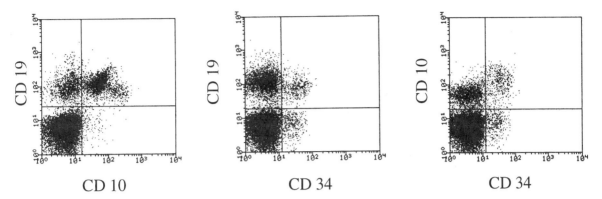

Figure 12.1.6 Triple labeling FACS analysis of normal child bone marrow for the B cell-specific marker CD19, the B cell-associated marker CD10, and the precursor cell marker CD34 (*left*). Among the double positive CD19$^+$/CD10$^+$ cells (which are the majority of B cells in BM) it is possible to identify a CD10^{high}/CD19^{low} population corresponding to the CD34$^+$/CD19$^+$ (*middle*) and CD34$^+$/CD10$^+$ (*right*) population, i.e. pro/pre-B-I cells (TdT$^+$).

The study of B cell development in humans began from the correlation between maturational stage and cell surface antigen expression of acute lymphoblastic leukemias (ALL) and lymphomas. Counterparts of many malignant cells have subsequently been identified in normal bone marrow (Uckun and Ledbetter 1988; Anderson et al 1984).

CD19 is expressed very early in B cell development and is known to be a pan-B cell marker, retained until the latest stages of cell activation and antibody secretion (Nadler et al 1983). The earliest B lineage cells are characterized by surface expression of CD19, CD10, CD34, and V$_{pre-B}$ and by the expression of TdT in the nucleus (Figs 12.1.6 and 12.1.7) (Larson and LeBien 1994; LeBien et al 1990; Guelpa-Fonlupt et al 1994). They do not yet express cytoplasmic μ heavy chains (cμ$^-$). These cells are referred to as pro/pre-B-I, on whose membranes the surrogate L chain (ΨLC) seems to be deposited before a functional μH rearrangement has occurred. A second population can be defined as being cytoplasmic μH$^+$, sIgM$^-$ (pre-B-II cells) which coexpresses CD19 and CD10 but has lost CD34 and TdT (Rehmann and LeBien 1994). About 10–15% of these cells still express V$_{pre-B}$ on the surface, and are large and actively cycling. The pre-B-II cells that are V$_{pre-B}^-$ (85–90%) can be still subdivided into 20% large, cycling cells and 80% small, resting

B cells. Most of the immature B cells still express CD19 and CD10 together with sIgM, in the absence of IgD (Fig. 12.1.5b) (Loken et al 1987). These cells are negative for the expression of ΨLC. By contrast, mature B cells, still CD19$^+$, lose the expression of CD10, and they acquire the expression of sIgD together with sIgM (Loken et al 1987).

Although rigorous cell surface marker studies have been performed by several groups,

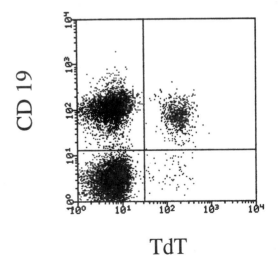

Figure 12.1.7 FACS analysis of BM cells permeabilized with FACS brand lysing solution and stained for CD19 on the surface and for TdT in the nucleus.

molecular studies of heavy- and light-chain loci during human B cell ontogeny are still lacking, owing to the more complicated genomic organization. This scenario of B cell development in humans closely resembles that observed in the mouse. Despite the similarities, B cell ontogeny in man and mice still differs partially in the expression of specific markers (c-kit, CD10, CD25, CD34), but it is clear that many marker molecules (CD19, CD40, surrogate L chain, and the V(D)J rearrangement machinery) have the same functions at the same time points during development.

Despite the possibility of dissecting the B cell development in both man and mice in discrete stages by the criteria described above, one should not forget that B cell development, like the development of the other hematopoietic lineages, is a dynamic process. This is best examplified by the finding that within 48 h all the noncycling, small, resting pre-B-II and immature B cells are labeled with bromodeoxyuridine *in vivo* (Forster and Rajewsky 1990).

Here we describe the methods that, in our laboratory, we use to dissect the different stages of B cell development (Rolink et al 1994a). Moreover, the method developed by us to grow mouse pro/pre-B-I cells for long periods *in vitro* is described in detail (Rolink et al 1991a).

References

Anderson KC, Bates MO, Slaughenhoupd BL, Pinkus GS, Schlossman SF, Nadler LM (1984) Expression of human B cell-associated antigens on leukemias and lymphomas: a model of human B cell differentiation. Blood 63: 1424.

Forster I, Rajewsky K (1990) The bulk of the peripheral B-cell pool in mice is stable and not rapidly renewed from the bone marrow. Proc Natl Acad Sci USA 87: 4781–4784.

Guelpa-Fonlupt V, Tonnelle C, Blaise D, Fougereau M, Fumoux F (1994) Discrete early pro-B and pre-B stages in normal human bone marrow as defined by surface pseudo-light chain expression. Eur J Immunol 24: 257–264.

Larson AW, LeBien TW (1994) Cross-linking CD40 on human B cell precursors inhibits or enhances growth depending on the stage of development and the IL costimulus. J Immunol 153: 584–594.

LeBien TW, Wormann B, Villablanca JG, et al (1990) Multiparameter flow cytometric analysis of human fetal bone marrow B cells. Leukemia 4: 354–358.

Loken MR, Shah VO, Dattilio KL, Civin CI (1987) Flow cytometric analysis of human bone marrow. II. Normal B lymphocyte development. Blood 70: 1316–1324.

Nadler LM, Anderson KC, Marti G, et al (1983) B4, a human B lymphocyte-associated antigen expressed on normal, mitogen-activated, and malignant B lymphocytes. J Immunol 131: 244.

Ogawa M, Matsutaki Y, Nishikawa S, et al (1991) Expression and function of c-kit in hemopoietic precursor cells. J Exp Med 174: 63–71.

Rehmann JA, LeBien TW (1994) Transforming growth factor-beta regulates normal human pre-B-cell differentiation. Int Immunol 6: 315–322.

Reininger L, Radaszkiewicz T, Kosco M, Melchers F, Rolink A (1992) Development of autoimmune disease in SCID mice populated with long-term 'in vitro' proliferating (NZB × NZW) F1 pre-B cells. J Exp Med 176: 1343–1353.

Rolink A. Melchers F (1991) Molecular and cellular origins of lymphocyte diversity. Cell 66: 1081–1094.

Rolink A, Kudo A, Karasuyama H, Kikuchi Y, Melchers F (1991a) Long-term proliferating early pre B cell lines and clones with the potential to develop to surface Ig-positive, mitogen reactive B cells *in vitro* and *in vivo*. EMBO J 10: 327–336.

Rolink A, Streb M, Nishikawa SI, Melchers F (1991b) The c-kit encoded tyrosine kinase regulates the proliferation of early pre B cells. Eur J Immunol 21: 2609.

Rolink A, Grawunder U, Haasner D, Strasser A, Melchers F (1993a) Immature surface Ig⁺ B cells can continue to rearrange kappa and lambda L chain gene loci. J Exp Med 178: 1263–1270.

Rolink A, Haasner D, Nishikawa S, Melchers F (1993b) Changes in frequencies of clonable pre-B cells during life in different lymphoid organs of mice. Blood 81: 2290–2300.

Rolink A, Grawunder U, Winkler TH, Karasuyama H, Melchers F (1994a) IL-2 receptor α chain (CD25, TAC) expression defines a crucial stage in pre-B cell development. Int Immunol 6: 1257–1264.

Rolink AG, Reininger L, Oka Y, Kalberer CP, Winkler TH, Melchers F (1994b) Repopulation of SCID mice with long-term *in vitro* proliferating pre-B-cell lines from normal and autoimmune disease-prone mice. Res Immunol 145: 353–356.

ten Boekel E, Melchers F, Rolink A (1995) The status of Ig loci rearrangements in single cells from different stages of B-cell development. Int Immunol 7: 1013–1019.

Uckun FM, Ledbetter JA (1988) Immunobiologic differences between normal and leukemic human B-cell precursors. Proc Natl Acad Sci USA 85: 8603.

Winkler TH, Melchers F, Rolink AG (1995a) Interleukin-3 and interleukin-7 are alternative growth factors for the same B-cell precursors in the mouse. Blood 85: 2045–2051.

Winkler TH, Rolink A, Melchers F, Karasuyama H (1995b) Precursor B cells of mouse bone marrow express two different complexes with the surrogate light chain on the surface. Eur J Immunol 25: 446–450.

FACS analysis of B lymphopoiesis in mouse and human bone marrow

12.2

Paolo Ghia
Ulf Grawunder
Thomas H. Winkler
Antonius Rolink

Basel Institute for Immunology, Basel, Switzerland

TABLE OF CONTENTS

Immunology Methods Manual
ISBN 0–12–442712–X

Monoclonal antibodies

B lymphopoiesis in bone marrow can be dissected into several stages based on the surface and intracellular expression of different markers (Fig. 12.1.1 in Chapter 12.1). A list of monoclonal antibodies (mAbs) that we routinely use for mouse bone marrow staining is given below. Most of these mAbs are commercially available from Pharmingen (San Diego, CA, USA) and/or the corresponding hybridomas can be obtained from ATCC.

Monoclonal antibodies

M41	rat (γ1) anti-mouse IgM (Leptin et al 1984)
187.1	rat (γ1) anti-mouse κL chain (Yelton et al 1981)
ACK-4	rat (γ2a) anti-mouse c-kit (Ogawa et al 1991)
7D4	rat (μ,κ) anti-mouse IL2Rα chain (CD25) (Malek et al 1983)
M5-114	rat (γ2b) anti-mouse MHC class II ($A^{b,d,q}$, $E^{b,d,k}$) (Bhattacharya et al 1981)
14.8	rat (γ2b) anti-mouse (CD45R) B220 (Kincade et al 1981)
LM 34	rat anti-mouse λ5 (Karasuyama et al 1993)
VP 245	rat anti-mouse V_{pre-B} (Karasuyama et al 1993)
NIM-R9	rat anti-mouse IgD (Parkhouse et al 1992)
S7	rat (γ2a) anti-mouse CD43 (Gulley et al 1988)
53-7.3	rat (γ2a) anti-mouse CD5 (Ly-1) (Ledbetter et al 1980)
B3B4	rat (γ2a) anti-mouse CD23 (IgE Fc receptor) (Rao et al 1987)
RA3-6B2	rat (γ2a) anti-mouse CD45R (B220) (Coffman and Weissman 1981)
M1/69	rat (γ2b) anti-mouse heat-stable-antigen (HSA) (Springer et al 1978)

6C3	rat (γ2a) anti-mouse BP-1 (Ramakrishnan et al 1990)
R26-46	rat (γ2a) anti-mouse λ1 + λ2 light chain (Pharmingen, San Diego, CA, USA, unpublished)

Polyclonal antibodies

Goat anti-mouse IgM-FITC	SBA, Birmingham, AL, USA

The large majority of the monoclonal and polyclonal antibodies which we use for human bone marrow staining are commercially available (Becton Dickinson, Caltag, Dako, Immunotech), FITC, PE, or biotin conjugated.

In the following, we list the most important antibodies used to study the B cell development in human bone marrow.

Monoclonal antibodies against:

CD 10	(Dakopatts, Glostrup, Denmark) (Caltag Laboratories, San Francisco, CA, USA)
CD 19	(Dakopatts, Glostrup, Denmark) (Coulter Immunology, Hialeah, FL, USA)
CD 25 (TAC)	(Dakopatts, Glostrup, Denmark) (Pharmingen, San Diego, CA, USA)
CD 24	(Boehringer Mannheim, GmbH, Germany)
CD 34	(Becton Dickinson, Mountain View, CA, USA)
CD 38	(Becton Dickinson, Mountain View, CA, USA)
TdT	(Immunotech, Marseille, France)

V$_{pre-B}$ (IgM di topo) (SL688)	(E. Sanz et al, manuscript in preparation)
CD 40	(EA-5)

Polyclonal antibodies

F(ab)$_2$ fragment of rabbit antibodies against:

μ Heavy chain	(Dakopatts, Glostrup, Denmark)
κ Light chain	(Dakopatts, Glostrup, Denmark)
λ Light chain	(Dakopatts, Glostrup, Denmark)

Goat antibodies against:

μ Heavy chain	(SBA, Birmingham, AL, USA)

δ Heavy chain	(SBA, Birmingham, AL)
κ Light chain	(Tago Diagnostics Inc., Burlingame, CA, USA)
λ Light chain	(Tago Diagnostics Inc., Burlingame, CA, USA)
κ Light chain	(SBA, Birmingham, AL, USA)
κ Light chain	(SBA, Birmingham, AL, USA)

F(ab)$_2$ fragment of goat antibodies against:

κ Light chain	(SBA, Birmingham, AL, USA)
λ Light chain	(SBA, Birmingham, AL, USA)

Purification of monoclonal antibodies using protein G

Preparatory work and handling of protein G sepharose columns

- 5 ml protein G sepharose beads (Pharmacia, LKB, Biotechnology AB, Uppsala, Sweden) are packed into a Bio-Rad Econo-column (Bio-Rad, Hercules, CA, USA) and are washed with 50–100 ml PBS to remove the 20% ethanol (used as a preservative) before being used for antibody purification.
- To obtain IgG-depleted fetal calf serum (FCS), up to 100 ml FCS is passed through a 5 ml protein G column.

- Protein bound to the protein G column is eluted with 0.1 M glycine-HCl, pH 2.2.
- The column is neutralized by washing with 50 ml PBS and in case of long-term storage is washed with 50 ml 20% ethanol.

Protein G columns used for FCS are always kept separate from columns intended for antibody purification. All procedures are performed at 4°C.

Purification of monoclonal antibodies

- A monoclonal antibody-producing hybridoma cell line is grown in

IMDM-based SF medium supplemented with 2% IgG-depleted FCS (see next section) until the cells begin to die.

- Culture supernatant (usually 1 liter) is harvested by centrifugation ($250g$, 10 min, 4°C) and is passed over a protein G column at 40–50 ml h^{-1}
- The column is washed with 20–30 ml PBS and bound antibodies are eluted with 0.1 M glycine-HCl, pH 2.2. Elution is monitored by UV absorption at $\lambda = 280$ nm.
- The eluted antibody is dialyzed three times against 1 liter of PBS at 4°C.
- The antibody concentration is determined by measuring the OD$_{280}$ (one OD$_{280}$ = 0.7 mg ml^{-1} protein); mAbs are stored at 4°C or at −20°C.

Labeling of monoclonal antibodies with biotin and fluorescein-isothiocyanate

- For biotinylation or FITC labeling, a mAb is dialyzed twice against 1 liter of 0.2 M NaHCO$_3$, pH 9.0.
- The concentration of the mAb is determined by measuring the OD$_{280}$; biotin-*N*-hydroxysuccinimide ester (Calbiochem-Novabiochem A.G., Lucerne, Switzerland) or fluorescein isothiocyanate Isomer I (Fluka A.G., Biochemica, Buchs, Switzerland) is dissolved in DMSO at 2.5 mg ml^{-1}.
- 80 µg of the biotin-NHS ester or FITC is added per mg of antibody to be labeled.
- The mixture is end-over-end rotated for 4 h at room temperature.
- The labeled mAb is dialyzed twice against 1 liter of PBS at 4°C and is stored at 4°C in PBS, 10 mM NaN$_3$.

Surface and intracellular staining for FACS analysis and cell sorting

Surface staining for FACS analysis

Surface staining of cells in suspension is carried out in U-bottom 96-well plates. All incubations are done on ice using ice-cooled antibody dilutions and FACS buffer. Whenever cells isolated from lymphoid organs are used for the stainings, cell clumps and tissue fragments are removed by filtration of the cell suspension through a nylon net.

When enough cells are available, 5×10^5 to 1×10^6 cells are regularly used for a single staining, but in rare cases, where cells are limited, stainings are done with as little as 2×10^4 cells. Monoclonal antibodies are used alone or in combination either directly FITC- or PE-labeled, or biotinylated antibodies, which have to be revealed by secondary staining with streptavidin-FITC, -PE, or -Tricolor. The appropriate concentration of each individual reagent for cell surface staining has to be determined in advance. All dilutions are prepared in PBS containing 2% FCS and 20 mM sodium azide (FACS buffer).

- Resuspend cells in FACS buffer at a concentration of about 1 to 2×10^7 cells ml^{-1}.
- Transfer 50 µl of the cell suspension into wells of a 96-well plate.
- Add 50 µl of the antibody solution ($2 \times$ concentrated).
- Incubate 30 min on ice.
- Centrifuge the plate 5 min at 250g, 4°C.

- Remove supernatant by gently flicking off the buffer.
- Resuspend cell pellets in 150 µl FACS buffer.
- Centrifuge, remove the supernatant, and resuspend in 150 µl FACS buffer.
- Centrifuge and remove the supernatant.
- Resuspend the cell pellets in 100 µl secondary reagent solution (streptavidin-FITC, streptavidin-PE, or streptavidin-Tricolor).
- Incubate 20–30 min on ice.
- Centrifuge twice, remove the supernatant, resuspend in 150 µl FACS buffer.
- Transfer into 1 ml FACS tubes (Micronic Systems, Lelystad, The Netherlands).

When cells are stained with directly FITC- or PE-labeled antibodies, incubation with the secondary reagent and the subsequent washing procedures are omitted. In case of a single-colour FITC staining, cells are occasionally resuspended in 150 µl FACS buffer containing 10 µg ml^{-1} propidium iodide after the final wash, in order to discriminate live from dead cells. Negative controls (secondary reagent only) and compensation controls (in case of double or triple stainings) are always included.

FACS analysis is carried out on a FACScan (Becton Dickinson, Mountain View, CA, USA) equipped with an argon laser tuned to 488 nm. Data are acquired and analyzed using the Lysis software package (Becton Dickinson).

Intracellular staining for FACS analysis

Cytoplasmic staining

(see Fig. 12.1.4)

- For fixation, 5×10^5 to 1×10^7 cells are resuspended in 1 ml ice-cold 4% paraformaldehyde (dissolved in PBS) and incubated for 10 min on ice.
- Fixed cells are washed twice with 1 ml ice-cold PBS (centrifugation at 250g, 5 min, 4°C).
- For permeabilization, cells are resuspended in 1 ml 0.2% Tween-20 (diluted in PBS) and incubated for 15 min at 37°C.
- Permeabilized cells are washed twice with 1 ml ice-cold FACS buffer (centrifugation at 250g, 5 min, 4°C).
- After the last wash, cells are resuspended in an appropriate volume of FACS buffer and transferred in 50 μl aliquots into a U-bottom 96-well plate.
- Subsequent staining and washing procedures, as well as FACS analysis are performed in the same way as described for surface staining of cells.

Nuclear staining (see Fig. 12.1.7)

After performing the surface staining, as described earlier, the samples can be analyzed for the expression of the nuclear enzyme terminal deoxynucleotidyl transferase (TdT). We currently use the procedure described by Drach et al (1991), which is cheap, fast, and reliable and still makes it possible to distinguish the cells by side–forward scattering profile. It is possible to use this method with a very limited amount of material, down to $1-2 \times 10^5$ cells per sample, which means that one can perform the

staining on small sorted populations. This method does not affect PE or Tricolor, making it ideal for three-color analysis.

- For fixation, the cells are resuspended in 1 ml of an ice-cold solution containing 1% paraformaldehyde (dissolved in PBS) and incubated for 2 min on ice.
- For permeabilization, without centrifugation, 1.5 volume of ice-cold absolute methanol is added and the cells are incubated for 20 min on ice.
- Permeabilized cells are washed twice with ice-cold FACS buffer (1500 rpm, 10 min, 4°C).
- After the last wash, cells are resuspended in an appropriate volume of fetal calf serum (as a blocking serum) for 15 min.
- Without centrifugation, a mixture of three FITC-conjugated anti-TdT monoclonal antibodies (Immunotech) is added at a final concentration of 1:20, for 30 min, on ice.
- Subsequent washing procedures and FACS analysis are performed in the same way as described for surface staining of cells.

Another very reliable and simple method utilizes FACS brand lysing solution (Becton Dickinson, San Jose, CA, USA) (Horvatinovich et al 1994); it simultaneously lyses red cells and permeabilizes white cells. As a consequence, with such a method, separation of mononuclear cells from whole blood or bone marrow is not necessary. In addition, the side light-scattering profile is excellent. The solution can also be used after performing gradient for separation of mononuclear cells.

- First the whole blood and bone marrow are stained with monoclonal antibodies, as described earlier.
- Then 2 ml of 1× FACS lysing solution is added to each tube and incubated

for 10 min at room temperature in the dark.
- The samples are centrifuged for 5 min at 300g at room temperature.
- After aspirating the supernatant, 1.0 ml of PBS with 0.1% azide is added and the tubes are gently vortexed.
- The samples are washed for 5 min at 300g at room temperature and the supernatant is aspirated as before.
- A mixture of FITC-conjugated anti-TdT antibodies is added at the right concentration and incubated for 30 min in the dark at 4°C.
- The samples are washed two times with PBS (0.1% azide) at 300g for 5 min and then 0.5 ml of 0.5% formaldehyde in PBS is added.
- FACS analysis is then performed as described before.

Surface staining for cell sorting

All staining and washing procedures are carried out on ice using ice-cold solutions.

If cells are sorted to enrich cell populations for subsequent cell culture, only PBS with 2% FCS is used as FACS buffer (i.e. NaN$_3$ is omitted) and all procedures are carried out under sterile conditions. Since cell numbers in sorting experiments (e.g. in the case of bone marrow samples) are of the order of about 1×10^7 to 1×10^8 cells per experiment, staining is carried out in a total volume of 3–4 ml FACS buffer. Antibody concentrations and washing conditions are identical to those for surface staining of cells for FACS analysis.

However, after the final wash, cell suspensions are always filtered through a nylon net to remove cell clumps which could easily clog the cell sorter. Cells for cell sorting are resuspended in FACS buffer at a concentration of about 5×10^6 cells ml^{-1}.

Cell sorting is performed on a Becton-Dickinson FACS-Star plus or a Becton-Dickinson 440 cell sorter (Becton-Dickinson, Mountain View, CA, USA). Cells are directly sorted into 0.5 ml medium containing 2% FCS (and IL-7, if pre-B cells are enriched).

DNA staining for cell cycle analysis

- Between 5×10^4 and 5×10^6 cells are fixed in about 5 ml 70% ethanol, at 4°C overnight, in FACS tubes. They can be kept longer at 4°C.
- The samples are then centrifuged at 2000 rpm for 5 min.
- The samples are resuspended in 500 μl of RNAase A (Sigma, Buchs, Switzerland) (0.5 mg ml^{-1} in 0.1 M Tris, pH 7.5, 0.1 M NaCl) and incubated for

30 min at 37°C.
- Without centrifuging, 500 μl of pepsin (Sigma, Buchs, Switzerland) (1 mg ml^{-1} in 0.4% HCl) is added and mixed and the samples are incubated for 15 min at 37°C.
- Without centrifuging, 1 ml of ethidium bromide (0.02 mg ml^{-1}, in 0.2 M Tris, pH 8–8.5, 0.5% BSA) is added and mixed, and the samples are incubated for another 15 min, at room temperature.

The tubes must be protected from light with aluminum foil.
- At the end of the incubation, the samples are kept on ice until analysis. Analysis is performed on FACScan, equipped with the DDM (Doublets Discriminating Module).

For samples with 5×10^4 cells the volumes can be scaled down by one-half.

In vivo BrdU labeling

The thymidine analogue 5-bromo-2'-deoxyuridine (BrdU) (Sigma, Buchs, Switzerland) is widely used to study lymphocyte population dynamics and lifespans. We routinely combine the cell cycle analysis described above and the *in vivo* BrdU labeling to address these two points. For *in vivo* BrdU labeling, mice are fed with $1 \, mg \, ml^{-1}$ BrdU in the drinking water for various times.

For testing the incorporation of BrdU we routinely first sort 2×10^4–10^5 of the cells of interest using the cell sorter.

The sorted cells are split into two fractions. One fraction is used as a control.

Cells are washed once in PBS and then fixed in $500 \, \mu l$ 70% ethanol for 20 min at room temperature.

After fixation, cells are washed twice with PBS (2000 rpm for 5 min) and then incubated with $500 \, \mu l$ of 3 M HCl–0.5% Tween 20 for 20 min to denature the DNA.

The cells are spun for 5 min at 2000 rpm and the pellet is resuspended in $250 \, \mu l$ of 0.01 M sodium tetraborate.

After 3 min incubation the tubes are filled up with PBS–0.5% Tween and centrifuged at 2000 rpm for 5 min.

PBS–0.5% Tween washing is repeated twice and $20 \, \mu l$ of FITC-labeled anti-BrdU antibody (Becton Dickinson) is added to the pellet.

After 20 min incubation, the cells are washed twice with PBS and subsequently analyzed on a FACScan..

References

Bhattacharya A, Dorf ME, Springer TA (1981) A shared alloantigenic determinant on Ia antigens encoded by the I-A and I-E subregions: Evidence for I region gene duplication. J Immunol 127: 2488.

Coffman RL, Weissman IL (1981) A monoclonal antibody that recognizes B cells and B cell precursors in mice. J Exp Med 153: 269–279.

Drach J, Gattringer C, Huber H (1991) Combined flow cytometric assessment of cell surface antigens and nuclear TdT for the detection of minimal residual disease in acute leukemia. Br J Haematol 77: 37–42.

Gulley ML, Ogata LC, Thorson JA, Dailey MO, Kemp JD (1988) Identification of a murine pan-T cell antigen which is also expressed during the terminal phases of B-cell differentiation. J Immunol 140: 3751–3757.

Horvatinovich JM, Sparks SD, Borowitz MJ (1994) Detection of terminal deoxynucleotidyl transferase by flow cytometry: a three color method. Cytometry 18: 228–230.

Karasuyama H, Rolink A, Melchers F (1993) A complex of glycoproteins is associated with V_{pre-B}/lambda 5 surrogate light chain on the surface of μ heavy-chain-negative early precursor B cell lines. J Exp Med 178: 469–478.

Kincade PW, Lee G, Watanabe T, Sun L, Scheid MP (1981) Antigens displayed on murine B lymphocyte precursors. J Immunol 127: 2262–2272.

Ledbetter JA, Rouse RV, Micklem HS (1980) T cell subsets defined by expression of Lyt-1, 2, 3 and Thy-1 antigens. Two-parameter immunofluorescence and cytotoxicity analysis with monoclonal antibodies modifies current views. J Exp Med 152: 280.

Leptin M, Potash MJ, Grutzmann R, et al (1984) Monoclonal antibodies specific for murine IgM. I. Characterization of antigenic determinants on the four constant domains of the μ heavy-chain. Eur J Immunol 14: 534.

Malek T, Robb R, Shevach E (1983) Identification and initial characterization of a rat monoclonal antibody reactive with the murine interleukin 2 receptor–ligand complex. Proc Natl Acad Sci USA 80: 5694.

Ogawa M, Matsuzaki Y, Nishikawa S, et al (1991) Expression and function of c-kit in hemopoietic precursor cells. J Exp Med 174: 63–71.

Parkhouse RME, Preece G, Sutton R, Cordell JL, Mason DY (1992) Relative expression of surface IgM, IgD and the Ig-associating α (mb-1) and β (B-29) polypeptide chains. Immunology 76: 535.

Ramakrishnan L, Wu Q, Yue A, Cooper MD, Rosenberg N (1990) BP-1/6C3 expression defines a differentiation stage of transformed pre-B-cells and is not related to malignant potential. J Immunol 145: 1603–1608.

Rao M, Lee W, Conrad D (1987) Characterization of a monoclonal antibody directed against the murine B lymphocyte receptor for IgE. J Immunol 138: 1845.

Springer TA, Galfre G, Secher DS, Milstein C (1978) Monoclonal xenogeneic antibodies to murine cell surface antigens: identification of novel leukocyte differentiation antigens. Eur J Immunol 8: 539–551.

Yelton DE, Desaymard C, Scharff MD (1981) Use of monoclonal anti-mouse immunoglobulin to detect mouse antibodies. Hybridoma 1: 5–11.

Determination of D_H-J_H rearrangements in precursor B cells by PCR

12.3

Dirk Haasner
Ulf Grawunder

Basel Institute for Immunology, Basel, Switzerland

TABLE OF CONTENTS

Immunology Methods Manual
ISBN 0–12–442712–X

Isolation of genomic DNA from tissue culture cells

To isolate genomic DNA, 5×10^5 cells are washed once with PBS (centrifugation in an Eppendorf Microfuge for 5–8 s), resuspended in 500 µl PBS, and subsequently lysed in a boiling water bath for 5 min. The lysed cells are treated with 0.2 mg ml^{-1} proteinase K at 55°C for 2 h. The preparations are boiled again (5 min) to inactivate the proteinase K and are extracted once with phenol–chloroform–isoamyl alcohol (25:24:1) and once with chloroform–isoamyl-alcohol (24:1). DNA is precipitated with 1 volume isopropanol and 0.1 volume 3 M sodium acetate, pH 5.2 at –20°C for 1 h. After centrifugation at 12,000 rpm in an Eppendorf Microfuge for 15 min at 4°C, the pellet is washed with 70% ethanol, air dried and dissolved in 500 µl 10 mM Tris–HCl pH 8.3.

Primers and PCR conditions used for detecting D_H-J_H rearrangements in mouse pre-B cells

The primer 3' of J_H has the following sequence:

5' GGGTCTAGACTCTCAGCCGGCTCCCT-CAGG 3';

the primer 5' of D_H has the sequence:

5' ACAAGCTTCAAAGCACAATGCCTGGCT 3'.

The $D_H J_H$ priming reaction detects all D_H-J_H rearrangements except those for $D_{FL16.2}$ and D_{Q52}.

The PCR is set up as follows:

5 µl DNA (equivalent of 5×10^3 cells)

6 µl PCR buffer (5×):

100 mM Tris–HCl, pH 8.3,

250 mM KCl

12.5 mM MgCl$_2$

3 µl nucleotide mix (10×) (2 mM each dNTP)

0.25 µM of each primer

0.1 µl Taq-polymerase (5 U µl^{-1}) (Roche Diagnostics Systems, Basel, Switzerland)

30 µl H$_2$O.

The reactions are set up on ice in 0.5 ml PCR tubes (Perkin Elmer, Norwalk, CT, USA) and the mixtures are finally overlaid with a drop of mineral oil to prevent evaporation during repeated heating and cooling during the PCR.

Primers and PCR conditions used for detecting D_H-J_H rearrangements

After short spinning (5–10 s) of the samples in an Eppendorf Microfuge, the PCR is carried out in a DNA Thermal Cycler (Perkin-Elmer Cetus), under the following conditions: 94°C, 40 s, 1 cycle (to initially denature the template DNA), followed by 35 cycles at 94°C, 20 s (denaturing), 65°C, 25 s and 72°C, 2 min.

Aliquots of 10–20 μl of the PCRs are mixed with 3–4 μl of 6× gel-loading mix (15% Ficoll-Paque, 0.25% bromophenol blue) and are analyzed on a 1–1.5% agarose gel containing 5 μg ml^{-1} ethidium bromide.

The approximate PCR product lengths are:

D_H-J_H1: 1700 bp
D_H-J_H2: 1450 bp
D_H-J_H3: 1100 bp
D_H-J_H4: 600 bp.

In vitro growth and differentiation of early mouse B cell precursors

12.4

Antonius Rolink
Fritz Melchers

Basel Institute for Immunology, Basel, Switzerland

TABLE OF CONTENTS

Immunology Methods Manual
ISBN 0–12–442712–X

Stromal cells and cytokine-producing cells

For all *in vitro* studies, IMDM-based serum-free (SF) medium, supplemented with 2% (w/v) FCS is used:

176.6 g IMDM-powder (Gibco, BRL, Paisley, UK)
30.24 g $NaHCO_3$ (Fluka, Buchs, Switzerland)
100 ml 100× nonessential amino acids (Gibco, BRL, Paisley, UK)
100 ml 100× Kanamycin (Gibco, BRL, Paisley, UK)
10 ml 5 mg ml^{-1} insulin solution (Sigma, St. Louis, MO, USA)
10 ml 50 mM 2-mercaptoethanol (Sigma, St. Louis, MO, USA)
30 ml 10% (w/v) primatone (ultrafiltered to exclude proteins > 10 kDa) (Quest International, Naarden, Naarden, NL, USA)
200 ml fetal calf serum (Gibco, BRL, Paisley, UK)
Make up to 10 liters triple-distilled H_2O

All cell cultures are kept in a humidified incubator at 37°C with 10% CO_2. The adherent stromal cells PA-6 (Kodama et al 1982) and ST-2 (Ogawa et al 1988) are routinely used as feeder cells for pre-B cell cultures. The stromal cells are grown in 175 cm^2 tissue-culture flasks and passaged every 3 days. Stromal cells are detached from the tissue culture plastic with 10 ml trypsin–EDTA (Tecnomara, Cramlington, UK) and are replated at 0.8–1 × 10^6 cells/175 cm^2 flask in 100 ml SF-medium.

Stromal cells, used as feeders for pre-B cells, are normally set up 3 days in

25 cm^2 flask, 7 × 10^4 cells in 7 ml medium

75 cm^2 flask, 2 × 10^5 cells in 25 ml medium
24-well plate, 5 × 10^3 cells ml^{-1} (1 ml/well)
96-well plate, 5 × 10^3 cells ml^{-1} (200 μl/well)

Directly before use as feeders, the stromal cell layers are γ-irradiated with 30 gy (3000 rad) to prevent further cell division during coculture with pre-B cells.

The myeloma cell line J588L transfected with the cDNA encoding for mouse IL-7 is used to produce IL-7 conditioned medium.

To determine the concentration of IL-7 in culture supernatants obtained from the transfected cell line, serial dilutions of the supernatants are applied to 5 × 10^3 of the IL-7-responsive cell line 5.7 in 96-well plates (Rolink et al 1991). In parallel, a cytokine standard with a known concentration is diluted in the same way and the cultures are incubated for 24 to 48 h at 37°C in a humidified incubator (depending on the cell density of the cultures). Thereafter, cultures were pulsed with 1 μCi of [*methyl*-^{3}H] thymidine (5 Ci mmol^{-1}) for 8–12 h at 37°C. After [*methyl*-^{3}H] thymidine incorporation, cells are frozen at −70°C and thawed at 37°C before they are harvested with an automated 96-well-harvester (1295-004, Betaplate, LKB-Wallac Oy, Turku, Finland) on to glass fiber filters (LKB-Wallac Oy). The filters are transferred into plastic bags and 5 ml of scintillation fluid (Beta Plate Scint, LKB-Wallac Oy) is added before the bags are sealed. Incorporated radioactivity is determined by means of a liquid scintillation counter (1205-Betaplate, LKB-Wallac Oy).

Culture supernatants of the IL-7-transfected J558 cells usually contain $1–2.5 \times 10^4\,U\,ml^{-1}$ of IL-7.

As an alternative, one can use commercially available human rIL-7 (Genzyme), which is also active on mouse cells.

Establishment of pre-B cell lines and clones from different organs of mice

Animals are killed with dry ice in a CO_2 chamber. All cell preparations are carried out with ice-cooled media, and cell suspensions are kept on ice.

Bone marrow cells are obtained by removing femur and tibia, opening of the bones at both ends, and flushing the marrow with 2–5 ml medium using a 5 ml disposable syringe and a 25- or 26-gauge needle.

Spleen, thymus, lymph nodes or fetal liver are removed under sterile conditions and put on a 200-mesh steel grid in a Petri dish with 5–10 ml medium. The organs are cut into pieces and gently squashed through the grid with the plunger of a 5 ml plastic syringe.

Cell suspensions are transferred into 15 ml or 50 ml conical tubes and clumps of cells are dissociated by pipetting.

Cells are spun down once (1000 rpm = 200g, 10 min, 4°C) and resuspended in 10–20 ml medium. The cell concentration is determined by diluting the samples in trypan blue solution using a Burker or Neubauer hemocytometer.

Cloning of pre-B cells is performed by limiting dilution in 96-well plates containing a semiconfluent, γ-irradiated, stromal cell layer.

Freshly isolated cells (e.g. from bone marrow or fetal liver) are appropriately diluted in medium containing $100–200\,U\,ml^{-1}$ of IL-7. After the medium has been removed from the stromal cells, 100 µl of IL-7-medium is added per well. By means of a multichannel pipette, serial 2-fold dilutions are carried out in such a way that cells are diluted below 1 cell/well. Finally, all wells are filled up to 200 µl with IL-7 medium and the plates are incubated for 6–7 days at 37°C, 10% CO_2.

Single wells are analyzed for growth of pre-B cell colonies using an inverted microscope.

Single pre-B cell colonies or pre-B cell lines are transferred into separate wells of a 24-well plate. Pre-B cell clones and lines are subsequently expanded to small (25 cm^2) and midsize (75 cm^2) tissue culture flasks coated with irradiated stromal cells.

Maintenance of stromal cell/IL-7-dependent pre-B cells

Fetal liver- and bone-marrow-derived pre-B cells are grown on a semi-confluent layer of γ-irradiated stromal cells in medium containing 100–200 U ml^{-1} mouse rIL-7.

Pre-B cell clones and lines grow in an indistinguishable manner on PA-6 and ST-2 cells.

Pre-B cells are removed from the adherent stromal cell layer by gentle clapping of tissue culture flasks with the flat palm or (when 24-well or 96-well plates are used) by aspirating the medium several times up and down.

Pre-B cell cultures are passaged every 3 days. Therefore, cultures have to be initiated at the following densities:

- 25 cm^2 flasks: 8×10^5 cells in 7 ml medium containing rIL-7
- 75 cm^2 flasks: 2.5×10^6 cells in 25 ml medium containing rIL-7
- 24-well plates: 5×10^4 cells ml^{-1} in 1 ml medium containing rIL-7
- 96-well plates 5×10^4 cells ml^{-1} in 0.2 ml medium containing rIL-7

After 3 days of culture, approximately 10 to 20 times the input cells can be harvested depending on the pre-B cell clone or line used for the expansion.

Differentiation of pre-B cell lines and clones *in vitro*

Differentiation of stromal cell/IL-7-dependent pre-B cell clones and lines is initiated by washing cells three times in IMDM-based SF-medium, 2% FCS (centrifugation at 250*g*, 5 min, 4°C), and seeding them in the same medium on a semiconfluent layer of γ-irradiated PA-6 or ST-2 cells at the following densities:

- 25 cm^2 flasks: 1×10^7 cells in 7 ml medium without rIL-7
- 75 cm^2 flasks: 3×10^7 cells in 25 ml medium without rIL-7

Cells are routinely harvested after 1, 2 or 3 days of culture in the absence of IL-7. At day 3 of culture the recovery of viable cells is about 5% for normal pre-B cell lines and clones and 80–90% for those expressing a bcl-2 transgene. Depending on the pre-B cell lines (but independently of whether bcl-2 transgene positive or negative)

between 5% and 20% of recovered viable cells express sIgM.

References

Kodama H, Amagai Y, Koyama H, Kasai S (1982) A new preadipose cell line derived from newborn mouse calvaria can promote the proliferation of pluripotent hemopoietic stem cells *in vitro*. J Cell Physiol 112: 89.

Ogawa M, Nishikawa S, Ikuta K, et al (1988) B cell ontogeny in murine embryo studied by a culture system with the monolayer of a stromal cell clone, ST-2: B cell progenitor develops first in the embryonal body rather than in the yolk sac. EMBO J 7: 1337–1343.

Rolink A, Kudo A, Karasuyama H, Kikuchi Y, Melchers F (1991) Long-term proliferating early pre-B cell lines and clones with the potential to develop to surface Ig-positive, mitogen reactive B cells *in vitro* and *in vivo*. EMBO J 10: 327–336.

In vitro culture of human B cell precursors

12.5

Paolo Ghia
Antonius Rolink
Fritz Melchers

Basel Institute for Immunology, Basel, Switzerland

TABLE OF CONTENTS

Immunology Methods Manual
ISBN 0–12–442712–X

Differentiation of human pre-B cells *in vitro*

In contrast to the difficulties of growing human B cell precursors in tissue culture, it is possible to follow *in vitro* the differentiation of pre-B cells into immature B cells, expressing IgM but not IgD, on the cell surface.

For all *in vitro* studies RPMI 1640 medium, supplemented with 10% FCS is used:

RPMI powder (Gibco, BRL, Paisley, UK)

plus

$2\,g\,l^{-1}$ $NaHCO_3$ (Fluka, Buchs, Switzerland)

2 mM L-glutamine (Gibco, BRL, Paisley, UK)

$100\,\mu g\,ml^{-1}$ $100\times$ kanamycin (Gibco, BRL, Paisley, UK)

10% fetal calf serum (Gibco, BRL, Paisley, UK)

Make up to 10 liters triple-distilled H_2O

All cell cultures were done in a humidified incubator at 37°C with 5% CO_2.

Isolation of cells from bone marrow

Human heparinized bone marrow is obtained by iliac crest aspiration, according to institutional guidelines, from adults and/or children healthy or free of hematological diseases.

Bone marrow mononuclear cells are isolated by Ficoll-paque gradient (density $1.077\,g\,ml^{-1}$) (Pharmacia, Uppsala, Sweden). The bone marrow sample is diluted in RPMI Medium (1:2) and stratified on Ficoll in a 2:1 ratio. The subsequent centrifugation is performed at 1500 rpm, 30 min, at room temperature. The mononuclear cells, taken from the interfaces between serum and Ficoll, are resuspended in RPMI medium and washed twice at 1500 rpm, 15 min.

Stroma cell lines

The adherent murine stromal cells ST-2 or human bone marrow-derived stroma cell lines are routinely used as feeder cells for pre-B cell cultures. To obtain cell lines from normal human bone marrow, the fractions of mononuclear cells are cultured in RPMI medium. Nonadherent cells are removed after 3–4 days of culture; the remaining adherent fraction reaches confluency in 3–4 weeks. The cells can then be detached from the plastic with trypsin–EDTA (Tecnomara, Cramlington, UK) and are replated at $0.8\text{–}1 \times 10^6$ cells/175 cm^2 flask in 100 ml RPMI medium. After about 10 passages these primary cells lose their growth capacity. These cells do not need to be irradiated when cultured together with pre-B lymphocytes.

It is possible to sort pre-B cells either using anti-CD19 or anti-CD10 or both antibodies, enriching for the cells which do not yet express surface immunoglobulins (κ^- and λ^-). After 1 day of *in vitro* culture, in medium only, 10–20% of these cells already express sIgM, being κ or λ positive.

However, when pre-B cells are cultured in medium only, they die rapidly by apoptosis and after 1 week only <1% of the cultured cells are still alive. The viability of the cells can be improved by seeding the precursors, from the beginning, on stroma cell layers either from human bone marrow or from murine cell lines (ST-2). It is then possible to obtain viable cells for analysis, even after 3 weeks of culture.

Outlook

Antonius Rolink
Fritz Melchers

Basel Institute for Immunology, Basel, Switzerland

TABLE OF CONTENTS

Immunology Methods Manual
ISBN 0–12–442712–X

Outlook

We have described here the methods that we use to analyze B lymphopoiesis in mouse and men. However others methods are used in different laboratories. For example Hardy's classification (Hardy et al 1991) of B cell subpopulations is widely used. This classification is based on the differential expression of CD43, CD24 (HSA), BP-1, IgM, and IgD on CD45R (B220) positive cells. In this classification mouse B cell development is subdivided into fractions A, B, C, C', D, E, and F. Fraction A, called pre-pro-B cells, expresses CD45R and CD43 but not yet CD24, BP-1, IgM, or IgD. This fraction is not included in our B cell development scheme since it does not express c-kit or CD25 and does not grow on stromal cells and IL-7. Recently we showed that a large part of this fraction A does not relate to B cell precursors but rather belongs to the NK cell lineage.

Fraction B is defined as CD45R (B220)$^+$, CD43$^+$, CD24 (HSA)$^+$, BP-1$^-$, IgM$^-$, and IgD$^-$, and called early pro-B cells. This fraction correlates with our pro/pre-B-I fraction since it expresses c-kit but not yet CD25 and moreover possesses the capacity to grow in the presence of stromal cells and IL-7.

The fractions C and C' are defined as CD45R (B220)$^+$, CD43$^+$, CD24 (HSA)$^+$, BP-1$^+$, IgM$^-$ and IgD$^-$, while C' has higher expression of CD24 than C.

Ehlich et al (1994) recently showed that the vast majority of cells in fraction C have no productive $V_H D J_H$ rearranged IgH chain loci, and thus are not able to give rise to functional B cells. We have not yet identified this cell type with our markers. Fraction C' is probably identical to our large CD25$^+$ SL/μH expressing pre-B-II cell (Winkler et al 1995).

Fraction D is defined as CD45R (B220)$^+$, CD43$^-$, CD24 (HSA)$^+$, BP-1$^+$, IgM$^-$ and IgD$^-$, called pre-B cells. This fraction consists of a mixture of what we call large pre-B-II cells that have lost the expression of the SL/μH pre-B cell receptor and small pre-B-II cells which are in the process of L chain rearrangement (ten Boekel et al 1995).

Fraction E corresponds to sIgM$^+$, sIgD$^-$ immature B cells while fraction F contains the sIgM$^+$, sIgD$^+$ mature B cells.

Thus, the classification of Hardy et al can be correlated with ours. However, both classifications have some drawbacks. As mentioned above, we are not yet able to detect the nonproductively VDJ rearranged fraction C. Hardy's fraction A, on the other hand, contains cells that do not belong to the B cell lineage. Moreover, with our classification we can split up Hardy fraction D into two subpopulations. It is also worth mentioning that the level of BP-1 expression is variable and dependent on the strain of mice used. Thus C57BL/6 mice express low levels of BP-1, while Balb/c and CBA/j mice express high levels. Thus far we have not observed this variability with c-kit and CD25 expression.

Also, Osmond and colleagues (Osmond 1990) and Nishikawa and colleagues (1988) have defined subpopulations of cells in the development of mature B cells from pluripotent stem cells and B lineage committed progenitors in mouse bone marrow. A correlation and comparison of their subpopulations with ours and those described by Hardy and colleagues can be found in Rolink et al (1994).

Concerning human B cell development, we tried to adapt our mouse classification to the data obtained from the analyses of human bone marrow in our as well as in others' laboratories. Despite the fact that molecular analyses of the rearrangement status of H chain and L chain gene loci during the different stages of development are still lacking and some markers are differently expressed, the overall appearance of human B cell ontogeny closely resembles that observed in the mouse. It is then likely that some molecules can play the same role at the same stages of differentiation. It should, however, be noted that there are still discrepancies among the data obtained in different laboratories in the expression patterns of some markers during human B cell development. In particular, the surrogate light-chain expression appears to differ widely.

Monoclonal antibodies have been raised against the triple complex of μH chain, V_{pre-B} and λ_5 protein, obtaining different antibodies with different reactivity (Guelpa-Fonlupt et al 1994; Lassoued et al 1993). Guelpa-Fonlupt et al have identified with their own reagent three discrete cell types: (1) a subpopulation expressing V_{pre-B} without μH chain on the surface; (2) a minor population coexpressing V_{pre-B} and μH without the conventional light chain; (3) a major subpopulation coexpressing V_{pre-B}, μH, κ or λ chains.

In contrast, Lassoued et al produced a monoclonal which identified only a relatively late stage of pre-B cell differentiation, consisting of cells not expressing CD34 antigen and TdT in the nucleus.

Moreover, the different capacities of mouse and human B cell precursors to grow *in vitro* remains the most striking difference between mouse and man. While mouse precursors can grow in different culture conditions (see above), there is not yet available a system capable of sustaining and expanding human B cell precursors at the same level. It is still to be clarified whether this is an inherent difference of human precursor B cells, or whether the right stroma cells and the right cytokines have yet to be found.

References

Ehlich A, Martin V, Muller W, Rajewsky K (1994) Analysis of the B-cell progenitor compartment at the level of single cells. Current Biology 4: 573–583.

Guelpa-Fonlupt V, Tonnelle C, Blaise D, Fougereau M, Fumoux F (1994) Discrete early pro-B and pre-B stages in normal human bone marrow as defined by surface pseudo-light chain expression. Eur J Immunol 24: 257–264.

Hardy RR, Carmack CE, Shinton SA, Kemp JD, Hayakawa K (1991) Resolution and characterization of pro-B and pre-pro-B cell stages in normal mouse bone marrow. J Exp Med 173: 1213–1225.

Lassoued K, CA N, Billips L, Kubagawa H, Monteiro RC, LeBien TW, Cooper MD (1993) Expression of surrogate light chain receptors is restricted to a late stage in pre-B cell differentiation. Cell 73: 73–86.

Nishikawa S-I, Ogawa M, Nishikawa S, Kunisada T, Kodama H (1988) B lymphopoiesis on stromal cell clone: stromal cell clones acting on different stages of B-cell differentation. Eur J Immunol 18: 1767–1771.

Osmond DG (1990) B cell development in bone marrow. Semin Immunol 2: 173–180.

Rolink A, Grawunder U, Winkler TH, Karasuyama H, Melchers F (1994) IL-2 receptor α chain (CD25, TAC) expression defines a crucial stage in pre-B cell development. Int Immunol 6: 1257–1264.

ten Boekel E, Melchers F, Rolink A (1995) The status of Ig loci rearrangements in single cells from different stages of B-cell development. Int Immunol 7: 1013–1019.

Winkler TH, Rolink A, Melchers F, Karasuyama H (1995) Precursor B cells of mouse bone marrow express two different complexes with the surrogate light chain on the surface. Eur J Immunol 25: 446–450.

Recommended reading

12.7

Anderson KC, Bates MO, Slaughenhoupd BL, Pinkus GS, Schlossman SF, Nadler LM (1984) Expression of human B cell-associated antigens on leukemias and lymphomas: a model of human B cell differentiation. Blood 63: 1424.

Bhattacharya A, Dorf ME, Springer TA (1981) A shared alloantigenic determinant on Ia antigens encoded by the I-A and I-E subregions: Evidence for I region gene duplication. J Immunol 127: 2488.

Coffman RL, Weissman IL (1981) A monoclonal antibody that recognizes B cells and B cell precursors in mice. J Exp Med 153: 269–279.

Coutinho A (1982) From the point of view of an immunologist: enemies from within or friends from long ago? Current Topics Microbiol Immunol 98: 113–126.

Dittel BN, LeBien TW (1995) The growth response to IL-7 during normal human B cell ontogeny is restricted to B-lineage cells expressing CD34. J Immunol 154: 58–67.

Drach J, Gattringer C, Huber H (1991) Combined flow cytometric assessment of cell surface antigens and nuclear TdT for the detection of minimal residual disease in acute leukemia. Br J Haematol 77: 37–42.

Ehlich A, Martin V, Muller W, Rajewsky K (1994) Analysis of the B-cell progenitor compartment at the level of single cells. Current Biology 4: 573–583.

Forster I, Rajewsky K (1990) The bulk of the peripheral B-cell pool in mice is stable and not rapidly renewed from the bone marrow. Proc Natl Acad Sci USA 87: 4781–4784.

Guelpa-Fonlupt V, Tonnelle C, Blaise D, Fougereau M, Fumoux F (1994) Discrete early pro-B and pre-B stages in normal human bone marrow as defined by surface pseudo-light chain expression. Eur J Immunol 24: 257–264.

Gulley ML, Ogata LC, Thorson JA, Dailey MO, Kemp JD (1988) Identification of a murine pan-T cell antigen which is also expressed during the terminal phases of B-cell differentiation. J Immunol 140: 3751–3757.

Hardy RR, Carmack CE, Shinton SA, Kemp JD, Hayakawa K (1991) Resolution and characterization of pro-B and pre-pro-B cell stages in normal mouse bone marrow. J Exp Med 173: 1213–1225.

Horvatinovich JM, Sparks SD, Borowitz MJ (1994) Detection of terminal deoxynucleotidyl transferase by flow cytometry: a three color method. Cytometry 18: 228–230.

Karasuyama H, Rolink A, Melchers F (1993) A complex of glycoproteins is associated with V_{pre-B}/lambda 5 surrogate light chain on the surface of μ heavy-chain-negative early precursor B cell lines. J Exp Med 178: 469–478.

Kincade PW, Lee G, Watanabe T, Sun L, Scheid MP (1981) Antigens displayed on murine B lymphocyte precursors. J Immunol 127: 2262–2272.

Kodama H, Amagai Y, Koyama H, Kasai S (1982) A new preadipose cell line derived from newborn mouse calvaria can promote the proliferation of pluripotent hemopoietic stem cells in vitro. J Cell Physiol 112: 89.

Larson AW, LeBien TW (1994) Cross-linking CD40 on human B cell precursors inhibits or enhances growth depending on the stage of development and the IL costimulus. J Immunol 153: 584–594.

Lassoued K, CA N, Billips L, Kubagawa H, Monteiro RC, LeBien TW, Cooper MD (1993) Expression of surrogate light chain receptors is restricted to a late stage in pre-B cell differentiation. Cell 73: 73–86.

Ledbetter JA, Rouse RV, Micklem HS (1980) T cell subsets defined by expression of Lyt-1, 2, 3 and Thy-1 antigens. Two-parameter immunofluorescence and cytotoxicity analysis with monoclonal antibodies modifies current views. J Exp Med 152: 280.

LeBien TW, Wormann B, Villablanca JG, et al (1990) Multiparameter flow cytometric analysis of human fetal bone marrow B cells. Leukemia 4: 354–358.

Leptin M, Potash MJ, Grutzmann R, et al (1984) Monoclonal antibodies specific for murine IgM. I. Characterization of antigenic determinants on the four constant domains of the μ heavy-chain. Eur J Immunol 14: 534.

Loken MR, Shah VO, Dattilio KL, Civin CI (1987) Flow cytometric analysis of human bone marrow. II. Normal B lymphocyte development. Blood 70: 1316–1324.

Malek T, Robb R, Shevach E (1983) Identification and initial characterization of a rat monoclonal antibody reactive with the murine interleukin 2 receptor–ligand complex. Proc Natl Acad Sci USA 80: 5694.

Moreau I, Duvert V, Banchereau J, Saeland S (1993) Culture of human fetal B-cell precursors on bone marrow stroma maintains highly proliferative CD20dim cells. Blood 81: 1170–1178.

Nadler LM, Anderson KC, Marti G, et al (1983) B4, a human B lymphocyte-associated antigen expressed on normal, mitogen-activated, and malignant B lymphocytes. J Immunol 131: 244.

Nishikawa S-I, Ogawa M, Nishikawa S, Kunisada T, Kodama H (1988) B lymphopoiesis on stromal cell clone: stromal cell clones acting on different stages of B-cell differentiation. Eur J Immunol 18: 1767–1771.

Ogawa M, Nishikawa S, Ikuta K, et al (1988) B cell ontogeny in murine embryo studied by a culture system with the monolayer of a stromal cell clone, ST-2: B cell progenitor develops first in the embryonal body rather than in the yolk sac. EMBO J 7: 1337–1343.

Ogawa M, Matsuzaki Y, Nishikawa S, et al (1991) Expression and function of c-kit in hemopoietic precursor cells. J Exp Med 174: 63–71.

Osmond DG (1990) B cell development in bone marrow. Semin Immunol 2: 173–180.

Parkhouse RME, Preece G, Sutton R, Cordell JL, Mason DY (1992) Relative expression of surface IgM, IgD and the Ig-associating α (mb-1) and β (B-29) polypeptide chains. Immunology 76: 535.

Ramakrishnan L, Wu Q, Yue A, Cooper MD, Rosenberg N (1990) BP-1/6C3 expression defines a differentiation stage of transformed pre-B-cells and is not related to malignant potential. J Immunol 145: 1603–1608.

Rao M, Lee W, Conrad D (1987) Characterization of a monoclonal antibody directed against the murine B lymphocyte receptor for IgE. J Immunol 138: 1845.

Rawlings DJ, Quan SG, Kato RM, Witte ON (1995) A long-term culture system for selective growth of human B cell progenitors. Proc Natl Acad Sci USA 92: 1570–1574.

Rehmann JA, LeBien TW (1994) Transforming growth factor-beta regulates normal human pre-B-cell differentiation. Int Immunol 6: 315–322.

Reininger L, Radaszkiewicz T, Kosco M, Melchers F, Rolink A (1992) Development of autoimmune disease in SCID mice populated with long-term 'in vitro' proliferating (NZB × NZW) F1 pre-B cells. J Exp Med 176: 1343–1353.

Renard N, Duvert V, Blanchard D, Banchereau J, Saeland S (1994) Activated CD4+ T cells induce CD40-dependent proliferation of human B-cell precursors. J Immunol 152: 1693–1701.

Rolink A, Melchers F (1991) Molecular and cellular origins of B lymphocyte diversity. Cell 66: 1081–1094.

Rolink A, Kudo A, Karasuyama H, Kikuchi Y, Melchers F (1991a) Long-term proliferating early pre-B cell lines and clones with the potential to develop to surface Ig-positive, mitogen reactive B cells in vitro and in vivo. EMBO J 10: 327–336.

Rolink A, Streb M, Nishikawa SI, Melchers F (1991b) The c-kit encoded tyrosine kinase regulates the proliferation of early pre-B cells. Eur J Immunol 21: 2609.

Rolink A, Grawunder U, Haasner D, Strasser A, Melchers F (1993a) Immature surface Ig+ B cells can continue to rearrange kappa and lambda L chain gene loci. J Exp Med 178: 1263–1270.

Rolink A, Haasner D, Nishikawa S, Melchers F (1993b) Changes in frequencies of clonable pre-B cells during life in different lymphoid organs of mice. Blood 81: 2290–2300.

Rolink A, Grawunder U, Winkler TH, Karasuyama H, Melchers F (1994a) IL-2 receptor α chain (CD25, TAC) expression defines a crucial stage in pre-B cell development. Int Immunol 6: 1257–1264.

Rolink AG, Reininger L, Oka Y, Kalberer CP, Winkler TH, Melchers F (1994b) Repopulation of SCID mice with long-term in vitro proliferating pre-B-cell lines from normal and autoimmune disease-prone mice. Res Immunol 145: 353–356.

Saeland S, Duvert V, Moreau I, Banchereau J (1993) Human B cell precursors proliferate and express CD23 after CD40 ligation. J Exp Med 178: 113–120.

Springer TA, Galfre G, Secher DS, Milstein C (1978) Monoclonal xenogeneic antibodies to murine cell surface antigens: identification of novel leukocyte differentiation antigens. Eur J Immunol 8: 539–551.

ten Boekel E, Melchers F, Rolink A (1995) The status of Ig loci rearrangements in single cells from different stages of B cell development. Int Immunol 7: 1013–1019.

Uckun FM, Ledbetter JA (1988) Immunobiologic differences between normal and leukemic human B-cell precursors. Proc Natl Acad Sci USA 85: 8603.

Winkler TH, Melchers F, Rolink AG (1995a) Interleukin-3 and interleukin-7 are alternative growth factors for the same B cell precursors in the mouse. Blood 85: 2045–2051.

Winkler TH, Rolink A, Melchers F, Karasuyama H (1995b) Precursor B cells of mouse bone marrow express two different complexes with the surrogate light chain on the surface. Eur J Immunol 25: 446–450.

Wolf ML, Buckley JA, Goldfarb A, Law CL, LeBien
  TW (1991) Development of a bone marrow culture
  for maintenance and growth of normal human B
  cell precursors. J Immunol 147: 3324–3330.

Yelton DE, Desaymard C, Scharff MD (1981) Use of
  monoclonal anti-mouse immunoglobulin to detect
  mouse antibodies. Hybridoma 1: 5–11.

Section 13

Modulation of the Humoral Immune Response and Its Measurement

Section Editor
Robbert Benner

LIST OF CHAPTERS

Introduction

13.1

Huub F.J. Savelkoul[1]
Eric Claassen[1,2]
Robbert Benner[1]

[1]*Department of Immunology, Erasmus University, Rotterdam, The Netherlands*
[2]*Department of Immunology and Infectious Diseases, TNO Prevention and Health, Leiden, The Netherlands*

In this Section we have selected a number of methods dealing with the induction and measurement of the humoral immune response, particularly geared at improving the successful generation of monoclonal antibodies. The individual chapters present a detailed theoretical background, a critical appraisal, and many practical pointers on crucial parameters for the successful application of the techniques.

The chapters are grouped into two parts. The first part deals with optimization of the generation of monoclonal antibodies by established methodology. To achieve this goal, essential information is necessary on the nature of the T and B cell epitopes of the relevant antigen (chapter 13.2). Extensive knowledge is required on properties of antigens and rules that govern antigen presentation to permit induction of a specific immune response in a highly efficient and specific fashion (chapter 13.3). Also new strategies are required to permit the generation of monoclonal antibodies specific for activation molecules expressed only transiently on stimulated cells. Exciting new technologies have

been developed to achieve this goal (chapter 13.4). Steering the isotype production in the immune response by cytokine treatment *in vivo* could potentially be important in obtaining monoclonal antibodies of preselected isotype. The alginate encapsulation technique permits *in vivo* cytokine treatment in a reliable and easy fashion (chapter 13.5).

Subsequently, the resulting immune response has to be evaluated in a quantitative way. We therefore include chapters on the detection and quantification of antibody-secreting cells (chapters 13.6 and 13.7), absolute quantification of antigen-specific antibodies (chapter 13.8) and the measurement of affinity of specific antibodies by several techniques, including ELISA and biosensors (chapter 13.9). Certainly, the newly developed biosensor technology is a major advancement in the absolute determination of association and dissociation rate constants, thereby permitting affinity determination of macromolecular interactions. We anticipate this technology becoming crucial in the selection of monoclonal

Immunology Methods Manual
ISBN 0–12–442712–X

antibodies based on functional characteristics.

Phage display techniques constitute another new development becoming increasingly important in the generation of monoclonal antibodies or their fragments (chapter 13.10). Using combinatorial phage display libraries, selection of antibody fragments can be achieved with proper specificity and high binding affinity.

The integration of the various techniques into the analysis of the humoral immune response *in vivo* and its proper measurement is discussed in chapter 13.11.

Epitope mapping by pepscan 13.2

Rob H. Meloen
Wouter C. Puijk
Wim M.M. Schaaper

DLO-Institute for Animal Science and Health, Lelystad, The Netherlands

TABLE OF CONTENTS

Immunology Methods Manual
ISBN 0–12–442712–X

Introduction

Design of vaccines and diagnostics, under-standing of the relationships between pathogen and their hosts (notably their interactions with the immune system) requires detailed knowledge of antigenic sites, i.e. the mapping of both B and T cell epitopes, often at the level of single amino acids.

Numerous methods have evolved to this end:

- Computer programs which predict epitopes from the primary structure of any given protein.
- Use of expression systems in which fragmented proteins are expressed separately and tested for their ability to bind a given antibody. Subsequent sequencing reveals the amino acid sequence of the Ab binding fragment.
- In the case of pathogens, notably viruses, mutants of viruses which survive in the presence of antibody are sequenced. Amino acid sequence changes are assumed to be directly or indirectly responsible for evading antibody recognition and these changes may thus be involved in antibody binding.
- The PEPSCAN. This is a systematic method based upon peptide synthesis on solid supports. The principle is shown in Fig. 13.2.1. In short it depends on the synthesis of multiple overlapping peptides coupled to solid supports. For a given sequence, all peptides of, for instance, 9 amino acids length, which overlap each other with eight amino acids, are used. The peptides are synthesized on to functionalized polyethylene pins formatted in such a way that they fit into microtiter wells (Plate 2). This allows easy assessment of antibody binding. Because the peptides are covalently bound to the solid supports, they can be cleaned from

noncovalently bound antibody and reused. In our hands, sets of peptides can be reused more than a hundred times over a period of at least 10 years. A typical example of a PEPSCAN result using a serum of an individual infected with HIV is shown in Fig. 13.2.2. This result was the first precise identification of the V3 epitope or the V3 loop (Goudsmit et al 1988).

In 1989 it was shown that it was possible to use PEPSCAN also for the precise determination of T cell epitopes when the peptides were removed after synthesis on the pin, allowing their use in functional T cell bioassays (Van der Zee et al 1989). Later it was shown that this latter variation can be used in almost any bioassay used to define bioactivity of receptor–

Table 13.2.1 Microorganisms and proteins subjected to epitope scanning in which the authors have been involved. In most cases these studies directly produced useful epitopes.

Microorganisms	Signal molecules
FMDV	LHRH
HIV	LH
Measles	FSH
TGEV	hCG
MHV	Inhibin
Pseudo rabies virus	Neuropeptides
Hog cholera	Cytokines (1, 2, 8, TNF-α)
Lelystad virus	Immunosuppressive peptides
Parvo viruses (CPV, PPV, B19)	CRH
SFV	
HPV	
Adeno virus	
FIV	
FeLV	
Rabies virus	
Malaria	
Treponema pallidum	
E. coli (Pilli, phoE, etc.)	
S. typhimurium	
BSE	
Mycobacterium leprae	

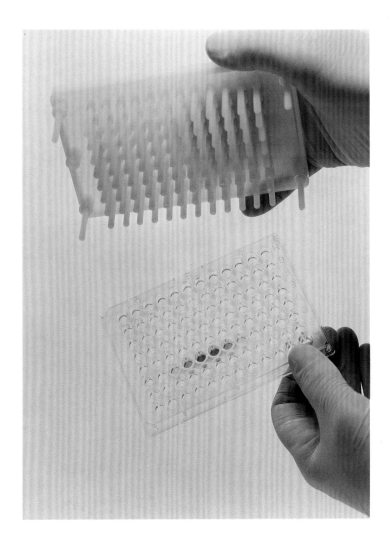

Plate 2 Physical format of the PEPSCAN. Peptides are synthesized on the top of polyethylene pins. The block of pins can be positioned over a standard microtiter tray, thus allowing each pin to fit into a separate well. In this example the last step of the PEPSCAN is shown: the incubation of the pins into the substrate. Pins that contain peptide-bound antibody are visualized using anti-antibody marked with peroxidase which binds only to pins already containing antibody, similarly to a normal ELISA.

ligand systems. Although PEPSCAN is as yet still limited in the mapping of conformational or discontinuous epitopes, it has been highly successful in the mapping, often in much detail, of B cell epitopes of more linear nature (Table 13.2.1). Obviously, mapping of T cell epitopes has been even more effective owing to the linear nature of these epitopes.

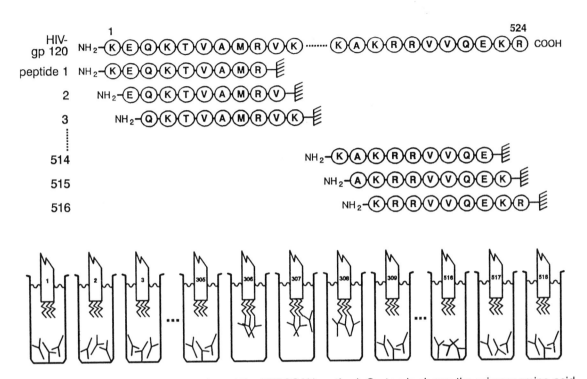

Figure 13.2.1 Schematic representation of the PEPSCAN method. On top is shown the primary amino acid sequence of gp120 of HIV-1, the major glycoprotein of HIV-1. Each circle represents an amino acid. For the amino acids the single-letter code is used (A = alanine, C = cysteine, D = aspartic acid, E = glutamic acid, F = phenylalanine, G = glycine, H = histidine, I = isoleucine, K = lysine, L = leucine, M = methionine, N = asparagine, P = proline, Q = glutamine, R = arginine, S = serine, T = threonine, V = valine, W = tryptophan, Y = tyrosine). The amino acid sequence of gp120 of HIV-1 is then divided into overlapping peptides, as indicated. Peptide no. 1 is the peptide which starts with amino acid no. 1 and ends with amino acid no. 9, peptide no. 2 is the peptide starting at amino acid no. 2 to amino acid no. 10, and so on. The peptides are synthesized on polyethylene rods, as shown in the lower part of the figure; the peptides are indicated as ʌʌʌ. (The rods are in such a configuration that they fit into microtiter wells; this greatly simplifies handling of the rods and data manipulation.) All rods with peptides are then contacted with the same antibody (indicated as —<). Some peptides will bind this antibody. After the rods are taken out of the antibody solution and have been washed, the antibody still present on the rods (bound to the peptide) can then be tested with anti-antibody conjugate for the presence of antibody. This directly produces the sequence of the peptide which has bound the antibody. After this process the antibody can be removed from the peptides and the peptides can be reused. One technician can easily test 2000 different peptides daily for reactivity with a given antibody. Over 2000 different peptides are being synthesized on a routine basis at our laboratory each month (from Meloen et al 1991, with permission).

Epitope mapping by pepscan

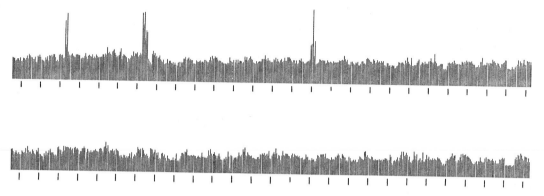

Figure 13.2.2 Results of antisera of two individuals; one infected with HIV-1 (top) and one not infected (bottom). The sera were tested with all overlapping nonapeptides of gp120 of HIV-1. The binding activities are shown as vertical lines proportional to the optical density of the ELISA. An optical density above the background indicates antibody binding activity. The positions of the peptides are shown on the horizontal axis: peptide 1 is at position 1 and represents the peptide which starts with amino acid 1 and ends with amino acid 9; peptide 2 starts at amino acid 2 and ends at amino acid 10, and so on. The antibodies in the serum of the infected individual (top) reacted with peptides around positions 57, 145, and 310. Reactivities against the latter peptides always occur in sera which neutralize the virus and this area of gp120 of HIV-1 has been named the 'principal neutralizing domain' (PND) (from Meloen et al 1991, with permission).

Methods

The Pepscan method has been amply described (Geysen et al 1984, 1985, 1987; Van der Zee et al 1989; Schaaper et al 1993; Meloen et al 1995).

In brief, each single peptide is synthesized on a polyethylene pin grafted with a polyacrylate- or a polystyrene-type resin using conventional peptide synthesis methods. For use in bio-assays in solution, as in the determination of T cell epitopes, appropriate linkers between resin and peptide are used to facilitate the cleavage from the pin at the end of the synthesis. Pin-bound peptides can be tested directly in 96-well microtiter plates (Greiner), using an ELISA.

Successful execution of these methods is highly dependent on experience and skill with respect to peptide chemistry and organic chemistry. Biochemists and biologists inexperienced in these disciplines who try to initiate PEPSCAN in their laboratories run the risk of ending up with unsolvable problems. The authors of this article offer a service which makes the PEPSCAN available at cost price for universities and nonprofit institutions (contact one of the authors at ID-DLO, PO Box 65, 8200 AB Lelystad, The Netherlands; tel. 31 320 238238; fax 31 320 238050).

In the meantime a whole variety of alternatives has evolved. The first was reported by Houghten (1985), who used standard peptide chemistry on conventional solid phase resins packed in sealed 'tea-bags'. His simultaneous multiple peptide synthesis (SMPS) allowed the simultaneous synthesis of a few hundred peptides at the most. Similar principles were used for the synthesis of peptides on polystyrene-grafted polyethylene films (Berg et al 1991), on paper (Frank et al 1992), or on cotton (Eichler et al 1991).

A different approach for the synthesis of large numbers of peptides involves (chemical) peptide libraries (for review see Scott and Graig 1994). In most of these libraries amino acid mixtures are used to replace single amino acid positions in a peptide, which results in very large numbers of peptides, but very low concentrations of the individual peptides. Although these methods are successful for the optimization of peptide–protein interactions, they are less suitable for the scanning of linear proteins as in PEPSCAN methods.

Practical aspects of the determination of B cell epitopes

First of all the length of the peptides used is important. Initially, hexapeptides were used (Geysen et al 1984). However, it was soon realized that this is far too short and a length of 9 was used subsequently. However, we currently use a standard length of 12 amino acids, because occasionally epitopes of this length are found which would be missed otherwise.

Second, the end groups are of importance. Charged groups need to be excluded, because epitopes are in general located within a protein chain and thus have no charged end groups. Therefore, the N-terminal is routinely acetylated. In our experience this end group is in general the best choice. The PEPSCAN has, in contrast to many other screening assays, a high signal-to-noise ratio. Therefore, antibody binding is observed even at very low affinities. The consequence is that (1) the affinity of positive bindings sometimes needs to be improved considerably before being applicable, for instance using systematic alteration (see below); and (2) sequences that bind antibody in PEPSCAN may do so because of some limited resemblance with the original but perhaps unknown epitopes (just two properly aligned amino acids spaced by a few others may be enough). Thus it is always necessary that the antibody binding amino acid sequence is confirmed using other systems; for instance using the peptide to induce in animals antipeptide antibodies with similar biological activities as the original antibody, or using the peptide to block the reaction of the original antibody and the original antigen. If confirmations using unrelated systems cannot be done, it is debatable how useful the sequence is.

Practical aspects of the determination of T cell epitopes

The length is important for T cell epitopes, less so for Th epitopes than for CTLs. Standard peptides for Th mapping have a length of 15 amino acids, those for CTLs a length of 8–11 amino acids (Falk et al 1992; Rötzschke et al 1992).

For CTLs the end group may also be important, so the N-terminal is not acetylated in these cases, as in the natural situation. Although the C-terminal should carry ideally a carboxy group, one can get away with using a carboxamide group. The ability of a peptide to react in functional T cell assays depends very much on the nature of the epitope, and thus on the concentration that can be applied. Strong epitopes are almost always found, detection of weaker ones depending on the amount of peptide that can be successfully split off the PEPSCAN pins. Normally this falls in the range of 100 μg to 1000 μg per pin. This is sufficient, in most assays, to measure peptides with significant activities up to 1 μmole.

Details of the mapped epitopes

When antibody binding peptides have been defined, it may be necessary to assess the minimum length (core) of the epitope and the role of the individual amino acids within the core and flanking the core. Using the vast number of permutations allowed by PEPSCAN, many 'algorithms' have been formulated to assess these aspects in a systematic fashion. The more useful ones are described here.

Core of the epitope

The core of the epitope can be readily assessed using a *deletion set* of peptides. A set of peptides is synthesized and in each peptide an amino acid is missing (Fig. 13.2.3). When these peptides are tested for antibody binding, the amino acids which cannot be deleted form the core of the epitope. Normally they fall into the middle of the epitope.

(a)

DELETION PEPTIDES

R I Q R G P G R A F V T I G

- I Q R G P G R A F V T I G

R - Q R G P G R A F V T I G

R I - R G P G R A F V T I G

R I Q - G P G R A F V T I G

R I Q R - P G R A F V T I G

etc.

(b)

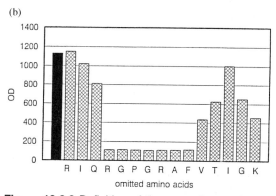

Figure 13.2.3 Definition of the core of an antigenic determinant using a deletion set of peptides. The amino acids of a reactive peptide are deleted, one at a time, the resulting peptides are tested with the antibody. If the deleted amino acid belongs to the core of the antigenic determinant, the reactivity of the peptide with the antibody is lost. (a) Example of a set of deletion peptides: A dash, the deleted amino acid. (b) A typical result with the peptides of the PND of HIV-1 and a neutralizing monoclonal antibody. The binding activity is shown as a bar proportional to the optical density in the ELISA. The solid bar represents the activity of the parent peptide RIQRGPGRAFVTIGK. The cross-hatched bars represent the reactivities of the deleted peptides. The amino acid below each bar represents the amino acid deleted from the parent sequence. The core of this antigenic determinant is RGPGRAF (from Meloen et al 1991, with permission).

(a)

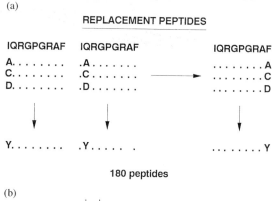

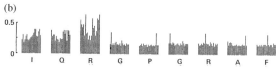

Figure 13.2.4 Definition of the irreplaceable amino acids of an antigenic determinant using a set of replacement peptides. (a) Replacement peptides are obtained by replacing each amino acid in turn with all other amino acids. For a nonapeptide this gives a set of 180 different peptide analogues. In this example the replacement set is shown for the peptide of the PND of HIV-1: IQRGPGRAF. (b) The set of peptides is tested against a polyclonal antibody; in this example results obtained with the serum of an individual infected with HIV are shown. The binding activity is shown as a vertical line proportional to the optical density in the ELISA. Each group of 20 lines corresponds to the complete set for one of the amino acids in the peptide. Within each group the left most line corresponds to substitution of the original amino acid by S, and the following lines to its substitution by amino acids GDNEQKRYHWFMLIV-PAT. This grouping accounts for the size and polarity of the side-chain and is derived from the circle arrangement described by Doolittle (1985), which places amino acids with similar properties near one another. To facilitate presentation we have 'cut' the circle between T and S; thus some amino acids with similar properties (i.e., small side-chains) are shown at the left- and right-hand sides. The sequence of the parent peptide is indicated. Six of the amino acids cannot be replaced (from Meloen et al 1991, with permission).

The role of the individual amino acids within the epitope (or within the core)

The role of the individual amino acids can be readily assessed using a set of *replacement* peptides in which individual amino acids, are replaced one at a time by, for instance, all other naturally occurring amino acids (Fig. 13.2.4). For each position this allows the assessment of whether that particular amino acid can be replaced; and if it can be replaced, whether by all or by a limited number of amino acids with similar properties, for instance. An often used limited version of this algorithm is the 'ala-scan', in which individual amino acids, one at a time, are replaced by alanine only.

The role of the flanking amino acids

This 'algorithm' reveals the minimum-length peptide which binds antibody and the contribution to binding made by the flanking amino acids. A limited PEPSCAN is made for the protein region which encompasses the epitope using all overlapping peptides of length 3 amino acids, 4 amino acids, and so on, up to 12 amino acids length. Antibody binding results show to what extent each individual amino acid addition contributes to binding (for instance, the extra amino acid of a tetrapeptide compared to each one of the two 3-amino acid peptides it contains).

T cell epitope

All 'algorithms', described for B cell epitopes, can be applied directly to study the details of T cell epitopes.

Peptide diversity libraries

Although PEPSCAN allows many permutations, the number of different peptides that can be used does not compare with those available in, for instance, phage display libraries.

Nevertheless, owing to the high signal-to-noise ratio of PEPSCAN, similar results can be obtained using small random libraries (up to 4550 peptides) or small motif libraries, i.e., libraries in which the number of permutations have been limited depending on the amino acid position. In our experience, numerous low-affinity leads are readily obtained, which can be subsequently upgraded in a few rounds to high-affinity peptides, applying one of the above-mentioned 'algorithms' (Van Amerongen et al 1992, 1994).

Conclusion

PEPSCAN can be readily used to define linear parts of B cell epitopes and most T cell epitopes systematically in much detail. At present it offers in addition the possibility to design mimics for conformational epitopes, applying small random peptides libraries.

Acknowledgment

We thank Mrs J. de Jager-Koning for the careful preparation of this manuscript.

References

Berg RH, et al (1991) Film-supported solid-phase peptide synthesis. In: Giralt E, Andreu D, eds. Peptides 1990. ESCOM, Leiden, pp. 149–150.

Doolittle RF (1985) Proteins. Scientific American 253: 699–703.

Eichler J, et al (1991) Multiple peptide synthesis on cotton carriers: Elucidation and characterization of an antibody binding site of CRF. In: Giralt E, Andreu D, eds. Peptides 1990. ESCOM, Leiden, pp. 156–157.

Falk K, et al (1992) Allele-specific motifs revealed by sequencing of self-peptides eluted from MHC molecules. Nature 351: 29.

Frank R (1992) Spot synthesis: an easy technique for the positionally addressable, parallel chemical synthesis on a membrane support. Tetrahedron 48: 9217–9232.

Geysen HM, et al (1984) Use of peptide synthesis to probe viral antigens for epitopes to a resolution of a single amino acid. Proc Natl Acad Sci USA 81: 3998–4002.

Geysen HM, et al (1985) Small peptides induce antibodies with a sequence and structural requirements for binding antigen comparable to antibodies raised against the native protein. Proc Natl Acad Sci USA 82: 178–182.

Geysen HM, et al (1987) Strategies for epitope analysis using peptide synthesis. J Immunol Methods 102: 259–274.

Goudsmit J, et al (1988) HIV type 1 neutralization epitope with conserved architecture elicits early type-specific antibodies in experimentally infected chimpanzees. Proc Natl Acad Sci USA 85: 4478–4482.

Houghten RA (1985) General method for the rapid solid-phase synthesis of large numbers of peptides: specificity of antigen–antibody interaction at the level of individual amino acids. Proc Natl Acad Sci USA 82: 5131–5135.

Meloen RH, et al (1995) PEPSCAN to determine T- and B-cell epitopes. In: Zegers N, Boersma W, Claassen E, eds. Immunological Recognition of Peptides in Medicine and Biology. CRC Press, Boca Raton, in press.

Rötzschke O, et al (1992) Peptide motifs of closely related HLA class I molecules encompass substantial differences. Eur J Immunol 22: 2453.

Schaaper WMM, et al (1993) Improvements in the synthesis of cleavable peptides on pins. In: Schneider CH, Eberle AN, eds. Peptides 1992. ESCOM, Leiden, pp. 312–313.

Scott JK, Craig L (1994) Random peptide libraries. Curr Opin Biotechnol 5: 40–48.

Van Amerongen A, et al (1992) Peptides reactive with a transmission blocking monoclonal antibody against *Plasmodium falciparum* Pfs25: 2000-fold affinity increase by PEPSCAN-based amino acid substitutions. Peptide Res 5: 269–274.

Van Amerongen A, et al (1994) Design of peptides with improved affinities for anti-human chorionic gonadotropin monoclonal antibodies. Peptide Res 7(2): 83–90.

Van der Zee R, et al (1989) Efficient mapping and characterization of a T cell epitope by the simultaneous synthesis of multiple peptides. Eur J Immunol 19: 43–47.

Antigens and antigen presentation

13.3

Marlies Leenaars[2,3]
Eric Claassen[1,2]
Wim J.A. Boersma[1]

[1]Division Immunological and Infectious Diseases, TNO Prevention and Health, Leiden, The Netherlands
[2]Department of Immunology, Erasmus University, Rotterdam, The Netherlands
[3]National Institute of Public Health the Environment (RIVM), Bilthoven, The Netherlands

TABLE OF CONTENTS

Immunology Methods Manual
ISBN 0–12–442712–X

Introduction

Antigens are those structures which the immune system recognizes as non-self. To achieve recognition and subsequent immune responsiveness, the antigen has to be presented to the immune system in such a way that it is recognized in a form most suitable for elimination of the antigen. An antigen in this context is any substance that is recognized by antigen-specific structures of the immune system: antibodies, T cell receptors. Antibodies can bind to unmodified antigens as well as to fragments and degradation products thereof. For recognition by T cells, antigens have to be degraded to peptides, which complexed with MHC molecules are recognized by T cell receptors. Recognition by the immune system is a prerequisite for an antigen to elicit a response. Molecular structures that can elicit an immune response are called immunogens. Although almost any substance can be recognized as an antigen, the immunogenicity of a substance is determined by the intrinsic properties of the antigen: source, nature, structure, complexity, and size (Table 13.3.1). Nevertheless, exactly what determines the capacity of a structure to induce an immune response (immunogenicity) and to be recognized by an antibody is a subject of discussion (Hopp 1986; Van Regenmortel 1986, 1989; Boersma et al 1993).

Here we will discuss antigens and their properties in *in vivo* immunization for the generation of immunodiagnostic reagents and in vaccine applications, with emphasis on induction of humoral immunity.

The recipient organism

In immunization and vaccination procedures most attention is usually focused on the properties of the antigen; however, the immune status of the immunized individual is also a decisive factor. The slow development of the immune system in young individuals, as the decline of the immune system in old individuals, determines the quality of an immune response. In newborns, circulating maternal antibodies which form complexes with the antigen may have enhancing as well as inhibiting effects on the immune responses dependent on antibody/antigen ratios. In young individuals, in general the response to (lipo)polysaccharides develops more slowly than responses to proteins; in a sense young individuals (mouse, man) are partly immunodeficient with respect to these antigens. In addition, protocols for intentional immunodeficiency induction (transplantation, malignancies) and natural immunodeficiency by superinfection (HIV) have a strong negative influence on immune responses.

Furthermore genetic factors such as sex, MHC haplotype, and intrinsic low or high responsiveness may play a role in the outcome of an immunization procedure.

Table 13.3.1 General features influencing immunogenicity of antigens

Parameter	Immunogenicity
Source	Xenogeneic > allogeneic > syngeneic > autologous
Chemical nature	Protein > polysaccharide > lipid
Size	High mol wt > low mol wt
Complexity[a]	Complex > simple

[a]Multiple antigenic/immunogenic molecules.

Antigen presentation

On the surface of cells, MHC molecules are expressed of which the function is to present self antigens and foreign antigens to potential responder cells with an immune surveillance function. Immunological antigen presentation is effected via two distinct routes: (1) For endogenous antigen presentation, intracellular proteins and therefore also viral proteins, for example, are expressed and processed in an active turnover process. As a consequence, peptides and fragments generated in this process are brought to the cell surface mainly bound to class I MHC molecules. There the complexed fragments are recognized by class I-responsive T cells (mainly CD8 T cells; cytolysis, suppression). (2) Exogenous antigens can be taken up by complex sequences of events leading to transfer of antigen to the cytoplasm (e.g., phagocytosis, pinocytosis) . Alternatively antigens which are recognized by surface immunoglobulins can be taken up directly in B cells. Antigens which are complexed with circulating antibodies can be bound via Fc receptors. In all cases an antigen processing pathway different from that for endogenous molecules is followed for digestion of antigens. After antigen processing, the antigen fragments are presented to the immune system in the context of MHC class II surface molecules. This form of presentation elicits a response mainly of CD4 T cells (T cell help, delayed type hypersensitivity). So-called professional antigen-presenting cells (APC) apart from B cells include dendritic cells and macrophages.

The main functions of activated T cells are: (1) cell mediated responses which are mainly direct cytotoxic responses to cells expressing specific antigens (in MHC bound form) and can be performed by CD4 and CD8 type T cells, (2) T cell help (Th) for B cells. Th1 and Th2 cells are each able to generate different sets of cytokines to modulate immune responsiveness of other cells. T and B cells interact not only via recognition of antigen by antigen receptors but also via mutual recognition of series of activation markers, i.e., receptors and ligands expressed on the surface, as discussed for CD40 and CD40 ligand by Laman and de Boer (see chapter 13.4). As a result, some antigens upon recognition will lead to production of other antibody isotypes from others. Also, the response of monocytes and macrophages in a DTH response is orchestrated by Th cells.

Primary immunization and memory

Primary immunization and vaccination aim at introduction of antigens so as to evoke an optimal immune response and to generate memory formation for secondary responsiveness. For antibody responses, priming generally leads to strong responses of IgM isotype and low levels of antibodies for which class switching is required (IgG, IgA, IgE). For class switching and memory formation, the activation of either Th1 or Th2 cells is necessary.

In general, expression of isotypes other than IgM in a primary immune response is low. After a second encounter with the same antigen, the isotype will be expressed mainly which is related to antigen and immunization conditions as discussed below. In the secondary or booster reaction, both the quantity (level) as well as the quality of the response is enhanced (higher affinity, specific isotype).

Specific antigenic properties may lead to specific isotype expression in immune responses (Table 13.3.2). In general, IgG will be elicited with (glyco)proteins, when the antigen is presented to the peripheral immune system. In human and mouse, the major isotype produced in response to proteins is IgG1. IgG3, apart from bacterial infections where polysaccharides play a role, is generally low in mice and in man. In man, in addition, IgG4 is relatively low. Isotype induction is dependent on nature of the antigen, the route of introduction, and the formulation in which the antigen is introduced. When similar antigens are presented to the immune system via the mucosal surface, the resulting isotype will be IgA. IgE is mainly induced by parasitic (surface) antigens

Table 13.3.2 General antigen categories and induced response

Antigens	MHC I pathway	MHC II pathway	Isotype
Soluble			
Protein	−	+	IgG1 > IgG2(a,b) >> IgG3 > (IgG4), IgE
Polysaccharide	−	±	IgM/IgG3 > IgG2a,b (mice) IgG2 (human)
Lipid	−	−	
Particulate/complex			
Dead			
Complex[a]	−	+	Antigen dependent
Micro-organism	−	+	Antigen dependent
Live			
Intracellular			
virus	+	+	IgG1 (mice) IgG1, IgG3 (human)
parasite	(+)	+, DTH	IgG, IgE (mice) IgG1, G3, G4, IgE (human)
bacteria	(+)	+	IgG1/IgG3 (mice) IgG1/IgG2 (human)
Extracellular			
virus	−	+	IgG1 (mice) IgG1, IgG3 (human)
parasite	−	+	IgG, IgE (mice) IgG, IgE (human)
bacteria	−	+	IgM, IgG2a/IgG3 (mice) IgM, IgG1/IgG2 (human)

[a]See Table 13.3.1.

which may act locally like allergens. Allergenic responses to other harmless antigens are also of the IgE isotype. Antigen–IgE complexes lead to activation of effector cells.

Delayed type hypersensitivity (DTH) reactions are in general developed to all protein antigens when introduced in low concentrations in the skin, where specific antigen-presenting cells (Langerhans cells, interdigitating cells) will take them up and migrate to the draining lymph nodes to activate specific T cells. The recruited effector cells are mainly monocytes and macrophages. DTH reactions can also be elicited with more complex and particulate antigens (such as xenogeneic erythrocytes, dead microorganisms).

Memory T cells play a pivotal role in the elicitation of secondary responses. In secondary responses mainly that isotype is expressed for which T help (Th1/Th2) is pro-vided. Memory T cells as well as memory B cells formed as a result of the priming immunization allow the immune system to respond more vigorously and more rapidly in a secondary or booster response.

T independent antigens

Binding of antigen fragments to class I or class II MHC molecules is observed for proteins. Other antigens, such as polysaccharides and lipids, do not follow the antigen presentation pathways described above. As a result they generally do not activate T cells and are therefore called T-independent antigens. The consequence is that these antigens in general do not lead to immune responses of other isotypes than IgM and in part to IgG3 (mouse) or IgG2 (man). T-independent activation of B cells in general occurs with polymeric antigens that are cross-linking surface immunoglobulins (for review see Laman and Claassen 1995; Van den Eertwegh et al 1992). Memory formation, affinity maturation as well as germinal center formation, is absent. To elicit IgG responses to this category of antigens, conjugation to T-dependent antigens (proteins) is necessary.

Isotypes of antibody and Fc receptors

Antibodies of different isotypes have specific roles in protective immune responses. Binding to Fc receptors may facilitate the opsonization of microorganisms and mediates antibody-dependent cellular cytolysis. Complement binding to the Fc part is of importance for activation of the complement system. Fc parts of IgG2a and IgG2b (in mouse) and IgG1 and IgG3 (human) have complement binding sites. These isotypes are therefore most effective in protection against viruses and other pathogens.

Antigens: source, chemical nature, size

The immune system of an organism reacts against any foreign compound (antigen) which crosses the defense lines of epithelia and skin and in general is tolerant (unable to react) towards its own body components, which may be immunogenic in other organisms. The intensity of an immune response is directly related to the degree of foreignness. The greater the phylogenetic distance between donor animal and the animal to be immunized, the better the immune response that is evoked (see Table 13.3.1). It is likely that in those cases peptides will bind to MHC molecules which are distinguishable from self peptides.

Antigens include substances as structurally diverse as proteins, polysaccharides, lipids, and nucleic acids. Most effective as immunogens are molecules which display diverse chemical and structural characteristics. Natural immune responses are directed mainly to proteins, glycoproteins, lipoproteins, lipopolysaccharides, and carbohydrates as part of the defense to infectious or invasive agents.

Proteins are synthesized as long, flexible polypeptide chains that assume a unique three-dimensional shape. Proteins are either rod-like fibrous (structural) proteins or globular proteins. Both structures are unfolded or denatured by extremes of pH and temperature and by high concentrations of dissociating agents (e.g., urea, guanidine, and sodium dodecyl sulfate), whereas more subtle alterations in conformation can be engendered by suitable modification of the amino acid side-chains. In general, these alterations in conformation, especially for structure-related determinants, are associated with changes in the antigenic reactivity.

Antigens with low molecular weight are in general not immunogenic. These so-called haptens may be small organic structures or short synthetic peptides which consist of a B cell epitope only. They may evoke immune response only when coupled to potent immunogens, e.g., proteins which provide the necessary T cell epitopes. The minimum molecular size required for immunogenicity can be debated. In principle the antigen needs at minimum one T cell epitope to be included, which may be expected in proteins/peptides which are of a size over 2–5 kDa.

(Glyco)proteins and (lipo)proteins dependent on the route of introduction and the adjuvant applied lead to T cell responses directed towards the protein backbone as well as antibody formation of isotypes other than IgM.

Polysaccharides exist in the capsules of bacteria, or may occur as lipopolysaccharides as part of the cell walls of Gram-negative bacteria. Purified polysaccharide antigens can serve as immunogens in certain species (e.g., mice and humans) but not in others (e.g., rabbits and guinea-pigs). Microbial polysaccharides are located on the cell surface and are, therefore, of importance in recognition and immune responses of a higher organism to microbacterial infection. Complex lipopolysaccharide antigens are found in a large variety of microorganisms, notably in Gram-negative bacteria. As they are largely T-independent, responses to (lipo)polysaccharides are mainly of the IgM isotype. Dependent on local antigen presentation, specific IgG isotypes (IgG3 in mouse) and IgA may be elicited. Pure lipids are not immunogenic. Coupled to proteins, they may act like haptens. Also, glycolipids such as cardiolipin may become immunogenic after binding to a carrier.

Role of physicochemical properties of antigens

In antibody responses the influence of physical parameters of the antigens is predicted in part on the basis of common-sense reasoning. Antigenic sites of intact proteins are accessible to large molecules, like antibodies, only when exposed on the outside of a protein. Hydrophilic sequences readily soluble in aqueous

solutions are thus most likely to be antigenic. Flexibility might enhance the fitting of antigen and antibody (Westhof et al 1984; Karplus and Schulz 1985; Fieser et al 1987), and hence segmental mobility of the epitope may enhance antigenicity (Novotny et al 1986). Loops of the protein may stick out from the globular form of a protein and as a result they may be more readily engaged in binding to a circulating antibody (Kyte and Doolittle 1982). Synthetic peptides forced into a loop were shown to have an enhanced affinity for preselected antibodies (Jemmerson and Hutchinson 1990).

Methods have been developed to describe the tendency to form secondary structures: α-helix or β-sheet or β-turns (Chou and Fasman 1978; Garnier et al 1978; Hopp and Woods 1981; Emini et al 1985). A relative value is attributed to each amino acid which represents its ranking in terms of hydrophilic behaviour (Hopp and Woods 1981; Kyte and Doolittle 1982). The 'surface seeking' tendency or affinity for the membrane interior was computed by Eisenberg et al (1984a,b). Normalized hydrophobicity values have been specially developed and adapted for the prediction of transmembrane sections of proteins which function as membrane-spanning transport proteins (Eisenberg et al 1984a,b). The mean value for a number of these physicochemical parameters, combined in an antigenic index, gives an estimate of the antigenic properties (index) of a protein segment (Wolf et al 1988). Amino acids exert mutual influences over short distances (Bangham 1988). In addition, amino acid sequences which, on a linear scale, are at great distance may interact by forming backfolding loops. It is of great importance to take this into account when choosing the length of peptides used for immunization (Jacob et al 1986; Horiuchi et al 1987).

Antigen complexity

Antigens may be divided into two groups: particulate complex multiantigens and single antigens. The group of particulate complex antigens includes intact microorganisms like bacteria, viruses, parasites, protozoans, and mammalian cells, but also artificial particles (e.g., poly(lactic acid) derivatives). Antigen presentation of particles generally follows the class II MHC pathway in specialized phagocytic cells. Polyclonal responses are directed to all constituents. Antigens which consist of one molecular species, which include proteins, peptides, polysaccharides, glycolipids, oligo-saccharides, haptens, and nucleic acids, evoke polyclonal responses to the various intra-anti-gen epitopes present. Immunodominance of specific determinants may lead to overrepre-sentation of specific responses.

Responses to dead microorganisms follow in general the profile for particulate antigens. For live microorganisms, antigen presentation is dependent on the character of the organism. Invasive microorganisms may become intra-cellular parasites to which in general DTH responses are directed (class II). More recently the involvement of γ/δ CD8 cells together with α/β CD4 Th1 and Th2 cells has been sub-stantiated (Kaufmann 1995). Viruses which infect cells are processed as class I MHC-restricted antigens.

Isotypes of antibody responses to micro-organisms are dependent on the structure of surface molecules and other constituents; e.g. for bacterium *Brucella abortus*, IgG2a, IgG3 (Snapper and Mond 1993). For nematode para-sites, IgG1- and IgE-restricted responses were observed (Snapper and Mond 1993).

When antigens are presented to the mucosal immune system, the uptake of soluble antigens takes place mainly via the endothelial cells of the mucosa, which then produce cytokines to enhance the specific immune response. Partic-ulate antigens are generally taken up by the microfold cells (M-cells) of the Peyer's patches and are presented there to the resting naive cells. Toxins and microorganisms may have specific receptors on cells along the mucosal lining. These mostly pathogenic agents encounter the immune system each in their specific way.

Antigen construction and preparation

The specificity of the immune response obtained (antiserum, T cell response) is dependent on the purity of the antigen applied. Minute impurities (<1%) may prove to be immunodominant, as is the case with many bacterial antigens. This is a problem encoun-tered with any antigen produced via recombi-nant DNA techniques and not marked with special tags for identification. In that case the selection of specific monoclonal antibodies may lead to specific reagents. Alternatively, the antigen may be synthesized. This, however, is only sensible for small peptides (<50 amino acids) and other low-molecular-weight substances.

Synthetic antigenic determinants have a wide variety of applications. The actual role of a determinant can best be judged from the results of immunizations, e.g., the antibody responses in sera, the recognition of antigenic determinants by antibodies in various immu-noassays, each with its typical microenvir-onmental conditions, and the specificity of the MAbs selected. Certainly for short peptides and haptens, polyclonal antibodies tend to be determinant-specific (monospecific) since they recognize one or a few more extensively over-lapping determinants in a short sequence. Such Pabs may differ slightly in epitope speci-ficity and thus may show an in-assay behavior similar to that of MAbs.

Haptens in general are small, rigid molecules which are recognized independently of the conditions of assay. In general, short peptides tend to elicit antibodies which recognize the denatured form of the protein only (Boersma et al 1988). Application of short peptides, there-fore, may be of advantage for the elicitation of antibodies to be used in assays where the antigen is encountered in denatured form, e.g., in SDS-PAGE, fixed tissue specimens (Van Denderen et al 1989, 1990). In contrast, Dyson et al (1985) showed that an immunogenic nonapeptide which showed a preferential conformation in aqueous solutions elicited anti-native antibodies to hemagglutinin of influenza virus. Ultrastructural analysis showed that longer peptides indeed tend to mimic the native structure of a protein much better since these peptides maintain a specific space-filling structure (Horiuchi et al 1987).

An important clue to the successful use of haptens and peptides (SP) as immunogens is the mode of presentation of the selected antigens to the immune system. Using these small antigens, the construct applied is usually an assembly consisting of a carrier protein, a hapten or peptide, and bridging coupling reagents (for review see Boersma et al 1993).

Antigen carrier

Antigens which in themselves do not contain elements for T cell activation (small peptides or polysaccharides) need a protein-carrier which provides these elements (T cell epitopes).

The properties of a carrier protein determine to a large extent the outcome of immunizations with the conjugated antigenic determinant, e.g., haptens and small peptides. Larger proteins (>60 kDa) are preferred as carriers because they contain a sufficient number of $-NH_2$, $-SH$, arginine, etc., groups for coupling to generate satisfactory determinant/protein ratios. In principle most proteins will suffice as carriers if derived from a non-self source. However, the more genetically distant a protein is from the animal to be used for immunization experiments, the better the chances for potent immunogenicity. Keyhole limpet hemocyanin (KLH), a large aggregated protein from a gastropod, is therefore often used as a carrier protein in mice, rabbits, goats, and sheep. Other widely used carriers are of bacterial origin: tetanus toxoid (TT), purified protein derivative of tuberculin (PPD), and diphtheria toxoid. Less efficient in mammals are ovalbumin (OVA) and bovine serum albumin (BSA) (Geerligs et al 1989), probably because of tolerance to these highly conserved proteins.

Chicken gamma-globulin (CGG) is rather immunogenic in mice.

However, coupling of relatively large numbers of determinants, depending on the protein, can have a negative influence on the properties of a conjugate. Overloading may lead to precipitation and decreased immunogenicity (Peeters et al 1989). Also, cross-linking agents such as diazo compounds and glutaraldehyde may drastically decrease the solubility and immunogenicity of a carrier–determinant conjugate.

For nonimmunogenic peptides, (e.g., B cell epitopes only) and haptens, coupling to a carrier protein to provide T cell help is required. This is not a matter of molecular mass since immunogenicity of peptides is in general not increased by homopolymerization with, for example, glutaraldehyde (Boersma, unpublished results). Polymerization using carbodiimides led to encouraging results, though in part based on the coupling of T-determinants (Borras-Cuesta et al 1988). Straightforward covalent elongation with a selected T cell determinant is an alternative option (Hackett et al 1985; Francis et al 1987; Zegers et al 1993).

A method has been developed for the synthesis of branched peptides or multiantigen peptides (MAP) (Tam and Zavala 1989). This multiple-antigen peptide method was developed for sensitive detection of anti-peptide antibodies, but in addition the multimeric peptides show enhanced immunogenicity compared to free peptides, peptide conjugates, or peptides still attached to their solid support (McLean et al 1991). MAP which include a specific T cell epitope may function as efficient immunogens.

Admixture of T cell epitopes and B cell epitopes under specific conditions may be sufficient for T cell and B cell activation. Such conditions can be found in water-in-oil emulsions and in liposomes. There the orientation of epitopes, dependent on their amphipathic character, may be such that they are presented so that T–B cell interaction is possible (Sarobe et al 1991; Partidos et al 1992; Prieto et al 1995).

Immunization procedures

Besides the immunogenicity of the antigen, a large number of parameters influence the result of each immunization: (1) the site at which the antigen is introduced; (2) the dose of antigen; (3) the scheduling of immunizations over time; (4) the type of adjuvant applied; and (5) the genetic make-up of the recipient (discussed above). These parameters determine which antigen-presenting cells the antigen meets first, how many and which cells will be involved in the response, as well as the quantity of the response (Tables 13.3.2 and 13.3.3).

The location at which the antigen is deposited in part determines the lymphoid organs activated and the isotype of the antibody response. For detailed description based on histocytochemical analyses, see Laman and Claassen (1995). Routine immunizations, intra-peritoneal (in mice) and subcutaneous (in rabbits and mice), rapidly deposit antigen mainly in spleen and peripheral lymph nodes. Soluble antigens injected intravenously (in the tail vein or retro-orbitally) also rapidly end up in the spleen.

Oral application in general leads to induction of tolerance. For effective immune responses (IgA) antigen presentation requires a vehicle which may either be a microorganism, a particle of specific size (1–10 μm), or a toxin. Application of antigen via the lungs (intratracheally) will in general follow the same rules as for oral application. Nasal application may efficiently lead to IgA induction.

Deposition of antigen at subcutaneous, intracutaneous, or intramuscular sites will in general lead to slow release of the antigen into the

Table 13.3.3 Immunomodulation by adjuvants

Function	Effect on antigen presentation	Adjuvant	Support MHC I presentation	Support MHC II presentation	Cytokines	Isotypes (mice)
Vehicle	Localization, enhanced	Inert beads	–	+		IgG1/IgE
		Mineral salts[a]	–	+, poor DTH	IL-3, IL-4, IL-5 II-6[i]	IgG1/IgE
	uptake	Oil emulsions[b]	–	+, DTH	IL-4	IgG1/IgE
Depot	Slow release	Mineral salts	–	+, poor DTH	IL-3, IL-4, IL-5, IL-6	IgG1/IgE
		Oil emulsions	–	+, DTH		IgG1/IgE
Micelle induction	Enhanced uptake	Liposomes[c]	+	+		IgG1/IgG2a,2b
		ISCOMS[d]	+	+		IgG1/IgG2a,2b
Macrophage activation	Enhanced uptake	NBP[e]	–	+, DTH		IgG2a,2b
		DDA[f]	–	+, DTH		IgG2a,2b
		Microbial products[g] Oil emulsions	–	+, DTH	IFN-γ, IL-4	IgG2a,2b IgG1/IgE
Direct B-cell activation		Microbial products	–	+, DTH	IFNγ, IL-4	IgG2a,2b
Growth/ differentiation of B cells		Cytokines[b] IL-1 IL-2 IFN-γ		+, DTH	IL-2 IFN-γ	IgM IgG2a

[a]Nicklas (1992); [b]Boersma et al (1992); [c]Buiting et al (1992); [d]Claassen and Osterhaus (1992); [e]Hilgers and Snippe (1992); [f]Verheul and Snippe (1992); [g]Warren et al (1986); [h]Heath and Playfair (1992); [i]Valensi et al (1994).

immune system. This slow release generally leads to continuous stimulation and strong immune responses. Leenaars et al (1994) studied the effect of subcutaneous injection of a volume of FCA/antigen (0.5 ml/rabbit) in one injection site compared to the same volume spread over 2 or 4 injection sites. Similar antibody responses were observed, for both immunization procedures, suggesting that dividing antigen over more sites does not improve the antibody response.

How the route of antigen administration controls the nature and intensity of the response is not known. Antigen-presenting cells (APC), different lymphoid tissues and the special characteristics of the regional immune system play an important role in the induction of the response. De Becker et al (1994) showed that the isotype and the amplitude of the B cell response can be regulated by the nature of the APC. The antigen presented by dendritic cells induces the production of Th1-dependent isotypes (IgG2a), whereas an antigen pulsed on peritoneal macrophages seems to induce a Th2-associated response (IgG1, IgE). The route of injection determines which APC comes into contact with the antigen. The route of injection is limited by the form of the antigen. Insoluble antigens cannot be immunized intravenously. In order to waste little of the antigen in the periphery, the antigen can be directly applied locally in the spleen either free or attached to a vehicle (nitrocellulose, Sepharose beads). Owing to reduction of antigen loss in the circulation, doses 50–100× lower than normal could be used on intrasplenic immunization (Spitz 1986; Hong et al 1989).

Certain immunogens, when injected parenterally, lead to the production of circulating antibody; when given intradermally, the same immunogen in relatively low doses may provoke delayed-type hypersensitivity in the absence of circulating antibody.

Peptides which appear to be low in immunogenicity upon routine intraperitoneal or intravenous immunization may be immunized locally to sensitize peripheral lymph node cells. In the mouse, immunization in the footpad or subcutaneously in the back may very efficiently activate regional lymph nodes when strong adjuvants are applied (Freund's).

Targeting of antigens to certain molecular structures on cells is a form of specific routing of antigen. For peptides, enhanced responsiveness has been shown to occur by targeting of these peptides to surface immunoglobulin or class I molecules (Casten 1988). For BSA, Lees et al (1990) showed that especially targeting with anti-IgD led to strong potentiation of antibody responses. Immunological targeting to, for example, MHC molecules acts as a replacement of adjuvant in that it targets the antigen and at the same time acts to activate the antigen-presenting cells. To this end, MHC-specific antibodies have been applied (Carayanniotis et al 1988; Carayanniotis and Barber 1990).

Dose of antigen

Immune responses are antigen dose-dependent. For most antigens an optimum is found in the dose–response relationship. A low dose induces the formation of small amounts of antibodies with relatively high affinity and specificity. Weak responses may be enhanced by coadministration of adjuvants (see below). The orders of optimal doses do not differ among various animals. For vaccination in humans, for most antigens 1–50 µg of antigen is administered. Similar optimal doses are found in cattle and small laboratory animals. For T-independent antigens, about 20 µg intravenously or intraperitoneally is sufficient in mice (Laman and Claassen 1995).

Studies of immunological tolerance illustrate that a very low or a very high dose of a foreign material can inhibit future responses to subsequent injection of an otherwise immunogenic dose.

Proteins with specific routes of entry into the immune system (e.g., toxins) may elicit immune responses in extremely low doses: 10^{-14} g for endotoxin in rodents. In general, doses of a few µg are sufficient for an efficient priming reaction. This shows that low antigen doses may efficiently stimulate T cells and that immunological memory may be induced by doses that do not produce detectable antibody responses. DTH reactions to specific antigens are evoked at lower doses than are needed for antibody responses to the same antigen.

Functions of adjuvants

When antigens are poorly immunogenic, the immune system needs a stimulus to evoke a response. Adjuvants are used for this purpose. Adjuvants are simple or complex admixtures (natural or synthetic) of compounds which upon administration to individuals lead to an aspecific immune stimulation (Tables 13.3.3 and 13.3.4).

The route of administration determines the localization. In addition, adjuvants often also function as the vehicle which takes care of localization of the antigen. Specific localization (intraperitoneal, intra- or subcutaneous, intrasplenal) of the antigen can be obtained by administration of antigen in water-in-oil or oil-in-water emulsions, antigen bound to or enclosed in liposomes, antigen bound or adsorbed on to amorphous or crystalline material as in alum hydroxide–phosphate gels, or with nitrocellulose or solid beads of various compositions. These immunizations have in common that the administration of antigen results in local immune responses. The surface area of the vehicle to which the antigen is attached determines the load and density of the antigen during antigen presentation. Vehicles convert soluble antigens into particulate material which is more readily ingested by antigen-

Table 13.3.4 Mode of action of adjuvants

Category	Examples	Mode of action
Mineral salts	Al(OH)$_3$, AlPO$_4$[a]	Vehicle Depot effect
Oil emulsions	FIA, montanide, specol[b]	Vehicle Depot effect Activation of macrophages
Microbacterial products	LPS, MDP, MPL, TDM[c]	Direct stimulation of B cells (LPS) Activation of macrophages Stimulation of T cells Activation of complement Enhanced antigen uptake
Saponins	Quil-A[d]	Facilitate cell–cell interaction Aggregation of antigen
Synthetic products	DDA[e]	Activation of macrophages Activation of complement
	Iscoms[f]	Facilitate cell–cell interaction Stimulation of T cells Aggregation of antigen
	Liposomes[g]	Vehicle Enhanced antigen uptake
	Non-ionic block polymer (NBP)[h]	Activation of macrophages Activation of complement
Cytokines	IL-2, IL-1-β, IFN-γ [i]	Growth and differentiation T and B cells Enhanced antigen presentation

[a]Nicklas (1992); [b]Boersma et al (1992); [c]Warren et al (1986); [d]Campbell and Peerbaye (1992); [e]Hilgers and Snippe (1992); [f]Claassen and Osterhaus (1992); [g]Buiting et al (1992); [h]Verheul and Snippe (1992); [i]Heath and Playfair (1992).

presenting cells such as macrophages. Some vehicles such as alum and oil emulsions have immune-stimulating (adjuvant) properties, while others, such as Sepharose beads and nitro-cellulose are relatively inert in an immunological sense.

Various adjuvants also offer a depot function. When deposited at a given site the antigen leaks slowly from the adjuvant compound and gradually becomes available for stimulation of the immune system.

Adjuvants are foreign to the body and therefore depending on the nature of the constituents, may in themselves be antigenic. If appropriately administered they lead to sterile inflammations that attract the various cells of the aspecific defense system — macrophages — which in turn produce *immunomodulating* factors that increase the level of immune surveillance. Macrophages as well as other antigen-presenting cells therefore play an important role in the generation of the adjuvant effect. Most adjuvants directly or indirectly stimulate the generation of interleukin-1 and also other factors that support growth and differentiation.

In summary, adjuvants contribute to improved antigen presentation and immuno-modulation; this is the true adjuvant function. The choice of the adjuvant is dependent on the type of antigen and on the desired immune response (cellular versus humoral). In vaccines adjuvants should preferentially enhance the immune response of all cell types involved in the process of generating protection. To this end, the stimulation of the antigen presentation, the T regulator cells (Th1 and Th2), the B cell and effector T cells is most important. For the generation of antibodies as immunore-agents, the requirements of immune stimulation are less stringent.

Properties of various adjuvants

The exact way in which adjuvants stimulate the immune system is not always known. The properties of various adjuvant products are summarized in Tables 13.3.3 and 13.3.4.

Mineral salts

Mineral salts consist of hydrated gels of aluminum hydroxide, aluminum phosphate, or calcium phosphate which form efficient vehicles to which proteins readily adsorb dependent on ionic strength and pH. The depot function is most prominent, whereas the immune stimulative capacity is weak. Mineral salts have the major drawback that they do not support cell-mediated immunity very well. The irritation caused suggests that these adjuvants function by local production of cytokines to activate antigen-presenting cells. Their relatively harmless character led to approval of these adjuvants in humans by the US Food and Drug Administration and the World Health Organization. Local (e.g., vaginal) application led to humoral response induction without harmful side-effects seen with other adjuvants. The effective depot function has led to combined use of mineral salts with other adjuvants such as microorganisms (*Bordetella pertussis*; Bomford 1980). Aluminum compounds may be suitable for antigens which are highly immuno-genic and available in large amounts, but they are not recommended when only weak immu-nogens (e.g., peptides, subunit vaccines) are used or when there is limited amount of antigen available. Compared with other adjuvants, mineral salts are rarely used for experimental immunization of laboratory animals.

Upon administration of aluminum salts in mice, Valensi et al (1994) detected IL-3, IL-4, IL-5, and IL-6 in serum 3–24 h after intramuscular injection, while IL-1α, IL-2, IFN-γ, and IL-10 were measured but not detectable in serum.

After primary immunization of mice with BSA in Al(OH)$_3$, Bomford (1980) found only antibodies of the IgG1 isotype, whereas during the secondary response an IgG1 and IgG2a response was stimulated by Al(OH)$_3$. This was confirmed by Byars et al (1991) who found, in contrast to several other adjuvants, that predominantly the IgG1 isotype and only little IgG2a and IgG2b was produced. Haaijman et al (1988) also found mainly IgG1 responses in BALB/c mice. It is suggested that this is of special importance when protective immunity is required, as antibodies of the IgG2a and IgG2b isotype act synergistically with complement and antibody-dependent effector cells. Aluminum adjuvants are very efficient in enhancing the synthesis of IgE antibodies in rabbits and rodents. $AlPO_4$ does not induce IgE antibodies (Allison and Byars, 1986).

Microorganisms and microbial products

Microorganisms and microbial products can exhibit strong adjuvant activity. In general the bacteria are applied in heat-killed form and coadministered with the immunogen under study. Mycobacteria cause severe side-effects (e.g., sterile inflammations and granulomas; Claassen et al 1992). Mycobacterial cell wall products have been studied to obtain fractions with similar immune-stimulating properties but without severe side-effects. Trehalose dimycolate (TDM, cord factor) was identified and found to have immunostimulating properties. The smallest subunit of the mycobacterial cell wall that still results in adjuvant activity is N-acetylmuramyl-L-alanyl-D-isoglutamine, abbreviated as muramyl dipeptide (MDP) (Azuma et al 1976).

Lipopolysaccharide (LPS) is the main surface structure of Gram-negative bacteria. The adjuvant activity of LPS was first described in 1956 by Johnson et al (1956). Its high toxicity precludes its clinical use. Low-toxicity analogues of LPS have been developed which retain their immunostimulatory properties. The lipid A moiety has received most attention in this respect, it can be prepared by chemical treatment of natural lipid A (Alving and Richardson 1984) or by chemical synthesis of lipid A (Yasuda et al 1982). Monophosphoryl lipid A (MLA) is an example of a lipid A derivative of LPS which retains all the immunostimulating properties of LPS (Gustafson and Rhodes 1992).

Microbial products may lead to mitogenic activation of B cells (LPS), and activation of macrophages, T cells and the complement system (reviewed by Warren et al 1986). Lipid A, as well as lipopolysaccharide, is hydro-phobic, while lipid A is also amphiphilic. Administration of this adjuvant separately or together with antigen leads to generation of cytokines. Most prominent is the production of IFN-γ by natural killer (NK) cells. This cytokine in turn activates antigen-presenting cells such as macrophages. Activated macrophages generate tumor necrosis factor-α. The activation of antigen presentation and production of cytokines following the administration of this adjuvant switches the humoral immune response in the mouse to IgG2a (Takayama et al 1991). In the promotion of protective immunity against bacteria and viruses, antibodies of the IgG2a isotype are considered to be superior owing to their complement fixation (Gustafson and Rhodes 1992). Only few nonpathogens can serve as adjuvant at the mucosal surfaces. Live *Lactobacillus* of various strains (*casei, plantarum*), can serve as adjuvants after intraperitoneal but also after oral administration (Boersma et al 1994). Their efficacy using peptide antigens is relatively low. Peptides need to be coupled or produced at the surface by *Lactobacillus* to generate acceptable levels of immune responsiveness (Claassen et al 1994, 1995; Boersma et al 1994).

Combination of lipid A or lipopolysaccharide with nonionic block polymers or alum leads to a shift to IgG1 antibodies. The same effect can be obtained with recombinant IFN-γ, which is much more expensive than LPS or lipid A and in addition cannot be applied to the sites where the local immune response is generated. LPS also stimulates cell-mediated immunity as measured by DTH.

Preparations consisting of a mixture of components from a diversity of microorganisms have been shown to influence severely the traffic of lymphocytes within the mucosal immune system, thereby enabling immunization on site to support responses in other mucosal locations also (Ruedl et al 1993).

The adjuvanticity of microbial products can be enhanced by mixing them with depot preparations of the immunogen (e.g., mineral salts, oil emulsions). The best-known example is Freund's complete adjuvant (FCA). The side-effects of FCA led to its being banned in various countries (see Claassen and Boersma 1992; Claassen et al 1992).

Toxins have been widely applied as immunological carriers with a high immune-stimulating potential. Heat-labile toxin of *E. coli*, tetanus toxoid as well as cholera toxin are the best-known examples (Vadjy and Lycke 1992). Simple coupling of peptides to these protein carriers leads to efficient immune response generation (Arnon 1991). Cholera toxin (CTX) is the most-investigated for its mucosal immunity-stimulating character.

Cholera toxin was shown to enhance the IL-1 production by macrophages. However, it inhibits T cell IL-2 production but enhances proliferation. *In vitro* enhanced IgA synthesis in lipopolysaccharide-stimulated B cells was demonstrated (Lycke et al 1989).

Other microbial products which have been tested as adjuvants include water-soluble adjuvants from *Nocardia* species, yeast glucans, various peptidoglycans, and certain bacterial exotoxins (Munoz 1964; White 1976).

Water-in-oil emulsions

Water-in-oil, reversed, and double emulsions with vaccines have been in use for a considerable time. Freund (1956) applied refined paraffin oil (Freund's incomplete adjuvant, FIA) with vaccines and since then this oil has been used as an adjuvant mainly for experimental purposes. Oil emulsions have a depot function, they can stimulate macrophages, and they promote uptake of antigen (Waksman 1979). The depot function of oil emulsions is enhanced by intraperitoneal, subcutaneous, intradermal, or intramuscular administration. The most potent adjuvant known for stimulating both humoral and cellular immunity is a combination of an emulsion of mineral oil and killed mycobacteria (FCA). Adjuvants like FCA stimulate the immune response very effectively but have dramatic side-effects. As a result not all adjuvants are permitted for human use or for application in veterinary medicine. In addition, in many countries laws on animal experimentation have limited the use of the most potent adjuvants for ethical reasons.

From the results of cytokine production, no skewing in either Th1 or Th2 response direction can be determined for this type of adjuvant. The mode of action relies on the enhancement of hydrophobicity of antigens, targeting to antigen-presenting cells, and on the presentation of antigens on a large surface area. The oil microspheres carry antigen to the surface of antigen-presenting cells. Endocytosis and presentation in the context of MHC class II is therefore expected (Boersma et al 1992).

The antigen to which an immune response has to be elicited is generally taken up into the hydrophilic phase (Bokhout et al 1981; Woodard and Jasman 1985; McKercher 1986; Woodard 1990).

Bokhout et al (1981) first described specol, a selected water-in-oil emulsion (oil-based adjuvant); its application has been evaluated in the prevention of future diseases by vaccination as well as in the enhancement of natural *in vivo* immune responses. This adjuvant was developed from their observation that vaccinated pigs were protected against pathogens not present in the vaccine. Specol certainly has advantageous properties with respect to animal care, ease and reproducibility of preparation and administration, and the use of constituents which generally are regarded as safe for use in animals.

Although proliferative T cell responses and DTH reactions could be demonstrated with use of specol, formal proof of the support of a cytolytic T cell response has yet to be produced. In contrast to most adjuvant admixtures, specol can be prepared easily and highly reproducibly since all constituents can be obtained from the suppliers in quality-controlled batches. The low viscosity of specol compared to many blends (FIA) facilitates administration by injection (Boersma et al 1992).

Water-in-oil emulsions activate macrophages, suggesting that the cytokine involved is most likely IL-1 (for review see Boersma et al 1992). Recent experiments suggest induction of IL-2, IL-4, and IFN-γ together with expression of the T cell activation marker gp39 (Zegers et al unpublished results). Most water-in-oil emulsions evoke the production of IgG1 responses (Allison and Byars 1991). With specol in BALB/c mice, IgG1 responses were frequently observed. In the mouse, IL-4 stimulates the generation of IgG1 and IgE while IFN-γ is involved in the IgG2a responses (Finkelman et al 1988a,b).

Liposomes

Liposome adjuvanticity is influenced by the charge, composition and the method of preparation. *In vivo*, liposomes act by facilitation of antigen uptake, ingestion, fragmentation, processing, and presentation, mainly by macrophages (for review see Claassen 1991). The antigen is encapsulated in the water or lipid phase or can be coupled to the surface (for review see Buiting et al 1992). Van Rooijen and Van Nieuwmegen (1980) showed that surface exposition of antigens is most efficient in antigen presentation. For peptides, elongation with a hydrophobic tail (lipopeptides) enhances the uptake and correct surface orientation of the determinants of interest in a given antigen. As such, a liposome can be envisaged as a hapten-vehicle with a large surface. The repetitive presence of the determinant may lead to B cell activation (IgM responses). T cell dependent responses are found only when a T cell determinant is included in the antigen. The fusion of liposomal and endosomal membranes delivers the antigen essentially in the MHC class II processing pathway (Germain 1986; Bevan 1987). Liposomes seem to support primarily CD4-dependent T cell responses (DTH, B cell help). Recently, activation of CD8 cells in cytolysis (class I dependent) after application of antigen with pH-sensitive liposomes has been documented. These liposomes introduce antigen in the cytoplasmic phase of antigen processing. However, questions with respect to the quantitative contribution of this route of processing based on liposomal delivery of antigen remain to be clarified.

The potency of liposomes can be further increased by the inclusion of other immune-stimulating agents such as bacterial lipopolysaccharide, lipid A, MDP, and saponins. Targeting of liposomes using antibodies anchored in the membrane is controversial (Buiting et al 1992). Liposome-associated proteins elicited IgG2a and IgG2b antibodies in the mouse (Phillips and Emili 1992). Whereas MDP preferentially stimulated IgG1 in their experiments, with a combination of liposomes and MDP a shift to IgG2 was observed. After oral application of pH-stable liposomes, Jackson et al (1990) observed secretory humoral immunity (IgA). Humoral responses elicited with antigens presented in liposomes in general show response profiles similar to responses elicited with water-in-oil emulsions.

Saponins

Saponins are plant-derived, rather complex, glycoside-containing compounds which are amphiphilic in nature and tend to form micelles. The best-known example of a saponin adjuvant is derived from the bark of *Quillaja saponaria* Molina. Quillaja saponin is a mixture of potent adjuvants. Crude saponin extracts are unreliable. Quil-A, one of the purified derivatives, is used as the basis for the formation of immune-stimulating complexes (ISCOMs) (Dalsgaard 1978; Morein 1990). Together with cholesterol (mostly also phospholipids) Quil-A tends to form particulate matrix structures with high adjuvanticity. Through hydrophobic interaction, proteins can strongly complex with these structures to form ISCOM (Lövgren and Morein 1988).

Saponins and ISCOM promote a strong immunogenic enhancement most probably by direct interaction with cellular membranes. The role of macrophages in the effectiveness of ISCOM has been described by Claassen et al (1995). Apart from humoral responses, saponins also seem to support CD8 (class-I restricted) mediated cellular cytotoxicity (Newman et al 1992; Campbell and Peerbaye 1992). CD8 T cell effected cytolysis has been demonstrated which might be dependent on IFN-γ production. The generation of IgG2a humoral responses in mice is indicative of this mode of action. It is not known how the induction of IFN-γ is established. Most saponins are toxic, and local, depot-like, administration may lead to necrosis. Oral immunizations of animals with ISCOM have been only partly successful. Relatively high doses of antigen (100 μg) were needed, but class I-restricted responses were also demonstrated (with the antigen ovalbumin). The support of mucosal immune responses has been documented upon intranasal application of antigens/vaccines (Jones et al 1988). Intravaginal responses have also been demonstrated but these required relatively high amounts of antigen (Thapar et al 1991). For application via the oral route, the induction of IFN-γ is thought to be most important. This is in agreement with the antibody isotypes observed (Finkelman et al 1988a). In mice, saponins stimulate production of IgG1, IgG2a, and IgG2b. The major part of the response is formed by IgG2a antibodies. IgE was not observed (Kensil et al 1991). Quil-A elicited mainly IgG1 in mice (Kenney et al 1989), whereas IgE responses have also been reported (Allison and Byars 1992). For those responses where elicitation of IgE has to be avoided (vaccination), the choice of adjuvant should be made carefully.

With MHC class II targeting, enhanced responsiveness in humoral immunity was mainly restricted to IgG1-restricted isotype (70%), whereas about 25% was of the IgG2a isotype in mouse (Skea and Barber 1993).

Synthetic adjuvants

Synthetic adjuvants are rather heterogeneous. The mode of action of *nonionic block* polymers (NBP) is not fully understood but the activation of the complement system as well as the activation of macrophages seems to play an important role. NBP have a low intrinsic toxicity. They are built from repetitive combinations of poly(oxyethylenes) and poly(oxypropylenes). NBP usually have amphipathic character. They stabilize water-in-oil emulsions and thereby may stabilize the tertiary structure of antigens as well. In contrast to most adjuvants studied, NBP have a strong adjuvant effect for poly- and oligosaccharides in particular, whereas for most proteins they are not as potent as many other adjuvants. The mode of action may be similar to that of lipid A since the predominant antibody response is IgG2a in the mouse, which is enhanced by IFN-γ production. The isotype distribution is influenced by the balance between the proportions of hydrophobic and hydrophilic moieties in the NBP

composition. CTL responses have not been documented, but it has been demonstrated that DTH responses are supported (for review see Verheul and Snippe 1992). Owing to their partly hydrophilic and partly hydrophobic nature, NBP are readily usable with liposomes and water-in-oil emulsions. With SAF, the combination adjuvant which includes a muramyl dipeptide analogue, a nonionic block polymer in a water-in-oil emulsion (Byars et al 1990), a skewing of the response to IgG2a antibodies was demonstrated in mice which was attributed to the induction of IFN-γ production (Th1 responses) (Kenney et al 1989; Byars 1991; Byars et al 1990).

Dimethyldioctadecylammonium bromide (DDA) is a representative of a specific category of adjuvants, the quaternary amines. The lipophilic character of DDA, which forms liposome-like structures in aqueous solutions, might be responsible for its capacity to enhance T cell responses. However, most responses documented are DTH responses, which is a CD4 T cell function, and CTL responses have not yet been described in detail. Adjuvant and antigen need to be administered together, following the same route, which suggests that in general binding of antigen and adjuvant is required to generate optimal responses. Binding of antigen to DDA serves the purpose of increasing the hydrophobicity of the antigen and neutralization of negatively charged moieties, but DDA is not an optimal adjuvant for eliciting antibody responses. For sufficient antibody response combination with alum or water-in-oil emulsions, for example, is required. The mild side-effects of DDA and some other quaternary amines make them promising adjuvants for veterinary application. For application in man, new less irritative analogues need to be developed (for review see Hilgers and Snippe 1992).

Lipopeptides are nontoxic and nonimmunogenic analogues of bacterial cell walls of Gram-negative bacteria (Wiesmuller et al 1983). They present a novel intrinsic adjuvant model for peptide immunization. Lipopeptides can not only serve efficiently to immunize peptides, in addition they can serve as a carrier with adjuvanticity (for review see Bessier and Jung

1992). They are amphipathic in nature, which seems to be a favorable property for adjuvancy. Lipopeptides coupled to low-molecular-weight haptens are able to elicit high antigen-specific antibody responses in mice and rabbits. *In vivo* as well as *in vitro* their ability to generate CTL responses even to single CTL epitopes has been demonstrated (Martinon et al 1992; Nair et al 1992). Conjugation as well as simultaneous and separate administration of adjuvant and antigen elicits immune responses. A drawback is that relatively high doses of antigen (10 times the dose used for regular immunization to generate antibody formation with adjuvant) are needed in most cases, but high antibody titers have been obtained. In a number of cases, valuable characteristics of lipopeptides were demonstrated. The toxicity of toxins decreased after conjugation without loss of immunogenicity. Other non- or poorly immunogenic substances showed enhanced immunogenicity (Bessler and Jung 1992). Lipopeptides are very easily applied and are commercially available. Recently the use of virus particles (HbsAg) as a carrier for peptides with a hydrophobic tail that dissolves in the outer membrane has been proposed as an alternative for both carrier and adjuvant (Neurath et al 1989).

As has been summarized above, most adjuvants which support cell-mediated immune responses (CMI), especially CTL responses, require an increase in the hydrophobicity of the antigen. It remains unclear what the mode of action might be. The chemical nature of most antigens which support CTL responses (micelles, specific liposomes, and ISCOM) suggests that binding or association with membranes of antigen-presenting cells is facilitated. With IFN-γ the conjugation of antigen enhanced the immune response (Heath and Playfair 1990).

Natural adjuvants

Cytokines are pleiotropic immunomodulatory agents which together form a partly degenerate system. The mode of action of various adjuvants has been unraveled to the level of immunomodulatory substances, among which are cytokines (Table 13.3.3). Cytokines mutually influence each other's function. Since cytokines are the primary immune response regulators, the analysis of adjuvant effects apart from antibody production, DTH responses, and cellular cytolysis is based on demonstration of cytokine production. Cytokines may mimic the actions that are generated with adjuvants. As a consequence, immunomodulation is aimed at the mimicking of adjuvant effects by administration of the cytokines themselves. These body-derived or physiological adjuvants can be administered or produced locally as a secondary effect of an immunization procedure using various other adjuvants (microorganisms, microbial membranes). Some cytokines have been evaluated for their adjuvant activity: interleukins-1 and -2, IFN-γ. These cytokines influence growth and differentiation of T and B cells as well as antigen presentation by various cells. However, because the mode of action of cytokines is complex, cytokine subunits with specific response-enhancing modalities were developed (IL-1). Fragments of cytokines have been selected which induce immune responsiveness without expressing the adverse side-effects. The application of the active component together with antigen has led to promising results for thymus-dependent as well as for thymus-independent antigens. Various modes of administration have been investigated: experimentally, intraperitoneal administration is most effective (Staruch and Wood 1983; Boraschi et al 1988; Tagliabue and Boraschi 1993).

Scheduling of immunization

T cell proliferation and production of response modifiers (cytokines) precedes the development of T-dependent antibody responses (class switch). For T cell responses *in vitro* in general the cells are isolated from lymphoid organs between 5 and 10 days after *in vivo* priming.

For optimal B cell responses it is helpful to allow resting periods of more than 4 weeks between the first and subsequent immunizations. The effective response is probably based on synchronization of resting B memory cells. Whether the decrease of primary antibody titers is a prerequisite for the second immunization to be given is a matter of debate. On one hand, the circulation of specific antibodies may lead to immune complex formation which may enhance antigen presentation; on the other, the reduction in antigen load by circulating antibodies, dependent on the ratio of antigen over antibody, determines whether immune complexes may lead either to stimulation (antigen excess) or to possible inhibition of responsiveness (antibody excess). For hybridoma production, a prefusion booster immunization was given after a long interval (a few months). The frequency of specific antibody-forming clones exceeded the level expected on the basis of the previous serum antibody responses. Repeated immunization with relatively low doses of antigen leads to stimulation and selection of clones that produce relatively high-affinity antibodies. This affinity maturation is used in the generation of strong polyclonal antisera.

Immunizations with peptide–carrier conjugates are given at 4-week intervals. In general the maximum titers are obtained after two to three subsequent immunizations. The best results following booster immunizations are obtained when priming titers have decreased. However, using KLH, a large aggregated protein which itself has adjuvant properties, anti-carrier titers will remain high for months. After priming with KLH as a carrier protein in mice, serum antibodies (IgG + IgM) increased gradually up to at least 21 days. Nevertheless, normal 4-week immunization schedules can be completed.

The timing of the immunization preceding spleen cell fusion depends on the routing: soluble antigens (conjugates) can be introduced intravenously 3 days before fusion. Less-soluble antigens are given intraperitoneally with FIA, as a water-in-oil emulsion or precipitated on aluminum hydroxide gel 4 days before an intended spleen cell fusion. Although we prefer spleen cells for fusion, for fusion of lymph nodes regional nodes are activated by immunization with adjuvant (as for intraperitoneal immunization). Mirza et al (1987) performed fusion of lymph node cells at day 14 after primary immunization for the production of anti-insulin MAbs.

ISCOM (immune-stimulating complexes) are generally applied intramuscularly or subcutaneously. In contrast to many other adjuvants, the administration of ISCOM-related antigen along these routes is most effective when the interval between two administrations is 6–8 weeks. Shorter intervals give low antibody titers (Lövgren et al 1990).

The timing of immunizations and responses is dependent on the mode of antigen presentation; the antigen preparation and formulation is important in this. Slow release of antigen from emulsions or crystalline deposits may introduce a delay.

Side-effects and safety of immunization procedures

In a Dutch Code of Practice (VHI 1993) the side-effects of immunization are scaled in arbitrary units of injury from I (low) to III (severe). Intravenous antigen application (when possible) scores I; intraperitoneal and subcutaneous injury score II. Intradermal and intramuscular application score injury level II, whereas intrasplenal immunization and immunization in the foot are level III.

The degree of injury is determined not only by routing but also by the nature of the adjuvant. Freund's complete adjuvant, liposomes, and ISCOM by various routes (e.g., intramuscularly) cause a high degree of injury and therefore are either better not used (FCA) or applied via routes that cause only a low degree of harm.

Snippe et al (1992) distinguish two categories of adjuvant applications: (i) includes adjuvant applications for which the efficacy aspect is more important than the safety/toxicology aspect and category (ii) applications where safety is more important than efficacy.

Group (i) consists of applications of adjuvants in experimental and food animals. All adjuvants seem to be applied in the former category and only four — water-in-oil and oil-in-water emulsions as well as aluminum salts and liposomes — in the latter category. For food animals, only FCA is considered not-allowed, whereas some doubts are expressed with respect to application of most other adjuvants. In this category the application of an adjuvant may be considered based on the presently available safety records of adjuvants. However, sufficient alternatives are available to eliminate FCA from animal experimentation in most fields of biomedical research.

In group (ii) applications of adjuvants in companion animals and humans are distinguished. For humans, only aluminum salts are allowed. Most other adjuvants are not considered for application except for specific purposes. Water-in-oil and oil-in-water as well as FCA are considered not to be applicable in humans at all. For companion animals, ISCOM are thought to be sufficiently safe enough for application. Water-in-oil emulsions as well as FCA are considered not to be applicable in companion animals. All other adjuvants are worth considering for application on the basis of the presently available safety records.

For application in humans, food animals, and pet animals the development of safe adjuvant formulations is still of great importance. The efficacy of present adjuvants seems in most cases to be in opposition to safety. For many vaccines an appropriate level of adjuvanticity is required to reach a sufficient level of vaccine efficacy. The ability to provide formulations which combine efficacy with safety is a challenge (see also Gupta et al 1993).

Conclusions and suggestions

Immunization, which aims at optimal antigen presentation, is a complex multivariate issue. The properties of the antigen, of the route of introduction, of the formulation, and of the recipient all contribute to the final result. Therefore, finding the optimum immunization result is a matter of experimentation and experience.

The tables included in this chapter may provide a guide to the choices to be made when specific immune responses to antigens have to be elicited.

The purity of an antigen determines the specificity of the responses. Subunits require better adjuvants than intact complex antigen. T cell determinants in general are MHC class-restricted; however, promiscuous epitopes are frequently encountered. Use of the latter epitopes with haptens produces efficient immunogens. Single epitopes (haptens) conjugated to an efficient carrier form efficient immunogens. The use of a 'foreign' protein as a carrier (KLH in mammals) is more efficient than a related carrier protein (BSA in mammals) (Boersma et al 1993). See also Table 13.3.1.

When generation of antibody for use in detection is the main aim of immunization, there is a rather broad choice of routes of introduction depending on the solubility of the antigen. Care should be taken to immunize more than once to obtain sera with relatively low titers of IgM antibodies, which in many immunoassays lead to background and cross-reactivity. Water-in-oil emulsions are simple to use and need no complex laboratory equipment. Subcutaneous or intraperitoneal application is easy to perform with a low injury index.

For production of monoclonal antibodies, cell fusion is best performed after a booster immunization (intravenously if possible) without adjuvant so as to avoid depot effects, and if necessary with alum-antigen precipitates as a vehicle (intraperitoneally) (Table 13.3.4).

When expression of specific isotypes is needed (complement fixation), the response can be skewed in that direction by the choice of adjuvant (lipopolysaccharides, liposomes, ISCOM, NBP) (Tables 13.3.2 and 13.3.3).

For DTH reactions the choices of adjuvants for specific antigens depend on the antigen properties. Mineral salts are not efficient. Water-in-oil emulsions, liposomes, various synthetic antigens, and ISCOM are very efficient (Tables 13.3.2– 13.3.4).

When cell mediated immunity on the level of class I-restricted responses is required for protection, then the choice of adjuvants is highly restricted. In fact, only ISCOM and to a lesser extent liposomes seem to be very efficient in elicitation of class I-restricted responses. However, this type of construction is needed only when we use selected fragments of antigens (isolated viral proteins, cell surface molecules) for the induction. When whole live viruses or cells are applied for immunization, class I-restricted responses are easily induced.

We have given above an outline of choices to be made with many variables. Although most of the data are from mouse and man, in general the picture is similar in other animals. For these other animals, however, the distribution of isotypes in immunoglobulins and the balance between Th1 and Th2 may be somewhat different. We hope that our guidelines will lead to successful immunizations.

New and exiting developments are moving rapidly from preliminary evidence to clinical application. Production and purification are difficult for many antigens. Recombinant DNA expression systems have therefore been developed for many antigens. An alternative route of influence over antigen presentation is the direct introduction of naked DNA (plasmids, vectors) into cells which bring the introduced sequence to expression. The antigens have proper glycosylation. This technique leads to antigen presentation via class I and possibly class II MHC molecules such that an immune response results. Not all easily accessible cells (skin, muscle) may be good antigen-presenting cells. However, when vectors are developed which, apart from sequences coding for antigens, in addition code for cytokines which induce the right molecular structures for antigen presentation, it seems that most problems can be resolved.

It is envisaged that this method of immunization will open new ways to treat malignancies (idiotypic vaccines) as well as chronic infections. The new 'DNA-vaccines' will be easily produced, purified, stored, and administered. The development of this alternative approach leads the immunization from expression products to the 'source' (genetic material). However, it will take some time before all regular vaccinations will follow this approach.

References

Allison AC, Byars NE (1986) An adjuvant formulation that selectively elicits the formation of antibodies of protective isotypes and cell mediated immunity. J Immunol Methods 95: 157–168.

Allison AC, Byars NE (1991) Immunological adjuvants: desirable properties and side-effects. Mol Immunol 28: 279–284.

Allison AC, Byars NE (1992) Immunological adjuvants and their mode of action. In: Ellis WR, ed. Vaccines: New Approaches to Immunological Problems. Butterworth–Heineman, Boston, pp. 431–449.

Alving CR, Richardson EC (1984) Mitogenic activities of lipid A and liposome associated lipid A: Effects of epitope density. Rev Infect Dis 6: 493–496.

Arnon R (1991) Synthetic peptides as the basis for vaccine design. Mol Immunol 28: 209.

Azuma I, et al (1976) Adjuvant activity of mycobacterial fractions: Immunological properties of synthetic n-acetylmuramyl dipeptide and related compounds. Infect Immunol 14: 18–27.

Bangham JA (1988) Data sieving hydrophobicity plots. Anal Biochem 174: 142–145.

Bessler WG, Jung G (1992) Synthetic lipopeptides as novel adjuvants. Res Immunol 143: 548–553.

Bevan MJ (1987) Antigen recognition. Class discrimination in the world of immunology. Nature 325: 192–194.

Boersma WJA, et al (1988) Antibodies to short synthetic peptides for specific recognition of partly denatured protein. Anal Chim Acta 213: 187–197.

Boersma WJA, et al (1992) Adjuvant properties of stable water-in-oil emulsions: evaluation of the experience with Specol. Res Immunol 143: 503–512.

Boersma WJA, et al (1993) Use of synthetic peptide determinants for the production of antibodies. In: Cuello AC, ed. Immunohistochemistry, 2nd ed. Wiley, New York, pp. 1–77.

Boersma WJA, et al (1994) Development of safe oral vaccines based on Lactobacillus as a vector with adjuvant activity. Proc ICHEM 2nd Int Congress, Biotech UK, Brighton.

Bokhout BA, et al (1981) A selected water-in-oil emulsion: composition and usefulness as an immunological adjuvant. Vet Immunol Immunopathol 2: 491–500.

Bomford R (1980) The comparative selectivity of adjuvants for humoral and cell mediated immunity. I. Effect on the antibody response to bovine serum albumin and sheep red blood cells of Freund's incomplete and complete adjuvants, Alhydrogel, Corynebacterium parvum, Bordetella pertussis, muramyl dipeptides and saponin. Clin Exp Immunol 39: 426–434.

Boraschi D, et al (1988) In vivo stimulation and restoration of the immune response by the non-inflammatory fragment 163–171 of human interleukin-1β. J Exp Med 169: 675–686.

Borras-Cuesta F, et al (1988) Enhancement of peptide immunogenicity by linear polymerization. Eur J Immunol 18: 199–202.

Buiting AMJ, et al (1992) Liposomes as antigen carriers and adjuvants in vivo. Res Immunol 143: 541–548.

Byars NE (1991) Improvement of hepatitis B vaccine by the use of a new adjuvant. Vaccine 9: 309–318.

Byars NE, et al (1990) Enhancement of antibody responses to influenza B virus haemagglutinin by use of a new adjuvant formulation. Vaccine 8: 49–56.

Campbell JB, Peerbaye YA (1992) Saponin. Res Immunol 143: 526–530.

Carayanniotis G, Barber BH (1990) Characterization of the adjuvant-free serological response to protein antigens coupled to antibodies specific for class II MHC determinants. Vaccine 8: 137–144.

Carayanniotis G, et al (1988) Delivery of synthetic peptides by anti class II monoclonal antibodies induces specific adjuvant free IgG response in vivo. Mol Immunol 25: 907–911.

Casten LA (1988) Enhanced T cell responses to antigenic peptides targeted to B cell surface immunoglobulin, Ia or class I molecules. J Exp Med 168: 171–181.

Chou PY, Fasman GD (1978) Prediction of the secondary structure of proteins from their amino acid sequences. Adv Enzymol 47: 46–148.

Claassen E (1991) The role of heterogeneous macrophage subsets in liposome targeting and immunomodulation. 41st Forum in Immunology. Res Immunol 143: 255–256.

Claassen E, Boersma WJA (1992) Characteristics and practical use of new generation adjuvants as an acceptable alternative for Freund's complete adjuvant. 44th Forum in Immunology. Res Immunol 143: 475–477.

Claassen E, et al (1992) Freund's complete adjuvant: an effective but disagreeable formula. Res Immunol 143: 478–483.

Claassen E, et al (1994) Use of Lactobacillus, a GRAS (generally recognized as safe) organism, as a base for a new generation of 'oral' live vaccines. In: Talwar GP and Kanury VSR, eds. Narosa Publishers, New Delhi, pp. 407–412.

Claassen E, et al (1995) New and safe 'oral' live vaccines based on *Lactobacillus*. In: Mestecky J, Russel MW, Jackson S, Michalek SM, Hogenova HT, Sterzl J, eds. Adv Exp Med Biol Advances in Mucosal Immunology, Part B. Plenum Press, New York, pp. 1553–1558.

Claassen I, Osterhaus A (1992) The ISCOM structure as an immune enhancing moiety: experience with viral systems. Res Immunol 143: 532–541.

Claassen IJTM, et al (1995) *In vivo* antigen detection after immunization with different presentation forms of rabies virus antigen: involvement of marginal metallophilic macrophages in the uptake of iscoms. Eur J Immunol 25: 1446–1452.

Dalsgaard K (1978) A study of the isolation and characterization of the saponin Quil-A. Acta Vet Scand 19: 1–40.

De Becker G, et al (1994) Immunoglobulin isotype regulation by antigen-presenting cells *in vivo*. Eur J Immunol 24: 1523–1528.

Dyson HJ, et al (1985) The immunodominant site of a synthetic immunogen has a conformational preference in water for a type II reverse turn. Nature 318: 480–483.

Eisenberg D, et al (1984a) Analysis of membrane and surface proteins sequences with a hydrophobic moment plot. J Mol Biol 179: 125–142.

Eisenberg D, et al (1984b) The hydrophobic moment detects periodicity in protein hydrophobicity. Proc Natl Acad Sci USA 81: 140–144.

Emini EA, et al (1985) Induction of hepatitis A virus neutralizing antibody by a virus specific synthetic peptide. J Virol 55: 836–839.

Fieser TM, et al (1987) Influence of protein flexibility and peptide conformation on reactivity of monoclonal antipeptide antibodies with a protein α-helix. Proc Natl Acad Sci USA 84: 8568–8572.

Finkelman FD, et al (1988a) Interferon-γ regulates the isotypes of immunoglobulin secreted during *in vivo* humoral immune responses. J Immunol 140: 1022–1027.

Finkelman FD, et al (1988b) IL-4 is required to generate and sustain *in vivo* IgE responses. J Immunol 141: 2335–2341.

Francis MJ, et al (1987) Non responsiveness to a foot and mouth disease virus peptide overcome by addition of foreign helper T cell determinants. Nature 330: 168–171.

Freund J (1956) The mode of action of immunologic adjuvants. Adv Tuberc Res 7: 130–148.

Garnier J, et al (1978) Analysis of accuracy and implications of simple methods for predicting the secondary structure of globular proteins. J Mol Biol 120: 97–120.

Geerligs HJ, et al (1989) The influence of different adjuvants on the immune respons to a synthetic peptide comprising amino acid residues 9–21 of herpes simplex virus type 1 glycoprotein D. J Immunol Methods 124: 95–102.

Germain RN (1986) The ins and outs of antigen processing and presentation. Nature 322: 687–689.

Gupta RK, et al (1993) Adjuvants — a balance between toxicity and adjuvanticity. Vaccine 11: 293–306.

Gustafson GL, Rhodes MJ (1992) Bacterial cell wall products as adjuvants: early interferon gamma as a marker for adjuvants that enhance protective immunity. Res Immunol 143: 483–488.

Haaijman JJ, et al (1988) Monoclonal antibodies directed against human immunoglobulins: preparation and evaluation procedures. In: Pal SB, ed. Reviews on Immunoassay Technology 1. MacMillan, London, pp. 59–93.

Hackett CJ, et al (1985) A synthetic decapeptide of influenza virus hemagglutinin elicits helper T cells with the same fine recognition specificities as occur in response to whole virus. J Immunol 135: 1391–1394.

Heath AW, Playfair JHL (1990) Conjugation of interferon gamma to antigen enhances its adjuvanticity. Immunology 71: 454–456.

Heath AW, Playfair JHL (1992) Cytokines as immunological adjuvants. Vaccine 10: 427–434.

Hilgers LATh, Snippe H (1992) DDA as an immunological adjuvant. Res Immunol 143: 494–503.

Hong TH, et al (1989) The production of polyclonal and monoclonal antibodies in mice using novel immunization methods. J Immunol Methods 120: 151–157.

Hopp TP (1986) Protein surface analysis. Methods for identifying antigenic determinants and other interaction sites. J Immunol Methods 88: 1–18.

Hopp TP, Woods KR (1981) Prediction of protein antigenic determinants from amino acid sequences. Proc Natl Acad Sci USA 78: 3821–3824.

Horiuchi N, et al (1987) Similarity of synthetic peptide from human tumor to parathyroid hormone *in vivo* and *in vitro*. Science 238: 1566–1568.

Jackson S, et al (1990) Liposomes containing anti-idiotypic antibodies: an oral vacine to induce protective secretory immune responses specific for pathogens of mucosal surfaces. Infect Immun 58: 1932–1936.

Jacob CO, et al (1986) Priming immune responses to cholera toxin induced by synthetic peptides. Eur J Immunol 16: 1057–1062.

Jemmerson R, Hutchinson RM (1990) Fine manipulation of antibody affinity for synthetic epitopes by altering peptide structure: antibody binding to looped peptides. Eur J Immunol 20: 579–585.

Johnson AG, et al (1956) Studies on the O antigen of *Salmonella typhosa*. V. Enhancement of antibody responses to protein antigens by the purified lipopolysaccharide. J Exp Med 103: 225–246.

Antigens and antigen presentation

Jones PD, et al (1988) Cellular immune responses in the murine lung to local immunization with influenza A virus glycoproteins in micelles and immunostimulatory complexes (ISCOM). Scand J Immunol 27: 645–652.

Karplus PA, Schulz GE (1985) Prediction of chain flexibility in proteins. Naturwissenschaften 72: 212–223.

Kaufmann SHE (1995) Immunity to intracellular microbial pathogens. Immunol Today 16: 338–343.

Kenney JS, et al (1989) Influence of adjuvants on the quantity, affinity, isotype and epitope specificity of murine antibodies. J Immunol Methods 121: 157–166.

Kensil CR, et al (1991) Separation and characterization of saponins with adjuvant activity form *Quillaja saponaria molina* cortex. J Immunol 146: 431–437.

Kyte J, Doolittle RF (1982) A simple method for displaying the hydropathic character of protein. J Mol Biol 157: 105–132.

Laman JD, Claassen E (1995) T cell independent and dependent humoral immunity. In: Snapper CM, ed. Cytokine Regulation of Immunoglobulin Synthesis and Class Switching. Wiley, Chichester, in press.

Leenaars PPAM, et al (1994) Evaluation of several adjuvants as alternatives to the use of Freund's adjuvant in rabbits. Vet Immunol Immunopathol 40: 225–241.

Lees A, et al (1990) Rapid stimulation of large specific antibody responses with conjugates of antigen and anti-IgD antibody. J Immunol 145: 3594–3600.

Lövgren K, Morein B (1988) The requirement of lipids for the formation of immunostimulating complexes (ISCOMS). Biotech Appl Biochem 10: 161–172.

Lövgren K, et al (1990) An experimental subunit vaccine (ISCOM) induced protective immunity to influenza virus infection in mice after a single intranasal administration. Clin Exp Immunol 82: 435–439.

Lycke N, et al (1989) Cellular basis of immunomodulation by cholera toxin *in vitro* with possible association to the adjuvant function *in vivo*. J Immunol 142: 20–27.

McKercher PD (1986) Oil adjuvants: their use in veterinary biologics. In: Nervig RM, Gough PM, Kaeberle ML, Whetstone CA, eds. Advances in Carriers and Adjuvants for Veterinary Biologists. Iowa State University Press, Ames, pp. 1–7.

McLean GW, et al (1991) Generation of antipeptide and anti-protein sera. Effect of peptide presentation on immunogenicity. J Immunol Methods 137: 149–157.

Martinon F, et al (1992) Immunization of mice with lipopeptides bypasses the prerequisite for adjuvant: immune response of Balb/c mice to human immunodeficiency virus envelope glycoprotein. J Immunol 149: 3416–3422.

Mirza IH, et al (1987) A comparison of spleen and lymphe node cells as fusion partners for the raising of monoclonal antibodies after different routes of immunization. J Immunol Methods 105: 235–243.

Morein B (1990) The ISCOM an immunostimulating system. Immunol Lett 25: 281–283.

Munoz J (1964) Effect of bacteria and abacterial products on antibody response. Adv Immunol 4: 397–440.

Nair S, et al (1992) Class I restricted CTL recognition of a soluble protein delivered by liposomes containing lipophilic polylysines. J Immunol Methods 152: 237–243.

Neurath AR, et al (1989) Hepatitis B virus surface antigen (HbsAg) as a carrier for synthetic peptides having an attached hydrophobic tail. Mol Immunol 26: 53–62.

Newman WJ, et al (1992) Saponin adjuvant induction of ovalbumin specific CD8 cytotoxic T-lymphocyte responses. J Immunol 148: 2357–2362.

Nicklas W (1992) Aluminium salts. Res Immunol 143: 489–494.

Novotny J, et al (1986) Antigenic determinants in proteins coincide with surface regions accessible to large probes (antibody domains). Proc Natl Acad Sci USA 83: 226–230.

Partidos CD, et al (1992) Antibody responses to non-immunogenic synthetic peptides induced by co-immunization with immunogenic peptides. Immunology 77: 262–268.

Peeters JM, et al (1989) Comparison on four bifunctional reagents for coupling peptides to proteins and the effect of the three moieties on the immunogenicity of the conjugates. J Immunol Methods 120: 133–143.

Phillips NC, Emili A (1992) Enhanced antibody response to liposome associated protein antigens: preferential stimulation of IgG2a/b production. Vaccine 10: 151–158.

Prieto I, et al (1995) Simple strategy to induce antibodies of distinct specificity: application to the mapping of gp120 and inhibition of HIV-infectivity. Eur J Immunol 25: 877–883.

Ruedl Ch, et al (1993) Oral administration of a bacterial immunomodulator enhances murine intestinal lamina propria and the Peyer's patch lymphocyte traffic to the lung: possible implications for infectious disease prophylaxis and therapy. Int Immunol 5: 29–36.

Sarobe P, et al (1991) Induction of antibodies against a peptide hapten and a class II presentable helper peptide. Eur J Immunol 21: 155–160.

Skea DL, Barber BH (1993) Studies of the adjuvant independent antibody response to immunotargeting. J Immunol 151: 3557–3568.

Snapper CM, Mond JJ (1993) Towards a comprehensive view of immunoglobulin class switching. Immunol Today 14(1): 15–17.

Snippe H, et al (1992) Characteristics and use of new-generation adjuvants. Res Immunol 143: 574–576.

Spitz M (1986) Single-shot intrasplenic immunization for the production of monoclonal antibodies. In: Methods Enzymol, 121: 33–41.

Staruch MJ, Wood DD (1983) The adjuvanticity of interleukin 1 *in vivo*. J Immunol 130: 2191–2194.

Tagliabue A, Boraschi D (1993) Cytokines as vaccine adjuvants: interleukin 1 and its synthetic peptide 163–171. Vaccine 11: 594–595.

Takayama K, et al (1991) Adjuvant activity of non-ionic block copolymers. V. Modulation of antibody isotype by lipopolysaccharides, lipid A and precursors. Vaccine 9: 257–265.

Tam JP, Zavala F (1989) Multiple antigen peptide. A novel approach to increase detection sensitivity of synthetic peptides in solid phase immuno-assays. J Immunol Methods 124: 53–61.

Thapar MA, et al (1991) Secretory immune responses in the mouse vagina after parenteral or intravaginal immunization with an immunostimulating complex (ISCOM). Vaccine 9(2): 129–133.

Vajdy M, Lycke NY (1992) Cholera toxin adjuvant promotes long-term immunological memory in the gut mucosa to unrelated immunogens after oral immunization. Immunology 75(3): 488–492.

Valensi JM, et al (1994) Systemic cytokine profiles in BALB/c mice immunized with trivalent vaccine containing MF59 oil emulsion and other advanced adjuvants. Immunology 153: 4029–4039.

Van Denderen J, et al (1989) Antibody recognition of the tumor-specific bcr-abl joining region in chronic myeloid leukemia. J Exp Med 169: 87–98.

Van Denderen J, et al (1990) Immunological characterization of the tumor specific bcr-abl junction in Philadelphia chromosome positive-acute lymphoblastic leukemia. Blood 76: 136–141.

Van den Eertwegh AJM, et al (1992) Immunological functions and in vivo cell–cell interactions of T cells in the spleen. Crit Rev Immunol 11(6): 337–380.

Van Regenmortel MHV (1986) Which structural features determine protein antigenicity. TIBS 11: 36–39.

Van Regenmortel MHV (1989) Structural and functional approaches to the study of antigenicity. Immunol Today 10: 266–272.

Van Rooijen N, Van Nieuwmegen R (1980) Liposomes in immunology: evidence that their adjuvant effect results from surface exposition of the antigens. Cell Immunol 49: 402–407.

Verheul AFM, Snippe H (1992) Non-ionic block polymer surfactants as immunological adjuvant. Res Immunol 143: 512–519.

VHI; Veterinary Public Health Inspectorate (1993) Code of Practice for immunization of laboratory animals (in Dutch), Rijswijk, The Netherlands, pp. 34.

Waksman BH (1979) Adjuvants and immune regulation by lymphoid cells. Springer Semin Immunopathol 2: 5–33.

Warren HS, et al (1986) Current status of immunological adjuvants. Annu Rev Immunol 4: 369–418.

Westhof E, et al (1984) Correlation between the segmental motility and the reaction of antigenic determinants in proteins. Nature 311: 123–126.

White RG (1976) The adjuvant effect of microbial products on the immune response. Annu Rev Microbiol 30: 579–600.

Wiesmuller KH, et al (1983) Synthesis of the mitogenic *S*-(2,3-bis(palmitoloxy)propyl)-*N*-palmitoyl pentapeptide from the *Escherichia coli* lipoprotein. Hoppe Seyler's Z Physiol Chem 364: 593–606.

Wolf H, et al (1988) An integrated family of amino acid sequence analysis programs. Comput Appl Biosci 4: 187–191.

Woodard LF (1990) Surface chemistry and classification of vaccine adjuvants and vehicles. In: Bacterial Vaccines. Alan Liss, New York, pp. 208–306.

Woodard LF, Jasman RL (1985) Stable oil-in-water emulsions:preparation and use as vaccine vehicles for lipophilic adjuvants. Vaccine 3: 137–144.

Yasuda T, et al (1982) Biological activity of chemically synthesized analogues of lipid A. Demonstration of adjuvant effect in hapten-sensitized liposomal system. Eur J Biochem 124: 405–407.

Zegers ND, et al (1993) Peptide induced memory (IgG) response, crossreactive with native proteins, requires covalent linkage of a specific B cell epitope with a T cell epitope. Eur J Immunol 23: 630–634.

Monoclonal antibodies for modulation of B cell activity

13.4

Jon D. Laman[1]
Mark de Boer[2]

[1]*Division of Immunological and Infectious Diseases, TNO Prevention and Health, Leiden, The Netherlands*
[2]*PanGenetics BV, Heemskerk, The Netherlands*

The work of J.D.L. is supported by grant 94–171 MS of the Netherlands Foundation for the support of MS-Research (Stichting Vrienden MS Research). Some of the methods described have been developed with the support of a TALENT stipend from the Netherlands Organization for Scientific Research (NWO) for a postdoctoral stay of J.D.L. in the laboratory of Dr R.J. Noelle (Dartmouth Medical School, Lebanon NH, USA).

TABLE OF CONTENTS

Introduction

This chapter is concerned with generation and use of monoclonal antibodies (MAb) against lymphocyte activation molecules involved in B cell activation. We here focus on two different receptor–ligand systems under the assumption that *mutatis mutandis* many of the rules and requirements found for these molecules apply to other activation molecules as well. These systems are B7-1/B7-2 with complementary ligands CD28 and CTLA4, and CD40 with its coreceptor CD40 ligand (CD40L, gp39). Both systems have been reviewed extensively in the recent past and we refer to these articles for more complete listings of primary references: for B7 molecules, Guinan et al (1994); Van Gool et al (1995); for CD40, Banchereau et al (1994), Durie et al (1994); Laman et al (1996).

T cells require two signals for activation and subsequent delivery of T cell help to B cells for specific antibody production. The first signal is provided by antigenic peptide presented to the clonally distributed T cell receptor in context of MHC class II. The second signal results from interaction of accessory molecules on antigen-presenting cells and on T cells, which are expressed upon antigenic stimulation and cellular interactions. The B7-1 (CD80) and B7-2 (CD86) molecules, whose expression can be induced on antigen-presenting cells (APC), interact with CD28 and CTLA-4 on T cells. CD28 is expressed constitutively, whereas CTLA-4 expression can be induced. Thus, two sets of two molecules cross-bind and signal each other. The crucial role of this receptor–ligand system in initiation of T cell function has sparked intense investigation into the mechanisms involved to allow its manipulation in vaccination and disease.

CD40, expressed on B cells, and CD40L (gp39), expressed on activated CD4$^+$ T cells, form a crucial receptor–ligand pair in regulation of humoral immunity. CD40–CD40L interaction provides an essential signal for B cell proliferation, expression of activation markers, immunoglobulin production and isotype switching. CD40–CD40L interaction is also required for formation of B memory cells and germinal centers, and signaling through CD40 prevents apoptosis of germinal center B cells. Defective expression of CD40L in humans leads to inability to produce isotypes other than IgM (hyper IgM syndrome) and to the absence of germinal centers. In addition, binding of CD40L to CD40 induced on monocytes can lead to monocyte effector mechanisms. Recent evidence indicates that the role of the CD40–CD40L axis in cellular interactions extends beyond antibody formation into antigen presentation in general, autoimmune disease and inflammation.

Since the introduction of the hybridoma technology by Kohler and Milstein (1975), a large number of monoclonal antibodies have been produced, which bind specifically to molecules expressed on the cell surface of lymphoid cells. These monoclonal antibodies have proved to be extremely powerful tools in immunological research. *In vitro* and *in vivo* studies using antibodies to cell surface-expressed activation molecules such as B7–CD28 and CD40L–CD40 have contributed significantly to the elucidation of the function of these molecules. Here, we discuss selected practical aspects of the use of monoclonal antibodies for manipulation of B cell activity. First, efficient and convenient methods to generate, select, and characterize monoclonal antibodies against cell surface molecules are described. Then, *in vitro* assays using these antibodies are described. Finally, general aspects of *in vivo* manipulation studies are highlighted. Special emphasis is put on practical pitfalls.

Generation of monoclonal antibodies against lymphocyte activation markers

Classical approaches to generation of monoclonal antibodies use purified protein in adjuvant for immunization (see chapter 13.3). Although effective for several applications, this approach has inherent drawbacks. As cell surface molecules have transmembrane and cytoplasmic regions, and often exist as oligomers on the membrane, generation of purified (recombinant) protein followed by mixing in adjuvant will result in significant denaturation, leading to loss of discontinuous B cell epitopes. Especially biologically active sites of receptor–ligand molecules tend to be of discontinuous nature. As a consequence, the purified protein–adjuvant immunization approach may strongly select against generation of functionally active antibodies that block receptor–ligand interaction or that have agonistic effects upon binding.

We therefore favor the use of an alternative approach to obtain monoclonal antibodies directed against cell surface molecules. Different eukaryotic cell lines such as Chinese hamster ovary cells, monkey kidney cells and insect cells can be genetically engineered to express foreign proteins, including glycosylation of the protein and expression of oligomers on the cell surface. Mice can subsequently be immunized with the cell lines engineered to express the molecule of interest. Immunization with intact xenogeneic cells generally results in a strong immune response and, therefore, the use of adjuvants is not required. This method does have some drawbacks. First, immune responses may occur to immunodominant epitopes (e.g., MHC antigens) that are not of interest. Second, when mice are immunized with whole cells, antibodies to a large number of different molecules are generated. It is therefore difficult to use the same cells to screen for specific antibody production by the hybridoma clones. Third, when the molecule of interest is expressed at low density, the frequency of mouse B cells specific for the antigen may be too low for the efficient isolation of appropriate hybridoma clones from the strong polyclonal immune response.

To circumvent these problems, we have developed a novel method using the high density expression of human cell surface antigens in insect cells for immunization (De Boer et al 1992). Using the baculovirus expression system (reviewed by Webb and Summers 1990; Luckow and Summers 1988; Maiorella et al 1988), cell surface proteins can be expressed at very high levels in insect cells (Webb et al 1989). When these cells are used for immunization, the high expression levels of the cloned proteins will increase the chance of obtaining specific antibodies. The major advantage of the methodology described below is that immunization of mice with insect cells expressing human cell surface molecules does not evoke antibodies cross-reactive with determinants on human cells other than those of interest. Therefore, human EBV-transformed B cells could be used for the screening of specific antibodies to human CD40 or human B7, which will be used as examples below to describe the technology.

The first step in this method is to obtain cDNA encoding the protein of interest. Since the introduction of PCR technology (Saiki et al 1985, 1988), it has become relatively simple to clone the cDNAs for proteins of which the coding DNA sequence has been published. One can use PCR primers spanning the complete coding region, and additionally incorporate restriction sites in these primers to facilitate cloning into expression vectors. For the generation of cDNA, RNA can be isolated from a suitable cell source using standard techniques (Chirgwin et al 1979). For CD40 and B7-1, we used an EBV-transformed human B cell line. As a result of transformation by EBV, B cell lines constitutively express these activation molecules.

- For RNA isolation, cells used as the source for the antigen to be cloned are washed twice with phosphate-buffered saline (PBS) and lysed in 5 M guanidinium thiocyanate in the presence of 0.7 M 2-mercaptoethanol. The cell lysate is then layered on a discontinuous CsCl gradient and centrifuged for 16 h at 26,000 rpm in a Beckman SW28 rotor. The RNA can be recovered by dissolving the pellet in DEPC-treated H_2O. The RNA is precipitated with ethanol once, resuspended in DEPC-treated H_2O, and can best be stored at −70°C.
- Alternatively, one can use one of the many commercially available RNA isolation kits.
- Total RNA (10 µg/reaction) is then converted to cDNA using random hexamer priming in 50 µl reaction buffer containing 500 units MLV-RT (Bethesda Research Laboratories, Bethesda, MD, USA), 5 µM random hexamer (Pharmacia, Piscataway, NJ, USA), 1 mM DTT, dNTP mix (0.5 mM each), 10 mM Tris–HCl pH 8.3, 50 mM KCl, 2.5 mM $MgCl_2$ and 0.1 mg ml^{-1} BSA (bovine serum albumin). After incubation at 37°C for 1 h, the samples should be boiled for 3 min and stored at −70°C.

The cDNA encoding the molecule of interest can be generated by PCR. For CD40 and B7-1 we used primers containing sequences having homology to known CD40 and B7-1 sequence. The primers also encoded restriction sites useful for cloning (*Bgl*II and *Kpn*I in the forward and backward primer, respectively). These primers were based on the published cDNA coding sequences for B7 and CD40 (Freeman et al 1989; Stamenkovic et al 1989).

- For PCR amplification, 1 µl of cDNA is mixed with 1 µl (10 pmol) of a forward primer, 1 µl (10 pmol) of a backward primer, and 47 µl of PCR mix. The PCR mix consists of 1.25 units *Taq* polymerase (Perkin-Elmer/Cetus, Norwalk, CT, USA), dNTP mix (0.2 mM each), 10 mM Tris–HCl pH 8.3, 50 mM KCl, 2.5 mM $MgCl_2$ and 0.1 mg ml^{-1} BSA. The 50 µl of PCR mixture is overlaid with 70 µl mineral oil and subjected to 25 cycles of amplification in a Perkin-Elmer/Cetus thermocycler (denaturation at 95°C for 30 s, primer annealing at 55°C for 30 s, and extension at 72°C for 1.5 min).
- The amplification products can then be digested with the restriction sites incorporated in the primers and isolated by size fractionation. Before expression in baculovirus, the DNA sequence of each fragment should be confirmed by sequence analysis to prevent the introduction of PCR-induced mutations.

The next step in this methodology is to express the cDNA encoding the protein of interest in the insect cells. For expression of human CD40 and B7-1, we used the baculovirus transfer vector pAcC8, digested with *Bgl*II and *Kpn*I. The amplified fragments were ligated to the linear pAcC8 vector at a ratio of insert to vector of 3:1. The ligation products were transfected into bacterial strain DH5α (Gibco/BRL, Gaithersburg, MD, USA) and recombinant pAcC8 vectors were selected on the basis of ampicillin resistance. Sequences encoding human CD40 and human B7 were recombined into the *Autographa californica* baculovirus (AcNPV) using the transfer vectors pAcCD40 (encoding the full-length CD40 molecule) and pAcB7 (encoding the full-length B7 molecule). The plasmids were cotransfected with wild type baculoviral DNA (2–10 pfu) (AcNPV) into Sf9 (*Spodoptera frugiperda*) cells at a density of 10^6 cells ml^{-1} (Summers and Smith 1987). Recombinant baculovirus-infected Sf9 cells were identified in soft agar and clonally purified (see Smith et al 1985; Summers and Smith 1987).

For optimal cell surface expression of the recombinant proteins, the cells can be harvested after 24–72 h of culture. If available, a positive control antibody can be used in ELISA or Western blot to confirm presence of the protein. If no antibody is available as a positive control, one can consider including a tag-sequence for which an antibody is at hand. Alternatively, one would have to rely on PCR or blot sequence information to infer that the molecule is present.

In the case of CD40 and B7-1, we checked the expression of the target molecules using a live cell ELISA.

- Sf9 insect cells infected with recombinant virus were cultured for 48 h in 24-well plates. After removal of the tissue culture medium, the plates with adherent insect cells were incubated for 45 min at room temperature (RT) with 0.25 ml of antibody in PBS with 1% BSA (PBS-BSA). After three washes with PBS-BSA, the plates were incubated for 35 min at RT with 250 µl of a 1/250 dilution of goat anti-(mouse total Ig) immunoglobulins conjugated to horseradish peroxidase (Zymed, South San Francisco, CA, USA) in PBS-BSA. Unbound peroxidase activity was removed by washing 5 times with PBS-BSA.
- Bound peroxidase activity was

revealed by the addition of an assay mixture prepared by diluting 0.5 ml of 2 mg ml^{-1} 3,3',5,5'-tetramethylbenzidine in ethanol to 10 ml with 10 mM sodium acetate, 10 mM EDTA buffer (pH 5.0) and adding 0.03% (v/v) H_2O_2. The reaction was stopped after 10 min by adding 100 µl of 1 M H_2SO_4. Plates were read at 450 nm using an automated ELISA reader.

This ELISA assay performed on live Sf9 cells showed that an established antibody to human CD40 reacted only with Sf9 cells infected with AcCD40 virus and not with AcB7 virus or mock-transfected insect cells.

After appropriate expression of the molecule of interest in the Sf9 insect cells has been confirmed, mice can subsequently be immunized intraperitoneally.

- For the generation of monoclonal antibodies to CD40 and B7-1, mice were injected at day 0 and day 14 with 5×10^6 live Sf9 cells infected with AcCD40 virus, AcB7 virus or AcCd3 virus (control virus). At day 21, 100 µl of serum was obtained to test for the presence of specific antibodies. After a rest period of at least 2 weeks, the mice received a final injection with 5×10^6 cells infected with AcCD40 or AcB7 virus.
- Three days after this last injection, the spleen cells were used for cell fusion with SP2/0 murine myeloma cells at a ratio of 10:1 using 50% poly(ethylene glycol) as previously described by De Boer et al (1988). The fused cells were resuspended in complete IMDM medium supplemented with hypoxanthine (0.1 mM), aminopterin (0.01 mM), thymidine (0.016 mM) and 0.5 ng ml^{-1} hIL-6 (Genzyme, Cambridge, MA, USA).

- The fused cells were then cultured in 96-well plates, so that approximately 60% of the wells would be expected to contain growing hybrids.
- After 10–14 days the supernatants of the hybridoma populations were screened for specific antibody production. To reduce the number of samples for screening, supernatants of 12 wells were pooled and used for fluorescent cell staining of EBV-transformed B cells.
- Subsequently, the supernatants of the positive pools were tested individually. Individual screening of wells instead of pooling of samples will prevent diluting out positive signals.

Positive hybridoma cells were cloned three times by limiting dilution in IMDM/FBS containing 0.5 ng ml^{-1} hIL-6. Four stable hybridomas producing anti-CD40 antibodies (5D12, 3A8, 3C6, and 5H7) and one stable hybridoma producing anti-B7-1 antibodies (B7-24) were isolated from two respective cell fusions. In addition, we have generated additional anti-B7-1 MAb and also anti-B7-2 MAb using this method.

In vitro modulation of lymphocyte activity

When studying activation of lymphocytes after providing stimulatory signals in the presence or absence of inhibitory of activating monoclonal antibodies, it is of utmost importance to perform detailed kinetic studies. This is essential to be able to discriminate between complete abolition of activation and delayed onset of activation as a result of addition of the inhibitory antibody. Ideally, analysis of the effects of modulating antibodies *in vitro* would use several measures such as proliferation [^{3}H]thymidine incorporation), functional assays (production of antibody isotypes and cytokines, by ELISA), and cell surface markers (the molecules of interest, but also general activation markers such as CD23, IL-2 receptor, high-affinity state of LFA-1, etc., by flow cytometry).

The anti-B7 and anti-CD40 antibodies described above have been used extensively in various *in vitro* experiments (De Boer et al 1993; Kwekkeboom et al 1994; Van Gool et al 1994a,b; Verwilghen et al 1994). Furthermore, the anti-B7 MAb B-24 is currently being tested in a clinical trial in steroid-resistant graft-versus-host disease.

The four anti-CD40 MAbs were shown to bind to a similar proportion of tonsillar B cells as does the established anti-CD40 MAb G28.5. Three of these monoclonal antibodies (5D12, 3A8, and 3C6) which were of the IgG2b subclass, were tested for their ability to deliver activation signals to human B cells in a B cell proliferation assay.

Human tonsillar B cells (4×10^4 per well) were cultured in 200 µl in microwells in the presence of anti-IgM coupled to Sepharose beads (5 µl ml^{-1}) or in the presence of anti-IgM plus rIL-2 (100 U ml^{-1}). Varying concentrations of the anti-CD40 MAb were added and [^{3}H]thymidine incorporation was measured at day 3 after 18 h pulsing.

None of the novel anti-CD40 MAb was able to significantly costimulate human B cell proliferation in the presence of immobilized anti-IgM or in the presence of immobilized anti-IgM and IL-2. In contrast, a control anti-CD40 MAb strongly costimulated human B cell proliferation in a concentration-dependent fashion. Unlike previously described MAb against CD40, these novel anti-CD40 MAb are not capable of stimulating human B cells. In contrast, the new MAb are able to efficiently block the proliferation of human B cells induced by a stimulatory anti-CD40 MAb. Addition of only 0.32 μg ml^{-1} of 5D12, 3C6, or 3A8 could almost completely block the stimulatory activity of a 4-fold excess of the stimulatory MAb. Thus, dependent on the epitope recognized, anti-CD40 antibodies can have strong stimulatory or inhibitory capacity.

It has been demonstrated that a mutant subclone of the murine thymoma EL-4, known as EL4B5, can strongly activate human B cells to proliferate and differentiate (Zubler et al 1985). This activation is cell–cell contact-dependent and, using the novel anti-CD40 MAbs described here, it has become clear that this activation is mediated via the CD40 molecule on the human B cells. Only 10 ng ml^{-1} of all three novel anti-CD40 MAbs is sufficient to completely block the activation of human B cells by the EL4B5 cells. Using a recombinant fusion of human CD40 and human IgM, it was confirmed that the EL4B5 cells express the CD40 ligand on the cell surface. Finally, the novel anti-CD40 MAbs are also able to inhibit the proliferation and differentiation of human B cells when stimulated with activated, CD40 ligand-expressing, human T cells. When purified T cells are stimulated through the TCR/CD3 complex and B7–CD28 interaction, the CD4$^+$ population of T cells express large amounts of the CD40 ligand. These T cells are very efficient in providing help for both IgM and IgG production by purified B cells. Again, the novel anti-CD40 MAbs are able to significantly inhibit the immunoglobulin production by human B cells: addition of only 40 ng ml^{-1} resulted in a 50% inhibition of both IgM and IgG in this culture system. However, addition of even 10 μg ml^{-1} of the antibodies did not result in more efficient inhibition. To exclude the possibility that the partial inhibition of antibody production by the human B cells was due to incomplete blocking of the CD40L–CD40 interaction, we performed a control experiment using CD40L-transfected mouse fibroblasts and demonstrated that B cell activation in this system, where the activation of the B cells is totally dependent on CD40L–CD40 interaction, could be blocked completely.

In vivo modulation of lymphocyte activity

General considerations

Although *in vitro* experiments using MAbs modulating B cell activity are highly informative and need to be done prior to *in vivo* studies, only the latter type of experiments will convincingly demonstrate physiological functions of the molecules under investigation. In addition, careful *in vivo* evaluation of MAbs is a prerequisite for clinical applications. Here, we discuss a number of practical aspects of such *in vivo* studies. *In vivo* experiments with hamster antibody against murine CD40L (antibody MR1: Noelle et al 1992) are used as an example.

MAb versus soluble ligand

When considering *in vivo* blocking of receptor–ligand interactions, the most obvious approaches are use of a specific MAb or of a recombinant soluble (truncated) form of the receptor/ligand. In the case of CD40–CD40L, modulation using recombinant CD40-Ig (Gray et al 1994) seems to be less efficient than using antibody MR1 against CD40L (Foy et al 1994). This is probably due to higher affinity of the antibody than of the soluble CD40-Ig molecule for CD40L. It is also possible that oligomerization of the soluble CD40-Ig molecule is required for efficient blocking. This need not be a strict rule, and soluble receptor molecules of sufficient affinity may be isolated or engineered.

Bioavailability and site of action

As for all therapeutics, one has to consider what concentrations have to be reached *in vivo* to efficiently block responses. The *in vitro* experiments will provide a good estimate for this. MR1 is a hamster antibody and, probably owing to its relatedness to murine antibody, its half-life in the mouse is fairly long (about 2 weeks). However, much depends on the access the modulating molecule has to its site of action. In the lymphoid organs, for example, with their high densities of lymphoid cells, effective concentrations may need to be higher than for modulation of interactions in cell suspensions or the peripheral blood. It is not surprising that anti-CD40L MAb administered intraperitoneally can gain access to the spleen to effectively block CD40–CD40L interactions in the spleen. But can anti-CD40L antibody access local inflammatory and autoimmune responses? Durie et al (1993) showed that collagen-induced arthritis of joints can be completely blocked. Similarly, the inflammatory response in the brain during peptide-induced EAE (experimental allergic encephalomyelitis) in mice can be blocked efficiently with anti-CD40L MAb, preventing disease (Gerritse et al 1996). Although formal proof is yet to be provided, these studies suggest that anti-CD40L MAb may gain access to target organs in disease without further manipulation.

In typical experiments, three doses of 250 μg anti-CD40L antibody i.p. per mouse are given at 2-day intervals (Foy et al 1993, 1994). These doses effectively blocked primary and secondary antibody responses as well as germinal center formation in the spleen. Titration experiments showed that 100 μg doses resulted in incomplete blocking of the plaque-forming cell response against T cell-dependent antigens. It is essential to estimate half-life of the antibody used for treatment; this can be done by a variety of methods, with detection by means of a species-specific conjugate against the modulating antibody on SDS-PAGE samples of serum being a straightforward approach. Assessment of circulating antibody should also take activity of the antibody into account and not merely its presence.

Expression of the molecule of interest *in vivo*

Modulation studies require good insight into the *in vivo* expression kinetics of the molecule under study. Time-course experiments *in vivo*, varying doses and scheduling of the activating agent (e.g., antigen) as well as doses and scheduling of the antibody, will be required for each new receptor–ligand pair. For example, the kinetics of CD40L expression *in vitro* and *in vivo* are quite distinct (discussed by Laman and Claassen 1996). Murine CD40L expression is induced *in vitro* by pharmacological activation of $CD4^+$ T cells with PMA and ionomycin after only 6 h, peaking at about 12–18 h, and returning to near-resting levels after 24 h, as evaluated by flow cytometry. In contrast, immunization of mice leads to emergence of CD40L-expressing cells in the spleen after 2–3 days, with a maximum at day 4 followed by a rapid decrease, as evaluated by immunocytochemistry (Van den Eertwegh et al 1993, 1994). A third pattern of CD40L expression was induced by stimulation of T cells with anti-CD3 in conjunction with B7-1 transfectants (de Boer et al 1993). In this system, CD40L expression peaked at 36–48 h and some expression could still be found at day 6 after activation. This system is more physiological than the rigorous trigger of PMA/ionomycin stimulation but, in contrast to the *in vivo* situation, no modulation of B7 expression occurs in this system.

In addition, secreted forms exist of many cell surface molecules, including CD40 and CD40L (see Banchereau et al 1994). Secreted forms can form complexes with the MAb used for modulation, leading to reduced effective concentrations. Also, the secreted molecules may obscure antibody binding sites on the complementary receptor, with the same result. Finally, CD40 can induce downregulation of CD40L gene expression, as well as endocytosis of the CD40L molecule.

Mechanism of action of antibody

The effects of the antibody when used *in vivo* should be evaluated. Does the antibody delete the cells expressing the complementary molecule, either directly or via ADCC? Does it induce functional incompetence of the cell, and if so, does this act through direct physical blocking of cellular interactions, through downregulation of expression, or by induction of anergy? For instance, it has been shown that treatment of mice with anti-CD40L antibody does not alter the number or the function of Th cells. Adoptive transfer of T cells from antibody-treated animals still transfers function. *In situ*, the frequencies of CD40L+ T cells producing IL-2, IL-4, and IFN-γ in response to immunization is similar in treated animals and controls, indicating that effector function is not impaired (Foy et al 1993; Van den Eertwegh et al 1994).

Anti-antibody responses

In vivo modulation studies should take into account that animals may generate anti-antibody responses against the modulating antibody. These responses can severely impair usefulness of experiments. Simple or more sophisticated ways around this problem are limiting treatment to one or few injections; using antibodies made in the same species; using antibodies of closely related species; and engineering the antibody to reduce immuno-

genicity. An interesting aspect of treatment with anti-CD40L antibody is that this antibody completely blocks the antibody response against TD antigens, and thereby also to itself as a foreign antigen.

Fc-receptor interactions

The possibility should be excluded that effects of modulating antibodies are due not to their specific binding but to interactions with Fc-receptors of the host. If such interactions are found to occur, use of F(ab')$_2$ fragments or molecular engineering of the antibody may offer alternative approaches.

Design of *in vivo* modulation analysis

For the design of *in vivo* modulation experiments, a number of parameters have to be considered in addition to the factors mentioned above.

First, the general kinetics of the B cell response to be studied (e.g., LPS stimulation, primary or secondary immunization with TD or TI antigens, infection, disease or challenge models) have to be known. The role the molecule of interest may have in initiation, maintenance, and/or termination of the response will determine scheduling of treatment.

Second, proper treatment of control animals will have to be identified, which is not necessarily simple. The optimal control treatment would be an isotype-matched antibody derived from the same species as the treatment antibody that binds to an unrelated cell surface molecule which is coexpressed with the molecule of interest. This is difficult to achieve, and it cannot be excluded that antibodies against the unrelated control molecule can affect the response. A complicating factor when using antibodies of hamster origin, such as MR1, is that the isotypes of hamster antibodies have not been identified, prohibiting selection of isotype-matched control antibodies. DEAE HPLC-purified hamster Ig (HIG) is then used as control antibody.

Third, appropriate assays for readout of manipulation *in vivo* have to be selected. We favor an integral approach in which evaluation of systemic responses (i.e., different isotypes and subclasses of antibodies produced, cyto-kines produced, evaluated by ELISA), by *ex vivo* functional experiments are performed in parallel with *in situ* events (antigen localization, antigen-specific antibody-forming cells, and cytokine producing cells in relation to the different anatomical compartments of lymphoid organs and other sites of cellular interactions of the immune system). Use of only peripheral, systemic readouts of the immune response and *in vitro* assays will obscure important informa-tion regarding cell–cell interactions. Two approaches will be discussed as examples.

Evaluation of B cell antibody responses

To evaluate the role of CD40L in antibody responses, *in vivo* treatment of mice with anti-CD40L antibody was evaluated using systemic readouts, adoptive transfers, and *in situ* evalua-tion (Foy et al 1993; Van den Eertwegh et al 1993).

- BALB/c mice were immunized with TD and TI-2 antigens (TNP–KLH/SRBC and TNP–ficoll, respectively). Typical doses were 100 µg of TD antigen and 20 µg for TI-2 antigen.
- For the *in situ* analysis, animals were immunized intravenously with the antigen in 200 µl of PBS, to elicit responses in the spleen.
- For memory responses, animals were immunized intraperitoneally with TD antigen in FCA.
- Antigen-specific ELISA was used to detect circulating titers of IgM, IgG1, IgG2a, IgG2b, IgG3, and IgE, using 1 mg ml^{-1} of KLH, TNP16-BSA, or TNP2-BSA to coat ELISA plates. Plaque assays were used to enumerate B cells secreting specific IgM. Quantification of intact anti-CD40L in the serum of mice was performed by SDS-PAGE of serum samples, using an enzyme-linked antibody against rat Ig cross-reactive with hamster but not mouse Ig (RG7–HRP). In addition, a flow cytometry assay using an anti-CD3 MAb-activated Th1 clone expressing CD40L was used to quantify circulating anti-CD40L antibody. Anti-CD40L antibody inhibited primary and secondary antibody responses against TD antigen in these experiments. However, the antibody response against TNP–ficoll was unaffected. Adoptive transfer experiments showed that anti-CD40L treatment does not delete or functionally affect SRBC-specific Th cells (Foy et al 1993).

In situ experiments are highly informative for analysis of events associated with antibody production *in vivo* and cellular interaction (Van den Eertwegh et al 1993). Antibody responses against TD and TI-2 antigens (TNP–KLH and TNP–ficoll) were evaluated immunocytochem-ically in tissue sections using hapten–enzyme conjugates (TNP–AP) and carrier–enzyme con-jugates. This allows detection of plasma cells producing hapten and carrier-specific antibod-ies. In addition, staining with antibody against CD40L (either using MR1–biotin/avidin–HRP or MR1 followed by RG7–HRP) allowed identifica-tion of CD40L-positive cells *in situ*. Using these methods, the numbers, anatomical localization, and kinetics of appearance of different cell types were assessed. Functional information was derived from immunocytochemical stain-ings for cells containing cytokines such as IFN-γ, IL-2, and IL-4. Double-staining proce-dures allowed visualization of interactions between cytokine-producing cells and antigen-specific antibody-forming cells, and between antibody-forming cells and CD40L$^+$ cells, and identification of cytokines produced by CD40L[+]/CD4[+] cells.

Collectively, the data showed that interactions between specific antibody-forming cells and CD40L[+]/CD4[+] T cells producing cytokines were restricted to the outer periarteriolar lymphocyte sheath (PALS) and around the terminal arterioles of the mouse spleen. The close juxtaposition of these cells suggests that cellular interactions are ongoing in regulating the antibody response. Claassen and Jeurissen (1996) have described detailed step-by-step procedures for the different immunocytochemical stainings mentioned. See also Van den Eertwegh et al (1991) and Claassen et al (1992). Recent methods for detection of apoptotic nuclei in tissue sections by *in situ* nick translation have further extended the repertoire of staining for functionally relevant parameters (see the Oncor manual for S7110 kit for extensive experimental detail and primary references).

Evaluation of B memory cell responses

To evaluate the role of cell surface molecules in generation of B cell memory, one can use systemic secondary antibody responses in adoptive transfer systems. In addition, the formation of germinal centers, the anatomical sites in secondary lymphoid organs where B memory cells are generated, can be evaluated.

Studying the role of CD40L in B memory cell responses Foy et al (1994), immunized CB17 (Igh[b]) mice intraperitoneally (i.p.), with 100 μg TNP–BSA in Freund's complete adjuvant (FCA) to induce formation of memory B cells. Four weeks later, B cells from these animals were adoptively transferred into irradiated (600 rad) congeneic BALB/c mice which had been primed with KLH (100 μg in FCA, i.p.). To assess memory B cell responses against the hapten TNP, mice were challenged with

10 μg soluble TNP–KLH at the time of transfer. On day 7, levels of systemic anti-TNP IgG1[b] were assessed using an antigen-specific allotype-specific ELISA. In a conventional ELISA approach (see chapter 13.8 for more detail), 5 mg ml[-1] of TNP6-OVA was absorbed on to microtiter plates and serum samples were incubated. Using biotin-conjugated anti-IgG1[b], followed by avidin–alkaline phosphatase (Zymed, South San Francisco, CA, USA), TNP-specific titers could be determined. Treatment of mice with anti-CD40L MAb (MR1) on days 0, 2, and 4 with 250 μg per injection, (i.p.), 4 weeks prior to adoptive transfer, led to abolishment of the memory response against TNP.

This indicated that CD40–CD40L interactions are required for memory B cell responses.

To confirm this finding *in situ* with respect to formation of germinal centers, *in vivo* treatment with TD antigen and anti-CD40L antibody was performed. Immunization with TD antigens leads to germinal center formation starting around day 3, peaking at days 10–14, followed by a gradual decline thereafter. Accurate kinetics have to be determined for each individual antigen separately.

Animals were immunized intraperitoneally with 100 μl of a 10% (v/v) solution of SRBC (Colorado Serum Co., Denver, CO, USA). SRBC are a very potent TD antigen and induce high numbers of germinal centers in murine spleen, typically 5–10 per single cross-section of tissue. Lower doses, such as 100 μl of 1% SRBC also suffice to induce a vigorous germinal center reaction.

Unpublished data suggest that for abolishment by anti-CD40L treatment of germinal centers that have already been formed, the lower dose of SRBC needs to be used. Hapten–carrier conjugates such as TNP–KLH and TNP–BSA can also be used in these experiments (100 μg, i.v.).

When animals were treated with anti-CD40L MAb on days 0, 2, and 4 after immunization, and sacrificed at days 9 or 11, total abolishment of germinal centers was found *in situ*. For this analysis, spleens were frozen in liquid nitrogen and 8 μm frozen sections were cut. After fixation with fresh acetone for 10 min, sections were stained with peanut agglutinin (PNA), anti-sIgD antibody (10.4.22), or anti-B220 antibody, all labeled with biotin. PNA is a lectin specifically binding to germinal center memory B cells. sIgD is present on resting B cells in primary follicles (follicles containing no germinal centers). B220 is a pan B cell marker. The secondary step for detection was avidin–HRP (1:500) (Sigma, St Louis, MO, USA), followed by staining with DAB (brown precipitate) or AEC (red precipitate) (for detailed step-by-step procedures of immunocytochemical stainings see chapter 13.7 and Claassen and Jeurissen 1996).

The results showed that treatment with anti-CD40L antibody abolished germinal center formation completely. In treated animals, only primary follicles were found, consisting of only sIgD$^+$ memory B cells but no PNA-binding memory B cells. In contrast, control animals treated with hamster Ig showed extensive germinal center formation as evidenced by foci of B cells intensely binding PNA, surrounded by a rim of sIgD$^+$ cells. As B220 is a pan B cell marker, no differences between treated animals and controls were found using anti-B220.

Although not performed in the study mentioned (Foy et al 1994), one can perform additional analysis of B memory cell responses *in vivo* by evaluating production of cytokines relevant to memory responses *in situ* (see above).

Concluding remarks and future perspectives

Monoclonal antibodies will remain valuable and indispensable tools for experimental dissection of the role of cell surface molecules in triggering lymphocyte activation and effector functions. In addition, a multitude of clinical applications of modulating MAb for treatment of autoimmune disease, rejection, and immunodeficiency can be envisioned. However, clinical applications inherently lag experimental concepts by years if not decades, even assuming that these applications easily surmount barriers of adverse side-effects and pharmacological constraints.

Several novel methods will improve and facilitate MAb-based approaches for analysis and manipulation of cell surface activation molecules. More detailed insight into cell types involved in antigen presentation and the role of cytokines in isotype switching and antibody production *in vivo* will contribute to more efficient generation of MAb, and to manipulation of the isotype/subclass of the antibody. DNA-immunization can in theory circumvent problems associated with production of sufficient quantities of purified antigen suited for immunization purposes. Genetic knockout mice can be used for generation of mouse antibodies against mouse proteins, because they are not tolerant to the protein under investigation.

Phage display techniques allow rapid screening and selection of MAb, or fragments derived from them with good binding capacity. Antibody engineering methods for improvement of antibody affinity have been described and are becoming increasingly sophisticated.

Surface plasmon resonance apparatus provides rapid and accurate assessment of association and dissociation rates, allowing selection of MAb based on desired functional characteristics.

References

Banchereau J, Bazan F, Blanchard D, et al (1994) The CD40 antigen and its ligand. Annu Rev Immunol 12: 881–922.

de Boer M, Ten Voorde GHJ, Ossendorp FA, Van Duijn G, Tager JM (1988) Requirements for the generation of memory B cells *in vivo* and their subsequent activation *in vitro* for the production of antigen-specific hybridomas. J Immunol Methods 113: 143–149.

de Boer M, Conroy L, Min HY, Kwekkeboom J (1992) Generation of monoclonal antibodies to human lymphocyte cell surface antigens using insect cells expressing recombinant proteins. J Immunol Methods 152: 15–23.

de Boer M, Kasran A, Kwekkeboom J, Walter H, Vandenberghe P, Ceuppens JL (1993) Ligation of B7 with CD28/CTLA-4 on T cells results in CD40 ligand expression, IL-4 secretion and efficient help for antibody production by B cells. Eur J Immunol 23: 3120–3125.

Chirgwin JM, Przybyla EA, MacDonald RJ, Rutter WJ (1979) Isolation of biologically active ribonucleic acid from sources enriched in ribonuclease. Biochemistry 18: 5294–5299.

Claassen E, Jeurissen SHM (1996) A step by step guide to *in situ* immune response analysis of lymphoid tissues by immunohistochemical methods. In: Weir DM, Blackwell C, Herzenberg L, Herzenberg L, eds. Handbook of Experimental Immunology, 5th edn. Blackwell Scientific.

Claassen E, Gerritse K, Laman JD, Boersma WJA (1992) New immunoenzyme-cytochemical stainings for the *in situ* detection of epitope specificity and isotype of antibody forming B cells in experimental and natural (auto)immune responses in animals and man. J Immunol Methods 150: 207–216.

Durie FH, Fava RA, Foy TM, Aruffo A, Ledbetter JA, Noelle RJ (1993) Prevention of collagen-induced arthritis with an antibody to gp39, the ligand for CD40. Science 261: 1328–1330.

Durie FH, Foy TM, Masters SR, Laman JD, Noelle RJ (1994) The role of CD40 in the regulation of humoral and cell-mediated immunity. Immunol Today 15: 406–411.

Foy TM, Laman JD, Ledbetter JA, Aruffo A, Claassen E, Noelle RJ (1994) gp39-CD40 interactions are essential for germinal center formation and the development of B cell memory. J Exp Med 180: 157–163.

Foy TM, Aruffo A, Ledbetter JA, Noelle RJ (1993) In vivo CD40–gp39 interactions are essential for thymus-dependent immunity. II. Prolonged in vivo Suppression of primary and secondary humoral immune responses by an antibody targeted to the CD40 ligand, gp39. J Exp Med 178: 1567–1575.

Freeman GJ, Freedman AS, Segil JM, Lee G, Whitman JF, Nadler LM (1989) B7, a new member of the Ig superfamily with unique expression on activated and neoplastic B cells. J Immunol 143: 2714–2722.

Gerritse K, Laman JD, Noelle RJ (1996) Functional and histological evidence forCD40–CD40-ligand interactions in multiple sclerosis.

Gray D, Dullforce P, Jainandunsing S (1994) Memory B cell development but not germinal center formation is impaired by *in vivo* blockade of CD40–CD40 ligand interaction. J Exp Med 180: 141–155.

Guinan EC, Gribben JC, Boussiotis VA, Freeman GJ, Nadler LM (1994) Pivotal role of the B7:CD28 pathway in transplantation tolerance and turner immunity. Blood 84: 3261–3282.

Köhler G, Milstein C (1975) Continuous culture of fused cells secreting antibodies of predefined specificity. Nature 256: 495–498.

Kwekkeboom J, De Rijk D, Van de Velde H, Kasran A, De Groot C, de Boer M (1994) Helper effector function of human T cells stimulated by anti-CD3 MAb can be enhanced by costimulatory signals and is partially dependent on CD40–CD40 ligand interaction. Eur J Immunol 24: 508–517.

Laman JD, Claassen E (1996) T cell independent and dependent humoral immunity. In: Snapper CM, ed. Cytokine Regulation of Immunoglobulin Synthesis and Class Switching. Wiley, Chichester, pp.23–72.

Laman JD, Claassen E, Noelle RJ (1996) Functions of CD40 and its ligand, gp39. Crit Rev Immunol, in press.

Luckow VA, Summers MD (1988) Trends in the development of baculovirus expression vectors. Bio/Technology 6: 47–55.

Maiorella B, Inlow D, Shanger A, Harano D (1988) Large scale insect cell culture for recombinant protein-production. Bio/Technology 6: 1406–1412.

Modulation of the humoral immune response

Noelle RJ, Roy M, Shepherd DM, Stamenkovic I, Ledbetter JA, Aruffo A (1992) A 39-kDa protein on activated helper T cells binds CD40 and transduces the signal for cognate activation of B cells. Proc Natl Acad Sci USA 89: 6550–6554.

Saiki RK, Scharf S, Faloona F, et al (1985) Enzymatic amplification of β-globin genomic sequences and restriction site analysis for diagnosis of sickle cell anemia. Science 230: 1350–1354.

Saiki RK, Gelfand DH, Staffel S, et al (1988) Primer-directed enzymatic amplification of DNA with a thermostable DNA polymerase. Science 239: 487–491.

Smith GE, Ju G, Ericson BL, et al (1985) Modification and secretion of human interleukin 2 produced in insect cells by a baculovirus expression vector. Proc Natl Acad Sci USA 82: 8404–8408.

Stamenkovic I, Clark EA, Seed B, et al (1989) A B cell activation molecule related to the nerve growth factor receptor and induced by cytokines in carcinomas. EMBO J: 8: 1403–1410.

Summers MD, Smith GE (1987) A manual of methods for baculovirus vectors and insect cell culture procedures. Texas Agricultural Experiment Station Bulletin No. 1555.

Van den Eertwegh AJM, Fasbender MJ, Schellekens MM, Van den Oudenaren A, Boersma WJA, Claessen E (1991) In vivo kinetics and characterization of IFN-γ-producing cells during a thymus-independent immune response. J Immunol 147: 439–446.

Van den Eertwegh AJM, Noelle RJ, Roy M et al (1993) In vivo CD40-gp39 interactions are essential for thymus-dependent humoral immunity. I. In vivo expression of CD40 ligand, cytokines, and antibody production delineates site of cognate T–B cell interaction. J Exp Med 178: 1555–1565.

Van den Eertwegh AJM, Laman JD, Noelle RJ, Boersma WJA, Claessen E (1994) In vivo T–B cell interactions and cytokine production in the spleen. In: CD40 and its ligand in the regulation of humoral immunity. Semin Immunol 6: 327–336.

Van Gool SW, Ceuppens JL, Walter H, de Boer M (1994a) Synergy between cyclosporin A and a monoclonal antibody to B7 in blocking alloantigen-induced T cell activation. Blood 83: 176–183.

Van Gool SW, de Boer M, Ceuppens JL (1994b) The combined effect of anti-B7 and cyclosporin A induces alloantigen-specific anergy. J Exp Med 179: 715–720.

Van Gool SW, Barcy S, Devos S, et al (1995) CD80 (B7-1) and CD86 (B7-2): potential targets for immunotherapy? Semin Immunol, 61st Forum in Immunology, CD28 and T-cell activation, pp. 183–196.

Verwilghen J, Lovis R, de Boer M, et al (1994) Expression of functional B7 and CTLA-4 on rheumatoid synovial T cells. J Immunol 153: 1378–1385.

Webb NR, Summers MD (1990) Expression of proteins using recombinant baculoviruses. Technique 2: 173–180.

Webb NR, Madoulet C, Tosi PF, et al (1989) Cell surface expression and purification of human CD4 produced in baculovirus-infected insect cells. Proc Natl Acad Sci USA 86: 7731–7735.

Zubler RH, Erard F, Less RK, et al (1985) Mutant EL-4 thymoma cells polyclonally activate murine and human B cells via direct cell interaction. J Immunol 134: 3662–3668.

Alginate encapsulation of cytokine gene-transfected cells

13.5

Huub F.J. Savelkoul

Department of Immunology, Erasmus University, Rotterdam, The Netherlands

TABLE OF CONTENTS

Principle of the method

Algin, or alginate, is a copolymer of β-D-manno-pyranosyluronate and α-L-gulopyranosyl uro-nate coupled by (1→4) glycosydic bonds. Alginate is extracted by the use of sodium hydroxide from the brown alga *Macrocystis pyrifera*. When exposed to divalent ions, like Ca^{2+}, an inert three-dimensional network is formed with large pores. This network is highly flexible at room temperature, but will jellify at 37°C. Cells are mixed with the alginate solution and subsequently squirted into a Ca^{2+} solution. As a result, all cells are caught in the alginate network, in which they stay because of the negative charge repulsion between the cell surface and the alginate. Owing to free diffusion of toxic waste products *out* of the capsule and nutrients *into* the capsule, the encapsulated cells stay alive for prolonged periods (e.g., weeks). Moreover, because of the capsule, activated cells during a potential cellular immune response in the host are unable to reach the encapsulated cells. In combination with a cell line that is stably transfected with a cytokine gene in a high copy number, the encapsulated cells will produce their main product continuously during the period facili-tated by the encapsulation procedure.

As many cell lines used for transfection are adherent growing cells, these can be grown efficiently on to Cytodex beads. The procedure described here facilitates the use of such fully grown beads for encapsulation. Encapsulated cytokine gene-transfected cells can be implanted *in vivo* in mice at several sites depending on the size of the capsules, number of capsules, and needle size. Moreover, these capsules can be used in *in vitro* culture systems to segregate contact-mediated versus factor-mediated signals (Fig. 13.5.1).

The method is based on the fast and efficient multiplication of cytokine gene-transfected cells in sufficient numbers to permit encapsula-tion and implantation *in vivo* by using rich culture conditions. Cytokine production by the cultured cells should be confirmed before encapsulation.

Materials and apparatus

- Cell lines of cells stably transfected with murine cytokine genes (e.g., CV-1/IL-4 cells)
- Cell culture media: RPMI or DMEM in combination with 10% (v/v) heat-inactivated (30 min at 56°C) fetal calf serum (FCS). The media are completed with 2 mM glutamine,

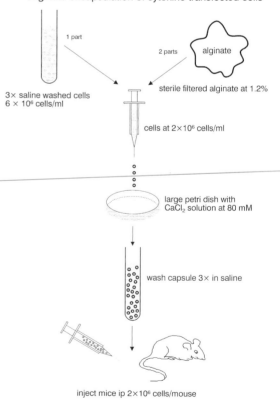

Alginate encapsulation of cytokine-transfected cells

1 part

2 parts   alginate

3× saline washed cells
6 × 10⁶ cells/ml

sterile filtered alginate at 1.2%

cells at 2×10⁶ cells/ml

large petri dish with
$CaCl_2$ solution at 80 mM

wash capsule 3× in saline

inject mice ip 2×10⁶ cells/mouse

Figure 13.5.1 Schematic representation of the encapsulation technique.

100 IU ml^{-1} penicillin, 50 µg ml^{-1} streptomycin, and 50 µM 2-mercaptoethanol.
- Cell culture facilities: laminar flow, incubator, centrifuge, microscope, sterile culture flasks and pipettes
- Alginate solution: 1.25% (w/v) filter-sterilized solution of algin in 20 mM Hepes buffer, filter sterilized, mycoplasm-free and endotoxin contamination <1 ng ml^{-1} (FMC Bioproducts, cat. no. 80020; distributor USA, FMC Rockland, ME; distributor Europe, Biozym, Landgraaf, The Netherlands). Bottles stored at room temperature; when in use store at 4°C.

- Cytodex 3 beads: Pharmacia, Uppsala, Sweden. Stock solution of 5 g resuspended in 250 ml 10 mM phosphate-buffered saline (PBS), pH 7.2, autoclaved 20 min at 121°C at 1.1 atm. The solution is stored at 4°C.
- Sterile stock solution of 1 M CaCl$_2$ in distilled water; store at 4°C.
- Saline (0.9 g l^{-1} distilled water), autoclaved, and stored at 4°C
- Tissue culture flasks T25 (Falcon #3081) and T75 (Falcon #3083)
- Petri dishes (10 cm^2)
- Syringes: 5 and 10 ml and needles 18–30 gauge

Culture of transfected cells

Adherent growing cells can be grown to a density of 1×10^5 cells ml^{-1} (10^7 cells per T75 culture flask). Cells can be trypsinized by adding 2.5 ml 0.25% trypsin per T75 flask and incubating for 1 min at 37°C in an incubator. Neutralize trypsin activity by addition of 7.5 ml of medium and washing the cells off the surface. This cell suspension can now be split 1:4 and recultured in new culture flasks. Alternatively, the cell suspension can be collected in a centrifuge tube. After centrifugation (5 min at 1500 rpm (400g) at 4°C) the supernatant is removed and the cell pellet is resuspended in 2 ml of medium. Meanwhile, to a new culture flask 8 ml of medium is added together with 4.5 ml of the resuspended solution of Cytodex beads. Add the cells to the culture flask and incubate for 24 to 48 h in an incubator.

Method 1: Nonadherent cells

1. Culture cells in the exponential phase of their growth.
2. Collect cells in centrifuge tube(s) and wash three times with sterile cold saline.
3. Bring cells up to 6×10^6 cells ml^{-1} with saline and estimate volume.
4. Add two volumes of alginate and mix by swinging the tube.
5. Attach to a syringe of appropriate size a 25- to 30-gauge needle, then

remove the plunger in a laminar flowhood. Transfer the mixture into the syringe and replace the plunger.

6. Squirt the mixture into a Petri dish containing a freshly prepared solution of 80 mM $CaCl_2$ in distilled water. Inject the mixture into the calcium solution at an angle of 45° while applying a moderate pressure. Continuously swirl the Petri dish with the calcium solution, to prevent clumping of the capsules. Transfer the capsules to a 50 ml centrifuge tube and leave 5 min to harden the capsules.

7. Wash the capsules three times with saline to remove the Ca^{2+} ions, by filling up the tube and allowing the capsules to settle. Remove supernatant to 7.5 ml.

8. After resuspension, 1 ml of capsules can be implanted intraperitoneally or subcutaneously using a 19-gauge needle and injecting very slowly into the mouse.

Check a small sample of capsules for their quality (Fig. 13.5.2).

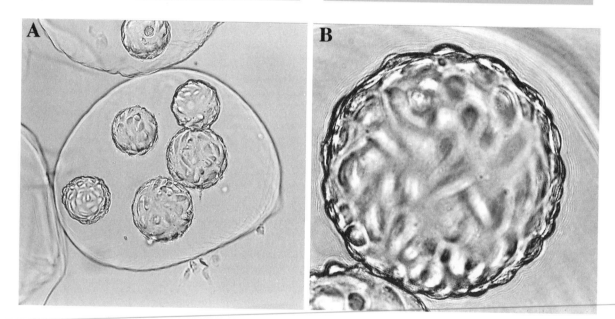

Figure 13.5.2 (a) Cytodex beads covered with IL-4 transfected cells encapsulated in alginate (40× magnification). (b) Detail of a Cytodex bead covered with CV-1/IL-4 (100× magnification).

Measurement of *in vivo* released cytokines from implanted capsules

The *in vivo* production of cytokines by encapsulated cytokine-transfected cells cannot be measured in the serum owing to the abundant presence of inhibitors, such as soluble receptors. Alternatively, despite the fact that cytokine is released continuously *in vivo*, at a particular time point not enough IL-4 may be present in the serum to be detected in the assays (ELISA or bioassays), owing to a low secretion rate. Therefore, we undertook analysis of mRNA expression by encapsulated CV-1/IL-4 washed out of the peritoneal cavity on various days after implantation. To ensure that equal amounts of cDNA were used in the PCR, a hypoxanthine phosphatidyl ribosyltransferase (HPRT) housekeeping gene-specific PCR was included for every sample. The results show that even up to 14 days after implantation IL-4 mRNA could still be detected. After day 14 the level of expression of both IL-4 mRNA and HPRT mRNA slowly declined, suggesting increasing cell death (Fig. 13.5.3).

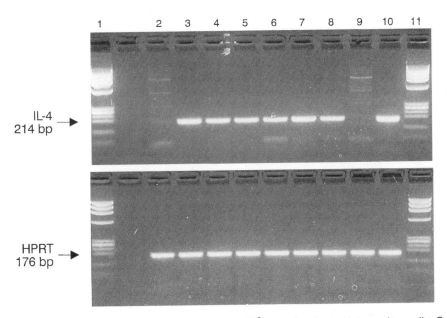

Figure 13.5.3 Alginate-encapsulated CV-1/IL-4 cells (2×10^6) were implanted intraperitoneally. Capsules were washed out of the peritoneum on days 0 (lane 3), 3 (lane 4), 5 (lane 5), 7 (lane 6), 10 (lane 7), 14 (lane 8), or 18 (lane 9). Total RNA was extracted, cDNA was prepared, and reverse-type PCR analysis was performed using an IL-4 primer set (upper) or HPRT primer set (lower). Lanes 1 and 11, base pair marker (Phi X174 RF DNA *Hae*III digest); lane 2, mock-transfected control; lane 10, positive control (CDC35 T cell clone).

Method 2: Adherent cells

1. Collect cells grown on Cytodex beads.
2. Wash the cells with sterile cold saline in centrifuge tubes by allowing the capsules to settle (2 min).
3. Take beads up in saline in twice the bead pellet volume.
4. Add two total volumes of alginate and mix.

5. Attach to a syringe of appropriate size a 21- or 23-gauge needle, then remove the plunger.
6. Transfer the mixture into the syringe and replace the plunger.

The rest of the procedure is identical to Method 1 using nonadherent cells.

Critical appraisal

- Beware of infection of the cell suspension with yeast or bacteria during cell culture, encapsulation, and implantation. It is therefore advisable to work in a laminar flow. Test the cell culture for possible mycoplasma infection prior to encapsulation.
- Ensure sufficiently high vitality (>90%) of the cultured cells. Perform trypan blue exclusion test before encapsulation.
- Wash the cells prior to encapsulation in saline to wash out any phosphate ions. After encapsulation wash excess $CaCl_2$ solution out with saline. Wash the cells with PBS or RPMI prior to implantation in vivo.
- Ensure that the culture of adherent growing cells is completely attached to Cytodex beads. Do not grow cells on beads for longer than 48 h as they tend to come loose.

- Too large a needle (23-gauge will produce capsules too large to be injected.
- When not enough force is being applied to the syringe or the angle between syringe and liquid surface is too small, the capsules will be too large to inject. If too much force is applied on the syringe, air bubbles are formed in the capsules, which is detrimental for survival of the cells.
- Capsules have to be allowed to harden for some time (e.g., 10 min) or they will break while injecting.
- Inject capsules with minimal force and using a large needle (18 or 19 gauge) to permit the capsules to deform and not break during injection.
- Carefully resuspend the capsules when filling the syringe for injection.

Applications

The immunoregulatory role of cytokines in immune responses is well established. Several cytokines are particularly powerful in directing B cell differentiation into specific isotypes; for example, IL-4 selectively induces isotype switching to IgE. A special interest has developed in the modulation of immune responses in experimental animal model systems by treatment with cytokine-neutralizing monoclonal antibodies or recombinant cytokines *in vivo*. The use of cytokines requires multiple injections with cytokines because of their short half-life, necessitating large amounts of purified cytokines. The availability of cell lines that are stably transfected with cytokine genes (Savelkoul et al 1991) enables the *in vivo* application of functional cytokines in doses sufficient to modulate specific immune responses. However, these cell lines will often be xenogeneic or allogeneic to the experimental animal and have to be encapsulated in such a way that no cellular response by the host will be induced. Moreover, the cytokine delivery system should ensure a constant systemic level of cytokine for prolonged periods. Alginate encapsulation (Savelkoul et al 1994) provides a simple, suitable method for immobilizing viable cells under mild conditions for *in vivo* implantation. The method is especially suited to studying the effects of cytokines on humoral responses (Van Oosterhout et al 1995; Van Ommen et al 1994a,b; Kraal et al 1994; Schilizzi et al 1995).

References

Kraal G, Schornagel K, Savelkoul HFJ, Maruyama T (1994) Activation of high endothelial venules in peripheral lymph nodes. The involvement of interferon-gamma. Int Immunol 6: 1195–1201.

Savelkoul HFJ, Seymour BWP, Sullivan L, Coffman RL (1991) IL-4 can correct the defective IgE production in SJA/9 mice. J Immunol 146: 1801–1805.

Savelkoul HFJ, Van Ommen R, Vossen ACTM, et al (1994) Modulation of systemic cytokine levels by implantation of alginate encapsulated cells. J Immunol Methods 170: 185–196.

Schilizzi BM, Savelkoul HFJ, de Jonge MWA, et al (1995) Antigen-specific IgG1 and IgE antibody-forming cells and hybridomas are decreased upon long-term IL-4 treatment *in vivo*. Scand J Immunol 41: 467–474.

Van Ommen R, Vredendaal AECM, Savelkoul HFM (1994a) Prolonged *in vivo* IL-4 treatment inhibits antigen-specific IgG1 and IgE formation. Scand J Immunol 40: 1–9.

Van Ommen R, Vredendaal AECM, Savelkoul HFM (1994b) IL-4 predominantly induces γ1ε-double positive B cells that are responsible for secondary IgE responses *in vivo*. Scand J Immunol 40: 491–501.

Van Oosterhout AJM, Van Ark I, Folkerts G, et al (1995) Anti-IL-5 inhibits virus-induced airway hyperresponsiveness but not bronchoalveolar eosinophilia in guinea pigs. Am Rev Respir Dis 147: 548–552.

Detection of immunoglobulin-secreting cells

13.6

Andre T.J. Bianchi
Guus Koch

DLO-Institute for Animal Science and Health, Lelystad, Netherlands

TABLE OF CONTENTS

Introduction

The enzyme-linked immunospot (ELISPOT) assay is the most appropriate assay for detecting and enumerating individual antibody or immunoglobulin-secreting cells *in vitro*. This assay is, as the name indicates, based on the well-known principle of the enzyme-linked immunosorbent assay. From its introduction, the ELISPOT assay (Sedgwick and Holt 1983; Czerkinszky et al 1983) appeared to be a worthy and more practical alternative for the well-known plaque-forming cell assay (Jerne and Nordin 1963). The ELISPOT assay is usually applied to enumerating antigen-specific, antibody-forming B cells. The solid phase is coated with a specific antigen to detect these cells. The ELISPOT assay can also be used to enumerate polyclonal B cell responses by coating the solid phase with antibodies against immunoglobulin isotypes of the species studied. This version of the ELISPOT assay is also called reversed enzyme-linked immunospot assay (Czerkinsky et al 1984) in analogy with the reversed (protein A) plaque assay (Gronowicz et al 1976).

The antibodies or immunoglobulins secreted by B lymphocytes are visualized as colored spots; the immunological fingerprints detected on the bottom of an antigen-coated or anti-immunoglobulin-coated microtiter plate. The spots are made visible by an insoluble colored product that is formed from a soluble substrate (and chromogen) by locally bound anti-Ig enzyme conjugates. The ELISPOT assay can be used easily to enumerate antibody- or immunoglobulin-producing cells of any species provided that immunoglobulin (isotype)-specific polyclonal or monoclonal antibodies are available (Bianchi et al 1990).

Antigen-specific responses against various soluble antigens, such as protein and polysaccharide antigens (Sedgwick and Czerkinsky 1992), and against particulate antigens (Franci et al 1986) have been studied with the ELISPOT technique. In fact, the use of the ELISPOT assay is not limited to the detection of antibody- or immunoglobulin-secreting cells; the assay can be constructed in such a way that any cell which secretes a specific product can be detected, for instance, to enumerate cytokine-producing cells. For more background information and for various other applications of the ELISPOT assay principle, the reader is referred to detailed reviews by Sedgwick and Holt (1986) and by Sedgwick and Czerkinsky (1992).

Materials

The following materials are needed for the standard ELISPOT assay.

- Polystyrene microtiter plates (high binding quality, e.g., Greiner)
- Coating material: (a) When polyclonal Ig responses are studied with the ELISPOT assay, antibodies specific for the Ig (subclass) of the species studied are used to coat plates. (b) When antibody-secreting B cell responses are studied with the ELISPOT assay, a specific antigen (e.g., ovalbumin, purified fimbriae of bacteria, LPS) is used to coat plates.
- Phosphate (0.02 M)-buffered saline (0.13 M NaCl) (PBS), pH 7.4
- 50 mM NaHCO$_3$ pH 9,6 or PBS pH 7.4 is used as coating buffer depending

on the material to be coated.
- PBS containing 2% BSA (Organon Technika) or 5% fetal bovine serum is used as blocking buffer.
- Washing solution: tap water containing 0.05% Tween
- Culture medium: RPMI 1640 containing 20 mM Hepes (Flow Laboratories) and supplemented with penicillin (200 units ml^{-1}) and streptomycin (0.2 mg ml^{-1})
- Primary antibody (monoclonal or polyclonal) specific for the antibodies or immunoglobulins of the species studied. This antibody can be conjugated either with alkaline phosphatase or with peroxidase.

  Note: When staining of the spots is weak using a primary antibody conjugate, amplification using a conjugated, secondary antibody is preferred. In this case an enzyme-conjugated, secondary antibody specific for the primary antibody is required.
- As an alternative to enzyme conjugates of primary or secondary antibodies, biotin conjugates of the various antibodies in combination with streptavidin–alkaline phosphatase or streptavidin–peroxidase conjugates (DAKO) are very useful for amplifying the staining of spots.
- *Substrate solution*

  (a) In combination with alkaline phosphatase conjugates: 2.3 mM 5-bromo-4-chloro-3-indolyl phosphate (Sigma) dissolved in 8 ml 0.625 M 2-amino-2-methyl-1-propanol working buffer pH 10.25 (containing 1 ml Triton X-405 per liter) (Sigma), and to stabilize the spots (optional), 2 ml 3% low-gelling agarose (Sigma or Serva).
  (b) In combination with peroxidase conjugates: 4 mg 3-amino-9-ethylcarbazole (Sigma)

dissolved in 250 μl N,N-dimethylformamide (Merck), added to 9.75 ml 0.05 M sodium acetate buffer pH 5.0. The solution is then filtered through a 0.22 μm filter. Just before use, 15 μl 30% H$_2$O$_2$ (Merck) is added.

The substrate volumes given are required for one 96-well microtiter plate.

Notes: (1) The staining solution for peroxidase conjugates is generally used in combination with nitrocellulose-bottomed 96-well microtiter plate (Millipore). (2) Air bubbles have to be prevented during addition of substrate solutions as they interfere with the enumeration of the spots.
- Incubator. Normally a 37°C incubator is used. In that case plates are placed in lockable, plastic boxes with wet paper to guarantee humidity during incubation. For ease, a 37°C tissue culture incubator is used with automatic CO$_2$ and humidity regulation.
- Multichannel pipette with disposable tips for 50–200 μl and micropipettes with disposable tips for 200–1000 μl and 20–200 μl.
- ELISA plate washer (optional)
- *Enumeration of spots*
  Normally an inverted microscope is used with a 40× magnification. For ease, an ELISPOT reader (see Fig. 13.6.1) is used. The spot assay reader shown in the figure has been constructed by the workshop of our institute and was based on a prototype from Logtenberg et al (Department of Immunology, University of Utrecht, The Netherlands). The reader produces a light projection of all spots in a well (30× magnification). The spots of these projections can be counted easily by eye. The essential parts of the reader are optic fiber illuminator (Schott), objective 10×

magnification and 0.3 numerical aperture (Nikon), *XY* stage for microtiter plates (Nikon).

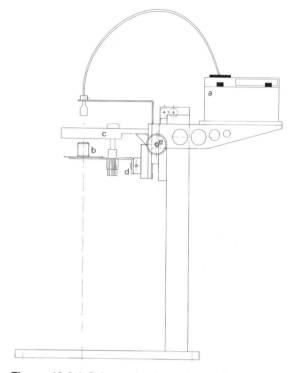

Figure 13.6.1 Schematic of an ELISPOT reader as constructed by the technical workshop of the DLO Institute for Animal Science and Health, Lelystad (scale 1:5): **a**, optic fiber illuminator; **b**, objective (10×, NA 0.3); **c**, *XY* stage for microtiter plate; **d**, focus adjustment knob; **e**, magnification knob.

Procedure

Coating of microtiter plates

- Optimal concentrations of antigens or anti-immunoglobulin antibodies in coating buffer are used to coat polystyrene microtiter plates overnight at 4°C or 3 h at 37°C. A 50 mM $NaHCO_3$ pH 9.6 buffer is often used for coating of proteins such as immunoglobulins (Voller 1976). The choice of the coating buffer depends on, amongst other things, the isoelectric point of the coating material.
- The optimal coating concentration is defined by checkerboard titration. Immunoglobulin concentrations of about $10 \, \mu g \, ml^{-1}$ purified Ig often appear to be optimal for coating of polystyrene plates for the ELISPOT assay. Coated plates can be often stored for months at −20°C before use.

Note: Virus-infected cell monolayers have also been used as 'antigen-coating' to enumerate virus-specific antibody secreting cells in an ELISPOT (Russell et al 1987).

Cell suspensions

- Cell suspensions from various origins are used in the ELISPOT assay, for example, cell suspensions prepared from lymphoid tissues, from peripheral blood, or harvested from in vitro cultures. Cell suspensions from solid tissue are prepared by mincing the tissue sample in culture medium with scissors and subsequently squeezing it through a 70 μm nylon gauze filter (Cell strainer from Becton Dickinson Labware).
- Nucleated cells are counted with a hemocytometer or are counted with an automatic cell counter (e.g., Coulter Counter, Coulter Electronics). The viability of the cell suspensions is determined by dye exclusion (e.g., nigrosine dye 0.2%). Three concentrations of lymphoid cells (between 10^2 and 10^6 cells per well) in culture medium are tested, usually in triplicate.

The spot-forming cell assay

1. Warm the microtiter plate at room temperature and remove the coating solution by flicking it off over a sink.
2. Wash the plate 3 times by hand with PBS. Per wash, the plate is filled with 100 μl PBS/well using a multichannel pipette and the solution is flicked off over a sink after 2 min.
3. Optional: Incubate the plate with blocking buffer for 30 min and flick off over a sink. Blocking can be useful to suppress background staining.
4. Incubate the plate with dilutions of various cell suspensions (100 μl per well) at 37°C in a humid atmosphere for 3–4 h.
   Note: Avoid any disturbance of the sedimented cells by vibration during incubation.
5. Wash the plate thoroughly 6 times with washing solution by hand or with

an automatic plate washer to remove the cells.

6. Incubate the plate with an optimal dilution of a primary antibody (100 µl/well) overnight at 4°C or for 3 h in a 37°C incubator.

7. When the primary antibody is not conjugated with alkaline phosphatase, an extra wash procedure followed by incubation with an optimal dilution of a secondary antibody conjugate is needed.

8. Wash the plate thoroughly 6 times with washing solution.

9. Incubate the plate with substrate solution (50 µl/well) for 2 h at room temperature.

10. The cell concentration that produces between 25 and 150 spots/well is used for calculation of the number/frequency of spot-forming cells.

Notes: (a) During the incubations the plates are covered with a lid. (b) Before incubation of the cells the plates are washed carefully, whereas after incubation of the cells the plates are washed intensively to get rid of cell remnants.

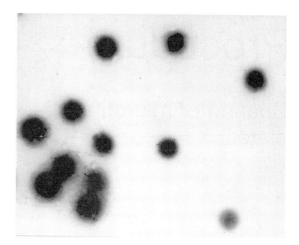

Figure 13.6.2 Spots of porcine spleen cells that secrete ovalbumin-specific IgG. The spots were formed on 0.2% glutaraldehyde-pretreated polystyrene plates that were subsequently coated with $10 \, \mu g \, ml^{-1}$ polymerized ovalbumin in PBS.

completely filling the wells with distilled water before counting.

Enumeration of the spots

- Spots are counted with an inverted microscope. A spot produced by an antibody or immunoglobulin-secreting cell has an intensely stained center and staining is more diffuse at the edge of the spot (see Fig. 13.6.2). When there is uncertainty about discrimination between real spots and grains of colored substrate or chromogen residues, one can check whether the 'spots' are diluted according to the cell dilutions.
- Undesired reflection of the light during counting can be prevented by

Trouble shooting

Microtiter plates for ELISPOT assays, especially when using small proteins, often require a higher coating of the solid-phase than is required for a comparable ELISA (Holt 1984). Coating of microtiter plates can be improved as follows:

1. Pretreatment of the polystyrene microtiter plates with freshly prepared 0.2% glutaraldehyde in distilled water overnight at 4°C (Logtenberg et al 1985) can improve the binding capacity of the plates for proteins.

2. Polyvinyl microtiter plates (Costar) may be used instead of polystyrene because of their higher binding capacity. However, polyvinyl plates are flexible and therefore less easy to handle. Moreover, polyvinyl plates do not have a flat bottom, making the

screening of the spots under the inverted microscope more difficult.

Nitrocellulose-bottomed 96-well microtiter plates (Moller and Borrebaeck 1985) have also been used for ELISPOT assays. Such plates have 100 times higher antigen-binding capacity than polyvinyl microtiter plates, because of the nitrocellulose matrix. The use of nitrocellulose plates is much more expensive and more complicated than that of standard polystyrene plates because the outer surface of the nitrocellulose membrane has to be carefully dried between each washing step to prevent leakage during the next incubation step. The screening of spots on the nontransparent nitrocellulose plates is also more difficult compared to the transparent polyvinyl and polystyrene plates.

3. Pretreatment of the coating material often improves the quality of the spots and increases the number of specific spots per well. A good example is given by Holt and coworkers (1984), who demonstrated that polymerization of the coating antigen, in their case ovalbumin polymerization with glutaraldehyde, improved the quality and the number of spots. For polymerization, glutaraldehyde was added dropwise to ovalbumin to achieve a final molar ratio of 25:1. This was incubated for 4 h and the pH was adjusted to 7.5.

Note: Since the conditions for coating of solid phases for ELISA and ELISPOT assays are essentially the same, the reader is referred to chapter 13.8 for detailed information on theoretical and practical aspects of solid phase coating.

Special version of the ELISPOT assay

Antibody-secreting cells of different isotypes can be detected after cell incubation in the same well with the ELISPOT technique. The following approaches are used.

1. Poly(L-lysine)-immobilized cell monolayers are repeatedly exposed to coverslips coated with antigen. The coverslips are subsequently incubated with different isotype-specific antibodies and the appropriate substrates to make the spots visible (Holt and Plozza 1986).

2. After cell incubation on nitrocellulose-bottomed micotiter plates, the plates are incubated simultaneously with an alkaline phosphatase- and peroxidase-conjugated antibody of different isotype specificity, and thereafter sequentially incubated with the different substrates (Czerkinszky et al 1988).

References

Bianchi ATJ, et al (1990) The use of monoclonal antibodies in an enzyme immunospot assay to detect isotype-specific antibody-secreting cells in pigs and chickens. Vet Immunol Immunopathol 24: 125–134.

Czerkinsky CC, et al (1983) A solid-phase enzyme-linked immunospot (ELISPOT) assay for enumeration of specific antibody-secreting cells. J Immunol Methods 65: 109–121.

Czerkinsky CC, et al (1984) Reversed enzyme-linked immunospot assay (RELISPOT) for the detection of cells secreting immunoreactive substances. J Immunol Methods 72: 489–496.

Czerkinsky CC, et al (1988) A novel two colour ELISPOT assay. 1. Simultaneous detection of distinct types of antibody-secreting cells. J Immunol Methods 115: 31–37.

Franci C, et al (1986) Further studies on the ELISA-spot technique. Its application to particulate antigens and a potential improvement in sensitivity. J Immunol Methods 88: 225–232.

Gronowicz EA, et al (1976) A plaque assay for all cells secreting Ig of a given type or class. Eur J Immunol 6: 588–590.

Holt PG, Plozza TM (1986) Enumeration of antibody-secreting cells by immunoprinting: sequential readout of different antibody isotypes on individual cell monolayers employing the ELISA plaque assay. J Immunol Methods 93: 167–169.

Holt PG, et al (1984) ELISA plaque assay for the detection of antibody secreting cells: observations on the nature of the solid phase and on variations in plaque diameter. J Immunol Methods 74: 1–7.

Logtenberg T, et al (1985) Enumeration of (auto)antibody producing cells in human using the 'SPOT-ELISA'. Immunol Lett 9: 343–347.

Jerne NK, Nordin AA (1963) Plaque formation in agar by single antibody-producing cells. Science 140: 405.

Moller SA, Borrebaeck CAK (1985) A filter immuno-plaque assay for the detection of antibody-secreting cells in vitro. J Immunol Methods 79: 195–204.

Russell PH, et al (1987) A rapid enzyme-linked semi-microwell assay for the enumeration of antibody-forming cells to viral and bacterial antigens in domestic animals. J Immunol Methods 101: 229–233.

Sedgwick JD, Czerkinsky C (1992) Detection of cell-surface molecules, secreted products of single cells and cellular proliferation by enzyme immunoassay. J Immunol Methods 150: 159–175.

Sedgwick JD, Holt PG (1983) A solid-phase immunoenzymatic technique for the enumeration of specific antibody-secreting cells. J Immunol Methods 57: 301–309.

Sedgwick JD, Holt PG (1986) The ELISA-plaque assay for the detection and enumeration of antibody-secreting cells. J Immunol Methods 87: 37–44.

Voller A, et al (1976) Enzyme immunoassays in diagnostic medicine. Theory and practice. Bull WHO 53: 53–65.

Immunocyto-chemical antigen-specific detection of antibodies

13.7

Eric Claassen[1,2]

[1]DLO Institute for Animal Science and Health, Lelystad
[2]Department of Immunology, Erasmus University Rotterdam, Rotterdam, The Netherlands

TABLE OF CONTENTS

Immunology Methods Manual
ISBN 0–12–442712–X

Introduction

In situ analysis of tissue sections by immuno-histochemical methods has elucidated essential actions taken by the immune system *in vivo* after encountering the antigen. When starting out on the path of *in situ* analysis, a few initial problems need to be overcome. First, the technical aspects concerning freezing, sectioning, storing, fixing, incubating, and staining the tissues can be an obstacle; these aspects are described in detail in chapter 32.7. Second, lymphoid organs are organized in discrete compartments, thus facilitating but also limiting cell–cell interactions. Knowledge of this organization is imperative for interpreting immunohistochemical results obtained by the techniques described in this chapter and has been described in detail before (cf. Claassen and Jeurissen 1996a).

Classically, primary antibodies are employed to specifically interact with conformational determinants (epitopes) present in tissue sections ('antigen' in Fig. 13.7.1a). Primary antibodies can be directly labeled with reporter molecules (e.g., fluorochromes, colloids, enzymes) to allow their revelation. Alternatively, they can be recognized by subsequently applied antibodies conjugated to a reporter molecule, or enzyme–anti-enzyme immune complexes, resulting in an amplification step (two-step procedure). Primary or secondary antibodies can also be derivatized with haptens (e.g., biotin, trinitrophenyl, digoxygenin) which can be specifically detected with their counter-structures (e.g., avidin or antibodies), resulting in more complex and usually more effective two- or three-step procedures. Basically, both antibody-based and DNA-probe-based immunohistochemistry can be performed as described in chapter 32.7, or with commercial kits. However, especially in a research setting, questions might arise which cannot be readily answered with standard reagents, and custom-made conjugates are then necessary. Here we describe conjugation and revelation procedures for the production and immunochemical use of such conjugates.

(a)

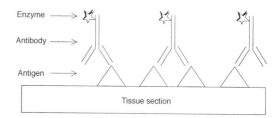

(b)

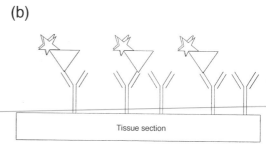

Figure 13.7.1 Schematic representation of the different approaches for immunochemical analysis of frozen tissue sections. (a) Revelation of antigen (e.g., cell-specific epitope, activation marker, cytokine, experimental antigen) by enzyme-labeled antibody. (b) demonstration of specific antibodies (e.g., in B cells or immune complexes) by antigen–enzyme conjugates for *in situ* humoral immune response analysis.

Antigen-specific antibody detection

Analysis of humoral immune responses is best performed by looking at the production of antigen-specific antibodies in combination with their isotype. In those cases where one might want to investigate antibodies in a tissue context (i.e., *in situ/in vivo*) in freshly frozen or archive material, there remains only one method of choice. This method makes use of the fact that lightly (acetone) fixed antibodies can still recognize and bind the epitope against which they are directed. By coupling the antigen to an enzyme and subsequent incubation on tissue sections, followed by routine immunohistochemistry, one can thus reveal antibodies based on their antigen specificity (Fab as shown in Fig. 13.7.1b and Fig. 13.7.2).

We have developed a direct immunoenzyme approach for the detection of antigen-specific antibody-forming B cells (AFC) *in situ*. In animal studies with model antigens, and human studies with viral antigens and autoantigens, specificity of the antibodies in tissue sections could be demonstrated after incubation with antigen–enzyme conjugates. Their isotype could be simultaneously determined using an anti-immunoglobulin (Fc chain-specific)–enzyme conjugate followed by appropriate (double) enzyme chemistry (cf. Van Rooijen and Claassen 1986). By means of hapten–enzyme and hapten–carrier conjugates made with haptens such as trinitrophenyl, penicilloyl, and arsonate, immune responses against thymus-dependent and T-independent and/or soluble versus particulate carriers could be studied. The method was adapted for use in other lymphoid tissues such as those associated with the gut or the lung. Antibodies visualized in this way can be both cytoplasmic in AFC or extracellular, the latter in the form, for example, of immune complexes (see Fig. 13.7.2 and color photomicrographs in: Laman et al 1990) or on memory B cells (Liu et al 1988).

The main strength in the method for detecting specific AFC lies in the fact that particularly those cells actively involved in any given (experimental or natural) *in vivo* immune response against a defined antigen (or even a single epitope thereof) can be identified. This is in sharp contrast to studies using only anti-Ig stainings, in which *all* immunoglobulin-forming cells are detected. In addition to this, the development and localization of the specific AFC can be studied in relation to the localization of the antigen, antigen-presenting cells, T-helper and T-suppressor cells, cytokine-producing cells (cf. Hoefakker et al 1995), and several other cell types and/or adhesion/activation markers involved in induction or regulation of the response (Claassen et al 1992; Claassen and Jeurissen 1996a). In some cases these double stainings may result in an intermediate

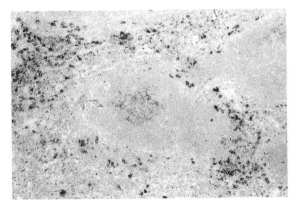

Figure 13.7.2 Specific antibody-forming B cells (AFC) directed against trinitrophenyl revealed (in a murine spleen 5 days after immunization with TNP–ficol) after incubation with TNP–alkaline phosphatase. Note antigen-specific immune complexes as web-like pattern in follicle, no AFC in surrounding inner-PALS (periarteriolar lymphocyte sheath; main T-cell area), or marginal zone, but mainly in outer-PALS and around terminal arterioles.

color, indicating that both markers under investigation reside at the same anatomical location. Furthermore, this can be done *in situ*, offering an integral picture, complete with cell–cell interactions *in vivo*, of the immune response (see also Laman et al 1991a; Van den Eertwegh et al 1992, 1994). We found that the *in situ* approach offered a number of advantages, such as the unequivocal detection of rare AFC ($1:10^7$), detection of multiple (cross-reactive or auto-) antibody specificities in a single sample, and the determination of tissue compartments, clone size, and distances between cells.

Antigen–enzyme conjugates can be effectively used for the detection of specific antibodies *in situ*, as described, but also with superior results in development of anti-Ig coated, antibody-catching ELISA (enzyme-linked immunosorbent assay) and the even more sensitive variant of the competition-ELISA. In a spot-ELISA, which is an improved modern version of the classical plaque-forming cell assay, these conjugates dramatically enhanced the sensitivity of the assay (10–20 times). These and other applications have been reviewed in chapter 13.6 and by Claassen et al (1992).

Principle of the method

Detection of antigen-specific antibody-forming cells is obviously dependent on the presence of active B cells in the tissue section. Basically this means that one has to start with animal or human material in which one expects these cells to be present. Furthermore, by definition, the method can only work when the antigen under investigation is coupled to an enzyme. Below we give coupling protocols for this purpose and for three different enzymes. The method does not work on formalin-fixed material and paraffin sections because of damage to the antibodies present in the tissue. The use and preparation of frozen tissue and cryosections is described in detail in chapter 32.7, as is the revelation of HRP and AP. For model studies in animals we make use of hapten-modified carrier molecules; below we give the protocol for preparation of such molecules and for trinitrophenylated alkaline phosphatase for detection of anti-TNP AFCs. Further details, when necessary can be found in Claassen and Jeurissen (1996a,b) and Van Rooijen and Claassen (1986).

Steps to be followed for detection of antigen-specific antibodies in tissue sections

In concordance with Murphy's law, remember 'when everything else fails: read the manual'. Read through this section *before* starting conjugation or immunochemistry.

1. Identify antigen to be studied (cf. Claassen et al 1991).
2. Obtain purified antigen (1–2 mg). Alternatively, synthesize B cell epitope peptide (cf. Zegers et al 1995)
3. Conjugate antigen to enzyme of choice.[a]. Protocols below can be used, but note that the molecular weight of the antibody is 160 kDa; protocols must be adapted for the molecular weight of the antigen.
4. Obtain and snap-freeze tissue.[b]

[a]Listed in this chapter;

5. Prepare frozen sections.[b]
6. Fix sections[b] (Laman et al 1991b).
7. Incubate overnight.[b]

8. Stain with appropriate substrate[a,b]

Selection and coupling of enzymes

Basically we have employed three enzymes, viz. alkaline phosphatase (AP), horseradish peroxidase (HRP), and β-galactosidase (GL).

- *Alkaline phosphatase*. AP was used most frequently, mainly because it can readily be modified with the dialysis glutaraldehyde method to provide simple coupling through ε-amino residues present in most proteins. The advantages of this modification of the classical one-step glutaraldehyde method are that no homopolymers are formed and much less antigen is needed.
- *Horseradish peroxidase*. In double staining the blue reaction product of AP we have found HRP in combination with the red immunodecoration by 3-amino-9-ethylcarbazole to yield best results. The more frequently used dye diaminobenzidine is not suited for double staining with AP substrates in our method.
- *β-Galactosidase*. This enzyme was chosen because its greenish-blue substrate provides an opportunity for cytochemical triple and quadruple stainings in combination with AP (blue + red) and HRP (red/brown; n.b., GL can be double stained even with diaminobenzidine). Furthermore, after acetone fixation of lymphoid tissue absolutely no background due to endogenous enzyme activity is observed. This latter fact provides the possibility for prolonged staining times (2–4 h) and we

have found assays with GL conjugates to be far more sensitive than those using PO or AP.

General aspects and caveats

- An essential step for detection of antigen-specific antibody-forming B cells and cytokine-producing cells is the preincubation step in a humid box to condition the freshly cut cryosections overnight. Such a box can be made by putting wet tissue on the bottom of a flat-bedded glass/plastic container. Ensure that droplets forming on the inside of the lid cannot fall onto the tissue sections (e.g., by placing container at slight angle).
- In most cases counterstaining is performed with hematoxylin because this nuclear staining does not interfere with membrane and cytoplasmic stainings as performed in most of the above protocols. Furthermore, it is non-water-soluble and can thus be combined with gelatin–glycerin embedding. When performing multiple stainings we usually do not find it necessary to employ a counterstain.
- Do not accidentally use Tris or glycine (or other amino-group-containing molecules) in coupling buffers (e.g., in buffer which

[b]Listed in chapter 32.7.

contains protein to be conjugated) where amino groups of the enzyme or carrier are involved.

- All conjugate coupling procedures are routinely performed in the dark; slides are always stored in a dark (to prevent fading of the stains) and relatively cool place. Results will remain visible for up to several years when proper embedding and storage is performed.
- Conjugates can be stored for prolonged periods (6–12 months) at 4°C when they are filter-sterilized and handled under microbiologically sterile (benchtop) conditions. Alternatively, conjugates can be diluted 1:1 with glycerin and kept at 20°C without freezing solid and without damage to the conjugate components.
- Always make fresh solutions, also for buffers, unless stated otherwise or proven to be effective.
- When double staining, always include single-stained sections for comparison in the same experiment/staining.
- Conjugate storage is best performed by mixing 1:1 with glycerin and placing at –20°C; no ice will form (keeping enzymes intact) and conjugates can be sampled with micropipette without 'thawing '.
- Gelatin–glycerin as water-miscible mounting medium: To 42 ml water add 8 g pure gelatin, 50 g glycerol, and 1 g phenol (to inhibit bacterial growth); stir under gentle heat until dissolved. Store at 4°C; before use bring to 37°C (can be stored and reheated ad infinitum).

Conjugation of trinitrophenyl to alkaline phosphatase

- Take 200 units AP (type VIIE, P-6774,1070 units mg^{-1}, Sigma) in 1 ml PBS.

- Add 1 mg 2,4,6-trinitrophenyl (TNP)-sulfonic acid.
- Mix for 30 min at room temperature in complete darkness.
- Dialyze extensively against PBS.
- Determine TNP substitution ratio (see below mol. wt AP = 100,000).
- Use conjugates which possess 9–12 TNP groups/AP molecule.
- Filter, sterilize, and store at 4°C or dilute 1:1 with glycerin and store at –20°C.

Spectrophotometric analysis for TNP substitution ratio

Measure extinction at 348 nm of protein before use (blank) and after trinitrophenylation (dilute when E is >>2.0!).

$$X = (\text{Dilution factor} \times E_{348-\text{TNP}}) - E_{348-\text{blank}}$$

X is divided by the molar extinction coefficient of 15,400 to give Y:

$$Y = \frac{X}{15,400}$$

The amount of protein used in (g l^{-1} = mg ml^{-1}) is divided by the molecular weight of the carrier to give Z. Then

$$\text{No. of TNP groups/mole carrier} = \frac{Y}{Z}$$

Note:

AFC specific for TNP can only be found after immunization of experimental animals with TNP- or DNP-carrier molecules. To prepare such TNP-carrier molecules, the same protocol as for TNP–AP can be followed. However, the amount of TNP-sulfonic acid added should be about 10 times greater on a molar basis. TNP-carrier molecules induce effective responses in the range of 30–45 TNP groups per 100 kDa carrier. When preparing TI-2 TNP molecules and starting out with neutral polysaccharides on which amino groups are introduced, one should retest the TI-2 characteristics of the molecule when more than 30 TNP groups per mole of carrier are introduced (Kraal et al 1989).

8. Purify by column chromatography preferably, or by dialysis against PBS when in a hurry.
9. Store at –20°C after mixing 1:1 with glycerin.

Conjugation to glutaraldehyde-activated alkaline phosphatase

Materials

Alkaline phosphatase (type VII-T, Sigma P-6774; other types and suppliers give inferior results; please compare first)

Phosphate-buffered saline (PBS), pH 7.2

25% glutaraldehyde (Baker's)

Antibody/antigen to conjugate (note molecular weight, see above)

0.2 M glycine–PBS

Glycerin

Dialysis tubing (membrane cutoff 40 kDa)

Method

1. Dilute 5 mg AP (Sigma, AP, type VII-T) in 1 ml PBS.
2. Dialyze extensively against PBS.
3. Add glutaraldehyde to the dialysis fluid (100 ml), to an end concentration of 0.2%; activate AP by overnight incubation at 4°C.
4. Remove excess glutaraldehyde by dialysis against PBS (2 × 1000 ml for 3 h), and transfer the activated AP to a test tube.
5. Add 2 mg antibody (in 1 ml PBS) to the activated AP.
6. Leave the reaction mixture overnight at 4°C.
7. Add 0.1 ml 0.2 M glycine–PBS and mix for 2 h at room temperature (to block

Conjugation to periodate-oxidized horseradish peroxidase

Materials

Horseradish peroxidase (HRP) (Sigma P-8375)

Sodium metaperiodate ($NaIO_4$) (Sigma P-1878)

MQ water or distilled water

Ethylene glycol (Merck)

Antibody/antigen to conjugate (note molecular weight, see above)

Sodium carbonate buffer, 1 M, pH 9.5

Sodium borohydride 98% (Aldrich no. 19807–2); working solution is 4 mg ml^{-1} MQ water

0.2 M glycine-PBS

Phosphate-buffered saline (PBS), pH 7.2

Glycerin

Dialysis tubing (cellulose membrane; Amicon or Ultrafree MC apparatus)

pH indicator paper

Method

1. Dissolve 4 mg HRP in 1 ml double-distilled (d-d) water; the color is brown.
2. Add 0.2 ml of a freshly prepared 0.05 M sodium periodate solution (11 mg

$NaIO_4$ in 1 ml d-d water); color should turn to green.

3. Incubate; stir gently (shield from light) for 20 min at room temperature.
4. Dialyze overnight against 0.001 M sodium acetate buffer (pH 4.4; dissolve 51 mg $AcONa \cdot 3H_2O$ in 1000 ml d-d water, and add 36 µl glacial acetic acid) at 4°C; transfer to test tube.

   Instead of removing periodate by dialysis one can inactivate it by adding 0.1 ml ethylene glycol to solution.
5. Add 1 ml with 8 mg antibody (antibody in PBS or culture medium).
6. Immediately add approximately 20 µl 0.2 M sodium carbonate buffer (pH 9.5; dissolve 13 mg $NaHCO_3$ and 16 mg Na_2CO_3 in 1 ml d-d water) to reach pH 9.5; check droplet on pH paper.
7. Stir well and incubate for 2 h at room temperature.
8. Add 0.1 ml fresh sodium borohydride solution (4 mg $NaBH_4$ in 1 ml d-d water).
9. Incubate for 2 h at 4°C.
10. Add 0.1 ml 0.2 M glycine–PBS and mix for 2 h at room temperature.
11. Purify by column chromatography preferably (Sephacryl S-200, 35 × 2.5 cm, PBS) or by dialysis against PBS when in a hurry.
12. Store at −20°C, preferably after mixing 1:1 with glycerin.

Conjugation to β-galactosidase with maleimidobenzoyl-succinimide

Materials

Antibody/antigen to conjugate (note molecular weight, see above)

N,N-Dimethylformamide (DMF; BDH Laboratory supplies)

MBS solution: 20 mg ml^{-1} in DMF (MBS, (m-maleimidobenzoyl-N-hydroxysuccinimide ester, #22310, Pierce; mol wt 314.2)

Phosphate-buffered saline (PBS), pH 7.2

β-Galactosidase (Boehringer Mannheim #745 731): 10 mg ml^{-1} in MQ or d-d water

Magnesium chloride ($MgCl_2$)

β-Mercaptoethanol

NaN_3

Glycerin

PD-10 column (Pharmacia #17-0851-01)

Ultrafiltration cell (Amicon)

Method

1. Take 10 mg antibody in 1 ml PBS and stir in a test tube on ice.
2. Add 157 µl maleimidobenzoylsuccinimide (MBS) solution (20 mg m-maleimidobenzoyl-N-hydroxysuccinimide ester in 1 ml DMF) dropwise to the antibody solution, every 5 min in 3 equal parts. In total 1 µmol MBS is added to 1 mg antibody.
3. Stir for 15 min. Meanwhile wash PD-10 column (Pharmacia) with at least 25 ml PBS.
4. Bring antibody–MBS solution to 2.5 ml precisely with PBS and centrifuge for 10 min at 3000 rpm to remove any precipitation.
5. Introduce the supernatant on to the PD-10 column and collect (later discard) the passage. Elute the column with 3.5 ml PBS, and collect the eluate.
6. Concentrate eluate to minimum 2–3 mg ml^{-1} with an Amicon ultrafiltration cell. Lower protein concentrations give inefficient

couplings, but quantities are meant as ratios: concentrate to the amount of protein you need.

7. Place antibody–MBS solution in test tube and add dropwise β-galactosidase solution (10 mg ml^{-1} d-d water). In total add 1 mg β-galactosidase per 1.25 mg antibody. After addition mix for 1 h at room temperature.

8. Add MgCl$_2$ and β-mercaptoethanol to antibody–MBS–β-galactosidase solution, both up to a final concentration of 10 mM.

9. Store the conjugate at 4°C with 0.1% NaN$_3$ or dilute the conjugate 1:1 with glycerin (add stabilizing proteins when indicated) and store at −20°C.

Revelation of HRP and AP

Proceed as described in chapter 32.7.

Revelation of β-galactosidase with X-galactopyranoside in turquoise

Materials

5-Bromo-4-chloro-3-indolyl-β-D-galactopyranoside (X-gal; Boehringer Mannheim #651 745)

N,N-Dimethylformamide (DMF; BDH Laboratory supplies)

MgCl$_2$·6H$_2$O, 1.1 mM

Potassium ferricyanide, 50 mM

Potassium ferrocyanide, 50 mM

Phosphate-buffered saline (PBS), pH 7.2

Hematoxylin (BDH Laboratory Supplies)

Glycerin/gelatin

Method

1. *Solution A*: Dissolve 10 mg X-galactopyranoside (5-bromo-4-chloro-3-indolyl-β-D-galactopyranoside) in 500 µl DMF. Solution A can be stored at 4°C for only 1 week.

2. *Solution B*: Add 7 ml of 1.1 mM MgCl$_2$·6H$_2$O to 500 µl 50 mM potassium ferricyanide and 500 µl 50 mM potassium ferrocyanide. Solution B can be stored at 4°C for some weeks and at least 8 months at −18°C.

3. Mix 12.5 µl of solution A with 500 µl of solution B (amounts are meant as ratios). Always use fresh working solution.

4. Incubate the slides horizontally in a closed incubation box under high humidity or vertically in coplin jars for 10–60 min at 37°C.

5. Reaction is ended by transferring the slides to PBS (rinse 2 × 5 min).

6. Slides are counterstained in haematoxylin solution for 1–5 s and rinsed with tapwater.

7. Slides are mounted in glycerin/gelatin with a cover slip.

References

Claaasen E, Jeurissen SHM (1996a) A step to step guide to *in situ* immune response analysis of lymphoid tissues by immunohistochemical methods. In: Weir DM, Blackwell C, Herzenberg L, eds. Handbook of Experimental Immunology, 5th edn. Blackwell Scientific, Oxford, in press.

Claassen E, Jeurissen SHM (1996b) Characterisation of antigens, antibodies and cells in lymphoid tissues by immunohistochemical methods. Current Protocols in Immunology, Wiley, London, chapter 5.9, in press.

Claassen E, Gerritse K, Laman JD, Boersma WJA (1992) New immuno-enzyme-cytochemical stainings for the *in situ* detection of epitope specificity and isotype of antibody forming B cells in experimental and natural (auto) immune responses in animals and man: A review. J Immunol Methods 150: 207–216.

Hoefakker S, Boersma WJ, Claassen E (1995) Detection of human cytokines *in situ* using antibody and probe based methods. J Immunol Methods 185: 149–175.

Kraal G, Ter Hart H, Meelhuizen C, Venneker G, Claasen E (1989) Marginal zone macrophages and their role in the immune response against T-independent type 2 antigens. Modulation of the cells with specific antibody. Eur J Immunol 19: 675–680.

Laman JD, Gerritse K, Fasbender M (1990) Double immunocytochemical staining for the *in vivo* detection of epitope specificity and isotype of antibody forming cells against synthetic peptides homologous to human immunodeficiency virus-1. J Histochem Cytochem 38: 457–462.

Laman JD, Van den Eertwegh AJM, Claassen E, Van Rooijer N (1991a) Cell–cell interactions, *in situ* studies of splenic humoral immune responses. In: Fornusek L, Vetvicka V, eds. Immune Accessory Cells. CRC Press, Boca Raton, pp. 201–231.

Laman JD, Kors N, Heeney JL, Boersma WJ, Claassen E (1991b) Fixation of cryo-sections under HIV-1 inactivating conditions, Integrity of antigen binding sites and cell surface antigens. Histochemistry 96: 177–183.

Liu Y-J, Oldfield S, MacLennan IC (1988) Memory B cells in T cell dependent antibody responses colonize the splenic marginal zones. Eur J Immunol 18: 355–360.

Van den Eertwegh AJM, Boersma WJ, Claassen E (1992) Immunological functions and *in vivo* cell–cell interactions of T lymphocytes in the spleen. Crit Rev Immunol 11: 337–380.

Van den Eertwegh AJM, Laman JD, Noelle RJ, et al (1994) *In vivo* T–B cell interactions and cytokine production in the spleen. In: CD40 and its ligand in the regulation of humoral immunity. Semin Immunol 6: 327–336.

Van Rooijen N, Claassen E (1986) Recent advances in thedetection and characterization of specific antibody forming cells in tissue sections. A review. Histochem J 18: 465–471.

Zegers ND, Boersma WJ, Claasen E (1995) Immunological Recognition of Peptides in Medicine and Biology. CRC Press, Boca Raton.

Quantification of immunoglobulin concentration by ELISA

13.8

Sulabha S. Pathak
Adri van Oudenaren
Huub F.J. Savelkoul

Department of Immunology, Erasmus University, Rotterdam, The Netherlands

TABLE OF CONTENTS

Introduction

Enzyme-linked immunosorbent assay (ELISA) was first described by Engvall and Perlmann (1971) and Schuurs and Van Weemen (1977) and provides a safe and simple method of measuring antigen-specific and total immuno-globulins (Ig). ELISA falls under the category of heterogeneous immunosorbent assays in which (a) the bound and free fractions of Ig (ligand) are physically separated by a washing procedure and (b) the antigen (Ig) to be detected or determined is either directly or indirectly physically attached to a solid phase.

Various types of ELISA have been developed and described in the literature. The commonly used ones can be divided into different categories depending upon the following criteria.

- The entity to be detected or quality to be determined — antigen- or antibody-specific, qualitative or (semi-)

quantitative ELISA.
- The setup of the assay — direct, indirect, competitive, inhibitory, noncompetitive, two-site (sandwich), or amplified assays.
- Nature of the substrate employed — fluorogenic, chromogenic, luminogenic.
- The amounts of reactants used — macro-ELISA performed in a tube, micro-ELISA employing a microtiter plate, or micro-ELISA needing micro quantities (5 µl) of sample performed in Terasaki trays, the Terasaki-ELISA.

Moreover, the solid phase used to adsorb the protein coat can be of various materials — glass, polypropylene, poly(vinyl chloride) (PVC), polystyrene, polycarbonate, nitrocellulose, silicone, etc., — though only polystyrene or PVC microtiter plates are commercially available.

Setup of the assay

The majority of ELISAs are performed in polystyrene microtiter plates using commercially available equipment, facilitating easy and automatic handling of the system. A serum sample of 100 µl is used to determine the concentration of Ig (IgM, IgA, IgG) in moderately sensitive ELISA. However, if simultaneous detection of a number of Ig classes (especially rare ones like IgD and IgE), autoantibodies, anti-idiotypic antibodies, etc., is required, a fairly large serum sample is needed. This problem is more pronounced in longitudinal studies on sera of mice or children, where it is often difficult to acquire the amount of serum needed to perform the assay. In the case of monoclonal antibody production, only very small quantities of supernatants are available for the assay, since testing for Ig production is desired as soon as

possible following fusion. We therefore adopted the Terasaki-ELISA performed in 60-well Terasaki trays using 5 µl of sample per well, thus reducing the amount of serum or supernatant needed. One further advantage of this method is that very small amounts of coating and detecting reagents are required. The Terasaki trays are made of polystyrene, so that the adsorption characteristics of these plates can be expected to be essentially similar to those of polystyrene microtiter plates. For the determination of total Ig, we use a two-site (sandwich) ELISA system in which the antigen (Ig) is characterized by two geometrically and molecularly distinct binding sites for the two specific antibodies employed. Thus, the Ig is sandwiched between the known anti-isotypic catching antibody adsorbed to a solid phase and a detecting antibody, in effect giving a much higher degree of specificity to the overall system. The catching as well as the detecting antibodies can be identical or different but they must be heterologous to the Ig determined. For antigen-specific assays, we use the antigen as the catching moiety and a heterologous detecting antibody. After adsorbing the antigen or catching antibody to the solid phase, all subsequent incubations are done until equilibrium is established and competition can no longer occur between the labeled and unlabeled ligand involved. Moreover, all other reagents, except the sample, are in slight excess, so that only the Ig that is bound from the sample is the limiting step. Thus, the signal obtained is directly related to the amount of bound Ig.

Other ELISA systems, whether competitive, homologous, one-site, homogeneous, or non-equilibrium-reaching, are well documented and will not be discussed here. The reader is referred to the available reviews (Wisdom 1976; Schuurs and Van Weemen 1977) and books (Voller et al 1981; Avrameas et al 1983; Bizollon 1984). Some relevant points to be considered before performing heterogeneous ELISA will be described here, after which essential parameters such as sensitivity, detection limit, and accuracy will be discussed.

Coating of the solid phase

Theoretical aspects

The coating of the solid phase can be done either by covalent coupling to the plastic matrix or to grafted plastic surfaces, or, as is done generally, by simple adsorption. Most proteins adsorb to plastic surfaces, probably due to hydrophobic interactions between the nonpolar (glyco)protein moieties with a net charge of zero and the nonpolar plastic matrix. The rate and the extent of adsorption (and therefore of coating) depend upon a number of different factors. The more important ones being the following.

- The nature of the molecule being adsorbed — the net charge, hydrophobicity, size, molecular weight, diffusion coefficient, etc.
- The concentration of the adsorbing molecule, which determines the occupancy rate.
- The nature of the coating buffer — its composition, molarity and pH — but also the absence of detergents, adsorbed spacer molecules for peptide binding (e.g., poly(L-lysine)), or agents permitting covalent coupling of highly charged molecules (e.g., glutaraldehyde).
- The solid phase. The chemical nature of the solid phase will obviously influence the extent of adsorption, as will the ratio of surface area to be coated to the volume of the coating solution (typically $1.5\,ng\,mm^{-2}$ or 6×10^9 molecules mm^{-2}, since a microtiter well is $33\,mm^2$ in surface this equals maximum immobilization of 50–150 ng/well). The shape of the wells, resulting in shearing or stress forces, can be critical. Large differences in the adsorbing properties of plates of different manufacturers and even among different batches from the same manufacturer have been observed.
- The temperature and duration of adsorption (typically overnight at 4°C, 3 h at 37°C or even 1 min in a microwave).

Langmuir proposed an equation to fit adsorption data which implies an equilibrium between protein molecules and a limited number of surface active sites on the solid phase. Both the equilibrium constant K_A and the maximal surface concentration of the adsorbed molecules can be calculated from the experimental data using the equation

$$\frac{1}{m} = \frac{1}{b} + \frac{1}{bK_Ac}$$

where c is the molar concentration of the protein in equilibrium with m grams of adsorbed protein per cm^2, and b and K_A are constants. The value of b corresponds to the maximum amount of protein that can be adsorbed to solid phase (saturation). K_A is the association constant corresponding to the reaction

$$P + S \underset{k_d}{\overset{k_a}{\rightleftharpoons}} PS ; \qquad K_A = \frac{k_a}{k_d}$$

where S represents surface active sites and P the protein to be adsorbed.

When only one kind of surface site is available and only these sites are occupied by the protein (no protein–protein association), a monolayer is formed according to this reaction. If two or more kinds of surface sites with different affinity for the protein are available, more complex reactions will take place.

The main criticism directed against this approach is that the equilibrium may not be a reversible one, as assumed by the Langmuir treatment. The basis for this criticism is that several authors could not detect protein desorption from plastic solid phases other than plates (Cantarero et al 1980). Morrisey (1977) and Lehtonen and Viljanen (1980), among others, on the other hand, have reported such desorption. Anything from 30% to 68% desorption has been reported. The underlying cause for this discrepancy might be that some proteins may indeed be irreversibly adsorbed or may undergo conformational change after being adsorbed, and thus escape detection by the enzyme used for the studies, viz., horseradish peroxidase. This question needs further study since desorption influences the reliability of ELISA. It should be noted that in sandwich techniques, since a large excess of capturing antibody exists on the solid phase, even an important loss may not influence the final result.

The adsorptive characteristics of polystyrene for different proteins like albumin and Ig show that the adsorptive behavior is not explainable only on the basis of simple charge differences and molecular weight and size. Adsorption is, however, dependent on the concentration of protein in the input solution and increases proportionally with temperature and incubation time. Beyond the range of concentrations where the proportion of protein adsorbed is independent of input, i.e., the region of independence, the kinetics of adsorption change and then seem to be inversely proportional to the protein size. The presence of competing proteins has no effect on the adsorptive characteristics of a given protein provided the total protein concentration of the mixture was within the range of independence. Up to a certain limit, a constant fraction of the proteins is adsorbed to plastic surfaces. For example, 80% of the IgM but only about 25% of BSA is adsorbed, with a limit of about 1.5 ng mm^{-2} for both proteins, independent of the input. According to Butler (1981), the protein molecules at this limit become equidistantly distributed on the surface, and the failure to exceed this coverage (about 1/3 of the surface) is due to steric hindrance. As these data were obtained from microtiter plates having an effective coating area of around 0.1 mm^2 μl^{-1}, they imply that at maximum 0.2 μg ml^{-1} of protein can be adsorbed before steric hindrance occurs. As a rule, under optimal working conditions a linear dose–response curve is obtained in ELISA. At higher inputs of analyte, deviation from this linearity can occur. This phenomenon is referred to as the 'hook effect' and was originally used synonymously with the 'prozone phenomenon' (Nakamura et al 1986). However, they are now recognized to be two independent phenomena. In ELISA, the 'hook effect' is commonly assumed to mean a release of coated antigen or antibody from the plate during the subsequent washing steps. However, in reality, it is the result of the release of coated antigen or catching antibody into the solution during sample incubation. As a result, both the coated (Ag^c) and the free (Ag^{sol}) molecules compete for binding sites on the Ig molecule (Ab). Therefore, two different equilibria can be established, leading to two different affinity constants (K_1 and K_2):

$$[Ag]^c + [Ag]^{sol} + [Ab] \rightleftharpoons [AgAb]^c + [AgAb]^{sol}$$

$$K_1 = \frac{[AgAb]^c}{[Ag]^c[Ab]} \; ; \quad K_2 = \frac{[AgAb]^{sol}}{[Ag]^{sol}[Ab]}$$

The ratio between these two affinity constants determines the extent of the hook effect. As a result the plateau value of ELISA titration curves may be lowered by increasing (but nonsaturating) concentrations of the coat (Pesce et al 1983).

The 'prozone phenomenon' is said to occur when a decreased instead of an increased signal is observed at higher levels of the analyte. It is presumed to be the result of monovalent binding between the detecting antibody or the conjugate and the analyte. It is thus a function of the affinity of the analyte and the detecting reagents, as well as the concentration, and is independent of the coat (Vos et al 1987). Subsequent washings dissociate weakly bound ligands from the plate and cause 'bleeding'. Bleeding refers to a progressive loss of antibodies bound to the immobilized ligand due to dissociation of immune complexes containing antibodies with low affinities.

Practical aspects

The usefulness of an ELISA depends upon the coating efficiency and the reproducibility of the coat. Thus, it is essential to ensure sufficient adsorption of the protein to the plastic surface. Various strategies can be employed to achieve this.

Use of coating buffers: Engvall and Perlmann (1971) have suggested the use of a high-pH–high-ionic-strength buffer (0.1 M Na_2CO_3, pH 9.6) for efficient coating of antibodies, which is perhaps the most widely used. Recent studies, however, have shown, that the method of passive adsorption of proteins — especially Ig — at this high pH leads to extensive protein denaturation. Only 10% of the adsorbed IgG

was found to retain its activity under these conditions (Butler et al 1992).

PBS (10 mM sodium or potassium phosphate buffer, pH 7.2–7.8) with 150 mM NaCl is also popular for coating of Ig, and is thought to cause much less protein denaturation. For antigen-specific ELISA, on the other hand, a large variation in the binding of various proteinic antigens to plastic surfaces can be expected. The choice of a proper buffer can help increase the binding. A variety of buffers can therefore be employed. Scott et al (1985), for example, found that 0.1 M acetic acid increased the efficiency of coating of dietary antigens. For a new assay, a proper investigation to establish the most suitable buffer is thus necessary. This may not, however, increase the coating efficiency sufficiently, and to further increase the binding of the proteins pretreatment of either the proteins or the polystyrene surface can be attempted.

Pretreatment of proteins: Partial denaturation and/or aggregation of proteins by heat treatment may help increase the binding of proteins. For several antigens (myoglobin, ovalbumin, albumin, transferrin, ferritin), an increase in sensitivity of ELISA was obtained after cross-linking by a carbodiimide-promoted condensation reaction. Pretreatment of the antibody with glutaraldehyde was found to increase its binding by 2- to 5-fold. Any kind of treatment of the proteins is likely to cause denaturation or partial unfolding of the molecule, and thereby change its antigenic properties. Thus, proper controls must be included to avoid erroneous results.

Pretreatment of the polystyrene surface: Activation of the polystyrene surface can be achieved by various means. Kemeny et al (1986) suggested ultraviolet irradiation of the plates. Boudet et al (1991) have successfully coated small synthetic peptides to microtiter plates by exposing them to UV radiation for up to 20 min. Pretreatment of the plate with a freshly diluted 2% solution of glutaraldehyde for 2 h has been found to increase antibody binding significantly.

Klasen and Rigutti (1983) have found that activation of polystyrene by 0.025% glutaraldehyde leads to reliable coating even by water-insoluble proteins dissolved in urea. The effect of glutaraldehyde is pH-dependent and is less polymerizing at low pH (5.0). Biotinylated microtiter plates are now also commercially available. These plates can be used to immobilize Ig via a streptavidin bridge. Ig thus immobilized have been found to retain a high degree of reactivity (Butler et al 1992). Alternatively, streptavidin can be passively adsorbed to the microtiter plate, and biotinylated proteins used for coating. Such methods are especially useful when dealing with small, synthetic peptides (Ivanov et al 1992).

For small antigens or haptens, it is advisable to precoat the plates with a polylysine or polyaspartate spacer or to use a hapten carrier conjugate for coating. Cyanogen bromide has likewise been used to increase the coating efficiency. Linking of antigen to the plastic via an antibody bridge decreases the variation due to differing adsorption of different antigens. Furthermore, it may also increase the sensitivity of the assay and is therefore used by many workers (Urbanek et al 1985).

To detect antibodies directed against cell-surface antigenic determinants, it is possible to coat the cells in monolayers on the solid phase using glutaraldehyde (0.02%, 15 min at room temperature), especially since such low concentrations of glutaraldehyde do not alter the conformation of the antigenic determinants.

One note of caution is necessary however. Pretreatment of the proteins or the polystyrene surfaces may lead to allosteric changes in the coated molecule, and thus decrease its immunoreactivity.

Use of various time–temperature combinations: Overnight coating at 4°C as suggested by Engvall and Perlmann (1971) is most frequently used. Various time–temperature combinations have been suggested in the literature. These vary from 1–2 h at 37°C to few hours at elevated temperatures followed by overnight incubation at 4°C.

All ELISA incubation times, including coating time, can be reduced by exposing the plates to 2.45 GHz microwaves combined with air blowing. Such usage, however, has to be carefully regulated, as otherwise denaturation of the proteins can occur. We have tried coating plates by allowing them to dry completely (37°C for 1 h 15 min with a fan) so as to ensure maximal coating. However, we did not find any increase in the overall sensitivity of the assay, perhaps because drying promotes protein denaturation (unpublished results). Ansari et al (1985) have also found that drying of plates following coating led to decreased activity of the solid-phase-bound antibody.

The concentration of proteins to be used in the coating solution has of course to be determined empirically. Excessively high protein concentrations should be avoided as they are likely to cause multilayers of proteins on the solid phase owing to protein–protein interaction. Also, competition between the immune serum and surface active sites for the antigen molecules may occur in ELISA. If this competition occurs, it will lead to affinity- and concentration-dependent desorption, leading to a nonlinear dose–response curve.

Engvall (1980) has suggested 1–10 μg ml^{-1} as the optimal concentration for coating of antigens and antibodies. This concentration can be used as a starting point. In our own experience with the microtiter system, we have observed that higher than optimal coating concentrations lead to a lowering of the signal. This could be explained by the formation of multiple protein–protein layers at higher coating concentrations, with consequent desorption during the ELISA.

The coated plates can be stored either wet in PBS with 1–2 mM NaN$_3$ at 4°C or dried in a desiccator for periods up to 4 weeks. In our experience, however, storing wet plates coated with monoclonal antibody at 4°C for more than 12–16 h leads to decreased sensitivity, probably because prolonged storage at a high pH leads to structural changes in the molecule or because of a time-related denaturation of the protein due to adsorption to plastic. However, we have observed that storing wet plates coated in PBS (pH 7.2) for up to 4 days has no undesirable effect on the ELISA.

After the coating has been completed, the trays are washed thoroughly, generally with the same running buffer as is used in the rest of the steps. PBS is commonly used, although Tris-buffered saline can also be employed.

The washing removes all the excess unadsorbed protein. Nonionic detergents such as

Tween 20 (TW20) are generally used in the wash buffer as they lower the surface tension and thus prevent formation of new hydrophobic interactions between the added protein and the solid phase. The effect of washes on dissociation of immune complexes is dependent on the affinity of the antibodies employed. When antibodies are immobilized on the solid phase at a concentration of 10^{-7} M (fairly typical for microtiter plates), low-affinity antibodies ($K_D = 10^5$–10^6 M) may be rapidly dissociated and underrepresented. With higher affinities ($K_D = 10^8$–10^9 M), however, activities are representative for concentrations.

All the remaining unoccupied binding sites are saturated with an unrelated, noninterfering protein to control nonspecific adsorption. An extra protein coating aimed at saturating the plastic support not occupied by the antigen has also been suggested. BSA (1–5%), gelatin (0.02–0.1%), or casein either in combination with TW20 or alone are in common use. Schonheyder and Andersen (1984) as well as Pruslin et al (1991) have shown that BSA post coating and addition of BSA to serum diluent decreases background readings and increases specific activity. However, batch-to-batch variation has often been found in BSA. The likelihood of the presence of contaminating bovine antibodies in BSA also cannot be ignored. Reservations about its general applicability have therefore been expressed (Smith et al 1993). TW20 was found to be a satisfactory blocking agent in experiments with murine monoclonal antibodies and high-titer animal antisera. In contrast, high nonspecific binding that was not abolished by TW20 was found to be a major problem in experiments with low-titer rabbit sera as well as in some assays with human antibodies from patient sera. Accordingly, the most suitable blocking agent will depend upon the individual system, that is, on the coat, the detecting reagent(s), and the nature of the sample, and will vary widely. Blocking of rat Ig immobilization, for example, confirms that small molecules (<10 kDa) block more effectively than larger molecules (>30 kDa). Ig were found to be able to penetrate blocking layers of BSA, giving high background levels and early failure of blocking with dilution.

Choice and preparation of the ligand

Theoretical aspects

Several characteristics of the enzyme to be conjugated to the detecting antibody need to be considered in choosing the enzyme (Table 13.8.1). These include the turnover number (i.e., the molecular activity), specific activity, detectability of the product, absence of substances likely to interfere with enzyme activity in the fluids to be measured, and so on.

Table 13.8.2 lists some of the enzymes in common use. The size of the antibody–enzyme conjugate can also influence the sensitivity of the ELISA system. The binding of the detection reagent at high concentrations of the primary antibody appears to be sterically inhibited in direct proportion to the size of the conjugate used, leading to a marked deviation from linearity in this region of the binding curve (Koertge and Butler 1985). Conjugation reactions involving only proteins (enzymes and antibodies) can be carried out easily using various bifunctional reagents, as extensively reviewed (Tijssen 1985).

Table 13.8.3 lists some of the most popular cross-linking methods. Generally speaking, use

Modulation of the humoral immune response

Table 13.8.1 Criteria for choice of enzyme label

1. Availability of low-cost purified homozygous enzyme preparations
2. High specific activity or turnover number
3. Presence of reactive residues through which the enzyme can be cross-linked to other molecules with minimal loss of enzyme and antibody activity
4. Capability of producing stable conjugates
5. Enzyme should be absent from biological test samples
6. Assay method that is simple, cheap, sensitive, precise, and not affected by factors present in the sample
7. Enzyme, substrates, cofactors, etc. should not pose potential health hazards

Table 13.8.2 Comparison of some enzymes available for ELISA

Enzyme	pH optimum	Specific activity (U mg^{-1})	Molecular weight (kDa)	Practical substrates	
				Visual	Fluorescent
Alkaline phosphatase	8–10	400	100	ONPP	MUF-P FM-P
Peroxidase	5–7	900	40	5-AS	NADH OPD HPA ABTS
β-Galactosidase	6–8	400	540	ONPG	MUF-G

Abbreviations: ABTS, 2,2-azino-di-(3 ethylbenzothiazolinsulfone-6) diammonium salt; 5-AS, aminosalicylic acid; FM-P, fluorescein methyl phosphate; HPA, hydroxphenylacetic acid; MUF-G, methylumbelliferylgalactose; MUF-P, methylumbelliferyl phosphate; NADH, nicotinamide–adenine dinucleotide; ONPG, *ortho*-nitrophenylgalactose; ONPP, *ortho*-nitrophenyl phosphate; OPD, *ortho*-phenylenediamine.

Table 13.8.3 Common methods used to prepare enzyme–antibody conjugates

Reactive group	Bifunctional reagent
Amino residues	Glutaraldehyde 1,4-Benzoquinone Succinimidyl-3-(2-pyridyldithiopropionate) (SPDP)
Sulfhydryl groups	*N, N'-o*-phenylenedimaleimide *m*-maleimidebenzoyl-*N*-hydroxysuccimimide ester (MBS)
Carbohydrate moieties	Periodate oxidation
Carboxyl groups	1-Ethyl-3-(3-dimethylaminopropyl) carbodiimide

of a heterobifunctional cross-linker should be preferred since this gives rise to conjugates free from polymeric products that are observed when using a homobifunctional linker. It has been suggested that the sensitivity of the ELISA is partially influenced by the method by which the enzyme is coupled to the detecting antibody. Tsang et al (1984) showed, using glutaraldehyde- and sodium periodate-coupled peroxidase–anti-human conjugates, that the former conjugates are more useful in quantitative assays where wide ranges are desirable. The latter conjugates, on the other hand, are more suited to qualitative assays, where sensitivity is more important.

After completion of the conjugation reaction, proper recovery of the enzyme–protein conjugate is important. Moreover, it is essential to ensure that the enzymatic activity resides only in the conjugated form and that the solution is free of contaminating native enzyme molecules. The specific activity of the enzyme can be measured by either end point or rate analysis. It is necessary to recognize that the optimal conditions of pH, substrate concentration, activators used, etc. for the conjugated enzyme can differ from those for the native enzyme. Thus, the experimental conditions for obtaining maximum activity of the conjugated enzyme in an ELISA may differ from those observed in the fluid phase. The proximity of the solid phase affects the microenvironment of the enzyme and may also lead to molecular conformational changes in the enzyme. It is thus not appropriate to optimize the conditions in a fluid phase and expect them to hold also in ELISA (Tijssen 1985). Furthermore, the antibody part of the conjugate must retain its binding affinity, and this has also to be confirmed in an assay similar to the one for which the conjugate has been prepared, i.e., an ELISA.

The efficiency of conjugation represents the amount of immunoreactive enzyme recovered. Factors that influence the efficiency are the type of reaction used for conjugation; the proportion of reactants, i.e., enzyme and antibody; and the heterogeneity/homogeneity of labeling sites for preparation of the immunogen and of the enzyme conjugate. Azimzadeh et al (1992) have suggested conjugating F(ab)' fragments instead of the whole antibody to the enzyme. This approach has the advantage that the enzyme and antibody fragments are present in the conjugate in a stoichiometric ratio of 1:1 and complications arising from monogamous bivalent binding of the antibody to the antigen (antibody) are eliminated (Azimzadeh and Van Regenmortel 1991).

The antiserum employed in the ELISA is crucial for the specificity of the assay and, by virtue of its affinity for the antigen to be detected in the assay, also for its sensitivity and detection limit. It is therefore evident that the antiserum preparation should be characterized as extensively as possible. This information is often not available for commercial antiserum preparations. Commonly, the only information available is that they are claimed to be affinity purified, i.e., to be adsorbed onto affinity columns bearing immobilized, potentially cross-reacting substances. Such preparations should therefore be tested elaborately before use. They must be tested in precipitation based assays as well as in inhibition ELISA. Displacement of a labeled antigen from the complex by competition with increasing concentrations of competing nonlabeled antigen reveals shared or cross-reactivity. Antibodies generally have lower affinities for cross-reactive antigens, so that higher concentrations of these antigens are necessary to obtain maximal inhibition. The degree of maximal inhibition is dependent on the degree of cross-reactivity between the various antigens. In inhibition-type ELISA a sufficient range of antibody concentrations should be tested on a panel of purified myeloma or hybridoma proteins. The panel should include specimens of every isotype and some of the isotypes should be represented by a lambda or a kappa light chain. Proteins of different idiotypes and allotypes should also be included, certainly when testing the isotype specificity of the antiserum.

This is all the more necessary when the preparation is a polyclonal antiserum induced by injecting heterologous animals with purified Ig of the appropriate (sub)class, since under these conditions a substantial part of the antiserum raised can be directed against the allotype, light-chain, or idiotype. When tested under such stringent criteria, not all commercially available sera turn out to be suitable for sensitive assays like ELISA.

The methodology for the induction and the purification of the IgG fraction of a specific antiserum by the use of monoclonal antibodies is essentially routine. Monoclonal antibodies are sensitive to purification procedures and the three-dimensional structure of the antibody can easily be distorted. Moreover, monoclonal antibody preparations often consist of incomplete chains. This could result in the use of partially denatured and/or incomplete Ig preparations in the raising of the antiserum. The quality of the antibody response can depend upon the immunization conditions and the type of adjuvant used. These factors can therefore affect the quality of the antiserum raised and should be considered while raising antisera for use in ELISA.

Apart from antibody–enzyme conjugates, other ligand–enzyme conjugates are in current use. The biotin–avidin system with a high affinity constant of $10^{15} M^{-1}$ is one such example (Bayer and Wilchek 1978). Thus, a system wherein the biotin is conjugated to the specific or detecting antibody or antigen and the avidin is linked to the end-product developing enzyme can be used, especially since biotin bound to a macromolecule retains its high affinity for avidin. Alternatively, biotin can be conjugated to both the detecting antibody and enzyme, and avidin then employed as a bridge between these biotinylated proteins. The use of biotin and avidin ensures that a high degree of enzyme is bound specifically to the Ig in the sample. In recent years the biotin–streptavidin system has found wide application. Biotinylated conjugates of polyclonal and monoclonal antibodies as well as enzymes conjugated to streptavidin are now commercially available, increasing the popularity of the system. Using this system we have achieved a sensitivity of $1–10 \, ng \, ml^{-1}$ for our isotype-specific microtiter ELISA.

S. aureus protein A and protein G of group G streptococci react with the Fc part of many mammalian species. It is especially useful as it can behave as a purified anti-antibody of restricted specificity with no species specificity.

Enzyme–protein A conjugates can therefore be used to detect Ig from samples in ELISA. A sandwich ELISA with enzyme-labeled protein A and unlabeled isotype-specific detecting antibody having a high affinity for protein A is currently in use. This method can be useful in assaying Ig with no binding affinity for protein A; e.g., rat monoclonal antibody directed against mouse cell-surface determinants or antigen specific as well as total human or mouse IgE antibody.

Practical aspects

In our fluorimetric Terasaki-ELISA assays, we have chosen the enzyme β-galactosidase from E. coli (EC 3.2.1.23) for a number of reasons. Although both alkaline phosphatase and β-galactosidase are widely used in combination with fluorigenic substrates, β-galactosidase is mostly used if high detectability is required. Also, alkaline phosphatase has the further disadvantage of producing reagent blanks which are 50-fold higher than those of β-galactosidase owing to the spontaneous degradation of the substrate. Moreover, β-galactosidase has a high specific activity ($400 \, U \, mg^{-1}$) and a high turnover number (12,500 substrate molecules transformed per mole of enzyme per min), and can be linked to the antibody by a number of chemical methods with a good degree of retention of its enzymatic activity without altering the immunological activity of the antibody. The conjugate is soluble and stable under the assay conditions and can be stored for prolonged time at 4°C. Last but not least, the enzyme itself, its substrate, or factors that interfere with enzyme activity, such as inhibitors, are all absent from the samples assayed.

The enzyme is conjugated to the antibody with the one-step glutaraldehyde method of Avrameas et al (1978, 1983). The antibody to be coupled is dialysed extensively against $0.1 \, M$ potassium phosphate buffer (pH 7.8) and a final protein concentration of $4 \, mg \, ml^{-1}$ is achieved. β-Galactosidase (specific activity $30 \, U \, mg^{-1}$;

Table 13.8.4 Physicochemical properties of β-galactosidase

Name	β-D-Galactoside galactohydrolase (EC 3.2.1.23)
Mass	456 kDa
Molecular form	Tetramer (16S)
	Monomers at pH3.5 or 11.5 or with mercurials
Stability	>30 min at 40°C with pH 6–8 in 100 mM 2-mercaptoethanol and 10 mM MgCl^2
Inhibitor	2-Mercaptoethanol alone
Isoelectric point	4.6
Specific activity	400 U mg^{-1}

Boehringer, Mannheim, FRG), 2 mg, is dissolved in 1 ml of the above buffer and a reaction mixture is prepared by mixing 1 ml of enzyme solution with 1 ml of antibody solution and 50 μl of 25% (v/v) glutaraldehyde solution (stored in the dark at 4°C; TAAB Laboratories Ltd, Reading, UK). The reaction is allowed to occur for 30 min at room temperature on an end-over-end mixer and thereafter blocked by the addition of 1 ml lysine (Sigma, St Louis, MO, USA) in the same phosphate buffer and stored overnight at 4°C. After extensive dialysis against PBS, additives such as 0.1% BSA, 0.2% gelatin, or 0.2% glycerol are added to give a 50% (v/v) final concentration. The conjugate is stored at 4°C. The additives make it possible to store the conjugate for a few months, while without them the shelf-life is reduced to few weeks. Ions can act as activators or inhibitors of enzymes. In the case of β-galactosidase, NH_4^+ and Mn^{2+} ions activate the enzyme, while high concentrations of Na^+ and heavy metal ions are inhibitory. Also β-mercaptoethanol can cause dissociation of the enzyme molecule in the absence of Mg^{2+}, while it is an excellent activator in its presence. These facts therefore need to be taken into account in choosing the conjugate and the substrate buffer (refer to Table 13.8.4).

Substrate reaction and product detection

Assays that utilize enzymatic markers offer the potential for a high degree of assay sensitivity and detectability. However, the detectability of enzyme immunoassays which utilize substrates that generate visibly colored molecules upon enzymatic activation is limited by the detection limit of these relatively low-energy markers. For example, while enzymes in common use have turnover rates of approximately 10^5 substrate molecules transformed per mole of enzyme per minute, the fact that some commonly used substrates require concentrations of 10^{-6} mol l^{-1} to be detectable by colorimetric instruments limits the usage of standard enzyme immunoassays to 10^{-11} mol l^{-1} (Ishikawa et al 1983). The current availability of a large number of substrates which generate fluorescent molecules upon enzymatic interaction offers the possibility of the development of an enzyme assay, which utilizes the fluorimetric measurement of the reaction end point. Moreover, fluorimetry measures a signal increase above relatively low background luminescence, while calorimetry measures a small decrease due to absorbance from a large amount of transmitted light. The detection limits of fluorimetry are frequently limited by background fluorescence such as that arising from serum or other biological samples. For this reason, fluorimetric ELISA methods have become practical only for heterogeneous systems, preferably when a solid-phase-bound antibody is used to facilitate washing.

This system thus offers the advantages both of the stability and quantitative nature of fluorescence measurement and of the catalytic magnification inherent in enzyme–substrate interaction. Since, as discussed above, the detection limit of a fluorigenic substrate can be orders of magnitude lower than that of a chromogenic substrate, this advantage can be used to increase the sensitivity of ELISA, thereby reducing sample sizes, using more dilute reagents, and/or shortening incubation times.

Most researchers using fluorimetric ELISA have employed conventional fluorimeters which require manual transfer of samples to cuvettes and measurement of their signals individually. Simply adding a flow cell and autosampler or sipper to a conventional fluorimeter can significantly decrease the labor involved.

Because it is emitted in all directions, there is more versatility in designing detectors for fluorescence than for absorbance. For example, reflectance fluorimetry has been used with thin-layer chromatography. A similar detector design could be used for fluorescence measurements directly from the wells of microtiter plates. With laser excitation, fluorescence can be measured from sub-microliter sample volumes. When choosing a fluorigenic substrate, there are several important considerations:

- The substrate should be readily available in a pure and stable form.
- It should be nonfluorescent under the detection conditions to be used.
- It must be efficiently converted by the enzyme label.
- The resultant product must be stable and highly fluorescent.
- There should be a large product Stokes shift, that is, a significant wavelength difference between the product's excitation and emission maxima.
- It is also preferable to have a long excitation wavelength for the product formed, as there will be lower background fluorescence at lower excitation energies (Milby 1985).

The most widely used substrates are those which generate the highly fluorescent ketone 4-methylumbelliferone (MUF) upon enzymatic cleavage, since this molecule can be measured at concentrations of 10^{-15} to 10^{-18} mol l^{-1}. We use MUF-G (4-methylumbelliferyl-β-D-galactoside; Sigma) as our substrate. This ELISA compares favorably with equivalent systems using chromogenic substrates like ortho-nitrophenyl-β-galactoside, which is detectable only at 100-fold higher concentration. The latter has a minimum detection limit of around 10 to 100 ng ml^{-1}. In the fluorigenic assay the minimum detection limit has been reported to be 10^{-15} mg ml^{-1} of IgG (Shalev et al 1980) and 10^{-10} mg ml^{-1} of IgE (Savelkoul et al 1989). Crowther et al (1990) have compared a number of enzyme–substrate combinations in ELISA. They have established that the combination of β-D-galactosidase and MUF-G resulted in maximum sensitivity. The substrate (0.25 mg ml^{-1}) is dissolved in 0.05 M potassium phosphate buffer, pH 7.8, with 10 mM MgCl$_2$ by rapid heating to 60°C.

Caution has to be exercised in the heating process as the substrate tends to disintegrate at temperatures higher than 80°C. Alternatively, the substrate can be dissolved to 0.5% in N,N-dimethylformamide and diluted to 0.1 mM with 10 mM sodium phosphate, pH 7.0, containing 0.1 M NaCl, 1 mM MgCl$_2$, 0.1% NaN$_3$, and 0.1% ovalbumin. Following solubilization, the substrate solution is cooled rapidly, aliquoted, and stored at −20°C. Such aliquots can be used for a few months provided they are reheated to 60°C and cooled rapidly to room temperature before use. In our experience, repeated freezing and thawing of the substrate leads to erroneous results. Also, freshly prepared substrate solution gives less interplate variation. We therefore prefer to prepare fresh substrate solution each day.

One of the major points to be considered when performing fluorescence-based ELISA is the dependence of the resultant fluorescence on the pH of the substrate solution and possible quenching by contaminating impurities. It is of crucial importance to regulate the pH in the substrate solution since, on changing the pH, the emission wavelength can shift up to 200 nm or fluorescence can be completely lost by formation of nonfluorescent ions. For MUF-G, the wavelengths used for excitation and emission analysis are 360 and 440 or 450 nm, respectively. A second consideration is that the

length of the substrate incubation step and the temperature (60 min, 37°C) are also critical for the fluorescent product formation. Hence, these parameters should be kept exactly constant in order to be able to compare the fluorescence measurements of various ELISAs performed on different days.

Expression and quantification of data

Untransformed data

The most common modes of expression of ELISA results are untransformed absorbance reading at a single sample dilution, end point antibody titer obtained by direct determination on calibration curves, and antibody activity unit from a calibrated dose–response curve. Such dose–response curves can be computed employing empirically transformed standard curves or by using equations derived from the law of mass action. Furthermore, for quantification the data can be expressed in a variety of ways: semilogarithmic plots, log–log plots, log–logit transformations, and so on. We shall briefly consider each of these. Any method that one chooses should fulfill two basic criteria: (1) stability of data under varying conditions, such as changes in assay design and minor variations in the experimental conditions; (2) linearity of response with dilutions of positive sera over a reasonably large interval. The optimal response and shape of dilution curves are both defined by a multitude of factors related to the main methodological choices made in establishing the ELISA structure as well as to within-method variables such as reagent characteristics and assay conditions which are subject to modifications and not always easy to predict and control. Assay conditions in the ELISA

affect the shape of the dilution curve (Malvano et al 1982). Even with identical assay conditions, the shape of the dose–response curves of different serum samples varies in some ELISAs (Gripenberg et al 1979). Also, it is known that antibody avidity and affinity differ between different sera, giving different kinds of dose–response curves (De Savigny and Voller 1980). Thus, useful information about antibodies may be lost if ELISA results are derived from a single serum dilution (Lehtonen and Eerola 1982).

Besides the dependence of signal intensity upon experimental conditions, the nonlinear correlation of signal with antibody activity in some regions of the dose–response curve sets limits to the use of nontransformed optical responses, despite its immediate availability and wide acceptance in practice. Nevertheless, a simple algorithm is proposed by Franco et al (1984) wherein multiple categorization of nontransformed absorbance values from ELISA plates is performed under microprocessor control. The printed output is a pictoral emulation of a 96-well plate with the color intensities represented for each reaction. This method is particularly suited for the screening of mouse hybridoma culture supernatants.

Quantification by the use of different equations derived from the law of mass action has been tried extensively, but, in general, no adaptation has been found suitable to the specific principles of ligand binding assays. Such efforts at obtaining an absolute concentration-dependent equation consisted in adapting the Langmuir adsorption isotherm and its consequence, the Scatchard equation. Further adaptations (e.g., the Sips equation) also could not be used sufficiently successfully.

End point titers

Titration of antibody is an established and comprehensible quantification method, even for ELISA. If standards calibrated in IU ml^{-1} are not available, titration is often the only generally accepted method.

The need for sequential dilutions of samples for definition and control of a cutoff value is an evident disadvantage intrinsic to titration. Other adverse features are less obvious: the range over which measurements can be made is rigidly defined by the detection system. The readings become increasingly insensitive at either end of this range. The assay precision relates to the sample dilution scale chosen and, when the greatest dilution still yielding a response above the cutoff value is taken as the titer, the measurement is made close to the detection limit, thus exaggerating the uncertainty in response. Moreover, the dilution at which the 'extinction' of response occurs depends upon the sensitivity of the assay, so that end point titers reflect variations of sensitivity level under different circumstances.

The expected parallelism between titers and antibody activity appears insufficient to counterbalance the unfavorable aspects, i.e., the high susceptibility to systematic and random variability, the amount of laboratory work involved, and the cost of the reagents. This suggests that caution is necessary in considering absorbance readings and end point titrations as suitable modes for expressing results. The results obtained by the titration method could at best be said to be semiquantitative.

Dopatka and Giesendorf (1992) suggested a novel method for antibody quantification in ELISA. In this, the sample is tested in a single dilution and the optical density value (OD) obtained is used in the equation

$$\log_{10}(\text{titer}) = \alpha OD^{\beta}$$

where α and β are constants to be determined for each batch of reagents. Since a single dilution is tested, the criticism above seems to hold for this method also. Further testing needs to be done before the method can find general application.

Reference standards

Since most quantitative ELISAs are based on the use of calibrators, it is prudent to realize the importance of the standard in determining the quality of the assay and allowing a comparison of the readings made with those obtained previously or in different circumstances. The recovery of the standard Ig is particularly dependent on the diluent used — a regular buffer or a similar matrix in which the sample Ig occurs (serum, liquor, etc.). Therefore, the sample and the standard should preferably be diluted in similar medium. Lyophilized reference samples are more prone to deterioration, denaturation, and accumulation of turbidity in comparison to fluid reference standards. It is therefore generally advisable to filter-sterilize, aliquot, and freeze the standard samples.

It is also a common practice to prepare standard samples for use in routine assays that are calibrated occasionally on internationally standardized reference preparations. Such routine Ig standards should contain highly purified (sub)class preparations that have been checked for intact Ig. Generally such standards are obtained by purifying myeloma or hybridoma proteins from which the recovery of intact material is all the more difficult. The Ig preparation should preferably contain members of the particular (sub)classes that carry the different light-chains and as many allotypes as possible. Also, in the case of hybridoma proteins, different specificities of the Ig produced should be included in the panel. Such sample mixtures can subsequently be tested in various assays for purity and reactivity against the antiserum preparations used in the ELISA. Next, the protein content in the reference sample can be established and a calibration graph can be constructed by running this standard in the ELISA. Also, a standard plot can be obtained in which the sample readings are plotted against the standard concentration. This plot can be the basis for further treatment of the data employing mathematical transformations.

Data transformation

The use of a reference scale eliminates systematic components of error and gives analytical consistency to the measurement. In this way, normalization between runs, between laboratories, and between methods is achieved. Availability of international reference sera greatly facilitates the use of calibration curves. In their absence, however, provisional arbitrary scales for antibody activity can be used. Such a procedure based upon interpolation on a calibration curve of a standard sample is, however, hampered by the weak reproducibility of the standard graph, the limited number of standard points used to draw the graphs, and the uncertainty of the sigmoid nature of the graph at the lower precision of the experimental results.

The calibration curves can be plotted in various ways. Commonly a calibration curve of OD versus log dilution of the standard is plotted. In our experience, however, probably because of differing affinities of the polyclonal Ig for a monoclonal coat, the standard and the sample curves often have different slopes, complicating interpretation of the data obtained. Butler et al (1985, 1992) considered logarithmic transformation of the mass law and arrived at the conclusion that the system can become independent of affinity within certain limits when log–log plots are used. Such log–log plots are also used in the 'multiple of normal activity' method (Ritchie et al 1983). This method employs an approximately parabolic relationship between the concentration of anti-

body and the absorbance (A_1 of sample 1 at dilution 1 and A_2 of sample 2 at dilution 2), which will result in a straight line on a log–log plot. The log–log plot will have a slope, m, and this factor is used for the calculation of the dilution factor. This factor can be defined as the number of times a serum should be diluted (D_2) over a reference serum (D_1) to obtain the same absorbance as the reference serum (= D_2/D_1):

$$\log \frac{D_2}{D_1} = -\log \frac{A_1/A_2}{m}$$

Characterization of the dose–response curve by approximating the dilution curve with a second-degree polynomial or a higher-degree spline function is also used. Gripenberg and Gripenberg (1983) approximate the function $y(x)$, which is the absorbance (or fluorescence) value obtained in ELISA when the serum dilution factor is $1/x$, by the polynomial

$$p(x) = a_0 + a_1 x + a_2 x^2$$

using the least-squares method, with equal weights at the discrete measurement values of x. Lehtonen and Viljanen (1982) have used the function $kx^{1/mx}$. Gordon et al (1985), on the other hand, have suggested the use of

$$y = a + \frac{b}{x} + \frac{c}{x^2}$$

Both are empirical equations where k, m and a, b, and c are constants to be arrived at by an iterative process. One advantage of using a polynomial lies in the simplicity of the calculation, since one has only to take a linear combination of the reading values. On the other hand, these polynomials lack monotonicity of the curves and are distorted by outlying values. Alternatively, the standard curve can be transformed into a four-parameter log–logit fit from which the sample Ig concentration can be determined.

Logistic transformation

For quantitative determination of the Ig concentration, the measured response is plotted in a semilogarithmic plot versus log dilution (i.e., absolute concentration) of the standard. Such dose–response curves are smooth, symmetrical, and sigmoidal in appearance. The measured response may be in terms of OD (chromogenic microtiter ELISA) or AFU (arbitrary fluorescence units; Terasaki-ELISA). The semilogarithmic curves may be described by the four-parameter logistic equation

$$y = \frac{a - d}{1 + \left(\dfrac{x}{c}\right)^b} + d$$

Here y represents the response variable, x is the arithmetic dose (absolute Ig concentration), a is the response at high dose (upper response plateau), d is the response at zero dose (lower response plateau), c is the dose resulting in a response halfway between a and d, and b is a slope factor that determines the steepness of the curve. In this equation, y can never exceed d. Therefore, the denominator must be greater than 1, and (x/c) is restricted to positive values.

The quantity y increases monotonically since the differential equation (of which the logistic function is the general solution)

$$\frac{dy}{dt} = my(a - y)$$

implies that $dy/dt > 0$. For $t \to -\infty$, y tends to zero, and for $t \to +\infty$, y tends to d. The growth starts slowly, then becomes faster and finally tapers off. Growth of the signal is fastest in the neighborhood of the point of inflection. To find its location, the second derivative must be equated to zero:

$$\frac{d^2y}{dt^2} = m(a-2y)\frac{dy}{dt}$$

This expression can only vanish if $a - 2y = 0$ or $y = a/2$; that is, the point of inflection is halfway between the lines $y = 0$ and $y = a$. To obtain the abscissa, let $y = a/2$ in the logistic equation and solve it with respect to t:

$$t = \frac{\ln k}{am}$$

This particular abscissa is positive or negative depending on whether $k < 1$ or $k > 1$.

The four-parameter logistic equation is not fitted to data directly to estimate all four parameters, rather a transformation of this equation is commonly used. A new variable Y is defined as $Y = (y-d)/(a-d)$, from which can be derived

$$\frac{Y}{1-Y} = \frac{1}{\left(\dfrac{x}{c}\right)^b}$$

By taking the natural logarithm, a linear transformation is derived:

$$\text{logit}(Y) = \ln \frac{Y}{1-Y} = -\ln\left(\frac{x}{c}\right)^b = \ln c^b - b \ln x$$

Before this linearized equation can be used, estimates of the parameters a and d must be obtained.

Standard linear regression is performed on the logit values of the response versus the log of the standard dilutions. The dilutions of known and unknown samples at which the logit is equal to zero are then compared to determine the concentration of the unknown. The latter plot is also useful to determine the dilution of the standard that gave a half-maximum reading (D_{st}). Given a constant amount of coated protein and constant affinity of the bound Ig versus the coat, the Ig concentration bound to the coat at D_{st} (C_{st}) should be equal to the Ig concentration in the sample (C_{sa}) that is bound to the coat at a certain dilution (D_{sa}). Thus, $C_{st} = C_{sa}$. From the sample response versus dilution factor, a similar transformed curve is prepared and the translation factor relative to the dilution factor of the standard that resulted in half-maximum response is determined. By comparison with the known Ig content in the standard, from this translation factor the absolute Ig content can be calculated:

$$C_{sa} = \frac{C_{st}D_{sa}}{D_{st}}$$

The data handling is generally completely microprocessor controlled.

The four-parameter log–logit transform takes into account the non-Gaussian error distribution of the multiple readings of each standard concentration tested. For each standard concentration tested, unequal weighted arithmetic means of the readings are to be taken, resulting in increasing standard deviations along the standard curve for higher concentrations of standard, as described by Rodbard (1971). This complicates considerably the linear regression analysis that has to be performed while constructing a standard graph. For this reason, microprocessor-based procedures have been designed to reduce the deviation of the fitted values from the actual readings.

Since it was found that essentially identical results could be obtained by using equal weighted means as well as by unequal weighting, a simplified two-parameter log–logit transform was proposed. This method, however, has some serious drawbacks, such as the strict dependency upon reaching equilibrium at each incubation step and its large systematic errors at either end of the curve. These effects can easily result in nonlinear log–logit plots.

Concentration determination

Controversy has arisen over whether the ELISA measures antibody concentration or antibody affinity. On the one hand, ELISA has been claimed to measure antibody concentration (Engvall and Perlmann 1971; Butler 1981; Butler et al 1985, 1992) and on the other hand to be a measure of antibody affinity or avidity. Moreover, as a discontinuous solid-phase assay comprising cycles of washes and a second antibody step during which the first antibody may dissociate, it would not be surprising from first principles if low-affinity antibodies were not detected in this system. It is all the more important to resolve this controversy because low-affinity antibodies may predominate early in an immune response (and hence play a role in the early diagnosis of infections) and may be important immunopathologically.

It appears that correlation of ELISA with affinity is associated with the form of data expression chosen. Those reports indicating that ELISA measures antibody concentration used the end point titration approach, whereas those suggesting that ELISA measures antibody affinity used the direct absorbance approach. From the work of Lehtonen and Eerola (1982) it can be concluded that ELISA absorbances read at low antibody dilution (i.e., antibody excess) preferentially reflect the binding by high-affinity antibodies in the preparation, whereas absorbances read at high antibody dilution (e.g., antigen excess) allow the detection of low-affinity antibody subpopulations as well. It has also been reported that high epitope density minimizes the influence of affinity and should be employed wherever possible. In our opinion, expressing the data in log–logit plots, along with the use of a standardized pure Ig as a reference standard, minimizes the influence of affinity and allows normalization between runs. The reference standard is obtained by mixing several purified hybridoma Ig of the same isotype. For antigen-specific ELISA the calibrator should consist of a mixture of monoclonals of the same antigenic specificity.

Important parameters

Some common characteristics of ligand-binding immunoassays determine the quality and the final applicability of such assays. These are the specificity, sensitivity, precision and reproducibility, and practicability.

The *specificity* of the assay depends to a large extent on the quality of the detecting antibody employed. Three points need to be borne in mind when selecting the detecting reagents:

- In our experience, polyclonal antibodies are better detecting reagents than monoclonals.
- Generally speaking, the use of enzyme-linked F(ab')$_2$ or, better still, Fab fragments is recommended.
- All the Ig employed in the assay should be tested explicitly for monospecificity.

As mentioned earlier, the detecting antibodies are purified from hyperimmune sera and need to be of high affinity to be of use as detecting reagents. For production of such high-affinity sera, some knowledge of affinity maturation is required. It was not until studies by Eisen and Siskind (1964), using the defined DNP–hapten system, that thermodynamically precise measurements of antibody affinity were made. These studies clearly illustrated that the affinity of antibody produced following immunization increases progressively and the rate of affinity maturation is related inversely to antigen dosage. However, affinities do not always mature, as indicated by the persistence of a low-affinity population throughout the antibody response, especially in the case of weak immunogens (polysaccharides, etc.). These facts should be borne in mind when raising hyperimmune sera.

The direct binding of the conjugate to the solid phase, which is only to some extent reduced by the use of detergents, should be avoided. The conjugate should thus be employed at the highest possible dilution. In a sandwich or two-site assay as described, a much better sensitivity is achieved by the double recognition of the Ig carrying at least two separate determinants, by the catching antibody, or by the antigen immobilized on the solid phase and the detecting antibody.

The *sensitivity* of an assay is usually defined as the lowest concentration giving a response which differs significantly from the zero-concentration response (detection limit). It would be more appropriate to define sensitivity as the change in response (dR) per unit amount of reactant (dC), dR/dC (Tijssen 1985). Therefore, on a sigmoid dose–response curve, sensitivity decreases at either end of the curve, resulting in an increasingly large error in the estimation of the analyte. *Detectability* reflects the confidence with which it can be stated that a certain response is larger than this error. Factors affecting specificity influence sensitivity to a large extent. Assay sensitivity is a function of the analyte concentration, detector sensitivity, and assay imprecision. Other assay conditions also have a role to play in determining the sensitivity. These include washing buffers and the performance of washing steps, the nature and molecular weight of the antigen (for antigen-specific assays), the ionic strength and pH of the diluent, the incubation period and temperature, and the molar ratio of enzyme label per detecting antibody molecule in the conjugate. Ekins (1990) has proposed that maximum sensitivity is dependent on Σ/K, where Σ is the experimental error of measurement and K is the affinity constant. Ekins (1990) has further developed mathematical arguments which show that 'nonspecific' binding is one of the important factors that limit the sensitivity of noncompetitive ELISA.

The optimal reagent dilutions will also greatly influence the sensitivity and detectability of the assay performed. First, the quantity of immobilized antigen or antibody is critically important. In all probability, at high concentrations of the coat protein multilayers are formed, so that the actual quantity of protein exposed to the antibodies is not increased. Moreover, the molecules may be packed too close together and therefore be antigenically ineffective owing to steric hindrance. On the other hand, when the concentration of the antigen or antibody immobilized is too low, the specific binding capacity is small and the sensitivity of the assay will be low. The optimal coating concentration for each assay system, as well as the optimal dilutions of all the other reagents to be employed in the assay, therefore need to be determined empirically by checkerboard titrations. The efficiency of the reading of the high-energy product formed determines the sensitivity of the assay. Most of the sensitive and quantitative assays described reach detection limits for antibody varying from 2 to 20×10^{-18} mole per ml. However, the sensitivity in terms of dR/dC, however, is rarely stated in such studies.

The precision and reproducibility of the assay are reflected in the slope of the log dose–response curve and the standard deviation of the individual reading at a certain dose of antigen. The steeper the log dose–response curve and the smaller the standard deviation of this line, the higher will be the precision of the assay. The general use of standard deviation (SD) is, strictly speaking, not valid since ELISA values for negative reference samples are distributed with a positive skew. Owing to this skewness, false positives occur with a significantly higher frequency than expected from a normal distribution. Nonparametric methods would be more accurate but are hardly used since they require larger sample sizes. SD is generally employed, but with a rather limited practical use. More replicates in the assay will only increase the reliability of the estimate for SD, but the spread of the individual data remains in principle unchanged. A more practical alternative is to estimate the 95% confidence interval of the arithmetic mean.

The use of ELISA in large-scale studies (e.g., epidemiological surveys) demands adequate assay precision, since in these situations it is not possible to evaluate all samples of interest in a single assay and between-assay variation must be minimized. Sources of between-assay variation include differential adsorptive characteristics of polystyrene (or PVC) plates, fluctuations in pH and temperature of buffers, errors in reagent preparation, reagent instability, variability in the washing procedure, timing of reactions, and volumetric pipetting errors.

Several methods are available for reducing between-assay variation in ELISA. These include expressing results as the ratio of OD of test sera and the OD of reference sera; expressing results as units read from a standard curve constructed from control samples tested on each plate; use of a plate correction factor (PCF), calculated as the ratio of the target OD to the actual OD of a single control serum or more than one control serum, to multiply the OD of the test sera.

Reproducibility of the ELISA method used can be further increased by defining a list of criteria for the standards on which quantification in the assay is based. Such a list should include stability, homogeneity, absence of contaminants, accurate division into smaller aliquots, moisture content, identity of properties with the test substance, and so on. Moreover, quality control samples should be included in the tests to monitor variations at different concentration levels and to ensure the quality of the tests. For regular ELISA users, a quality control chart plotting the results of the control samples over time can provide a simple means of monitoring trends and hence a drift in assay parameters.

The *practicability* of the assay refers not only to the speed and the ease of performance of the assay but also to the possibility of automation of monotonous steps like washing, addition of reagents (whether sample, conjugate, substrate etc.), and especially the measurement of the signal and calculation and computation of the results. A range of automated washing as well as measuring systems, along with software packages for computation of results, are available for the microtiter system. For Terasaki-ELISA, on the other hand, little automated equipment is available. All these parameters must be taken into account while choosing and performing ELISA.

References

Ansari AA, et al (1985) ELISA solid phase: stability and binding characteristics. J Immunol Methods 84: 117–124.

Avrameas SA, et al (1978) Coupling of enzymes to antibodies and antigens. Scand J Immunol 8: 7–23.

Avrameas SA, et al (1983) Immunoenzymatic Techniques. Elsevier, Amsterdam.

Azimzadeh A, Van Regenmortel MHV (1991) Measurement of affinity of viral monoclonal antibodies by ELISA titration of free antibody in equilibrium mixtures. J Immunol Methods 141: 199–208.

Azimzadeh A, et al (1992) Measurement of affinity of viral monoclonal antibodies using Fab'-peroxidase conjugate. Influence of antibody concentration on apparent affinity. Mol Immunol 29: 601–608.

Bayer EA, Wilchek M (1978) The avidin–biotin complex as tool in molecular biology. Trends Biochem Sci 3: 257–259.

Bizollon CA (1984) Monoclonal Antibodies and New Trends in Immunoassay. Elsevier Science, Amsterdam.

Boudet F, et al (1991) UV-treated polystyrene microtiter plates for use in ELISA to measure antibodies against synthetic peptides. J Immunol Methods 142: 73–82.

Butler JE (1981) The amplified ELISA: principles of and application for the comparative quantitation of class and subclass specific antibodies and the distribution of antibodies and antigens in biochemical separates. Methods Enzymol 73: 482–523.

Butler JE, Ni L, Nessler R et al (1985) The amplified immunosorbent assay (a-ELISA). In: Ngo TT, Lenhoff HM, eds. Enzyme Mediated Immunoassay. Plenum, New York, pp. 241–276.

Butler JE, et al (1992) The physical and functional behavior of capture antibodies adsorbed on polystyrene. J Immunol Methods 150: 77–90.

Cantarero LA, Busher JE, Osbourne JW (1980) The adsorptive characteristics of proteins for polystyrene and their significance in solid phase immunoassay. Anal Biochem 105: 375–382.

Crowther JR, Angarita L, Anderson J (1990) Evaluation of the use of chromogenic and fluorogenic substrates in solid-phase enzyme linked immunosorbent assays (ELISA). Biologicals 18: 331–336.

De Savigny D, Voller A (1980) The communication of ELISA data from laboratory to clinician. J Immunoassay 1: 105–128.

Dopatka HD, Giesendorf B (1992) Single point quantification of antibody in ELISA without need of a reference curve. J Clin Lab Anal 6: 417–422.

Eisen HN, Siskind GW (1964) Variations in affinities of antibodies during the immune response. Biochemistry 3: 996–1008.

Ekins R (1990) Merits and disadvantages of different labels and methods of immunoassay. In: Price CP, Newman DJ, eds. Principles and Practice of Immunoassay. Stockton Press, New York, pp. 96–123.

Engvall E (1980) Enzyme immunoassay ELISA and EMIT. Methods Enzymol 70: 419–439.

Engvall E, Perlmann P (1971) Enzyme-linked immunosorbent assay (ELISA) quantitative assay of immunoglobulin G. Immunochemistry 8: 871–874.

Franco EL, et al (1984) Computer assisted multiple categorization of absorbance values in ELISA in pictorial emulation of 96-well plates. J Immunol Methods 70: 45–52.

Gordon J, et al (1985) Immunoblotting and dot immunobinding. In: Habermahl KO, ed. Rapid Methods and Automation in Microbiology and Immunology. Springer-Verlag, Berlin, pp. 103–114.

Gripenberg M, Gripenberg G (1983) Expression of antibody activity measured by ELISA, anti-DNA antibody activity characterized by the shape of the dose–response curve. J Immunol Methods 62: 315–323.

Gripenberg M, et al (1979) Demonstration of antibodies against Yersinia enterocolitica lipopolysaccharide in human sera by enzyme-linked immunosorbent assay. J Clin Microbiol 10: 279–284.

Ishikawa E, et al (1983) Ultrasensitive enzyme immunoassay using fluorogenic, luminogenic, radioactive and related substrates and factors to limit sensitivity. In: Avrameas SA, Druet P, Masseyeff R, Feldmann G, eds. Immunoenzymatic Techniques. Elsevier, Amsterdam, pp. 219–232.

Ivanov VS, et al (1992) Effective method for synthetic peptide immobilization that increases the sensitivity and specificity of ELISA procedures. J Immunol Methods 153: 229–233.

Kemeny DN, et al (1986) The use of monoclonal and polyspecific antibodies in the IgE sandwich ELISA. J Immunol Methods 87: 45–50.

Klasen E, Rigutti A (1983) A solid phase EIA which allows coating of peptides and water-insoluble protein showing no 'hook effect' immunoreactivity of haemoglobin. In: Avrameas SA, Druet P, Masseyeff E, Feldmann G, eds. Immunoenzymatic Techniques. Elsevier, Amsterdam, pp. 159–161.

Koertge TE, Butler JE (1985) The relationship between the binding of primary antibody to solid-phase antigen in microtitre plates and its detection by the ELISA. J Immunol Methods 83: 283–299.

Lehtonen O-P, Eerola E (1982) The effect of different antibody affinities on ELISA absorbance and titre. J Immunol Methods 54: 233–240.

Lehtonen O-P, Viljanen MK (1980) Antigen attachment in ELISA. J Immunol Methods 34: 61–70.

Lehtonen O-P, Viljanen MK (1982) A binding function for curve-fitting in enzyme-linked immunosorbent assay (ELISA) and its use in estimating the amounts of total and high affinity antibodies. Int J Bio-Med Comput 13: 471–479.

Malvano R, et al (1982) ELISA for antibody measurement: aspects related to data expression. J Immunol Methods 48: 51–60.

Milby KH (1985) Fluorometric measurements in enzyme immunoassays. In: Ngo TT, Lenhoff HM, eds. Enzyme Mediated Immunoassay. Plenum, New York, pp. 325–341.

Morrisey BW (1977) The adsorption and confirmation of plasma protein: a physical approach. Ann NY Acad Sci 283: 50–64.

Nakamura RM, et al (1986) Enzyme immunoassays: heterogeneous and homogeneous systems. In: Weir DM, ed. Handbook of Experimental Immunology. Blackwell Scientific, Oxford, pp. 27.1–27.20.

Pesce A, et al (1983) Theories of immunoassay employing labeled reagents with emphasis on heterogeneous enzyme immunoassay. In: Avrameas SA, Druet P, Masseyeff R, Feldmann G, eds. Immunoenzymatic Techniques. Elsevier, Amsterdam, pp. 127–138.

Pruslin FH, et al (1991) Caveats and suggestions for the ELISA. J Immunol Methods 137: 27–35.

Ritchie DG, et al (1983) Two simple programs for the analysis of data from enzyme-linked immunosorbent assay (ELISA) on a programmable desk-top calculator. Methods Enzymol, 92: 577–588.

Rodbard D (1971) Statistical aspects of radioimmunoassays. In: Odell WD, Daughaday WH, eds. Principles of Competitive Protein-binding Assays. Lippincott, Philadelphia, pp. 204–239.

Savelkoul HFJ, et al (1989) Terasaki-ELISA for murine IgE-antibodies. II. Quantitation of absolute concentration of antigen-specific and total IgE. J Immunol Methods 116: 277–285.

Schonheyder H, Andersen P (1984) Effects of bovine serum albumin on antibody determination by the enzyme-linked immunosorbent assay. J Immunol Methods 72: 251–259.

Schuurs WHM, Van Weemen BK (1977) Enzyme immunoassay. Clin Chem Acta 81: 1–40.

Scott H, et al (1985) Performance testing of antigen-coated polystyrene microplate for measurements of serum antibodies to bacterial and dietary antigens. Acta Pathol Microbiol Immunol Scand Sect C-93: 117–123.

Shalev A, et al (1980) Detection of attograms of antigen by a high sensitivity enzyme-linked immunosorbent assay (HS-ELISA) using a fluorometric substrate. J Immunol Methods 38: 125–139.

Smith SC, et al (1993) Pitfalls in the use of ELISA to screen for monoclonal antibodies raised against small peptides. J Immunol Methods 158: 151–160.

Tijssen P (1985) Practice and theory of enzyme immunoassays. In: Burdon RH, Van Knippenberg, eds. Laboratory Techniques in Biochemistry and Molecular Biology. Elsevier Science, Amsterdam, pp. 1–549.

Tsang VCW, et al (1984) Quantitative capacities of glutaraldehyde and sodium m-periodate coupled peroxidase–anti-human IgG conjugates in enzyme-linked immunoassays. J Immunol Methods 70: 91–100.

Urbanek R, et al (1985) Use of the enzyme-linked immunosorbent assay for measurement of allergen-specific antibodies. J Immunol Methods 79: 123–131.

Voller A, et al (1981) (eds) Immunoassays for 80's. University Park Press, Baltimore, Maryland.

Vos Q, et al (1987) The effect of divalent and univalent binding on antibody titration curves in solid phase ELISA. J Immunol Methods 103: 47–53.

Wisdom BG (1976) Enzyme immunoassay. Clin Chem 22: 1243–1255.

ELISA quantification of Ig concentration

Determination of antibody affinity and affinity distributions

13.9

Sulabha S. Pathak
G. John M. Tibbe
Huub F.J. Savelkoul

Department of Immunology, Erasmus University, Rotterdam, The Netherlands

TABLE OF CONTENTS

Immunology Methods Manual
ISBN 0–12–442712–X

Introduction

Like any other biochemical reaction, antibody–antigen reactions are reversible and can be characterized by measuring both association and dissociation rate constants. In the case of a monovalent antibody site reacting with a monovalent determinant, the equilibrium constant is conventionally defined as the 'affinity' of the antibody. It is therefore the ratio of the molecular association and dissociation rates. Thus, if we consider the equation

$$Ag + Ab \overset{k_1}{\underset{k_2}{\rightleftharpoons}} Ag.Ab$$

then,

Affinity = equilibrium constant =

$$\frac{k_2}{k_1} = K_d = \frac{1}{K_a}$$

When both the antibody and the antigen are multivalent, the equilibrium constant has been termed the 'avidity' of the antibody. Avidity, thus, depends upon affinity but also involves other contributing factors such as valency, the method of measurement, and so on, that are not necessarily concerned with the primary antibody–antigen reaction. In a polyclonal situation, such as the estimation of the affinity of antisera, the situation is further complicated by the fact that the serum is a mixture of antibodies of different affinities. Thus, the affinity of the serum is a kind of 'average affinity' or the 'functional affinity'. The functional affinity is influenced by the affinity distribution of the various antibodies capable of reacting to the same epitope in the serum as well as the method of determination.

The experimental determination of the affinity of a monoclonal antibody for its antigen is of considerable importance. It is the basic experimental parameter in a variety of studies — for example, the use of monoclonal antibodies as conformational probes, the thermodynamic approach to the study of the molecular basis of the antigen–antibody interaction, and so on (Goldberg and Djavadi-Ohaniance 1993). Antibody screening as well as ranking of antibodies with regard to their reactivity to a tumor-associated antigen is a common assay procedure in the production of monoclonal antibodies for diagnosis and therapy. Knowledge of the functional affinity as well as the affinity distribution patterns of the various antibody populations in the serum is important in the study of ongoing immune responses and especially the phenomenon of affinity maturation.

The possibility that variation in the binding affinity of antibodies can influence the biological properties of antisera was recognized several years ago. A range of factors are known to influence antibody affinity. These include the dose and nature of the immunogenic stimulus, the immunization scheme, genetic factors, qualitative and quantitative aspects of lymphocyte function, dietary and hormonal factors, reticulo-endothelial function, and the effects of free antibody and antigen–antibody complexes in the microenvironment. The mechanisms by which these factors affect affinity are not clear, particularly since the cellular basis of the control of affinity has not been fully characterized.

Methods of affinity determination

It is clear from the equation above that the determination of antibody affinity requires the measurement, at equilibrium, of antibody-bound and free antigen or the measurement of antigen–bound and free antibody. The methods used obviously must not disturb the equilibrium and should not preferentially detect antibodies of a particular affinity. In practical terms the analysis requires the availability of large amounts of purified labeled (radio-, enzyme- or fluorescence-labeled) ligand, undamaged by the labeling procedures.

A variety of methods have been used to measure antibody avidity or average affinity. The most widely accepted among them is equilibrium dialysis, in which the reactants are separated by a semipermeable membrane which allows movement of one of the reactants (the labeled antigen), but not the other (antibody). The amount of label on either side of the membrane at equilibrium is then quantified. Affinity can then be calculated by the Scatchard equation. However, the obvious disadvantage of this method — the strict requirement for small haptens — limits its usage.

A number of different methods have therefore been employed to generate the data required for the Scatchard analysis. These include precipitation by ammonium sulfate or poly(ethylene glycol), electrophoresis in agarose gels, radioimmunoassay and, recently, ELISA (Gaze et al 1973; Nimmo et al 1984; Friguet et al 1985; Heegaard and Bjerrum 1991). However, strictly speaking, the K_d can be rigorously determined only for an equilibrium in homogeneous solution and no straightforward thermodynamic theory can describe the equilibrium in heterogeneous phase systems.

A different approach to the determination of antibody avidity is the use of agents that disrupt the already formed antigen–antibody complexes, e.g., temperature, chaotropic ions such as thiocyanate, perchlorate, iodide, low-pH buffers (Lee et al 1974; Germuth et al 1979). Chaotropic ions in particular exhibit a remarkable capacity to modify ionic forces responsible for stabilization of the tertiary structure of many molecules and can disrupt or modify the intermolecular interactions between antigen and antibody that are mediated by such forces and which have been recognized as promising for the study of antibody affinity.

Lately, biosensor technology has made it possible to measure antigen–antibody reactions in real time as they are occurring (see later).

Affinity determination

Calculations

In calculating the affinity constant for polyclonal serum antibodies and more recently for monoclonal antibodies, the most frequently used method has been that of Scatchard (Scatchard 1949; Soos and Siddle 1982; Steward and Lew 1985). Scatchard analysis allows estimation of antibody affinity by the regression analysis of bound over free ligand (antigen) concentration versus the bound ligand concentration, under circumstances where the ligand concentration is variable and the antibody concentration is constant. From the application of the law of mass action to antigen–antibody interaction, the following form of the Langmuir adsorption isotherm may be derived:

$$\frac{[Ab.Ag]}{[Ab]} = r = \frac{nK_a[Ag]}{1+K_a[Ag]}$$

where $r =$ moles antigen bound per mole of antibody; $[Ab \cdot Ag] =$ bound antigen concentration; $[Ab] =$ free antibody concentration; $[Ag] =$ free antigen concentration (at equilibrium); $n =$ antibody valence; and $K_a =$ association constant or affinity. From this equation we arrive at

$$\frac{r}{[Ag]} = nKa - rK_a = K_a(n-r)$$

which is the Scatchard equation. Hence, a plot of $r/[Ag]$ versus r over a range of free antigen concentrations allows values of K_a and n to be assessed. When half the divalent ($n = 2$) antibody binding sites are bound ($r = 1$) then

$$\frac{1}{[Ag]} = 2K_a - K_a = K_a$$

and thus K_a equals the reciprocal of the free concentration at equilibrium when half the antibody sites are antigen-bound. Alternatively, the free antigen ($[Ag]$) at equilibrium is determined by subtracting the bound antigen from the total antigen. The reciprocals of bound antigen ($1/[Ab–Ag]$) and free antigen ($1/[Ag]$) are plotted according to the Langmuir equation

$$\frac{1}{[Ab–Ag]} = \frac{1}{K_a[Ag][Ab_t]} + \frac{1}{[Ab_t]}$$

to obtain a value for the total amount of antibody present, expressed as total antigen binding sites Ab_t. Thus, when $1/[Ag] = 0$ then ($1/[Ab–Ag] = (1/Ab_t)$); Ab_t may be determined by extrapolation of the linear portion of the plot directly by determining the value of $1/[Ag]$ when 50% of the total antigen binding sites are

bound. The relationship must be linear over a range of ligand concentration, at least one order of magnitude on either side of the K_a value. If linearity is not observed over this range, the assumptions made in deriving the Scatchard formula do not apply and the calculated K_a value is not meaningful. Frequently a portion of the plot, often covering only a narrow range of antigen concentration, is found to be roughly linear and is arbitrarily chosen for the calculation. Furthermore, even when using isolated and purified anti-hapten antibodies, these plots frequently deviate from linearity owing to the existence of heterogeneity of antibody affinities within an antibody population. This heterogeneity makes it necessary for the equilibrium concentrations of bound and free antigen (or antibody) to be determined over a range of free antigen (or antibody) concentrations. The interpretation of the results is thus based on the overall affinity estimation and the value is therefore often called the 'average' affinity, although the value estimated empirically may not represent the mean, mode, or median.

Affinity determination in ELISA

Though many studies have been conducted on the effect of antibody affinity on ELISA measurements, application of noncompetitive ELISA for affinity measurement has been limited (Nimmo et al 1984; Steward and Lew 1985). ELISAs have been mainly used for quantification. Friguet et al (1985) have used this technique to determine the amount of unreacted antibody that is left in the reaction mixture after the antigen–antibody reaction in the fluid phase and performed Scatchard analysis on the data obtained. Alternatively, a mathematical approach involving curve-fitting procedures for estimating avidity from ELISA dilution curve data has been explored. Binding constants have also been estimated from adsorption kinetics data (Matikainen and Lehtonen 1984; Li 1985). Recently, Beatty et al (1987a,b) have used the approach of determining relative affinities by varying the input concentration in the coating solution. Competitive EIA techniques have also been employed (Nieto et al 1984). These methods, however, require special methodology and are not universally applicable, especially where the antibody is present in minute quantities, as is the case with IgE ($<1\ \mu g\ ml^{-1}$ in normal mouse serum). Moreover, some of these techniques are laborious and often require partial purification and concentration determination of the Ig involved.

The major drawback of applying general ELISA techniques for affinity measurement is the fact that though it is possible to determine the input concentration of the ligand in the coating solution, it is almost impossible to determine the amount of ligand that has been actually adsorbed to the wells, even if one assumes such adsorption to be uniform all over the plate and therefore the concentration of ligand per well to be identical. The approach of Scatchard requiring the determination of bound over free antigen or antibody ratios cannot easily be adapted to the ELISA system since the molar concentration of antibody that complexes with the antigen cannot easily be computed. The reaction developed is proportional to the complex formed, so that it is not possible to determine the amount of free antibody (antigen) on the solid phase. Moreover, low-affinity antibodies are said to be preferentially underestimated in ELISA. Also, independent of its affinity, subpopulations of <10% concentrations could not be detected in ELISA. It is especially difficult to measure the affinity constant of a reaction where both the reactants are Ig. Nevertheless, because of its ease and simplicity, ELISA remains a popular method of quantification and affinity determination.

We have proposed the use of a simple binding analysis — sequential equilibrium binding analysis to determine the 'functional affinity' of Ig without prior purification and even when present in minute quantities (Pathak et al 1989). An apparent equilibrium constant, K_{rel} is measured by fitting the ELISA data to two different formulas derived from the law of mass action. The relation between the K_{rel} and K_d, the conversion factor, is then established. Once established, the relation can be employed for computing K_d from K_{rel} provided the detection system remains identical.

The quantitative relationship of the interaction between antigen and antibody at equilibrium is governed by the law of mass action (Fazekas De St Groth 1979). It is represented in various forms, one of them being

$$\frac{[nC-x][P-x]}{[x]} = K_d \tag{1}$$

where [C] is the original concentration of antigen molecules with n epitopes/molecule; [P] is the original concentration of paratopes; [x] is the concentration of complex formed at equilibrium; K_d is the equilibrium constant $= 1/K_a$. All the concentrations are expressed in moles/liter. K_d has the dimension of moles/liter.

Thus, when the concentration of [nC] is constant, but the concentration of [P] varies, we arrive at

$$\frac{[nC-x_1][P_1-x_1]}{[x_1]} = \frac{[nC-x_2][P_2-x_2]}{[x_2]} = K_d \tag{2}$$

$[x_1], [x_2]$ are the complexes formed at concentrations $[P_1]$ and $[P_2]$ of the paratopes, respectively.

Upon imposing the condition that $[P_2] = [P_1-x_1]$, eq. (2) can be rewritten as

$$\frac{[nC-x_1][P_1-x_1]}{[x_1]} = \frac{[nC-x_2][P_1-x_1-x_2]}{[x_2]} = K_d \tag{3}$$

or

$$[P_1-x_1] = \frac{K_d[x_1]}{[nC-x_1]} \tag{4}$$

and

$$[P_1-x_1-x_2] = \frac{K_d[x_2]}{[nC-x_2]} \tag{5}$$

Subtracting eq. (5) from eq. (4) results in

$$[x_2] = \frac{K_d[x_1]}{[nC-x_1]} - \frac{K_d[x_2]}{[nC-x_2]} \tag{6}$$

or

$$[nC-x_1][nC-x_2] = K_d[nC]\frac{[x_1-x_2]}{[x_2]} \quad (7)$$

This equation can be employed in practice within certain limits. Although theoretically $[nC]$ will never be equal to $[x_1]$, practically $[nC]$ tends towards $[x_1]$ when $[nC]$ is limiting and $[x_1]$ and/or $[x_2]$ tend to zero at the upper and the lower plateaus of the dilution curve, respectively. We thus have a range within which this equation can be used.

Also, reconsidering eq. (2), and extending the condition

$$[P_1-x_1] = [P_2]$$

$$[P_2-x_2] = [P_3]$$

$$[P_3-x_3] = [P_4] \dots [P_n]$$

where $[x_n] \to 0$, we arrive at

$$[P_1-x_1] = [x_2] + [x_3] + \dots + [x_n] \quad (7a)$$

$[x_2]$, $[x_3]$, ..., $[x_n]$ being the complexes formed at successive equilibrium steps, with the original concentration of paratopes $[P_1]$, and the total concentration of epitopes $[nC]$ kept constant at each step. Therefore eq. (2) can be rewritten as

$$\frac{[nC-x_1][x_2+x_2+\dots]}{[x_1]} = K_d$$

or

$$\frac{[nC]}{K_d} = \frac{[x_1]}{K_d} = \frac{[x_1]}{[x_2+x_3+\dots]} \quad (8)$$

Both eq. (7) and (8) give a straight line in the range under consideration. They can be used to determine the value of K_d. Thus the slope of the plot of $[x_1-x_2]/x_2$ versus $[nC-x_1][nC-x_2]$ will be $[nC]K_d$. The inverse of the slope of the plot of $[x_1]/\Sigma[x_2]$ versus $[x_1]$ will give the K_d, the intercept on the y axis can be used to arrive at the value of $[nC]$ (Fig. 13.9.1). The first condition of eq. (2) can be met in an ELISA by transferring the liquid in the wells after equilibrium has been reached to a fresh plate previously coated and blocked as the initial one and incubating the second plate until equilibrium is once again established.

It is essential to determine the value of $[nC]$, the maximum number of epitopes that are available for reaction with the paratopes in each well in order to allow the application of eq. (7). When $[nC]$ becomes limiting, i.e., in the plateau region of the dilution curve, $[x_1]$ tends towards $[nC]$, and the values of $[x_1]$ and $[x_2]$ are

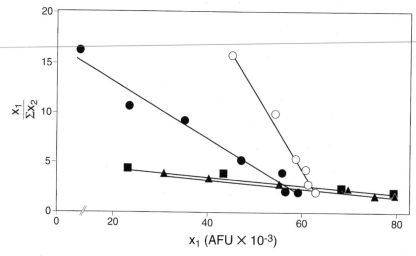

Figure 13.9.1 Sequential equilibrium binding analysis for affinity ranking in ELISA. Four individual monoclonal anti-TNP antibody preparations were tested on plates coated with the antigen. The product developed in each plate (OD or AFU) was fitted to eq. (8). Affinities can be ranked in decreasing order (○, ●, ■, ▲).

approximately equal. These values of $[x_1]$ remain unchanged over a range of decreasing concentrations of $[P]$ while the value of $[x_2]$ decreases. The response value of $[x_1]$ in this region is approximately equal to $[nC]$. The value of $[x_1]$ in the plateau region of the dilution curve (at the highest concentration of Ig) is sometimes lower than that obtained at lower Ig concentrations (the steep part of the curve). Similarly, $[x_2]$ is sometimes found to be greater than $[x_1]$ in this region.

On purely theoretical grounds, two explanations are possible. First, the plateau region of the dilution curve is the antibody excess zone so that only one antibody site can complex with the antigen, leading to a loose binding. Some of this Ig will be lost in the subsequent washing steps giving a reduced signal. Alternatively, the detecting reagents either cannot bind or bind loosely to the Ig due to steric hindrance, leading to a deviation from linearity in the dose–response curve on the first plate.

The process of liquid transfer can be repeated a number of times to get the values required for eq. (8). However, the repeated transfer of the liquid is a tedious and laborious process. The systematic error in the measurement will also be amplified and loss of minute amounts of the liquid in each successive transfer will further distort the signal obtained.

We therefore suggest the use of an empirical equation that can be used to predict the values of $[x_3]$, $[x_4]$, etc., once the values of $[x_1]$ and $[x_2]$ are established. A close agreement of the predicted and the experimentally determined values is generally observed (Fig. 13.9.2).

$$[x_1] = m \frac{[x_1 - x_2]}{[x_2]} + C \qquad (9)$$

All the variables in eqs (8) and (9) are expressed in terms of response (e.g., optical density). The K_{rel} value so determined will therefore also be in these terms and a conversion factor will have to be determined to arrive at K_d in moles/liter. This conversion factor is dependent on the detecting and measuring system used and will be unique for a particular system.

A number of assumptions have to be made in applying the above equations to ELISA.

1. The antigen–antibody interaction wherein the antigen is bound to the solid phase and the antibody is in the liquid phase is governed by the law of mass action.
2. The reaction reaches an equilibrium and

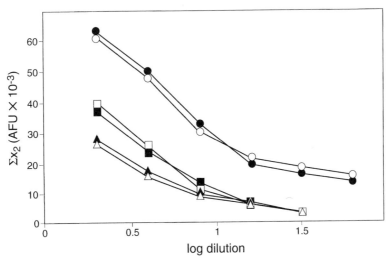

Figure 13.9.2 Comparison of the actual and predicted $\Sigma[x_2]$ values. Sequential equilibria were established between a coat and three different monoclonal antibodies. A summation of the response values (OD or AFU) of plates 2,3,4,..., n gave the experimental $\Sigma[x_2]$ (closed symbols). The response values of the first and second plates were fitted in eq.(9) to arrive at the predicted $\Sigma[x_2]$ values (open symbols). A plot of $\Sigma[x_2]$ versus log dilution was then constructed.

Antibody affinity and affinity distributions

this equilibrium is not substantially affected by the binding of the antibody to the immobilized antigen and its consequential removal from the fluid phase.

3. The antigenic epitopes remain essentially unaltered after adsorption to the solid phase, so that the behavior of the paratopes and therefore the K_d remains unaffected in this system.

4. The epitopes bound to the solid phase behave independently of each other and there is no cooperative binding between the two identical binding sites of the antibody.

5. The coating of the solid phase is uniform, i.e., the average number of epitopes that can react with the paratopes per well is the same for a given antigenic input in the coating solution.

It is possible to determine the K_{rel} (and K_d) of reactions between antigen and antibody as well as between anti-Ig antibodies and Ig by the sequential equilibrium binding analysis. A large conversion factor is required to arrive at the K_d from K_{rel}. However, the use of this large factor is likely to introduce/amplify errors and we restrict the usage to relative ranking only. We have used this system extensively for evaluating immunization protocols as well as relative ranking of monoclonal antibodies. In our experience, the system works best for monoclonal antibodies and for polyclonal sera consisting of relatively simple affinity distributions. When the serum consists of a wide range of affinities, e.g., late secondary or tertiary responses, the K_{rel} can only be determined over a very small range of dilutions, and caution needs to be exercised in evaluating the results.

Determination of affinity distributions

It is known that the binding of antibodies produced by the immune system in response to

'foreign' antigens is heterogeneous and that the distribution of association constants of the antigen–antibody reaction evolves in time under various circumstances (Eisen and Siskind 1964). The study of the antibody distribution pattern as a function of time after immunization is useful for the quantification of the immune memory (Yee 1991).

The affinity constants of polyclonal serum antibodies represent the overall binding properties of a normally heterogeneous population of antigen-specific antibodies and have limited usage (Bruderer et al 1992). An accurate assessment of affinity distributions within polyclonal populations of antigen–specific antibodies depends either on the isolation of antibody populations according to differences in affinity, or on the detectability of these populations in the presence of additional populations with the same specificity. Such isolation and quantification of antigen–specific antibodies requires large samples and is difficult and labor-intensive. Moreover, an accurate detection of the affinity populations in polyclonal samples is problematic owing to the complexity of the populations to be analyzed and the multiplicity of the possible interactions.

Several ELISA-based assays have been described for the measurement of affinity distributions (Nieto et al 1984; Rath et al 1988; Van Dam et al 1989). We have adapted a simple competitive inhibition ELISA method for the relative determination of affinity distribution patterns in a polyclonal serum using a previously described protocol (Rizzo et al 1992). Increasing amounts of the hapten are added to serum aliquots and incubated overnight. The mixture is then run in a regular ELISA using the same hapten–protein conjugate as the coat. In this type of an assay, the high-affinity antibody is inhibited from binding to the ELISA plates by the low concentrations of the free hapten. As the free hapten concentration is incrementally increased, lower-affinity antibody is inhibited, until at sufficiently high hapten concentrations all hapten-specific antibody is inhibited. The results are expressed in a histogram (Fig. 13.9.3).

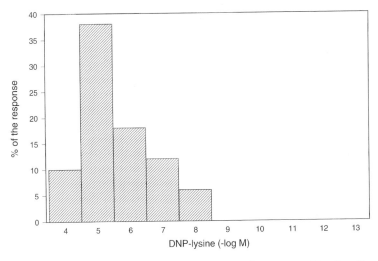

Figure 13.9.3 Histogram of the affinity distributions of DNP-specific serum antibodies. Sera were collected at certain days after immunization and run in a competitive inhibition ELISA. The abscissa indicates the concentration of free DNP-lys (–log M) used for inhibition. Affinity increases to the right. The ordinate shows the percentage of antibody present in each affinity subgroup.

Application of biosensors to interaction kinetics

For the understanding of structure–function relationships in biomolecular interactions and for the technical use of these biomolecules, the characteristic properties of the interaction must be known. Ideally, it should be possible to characterize the reaction between unlabeled reactants, but this has proved difficult by traditional methods (Karlsson et al 1991). Evanescent optical sensing techniques are finding increasing applications for the monitoring of biomolecular interactions. They offer the ability to perform real-time quantitative assays allowing for the kinetic analysis of interactions, and are generic techniques applicable to a wide range of assays. They exploit the ability to perform assays at solid surfaces and use evanescent fields to probe changes in refractive index, layer thickness, absorption, or fluorescence caused by reactions occurring within a few hundred nanometers of the sensor surface.

Two main types of optical evanescent sensors are commercially available: those based on surface plasmon resonance (SPR) and those based on waveguiding techniques (resonant mirror). Here we focus on the resonant mirror biosensor. The basic principle of all evanescent sensors is similar. When light travels through an optically denser medium (e.g., a glass prism) it is totally reflected back into the prism when reaching an interface with an optically less dense medium, provided that the angle of incidence is larger than the critical angle. This phenomenon is known as total internal reflection. Although the light is said to be totally reflected, a component of the incident light, called the evanescent wave, penetrates a distance of the order of one wavelength into the less-dense medium to give the evanescent field. Both of the commercially available optical biosensor types (IAsys™ biosensor, Fisons plc, Applied Sensor Technology, Cambridge, UK;

Antibody affinity and affinity distributions

BIAcore™ biosensor, Pharmacia Biosensor AB, Uppsala, Sweden) exploit this phenomenon of evanescent wave generation.

The resonant mirror biosensors

In the waveguiding technique devices, the sensing layer is placed in the evanescent region of a guided mode propagating in a dielectric waveguide structure. The resonant mirror biosensor is essentially similar in structure to an SPR device. Light is totally internally reflected from the sensing surface by means of a prism. At the sensing surface, there is present a dielectric resonant layer of high refractive index that is separated from the prism by a layer of low refractive index. This low-refractive-index layer is sufficiently thin (around 1 μm) that light may couple into the resonant layer via the evanescent field. Efficient coupling occurs only for certain resonant angles, where phase matching between the incident beam and the resonant modes of the high-index layer is achieved. At the resonant point, light couples into the high-index layer and propagates some distance along the sensing interface before coupling back in the prism. The angle of excitation of resonance is sensitive to changes at the sensing interface. By monitoring shifts in the resonance angle, changes occurring as a result of binding or dissociation of biomolecules can be measured in real time. The IAsys biosensor is a cuvette system based on the resonant mirror waveguiding technique (Fig. 13.9.4). It includes a sensor device with integrated optics incorporated in a reinsertable microcuvette sample cell, a resonance angle detector, and a microprocessor-based system controller. A micro-stirrer in the instrument ensures efficient transport of reagents and analytes to the sensor surface for kinetic analysis on the instrument. The inner surface of the microcuvette, i.e., the sensor surface, is coated with carboxymethylated dextran (CMD). This provides a hydrophilic and flexible matrix suitable for studies of biomolecular interactions and for efficient immobilization of ligands through covalent linkage of target molecules. Alternatively, cuvettes coated with silane are also available. The cuvette design minimizes sample requirements and offers the possibilities of sample recovery and reinsertion of the

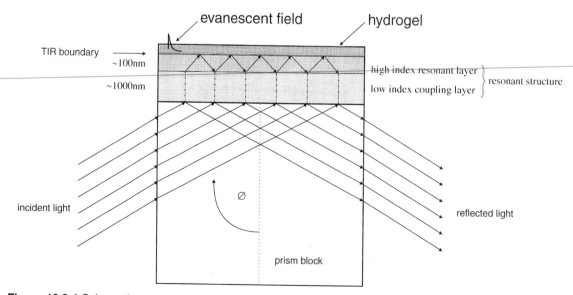

Figure 13.9.4 Schematic representation of the IAsys resonant mirror sensing device built in to a disposable cuvette.

sample cells. The coated sample cells can be stored at refrigeration temperatures between runs. This design also permits the operator to use a range of ligate volumes (50 to 200 µl). The change in the angle of resonance can be directly related to the presence and concentration of biomolecules in real time. The instrument records this change in resonance angle in terms of arc-seconds.

Immobilization of the ligand

For the biosensor, it is possible to immobilize various molecules to the CMD layer by conventional chemical methods. The ligand may be one of the reactants in the interaction of interest or a capture antibody having a high affinity for one of the reactants. In order to optimize the immobilization procedure, the electrostatic uptake characteristics of the ligand should be evaluated with respect to concentration and immobilization pH. This allows sufficient concentration of protein available for immobilization on to the matrix. Buffer selection is dependent on the isoelectric point of protein to be immobilized. Generally, proteins are coupled most efficiently at pH close to their isoelectric point (pI). Proteins or peptides of interest can be immobilized using simple carbodiimide chemistry (Lofas et al 1991). The dextran surface is activated by adding a mixture of N-ethyl-N'-(dimethylaminopropyl)carbodiimide and N-hydroxysuccinimide. Next, the protein/peptide of interest is added to the dextran layer. A covalent bond is formed between the free amines (e.g., lysine, arginine, histidine) or N-terminus on the peptide and the reactive ester group. The remaining activated groups are then deactivated by reaction with ethanolamine. A final wash with acid/alkali removes any unreacted but nonspecifically bound molecules and the sensor is ready for studying biomolecular interactions. The immobilization step itself can be monitored in real-time, allowing quantification of the surface concentration of the immobilized ligand (Fig. 13.9.5).

Ligand purity and homogeneity, proper choice of immobilization buffer, the extent of the electrostatic uptake, concentration of the ligand and the time of incubation of the various immobilization reagents are all critical for the successful application of the biosensor and have to be carefully standardized. The immobilized cuvette can be used for multiple runs. An appropriate reagent (acid/alkali/urea) can be used for regeneration. The measurement of the resonance signal is volume-dependent up to a volume of 50 µl per cuvette, after which the measurement becomes volume-independent. We therefore consider it essential to consistently use a constant volume in the cuvette. Moreover, a total volume exchange of running buffer with sample has to be avoided as it causes extensive baseline shifts.

The binding stoichiometry is calculated from:

Analyte : ligand =

$$\frac{\text{analyte response}}{\text{analyte MW}} : \frac{\text{ligand response}}{\text{ligand MW}}$$

Numerous parameters are important in studying interaction analysis with the biosensor.

- *The quality of the ligand*. Apart from factors such as purity and homogeneity, the sensitivity of the ligand to the process of immobilization will affect the quality of the assay. The extent of denaturation and/or alteration in conformation may result in nonhomogeneous ligand sites.
- *Immobilization*. The gentleness and randomness of immobilization will affect the quality of the analysis. The level of immobilization is also important and will be dictated by the ligand–ligate system to be studied as well as the desired application.
- *The quality of the ligate*. Purity of the ligate is not essential but is desirable. Various proteins may react nonspecifically with the coat, and the absence of such reaction must be

ensured. If the preparation contains aggregates or fragments, the ligate may display a heterogeneity with respect to binding properties and complicate data interpretation. Nonspecific binding is of special importance when the ligate is present in a complex protein mixture such as serum or supernatant.

- *Controls*. Proper positive and negative controls must be included in the experiments. These can be used for correction of the nonspecific binding as well as ensuring the quality of the immobilized cuvette even after multiple runs.
- *Method of regeneration*. Ideally regeneration should be as gentle as possible, but at the same time the agent must regenerate the entire surface. Repeated regenerations may affect the

quality of the immobilized ligand. Often, for complex protein mixtures, one reagent may not be sufficient, and a combination of reagents should be investigated (HCl, EDTA, urea, high-salt buffers, etc.).

- *Method of data analysis*. It is possible to use either linear or nonlinear regression for data analysis. However, the reaction between the immobilized ligand and the solution phase ligate cannot be described by a simple linear equation and hence nonlinear regression analysis is to be preferred.

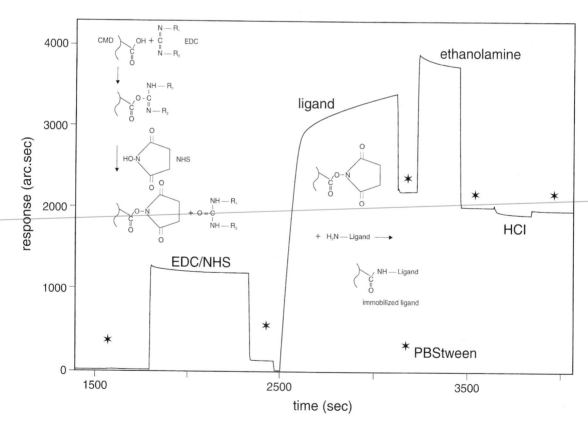

Figure 13.9.5 Profile of immobilization of ligand (protein) to CMD-coated cuvette. The CMD is activated by the addition of EDC and NHS. The primary amino groups in the protein bind covalently to the carboxyl groups on the CMD. The surface is deactivated with ethanolamine and regenerated with HCl.

Interaction kinetics

Antigen–antibody interactions have traditionally been measured using a variety of methods such as radioimmunoassay, ELISA, fluorescence quenching, and so on. All these approaches have many theoretical and practical drawbacks (Nygren et al 1987; Malmborg et al 1992). When information regarding the kinetics of the reaction *per se* is required, for studying affinity maturation, the binding characteristics of recombinant antibody fragments, the selection of monoclonal antibodies, and so on, rapid kinetic measurements are desired and the traditional methods are not suitable (Lofas 1991). Biosensors have been used successfully for such measurements (reviewed in Malmqvist 1993). Various dilutions of the ligate are added to the ligand-immobilized cuvette and the association kinetics are followed in time. Replacing the ligate with buffer allows the dissociation kinetics to be studied in real time as well. Fig. 13.9.6 shows a typical dose–response curve obtained on a biosensor. This primary biosensor data can be treated in various ways to arrive at the kinetic constants.

Most of the published studies use the traditional method involving linear transformations of the primary biosensor data for the determination of kinetic constants. However, linear transforms also transform the parameter-associated errors (Malmqvist 1993). Accordingly, the use of nonlinear regression is increasing (O'Shannessy et al 1993; Edwards et al 1995).

Theoretical considerations

A detailed theoretical consideration of kinetic analysis is given elsewhere (Karlsson et al 1991; O'Shannessy et al 1993). Here, only the linear and nonlinear regression analysis methods are briefly outlined. The monovalent homogeneous binding between monoclonal antibody (B) and its antigen (A) to form the complex (AB) can be expressed by the equation

$$[A]\ [B] \underset{k_{diss}}{\overset{k_{ass}}{\rightleftharpoons}} [AB] \qquad (10)$$

The net rate of complex formation is given by

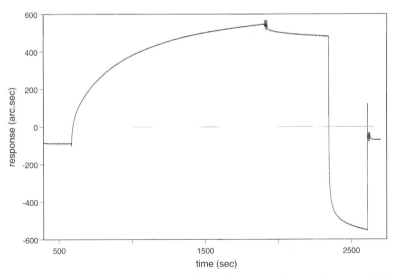

Figure 13.9.6 A typical dose–response curve obtained on a biosensor. Goat anti-mouse-IgG was immobilized to the sensor surface. After establishing a base line, $50\,\mu g\,ml^{-1}$ of mouse-IgG was added to it. The sample was replaced with PBS and allowed to dissociate for 4 min. The sensor surface was regenerated with an acid wash.

$$\frac{d[AB]}{dt} = k_{ass}[A]_t[B]_t - k_{diss}[AB]_t \qquad (11)$$

When A is immobilized on the sensor surface and the concentration of B is in large excess and continuously replenished at the boundary layer, a pseudo-first-order reaction results. The concentration of complex AB is monitored by the biosensor directly as the change in response against time, and the rate of change of response (R) with time is given by

$$\frac{dR}{dt} = k_{ass}[B](R_{max} - R_t) - k_{diss}R_t \qquad (12)$$

where [B] is the concentration of the ligate in the free solution; R_{max} is the total number of binding sites for B on the immobilized ligand A, expressed in the terms of the biosensor response R; R_t is the number of binding sites on A occupied at time t, expressed as in terms of biosensor response; dR/dt is the rate of formation of the complex AB expressed as response R per second.

Rearrangement of eq.(12) gives

$$\frac{dR}{dt} = k_{ass}[B]R_{max} - k_{ass}[B]R_t - k_{diss}R_t \qquad (13)$$

When the response R is measured for varying dilutions of the sample (and thus varying [B]), dR/dt can be calculated. The slope of a plot of dR/dt versus R gives $K' = k_{ass}[B] + k_{diss}$. When these K' values are plotted against [B], a straight line results. The slope of this line is k_{ass} and the intercept is k_{diss}. The units of the association constant k_{ass} are $M^{-1}s^{-1}$. In general, when the k_{diss} is expected to be low (as is the case for antigen–antibody reactions) the intercept value is too close to zero and the error on the value is too large to give a true k_{diss}. The dissociation of the complex AB is observed as a decrease in response when sample is replaced by a buffer. At this point the concentration of the free molecule B is zero, and from the eq. (13) we arrive at

$$\frac{dR}{dt} = -k_{diss}R_t \qquad (14)$$

When the conditions at the sensor are such that it can be assumed that there is no reassociation of the complex AB from the released B:

$$\ln\frac{R_{t1}}{R_{tn}} = k_{diss}(t_n - t_1) \qquad (15)$$

Rt_1 is the response at starting time t_1, and Rt_n is the response at time t_n. The slope of a plot of decrease in response versus time yields k_{diss} in s^{-1}. The association constant can be calculated as $K_a = k_{ass}/k_{diss}$ in M^{-1}.

As noted earlier, linear transforms, such as those described above, also transform the errors in the primary data. In addition, little information is obtained to define the reproducibility or the certainty of the derived parameters. Nonlinear regression helps circumvent these problems. It is not necessary to rearrange the original experimental results prior to analysis and so there is no distortion of the original error distributions (Leatherbarrow 1990).

For nonlinear regression analysis, the integrated form of the rate equation has to be used. In the cuvette [B] can be considered constant as B is added in large excess, and efficient stirring ensures that the concentration remains constant throughout the cuvette. Then the concentration of antigen at time $t([A]_t)$ is given by

$$[A]_t = [A]_0 - [AB]_t \qquad (16)$$

$$\frac{d[AB]}{dt} = k_{ass}[B]([A]_0 - [AB]_t) - k_{diss}[AB]_t \qquad (17)$$

$$\frac{d[AB]}{dt} = k_{ass}[B][A]_0 - [AB]_t(k_{ass}[B] + k_{diss}) \qquad (18)$$

Separating the terms in $[AB]_t$ and t and integration yields

$$\int \frac{[AB]t}{k_{ass}[B][A]_0 - [AB]_t(k_{ass}[B] + k_{diss})} = \int dt \qquad (19)$$

$$\frac{\ln(k_{ass}[B][A]_0 - [A][B]_t(k_{ass}[B] + k_{diss}))}{-(k_{ass}[B] + k_{diss})} = t + c \qquad (20)$$

At $t=0$, [AB]=0, therefore

$$c = \frac{\ln(k_{ass}[B][A]_0)}{-(k_{ass}[B]+k_{diss})} \quad (21)$$

$$\frac{\ln(k_{ass}[B][A]_0-[A][B]_t(k_{ass}[B]+k_{diss}))}{-(k_{ass}[B]+k_{diss})} =$$

$$t - \frac{\ln(k_{ass}[B][A]_0)}{k_{ass}[B]+k_{diss}} \quad (22)$$

or

$$\frac{k_{ass}[B][A]_0-[A][B]_t(k_{ass}[B]+k_{diss})}{k_{ass}[B][A]_0} =$$

$$e^{-(k_{ass}[B]+k_{diss})t} \quad (23)$$

Hence

$$[AB]_t = \frac{k_{ass}[B][A]_0}{k_{ass}[B]+k_{diss}}(1-e^{-(k_{ass}[B]+k_{diss})t}) \quad (24)$$

This can be rearranged as

$$\frac{k_{ass}[B][A]_0}{k_{ass}[B]+k_{diss}} = \frac{[B][A]_0}{[B]+\left(\dfrac{k_{diss}}{k_{ass}}\right)} = \frac{[B][A]_0}{[B]+K_d} \quad (25)$$

This is equivalent to the Michaelis–Menten equation, and gives the [AB] at equilibrium ([AB]$_\infty$) as

$$[AB]_t = [AB]_\infty(1-e^{-k_{on}t}) \quad (26)$$

where k_{on} is the observed pseudo-first-order constant.

[AB] at equilibrium can be termed the extent (E) of the reaction. Thus the association of the antibody with the immobilized antigen can be described by the following pseudo-first-order equation:

$$R_t = R_0+E(1-e^{kt}) \quad (27)$$

where R_0 is the initial response and E is the extent of the change in response.

However, the instrument response does not always increase in a single exponential manner, and frequently at least two distinguishable phases are observed. This biphasic response is better described by

$$R_t = R_0+E_1(1-e^{k_{on(1)}t}) + E_2(1-e^{k_{on(2)}t}) \quad (28)$$

where E_1 is the extent of the first association phase characterized by the rate constant $k_{on(1)}$ and E_2 is the extent of the second association phase characterized by the rate constant $k_{on(2)}$.

To facilitate the calculations, the biosensor is supplied with the necessary software packages that allow the use of nonlinear regression analysis to fit the association phase of the interaction profiles to the equations above, eliminating the need for equilibrium binding to occur. The appropriate association phase data are selected and monophasic and biphasic fits are performed to arrive at an apparent on rate (k_{on}). The residual error plot, also produced by the program, allows the selection of the k_{on}. By determining the k_{on} for various concentrations of the ligate (minimum 5), a linear regression plot of k_{on} versus [B] is constructed. The slope of this plot is k_{ass} as described by the equation

$$k_{on} = k_{ass}[B] + k_{diss} \quad (29)$$

The intercept of the plot is k_{diss}. However, as suggested earlier, calculation of k_{diss} from such plots is likely to be erroneous. Data transformation needed in the linear regression analysis also transforms the error in the primary data. It is therefore better to calculate k_{diss} by nonlinear regression analysis using the integrated rate equation

$$R_t = Ee^{-k_{diss}t} + R_f \quad (30)$$

where R_t is the instrument response at time t after replacing ligate with buffer; R_f is the final response after dissociation is complete; and E is the extent of change in response.

Edwards et al (1995) have investigated in detail the application of biosensors to biomolecular interaction analysis. They have investigated the kinetic behavior of biomolecules in biosensors and compared them to the behavior in liquids. It is suggested that the biphasic responses observed in biosensors seem to be intrinsic to CMD surfaces (Malmqvist 1993). The observed biphasic interaction is most probably due to steric hindrance caused by ligate binding to ligand immobilized to the

CMD on the sensor surface. In such cases, the fast phase of the binding profiles should be directly comparable to the situation in free solution.

A number of precautions need to be taken while determining kinetic constants in biosensors. However, with an understanding of the possible artifacts that can interfere with biosensor measurements, it is possible to obtain relevant kinetic information even with unpurified material. Association rate constants in the range 10^3 to $10^6\,\mathrm{M}^{-1}\,\mathrm{s}^{-1}$ and dissociation rates of 10^{-5} to $10^{-2}\,\mathrm{s}^{-1}$ are within the instrument's capability, whereas determining equilibrium constants of molecules with high dissociation rates is especially difficult with conventional techniques (Nygren et al 1987).

Conclusions

Biosensors can be defined as devices that combine a biological recognition mechanism with a suitable transducer which generates a measurable signal in response to changes in concentration of a given biomolecule at the detector surface. They offer the possibility of characterizing biomolecular interactions in real time without the need to label either of the reactants.

Furthermore, owing to the possibility of visualizing each step, complex assays can be performed and controlled relatively easily. In conventional techniques, the quality of such assays can only be judged by the final step, and the defects, if any, in the intermediate steps cannot be identified easily. With adequate precautions and proper knowledge of the possible artifacts, this technology promises to open avenues of research that were hitherto inaccessible or could be approached only very indirectly.

For the determination of functional affinity and avidity of Ig specific for macromolecular antigens, both ELISA and biosensor technology are available. Under these conditions, conventional affinity determination by equilibrium dialysis is not feasible. ELISA-based affinity measurement is particularly helpful when ranking large series of, for example, monoclonal antibodies for their affinity for the same antigen. This method is less suitable for functional affinity determination because of the need for a generally large conversion factor. This conversion factor is based upon a comparison between the K_{rel} arrived at in ELISA with that measured in equilibrium dialysis or some such unequivocal method. Again this is not feasible for macromolecular antigens. The advent of biosensor technology has enabled absolute affinity determination by permitting kinetic binding analysis by highly sensitive measurement of binding events on a sensor surface. This, combined with the development of software enabling nonlinear regression analysis of binding data, has made possible measurements of association and dissociation rate constants. We see numerous applications of this methodology in the typing of monoclonal antibodies, following ongoing immune responses, characterization of receptor–ligand interactions and the binding of molecules to living cells, and so on.

References

Beatty JD, et al (1987a) Method of analysis of non-competitive enzyme-immunoassays for antibody quantification. J Immunol Methods 100: 161–172.

Beatty JD, et al (1987b) Measurement of monoclonal antibody affinity by non-competitive enzyme immunoassay. J Immunol Methods 100: 173–179.

Bruderer U, et al (1992) Analyses of affinity distributions within polyclonal populations of antigen-specific antibodies. J Immunol Methods 151: 157–164.

Edwards PR, et al (1995) Kinetics of protein–protein interactions at the surface of an optical biosensor. Anal Biochem, in press.

Eisen HN, Siskind GW (1964) Variations in affinities of antibodies during the immune response. Biochemistry 3: 996–1008.

Fazekas De St Groth S (1979) Quality of antibodies and cellular receptors. In: vol. I. Lefkovits I, Pernis B, eds. Immunological Methods. Academic Press, New York, pp. 1–42.

Friguet B, et al (1985) Measurements of true affinity constant in solutionof antigen–antibody complexes by enzyme-linked immunosorbent assays. J Immunol Methods 77: 305–319.

Gaze SN, et al (1973) The use of a double-isotope method in the determination of antibody affinity. J Immunol Methods 3: 357–364.

Germuth FG, et al (1979) Passive immune complex glomerulonephritis in mice models for various lesions found in human disease. 1. High avidity complexes and mesangiopathic glomerulonephritis. Lab Invest 41: 360–365.

Goldberg ME, Djavadi-Ohaniance L (1993) Methods for measurementof antibody/antigen affinity based on ELISA and RIA. Curr Opin Immunol 5: 278–281.

Heegaard NHH, Bjerrum OJ (1991) Affinity electrophoresis for determination of binding constants for antibody–antigen reactions. Anal Biochem 195: 319–326.

Karlsson R, et al (1991) Kinetic analysis of monoclonal antibody–antigen interactions with a new biosensor based analytical system. J Immunol Methods 145: 229–240.

Leatherbarrow RJ (1990) Using linear and non-linear regression to fit biochemical data. Trends Biochem Sci 15: 455–458.

Lee RG, et al (1974) Competitive adsorption of plasma proteins on to polymer surfaces. Thromb Res 4: 485–490.

Li CK (1985) ELISA-based determination of immunological binding constants. Mol Immunol 22: 321–327.

Lofas S (1991) Bioanalysis with surface plasmon resonance. Sensors Actuators 5: 79–84.

Malmborg A-C, Michaelson A, Ohlin M, et al (1992) Real-time analysis of antibody–antigen reaction kinetics. Scand J Immunol 35: 643–650.

Malmqvist M (1993) Surface plasmon resonance for detection and measurement of antibody–antigen affinity and kinetics. Curr Opin Immunol 5: 282–286.

Matikainen MT, Lehtonen O-P (1984) Relation between avidity and specificity of monoclonal anti-chlamydial antibodies in culture supernatants and ascitic fluids determined by enzyme immunoassay. J Immunol Methods 72: 341–347.

Nieto A, et al (1984) Direct measurement of antibody affinity distribution by hapten-inhibition enzyme immunoassay. Mol Immunol 21: 537–542.

Nimmo GR, et al (1984) Influence of antibody affinity on the performance of different antibody assays. J Immunol Methods 72: 177–187.

Nygren H, et al (1987) Kinetics of antibody binding to solid-phase-immobilized antigen. Effect of diffusion rate limitation and steric interaction. J Immunol Methods 101: 63–71.

O'Shannessy DJ, et al (1993) Determination of rate and equilibrium binding constants for macromolecular interactions using surface plasmon resonance: Use of non-linear least square analysis methods. Anal Biochem 212: 457–468.

Pathak SS, et al (1989) Terasaki-ELISA for murine IgE: III. Determination of absolute concentration and functional affinity sequential equilibrium binding analysis. J Immunol Methods 123: 71–81.

Rath S, et al (1988) An inhibition enzyme immunoassay for estimating relative antibody affinity and affinity heterogeneity. J Immunol Methods 106: 245–249.

Rizzo LV, et al (1992) Generation of B cell memory and affinity maturation. Induction with Th1 and Th2 T cell clones. J Immunol 148: 3733–3739.

Scatchard G (1949) The attraction of proteins to small molecules and ions. Ann NY Acad Sci 51: 660–672.

Soos M, Siddle K (1982) Characterization of monoclonal antibodies directed against human thyroid-stimulating hormone. J Immunol Methods 51: 57–68.

Steward MW, Lew AM (1985) The importance of antibody affinity in the performance of immunoassays for antibody. J Immunol Methods 78: 173–190.

Van Dam GJ, et al (1989) Estimation of the avidity of antibodies in polyclonal antisera against *Streptococcus pneumoniae* type 3 by inhibition ELISA. Mol Immunol 26: 269–274.

Yee E (1991) Reconstruction of the antibody affinity distribution from the experimental data by a minimum cross-entropy procedure. J Theor Biol 153: 205–227.

Analysis of the humoral immune response using combinatorial phage display libraries

13.10

Jessica van der Heijden[1,2]
John de Kruif[1]
Tjerk W.A. de Bruin[2]
Ton Logtenberg[1]

Departments of [1]*Immunology and*
[2]*Internal Medicine, University Hospital, Utrecht, The Netherlands*

TABLE OF CONTENTS

Immunology Methods Manual
ISBN 0–12–442712–X

Introduction

The analysis of antibody responses to auto- and xenoantigens has been pursued by immunologists interested in a diverse array of scientific questions related to the mechanisms of antibody repertoire diversification, the protective and pathogenic actions of antibodies, the molecular changes in the genes encoding antibodies, and the dissection of the epitopes recognized by cells of the immune system. In the last two decades, immortalization of the antibody-producing cells by somatic hybridization or viral transformation has frequently been the method of choice to obtain monoclonal cell lines secreting a single species of antibody. Though extremely useful, this approach permits only a limited sampling of the immune repertoire and has posed severe technical limitations in organisms other than the mouse. These notions have spurred the development of molecular approaches that obviate the need for cell immortalization to obtain monoclonal antibodies with desired specificities. This approach is based on 'immortalization' of the genes encoding the antibodies rather than the cell lines producing them. The cloning of antibody variable (V) region genes ('repertoire cloning'), the expression of these genes as fragments of antibody molecules on the surface of bacteriophages ('phage antibody display'), and the subsequent selection of desired antibody specificities by binding of phages to antigen represents a recombinant DNA-based approach that mirrors many of the features of the normal immune system. Here, we discuss the possibilities and limitations of repertoire cloning for the study of antibodies and their targets, with particular focus on the human immune response.

General principles of repertoire cloning and expression

The first step of repertoire cloning consists of reverse transcription and PCR amplification of the heavy and light chain variable region genes (V_H and V_L) expressed in the B lineage cells of an individual selected to produce the antibody specificity of interest as a result of immunization or as a result of viral infection, autoimmune disease, or allergic conditions. The collections of amplified V regions are cloned into phage or phagemid expression vectors such that random pairing of V_H and V_L regions occurs. After introduction of the V_H/V_L combinations into bacterial cells and production of phage particles, the antibody fragments are expressed as fusion proteins on the surface of bacteriophages. To select antibodies of interest, the library is incubated with the target antigen to facilitate binding of phages expressing the correct specificity (panning). Nonbinding phages are removed by washing, whereas binding phages are eluted from the target antigen and propagated in bacterial cells. The ensuing library is used for subsequent rounds of panning, resulting in further enrichment of specific phages. After multiple rounds of panning, monoclonal phage antibodies are produced and tested for binding to the target antigen. Finally, antibodies may be produced as scFv or Fab fragments or modified to meet specific requirements.

Construction of combinatorial libraries

Vectors for construction of combinatorial libraries

Combinatorial libraries may be constructed to contain Fab or single-chain Fv (scFv) antibody fragments. Fab fragments faithfully represent the original, albeit monovalent, antibody binding moiety of natural antibodies. In monovalent scFv antibody fragments, the V_H and V_L domains are linked through a glycine-rich universal peptide linker, designed to interfere minimally with the antigen-binding portion of the molecule. Although a number of different vectors for phage antibody display have been constructed, phagemid vectors appear to offer the most attractive and versatile system.

For most cloning purposes, phagemids can be manipulated as plasmid molecules while, during production of phage particles, super-infection of bacterial cells with helper phage permits packaging of phagemid DNA in a fashion identical to that in the phage itself. The Fab or scFv antibody fragments may be expressed on the surface of bacteriophage as fusion proteins to the coat proteins g3p (3–5 copies per virion) or g8p (2800 copies/virion) or truncated forms thereof. Expression of antibody fragments as g8p fusion proteins results in multivalency and results in the selection of both low-affinity ($<10^6\,M^{-1}$) and higher-affinity antibodies. Expression of antibody fragments as g3p fusion proteins employing a phagemid vector results in monovalency on the majority of phage particles and is the method of choice for repertoire cloning when selection of higher-affinity antibodies is desirable. The pComb3

vector, designed for display of Fab fragments, is the prototype of a phagemid vector in which the amplified V_H genes are cloned as gene fusions to truncated g3p (Barbas et al 1991). The amplified V_L genes are introduced into the V_H gene-containing phagemids in a separate cloning step. Upon expression and protein synthesis, the V_H and V_L peptide chains are directed to the periplasmic space by a pelB leader sequence. The V_H fusion peptide is anchored to the inner membrane by the g3p moiety, whereas the V_L peptide is secreted in the periplasmic space and associates with the V_H chain to form a Fab fragment. Super-infection with helper phage finally results in the production of phage particles expressing Fab fragments. The pHEN 1 vector is a prototype of a phagemid vector for display of scFv antibody fragments as g3p fusion proteins (Marks et al 1991). In the originally described approach, the repertoires of V_H and V_L genes and the linker are assembled in a three-step PCR prior to cloning in the vector (Clackson et al 1991), which technically proves to be a tedious step in the efficient generation of combinatorial libraries. Improved vectors that allow the separate cloning of V_H and V_L repertoires are currently available (Engelhardt et al 1994; J. van der Heijden, unpublished results).

The tissue source of V regions

Several parameters have to be considered in determining the choice of tissue for cloning of the V regions that constitute the building blocks of phage display libraries. High serum titers of antibodies to the antigen of interest are considered to reflect an immune system with actively secreting and clonally expanded B lineage cells. RNA extracted from tissues that harbor these cells will be enriched for the genes encoding the antibodies of interest, thus considerably reducing the number of recombinants that needs to be generated and screened and increasing the chance of finding combinations of V_H and V_L regions originally present in a B cell clone. The richest source of antibody-secreting cells is the bone marrow, while spleen and tonsil have also been used successfully to construct combinatorial libraries. In organ-specific autoimmune disease, the antibody-secreting cells are often found in the target organ or draining lymph nodes (Rapoport et al 1995). The library can be constructed from whole-tissue mRNA or from purified, tissue-infiltrating B cells. When lymphoid organs are not available, peripheral blood lymphocytes may be used for RNA extraction, although the pool of peripheral blood B cells only temporarily contains high numbers of specific antibody-secreting cells, shortly after secondary contact with antigen (UytdeHaag et al 1985).

Amplification and cloning of V regions

Standard procedures for extraction of RNA, cDNA synthesis, and PCR amplification are generally applicable for V_H and V_L gene cloning from a variety of tissues. The size of the library is critically dependent on the amount and quality of amplified material. In our hands, purification of mRNA using oligo(dT) Dynabeads (Dynal, Norway) according to the manufacturer's recommendations, in combination with AMV reverse transcriptase supplied by Boehringer, gives reproducible yields and high-quality amplified fragments. Sets of primers required for amplification of human V_H and V_L genes have been designed to hybridize to 'all' members of the different families and avoid primer-based selective amplification of individual V genes (reviewed in Rapoport et al 1995; Soderlind et al 1992; Winter et al 1994). The choice of immunoglobulin constant region primer used for first-strand cDNA synthesis further permits the selective amplification of antibodies of a particular isotype.

Selection of phages

Selection of phages is performed by panning against antigen. In the standard approach, purified target protein is coated to a solid phase such as a well of a microtiter plate, a tube, or the beads commonly used for affinity chromatography. Buffers that work well in ELISA can be used to coat antigen to plates and tubes. Alternatively, antigen may be biotinylated, allowing phage binding in solution followed by capture in streptavidin beads (Hawkins et al 1995). Nonbound phages are removed by extensive washing. The coating and washing procedures may be manipulated to select antibodies with desired properties. For example, low antigen densities and extended washing times promote the selection of high-affinity binders (Marks et al 1992). Bound phages are eluted with high- or low-pH buffers, or by competition with antigen or antibody (Winter et al 1994; Burton and Barbas 1994; Meulemans et al 1994).

Structural alteration of antigens as a result of coating or purification procedures may result in the selection of phage antibodies that fail to bind to the native antigen. Alternative strategies have been successfully employed to circumvent this problem. Phage antibodies have been obtained by direct selection on antigens expressed on intact bacteria and viral particles, prokaryotic and eukaryotic cells, and bacteria expressing the target antigen as a recombinant fusion protein (Winter et al 1994; Burton and Barbas 1994; Marks et al 1993; Portolano et al 1993; De Kruif et al 1995; Siegel and Silberstein 1994; Bradbury et al 1995). Phage panning on heterogeneous mixtures of cells and flow cyto-metry-based cell sorting of the target population has proved extremely powerful in obtaining cell subpopulation-specific antibodies (De Kruif et al 1995). In this procedure, the nonselected cells in the mixture act as absorber cells of unwanted specificities, whereas the flow-cyto-metric isolation provides a very rigorous wash-ing procedure. Nontransfected cells may be used to preabsorb phage libraries, followed by selection of phages on cells transfected with a gene encoding the protein of interest (Portolano et al 1993). Finally, we have successfully used whole tissue fragments for phage selection (unpublished results).

The relationship between natural antibodies and antibodies from combinatorial libraries

In combinatorial libraries, collections of V_H and V_L genes are randomly combined to produce Fab or scFv fragments. For analysis of the humoral immune response against an (auto)antigen, it is imperative to determine whether the V_H/V_L combinations found in phage antibodies selected from libraries faithfully represent those expressed in the original B lymphocytes. Combinatorial libraries have been constructed from V regions cloned from immunized individuals and individuals suffering from viral infections or autoimmune disease. In two studies in immunized mice, a direct comparison has been made between the V regions utilized in library- and hybridoma-derived antibodies with similar specificities. The results of these experiments led to opposing conclusions regarding the faithful representation of naturally occurring V_H/V_L combinations in the library (Caton and Koprowski 1990; Gherardi and Milstein 1995).

Although a bias towards naturally occurring V_H/V_L combinations may be achieved through the choice of V regions used for constructing the library (for example, the target organ in organ-specific autoimmune disease), in some cases combinations of a single V_H gene with multiple V_L genes may yield antibodies with the same or similar apparent specificities and affinities. In fact, an approach termed 'light chain shuffling' has been employed in which a single V_H gene is combined with a library of V_L genes in order to select new combinations with increased affinity for the antigen (Marks et al 1992). The dominance of particular V_H genes in establishing antigen specificity has been observed in both recombinant and natural antibodies (Burton and Barbas 1994; Radic and Weigert 1995; Schutte et al 1995 and references therein). At the other end of the spectrum, both V_H and V_L may be required for high-affinity binding to antigen, limiting the set of V_L genes with which V_H can associate under the constraint of preserving specificity and affinity (Costante et al 1994). Finally, identical light chains have been found in combination with widely varying heavy chains in antibodies of different specificities (Kabat and Wu 1995). In these cases, it has been suggested that V_H dominates the specificity while the particular V_L gene is a suitable partner for many different V_H genes (light chain 'plasticity'; Burton and Barbas 1994). Based on these considerations, it should be noted that antibodies with promiscuous heavy chains and plastic light chains will be recovered from combinatorial libraries with greater frequency than their representation in the natural response. Skewing may be further enhanced as a result of the procedures used for constructing combinatorial libraries, i.e., amplification, cloning, and expression of V_H regions.

On the basis of an extensive body of data on natural immune responses, other criteria may be formulated to assess how faithfully library-derived antibodies report on the *in vivo* immune response. Upon interaction with antigen, B lymphocytes clonally expand and activate their somatic hypermutation mechanism. The net result of this process is the generation of large populations of a restricted set of clonally related B cells that express somatically diversified Ig receptors with high affinity to the inciting antigen. As a corollary, when multiple different antibodies are obtained from a single combinatorial library, it may be expected that imprints of these processes will be found in the corresponding V regions. Indeed, detailed analysis of the V_H and V_L nucleotide sequences of a large panel of recombinant antibodies against the HIV gp120 protein from a single library showed clear evidence for the expression of a restricted set of clonally-related V regions (Barbas et al 1993). Comparison of recombinant antibodies from libraries of different individuals mounting an immune response to the same antigen may provide further information. Recurrent utilization of particular V_H and/or V_L genes and perhaps similarity in the length and composition of heavy-chain CDR3 regions may be anticipated in some instances.

The relationship between antibodies from combinatorial libraries and natural antibodies may be further evaluated by a variety of additional criteria related to epitope specificity and functional activity of the antibodies. For these purposes, it is recommended that the monovalent recombinant Fab or scFv antibodies are converted to bivalent whole antibody molecules of the correct isotype. Recombinant and serum-derived antibodies should have similar affinities and epitope specificity. Neutralizing capacity of natural anti-viral antibodies or stimulatory activity of thyroid cells of natural anti-TSH receptor autoantibodies are examples of functional activities that should be recapitulated in antibodies from combinatorial libraries.

Concluding remarks

Combinatorial libraries represent an exciting new tool for studying the humoral immune response with respects to V gene utilization, epitope specificity, and functional capacity of antibody molecules evoked as a result of immunization, infection, allergy, or autoimmune disease. Properties of individual V_H and V_L regions (heavy chain dominance, light chain plasticity) as well as amplification and cloning procedures may skew the repertoire of recombinant antibodies in a combinatorial library. Rigorous comparison of the genetic makeup, epitope specificity, and functional properties of recombinant antibodies from libraries and antibodies secreted by immortalized cell lines generated from the same tissue will establish to what extend recombinatorial libraries provide an accurate tool for study of the natural humoral immune response.

References

Barbas CF, Kang AS, Lerner RA, Benkovic SJ (1991) Assembly of combinatorial antibody libraries on phage surfaces: the gene III site. Proc Natl Acad Sci USA 88: 7978–7982.

Barbas CF, Collet TA, Amberg W, et al (1993) Molecular profile of an antibody response to HIV-1 as probed by combinatorial libraries. J Mol Biol 230: 812–823.

Bradbury A, Persic L, Werge T, Cattaneo A (1995) Use of living columns to select specific phage antibodies. Bio/Technology 11: 1565–1569.

Burton DR, Barbas CF, III (1994) Human antibodies from combinatorial libraries. Adv Immunol 57: 191–280.

Caton AJ, Koprowski H (1990) Influenza virus hemagglutinin-specific antibodies isolated from a combinatorial expression library are closely related to the immune response of the donor. Proc Natl Acad Sci USA 87: 6450–6454.

Clackson T, Hoogenboom HR, Griffiths AD, Winter G (1991) Making antibody fragments using phage display libraries. Nature 352: 624–628.

Costante G, Portolano S, Nishikawa T, et al (1994) Recombinant thyroid peroxidase-specific autoantibodies. II. Role of individual heavy and light chains in determining epitope recognition. Endocrinology 135: 25–30.

De Kruif J, Terstappen L, Boel E, Logtenberg T (1995) Rapid selection of cell subpopulation-specific human monoclonal antibodies from a synthetic phage antibody library. Proc Natl Acad Sci USA 92: 3938–3942.

Engelhardt O, Grabheer R, Himmler G, Tuker F (1994) Two-step cloning of antibody variable genes domains in a phage display vector. Biotechniques 17: 44–46.

Gherardi E, Milstein C (1995) Original and artificial antibodies. Nature 357: 201–202.

Hawkins RE, Russell SJ, Winter G (1995) Selection of phage antibodies by binding affinity: mimicking affinity maturation. J Mol Biol 226: 889–896.

Kabat EA, Wu TT (1995) Identical V region amino acid sequences and segments of sequences in antibodies of different specificities: relative contributions of V_H and V_L genes, minigenes, and complementarity-determining regions to binding of antibody-combining sites. J Immunol 147: 1709–1719.

Marks JD, Hoogenboom HR, Bonnert TP, et al (1991) By-passing immunization. Human antibodies from V-gene libraries displayed on phage. J Mol Biol 222: 581–597.

Marks JD, Griffith AD, Malmquist M et al (1992) By-passing immunization: building high affinity human antibodies by chain shuffling. Bio/Technology 10: 779–783.

Marks JD, Onwehand WH, Bye JM, et al (1993) Human antibody fragments specific for human blood group antigens from a phage display library. Bio/Technology 11: 1145–1149.

Meulemans EV, Slobbe R, Wasterval P, et al (1994) Selection of phage-displayed antibodies specific for a cytoskeletal antigen by competitive elution with a monoclonal antibody. J Mol Biol 244: 353–360.

Portolano S, McLachlan SM, Rapoport B (1993) High affinity, thyroid-specific human autoantibodies displayed on the surface of filamentous phage use V genes similar to other autoantibodies. J Immunol 151: 2839–2851.

Radic MZ, Weigert M (1995) Genetic and structural evidence for antigen selection of anti-DNA antibodies. Annu Rev Immunol 12: 487–520.

Rapoport B, Portolano S, McClachlan SM (1995) Combinatorial libraries: new insights into human organ-specific autoantibodies. Immunol Today 16: 43–49.

Schutte MEM, van Es JH, Silberstein LE, Logtenberg T (1995) $V_H4.21$-encoded natural autoantibodies with anti-i specificity mirror those associated with cold hemagglutinin disease. J Immunol 151: 6569–6576.

Siegel DL, Silberstein LE (1994) Expression and characterization of recombinant anti-Rh(D) antibodies on filamentous phage: a model system for isolating human red blood cell antibodies by repertoire cloning. Blood 83: 2334–2344.

Soderlind E, Simonsson AC, Borrebaeck CAK (1992) Phage display technology in antibody engineering: Design of phagemid vectors and in vitro maturation systems. Immunol Rev 130: 109–125.

UytdeHaag FGCM, Loggen HG, Logtenberg T et al (1985) Human peripheral blood lymphocytes from recently vaccinated individuals produce both homotypic and heterotypic antibody upon in vitro stimulation with one type of poliovirus. J Immunol 135: 3094–3101.

Winter G, Griffiths AD, Hawkins RE, Hoogenboom HR (1994) Making antibodies by phage display technology. Annu Rev Immunol 12: 433–455.

Outlook

13.11

Huub F.J. Savelkoul[1]
Eric Claassen[1,2]
Robbert Benner[1]

[1]Department of Immunology, Erasmus University, Rotterdam, The Netherlands
[2]Department of Immunology and Infectious Diseases, TNO Prevention and Health, Leiden, The Netherlands

TABLE OF CONTENTS

Immunology Methods Manual
ISBN 0–12–442712–X

Induction of the humoral immune response

The highly efficient induction of a specific immune response can now be based on detailed knowledge of the relevant T cell and B cell epitopes of the antigen of choice as determined, for example, by the Pepscan method (chapter 13.2) in the case of peptides. Moreover, much has been learned over recent years about rules that govern efficient antigen presentation in relation to the complexity and the physicochemical nature of the antigenic molecule itself (chapter 13.3). This, in combination with dose, carrier proteins, and adjuvants can now lead to highly efficient immunization procedures. Particularly, the type of adjuvant determines to a large extent the features of the subsequent immune response. Although the adjuvants are also thought to determine the isotype specificity of the response, this capacity is also dependent on the carrier protein used. For this reason, the choice of carrier protein when using (synthetic) peptides is crucial. Newly developed methods of immunization, such as recombinant DNA expression systems or naked DNA, based on solid understanding and optimization of the relevant parameters, can increase the successful generation of monoclonal antibodies and eventually the generation of vaccines.

Apart from generating monoclonal antibodies against soluble antigens, activation markers transiently expressed on target cells have also become important targets. As outlined in chapter 13.4, conventional immunization schemes are not appropriate for obtaining useful monoclonal antibodies for such antigens. Alternative methods, such as immunization with insect cells transfected with baculovirus containing the cDNA encoding the protein of interest, has been shown to be very efficient for the generation of several monoclonal antibodies.

To facilitate efficient induction of immune responses essential for the successful generation of monoclonal antibodies, it is important to shift the balance of cytokines present in vivo. Moreover, in this way specific antibodies of predetermined isotype can be induced as several cytokines are capable of inducing switching to selective isotypes. A simple method, based on implantation of stable cytokine gene-transfected cells encapsulated in alginate (chapter 13.5) has been developed that enables effective delivery of cytokines in vivo in a constant fashion.

It can be anticipated that many new approaches will be followed that will improve the generation of monoclonal antibodies that can be applied widely in current (immunological) research. Methods such as DNA immunization, immunization in gene-targeted mice, immunization with transfected xenogeneic cells, and so on, will facilitate the generation of monoclonal antibodies specific for targets that are only transiently expressed or too well conserved to be immunogenic.

Measurement of the immune response

To evaluate the efficacy of the induced immune response, it is essential to quantify the immune parameters of isotype production, antigen-specific responses, and affinity determination. These parameters can only be evaluated with a solid theoretical background and advanced technology. Quantitative determination of immunoglobulin concentrations can be achieved by immunoassays (ELISA) when performed under optimal conditions (chapter 13.8). Moreover, essential data reduction using four-parameter logistic transformation of the data is generally considered the most reliable procedure, particularly at lower readings when dilute sample concentrations are present.

The same parameters that determine the optimization of an ELISA procedure also apply to the development of ELISPOT assays (chapter 13.6). This methodology is particularly useful in determining the frequency of antibody-secreting cells and the distribution pattern of such antibody-secreting cells over the various lymphoid organs during ongoing immune responses. Moreover, a reliable ELISPOT assay can form the basis of limiting dilution assays for determination of the precursor frequency of antibody-secreting cells and clonal sizes of activated precursors. Important additional information can be obtained when combining data from ELISPOT assays on cultured cell suspensions and a quantitative ELISA on the culture supernatant of these cells. In this way, secretion rates (molecules secreted per cell per second) of antibody-secreting cells can be obtained *in vitro*. Apart from ELISPOT assays specific for antibody-secreting cells, similar types of assays have been designed for cytokine-secreting cells.

For full appreciation of the mechanisms involved in immune responses *in vivo*, immunohistochemical methods are powerful in analyzing the structural organization and cellular interactions *in situ* (chapter 13.7). The combination of antigen-specific staining with an antibody- or cell-targeted staining has proved to be extremely informative. Cell-targeted staining can be based on subset- or differentiation stage-specific monoclonal antibodies. In combination with the other methods discussed, this provides a relatively complete picture, permitting quantification of the nature and extent of immune responses *in vivo*.

The affinity of monoclonal antibodies to a large extent determines the detection limit and the speed of assay based on equilibrium measurements (chapter 13.9). Therefore, characterization of monoclonal antibody affinity and kinetics is important in the selection of appropriate reagents for immunoassays. More fundamentally, changes in association and dissociation rate constants and the resulting affinity can be used to probe structure and function in the reactants. This can be important when antibodies or antigens are modified and in the characterization of recombinant proteins. Moreover, affinity measurement is important in defining the efficacy of affinity maturation during primary and memory immune responses.

Recent advances in the technology of repertoire cloning and the expression of these genes as antibody fragments on bacteriophages ('phage display techniques', chapter 13.10) permit the rapid screening and selection of monoclonal antibodies of predefined specificity. The tissue source used for construction of a combinatorial library determines the limits of the specificities of antibody fragments that can potentially be selected. As outlined in this Section, the phage display technique permits the generation of useful reagents with specificity for (auto)antigens or highly conserved antigenic structures.

Section 14

Signaling Pathways and Lymphocyte Cell Repertoires

Section Editor
Carlos Martinez-A.

Introduction

14.1

Carlos Martinez-A.

Centro Nacional de Biotecnología, Universidad Autónoma, Madrid, Spain

The analysis of the spatio-temporal pattern of gene expression is one of the steps for the understanding of its biological function. Whole mount *in situ* hybridization is a rapid and convenient method, introduced in *Drosophila*. The initial procedure has been adapted by a number of laboratories to other experimental systems such as *Xenopus*, chick, and mouse. The potentiality of whole mount *in situ* hybridization is not restricted to developmental studies but may also be used to analyze the expression of molecules involved in the commitment, differentiation, and function of the immune system. In mouse embryos, the procedure can be applied ideally between 7.5 and 10.5 days post conception (dpc). The overall pattern of hybridization is immediately apparent and small groups of cells that express the gene of interest can easily be identified. We present a nonradioactive whole mount *in situ* hybridization method that we have successfully used for post-implantation mouse embryos and organs.

Hematopoietic cell development begins in the mouse embryo at approximately day 7 of gestation and is thought to originate in the yolk sac and proceed to the fetal liver before becoming resident in the adult bone marrow. The studies of hematopoietic progenitor activity in post-implantation and early mid-gestation embryos are sparse either because the techniques for the identification of such cells *in vivo* are technically difficult or because such activity is limited. The generation of hematopoietic chimeras has provided a powerful tool for the study of the development and maturation of the immune system. This section describes a protocol for the engraftment of hematopoietic donor cells into the fetal liver of immunodeficient mouse embryos at early stages of fetal lymphocyte development. This system avoids the immunological defenses of the recipient without the use of toxic conditioning and can also permit engraftment at a time point before irreversible degeneration of the lymphopoietic compartment occurs.

The development of techniques for the transfer and expression of exogenous genes into eukaryotic cells has allowed a rapid advance in regulation studies on gene expression. Although different methods have been used to transfer DNA to hematopoietic cells, retroviral-mediated gene transfer seems to be the most efficient. The retroviral vectors provide several advantages, including the high infection rate of the target cells and the stability of the transduced sequences. Furthermore, the integration of a reduced copy number (usually 1) of the gene in the recipient genome allows us to trace the progeny of individual clones using the integration site as a genetic marker. We describe in this Section an efficient protocol (IES) for retroviral gene transfer into the mouse hematopoietic system using infective supernatants combined with *ex vivo* expansion and selection of bone marrow transduced populations.

During development as well as during adult life cell numbers are maintained through a continuous process of cell growth and cell death. Physiologically most cells die by programmed cell death (PCD), or apoptosis. Here we describe several methods for the cytofluorometric detection of apoptosis and apoptosis-associated phenomena. Most of the systems used for the detection of apoptosis *in vivo* require *in vitro* cell cultures. The necessity of culturing cells *in vitro* to reveal DNA fragmentation has been observed in animals and patients treated with lymphocyte-depleting cytostatic drugs, as well as alterations that are closely associated with apoptosis in HIV-infected donors. We have developed methods, described here, to detect apoptosis before nuclear DNA degradation is observed, which can be used to obtain information on PCD that is ongoing *in vivo*. Such pre-apoptotic alterations concern certain aspects of mitochondrial function, namely, a reduction in mitochondrial transmembrane potential and an increased generation of reactive oxygen species that causes damage to the mitochondrial membrane.

Both cell growth and the induction of apoptosis require cell activation. Protein kinases regulate the early signals of most cellular activation processes. Lymphocytes are no exception and, in fact, the number is increasing of protein kinases being described that regulate not only proliferative responses of B and T lymphocytes but also the early differentiation processes and specific effector functions. One of the methods described in this Section addresses the procedures involved in the detection of the protein kinases as well as the consequences of protein kinase activation.

The development and function of leukocytes are controlled by a large number of cytokines. These mediators are expressed by many different cell types and can exert restricted and pleiotropic effects on various cell lineages. The ability to detect low levels of expression of a cytokine or its receptors is critical given that some of these molecules may require only a few ligand molecules to exert full biological function on a target cell. Chapter 14.7 describes a very sensitive and quantitative method for the measurement of mRNA molecules encoding cytokines based on competitive PCR. This method should, however, be generally applicable to any new PCR-based system.

Detection of apoptosis and apoptosis-associated alterations

14.2

Guido Kroemer[1]
Lisardo Bosca[2]
Naoufal Zamzami[1]
Philippe Marchetti[1]
Sonsoles Hortelano[3]

Carlos Martinez-A[3]

[1]CNRS-UPR420, Villejuif, France
[2]Instituto de Bioquímica, Universidad Complutense, Madrid, Spain
[3]Centro Nacional de Biotecnología, Universidad Autónoma, Madrid, Spain

TABLE OF CONTENTS

Immunology Methods Manual
ISBN 0–12–442712–X

Introduction

Apoptosis, or programmed cell death (PCD), is a physiological process that probably constitutes the natural fate of end-stage differentiated cells, including immunologically relevant cells (Duvall and Wyllie 1986; Kroemer 1995; Kroemer and Martinez-A 1995; Møller, 1994). In contrast to necrosis, apoptosis involves the disruption of nuclear and cytoplasmic structures prior to plasma membrane permeabilization. Depending on the context, the absence of a survival signal, the presence of a death-inducing signal, or the combination of contradictory signals can induce physiological apoptosis. Apoptosis also results from cellular responses to subnecrotic levels of physical, chemical, or osmotic damage. End-stage apoptosis is commonly characterized by nuclear condensation and fragmentation, a stepwise, ordered degradation of nuclear DNA, and the maintenance of nearly intact internal and external cell membranes. This is the reason why most systems for the quantification of apoptosis are based on the detection of nuclear DNA degradation. Advanced DNA degradation is mostly accompanied by alterations in membrane permeability and a loss of phospholipid bilayer dissymmetry, both of which can be measured by single-step colorations.

End-stage apoptosis is difficult to detect in freshly isolated peripheral lymphocytes from the spleen, the lymph nodes, or the circulation, even during treatments that cause massive cell depletion *in vivo* and are known to cause apoptosis *in vitro* (Begleiter et al 1994; Gonzalo et al 1993; Johnston et al 1992; Kawabe and Ochi 1991; Meyaard et al 1992). Thus, to reveal DNA alterations indicative of apoptosis in peripheral lymphocytes, it is indispensable to culture cells for a few hours *in vitro*. This phenomenon can be attributed to the sequestration or phagocytic removal of cells undergoing apoptosis *in vivo* (Huang et al 1994; Savill et al 1993). The necessity of culturing cells *in vitro* to reveal DNA fragmentation has been observed in animals and patients treated with lymphocyte-depleting cytostatic drugs, as well as in HIV-infected donors (Begleiter et al 1994; Gonzalo et al 1993; Johnston et al 1992; Kawabe and Ochi 1991; Meyaard et al 1992). In these cases, alterations that are closely associated with apoptosis, yet precede nuclear DNA damage, can be exploited to obtain information on PCD that is ongoing *in vivo*. Such pre-apoptotic alterations concern certain aspects of mitochondrial function, namely, a reduction in mitochondrial transmembrane potential and an increased generation or reactive oxygen species causing damage to the mitochondrial membrane (Zamzami et al 1995a,b).

Here, several methods for the cytofluorometric detection of apoptosis and apoptosis-associated phenomena are described. None of these methods allows for unequivocal distinction between apoptosis and necrosis, given that apoptosis is typically followed by secondary necrosis. To discriminate the types of cell death, it is imperative either to assess cellular morphology by electron microscopy or to determine the electrophoretic pattern of DNA fragmentation.

Cytofluorometric methods for detection of apoptosis

Any laboratory interested in the quantification of apoptosis has to be equipped with a standard cytofluorometer. Protocols of apoptosis quantification based on light-microscopic or fluorescence-microscopic examination of Mayer-Grünwald or acridine orange-stained cell preparations are time-consuming and quite subjective. Techniques based on the electrophoretic determination of DNA fragmentation are not suitable for quantification. The protocols detailed in this section are designed for conventional single-laser cytofluorometers allowing for excitation within the visible spectrum (e.g., Elite 2 from Coulter, Facscan from Becton Dickinson). The fluorochromes used to determine apoptotic changes in cellular physiology are listed in Table 14.2.1.

Determination of nuclear DNA degradation

Two strategies may be followed for the cytofluorometric assessment of apoptotic DNA degradation. In the first case, chromatinolysis causes an apparent reduction in nuclear DNA content. In consequence, staining with a DNA-intercalating dye will allow for the detection of a cell population containing less DNA than normal (euploid or diploid) G_0/G_1 cells: 'hypoploid' or 'subdiploid' cells. This method is simple and cheap, but is afflicted with the problem that the strong fluorescence of the intercalating dye renders difficult the simultaneous assessment of cell surface markers. Alternatively, the presence of DNA strand breaks can be determined by enzymatic methods. DNA breaks create potential acceptor sites for enzymes involved in DNA synthesis. Addition of such enzymes (terminal deoxyribonucleotidyltransferase, DNA

polymerase, etc.), together with fluorescent deoxytrinucleotide derivatives, thus positively reveals DNA fragmentation. This method is comparatively expensive but it has several advantages: the positive identification of apoptotic events and the possibility of combination with phenotypic analysis of the cell surface.

Assessment of DNA hypoploidy

Materials

- *Propidium iodide stock solution:* 1 mg ml^{-1} propidium iodide (Molecular Probes) in water, to be stored in the dark at 4°C
- *RNAase A stock solution:* 10 mg ml^{-1} RNAase A (Sigma) diluted in water and stored in aliquots at −20°C
- *Sample buffer:* PBS (pH 7.2) containing 1 g l^{-1} glucose. Solution should be passed through a 0.22 μm filter
- *Staining solution:* 18 volumes sample buffer + 1 volume propidium iodide stock solution + 1 volume RNAase A stock solution
- Ethanol for fixation of cells: 70% ethanol kept at −20°C

Permeabilization, fixation, and staining procedure

1. For ethanol permeabilization, adjust cells in 5 ml plastic tubes to a number of 2×10^6 cells/tube. Wash cells twice in ice-cold PBS. It is critical that each sample contains roughly the same number of cells. Centrifuge (600*g*,

Table 14.2.1 Fluorochromes used to detect PCD-associated alterations

Fluorochrome	Alterations detected during PCD	Reference
Determination of nuclear DNA degradation		
Propidium iodine (PI)	Intercalates into DNA and may be used to assess DNA content in ethanol-permeabilized cells. Reduced staining indicates subdiploidy.	Nicoletti et al (1991)
FITC-dUTP biotin–dUTP	Incorporation into DNA strand breaks by DNA polymerase or terminal deoxyribonucleotidyltransferase	Meyaard et al (1992) Gavrieli et al (1992)
Mitochondrial aberrations associated with apoptosis		
Rhodamine 123 DiOC$_6$(3)[a] JC-1[b]	Incorporate into mitochondria depending on electric transmembrane potential ($\Delta\Psi_m$). Reduced uptake in pre-apoptotic cells.	Zamzami et al (1995b) Zamzami et al (1995a)
Dihydroethidine (HE)	Reacts with reactive oxygen species (ROS) including superoxide anion to form ethidium bromide. Increased fluorescence indicates elevated ROS generation.	Carter et al (1994)
Nonyl acridine orange (NAO)	Interacts stoichiometrically with cardiolipin of the inner mitochondrial membrane. Decreased labeling indicates oxidation of cardiolipin by locally produced ROS.	Petit et al (1992)
Apoptosis-associated membrane alterations		
Merocyanine 450 (MC450)	Determines loosened packaging of phospholipids in the outer layer of the membrane. Increased fluorescence indicates loss of transbilayer phospholipid asymmetry associated with apoptosis.	Fadok et al (1992)
Propidium iodine (PI)	Increased diffusion into cells with leaky plasma membranes. Increased uptake into still intact cells undergoing apoptosis (intermediate population).	Nicoletti et al (1991)

[a]3,3'-Dihexyloxacarbocyanine iodide.
[b]5,5',6,6'-Tetrachloro-1,1',3,3'-tetraethylbenzimidazolcarbocyanine iodide.

7 min) and discard supernatant. Vigorously vortex the pellet in the remaining fluid for 10 s. Continue vortexing cells and slowly add 1 ml of cold 70°C ethanol dropwise.

2. Samples should be fixed at least overnight at 4°C in the dark. However, cells can be stored for months.

3. For staining, briefly vortex the sample and centrifuge at high speed ($2000g$ 5 min), pour off supernatant, gently vortex to resuspend cells in residual ethanol and add 200 to 1000 µl freshly prepared staining solution. The final cell concentration should be 10^6 cells ml^{-1}. Incubate for at least 30 min at room temperature while gently rocking.

4. Analyze samples within 24 h in a cytofluorometer (excitation at 488 nm and emission at 617 nm).

Dual staining for the simultaneous determination of nuclear DNA loss and cell viability

In this procedure two dyes are used: one to characterize the necrotic population on the basis of its loss of membrane integrity and a second dye to stain the DNA of cells with still-intact membranes.

- In a first step, cells are resuspended in PBS containing 0.005% propidium iodide. This allows for the staining of nonviable cells.
- The cell suspension is washed twice with PBS to remove the excess of propidium iodide, and permeabilized with 70% ethanol in PBS.
- The cell pellet is resuspended in PBS containing 0.1% Hoechst 33342 and the red (propidium iodide) and blue (Hoechst 33342) fluorescence is assessed. The staining with each dye is proportional to the viability of the cells and allows the determination of the percentage of dead (red fluorescence) and viable (blue fluorescence) cells, as well as for the determination of DNA content in the viable and nonviable fractions. Note that this technique requires excitation in the UV spectrum.

Enzymatic determination of DNA degradation

Two protocols are available for the determination of DNA fragmentation. Fluorescent trinucleotide deoxynucleotides such as fluorescein-12–2'deoxyuridine-5'-triphosphate are incorporated into DNA acceptor sites created by DNA fragmentation. The enzymes capable of transferring the fluorochrome–nucleotide on to such acceptor sites may be either DNA polymerase or terminal deoxyribonucleotidyltransferase (TdT). It appears that the type of DNA fragmentation associated with apoptosis, as compared to necrosis, creates relatively more acceptor sites for the action of TdT (5'-OH termini on free ends of DNA) than for DNA polymerase (single-strand breaks) (Gold et al 1994). In consequence, only the protocol for labeling with TdT, also termed 'end-labeling' or 'tailing' is outlined below.

Materials

- 1 mM fluorescein-12–2'deoxyuridine-5'-triphosphate solution (Boehringer Mannheim); aliquot and store at –20°C protected from light.
- 1 mM adenosine triphosphate (ATP); aliquot and store at –20°C
- Terminal deoxyribonucleotidyl-transferase (TdT; Boehringer Mannheim) in glycerol. Store at –20°C. Remove the quantity needed for the assay while preparing the reaction solution.
- 5× concentrated reaction buffer (1 M potassium cacodylate, 125 mM Tris-HCl, 1.25 mg ml^{-1} bovine serum albumin; pH 6.6)
- 25 mM cobalt chloride solution
- *Lysis buffer:* 0.1% Triton X-100 (Sigma), 3.4 mM sodium citrate; 0.5 mM Tris–HCl (pH 7.6) to be prepared freshly and to be kept on ice.

Procedure

1. Fix cells in 2% paraformaldehyde–PBS solution for a minimum of 2 h. Cells can be labeled prior to fixation with suitable antibodies revealed by phycoerythrin and can be left in the fixing solution for several days.
2. Wash cells twice in PBS.
3. Resuspend in 100 µl lysis buffer and incubate 2 min at 4°C or on ice.
4. Wash cells twice in PBS.
5. Resuspend cells in a total volume of 25 µl of the reaction solution containing: 0.15 nM FITC-dUTP, 2.5 µl 25 mM CaCl$_2$, 5 µl TdT buffer (5×), 1.5 nmol ATP, 12.5 U TdT. This solution should be prepared freshly on ice and TdT should be the last reagent to be added.
6. Incubate for 2 h at 37°C.
7. Wash 3× in PBS.
8. Resuspend in 500 µl PBS and analyze by cytofluorometry.

Note: This technique may be combined with phenotypic characterization of surface markers (Sgonc et al 1994). Immunofluorescence labeling should be performed with the help of phycoerythrin- or cytochrome-labeled antibodies, followed by fixation of cells and subsequent enzymatic determination of DNA fragmentation.

Special application: simultaneous assessment of cell cycle and DNA fragmentation

In some experimental designs it is of interest to assess simultaneously cell cycle and nuclear signs of apoptosis. Analysis of DNA content (above) provides information on ongoing S phases and M/G_2 events, as well as hypoploidy. However, this type of analysis does not show from which phase of the cell cycle apoptotic (hypoploid) cells are recruited. One possibility for addressing this problem is to induce apoptosis in cells exposed to 5-bromo-2-deoxyuridine (BrdUrd). BrdUrd is a thymidine analogue that is incorporated into the DNA of cells during the S phase of cell cycle and can be detected with the aid of specific monoclonal antibodies. Two possibilities are available. In one method, agarose DNA gel electrophoresis of nuclear DNA can be subjected to Southern blotting, followed by immunochemical determination of BrdUrd in the intact and/or fragmented DNA (Gonzalo et al 1994a). In the other, it is possible to perform a dual fluorescence assay with the end-labeling method (i.e., to incorporate FITC–dUrd into fragmented DNA, as described above) followed by dihydroethidine-developed detection of BrdUrd incorporation into cells (Gonzalo et al 1994b). This method is described below.

- Cells should be labeled prior to induction of apoptosis using 5-bromo-2-deoxyuridine (BrdUrd, Sigma) employing suitable protocols of *in vitro* or *in vivo* incubation with the reagent.

- Perform end-labeling as described above.
- Incubate cells 30 min with DNAase I (35 U in 35 µl, room temperature) to partially denature DNA, and wash twice in PBS + 2% FCS.
- Incubate cells with 0.4 µg monoclonal antibody specific for BrdUrd (BMC 9318 from Boehringer Mannheim); wash three times.
- Incubate with goat anti-mouse IgG1–phycoerythrin conjugate (Southern Biotechnology); wash twice. During analysis, gate on normal-sized cells only. Isotype-matched, nonspecific conjugates and cells that have been incubated in the absence of BrdUrd serve as negative controls.

Cytofluorometric determination of mitochondrial aberrations associated with early apoptosis

Physiological PCD, that is, PCD that is not induced by direct genotoxic insults, is characterized by alterations of mitochondrial parameters that clearly precede nuclear DNA degradation. A significant reduction in mitochondrial transmembrane potential ($\Delta\Psi_m$) precedes morphological alterations of the nucleus, cell shrinkage, as well as oligonucleosomal DNA fragmentation in many different systems of apoptosis induction: glucocorticoid-induced depletion of lymphocytes *in vivo*, activation-induced apoptosis of T and B cells, tumor necrosis factor-induced PCD of U937 cells, and so on (Zamzami et al 1995a,b). Those cells that exhibit a reduced $\Delta\Psi_m$ can be subdivided into two subpopulations, a first population that still behaves normally with respect to the generation of reactive oxygen species, and a second population that hyperproduces superoxide anions in uncoupled mitochondria (Zamzami et al 1995a). The size of this latter population is a sensitive marker of ongoing apoptotic depletion *in vivo*, given that it represents a more advanced stage of the apoptotic process.

Determination of the mitochondrial transmembrane potential ($\Delta\Psi_m$)

Several different cationic fluorochromes can be employed to measure mitochondrial trans-membrane potentials. These markers include rhodamine 123 (Rh123), 3,3'-dihexyloxacarbocyanine iodide (DiOC$_6$(3)) (fluorescence in green) (Petit et al 1990) and 5,5',6,6'-tetrachloro-1,1',3,3'-tetraethylbenzimidazolcarbocyanine iodide (JC-1) (fluorescence in red and green) (Johnson et al 1980; Smiley, 1991). DiOC$_6$(3) allows for double staining with phycoerythrin labeled antibodies (Zamzami et al 1995a,b). Compared to Rh123, DiOC$_6$(3) offers the important advantage that it does not show major quenching effects. JC-1 incorporates into mitochondria, where it either forms monomers (fluorescence in green, 527 nm) or, at high transmembrane potentials, aggregates (fluorescence in red, 590 nm) (Smiley, 1991). Thus, the quotient between green and red JC-1 fluorescence provides an estimate of $\Delta\Psi_m$ that is independent of mitochondrial mass.

Materials

- *Stock solutions of fluorochromes:* Rh123 should be diluted to 10 mM in ethanol, DiOC$_6$(3) to 40 μM in DMSO, and JC-1 to 10 mM in DMSO. All three fluorochromes can be purchased from Molecular Probes and should be stored, once diluted, at –20°C in the dark.
- *Working solutions:* Dilute Rh123 to 100 μM (5 μl stock solution + 500 μl PBS), DiOC$_6$(3) to 400 nM (10 μl stock solution + 1 ml PBS), and JC-1 to 20 μM (2 μl stock solution + 1 ml PBS). These solutions should be prepared freshly for each series of stainings.

Staining protocol

1. Cells (5–10 × 10^6 in 0.5 ml PBS) should be kept on ice until the staining. If necessary, cells can be labeled with specific antibodies conjugated to compatible fluorochromes (phycoerythrin for DiOC$_6$(3), fluorescein isothiocyanate for Rh123) before determination of mitochondrial potential.

2. For staining, add the following amounts of working solutions to 0.5 ml of cell suspension: 25 μl Rh123 (final concentration 2.5 μM), 25 μl DiOC$_6$(3) (final concentration 20 nM), or 25 μl JC-1 (final concentration 1 μM) and transfer tubes to a water bath kept at 37°C. After 15–20 min of incubation, return cells to ice. Do not wash the cells.

3. Perform cytofluorometric analysis within 10 min, while gating the forward and side scattering on viable, normal-sized cells. When large series of tubes are to be analyzed (>10 tubes), the interval between labeling and cytofluorometric analysis should be kept constant. When using an Epics Profile cytofluorometer (Coulter), Rh123 should be monitored in FL1 (excitation 488 nm; emission 501 nm), DiOC$_6$(3) in FL1 (excitation 488 nm; emission 500 nm), and JC-1 in FL1 versus FL2 (excitation 488 nm; emission at 527 and 590 nm. The following compensations are recommended for JC-1: 10% of FL2 in FL1, and 21% of FL1 in FL2 (indicative values). DiOC$_6$(3) can be combined with dihydroethidine staining (see below).

Mitochondrial generation of reactive oxygen species (ROS) and local ROS effects

It is possible to determine the production of reactive oxygen species (ROS) by dihydroethidine (HE), a substance that is oxidized by superoxide anion to become ethidium bromide (EthBr) and to emit red fluorescence (Rothe and Valet 1990). HE is more sensitive to superoxide anion than 2',7'-dichlorofluorescein diacetate, which measures H_2O_2 formation (Carter et al 1994; Rothe and Valet 1990). Thus, enhanced HER→EthBr conversion can be observed in cells that do not label with 2',7'-dichlorofluorescein diacetate. Alternatively, the damage produced by reactive oxygen species in mitochondria can be determined indirectly by assessing the oxidation state of cardiolipin, a molecule restricted to the inner mitochondrial membrane. Nonyl acridine orange (NAO) interacts stoichiometrically with intact, nonoxidized cardiolipin (Petit JM et al 1992). In consequence, a reduction in NAO fluorescence indicates a decrease in cardiolipin content.

Materials

- *Stock solutions of fluorochromes:* 4.73 mg ml^{-1} (10 mM) HE should be stored at –20°C. 3.15 mg ml^{-1} (10 mM) NAO in DMSO should be stored at 4°C. Both fluorochromes, available commercially from Molecular Probes, are light-sensitive.
- *Working solutions:* Dilute 5 µl of either of the stock solutions in 5 ml PBS (final concentration 10 µM). Prepare freshly for each series of stainings.

Staining protocol

1. Cells (5–10×10^6 in 1 ml PBS) should be kept on ice until the staining. Before HE staining, cells may be labeled with specific antibodies conjugated to either fluorescein isothiocyanate or phycoerythrin.

2. For staining, add the following amounts of working solutions to 1 ml of cell suspension: 25 µl HE (final concentration 2.5 µM), 10 µl NAO (final concentration 100 nM), or 25 µl JC-1 (final concentration 1 µM) and transfer tubes to a water bath kept at 37°C. After 15–20 min of incubation, return cells to ice. Do not wash cells.

3. Perform cytofluorometric analysis within 10 min, while gating forward and side scattering on viable, normal-sized cells. When large series of tubes are to be analyzed (>10 tubes), the interval between labeling and cytofluorometric analysis should be kept constant. When using an Epics Profile cytofluorometer (Coulter), HE should be monitored in FL3 and NAO in FL1 or FL3. For double stainings, compensations have to be made in accordance with the apparatus. If HE is combined with NAO, we recommend compensating FL3 – FL1 = 27% and FL1 – FL3 = 1%. In this case, NAO is measured in FL3. For double staining with HE and DiOC$_6$(3) the compensation should be approximately FL1 – FL3 = 5% and FL3 – FL1 = 2%.

Cytofluorometric determination of apoptosis-associated cell membrane alterations

Normal plasma membranes exhibit a marked phospholipid asymmetry, with phosphatidylcholine and sphingomyelin predominantly on the external layer, and most of the phosphatidylethanolamine and phosphatidylserine on the inner layer. This asymmetry is maintained by an ATP- and Mg^{2+}-dependent enzyme mediating the inward transport of negatively charged phospholipids and/or association of phosphatidylserine with membrane skeletal proteins. Apoptosis is accompanied by a loss of phospholipid asymmetry, with exposure of negatively charged phosphatidylserine on the outer leaflet. The resulting alteration of lipid packaging gives rise to an enhanced fluorescence with merocyanine 540 (MC540) (Fadok et al 1992). In addition, apoptosis is accompanied by an enhanced membrane permeability. Since cells undergoing apoptosis are less permeable than lysed cells, they have a characteristic intermediate phenotype with low-molecular-weight vital dyes such as propidium iodide (PI) (Nicoletti et al 1991).

Determination of alterations in membrane lipid packaging accompanying apoptosis

Materials

- *Stock solution:* 50 mg ml^{-1} MC540 (Molecular Probes) in ethanol, to be kept in the dark at –20°C.
- *Working solution:* Dilute 10 μg of stock solution in 500 μl PBS (final concentration 1 mg ml^{-1}) before use.

Staining

1. Add 2 μl working solution to 100 μl PBS containing 0.5–1 × 10^6 cells. After 3 min of incubation at room temperature, add 900 μl PBS to tubes and analyze immediately.
2. MC450 emission can be detected both in FL1 and in FL2. It is compatible with phycoerythrin-labeling of surface markers.

Quantification of cell viability with propidium iodide (PI)

During end-stage apoptosis, cells progressively lose their capacity for vital dye exclusion. Classically, cells undergoing apoptosis first show an intermediate incorporation of PI and finally display a complete loss of membrane permeability and become PI^{high} (Nicoletti et al 1991).

Materials

- *Stock solution:* 1 mg ml^{-1} PI (Sigma) in distilled water. Store at 4°C in the dark.
- *Working dilution:* Predilute 10 μl PI with 90 μl PBS.

Staining

1. Add 20 μl PI working solution (final concentration 2 μg ml^{-1}) to 1 ml cell suspension (0.5–1 × 10^6 cells ml^{-1}). Cell should be analyzed after 5 min of incubation at room temperature.
2. Analyze in FL3. Cells exhibiting intermediate labeling are likely to be on the way to undergoing apoptosis. PI staining is compatible with labeling with DiOC$_6$(3) or FITC.

Assessment of DNA fragmentation

Two alternative protocols are described in this subsection. DNA fragments arising from the activation of endogenous nucleases can be determined either in the whole cell or in the cytoplasm only.

Assessment of oligonucleosomal fragmentation of nuclear DNA by agarose gel electrophoresis

A regular 'ladder-type' pattern of DNA fragmentation is commonly considered to define apoptosis, whereas random degradation of DNA giving rise to a smear is thought to indicate necrosis as the mechanism of cell death. The regular pattern of DNA fragmentation is due to endonclease-driven digestion of DNA into mono- or oligomers of oligonucleosomal DNA stretches, i.e., 200 bp and multiples thereof. The oligonucleosomal DNA fragmentation can easily be detected by agarose gel electrophoresis of nuclear DNA. It has to be understood, however, that this technique is not suitable for the *quantification* of apoptosis.

Materials

Stock solutions

- 0.5 M EDTA (pH 8.0) to be stored at room temperature.
- 10% SDS to be stored at RT.
- 50 mM Tris–HCl (pH 7) to be stored at room temperature.
- 20 mg ml^{-1} proteinase K (Boehringer Mannheim) to be aliquoted and frozen at –20°C.
- 1 mg ml^{-1} RNAase (Sigma) to be aliquoted and frozen at –20°C. RNAase should be heat-inactivated to remove traces of contaminating DNAase (15 min at 100°C).
- *Gel loading buffer:* 0.25% bromophenol blue, 0.25% xylene cyanol, 40% (w/v) sucrose in water. To be stored at 4°C.
- *TAE (Tris acetate) buffer 50×*: 242 g Tris base, 57.1 ml glacial acetic acid plus 100 ml 0.5 M EDTA (pH 8.0) to be dissolved in 1 liter of H$_2$O. To be stored at room temperature.

Working solutions

- *Working solution A:* 20 μl EDTA (final concentration 10 mM) + 50 μl SDS (final concentration 0.5%) + 900 μl Tris–HCl (50 mM) + 25 μl of prediluted proteinase K (1 μl stock diluted in 1 ml of Tris–HCl; final concentration 0.5 μg ml^{-1})
- *Working solution B:* 10 μg RNAase + 1 ml 50 mM Tris–HCl

DNA extraction

Cells (1–2 × 10^6 cells) should be contained in E-tubes, preferably as pellets without any supernatant. Dry cell pellets can be frozen at –20°C until DNA extraction. It is important that each vial contain the same number of cells. The protocol of DNA extraction is explained in the following flow diagram:

Add 20 µl of working solution A to cell pellet, gently resuspend

⇓

Incubate 1 h at 50°C

⇓

Heat-inactive proteinase K contained in solution A by incubating tube for 10 min at 70°C

⇓

Add 10 µl working solution B

⇓

Incubate 1 h at 37°C

⇓

Samples may be stored overnight at 4°C or may be analyzed immediately by electrophoresis

Gel electrophoresis

Prepare horizontal agarose gel as described by Maniatis and colleagues (Sambrook et al 1989). It is recommended to employ 1% agarose gels prepared in 1× TAE buffer supplemented with ethidium bromide to a final concentration of 0.5 µg ml^{-1}. Gels should be at least 6 mm thick and slots should measure 2 × 5 × 5 mm. Mount gel in the electrophoretic tank, and add 1× TAE buffer so that the upper level of the buffer and the gel reach exactly the same level; in this way the slots produced by the comb remain dry. Add 10 µl of gel loading buffer to the sample and load the entire amount into the appropriate slot using disposable 50–250 µl tips cut with a pair of scissors to about two-thirds of the original length. This allows for pipetting of the viscous DNA-containing sample. Size markers in the range 100 to 2000 bp should be included.

Application of a field of 5 V cm^{-1} over 15 min causes DNA in the slots to enter the gel. Thereafter, the gel can be covered with TAE buffer. Electrophorese at 5 V cm^{-1} for 2.5 h. Examine in a UV transilluminator and photograph. Densitometric analysis may provide approximate information on the relative content of oligonucleosomal versus intact DNA.

Determination of DNA fragments in the cytosol

During the apoptotic process, nucleosomes and oligonucleosomes produced by internucleosomal DNA fragmentation are released from the nucleus to the cytoplasm, where these molecules remain stable for several hours. The presence of nucleosomal DNA in the cytosol requires the cleavage of the nuclear chromosomal DNA to fragments that may cross the nuclear membrane; therefore, this process is delayed in time with respect to the nuclear fragmentation.

The protocol allows for the determination of oligonucleosomal DNA in the cytosol of a wide number of cells from different origins. Kinetic analysis of the appearance of fragments in the cytosol is recommended to optimize sampling time points.

Additional materials

- *Lysis buffer:*

    20 mM EDTA (pH 8.0)

    0.5% Triton X-100

    5 mM Tris–HCl, pH 8.0

- *Phenol–chloroform:*

    Prepare 250 g of phenol in 30 ml of 1 mM Tris–HCl, pH 8.0.

    Combine 1 volume of phenol equilibrated at pH 8.0 with 1 volume of 24:1 chloroform–isoamyl alcohol.

Procedure

1. Cultured cells attached to a matrix or to plastic dishes, or in suspension are

washed twice with an excess of ice-cold PBS, and the cell pellet or the cell layer is filled with ice-cold lysis buffer (0.2–0.4 ml per 10^6 cells). After gentle shaking for 15 min, check under the microscope for cell disruption. Immediately centrifuge in an Eppendorf tube for 10 min and transfer the supernatant.

2. To deplete the extract of RNA and proteins, treat for 30 min at 37°C with heat-inactivated RNAase (0.1 mg ml^{-1} of homogenate) and for 2 h at 56°C with proteinase K (0.1 mg ml^{-1} of homogenate). Samples are centrifuged for 5 min in an Eppendorf tube and the supernatant is mixed with 1 volume of standard phenol–chloroform mixture. After thorough mixing for 30 s let settle for 15 min at 4°C. Centrifuge for 5 min in the Eppendorf and remove an exact amount of the overlaying aqueous solution, which should be clear. If it has a cloudy aspect, proceed to a second phenol–chloroform extraction.

3. DNA is precipitated with at least 2 volumes of absolute ethanol and 0.1 volume of 2 M ammonium acetate, and stored for 24 to 48 h at –20°C. The tubes are centrifuged for 15 min in an Eppendorf and the pellet is washed with 75% ethanol solution at –20°C. The DNA pellet is solubilized in Tris–EDTA buffer (pH 8.0), and the DNA content is quantified by spectrophotometric reading at 260 and 280 nm. DNA is subjected to electrophoretic analysis as described above.

Note: Phenol–chloroform extractions can be omitted in cells with low rRNA content. In this case, after treatment with RNAase and proteinase K, DNA can be directly precipitated with ethanol.

Concluding remarks

The optimal method to be employed for quantification of apoptosis is dictated by the nature of the cell type investigated. If experiments are performed on homogeneous cell populations (e.g., cell lines), the assessment of DNA hypoploidy or loss of viability is largely sufficient to provide information on the kinetics of apoptosis. In contrast, if the cell population that is being investigated is rather heterogeneous, simultaneous determinations of cell phenotype and of DNA degradation using the 'tailing' method are recommended. Determination of mitochondrial transmembrane potentials and ROS generation may be particularly useful for the assessment of lymphoid cell function *ex vivo*. To prove that cell death is due to apoptosis rather than necrosis, it is useful to perform at least one series of control experiments and to demonstrate, by agarose gel electrophoresis, the regular, oligonucleosomal DNA fragmentation that occurs prior to cytolysis. Similarly, morphological (ideally electron-microscopic) analysis provides objective criteria for the discrimination of apoptosis and necrosis. Nonetheless, in some experimental systems DNA fragmentation and apoptosis have been dissociated. Indeed, cytoblasts (enucleated cells) can be induced to undergo apoptosis (Jacobson et al 1994), a finding that emphasizes that the nucleus is only one of the targets of the apoptotic process and that current investigation of programmed cell death should not neglect apoptosis-associated cytoplasmic alterations.

References

Begleiter A, Lee K, Israles LG, Mowat MRA, Johnston JB (1994) Chlorambucil induced apoptosis in chronic lymphocytic leukemia (CLL) and its relationship to clinical efficacy. Leukemia 8: S103–S106.

Carter WO, Narayanan PK, Robinson JP (1994) Intracellular hydrogen peroxide and superoxide anion detection in endothelial cells. J Leukocyte Biol 55: 253–258.

Duvall E, Wyllie AH (1986) Death and the cell. Immunol Today 7: 115–119.

Fadok VA, Voelker DR, Campbell PA, Cohen JJ, Bratton DL, Henson PM (1992) Exposure of phophatidylserine on the surface of apoptotic lymphocytes triggers specific recognition and removal by macrophages. J Immunol 148: 2207–2216.

Gavrieli Y, Sherman Y, Ben-Sasson SA (1992) Identification of programmed cell death in situ via specific labeling of nuclear DNA fragmentation. J Cell Biol 119: 493–498.

Gold R, Schmied M, Giegerich G, et al (1994) Differentiation between cellular apoptosis and necrosis by the combined use of in situ tailing and nick translation techniques. Lab Invest 71: 219–225.

Gonzalo JA, Gonzalez-Garcia A, Martinez-A C, Kroemer G (1993) Glucocorticoid-mediated control of the clonal deletion and activation of peripheral T cells in vivo. J Exp Med 177: 1239–1246.

Gonzalo JA, Baixeras E, Gonzalez-Garcia A, et al (1994a) Differential in-vivo-effects of a superantigen and an antibody targeted to the same T cell receptor: activation-induced cell death versus passive macrophage-dependent deletion. J Immunol 152: 1597–1608.

Gonzalo JA, Gonzalez-Garcia A, Baixeras E, et al (1994b) Pertussis toxin interferes with superantigen-induced deletion of peripheral T cells without affecting T cell activation in vivo. Inhibition of deletion and associated programmed cell death depends on ADP-ribosyltransferase activity. J Immunol 152: 4291–4299.

Huang L, Soldevila G, Leeker M, Flavell R, Crispe IN (1994) The liver eliminates T cells undergoing antigen-triggered apoptosis in vivo. Immunity 1: 741–749.

Jacobson MD, Burne JF, Raff MC (1994) Programmed cell death and Bc1-2 protection in the absence of a nucleus. EMBO J 13: 1899–1910.

Johnson LV, Walsh ML, Chen LB (1980) Localization of mitochondria in living cells with rhodamine 123. Proc Natl Acad Sci USA 77: 990–994.

Johnston JB, Lee K, Verburg L, et al (1992) Induction of apoptosis in CD4+ prolymphocytic leukemia by deoxyadenosine and 2'-deoxyformycin. Leuk Res 16: 781–788.

Kawabe Y, Ochi A (1991) Programmed cell death and extrathymic reduction of Vβ8+ CD4+ T cells in mice tolerant to Staphylococcus aureus enterotoxin B. Nature 349: 245–248.

Kroemer G (1995) The pharmacology of T lymphocyte apoptosis. Adv Immunol, in press.

Kroemer G, Martinez-A C (eds) (1995) Apoptosis of immune cells. Curr Top Microbiol Immunol 200: 175.

Meyaard L, Otto SA, Jonker RR, Mijnster MJ, Keet RPM, Miedema F (1992) Programmed cell death in HIV-1 infection. Science 257: 217–219.

Møller G (1994) Apoptosis in the immune system. Immunol Rev 142:

Nicoletti I, Migliorati G, Pagliacci MC, Riccardi C (1991) A rapid simple method for measuring thymocyte apoptosis by propidium iodide staining and flow cytometry. J Immunol Methods 139: 271–280.

Petit JM, Maftah A, Ratinaud MH, Julien R (1992) 10 N-nonyclacridine orange interacts with cardiolipin and allows for the quantification of phospholipids in isolated mitochondria. Eur J Biochem 209: 267–273.

Petit PX, O'Connor JE, Grunwald D, Brown SC (1990) Analysis of the membrane potential of rat- and mouse-liver mitochondria by flow cytometry and possible applications. Eur J Biochem 389–397.

Rothe G, Valet G (1990) Flow cytometric analysis of respiratory burst activity in phagocytes with hydroethidine and 2',7'-dichlorofluorescein. J Leukocyte Biol 47: 440–448.

Sambrook J, Fritsch ER, Maniatis T (1989) Molecular Cloning: A Laboratory Manual. Cold Spring Harbor Laboratory Press, Cold Spring Harbor, NY.

Savill J, Fadok V, Henson P, Haslett C (1993) Phagocyte recognition of cells undergoing apoptosis. Immunol Today 14: 131–136.

Sgonc R, Boeck G, Dietrich H, Gruber J, Recheis H, Wick G (1994) Simultaneous determination of cell surface antigens and apoptosis. Trends Genet 10: 41–42.

Smiley ST (1991) Intracellular heterogeneity in mitochondrial membrane potential revealed by a J-aggregate-forming lipophilic cation JC-1. Proc Natl Acad Sci USA 88: 3671–3675.

Zamzami N, Marchetti P, Castedo M, et al (1995a) Sequential reduction of mitochondrial transmembrane potential and generation of reactive oxygen species in early programmed cell death, submitted.

Chapter 14.2

Zamzami N, Marchetti P, Castedo M, et al (1995b) Reduction in mitochondrial potential constitutes an early irreversible step of programmed lymphocyte death *in vivo*. J Exp Med, in press.

Use of retroviral vectors in lymphohemato-poietic lineage analysis

14.3

Florencio Varas[1]
Carlos Martinez-A[2]
Antonio Bernad[2]

[1]Unidad de Biologia Molecular y Celular, CIEMAT, Madrid, Spain
[2]Centro Nacional de Biotecnología, Universidad Autónoma, Madrid, Spain

TABLE OF CONTENTS

Immunology Methods Manual
ISBN 0–12–442712–X

Abstract

Retroviral-mediated gene transfer nowadays represents the finest methodology available for achieving a persistent and stable phenotypic modification of the lymphohematopoietic system. Although several procedures have been reported, this chapter is mainly focused on the description of an efficient protocol (IES) for transducing long-term repopulating progenitors. Freshly harvested adult murine bone marrow (BM), enriched in proliferatively activated precursors by *in vivo* 5-FU treatment, is infected with pXT1-retroviral supernatants. Afterwards, BM cells are cultured with IL-3/SCF followed by a G418 selection step of the transduced cells. This strategy enables the collection of a homogeneously marked cell graft with a highly increased content in committed and pluripotent progenitors which play an essential role in accelerating the hematopoietic recovery in myeloablated recipients. Furthermore, these infected, expanded, and selected grafts reconstitute, in the long term, most of the lymphohematopoietic system of lethally irradiated and transplanted animals. This transduction protocol, in combination with PCR-recognizable retroviral vectors and cell sorting, may become a valuable tool for studies on the lymphohematopoietic lineage development and functional relationships.

Introduction

The development of techniques for the transfer and expression of exogenous genes into eukaryotic cells has allowed a rapid advance in regulation studies on gene expression, as well as in the description of important regulatory factors. The lymphohematopoietic system has been the focus of most efforts for gene transfer owing to the experience accumulated in manipulation and successful transplantation, and the availability of both *in vivo* and *in vitro* techniques for assaying a quite high number of progenitors.

Although different methods have been used to transfer DNA to hematopoietic cells, retroviral-mediated gene transfer seems to be the most efficient. The retroviral vectors provide several advantages including (a) high infection rate of the target cells, (b) stability of the transduced sequences after their integration as provirus, and (c) transfer of a reduced copy number (usually 1) of the gene into the recipient genome. Retroviral vectors have aided in unraveling biological roles played by genes after expression in murine bone marrow cells. For example, the deregulated expression of different oncogenes (Keller and Wagner 1989; von Ruden et al 1992) and hematopoietic growth factors (Johnson et al 1989; Hawley et al 1992; Fraser et al 1993) has resulted mainly in myeloproliferative syndromes. However, for application to somatic gene therapy strategies, achievement of a persistent modification of the hematopoietic system depends on stable transfer into pluripotent hematopoietic stem cells (PHSC), which currently remains an open field.

In another sense, because proviral integration usually occurs once and is essentially random, it is feasible to trace the progeny of individual clones using the integration site as a genetic marker (Lemischka et al 1986; Dick et al 1985). Before retroviral marking was developed, the generation of cytogenetic chromosomal markers induced by sublethal irradiation served to demonstrate the existence of PHSC contributing to all blood lineages (Wu et al

1965; Nowell et al 1970; Abramson et al 1977) and for the establishment of the clonal origin of the CFU-S (colony forming unit in spleen) (Becker et al 1963; Wu et al 1967). However, the low efficiency of this technique in inducing the genetic markers hindered the dynamic analysis of these populations. Using retroviral marking, these preliminary results have been confirmed and implemented. Several authors have proposed the existence of certain hetero-geneity within the stem cell compartment based on the self-renewal and developmental potential of different cell populations (Dick et al 1985; Keller et al 1985; Lemischka et al 1986), distinguishing totipotent stem cells from line-age-committed ones. Nevertheless, it is not clear whether these restrictions may also be the

result of stochastic mechanisms acting on real totipotent stem cells (Jordan et al 1990).

The analysis of individual cell lineages would be greatly improved if the lymphohemato-poietic system of transplanted animals were mostly reconstituted by transduced cells. To achieve this goal, the IES retroviral gene trans-fer protocol may represent a useful tool because reduced grafts (10^5–10^6 transduced cells) repopulate predominantly the hemato-poietic organs of long-term surviving mice. The combination of this protocol with a collection of retroviruses easily identifiable by PCR (Cepko et al 1993) and cell sorting might constitute a powerful experimental system for studies on specific lineages or even at the clonal level.

Retroviral gene transfer to murine lymphohematopoietic stem cells

Gene transfer by retroviral vectors is based on replication-deficient but infection-competent pseudoviruses. A large number of retroviral vectors have been developed, based on the similar genetic structures of several murine viruses and employing different strategies for the expression of exogenous genes. These retrovirus-like particles are produced by pack-aging cell lines, usually constructed from mur-ine fibroblasts. These cell lines supply in trans all the protein products required to encapsidate the retroviral vector-encoded RNA genome. On the other hand, retroviral vectors include sequences required in cis for packaging and reverse transcription, allowing their proviral integration into the genomic DNA of the trans-duced target cells.

In the present chapter we focus on the careful presentation of the methodology for efficient retroviral gene transfer to BM stem cells from adult mice, pointing out the topics related to optimal BM manipulation and evalua-tion of transduction efficiency. For more detailed information about retroviral vectors and packaging cell lines, we suggest a review of specialized literature (e.g., Chang and John-son 1989; Kriegler 1990; Miller et al 1993).

Retroviral vectors and packaging cell lines

These vectors usually contain a positively selectable gene under the transcriptional control of the LTR sequences or a second internal promoter. The most common are genes confering drug resistance (neomycin, methotrexate, puromycin, hygromycin), although genes encoding proteins expressed on the cell surface (MDR, CD24) have also been reported. Some vectors are also suitable for the cloning and potential expression of another exogenous gene. Throughout the protocol to be described for the transduction of lymphohematopoietic stem cells, we used the pXT1 retroviral vector (Boulter and Wagner 1987) which transcribes the neomycin phosphotransferase gene (neo^r) from the Moloney murine leukemia virus LTR. In this method, the positive selection of cell populations expressing the indicator gene seems to be one of the critical steps. The use of vectors expressing different selectable genes will imply the corresponding modifications (see later).

With the tropism of the viral particles, the main point to be considered is the nature of the envelope protein included in the particles, which will guide the primary interaction with the cellular receptor. This tropism is provided by the particular packaging cell line used. The most common, $\Psi2$ and PA317 (Mann et al 1983; Miller and Buttimore 1986), allow one to obtain infectious particles restricted to rodent cells (ecotropic) or with a wide host range (amphotropic), respectively. Recently other packaging cell lines, GP+E–86 (Markowitz et al 1988a) and its amphotropic version GP+env-AM12 (Markowitz et al 1988b), which present a fragmented helper retroviral genome, have been developed to reduce risk of the generation of replication-competent viruses by recombination with vector sequences. However, lower retroviral titers are usually obtained.

Production of supernatant containing defective retroviral particles

The first step consists of the generation of stable retrovirus-producing packaging cell lines. We have developed the IES protocol (Bernad et al 1994) for the transduction of murine hematopoietic stem cells by means of pXT1-derived retroviral particles packaged in the ecotropic $\Psi2$ cell line. The use of other cell lines may require minor changes. To obtain high-titer producer lines, it is usually recommended to infect the packaging cell line rather than to transfect by conventional methods, as the former procedure yields better results. However, ecotropic virus producers are resistant to infection by ecotropic viruses, as amphotropic producers are resistant to amphotropic virus infection. The protocol therefore includes the transfection of the amphotropic cell line (PA317) and the use of its conditioned medium to infect the ecotropic producer $\Psi2$.

Isolation of retrovirus-producing packaging cell lines

Materials

- PA317 and $\Psi2$ cell lines (ECACC) are routinely cultured in Dulbecco's Minimal Essential Medium (DMEM; Gibco Laboratories) supplemented with 75 IU ml^{-1} penicillin (Sigma), 75 mg ml^{-1} streptomycin sulfate (Sigma), and 10% heat-inactivated fetal bovine serum (FBS; Gibco Laboratories) at 37°C in a fully-humidified atmosphere with 5% CO_2 in air.

- 60 mm diameter and 24-well culture plates (Nunc).
- *Trypan blue solution:* 0.4% (w/v) trypan blue (Merck) in 139 mM NaCl, and 49 mM KH_2PO_4. The solution is adjusted to pH 7.0–7.3.
- *Trypsin solution:* 0.05% (w/v) trypsin (Sigma) in 5 mM glucose, 0.5 mM EDTA, 5 mM KCl, 137 mM NaCl, 7 mM $NaHCO_3$, and 0.015% (w/v) phenol red (Sigma). The solution is adjusted to pH 7.0–7.3.
- *Polybrene stock solution:* 5 mg ml^{-1} polybrene (Aldrich Chemical Company) in PBS, filtered through 0.22 μm and stored in aliquots at –20°C.
- Freshly prepared plasmidic retroviral vector DNA.
- *Geneticin (G418) stock solution:* 100 mg ml^{-1} G418 (Gibco Laboratories) — a neomycin analogue — in DMEM, neutralized with NaOH to pH 7.0, filtered through 0.22 μm, and stored in aliquots at –20°C. G418 dosage is always given as the equivalent weight of active product.

Method

1. One day before transfection, 2×10^5 PA317 cells per 60 mm culture plate are seeded and incubated at 37°C overnight.
2. Transfection is performed by the method of choice. We have obtained good results with the conventional calcium phosphate method using 10 μg of plasmidic DNA purified on a CsCl–ethidium bromide gradient (Sambrook et al 1989).
3. Forty-eight hours post-transfection the supernatant is collected, clarified by centrifugation (400g, 5 min, 4°C) and filtered through a 0.8 μm filter. This supernatant (PA317-conditioned medium) is stored at –80°C.
4. One day before infection, 10^5 Ψ2 cells are seeded per 60 mm culture plate and incubated overnight at 37°C.
5. The culture medium from the Ψ2 plates is removed and 1 ml of the PA317 viral supernatant supplemented with polybrene to a final concentration of 8 μg ml^{-1} is added. Plates are incubated at 37°C for 1.5 h with occasional gentle shaking. This solution is then replaced with 5 ml of fresh culture medium and incubated at 37°C.
6. 48 h post-infection, the cells are trypsinized and split to 1/10, 1/20, and 1/40, adding G418 to the culture medium up to 1 mg ml^{-1}. Cells expressing the resistance gene will proliferate and form colonies, whereas the rest will die within 5–14 days. During this period, medium containing dead cells is removed every 3 days and replaced with fresh culture medium containing 1 mg ml^{-1} G418.
7. At this point, depending on the purpose of the experiment, it is possible to harvest a pool of retroviral-producing cells, or to pick individual colonies to select a very high-titer producer clone. In the former case, select a plate or plates containing a sufficient number of colonies, trypsinize, and expand the resulting pool. Alternatively, to pick individual colonies, select plates with isolated clones. Remove the culture medium and add 5–10 μl of trypsin solution on to the colony. After 2–3 min at room temperature, recover the suspended cells with a Gilson pipetteman and transfer them to 1 ml culture medium in a 24-well plate. When cells have grown to confluence, they are expanded for freezing and virus titration assays. The higher-titer producer clones must also be tested for the absence of replication-competent helper viruses.

Titration of virus producer clones on NIH/3T3 cells

Materials

- NIH/3T3 (ECACC) and retroviral producer cells are routinely cultured in Dulbecco's Minimal Essential Medium (DMEM; Gibco Laboratories) supplemented with 75 IU ml^{-1} penicillin (Sigma), 75 mg ml^{-1} streptomycin sulfate (Sigma), and 10% heat-inactivated fetal bovine serum (FBS; Gibco Laboratories) at 37°C in a fully-humidified atmosphere with 5% CO_2 in air.
- 60 mm diameter plates (Nunc).
- *Polybrene stock solution:* 5 mg ml^{-1} polybrene (Aldrich Chemical Company) in PBS, filtered through 0.22 μm and stored in aliquots at –20°C.
- *Geneticin (G418) stock solution:* 100 mg ml^{-1} G418 (Gibco Laboratories) — a neomycin analogue — in DMEM, neutralized with NaOH to pH 7.0, filtered through 0.22 μm, and stored in aliquots at –20°C. G418 dosage is always given as the equivalent weight of active product.
- *PBS:* 8 mM Na_2HPO_4, 1.5 mM KH_2PO_4, 137 mM NaCl, 2.7 mM KCl.
- *Crystal violet solution:* 0.2% (w/v) crystal violet (Merck) in water. Stored at room temperature.
- Methanol (Merck)

Method

1. Virus producer cells are grown and the day before confluence is reached the culture medium is changed. Twenty-four hours later, the supernatant is collected and stored as described in the preceding protocol.
2. The day before the titration assay, 10^5 NIH/3T3 cells are seeded per 60 mm plate in culture medium and incubated overnight at 37°C. Prepare enough plates to test three serial dilutions of each viral supernatant, leaving additional plates for uninfected and positive controls.
3. On the day of assay, 10^{-3}, 10^{-4}, and 10^{-5} dilutions of each supernatant are prepared in culture medium supplemented with polybrene to a final concentration of 8 μg ml^{-1}.
4. The culture medium from the NIH/3T3 plates is removed and 1 ml of the corresponding supernatant dilution is added. The plates are incubated at 37°C for 1.5 h with gentle shaking every 10 min. The infection medium is then replaced with 5 ml of fresh culture medium and the incubation is continued at 37°C for another 10–12 days.
5. On the third day after infection, G418 is added to a final concentration of 1 mg ml^{-1}. During the following days, the culture medium supplemented with G418 (1 mg ml^{-1}) is changed every 2–3 days.
6. Ten to twelve days after infection, when the colonies are clearly recognizable under the microscope, the plates are stained according to the following protocol:

   - Remove the culture medium.
   - Wash the plates twice with PBS.
   - Fix the cells, adding 1–2 ml of methanol at room temperature for 5 min.
   - Remove the methanol and stain the plates with 2 ml of crystal violet solution. Incubate at room temperature for 5–10 min.
   - Wash extensively with water. Colonies must be clearly stained over a dull background.

7. The viral titer per ml of supernatant is defined as the number of scored colonies multiplied by the inverse of the dilution plated, and is expressed in colony-forming units *neo^r* ml^{-1} (CFU *neo^r* ml^{-1}).

Detection of replication-competent viruses (helper viruses)

Prior to selection of a high-titer producer clone for further experimentation, it must be tested for the absence of competent helper virus generation. This can be analyzed using a genome rescue assay (Cornetta et al 1991). For this purpose we obtained a test-line, 3T3-A, derived after NIH/3T3 infection with the retroviral vector pXT1 and subsequent G418 selection. Hence, this line harbors a replication-defective genome containing a selectable marker (neo^r), but is unable to package retroviral particles. After the infection of this cell line with a viral supernatant, the presence of replication-competent viruses would promote the encapsidation and amplification of the defective viral genomes, which is easily evidenced by their ability to form neo^r colonies in NIH/3T3 cells.

Materials

- NIH/3T3 (ECACC) and 3T3-A cells are routinely cultured in Dulbecco's Minimal Essential Medium (DMEM; Gibco Laboratories) supplemented with 75 IU ml^{-1} penicillin (Sigma), 75 mg ml^{-1} streptomycin sulfate (Sigma), and 10% heat-inactivated fetal bovine serum (FBS; Gibco Laboratories) at 37°C in a fully humidified atmosphere with 5% CO_2 in air.
- 60 mm diameter plates (Nunc).
- *Polybrene stock solution:* 5 mg ml^{-1} polybrene (Aldrich Chemical Company) in PBS, filtered through 0.22 μm and stored in aliquots at −20°C.
- Trypsin solution: 0.05% (w/v) trypsin (Sigma) in 5 mM glucose, 0.5 mM EDTA, 5 mM KCl, 137 mM NaCl, 7 mM $NaHCO_3$, and 0.015% (w/v) phenol red (Sigma). Solution is adjusted to pH 7.0–7.3

Method

1. The day before assay, 10^5 3T3-A cells are seeded per 60 mm plate in culture medium and incubated overnight at 37°C.
2. The culture medium from the 3T3-A plates is removed and replaced by 1 ml of the supernatant to be tested, supplemented with polybrene at a final concentration of 8 μg ml^{-1}. The plates are incubated at 37°C for 1.5 h with gentle shaking every 10 min. The infection medium is then replaced with 5 ml of fresh culture medium and incubation is continued at 37°C for another 2–3 weeks.
3. During this period, when confluence is reached, the cells are split to 1/10–1/20 adding 1 ml of the previous supernatant to the fresh culture medium. This procedure is repeated 3–4 times to allow the spread of potential helper viruses.
4. The supernatant collected (as described under 'Isolation of Retrovirus-producing packaging cell lines') from the last cell passage is tested for the presence of CFU neo^r in NIH/3T3 (see preceding protocol).

Production of infective supernatant for the transduction of murine bone marrow populations

In accordance with the BM transduction protocol proposed in the next subsection, it is necessary to produce a batch of high-titer infective supernatant (≥ 1–2×10^6 CFU neo^r ml^{-1}) which is helper-virus-free (see earlier in this subsection).

Materials

- High-titer retroviral producer cells are routinely cultured in Dulbecco's

Minimal Essential Medium (DMEM; Gibco Laboratories) supplemented with $75\,IU\,ml^{-1}$ penicillin (Sigma), $75\,mg\,ml^{-1}$ streptomycin sulfate (Sigma), and 10% heat-inactivated fetal bovine serum (FBS, Gibco Laboratories) at 37°C in a fully humidified atmosphere with 5% CO_2 in air. When indicated, cells are grown with Iscove's modification of Dulbecco's medium (IMDM; Gibco Laboratories) supplemented with 5% FBS, maintaining the other culture conditions as above.

- *Gelatin stock solution:* 10% gelatin (BioRad) solution in PBS is prepared, autoclaved, and stored at 4°C.
- $75\,cm^2$ culture flasks (Falcon)

Methods

1. The selected high-titer and helper virus-free Ψ2–pXT1 clone is grown and expanded in 1% gelatin-pretreated flasks to avoid massive detachment during supernatant harvest.
2. When the monolayers are 70–80% confluent, cells receive 20 ml per flask of fresh IMDM supplemented with 5% FBS. Twenty-four hours later the supernatants are harvested, clarified, filtered and stored as indicated in an earlier protocol. The titer of frozen supernatant batches is stable for long periods.

Transduction of murine bone marrow hematopoietic stem cells using the IES protocol

Gene transfer by retroviral vectors occurs only in cells which are actively proliferating during the infection (Miller et al 1990). Because hema-topoietic stem cells are mostly quiescent in adult bone marrow, this cell population must be recruited into cycle to be suitable for transduction. To achieve this, a widely used experimental strategy consists of the *in vivo* administration of 5-fluorouracil (5-FU) to donor mice. This cell cycle-dependent drug, by killing more committed progenitors, enables the enrichment in hematopoietic stem cells and promotes their proliferative activation (Lerner and Harrison 1990; Harrison and Lerner 1991). As other authors have reported previously, this cell population is an adequate substrate for retroviral gene transfer. Taking advantage of this strategy and exploring the optimal combination of different parameters, we have developed a very efficient protocol — IES (infection/expansion/selection) — to transduce hematopoietic stem cells (Bernad et al 1994). The first step is an infection procedure in which BM cells freshly harvested 4–5 days after a single 5-FU treatment ($150\,mg\,kg^{-1}$ body weight) are mixed with a supernatant containing a high titer of pXT1-derived retroviral particles. Subsequently, target cells are subjected to an expansion period with a combination including interleukin 3 (IL-3) and stem cell factor (SCF). These hematopoietic growth factors have been reported to preserve the *in vitro* viability of stem cells (Katayama et al 1993) and to promote retroviral infection in primitive hematopoietic stem cells (Luskey et al 1992). This expansion also enables the collection of a graft enriched in committed and pluripotent hematopoietic progenitors (Muench et al 1992; Varas et al 1994a,b), which should accelerate hematopoietic recovery in transplanted mice (Muench and Moore 1992; Serrano et al 1994). Finally, transduced cells are positively selected because of their exogenously conferred resistance to geneticin, which maximizes their long-term repopulating contribution into lethally irradiated recipients (Bernad et al 1994). This transduction protocol is summarized in Fig. 14.3.1.

Materials

- IMDM (Iscove's modification of Dulbecco's medium) (Gibco Laboratories) supplemented with 75 IU ml^{-1} penicillin (Sigma) and 75 mg ml^{-1} streptomycin sulfate (Sigma)
- WEHI-3b conditioned medium (WCM).

Conditioned medium from WEHI-3b cell line is used as a source of interleukin 3 (IL-3)

- 5-Fluorouracil (5-FU) (Roche). Supplied as a stock liquid solution at 50 mg ml^{-1}. This drug is injected intravenously into mice via a lateral tail vein at a dosage of 150 mg kg^{-1} body weight in a final volume of 200 μl. For

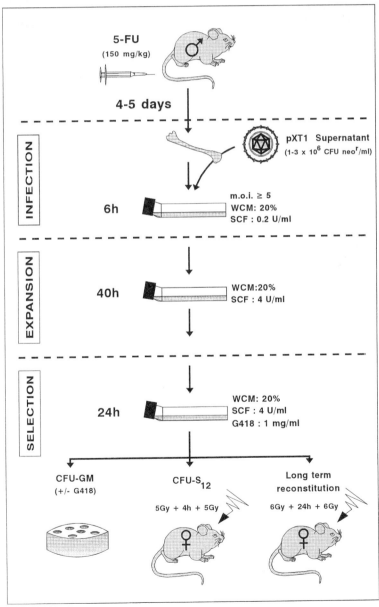

Figure 14.3.1 Schematic diagram of the infection/expansion/selection (IES) protocol for retroviral gene transfer to hematopoietic precursors.

this purpose, the stock solution may be diluted with IMDM.

- Stem cell factor (SCF) (Genetics Institute). Provided as recombinant soluble murine SCF. One unit of activity is the half-maximal activity as measured by the MO-7 proliferation assay
- *Polybrene stock solution:* 5 mg ml^{-1} polybrene (Aldrich Chemical Company) in PBS, filtered through 0.22 μm and stored in aliquots at –20°C
- Culture flasks (Falcon)
- *Turk solution:* composed of 20% acetic acid (Merck) and 0.01% methylene blue (Sigma). Bone marrow cell suspensions are diluted 1/10 in this solution for nucleated cell counting.

Transduction protocol

1. Four to five days after a single 5-FU treatment in 8–12-week-old male donor mice, their hind limbs (tibiae and femora) are perfused at room temperature under sterile conditions with IMDM. Cellular aggregates are dispersed by repeated passages through a 25G needle and the resulting cell suspension is not subjected to hemolysis.
2. Donor bone marrow is mixed with pXT1-derived infective supernatants ($1-3 \times 10^6$ CFU *neo*^{r} ml^{-1}) in culture flasks to a final concentration of $1-2 \times 10^5$ cells ml^{-1} and supplemented with 20% FBS, 20% WCM, 0.2 U ml^{-1} SCF and 5 μg ml^{-1} polybrene. The resulting multiplicity of infection (m.o.i) value must be above 5.

3. After a 6 h infection period at 37°C in a fully-humidified atmosphere with 5% CO_2 in air, culture flasks are scraped and cells are collected after centrifugation at 400g for 5 min.

4. The cell pellet is resuspended and adjusted to 2×10^6 cells ml^{-1} in fresh IMDM, supplemented as previously described but increasing SCF to a final concentration of 4 U ml^{-1} and eliminating polybrene.

5. Forty hours later, G418 is added to the culture at a 1 mg ml^{-1} final concentration and this selection step is extended for 24 h.

6. Culture flasks are scraped again and cells are collected (as described in step 3), washed once, and resuspended in fresh IMDM.

7. The infection efficiency is estimated by seeding different cell numbers into clonogenic assays both *in vitro* (CFU-GM assay) and *in vivo* (CFU-S assay). Transduced cells are also injected into lethally irradiated female mice to test the ability of the transplanted cells to reconstitute their lymphohematopoietic system. These assays are described below.

Long-term analysis of lymphohematopoietic reconstituted mice

At least 4 months after being transplanted with the transduced graft, surviving animals are killed and their hematopoietic organs (bone marrow, spleen, and thymus) are removed and processed for DNA extraction. Recipient bone marrow can be also assayed to determine its content in committed CFU-GM progenitors, as well as transplanted into secondary recipients to quantify the extent of transduction into the pluripotent CFU-S population and the preservation of the long-term repopulating ability.

Materials

- Irradiation equipment (Philips MG 324 x-ray unit or equivalent)
- IMDM (Iscove's modification of Dulbecco's medium) (Gibco Laboratories) supplemented with 75 IU ml^{-1} penicillin (Sigma) and 75 mg ml^{-1} streptomycin sulfate (Sigma)
- Horse serum (HS) (Gibco Laboratories), heat-inactivated at 56°C for 30 min
- WEHI-3b conditioned medium (WCM). Conditioned medium from WEHI-3b cell line is used as source of interleukin 3 (IL-3)
- Bacto-agar (Difco Laboratories). Stock solution is prepared at 3% (w/v) in hot distilled H$_2$O, previously sterilized by autoclaving. Solidified suspension must be dissolved in a hot water bath before use.
- 35 mm diameter plastic culture dishes (Nunc)
- *Geneticin (G418) stock solution:* 100 mg ml^{-1} G418 (Gibco Laboratories) — a neomycin analogue — in DMEM, neutralized with NaOH to pH 7.0, filtered through 0.22 μm and stored in aliquots at −20°C. G418 dosage is always given as the equivalent weight of active product.
- *Tellesniczky's solution:* Fixing solution composed of 44% ethanol, 31% acetic acid and 2.3% formaldehyde. After a few minutes immersion in this solution, spleen colonies are easily visualized because of their white color over the brown splenic background.
- 1 M pH 7.5 and 1 M pH 8.0 Tris–HCl (Sigma)
- 10% sodium dodecyl sulfate (SDS) (Sigma)
- 0.5 M ethylenediaminetetraacetic acid (EDTA) pH 8 (Merck)
- 5 M NaCl (Merck)
- Phenol (Merck) equilibrated to pH 8.0 with Tris–HCl
- Chloroform (Merck)
- Phenol–chloroform. Mix 1:1 phenol pH 8.0 and chloroform
- 3 M pH 5.0 ammonium acetate (Merck)
- Dextran sulfate (Sigma)
- Proteinase K (Boehringer Mannheim). The stock solution is prepared at 20 mg ml^{-1} dissolved in water.
- RNAase A (Boehringer Mannheim): 10 mg ml^{-1} in 10 mM Tris–HCl pH 7.5 and 15 mM NaCl. Heat to 100°C for 15 min and allowed to cool slowly to room temperature.

- *Carrier DNA stock solution:* Sheared salmon sperm DNA (Sigma) 10 mg ml^{-1} in water
- *20× SSC stock solution:* 3 M NaCl, 300 mM sodium citrate pH 7.0
- *50× Denhardt's reagent stock solution:* 50% (w/v) Ficoll (Pharmacia), 50% (w/v) polyvinylpyrrolidone (Sigma), 50% (w/v) bovine serum albumin (Sigma, fraction V)
- Cold 100% ethanol (Merck)
- Cold 70% ethanol (Merck)
- *TE:* 10 mM Tris–HCl, 1 mM EDTA pH 8.0
- x-Ray films (Amersham International)
- N-Hybond+ membranes (Amersham International).

Methods

CFU-GM assay

1. Different cell numbers are resuspended in IMDM supplemented with 25% HS, 10% WCM and mixed with Bacto-agar to a 0.3% final concentration. This suspension should be brought to a final volume of 3.5 ml.
2. Aliquots of 1 ml are poured into each of three 35 mm plastic culture dishes. After the seeding, the plates are stored for 5 min at 4°C to facilitate solidification.
3. Plates are cultured at 37°C in a fully-humidified atmosphere with 5% CO_2 in air.
4. Seven days later, colonies containing 50 or more cells are scored under a dissecting microscope.

The percentage of CFU-GM progenitors expressing resistance to G418 is determined from the number of colonies growing in the presence or absence of 0.5 mg ml^{-1} G418.

CFU-S assay

1. Female mice (12–24 weeks old) are conditioned by total body irradiation (in our hands, a 5 Gy + 4 h + 5 Gy x-ray irradiation protocol with a dose rate of 1.03 Gy min^{-1} works for various mouse strains).
2. Within the first 4 h after the completion of the irradiation conditioning, groups of 8–10 recipient mice are independently injected via a lateral tail vein with a 200–500 µl volume of different cell numbers from the suspension to be assayed.
3. Transplanted mice are housed individually and given sterilized food and acidified water (pH 2.8) *ad libitum.*
4. Twelve days later, surviving animals are killed and their spleens are removed and fixed in Tellesniczky's solution to facilitate the quantification of the splenic colonies under a dissecting microscope.

When molecular analysis is required, splenic colonies are individually dissected without fixation and processed for DNA extraction, as described below.

Long-term repopulating ability assay

1. Female mice (12 weeks old) are conditioned by total body irradiation (in our hands, a 6 Gy + 4 h + 6 Gy x-ray irradiation protocol with a dose rate of 1.03 Gy min^{-1} works for several mouse strains).
2. Within the first 4 h after the completion of the irradiation conditioning, groups

of recipient mice are independently injected via a lateral tail vein with a 200–500 μl volume containing appropriate cell numbers from the suspension to be assayed.

3. Especially during the first month post-irradiation, transplanted mice are housed individually and given sterilized food and acidified water (pH 2.8) *ad libitum*.

4. Animals surviving for the first 4 months post-transplantation can be killed and their hematopoietic organs removed and processed for DNA extraction or clonogenic assays.

DNA extraction

This protocol is performed essentially as described by Laird et al (1991).

1. Splenic colonies, spleens, and thymuses are homogenized in IMDM with the aid of a Potter tissue homogenizer. As for perfused bone marrow, cell clumps are disaggregated by serial passages through a 25G needle.

2. Cells are collected after centrifugation at 400g for 5 min. Cell pellets are resuspended in 10 mM Tris pH 7.5 with 500 μg ml^{-1} protease K, 0.5% SDS, 10 mM EDTA pH 8.0, and 150 mM NaCl.

3. Biological samples are digested overnight at 55°C with vigorous shaking.

4. RNAase A is added to a final concentration of 50 μg ml^{-1}. Samples are then incubated 1–2 h at 37°C with gentle shaking.

5. Genomic DNA is extracted 1:1 (v/v) with phenol, phenol–chloroform, and chloroform. DNA is precipitated for 1 h at −70°C with 0.3 M ammonium acetate and 2 volumes of cold 100% ethanol.

After the centrifugation (10 min, 10,000g, 4°C), the DNA pellet is washed once with cold 70% ethanol, dried at 37°C and resuspended in 0.2× TE.

6. DNA concentration is calculated by determining the OD_{260}.

Exogenous contribution to the lymphohematopoietic system from long-term repopulated mice

1. Genomic DNA samples (10 μg) are digested overnight with restriction enzymes cutting outside (i.e., *Bam*HI) or inside (i.e., *Dra*I) the proviral genome of the pXT1-retroviral vector.

2. Digested samples are electrophoresed on a 0.8% agarose gel in 1× TBE at a low voltage (1–2 V cm^{-1}).

3. DNA fragments are transferred from agarose gels to nylon membranes using the alkaline method (Sambrook et al 1989).

4. The membranes are prehybridized for 3 h at 42°C in 10 mM Tris 7.5, 50% formamide, 5× SSC, 5× Denhardt's reagent, and 0.5% SDS.

5. Membranes are hybridized overnight with the proper probe (see below) in the same conditions as for the prehybridization but supplemented with 10% dextran sulfate and 50 μg ml^{-1} denatured carrier DNA.

To detect integrated pXT1 retroviral sequences, membranes are developed with a probe for the *neo*^{r} gene, while a probe specific for Y-chromosome mouse sequences is used to test the exogenous contribution in the female recipients. To verify DNA load, membranes are hybridized with a probe for a single-copy gene (i.e., M-CSF).

6. Membranes are washed three times, 1 h each, at 56°C in 0.1× SSC and 0.5% SDS, followed by one wash in 2× SSC.

7. Membranes are exposed wet to x-ray films.

Discussion

We have developed an efficient protocol (IES) for retroviral gene transfer into the mouse hematopoietic system using infective supernatants combined with ex vivo expansion and selection of BM transduced populations. Bone marrow cocultivation with a packaging cell line, although it has been widely recognized as an efficient procedure for gene transfer, presents several limitations such as the preferential loss by attachment of very primitive precursors (Einerhand et al 1993) and the intrinsic variability of the cocultures. Transduction mediated by supernatants, although yielding a lower efficiency (Bodine et al 1991; Bordignon et al 1989), greatly improves the reproducibility of BM infection by using defined batches of helper-free retroviral vectors.

We have evaluated several conditions for the in vitro manipulation of BM cells from mice treated with 5-FU to increase both the proportion and the proliferative rate of PHSC. The in vitro incubation of BM cells with IL-3/SCF significantly expanded the CFU-GM and CFU-S$_{12}$ progenitors after 3 days in culture (Bernad et al 1994) without exhausting the stem cell function (Varas et al 1994b). The expansion of hematopoietic precursors was shown to enhance the survival rate and facilitate the recovery of transplanted animals, as reported for other hematopoietic growth factor combinations (Serrano et al 1994; Muench et al 1992; Muench and Moore 1992). Under these culture conditions, the efficiency of infection was determined to be optimal using freshly harvested 5-FU-treated BM. The prolonged incubation in IL-3/SCF, although it promoted the

exponential growth of CFU-GM progenitors, significantly reduced their susceptibility to retroviral infection (Bernad et al 1994). This result suggests that the proliferation rate is not the bottleneck in this procedure and that there must be other limiting factors. Finally, the culture conditions described were also shown to be suitable for the efficient selection, on the basis of the G418 resistance conferred by the retroviral construct, of both committed and stem cell precursors (Bernad et al 1994).

In spite of a relatively low efficiency of infection, this protocol enables the collection of grafts highly enriched in homogeneously labeled hematopoietic precursors. The analysis of the hematopoietic organs from lethally irradiated animals reconstituted with BM cells transduced using the IES protocol demonstrates the efficiency of the method for the transduction of very primitive precursors. The pattern of junction fragments generated in Southern analysis showed the mono- or oligoclonal reconstitution of the recipients in the long term, consistent with results obtained previously by other authors (Lemischka et al 1986; Dick et al 1985). In addition, a high proportion of femoral CFU-GM from long-term transplanted mice are resistant to G418. This may be a consequence not only of the extinction of untransduced cells, but also of the in vitro selection of transduced PHSC with the highest LTR-driven expression of the neo^r gene, suggesting that a selection step could be critical for achieving a lasting and uniform expression of exogenous genes.

Concluding remarks

The TES protocol has proved an efficient procedure for the insertion of genetic markers into a large proportion of the mouse hematopoietic system, based on the combination of the *in vivo* proliferative activation of PHSC with the *in vitro* expansion and selection of the transduced bone marrow cells. The method renders chimeric animals reconstituted predominantly by transduced cells for long periods after transplantation. The combination of this transduction protocol with cell sorting and PCR-recognizable retroviral vectors will be a valuable tool for the investigation of PHSC biology and lymphohematopoietic lineage development.

References

Abramson S, Miller RG, Phillips RA (1977) The identification in adult bone marrow of pluripotent and restricted stem cells of the myeloid and lymphoid systems. J Exp Med 145: 1567–1579.

Becker AJ, McCulloch EA, Till JE (1963) Cytological demonstration of the clonal nature of spleen colonies derived from transplanted mouse marrow cells. Nature 197: 452–454.

Bernad A, Varas F, Gallego JM, Almendral JM, Bueren JA (1994) *Ex vivo* expansion and selection of retrovirally transduced bone marrow: an efficient methodology for gene-transfer to murine lympho-haematopoietic stem cells. Br J Haematol 87: 6–17.

Bodine DM, McDonagh KT, Seidel NE, Nienhuis AW (1991) Survival and retrovirus infection of murine hematopoietic stem cells *in vitro*: effects of 5-FU and method of infection. Exp Hematol 19: 206–212.

Bordignon C, Yu S, Smith CA, et al (1989) Retroviral vector mediated high-efficiency expression of adenosine deaminase (ADA) in hematopoietic long-term cultures of ADA-deficient marrow cells. Proc Natl Acad Sci USA 86: 6748–6752.

Boulter CA, Wagner EF (1987) A universal retroviral vector for efficient constitutive expression of exogenous genes. Nucleic Acids Res 15: 7194.

Cepko CL, Ryder EF, Austin CP, Walsh C, Fekate DM (1993) Lineages analysis using retrovirus vectors. Methods Enzymol 225: 933–960.

Chang JM, Johnson G (1989) Gene transfer into hemopoietic stem cells using retroviral vectors. Int J Cell Cloning 7: 264–280.

Cornetta K, Morgan RA, Anderson WF (1991) Safety issues related to retroviral-mediated gene transfer in humans. Human Gene Ther 2: 5–9.

Dick JE, Magli MC, Huszar D, Phillips RA, Bernstein A (1985) Introduction of a selectable gene into primitive stem cells capable of long-term reconstitution of the hemopoietic system of W/W^v mice. Cell 42: 71–79.

Einerhand MPW, Bakx TA, Kukler A, Valerio D (1993) Factors affecting the transduction of pluripotent hematopoietic stem cells: long-term expression of a human adenosine deaminase gene in mice. Blood 81: 254–263.

Fraser CC, Thacker JD, Hogge DE, Fatur-Saunders D, Takei F, Humphries RK (1993) Alterations in lymphopoiesis after hematopoietic reconstitution with TL-7 virus-infected bone marrow. J Immunol 151: 2409–2418.

Harrison DE, Lerner CP (1991) Most primitive hematopoietic stem cells are stimulated to cycle rapidly after treatment with 5-fluorouracil. Blood 78: 1237–1240.

Hawley RG, Fong AZC, Burns BF, Hawley TS (1992) Transplantable myeloproliferative disease induced in mice by an interleukin 6 retrovirus. J Exp Med 176: 1149–1163.

Johnson GR, Gonda TJ, Metcalf D, Hariharah IK, Cory S (1989) A lethal myeloproliferative syndrome in mice transplanted with bone marrow cells infected with a retrovirus expressing granulocyte-macrophage colony stimulating factor. EMBO J 8: 441–448.

Jordan CT, McKearn JP, Lemischka IR (1990) Cellular and developmental properties of fetal hematopoietic stem cells. Cell 61: 953–963.

Katayama N, Clark SC, Ogawa M (1993) Growth factor requirement for survival in cell-cycle dormancy of primitive murine lymphohematopoietic progenitors, Blood 81: 610–616.

Keller G, Wagner EF (1989) Expression of v-src induces a myeloproliferative disease in bone-marrow-reconstituted mice. Genes Dev 3: 827–837.

Keller G, Paige C, Gilboa E, Wagner EF (1985) Expression of a foreign gene in myeloid and lymphoid cells derived from multipotent haematopoietic precursors. Nature (London) 318: 149–154.

Kriegler M (1990) Gene Transfer and Expression: A Laboratory Manual. Stockholm Press, New York.

Laird PW, Zijderveld A, Lindeus K, Rudniki MA, Jaenisch R, Berns A (1991) Simplified mammalian DNA isolation procedure. Nucleic Acid Res 19: 4293.

Lemischka IR, Raulet DH, Mulligan RC (1986) Developmental potential and dynamic behavior of hematopoietic stem cells. Cell 45: 917–927.

Lerner C, Harrison DE (1990) 5-Fluorouracil spares hemopoietic stem cells responsible for long-term repopulation. Exp Hematol 18: 114–118.

Luskey BD, Rosenblatt M, Zsebo K, Williams DA (1992) Stem cell factor, interleukin-3, and interleukin-6 promote retroviral-mediated gene transfer into murine hematopoietic stem cells. Blood 80: 396–402.

Mann R, Mulligan RC, Baltimore D (1983) Construction of a retrovirus packaging mutant and its use to produce helper-free defective retrovirus. Cell 33: 153–159.

Markowitz D, Goff S, Bank A (1988a) A safe packaging line for gene transfer: separating viral genes on two different plasmids. J Virol 62: 1120–1124.

Markowitz D, Goff S, Bank A (1988b) Construction and use of a safe and efficient amphotropic packaging cell line. Virology 167: 167–400.

Miller AD, Buttimore C (1986) Redesign of retrovirus packaging cell lines to avoid recombination leading to helper virus production. Mol Cell Biol 6: 2895–2902.

Miller AD, Miller DG, Garcia JV, Lynch CM (1993) Use of retroviral vectors for gene transfer and expression. Methods Enzymol 217: 581–599.

Miller DG, Mohammed AA, Miller AD (1990) Gene transfer by retrovirus vectors occurs only in cells that are actively replicating at the time of infection. Mol Cell Biol 10: 4239–4242.

Muench MO, Moore MAS (1992) Accelerated recovery of peripheral blood cell counts in mice transplanted with in vitro cytokine-expanded hematopoietic progenitors. Exp Hematol 20: 611–618.

Muench MO, Schneider JG, Moore MAS (1992) Interactions among colony-stimulating factors, IL-1β, IL-6 and Kit-ligand in the regulation of primitive murine hematopoietic cells. Exp Hematol 20: 339–349.

Nowell PC, Hirsch BE, Fox DH, Wilson DB (1970) Evidence for the existence of multipotential lymphohematopoietic stem cells in the adult rat. J Cell Physiol 75: 151–158.

Sambrook J, Fritsch EF, Maniatis T (1989) Molecular Cloning: A Laboratory Manual. Cold Spring Harbor Laboratory Press. Cold Spring Harbor, New York.

Serrano F, Varas F, Bernad A, Bueren JA (1994) Accelerated and long-term hematopoietic engraftment in mice transplanted with ex vivo expanded bone marrow. Bone Marrow Transpl 4: 855–862.

Varas F, Bernad A, Bueren JA (1994a) Analysis of the differentiation rate of 5-FU treated murine bone marrow under ex vivo expansion with ILs 3+6 and IL3+SCF (abstract). Exp Hematol 22: 786.

Varas F, Bernad A, Bueren JA (1994b) Ex vivo expansion of 5-FU treated murine bone marrow with IL-3 and stem cell factor (abstract). Bone Marrow Transpl 15 (suppl 3): 539.

von Ruden T, Kandels S, Radaszkiewicz T, Ullrich A, Wagner EF, (1992) Development of a lethal mast cell disease in mice reconstituted with bone marrow cells expressing the v-erbB oncogene. Blood 79: 3145–3158.

Wu AM, Till JE, Siminovitch L, McCulloch EA (1965) Cytological evidence for a relationship between normal hematopoietic colony-forming cells and cells of the lymphoid system. J Exp Med 127: 455–464.

Wu AM, Till JE, Siminovitch L, McCulloch EA (1967) A cytological study of the capacity for differentiation of normal hemopoietic colony-forming cells. J Cell Physiol 62: 327–333.

Characterization of antigen–antibody and ligand–receptor interactions

14.4

Mario Mellado
Leonor Kremer
Santos Manes
Carlos Martinez-A
J. M. Rodriguez Frade

Centro Nacional de Biotecnología, Universidad Autónoma, Madrid, Spain

TABLE OF CONTENTS

Immunology Methods Manual
ISBN 0–12–442712–X

Abstract

Immunological techniques have been used to characterize the ligand–receptor binding through the recognition by antibodies of ligand or receptor or the complex. Here we describe the use of enzyme-linked immunosorbent assays, radioimmunoassays, and real-time bio- specific interactions to characterize the binding between growth hormone and its receptor as a model, placing special emphasis on protocols, advantages, limitations, and conclusions derived from each assay to give an overview of the system.

Introduction

The study of receptor–ligand interactions is essential for understanding the molecular mechanisms involved in biological processes. Recognition followed by association is the consequence of the complementarity of individual residues ordered in a specific conformation on the molecular interface of interacting molecules.

Such complementarity determines the specificity and stability of the interaction. Several immunological techniques, such as enzyme-linked immunosorbent assay (ELISA) or radioimmunoassay (RIA), have been widely used for measurement of affinity and specificity analysis of protein–protein interactions. Recently, a new area in the study of biospecific interactions consisting of the use of biosensors to monitor interactions directly in real time has been developed.

We will describe these techniques using the interaction between growth hormone (GH) and its receptor as a model, placing special emphasis on protocols, advantages, limitations, and the different but complementary conclusions derived from each assay, to give an overview of the system.

The GH receptor (GHR) belongs to a large family of single transmembrane receptors which includes the prolactin and cytokine receptor (Kelly et al 1991). It is a glycoprotein of 620 amino acids with an extracellular hormone-binding domain rich in cysteines, a single transmembrane region, and a cytoplasmic domain which shares no homology with the other members of the family (Bazan 1989). It does not show any obvious homology to other receptors which have domains corresponding to known signaling mechanisms (Bazan 1989). The GHR is heavily N-glycosylated and highly susceptible to proteolysis (Leung et al 1987). In fact, a serum protein that binds GH has been identified (Baumann et al 1986) (GH binding protein or GHBP) and corresponds to the extracellular domain of human GHR. GH-induced receptor homodimerization is essential for hormone activity (Cunningham et al 1991).

In order to characterize this hormone–receptor interaction, we have derived monoclonal antibodies that have been screened in different assays, and which are described below together with advantages and disadvantages.

Abbreviations

Ag, antigen; Ab, antibody; BSA, bovine serum albumin; DMF, N,N-dimethylformamide; DMSO, dimethylsulfoxide; EIA, enzyme-linked immunoassay; GAM, goat anti-mouse immunoglobulin antibody; GH, growth hormone; GHBP, growth hormone binding protein (extracellular part of the GH receptor); mAb, monoclonal antibody; NMS, normal mouse serum; OD, optical density; OPD, o-phenylenediamine; PEG, poly(ethylene glycol); PO, peroxidase;

Screening in ELISA

General strategies

Antigen–antibody interactions are the basis for all immunochemical techniques. The nature of this interaction, which involves not only steric factors but also ionic, hydrophobic and hydrogen bonding, and Van der Waal's forces, directly affects the specificity, cross-reactivity and sensitivity of the different antibodies (Steward 1977). These immunoassays, including Western blot, EIA, RIA, and others, employ monoclonal antibodies as specific reagents for the follow-up of a given antigen. They are chosen because of their monospecificity and homogeneity, but their interaction with the corresponding antigen may vary greatly from one assay to another. This is why the selection of the appropriate monoclonal antibodies (mAbs) must rely not only on their characteristics under native conditions (affinity, specificity, and so on) but also on the conditions under which the mAb will be used. Thus, a mAb may be selected to recognize labeled antigens (Ag) (for use in RIA), Ag adsorbed to a solid-phase (EIA), denatured antigens (Western blot) or Ag present on cell surfaces. The behavior of a given mAb upon labeling, adsorption to a solid-phase, etc., also should be tested. When mAbs are used to follow up ligand–receptor interactions, the interaction of the mAb with either ligand or receptor must be considered, as it may affect this binding and the events following binding.

Materials

- Polystyrene 96-well microtiter plates (Maxisorb, Nunc, Denmark)
- Phosphate-buffered saline (PBS) (0.008 M Na_2HPO_4, 0.0018 M NaH_2PO_4, 0.15 M NaCl)

- PBS with 0.5% bovine serum albumin (Sigma Chemical Co., St Louis, MO, USA). If used routinely, storage of ready-to-use frozen aliquots is recomended.
- Distilled water
- o-Phenylenediamine dihydrochloride (OPD, 4 mg ml^{-1} in 0.15 M sodium citrate buffer, pH 5.0). Preweighed tablets are available from Sigma. As it is a highly carcinogenic product, avoid handling as much as possible. Storage of frozen aliquots is recomended.
- Hydrogen peroxide 33% (v/v)
- Peroxidase-labeled goat anti-mouse immunoglobulins (H+L chains) (Tago Inc., Burlingame, CA, USA; ICN Biomedicals Inc., Costa Mesa, CA, USA)
- Streptavidin (Sigma)
- Streptavidin–PO (Sigma). A 0.5 mg ml^{-1} solution in PBS–0.5%BSA is recommended, further diluted 1:2 with glycerol and stored at –20°C. In these conditions, it is active for more than 6 months.
- Affinity-purified GAM prepared in goats immunized with mouse IgGs and purified against mIgGs coupled to Sepharose. Sera are available commercially (ICN Biomedicals).

Procedures

Antibody-capture EIA

Antigen is immobilized on a solid phase, and the antibody is allowed to bind. The antibody can be labeled directly or can be detected using a labeled secondary reagent that will recognize the antibody specifically. The amount of antibody that is bound determines the strength of the signal. Negative controls are wells coated with the same antigen but using nonrelated antibodies as second step. The reactions are stopped when positive controls are optimally developed (Fig. 14.4.1a).

Procedure

1. Bind the corresponding antigen to the wells by adding 100 μl of antigen solution to each well (2.5 μg ml⁻¹ in PBS).

   Note: Titrate the antigen (synthetic peptide or protein) in order to obtain maximum adsorption. Concentrations ranging from 0.5 to 5 μg ml⁻¹ work well for most proteins. Impure antigen preparations (<10% purity) should not be used, as minimal adsorption of the protein of interest will result. Avoid detergents which will compete with the antigen for binding to the plate. Other buffers (carbonate buffer 0.1 M Na_2CO_3, pH 9.6) also give good results.

2. Incubate overnight at 4°C.

   Note: Shorter incubation periods should be used at 37°C; in this case, 90–120 min of incubation suffices.

3. Wash plates three times with distilled water.

   Note: PBS or PBS–0.05% Tween 20 are equally effective.

4. Saturate protein binding on the plate

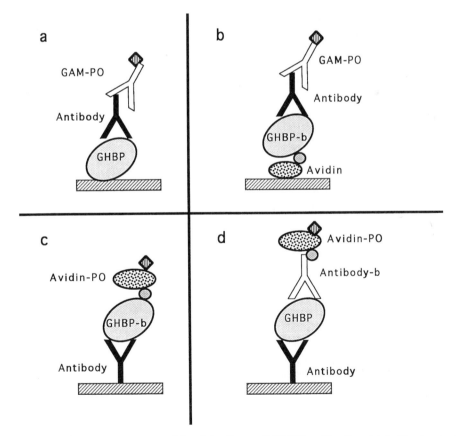

Figure 14.4.1 Enzyme immunoassays used for detecting anti-GHBP antibodies: (a) Antibody-capture EIA; (b) antibody-capture EIA using avidin as spacer; (c) antigen-capture EIA; (d) Two-site antigen-capture EIA.

Ag–Ab and ligand–receptor interactions

by incubating with 200 µl per well of PBS–0.5% BSA. Incubate for 1 h at 37°C.

Note: Other suitable blocking solutions include 10% nonfat dried milk or 1% gelatin. Depending on the protein adsorbed, plates can be stored at 4°C, preferably dried and vacuum-closed.

5. Wash the plates three times with distilled water.
6. Add 100 µl of antibody (cell-culture supernatant, diluted sera or ascites fluid). Incubate for 1 h at 37°C. If necessary, dilute samples in blocking buffer.
7. Remove the unbound antibody by three washes with distilled water.
8. Add the previously titrated labeled secondary antibody (GAM-PO). Use 100 µl per well. Incubate for 1 h at 37°C.

Note: Avoid the use of azide as preservative with horseradish peroxidase detection systems. Prior to assay, prepare and titer the secondary reagent (anti-immunoglobulin antibodies, enzyme-labeled), taking into account the assay limits, that is, the amount of Ig present in the supernatant, Ig subclass, etc. Although manufacturers offer instructions for the use of these reagents, be sure to check under your assay conditions. In some cases, the use of a biotinylated second antibody, followed by PO-labeled streptavidin may enhance the sensitivity of the assay.

9. Remove the unbound secondary reagent by washing 3 times with distilled water.
10. Add the enzyme substrate. Use 100 µl per well of substrate solution (OPD containing 1/2000 H_2O_2).
11. Wait until the positive controls are developed optimally and stop the reaction by adding 50 µl per well of 3 N H_2SO_4. Maximum OD is achieved at 492 nm.

Note: All the steps can be carried out using only 50 µl of the different reagents. Be sure to check all materials, reagents, buffers, etc., in advance. Most trouble shooting can be avoided by using the proper microtiter plate, second antibody, substrate, washing buffer, blocking protein, etc.

Comments and trouble shooting

Although adsorption of proteins on the plastic is supposed to be a random process, some proteins bind preferentially to given surface regions. This, together with conformational changes which may occur on proteins upon binding to the solid phase, can mask epitopes of interest in this type of assay.

To avoid this problem, the binding of proteins to the solid phase via a spacer conserves the protein's native conformation. A general method is the use of biotin-labeled proteins which are bound to plates coated with avidin or streptavidin. In this case, it is important to consider that the labeling process also may alter the protein structure or mask some epitope as a result of the presence of biotin (Fig. 14.4.1b).

Avidin capture EIA

Procedure

1. Bind avidin or streptavidin to the wells by adding 100 µl of protein solution (2.5 µg ml^{-1}).
2. Incubate overnight at 4°C or 2 h at 37°C.
3. Wash the plate three times with distilled water.
4. Block free sites for protein binding on the plate with PBS–0.5% BSA. Incubate 1 h at 37°C.
5. Wash the plates three times with distilled water.

6. Add 100 µl of the biotin-labeled antigen to each well. Prepare dilutions in blocking buffer. Incubate 1 h at 37°C.

Note: Pretitrate the antigen to obtain an optimal working dilution.

Continue the assay from step 6 of the preceding protocol.

Biotin labeling

Biotinylation of proteins offers a unified approach because a single biotin-labeled binder can be used with avidin conjugated to different probes (enzymes, radiolabels, fluorescent agents, chromophores, chemoluminescent agents) depending on the method to be employed.

Materials

A. 0.1 M sodium bicarbonate ($NaHCO_3$) pH 9 containing 0.15 M NaCl

B. 1 mg ml^{-1} *N*-hydroxysuccinimidobiotin (Sigma Chemical Co., St Louis, MO, USA), prepared just prior to use in DMSO

Procedure

1. Dialyze the protein against buffer A. The final concentration must be around 1 mg ml^{-1}.
2. Add 0.1 ml of solution B for each milligram of protein.
3. Mix the contents gently and allow the reaction to proceed in darkness for 2 h at room temperature.
4. Dialyze the sample against your assay buffer.
5. Dilute the conjugate with an equal volume of glycerol, separate into aliquots, and store at –20°C.

Note: Some aliquots of the biotin-labeled protein should be stored at –20°C or –80°C without glycerol to

avoid the interference of this product in some assays.

For most applications, biotinylation of proteins via amino groups is the preferred approach. Other group-specific reagents, such as tyrosyl and histidyl amino acid side-chains (*p*-diazobenzoylbiocytin), sulfydryl groups (3-(*N*-maleimidopropionyl)biocytin) or sugar residues and carboxyl-groups (biocytin hydrazide), have been synthesized as potential substitutes for amine-specific reagents.

Antigen-capture EIA

Although this type of assay is used to quantify antigens, we have also used it to detect specific antibodies. Unlabeled antibody is bound to the solid phase directly or through a capture protein (protein A, anti-immunoglobulin antibody). The antigen is labeled and titrated. Unbound proteins are removed by washing and the reaction is developed using the label agent.

The sensitivity of the assay depends on the number of antibodies that bind to the solid phase and the specific activity of the labeled antigen. Check for the best labeling method, in order to minimize protein alterations (Fig. 14.4.1c.)

Procedure

Prior to the assay, the antigen is labeled with biotin or other labeling agents.

1. Bind the anti-immunoglobulin antibody to the plates by adding 100 µl to each well (2.5 µg ml^{-1} in PBS). Incubate overnight at 4°C or 2 h at 37°C.
2. Wash the plate three times with distilled water.
3. Block the remaining sites for protein binding on the plates by adding 200 µl per well of PBS–0.5%BSA. Incubate 1 h at 37°C.

4. Wash three times with distilled water.
5. Add 100 µl of the labeled GHBP at the predetermined dilution. Dilutions should be made with blocking buffer.
6. Incubate 90 min at 37°C.
7. Wash the plate three times with distilled water.
8. If the antigen is iodine-labeled, read the counts bound directly and compare with the counts bound to an unrelated antibody (negative control). When using biotinylated antigens, add 100 µl per well of peroxidase-labeled streptavidin (0.5 ng ml^{-1} in blocking buffer). Incubate 30 min at 37°C. Wash three times with distilled water and develop the reaction using OPD/H$_2$O$_2$ as substrate.

Radioiodination procedures

Radioiodination is an easy method that yields highly sensitive, accurate measurement of very low amounts of protein. The different labeling methods vary with the enzymatic (lactoperoxidase) or chemical (Iodo-Gen® and chloramine T) nature of the oxidizing agent, and with the site of iodine incorporation: tyrosine (chloramine T, Iodo-Gen®, and lactoperoxidase), lysines (Bolton-Hunter).

Chloramine T

Materials

A. 0.5 M sodium phosphate buffer pH 7.6
B. 0.05 M sodium phosphate buffer pH 7.6
C. 0.01 M sodium phosphate pH 7.6 containing 0.15 M sodium chloride, 0.01 M EDTA, and 0.25 g/100 ml bovine serum albumin
D. 25 mg ml^{-1} chloramine T prepared just prior to use in 0.05 M sodium phosphate buffer pH 7.6

E. 0.7 mg ml^{-1} sodium metabisulfite prepared just prior to use in sodium phosphate buffer pH 7.6 containing 0.25 g/100 ml bovine serum albumin

F. 0.05 M sodium phosphate buffer pH 7.6 containing 1 g/100 ml potassium iodide and 8 g/100 ml saccharose

G. Buffer C without bovine serum albumin

Procedure

1. 5 µg antigen in 20 µl buffer B plus 10 µl buffer A and 5 µl buffer D.
2. Add 5 µl of Na ^{125}I (100 mCi ml^{-1}).
3. Mix the contents gently.

   Note: Poor mixing is probably the commonest cause of low yield of labeled protein by this method.

4. Allow reaction to proceed for 10 min.
5. Add 50 µl of buffer E and 200 µl of buffer F to stop the reaction.
6. Separate the radioactively labeled protein from the remaining radioactive iodide using a Sephadex® G-25 M column equilibrated with buffer C prior to use.

   Note: Prepacked columns are available from Pharmacia, Uppsala, Sweden.

7. Use buffer G to elute.

Iodo-Gen®

Materials

A. 0.5 M sodium phosphate buffer pH 7.6
B. 0.05 M sodium phosphate buffer pH 7.6
C. 0.01 M sodium phosphate buffer pH 7.6 containing 0.15 M sodium chloride and 0.01 M EDTA
D. Buffer C containing 0.25 g/100 ml bovine serum albumin.

Procedure

1. Dissolve 0.1 mg Iodo-Gen® in 1 ml dichloromethane.
2. Add 50 µl of Iodo-Gen® solution to a small plastic tube. Flush the tube with nitrogen gas until the solvent has evaporated. The tube can be stored at –20°C until required.
3. Cool the tube in ice water and add 15 µl of buffer A, 2.5 µl of the protein (1 mg ml^{-1}) to be labeled, and 2.5 µl of Na^{125}I (100 mCi ml^{-1}).
4. Gently shake the tube and allow the reaction to proceed for 10 minutes.
5. Stop iodination by transferring the reaction mixture from the coated tube to an uncoated plastic tube containing 300 µl of buffer C.
6. Leave for 15 min and then add 400 µl of buffer C.
7. Separate the radioactively labeled protein from the residual radioactive iodide using a Sephadex® G-25 M column equilibrated with buffer D prior to use.
8. Use buffer C to elute.

Lactoperoxidase

Materials

A. 0.01 M sodium phosphate buffer pH 7.6 containing 0.15 M sodium chloride and 0.01 M EDTA

B. Buffer A plus 0.25 g/100 ml bovine serum albumin

C. 0.05 M sodium phosphate buffer pH 7.6

D. 0.5 M sodium phosphate buffer pH 7.5

Procedure

1. Prepare stock solution of 0.2 mg ml^{-1} lactoperoxidase in buffer C.
2. Add 10 µl of 1/25,000 diluted hydrogen peroxide solution to a small plastic tube containing 15 µl of buffer D, 15 µl of the protein (2.5 µg) to be labeled, 10 µl of 0.2 mg ml^{-1} lactoperoxidase, and 2.5 µl of Na^{125}I (100 mCi ml^{-1}).
3. Gently shake the tube and allow the reaction to proceed for 10 min. During this time, 5 µl of 0.2 mg ml^{-1} lactoperoxidase and 5 µl of 1/25,000 diluted hydrogen peroxide are added.
4. Stop iodination by adding 0.5 ml of buffer A.
5. Separate the radioactively labeled protein from the residual radioactive iodide using a Sephadex® G-25 M column equilibrated with buffer B prior to use.
6. Use buffer A to elute.

Bolton–Hunter reagent

Materials

A. 0.01 M sodium phosphate buffer pH 7.6 containing 0.15 M sodium chloride and 0.01 M EDTA

B. Buffer A plus 0.25 g/100 ml gelatin

C. 0.1 M sodium borate buffer pH 8.5

D. 0.2 M glycine in buffer C.

Procedure

1. Evaporate the benzene–DMF solution of ^{125}I Bolton–Hunter reagent to dryness.
2. Cool all reagents in ice water.
3. Add 5 µl protein solution (1 mg ml^{-1} in buffer C) to the dry reagent.
4. Gently shake the tube and allow conjugation to proceed for 15 min at 4°C.
5. Add 0.5 ml of buffer D, mix and leave for 5 min at 4°C.
6. Separate the radioactively labeled protein from the residual radioactive iodide using a Sephadex® G-25 M

column equilibrated with buffer A prior to use.
7. Use buffer D to elute.

Comments and trouble shooting

In choosing an enzymatic or radioactive label, several factors have to be taken into account.

- Label must not alter those residues which are essential for the structural stability or biological activity of the protein of interest (both iodine and biotin labeling may alter the protein, depending on the degree of incorporation of label and the specific group to which it is bound).
- Half-life of the label (biotin can be stored for years without loss of specific activity).
- Need for special equipment.
- Health hazard.
- Sensitivity of the assay and possibility of quantification (enzyme labels are not homogeneous and quantification is complex).

Two-site antigen-capture assay

This commonly used sandwich-type assay is modified for use in screening assays using culture supernatants as a source of capture antibodies. The assay is similar to the antigen-capture assay, except that in this case we use an unlabeled antigen and a second antigen-specific biotin-labeled antibody. Its main advantage is that crude preparations can be used as the source of antigen. Its main disadvantages are the need for a good labeled second antibody and nondetection of antibodies that may be directed against the same region as the labeled antibody (Fig. 14.4.1d).

Procedure

Follow steps 1 to 5 exactly as for antigen-capture assay (p. 1149).

6. Add antigen in dilution buffer.

Note: If possible, titrate the antigen before use.

7. Incubate 60 min at 37°C.
8. Wash three times with distilled water.
9. Add biotin-labeled second antibody and incubate 60 min at 37°C.

Continue the assay with addition of streptavidin–PO and OPD as in the antigen-capture assays.

Comments and trouble shooting

When the antibodies to be bound to the solid phase and the labeled second antibody are derived from the same or related species, a blocking step should be included to avoid nonspecific signals. In our case, where both the first and second antibodies are of mouse origin, we include a blocking step with 3% normal mouse serum (NMS) after the addition of supernatants. All subsequent steps are carried out in buffer containing the same concentration of NMS.

A flow-chart summarizing these types of assays is shown in Fig. 14.4.2.

Radioimmunoassay for soluble antigens

Radioimmunoassays (RIA) are used as a sensitive method for measuring the primary interaction between antigen and antibody. The capacity of RIA to detect ever-smaller amounts of analyte increases as the operating amount of antibody is reduced, the limit being dictated in practice by the ability of the signal and the counter to distinguish the bound labeled antigen from background and by the equilibrium constant of the reaction between analyte and antibody. The labeling process should not alter the antigen structure or any other characteristics of interest. This is particularly important if we are trying to study later biological activities of the antigen, for example, interactions between receptors and ligands.

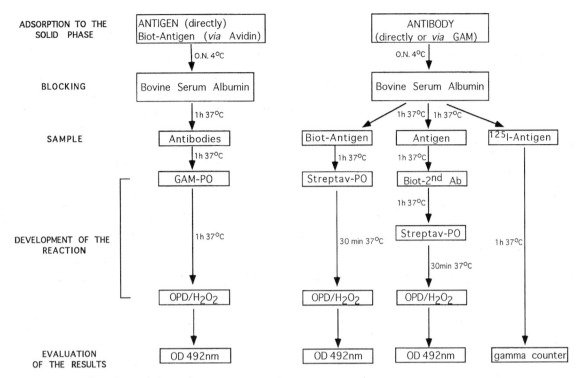

Figure 14.4.2 Flowchart of the various assays used.

Materials

- Radioiodinated antigen
- RIA buffer (0.01 M EDTA containing 0.001 M NaH_2PO_4, 0.0088 M Na_2HPO_4 and 0.15 M NaCl pH 7.6)
- RIA buffer containing 0.25 g/100 ml BSA
- Goat anti-mouse immunoglobulin antiserum
- PEG 6000

Procedure

1. Add 200 µl of NMS 1% to each polystyrene tube, plus 100 µl of the monoclonal antibody supernatant or diluted polyclonal serum, plus 100 µl of the iodinated antigen (20,000–40,000 cpm).
2. Incubate overnight at 4°C.
3. Add 200 µl/tube of the precipitating antibody. We use 5% goat anti-mouse immunoglobulin antibody. Add 200 µl/tube of 15% PEG 6000.
4. Incubate 60 min at room temperature.
5. Centrifuge tubes at 3000 rpm, 25 min at room temperature. Discard the supernatant and count the pellet.

Use of BIAcore™

General strategies

Biosensors allow direct monitoring of interactions in real time, without labeling any of the interactants, and kinetic information is readily derived. The biosensor combines a biological recognition mechanism with a sensing device or transducer. Recently, an instrument has been commercialized for biomolecular interaction analysis (BIA) based on surface plasmon resonance under the name of BIAcore™ (BIAcore™ System Manual 1992; Fagerstam et al 1990).

Surface plasmon resonance (SPR) is an optical phenomenon generated by direct obliquely incident light on the base of a prism which is coupled optically to a thin, semi-transparent gold film modified with carboxy-methylated dextran. Under conditions of total internal reflection, the electronic oscillations give rise to a nonpropagating evanescent wave that extends approximately 100 nm out from the gold surface into the sample solution and decays exponentially as a function of distance. The formation of a ligand–ligate complex immobilized on to the dextran changes the refractive index close to the gold layer in proportion to the mass of the interacting molecule bound to the sensor surface (Löfas and Johnsson 1990). It allows measurement of biospecific binding of an analyte in continuous flow to an immobilized ligand in real time, without labeling either the ligand or the analyte.

Applications of real-time interaction analysis using BIAcore™ have different uses, including the determination of analyte concentrations, estimation of affinity and kinetic rates, epitope mapping, and, in general, analysis of protein–protein, DNA–protein and DNA–DNA interactions (BIAcore™ System Manual 1992). We will focus on the kinetic analysis of ligand–receptor interactions.

BIAcore is controlled from a personal computer running BIAlogue™ software under Microsoft® Windows. Automatic operation in BIAcore is controlled by the Method Definition Language. The commands and their functions are explained in the Instrument System Manual (BIAcore System Manual, 1992). To plan and evaluate experimental work with BIAcore, several considerations must be taken into account. One of the molecules, the ligand, is covalently immobilized on the dextran surface of the biosensor. This surface forms one wall of a microflow cell which is in contact with a continuous flow of buffer controlled by an integrated liquid handling system, the microfluidic cartridge (IFC), which controls both the injection of small amounts of the interacting molecules (1–50 μl) and the flow of the buffer.

After immobilization, specific interactions between the injected analyte and the adsorbed ligand produce an increased matrix protein concentration (if the interaction in fact occurs) and consequently a change in the SPR signal. These changes in SPR signal are recorded as resonance units (RU) and are displayed on a sensorgram with reference to time along the y-axis. Thus, analyte–ligand association and dissociation can be followed as they take place.

The interacting analyte can be eluted from the immobilized ligand by washing with a chaotropic, acidic, or basic solution. In this way, the biosensor matrix is regenerated after each interaction cycle, allowing the analysis of new interactions on the same sensor. SPR instrumentation is very sensitive and yields reproducible results in the quantification of macromolecular interactions. The dextran matrix exhibits minimal nonspecific binding, so analytes can be measured in crude samples. The running buffer defines the baseline and all responses are expressed relative to this level.

An *E. coli*-produced extracellular part of the growth hormone receptor (GHBP), that binds to GH with the same affinity as the entire receptor, is used to analyze GH–GHR interactions. Study of the crystal structure of the complex shows that one GH can bind to two GHBP molecules (De Vos et al 1992). We describe below the kinetic analysis of this interaction using SPR technology.

Analysis of kinetic parameters in ligand–receptor interactions

Quantitative analysis of macromolecular interactions is important not only for the study of protein–protein interactions, but also because it can provide valuable information for the design of therapeutically relevant agonists or antagonists.

The strength of binding can be defined as the apparent affinity constant (K_a), which is the ratio between association and dissociation rate constants. When two proteins, A and B, interact in solution to form a complex AB, the rate of formation of the AB complexes with time may be described by the equation

$$\frac{d[AB]}{dt} = k_{ass}[A][B] - k_{diss}[AB] \qquad (1)$$

where k_{ass} and k_{diss} are the association and dissociation rate constants, respectively.

Equation (1) can be rewritten for BIAcore™ (Karlsson et al 1991; Altschuh et al 1992; Fagerstam et al 1992) as

$$\frac{dR}{dt} = k_{ass} C R_{max} - (k_{ass} C + k_{diss}) R \qquad (2)$$

where dR/dt is the rate of formation of surface complexes, C is the concentration of analyte in free solution and R_{max} is the total number of binding sites of the immobilized ligand expressed as SPR response.

The k_{ass} is determined from the concentration-dependent rate of association in pseudo-first-order conditions, i.e., in excess of the liquid-phase partner in the interaction. Thus, k_{ass} can be calculated from the straight line obtained by plotting dR/dt versus R. The biosensor calculates dR/dt at each report point. The derivative is obtained using least-squares adjustment of the values with a report point interval of 10 s. If the dR/dt vs R plots are performed at different concentrations of the free analyte, each of the lines obtained will have a slope k_s corresponding to

$$k_s = k_{ass} C + k_{diss} \qquad (3)$$

When the k_s values are plotted versus the concentration C, the slope of the new line obtained will be the k_{ass} and the intercept with the k_s axis will correspond to k_{diss}. Using this evaluation method, the interaction does not need to reach equilibrium.

When k_{diss} is calculated as described above, the long extrapolation to zero concentration usually results in a value of low accuracy. The k_{diss} is normally calculated when the sample pulse has passed the surface and the concentration C of free analyte drops to zero, using the equation

$$\ln(R_0/R_n) = k_{diss}t \qquad (4)$$

where R_0 and R_n correspond to the SPR signal at times $t=0$ and $t=n$, respectively. The k_{diss} will be calculated from the slope of the plot (R_0/R_n) versus $(t_n - t_0)$. Usually, analysis of data in this way yields a linear plot. However, in some instances, the data better fits a polynomial expression, indicating dual dissociation rate constants.

Experimental procedures

Materials and instruments

- Real-time BIA measurements were made using a BIAcore™ System (Pharmacia Biosensor AB, Sweden).
- BIAcore Sensor Chip CM5. Store at 4–8°C. (Pharmacia)

- Surfactant P20 (Pharmacia)
- *Amino coupling kit for ligand immobilization:* 0.1 M N-hydroxysuccinimide (NHS), 0.4 M (N-ethyl-N'-(3-diethylaminopropyl)-carbodiimide (EDC), and 1 M ethanolamine hydrochloride, pH 8.5 (Pharmacia)
- 40% (w/w) glycerol (Sigma, USA).
- *HBS buffer (running buffer):* 10 mM Hepes, pH 7.4, containing 150 mM NaCl, 3.4 mM EDTA, and 0.05% surfactant P20
- 10 mM sodium acetate buffer, pH 4.5 (Sigma, USA)
- Recombinant human GH (Pharmacia) produced in *Escherichia coli*.
- Recombinant human GHBP (Pharmacia) produced in *Escherichia coli*.

All solutions and buffers used should be filtered through a 0.22 μm filter. The running buffer always should be tempered and degassed prior to use.

Procedures

Immobilization of the ligand

As mentioned above, covalent coupling of the ligand to the sensor chip has to be done in a way that does not disturb analyte binding capacity. Routine protocols for ligand immobilization and sensor surface regeneration are described in the Instrument System Manual. However, for each new analyte the optimal conditions should be revised to insure that the interaction studied can be dissociated without damaging the covalently attached ligand. This requires attention to ligand stability, the retention of native conformation during consecutive cycles of binding and regeneration, the reproducibility of the SPR signal, and the ability to quantify soluble analytes in both purified and unpurified preparations.

The instrument has to be switched on approximately 1 h before use to temper the unit. To confirm the normal functioning of the apparatus, several steps must be covered before coupling proteins to a new sensor chip. These steps, such as docking of the optical system to the sensor or initiating the flow network, are part of routine daily work with BIAcore™ and are described extensively in the Instrument System Manual. Coupling of the ligand is performed after equilibration of the system with HBS using automatic injections executed by a robotic unit incorporated into the instrument. A procedure suitable for most of the proteins uses amine coupling in which the activated dextran surface reacts with uncharged primary amine groups of the ligand (lysine residues and amino terminal). During this process the buffers should not include components containing amino groups as they may suppress ligand immobilization.

1. Prior to use, allow the sensor chip to reach room temperature.
2. Insert the sensor chip into BIAcore using the DOCK command in BIAcore.
3. Initiate the system with HBS using the INITIATE command of the BIAcore User Working Tools.
4. Normalize the response of the photodiode array detectors with 40% glycerol solution using the command NORMALIZE of the BIAcore User Working Tools.
5. Perform a Dip Check to insure the integrity of the sensor chip.
6. Keep a continuous flow of the running buffer (HBS) passing over the sensor surface at 5 μl min⁻¹.

   Note: The choice of running buffer may vary with the properties of the interactants. HBS is the buffer most used in the experiments described in the bibliography. During the entire immobilization procedure, flow is maintained at 5 μl min⁻¹.

7. Inject the ligand in the appropriate buffer and determine the lowest ligand concentration that produces maximum electrostatic adsorption (preconcentration). Choose the

concentration of ligand for immobilization (usually 10 to 100 μg ml⁻¹).

Note: For immobilization of a ligand, a preconcentration effect is used to concentrate the ligand in the sensor chip matrix, which is based on the electrostatic attraction between ligand (positively charged) and surface (negatively charged carboxylic groups in the dextran). This preconcentration enhances the capacity of the coupling procedure. The ionic strength of the coupling buffer has to be low, less than 10 mM. Proteins are prepared at a pH below their isoelectric point and the pH of this buffer is changed for the immobilization of different substances.

8. Inject 10–35 μl of a freshly prepared solution containing 0.2 M EDC and 0.05 M NHS to activate the carboxylated dextran matrix.

   Note: A 7 min activation results in 30–40% of the carboxyl groups being EDC/NHS-activated, with the remaining groups negatively charged to interact electrostatically with the protein to be immobilized.

9. Inject 5 to 35 μl of the ligand into the selected buffer.

   Note: The quality and purity of the immobilized ligand determines the specificity and effectiveness of analyte binding. The purity of the ligand is therefore of vital importance for experimental results. The recommended purity of the ligand is >90%, and, ideally, affinity-purified reagents should be used.

10. Inject 35 μl of ethanolamine to block the remaining NHS ester groups.

    Note: Most of the noncovalently bound ligand is eluted during this step.

11. Inject 4–20 μl of the regeneration solution to remove fully any remaining noncovalently bound ligand once the

sample injection is completed and a stable HBS reading is obtained.

Note: Select the regeneration conditions to insure a constant binding capacity for at least 50 cycles.

A representative sensorgram for the immobilization of GH is shown in Fig. 14.4.3. In these conditions (Table 14.4.1) a level of approximately 360 RU of GH bound to the sensor is obtained, corresponding to about 0.36 ng mm⁻² of GH. This low amount of GH bound to the sensor was selected to avoid limited mass transport of GHBP during kinetic assays.

Interaction of the analyte with the immobilized ligand

Preliminary binding experiments must be performed to determine the useful concentration range for kinetic measurements, to optimize buffer solutions and flow rates, and to establish conditions for regeneration that elute bound protein without significant loss of binding capacity. To perform kinetic experiments, the level of immobilized ligand on the sensor surface and the maximum level of resonance that can be obtained after analyte binding (R_{max}) must be optimized. Insure that the resonance changes reflect interaction kinetics rather than mass transport, and that equilibrium is not reached very rapidly. For this verification, the Simulation part of the Kinetic Evaluation software can be used. Moreover, the optimal amount of immobilized protein can be determined empirically: if the binding rate kinetics are too fast with a given level of immobilized protein, then coupling should be repeated with a shorter activation time; if binding rates are initially limited by mass transport, then the amount of immobilized ligand should be changed.

For kinetic determinations, we used the extracellular part of the GH receptor at different concentrations (7.8–1000 nM in HBS) injected at a flow rate of 5 ml min⁻¹. At the end of the injection buffer, flow was continued for 30 min

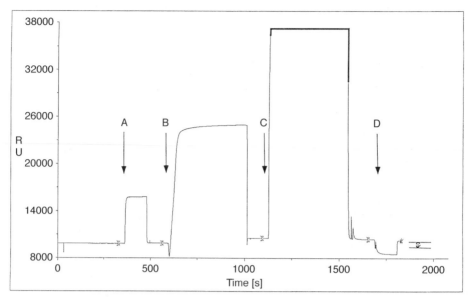

Figure 14.4.3 Sensorgram obtained on immobilization of GH on sensor chip CM5. During the immobilization protocol, a constant flow of HBS of $5 \mu l \, min^{-1}$ was maintained. At (A), $10 \mu l$ of EDC–NHS solution was injected. At (B), $35 \mu l$ of $50 \mu g \, ml^{-1}$ GH in $10 \, mM$ sodium acetate buffer, pH 4.5, was injected. At (C), $35 \mu l$ of ethanolamine solution was injected to react with residual NHS-esters on the sensor chip surface. At (D), $10 \mu l$ of $30 \, mM$ HCl was injected to wash noncovalently bound GH. The sensorgram represents the immobilization of GH detected in real time. The degree of immobilization, 360 RU in this example, is taken as the difference in response between the unaltered dextran surface and the response after HCl washing, as indicated by the arrows.

Table 14.4.1 Protocol for immobilizing GH on sensor chip CM5. Buffer flow is maintained at $5 \mu l \, min^{-1}$ throughout the immobilization procedure.

BIAcore™ immobilization protocol time(s)	Event	Comments
0	HBS flow 5 μl min⁻¹	Start cycle
200	Mix NHS + EDC 1:1	
360	Inject 10 μl NHS–EDC	Activate surface
600	Inject 35 μl of GH	Couple GH
1120	Inject 35 μl ethanolamine	Deactivate excess reactive groups
1670	Inject 10 μl HCl	Regenerate surface
1830		End cycle

and dissociation was measured. The immobilized GH was stable for 50 cycles using the regeneration procedure described above.

1. Use a recently coated sensor chip.

   Note: Test that the sensor chip is active, that it can bind specifically the analyte, and that the regeneration conditions are optimized.

2. Allow running buffer to flow continuously through the flow cell before injection of the analyte (baseline).

   Note: Use a flow of 2–5 μl min⁻¹ to avoid excessive consumption of analyte.

3. Inject 20–30 μl of serial analyte dilutions.

   Note: Use purified analyte if available. However, nonpurified substances can be used if they do not interact with any other protein in the sample.

4. Immediately after injection, allow the

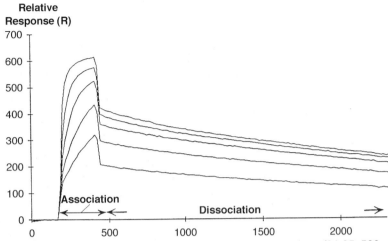

Figure 14.4.4 A set of sensorgrams for binding of a series of concentrations (31.25–500 nM) of *E. coli* GHBP to GH immobilized on the sensor surface. Aliquots of GHBP (20 μl) were passed sequentially over a flow cell containing GH and binding was monitored as resonance units (RU) on the sensorgram. Association was monitored over this period. At the end of the injection pulse, the buffer flow was continued and dissociation was measured. At the end of the cycle, the sensor chip was regenerated with 30 mM HCl.

Table 14.4.2 Kinetic analysis of the association phase for the interaction between GHBP and GH

GHBP concentration (nM)	k_s (s^{-1})	Correlation coefficient for k_s
31.25	0.003496	0.872
62.5	0.008224	0.985
125	0.018382	0.999
250	0.034744	0.999
500	0.053958	0.994
1000	0.097727	0.993

GH was immobilized on the sensor chip (0.36 ng mm^{-2}) and various concentrations of GHBP were injected on the sensor surface. k_s values were from the slopes of dR/dt vs R plots. In each experiment, report points of dR/dt and R values were recorded about every 10 s over the region 30–240 s following the injection of GHBP. Other conditions were as described for Fig. 14.4.5. The results of one representative experiment are shown.

running buffer to flow through the flow cell during the dissociation.

Note: The net difference between the baseline signal and the signal after completion of analyte injection represents the binding value of that particular sample. High flow (10–20 μl min^{-1}) will minimize further reassociation of the analyte during the dissociation phase.

5. Inject 4–15 μl of the regeneration solution to elute the remaining bound analyte.

Table 14.4.3 Apparent kinetic constants for the interaction of recombinant human GHBP with immobilized human GH

	k_{ass} (mean ± SD) ($M^{-1}\,s^{-1}$)	k_{diss} (mean ± SD) (s^{-1})	K_{aff} (k_{ass}/k_{diss}) (mean ± SD) (M^{-1})
E.coli GHBP	95,630 ± 635	0.00039 ± 0.000015	$2.42 \times 10^8 \pm 0.116$

The initial k_{diss} (30–180 s) was calculated for 500 nM GHBP. Results are given as mean ± standard deviation (SD) of three different experiments.

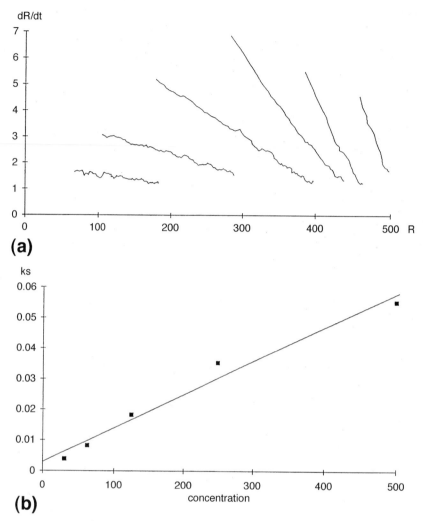

Figure 14.4.5 (a) Plot of dR/dt vs R values from Fig.14.4.2 and (b) plot of slopes versus GHBP concentration. (a) Example of the dR/dt vs R plot for the association part of the sensorgram at 31.25, 62.5, 125, 250, 500, and 1000 nM (Fig.14.4.2). A total of 360 RU of GH was immobilized on the sensor surface and the recombinant GHBP produced in *E. coli* (20 μl) was injected at a flow rate of 5 μl min^{-1}. Report points were taken every 10 s between 30 and 240 s after GHBP injection. Values of dR/dt < 0 were discarded. (b) Determination of the association rate constant for GHBP and GH. The slopes (k_s) of the dR/dt vs R plots from (a) were plotted against GHBP concentration. The slope of this plot yields the apparent association rate constant, k_{ass}, assuming unimolecular binding kinetics.

Rate constant determination

Kinetic experiments can be evaluated using the BIAlogue™ Kinetics Evaluation Software provided with the BIAcore™ instrument.

Typical sensorgrams showing the association and dissociation rate constants for GH–GHBP interaction are shown in Fig. 14.4.4. Data for determining the association between immobilized GH and soluble GHBP were analyzed using BIAlogue Kinetics Evaluation Software during an association phase over the 240 s injection period.

The initial 30 s period was ignored, because much of the resonance is due to a change in the refractive index of the sample injected. Data obtained at the end of the injection time, when the system approaches the equilibrium, were also discarded. During this period, different dR/dt vs R plots were obtained (Fig. 14.4.5a).

All of these lines gave a good fit to the linear equation (3), with the resulting K_s values being linearly dependent on concentration (Fig. 14.4.5b; Table 14.4.2).

Dissociation of the GH–GHBP complex was followed up to 30 min with continuous flow of HBS buffer over the sensor surface. Plotting the data according to equation (4), the line was observed to curve slightly with a faster initial rate of dissociation. Possible explanations for this early dissociation phase include reassociation of the GHBP as it flows over the immobilized GH; heterogeneity in binding to GH binding sites 1 and 2; and molecular heterogeneity. Only data obtained after the 30 s period were used to calculate k_{diss} owing to the change in the refractive index after injection of the sample. Using this scheme the dissociation phase fits a linear equation, showing independence of the GHBP concentration used (Fig. 14.4.6; Table 14.4.3).

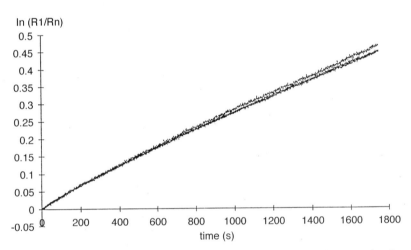

Figure 14.4.6 Determination of the dissociation rate constant for the GHBP–GH complex from the data in Fig. 14.4.5. Report points for R were taken at 10 s intervals between 30–1800 s after termination of GHBP injection for the dissociation of a series of concentrations (125–500 nM). Rate constants for the initial (30–180 s) and late (180–1000 s) dissociation were obtained from the slope of the $\ln(R_1/R_n)$ versus ($t_1 - t_n$) plots with R_1 and R_n referring to R at times t_1 and t_n, respectively.

References

Altschuh D, Dubs M, Weiss E, Zeder-Lutz G, Van Regenmortel M (1992) Determination of kinetic constants for the interaction between a monoclonal antibody and peptides using surface plasmon resonance. Biochemistry 31: 6298–6304.

Baumann G, Stolar MW, Amburn K, Barsano CP, DeVries BC (1986). A specific growth hormone-binding protein in human plasma: initial characterization. Endocrinol Metab 62: 134–141.

Bazan JF (1989) A novel family of growth factor receptor: a common binding domain in the growth hormone, prolactin, erythropoietin and IL-6 receptor, and the p75 IL-2 receptor b-chain. Biochem Biophys Res Commun 164: 788–795.

BIAcore™ System Manual (1992) Pharmacia Biosensor AB, Sweden.

Cunningham BC, Ultsch M, De Vos AM, Mulkerrin MG, Clauser KR, Wells JA (1991) Dimerization of the extracellular domain of the human growth hormone receptor by a single hormone molecule. Science 254: 821–825.

De Vos AM, Ultsch M, Kossiakoff AA (1992) Human growth hormone and extracellular domain of its receptor: crystal structure of the complex. Science 255: 306–312.

Fagerstam L, Frostell A, Karlsson R, et al (1990) Detection of antigen–antibody interactions by surface plasmon resonance. J Mol Recogn 3: 208–214.

Fagerstam L, Frostell A, Karlsson R, Persson B, Ronnberg I (1992) Biospecific interaction analysis using surface plasmon resonance detection applied to kinetic, binding site and concentration analysis. J Chromatography 597: 397–410.

Karlsson R, Michaelsson A, Mattsson L (1991) Kinetic analysis of monoclonal antibody–antigen interactions with a new biosensor based analytical system. J Immunol Methods 145: 229–240.

Kelly PA, Djiane J, Postel-Vinay M-C, Edery M (1991) The prolactin/growth hormone receptor family. Endocrinol Rev 12: 235–251.

Leung DW, Spencer SA, Cachianes G, et al (1987) Growth hormone receptor and serum binding protein: purification, cloning and expression. Nature 330: 537–543.

Löfas, S, Johnsson B (1990) A novel hydrogel matrix on gold surfaces in surface plasmon resonance sensors for fast and efficient covalent immobilizations of ligands. J Chem Soc Chem Commun 21: 1526–1528.

Steward MW (1977) Affinity of the antibody–antigen reaction and its biological significance. In Glynn LE, Steward MW eds. Immunochemistry: An Advanced Textbook. Wiley, London, pp. 223–262.

Lymphoid kinase detection and activation

14.5

Ana C. Carrera
Carlos Martinez-A.

Centro Nacional de Biotecnologia, Universidad Autonoma, Madrid, Spain

TABLE OF CONTENTS

Immunology Methods Manual
ISBN 0–12–442712–X

Introduction

Protein kinases regulate the early signals of most cellular activation processes. Lymphocytes are not an exception and, in fact, an increasing number of protein kinases are being described that regulate not only proliferative responses of B and T lymphocytes but also the early differentiation processes and specific effector functions. Protein kinases consist of a catalytic conserved core (responsible for transferring phosphate), flanked by regulatory regions important for functions such as regulation and subcellular localization. The catalytic core contains highly conserved sequence stretches, but the regulatory region is also usually composed of sequence motifs present in different signal transduction molecules (such as SH2 and SH3 domains) that allow the prediction of certain features in the role of the kinase.

Protein kinases can be studied from two different and complementary views: biochemical approaches and genetic approaches. This chapter is dedicated mainly to biochemical approaches, since genetic approaches for the study of protein kinases will be similar to those for other proteins. The first set of methods described will refer to the detection of protein kinases. These will be followed by a more kinase-specific treatment referring to protein kinase activity detection, methods that are to a large extent based on the ability of protein kinases to transfer the γ-phosphate of the ATP to an amino acid side-chain. Finally, given that the limitation of some of the methods is the small amount of protein kinase expressed naturally in cells, there will be a discussion of the methods for overexpression of protein kinases.

Protein kinase detection

General strategies

Detection of protein kinases when a specific antibody is available can be achieved by flow cytometry, Western blotting, and metabolic labeling. Western blotting and intracellular flow cytometric staining are the two simplest. However, metabolic labeling allows subsequent tryptic peptide analysis (see below) and ^{32}P metabolic labeling gives information on the phosphorylation state of the kinase (kinases often regulate by intra- or intermolecular phosphorylation). Detection of kinases upon metabolic labeling is also mandatory when the available antibody does not recognize the denatured kinase (in Western blot) or when

determination of the sites of phosphorylation is required. For the analysis of protein kinase detection upon metabolic labeling, as for the detection of kinase activity *in vitro*, a common step is the extraction of cellular lysates and immunoprecipitation of the kinase. A basic protocol for cell lysis and kinase immunoprecipitation is included in this chapter. Detection of protein kinases can also be made by kinase reaction in immunoprecipitates (see below). For all this analysis computer technology is recommended to quantify the signal corresponding to the kinase.

Intracellular flow cytometric staining

Materials

Staining buffer

Digitonin

Specific Ab

Phycoerythrin- or fluorescein-conjugated second antibody

Flow cytometer

Procedure

1. Collect 10^5–2×10^5 cells in an Eppendorf tube and wash them in staining buffer.
2. Resuspend the cell pellet in 50 μl of 1:50 (v/v) dilution of digitonin solution in staining buffer at room temperature (cell permeabilization, only required for intracellular kinases).
3. Incubate 4 min at room temperature.
4. Wash the cells twice in 1.5 ml of ice-cold staining buffer.
5. Resuspend the cells in 30–50 μl of staining buffer containing the specific antibody (10 ng to 1 μg, previously optimized); incubate 20 min at 4°C.
6. Wash twice in staining buffer.
7. Resuspend the cells in 30–50 μl of staining buffer and add the appropriate second antibody coupled to phycoerythrin or fluorescein; incubate 20 min at 4°C.
8. Wash twice in staining buffer.
9. Resuspend in 100 μl of staining buffer and analyze by flow cytometry immediately (or resuspend in 4% paraformaldehyde in PBS for analysis in the following 24 h).

Solutions

- *Digitonin solution*: 0.25% digitonin in H_2O
- *Staining buffer*: 1× PBS, 0.5% BSA, 1% FCS, 0.65 g l^{-1} NaN$_3$.
- *PBS*: 8.0 g NaCl, 0.2 g KCl, 1.44 g Na$_2$HPO$_4$ and 0.24 g KH$_2$PO$_4$, adjust to pH 7.2, make up to 1 l with ddH$_2$O.

Trouble shooting

- If permeabilization is partial, increase the incubation time of permeabilization or decrease the number of cells.
- If permeabilization is efficient but the staining is low, increase the concentration of the first antibody.
- If negative control gives a high signal, wash the sample in higher volumes; pass the samples through the cytometer slowly.

Western blotting

Materials

Transfer apparatus

Whatman 3MM

Nitrocellulose or Immobilon

Nonfat powdered milk

Gelatin or BSA

TBS

TBST

Conjugated second antibody

Developing kit

Specific antibody

Transfer buffer

Procedure

1. Run a previously immunoprecipitated protein sample or a total cellular extract (amount of cell extract corresponding to 50–100 µg of protein) in SDS-PAGE.
2. Transfer the gel immediately after electrophoresis. Assemble the transfer sandwich: (i) plastic cassette, (ii) sponge, (iii) two pieces of Whatman 3MM, (iv) gel, (v) nitrocellulose (prewetted in transfer buffer) or Immobilon (prewetted in methanol), (vi) one piece of Whatman 3MM, (vii) sponge, (viii) plastic cassette. When the cassette is placed in the transfer tank the nitrocellulose or Immobilon should be closest to the positive electrode. Transfer gel for 4 h at 0.4 mA at 4°C.
3. Block residual binding sites on the membrane by incubating it 1 h at room temperature in 10% nonfat powdered milk in TBS (or 2% gelatin or 2% BSA).
4. Rinse in TBST.
5. Probe the filter by incubating it at room temperature for 1 h with the primary antibody diluted in the minimal volume of TBS (for 15 ml: 250 µl of culture supernatant, 10 µl of polyclonal antibody or 2–4 µl of ascites, approximately 20 µg of antibody).
6. Rinse vigorously twice for 15 min with 50 ml of TBST–1% milk.
7. Incubate with the minimal volume of TBST–1% milk containing the second antibody conjugated to horseradish peroxidase (or alkaline phosphatase, depending on the developing method). The exact amount of antibody to be used should previously be optimized.
8. Rinse vigorously in 50 ml of TBST–1% milk for 15 min, change the TBST–1% milk and incubate for another 15 min. Change the TBST–1% milk again and incubate for another 30 min at room temperature.
9. Develop the blot using Amersham developing kit for horseradish peroxidase or Promega NBT-BCIP developing kit (for alkaline phosphatase).

Solutions

- *Transfer buffer*: 12.1 g Tris-base, 57.67 g glycine, 800 ml methanol, make up to 4 liters with H_2O.
- *TBS*: 10 mM Tris pH 8.0, 150 mM NaCl
- *TBST*: 10 mM Tris pH 8.0, 150 mM NaCl, 0.1% Tween 20.

Trouble shooting

- If the background is high, increase the number of washes, the volume used to wash, the speed of the shaker and, if required, the percentage of Tween 20 in the TBST.
- If the specific signal is low, increase the concentration of the first and/or second antibody and increase the incubation periods.

Metabolic labeling of lymphoid cells

General strategies

Labeling of cellular proteins from normal or transformed lymphoid cells can be achieved by growing the cells in the presence of a medium containing radioactive amino acids or radioactive orthophosphate. In addition, transmembrane proteins can also be labeled in intact cells by membrane iodination (Fraker and Speck 1978). The majority of specific lymphoid protein kinases are intracellular; thus, we would only include the methods for intracellular detection. The conditions of labeling vary depending on the rate of division of the cell, sensitivity to radiation, and ability of the cells to survive in conditioned media. Optimal labeling of transformed cells is normally achieved at 3–4 h, while for resting peripheral cells the time is approximately 16 h. Conditions in each particular system should be optimized in small-scale assays to balance cell survival and labeling efficiency.

Materials

Phosphate-free RPMI

[^{32}P]orthophosphate (6000 Ci mmol^{-1})

Shielding material

FCS dialyzed against 150 mM NaCl

Methionine-free RPMI

[^{35}S]methionine (1000 Ci mmol^{-1}).

Procedures

^{32}P Labeling of Jurkat cells

1. Make sure that the cells are healthy and growing exponentially at the time of labeling. Prepare shielding material.
2. Wash 5×10^7 cells in phosphate-free RPMI, incubate 1 h at 37°C.
3. Spin down (400g 10 min) and resuspend in 20 ml of phosphate-free RPMI containing 1% glutamine, 1% phosphate-free FCS (dialyzed against 150 mM NaCl) and 5 mCi [^{32}P]orthophosphate (6000 Ci mmol^{-1}).
4. Incubate 4 h at 37°C, 5% CO$_2$.
5. Cells are ready to be activated and/or to be lysed (see below).

^{32}P Labeling of normal peripheral human T cells

1. Prepare 5×10^7 purified T cells (Carrera, 1987). Prepare shielding material.

2. Wash the cells in phosphate-free RPMI, incubate for 1 h at 37°C. Repeat this step.
3. Spin down and resuspend in 25 ml of phosphate-free RPMI containing 1% glutamine, 1% phosphate-free FCS (dialyzed against 150 mM NaCl) and 2.5 mCi [^{32}P]orthophosphate (6000 Ci mmol^{-1}).
4. Incubate 16 h at 37°C, 5% CO$_2$.
5. Cells are ready to be activated and/or to be lysed (see below).

Metabolic labeling

1. Cells are prepared and incubated in methionine-free RPMI as above.
2. Spin down and resuspend at 2×10^6 to 1×10^7 cells ml^{-1} in methionine-free RPMI containing 1% glutamine, 1% S-free-FCS (dialyzed against 150 mM NaCl) and 0.2 to 0.5 mCi ml^{-1} [^{35}S]methionine (1000 Ci mmol^{-1}).
3. Incubate for 16 h at 37°C, 5% CO$_2$ (if slowly-growing cells, like peripheral normal lymphocytes) or for 3.5 h at 37°C 5% CO$_2$ (if rapidly growing cells).
4. Cells are ready to be activated and/or to be lysed (see below).

Trouble shooting

- If the cell viability after labeling is low, try decreasing the specific activity of the radioisotope or decreasing the labeling incubation time.
- If labeling of proteins is not efficient, try increasing the specific activity of the radioisotope or increasing the incubation time.
- A high background signal in the autoradiography of the SDS-PAGE indicates that the samples have contaminated the running buffer, or that radioisotope migrating ahead of the SDS-PAGE front has contaminated the gel. Load more carefully, handle the gel after running carefully to avoid carrying

the protein A–Sepharose beads from the loading wells (often contaminated) and let the front of the SDS-PAGE migrate out of the gel before stopping the electrophoresis.

Lysis, immunoprecipitation, and reprecipitation

Materials

PBS

Lysis buffer

Protease and phosphatase inhibitors

Microfuge

Ice bucket

Glycerol

Eppendorfs

Lysis buffers

Protein A–Sepharose and/or protein G–Sepharose

Washing solutions (see solutions below)

GSD

Procedures

Cell lysis

1. Prior to lysis, wash the cells in $1\times$ PBS (spin $400g$, 10 min).
2. Resuspend the cells in the appropriate volume ($0.25–1 \times 10^8$ cells ml^{-1}) of ice-cold lysis buffers (see Note a) (RIPA lysis buffer, 1% NP-40/Triton X-100 lysis buffer or 1% digitonin/Brij 96 lysis buffer), containing protease and phosphatase inhibitors (see recipes below).

3. Incubate for 30 min at 4°C in an end-over-end rotor.
4. Spin at $1.6 \times 10^3 g$ for 30 min.
5. Collect supernatant and use immediately if possible, or add 10% glycerol and store at –80°C.

Note
(a) To select the lysis buffer, consider that RIPA buffer would break noncovalent protein–protein interactions and disrupt nuclei. 1% NP-40 or Triton X-100 lysis buffer would not lyse nuclei and does not break strong ionic protein–protein interactions. 1% Digitonin or 1% Brij 96 lysis buffer would not be 100% efficient in solubilizing cell membranes but would maintain ionic and hydrophobic noncovalent protein–protein interaction.

Immunoprecipitation of kinases

1. Incubate 0.5 ml of lysate (containing approximately 2.5×10^7 cell equivalents corresponding to 200 µg–1 mg of protein depending on cell size) for 30 min at 4°C with 50 µl of preclearing solution (lysis buffer containing 300 µl protein A ml^{-1}; see 'trouble shooting' for more stringent preclearing conditions).
2. Collect supernatant, and repeat step 1.
3. Collect supernatant and add the specific antibody (see Note a). Incubate 3 h at 4°C.
4. Add 30 µl of 50% protein A–Sepharose in lysis buffer (preincubated for 1 h with secondary Ab if required) or 30 µl of 50% protein G–Sepharose (see Note b).
5. Wash the immunoprecipitates twice in lysis buffer. To wash, add 1 ml of the washing buffer, vortex, spin in microfuge at low speed, and discard supernatant.

6. Wash the immunoprecipitates three times in 0.5 M LiCl, 50 mM Tris pH 7.5 (strong wash (see Note c), to decrease the number of non-covalently associated proteins), or wash three times in 0.15 M NaCl, 50 mM Tris pH 7.5 (mild wash, to preserve associated proteins).
7. Wash the immunoprecipitate in kinase buffer for subsequent kinase assay or boil in GSD if the immunoprecipitate is ready to be resolved by SDS-PAGE (Laemmli, 1970).

Notes
(a) 5–10 μl of polyclonal sera or 5 μg of purified mAb or 1–2 μl of ascites or 100–200 μl of culture supernatant.
(b) Particularly if primary Ab is rat, mouse IgG1 or human IgG3 and no secondary antibody is available.
(c) Strong washes can be complemented by washing the immunoprecipitates in 50 mM Tris pH 7.5 containing 0.1% SDS.

Detection of kinase-associated proteins by reprecipitation

1. Perform the first immunoprecipitation (from no less than 300 μg of protein; see Note a).
2. Perform an *in vitro* kinase assay (see below) to label kinase-associated proteins (see Note b).
3. Add 400 μl of reprecipitation solution (0.4% SDS, 50 mM triethanolamine, 100 mM NaCl, 2 mM EDTA, 2 mM βME) to the radiolabeled immunoprecipitated proteins and boil for 5 min.
4. Collect the supernatant by centrifugation, place on ice for 5 min.
5. Add 100 μl of 10% Triton X-100 and 10 mM iodoacetamide. Dilute solubilized phosphoproteins 1:2 in lysis buffer.
6. Samples can be stored at –80°C.

7. Set up a second immunoprecipitation following the same protocol as for the first immunoprecipitation (see above). Preclearing and strong washes are particularly recommended for second immunoprecipitations.

Notes
(a) Obtained from lysing cells in mild detergent and washing the immunoprecipitate under mild washing conditions.
(b) Alternatively, reprecipitations can be carried out from immunoprecipitates of lysates obtained from metabolically labeled cells; in this case, associated proteins are radiolabeled during the labeling and do not require *in vitro* kinase assay.

Solutions

- *RIPA buffer*: 1% Triton X-100, 1% sodium deoxycholate, 0.1% SDS, 150 mM NaCl, 10 mM Tris pH 7.2, 1 mM EGTA
- *Triton X-100 (or 1% NP-40) lysis buffer*: 1% Triton X-100 (or 1% NP-40), 50 mM Hepes pH 7.5, 150 mM NaCl. *Optional*: 1 mM EDTA.
- *1% digitonin (or 1% Brij 96) lysis buffer*: Prepare the detergent at 2% (commercial digitonin is usually only 50% pure so 4% of product should be prepared) in H$_2$O; warming will be required for solution (to 100°C in the case of digitonin); adjust to pH 8.0 with NaOH. Filter and store aliquots at –20°C. To use, mix 1:1 (v/v) with a solution of 20 mM triethanolamine, 2 mM EDTA, 300 mM NaCl pH 8.0, 20% glycerol.
- Protease and phosphatase inhibitors: For 1 ml of lysis buffer add 10 μl of 200 mM PMSF in acetone (store at room temperature), 10 μl of 0.1 M sodium orthovanadate in H$_2$O (store at

4°C), 1 µl of 10 mg ml^{-1} leupeptin in H$_2$O (store at −20°C), 1 µl of 10 mg ml^{-1} (or 30 units ml^{-1}) aprotinin in H$_2$O (store at 4°C), 1 µl of 1 mg ml^{-1} pepstastin in methanol (store at −20°C). *Optional*: 50 mM Na$_4$P$_2$O$_7$·10H$_2$O.
  * *GSD*: 10 ml glycerol, 30 ml 10% SDS, 5 ml 0.5 M Tris pH 6.8, 10 mg bromophenol blue, 55 ml H$_2$O.

Trouble shooting

  * If secondary antibody is to be used for the immunoprecipitation, it should also be added to the preclearing mixture. Similarly, if preimmune serum is available, it is recommended to include it during the preclearing. If preliminary results give high background problems, include in the preclearing 0.5 M NaCl.
  * If the immunoprecipitated protein is low, increase the lysate amount used to immunoprecipitate, increase the concentration of antibody, and increase the incubation time (no more than 4 h if a kinase assay is to be performed).
  * Reprecipitation is only 10% efficient (this is easy to test by doing the second immunoprecipitate with the same antibody as the first). Therefore, if the signal is too low, try increasing the starting material. Use a phosphoImager to enhance the signal and heavily phosphorylated starting material. Make sure that the antibody used for the second immunoprecipitate recognizes denatured protein.
  * If the protein is degraded or dephosphorylated, try using tested protease and phosphatase inhibitors, and consider adding some other inhibitors. Try all the steps sequentially; work at 4°C.

Protein kinase activation

General strategies

Methods for evaluation of protein kinase activity are based on the ability of these enzymes to transfer phosphate. We give below the protocols for cell starvation that should precede cell induction via specific transmembrane receptors (for example, TcR, BcR, CD4, CD8, CD19, interleukin receptors, etc.). Once the cells have been induced, the activation of protein kinases can be detected *in vitro* or *in vivo* by their ability to autophosphorylate and to transfer phosphate to exogenous substrates. Finally, most activated protein kinases affect cellular metabolism by phosphorylating specific amino acids in specific substrates. Protocols for analysis of amino acid specificity and substrate specificity are given.

Cell triggering for kinase induction

Commentary

Transformed lymphoid cells, used for many of the studies of lymphoid kinases, are actively growing cells and often behave as already triggered cells. Even freshly isolated lymphocytes (obtained from lymphoid organs or peripheral blood) are not by all criteria unstimulated cells. Therefore, unless the study refers to the comparison of freshly isolated lymphocytes of differently treated animals, it is always appropriate to utilize a starvation method before the cells are induced using Abs or ligands to specific cellular receptors.

Materials

BSA

RPMI

Induction reagent (lymphokine or specific antibodies)

PBS.

Procedure for starvation and induction of lymphoid cells

1. Lymphokine-dependent cell lines are collected and washed three times in RPMI containing 2 mM L-glutamine, 1% antibiotics (Flow) and 1% BSA (without lymphokine or FCS). Cells are then resuspended at 10^7 cells ml^{-1} in the same medium and incubated at 37°C, 5% CO_2 for 4 h (longer periods of time can be tested). Normal lymphocytes and transformed cells (lymphokine-independent) require only 1 wash and 1 h starvation period.
2. For induction, cells can be directly incubated at 37°C, 5% CO_2 for the selected period of time in the presence of the appropriate lymphokine or Ab. Alternatively, cells can be preincubated in the cold in the presence of the induction reagent and then triggered at 37°C.
3. Induction should be stopped rapidly (dilution in cold medium or PBS or a quick spin can be used).
4. Cell pellet is ready for lysis (washing in PBS before lysis is recommended if analysis of total lysates is to be performed) and analysis.

Assay for *in vitro* kinase activity

Commentary

Below we include an example of an *in vitro* kinase assay for kinases homologous to pp60^{src}. For other kinases, optimizations should be performed varying incubation time, enzyme and substrate concentration, ATP specific activity, inclusion of Mn^{2+} or Mg^{2+} and other active reagents such as EGTA, DTT, and inhibitors of phosphatases such as β-glycerol phosphate or NaF.

Materials

Kinase buffer,

[^{32}P]ATP 3000 Ci mmol^{-1}

Exogenous substrate (if required)

EDTA

Gel electrophoresis apparatus

β counter and/or phospholmager

Procedures

In vitro kinase assay of src kinases

1. Enzymes purified by immunoprecipitation (see above) and enzymes purified to homogeneity are appropriate for this assay. Place 50 ng of the purified enzyme or the immunoprecipitate in 10 μl of 10 mM Tris–HCl pH 7.4. Incubate 1 min at 25°C.
2. Add 20 μl of 2× kinase buffer. Incubate 1 min at 25°C.
3. If exogenous substrate phosphorylation is to be tested, add 10 μl containing 10 μg of acid-denatured enolase or the desired prepurified substrate in an excess of

at least 100 times with respect to enzyme. When peptides are used as substrates, 1.5 mM final concentration is recommended (if no substrate is added, use 1× kinase buffer).

4. Incubate the reaction 2 min at 30°C (to analyze enzyme activation) or 15 min at 25°C (when an efficient phosphorylation of the exogenous substrate is desired).

5. Stop the reaction by adding of 10 μl 100 mM EDTA pH 8.0.

6. Kinase reactions are usually resolved by SDS-PAGE (Laemmli 1970). In the case of phosphorylated peptides, follow the procedure of Casnellie (1991).

7. Phosphate incorporated into enzyme or substrate can be quantified by liquid scintillation counting or using the phospholmager.

Quantitative in vitro kinase assay

1. Purified enzymes are required to determine enzymatic parameters. The amount of enzyme used should be low and appropriate to keep the incorporation of phosphate into substrate in the linear range. As a first approximation, 10–50 ng of enzyme can be tested.

2. Follow steps 2 to 7 of the previous protocol, trying to carry out the reaction and addition of ingredients in the minimal volume and for an incubation period short enough to keep the incorporation of phosphate into enzyme or substrate in the exponential range (1 to 3 min). For enzymatic analysis, substrate should be purified and used in at least 500-fold excess.

3. To obtain the K_m for ATP, fix the concentration of substrate and vary the concentration of ATP to obtain the saturation curve. To obtain the K_m for substrate and the apparent V_{max} of

substrate phosphorylation, fix the concentration of ATP and vary the concentration of substrate. Analyze the data (moles of phosphate incorporated into substrate) following the procedure of Eisenthal and Cornish-Bowden (1974) and Lineweaver and Burk (1934). Once approximate values of apparent K_m and V_{max} are obtained, repeat the measurements varying the concentrations of substrate and ATP around 0.1 to 10 times the values of K_m.

Solutions

- 2× Kinase buffer:

    50 mM Tris–HCl pH 7.4,

    20 mM MnCl$_2$

    1 μM ATP

    1 μCi μl^{-1}[γ-^{32}P]ATP 3000 Ci mmol^{-1}.

Trouble shooting

- If the phosphorylation of substrate is low, try increasing the specific activity of the [^{32}P]ATP in the kinase buffer. Make sure that the substrate is not aggregated; increase the incubation period for the kinase reaction.

- If the background is high, let the SDS-PAGE front come out of the gel before stopping the electrophoresis. Fix the gel in large volumes.

- In kinetic analysis, if the saturation curve is not exponential, the kinase is not under optimal enzymatic conditions. Increase the concentration of protein substrate and ATP; decrease the concentration of enzyme; decrease the incubation time for the kinase reaction to improve the values of initial velocity. Make sure that your preparation does not contain a fraction of inactive enzyme.

In vivo analysis of kinase activation

General strategies

The ability of a kinase to transfer phosphate to protein substrates in intact cells is not strictly measurable. However, several approaches allow one to obtain information on the activation state of a kinase in intact cells. First, many kinases autophosphorylate upon activation. As a consequence of the process of autophosphorylation, some kinases undergo gel mobility retardation upon activation (for example c-Raf; Morrison et al 1988). This mobility retardation has been used as an indicator for enzyme activation. In addition, peptide analysis shows whether or not a kinase has been modified at the autophosphorylation site or at a different site (by other cellular kinases). To analyze phosphorylated peptides, cells should be ^{32}P-metabolically labeled, the radiolabeled kinase immunopurified and resolved by SDS-PAGE, and this immunopurified kinase fragmented into peptides. Resolution of these phosphorylated peptides yields a peptide map, which, when characterized (see below), indicates which sites of the kinase have been modified. Another method for discovering whether a kinase has been activated is the analysis of specific substrates. This requires that the kinase have a well-characterized specific substrate (for example, S6 kinases; Strugill 1988). We include here a basic protocol for peptide preparation and resolution.

Materials

V8 protease

V8 reaction buffer

Electrophoresis apparatus

Cooling unit for TLE

Trypsin

MeOH

Speed vacuum

Glycerol

DTT

Bromophenol blue

TLE buffers

Dye mix

Whatman paper.

Procedures

V8 peptide preparation

1. Immunopurified kinases extracted from lysates of labeled cells and resolved by SDS-PAGE constitute appropriate samples for V8 treatment. Run the preliminary SDS-PAGE as usual (Laemmli, 1970), rinse the gel in ddH$_2$O, and expose without fixing in a plastic bag. Excise the band of interest and Čerenkov count.
2. Prepare a 12–15% second SDS-PAGE, including 1 mM EDTA in all the gel mixes. It is recommended that the second gel be thicker than the slice to proteolyze.
3. To load the gel, put it in the gel tank and add running buffer to the bottom chamber. Fill the stacking gel with the V8 reaction buffer (0.1% SDS, 1 mM EDTA, 2.5 mM DTT and 0.125 mM Tris, pH 6.8) and put the gel slices into the wells. (Gel can also be dried without fixing, and the dry slices put in the wells after removing the backing paper).
4. Incubate 10 min and put the running buffer into the top chamber. Overlay with 10 µl of the V8 reaction buffer containing 10% glycerol, 0.001% bromophenol blue, and between 50 ng and 5 µg of V8 protease.
5. Let the samples enter the stacking gel.

Once inside, stop the electrophoresis and incubate 30 min.

6. Run the gel as usual, dry, and analyze by autoradiography.

Tryptic peptide preparation

1. For sample preparation, immunoprecipitated labeled proteins (obtained by in vitro kinase assay or from lysates of metabolically labeled cells) are resolved by SDS-PAGE. Gels are dried and exposed, and the bands of interest are excised from the gel and Čerenkov counted.

2. Incubate the excised gel band in 3 ml of 30% MeOH in ddH$_2$O, at 37°C for 1 h (repeat 3 times). Incubate again overnight.

3. Scrape the paper and transfer the gel to an Eppendorf tube. Dry in the speed vacuum for 1 h.

4. Make fresh trypsin solution: 50 μg ml^{-1} in 50 mM ammonium bicarbonate (for 10 ml 0.04 g NH$_4$HCO$_3$, 0.5–1 mg trypsin in ddH$_2$O). Other proteases can be utilized similarly.

5. Add 0.5 ml trypsin solution per slice, incubate at 37°C 20 h. Spin 10 min at 400g; collect supernatant; supplement it with 0.5 ml of fresh trypsin solution and incubate at 37°C for 6 h.

6. Spin the tubes in a microfuge for 10 min. Transfer supernatant to a fresh tube; spin another 10 min; transfer supernatant to another fresh tube; dry the sample.

7. Wash the dried sample sequentially in decreasing volumes of ddH$_2$O. First add 1 ml, vortex, and dry in the speed vacuum. Repeat the process with 0.5, 0.3, 0.2, and 0.1 ml of ddH$_2$O. Čerenkov count the sample.

8. Resuspend it in 1–2 μl of 4 mM βME in H$_2$O (5 μl of 13 M βME in 15 ml H$_2$O). Load in the TLE at least 10^3 cpm (in 0.5 μl).

Thin-layer electrophoresis and chromatography for tryptic peptides

1. TLE in the first dimension is tried at pH 1.9 and pH 8.9 (see buffers, below) to optimize separation of the tryptic peptides for a given protein.

2. Use plastic-backed cellulose TLE plates, 20 cm × 20 cm. Label the plates using a blunt pencil (to avoid removing cellulose). Draw a line through the middle of the plate and mark two spots (on both sides of the central pencil-line, 1.5 cm away from it and 1.5 cm away from the edge of the plate).

3. Cut the plate along the central line and load on the marked spot 0.5 μl of dye mix.

4. Load similarly the 0.5 μl containing the peptides in βME. To load, avoid touching the cellulose with the loading capillary (or 1–10 μl pipette); instead, carefully let the drop be absorbed by the cellulose, making the spot the smallest possible. Let it dry and cut the plate along the central line.

5. Electrophoresis: Fill the chambers with the appropriate buffer (pH 1.9 or pH 8.9). Connect the apparatus to a water supply cooling unit. Spread 5 ml of the electrophoresis buffer on the surface of the cooling unit, then place the dry TLE plate. Wet the TLE plate slowly using a piece of Whatman prewetted in the electrophoresis buffer. Start from the periphery of the TLE plate, to finally wet the area of the spot, letting the buffer move in towards the spot from all directions at the same time to avoid sample streaking.

6. Prepare the wicks: cut and wet in buffer two 20 cm × 10 cm pieces of Whatman paper to connect the plate and the two chambers. Connect the power supply and run 27 min at 1000 V (pH 8.9) or 25 min at 1000 V (pH 1.9). Allow to dry completely afterwards.

7. Second-dimension chromatography should be run in a chemical hood. Make the chromatography buffer the day before and place 100 ml in the chromatography tank to equilibrate.

8. Place each half of the TLE plate resting on the bottom of the tank with the dyes at the bottom (avoid direct contact of the sample and the buffer). Let the buffer and the dye move up. Run the chromatograph until the front is about 0.5 cm from the top.

9. Remove the plate from the tank, dry in the hood (you can use a hair dryer), and expose it. If starting with 1000 cpm, expose for two days or use the phosphoimager. To know which peptide is the tryptic peptide that includes the autophosphorylation site (if not previously described) you can extract the heavily phosphorylated peptides from the TLE and sequence, or analyze their comigration in TLE (or coelution in HPLC) of the corresponding peptides synthesized *in vitro*.

Solutions

- *V8 reaction buffer*: 0.1% SDS, 1 mM EDTA, 2.5 mM DTT, and 0.125 mM Tris, pH 6.8
- *TLE pH 8.9 buffer*: 1% $(NH_4)_2CO_3$ HPLC grade in ddH_2O, adjust to pH 8.9 with NH_4OH
- *TLE pH 1.9 buffer*: 150 ml HCl, 50 ml formic acid, 800 ml ddH_2O
- *Chromatography buffer*: For 100 ml: 37.5 ml n-butanol, 25 ml pyridine, 7.5 ml HCl, 30 ml ddH_2O
- *Dye mix*: 0.3% bromophenol blue, 0.3% xylene cyanol, 0.3% orange G, and 0.3% acid fuchsin.

Trouble shooting

- If peptide maps are affected by the presence of residual salts in the peptide sample, increase the number of washes of the sample.
- Load the TLE carefully to avoid removing cellulose. Make sure that the cooling unit works perfectly to avoid sample burning.

Protein kinase specificity analysis

General strategies

Most protein kinases affect cell metabolism by phosphorylating specific substrates in Ser/Thr or Tyr. We include below the protocols for analyzing the amino acid specificity and substrate specificity of protein kinase. For analysing the amino acid specificity of a protein kinase, the starting material can be *in vitro* autophosphorylated enzyme (from *in vitro* kinase assay), or phosphorylated specific substrate (from *in vitro* kinase assay), or either or both phosphorylated *in vivo* (from metabolic labeling assays).

For analysis of substrate specificity of a protein kinase the best preliminary assay is the *in vitro* kinase assay of the enzyme in the presence of the putative substrate (see above). After demonstrating that the kinase is able to phosphorylate the enzyme *in vitro*, the ability of the enzyme to phosphorylate substrate *in vivo* should be proved. The kinase should be 'activated' by triggering the appropriate receptors or, if possible, by introducing a mutation in it that renders the kinase constitutively active (see Cooper 1990). The substrate should appear in its phosphorylated form in cells only when the enzyme is active. This can be analyzed by immunoprecipitation from metabolically labeled cells. In addition, a physiologically specific substrate should be phosphorylated *in vitro* in the same site as *in vivo*. This can be analyzed by comparison of the peptide maps of *in vivo* and *in vitro* phosphorylated substrate (see above).

Materials

- HCl
- Cellulose TLE plates
- Phosphoamino acid loading buffer
- Ninhydrin
- Whatman paper
- Phosphoamino acid electrophoresis buffer

Procedures

Phosphoamino acid analysis in one dimension

1. Samples resolved by SDS-PAGE are excised from the gel.
2. Spots are first trypsinized as described for two-dimensional peptide mapping. The supernatants are lyophilized, resuspended in 100 μl of ddH$_2$O and re-lyophilized in microfuge tubes.

   Note:
   Hydrolysis can also be done directly in Immobilon for convenience (see Luo et al (1990) for details). Immobilon should not be preincubated with any blocking reagent or Abs. Immobilon (the appropriate excised fragment) should be wetted in 0.5 ml methanol for 0.5 min, in 0.5 ml of ddH$_2$O for 0.5 min, and in 200 μl of constant-boiling HCl for 90 min at 100°C. The subsequent steps are common.

3. Peptides are incubated with 100 μl of constant-boiling HCl at 100°C for 90 min.
4. HCl is subsequently removed by lyophilization followed by four cycles of vacuum centrifugation of the sample in the presence of 500 μl, 50 μl, and 25 μl of ddH$_2$O (at room temperature).
5. Samples are resuspended in 10 μl of one-dimensional phosphoamino acid loading buffer.
6. 1 μl of each sample is spotted on to a 20 cm × 20 cm plastic-backed cellulose thin-layer chromatography plate 3 cm from the edge.
7. The plate is moistened with wet Whatman 3MM dipped in pH 3.5 electrophoresis buffer. To wet the sample spot, the buffer is applied in a circle around the spot and allowed to diffuse evenly into the spot.
8. Electrophoresis is performed for 75 min at 800 V in a cooled TLE apparatus.
9. Dry the plate; visualize the standards with 0.2% ninhydrin in acetone; dry and expose.

Two-dimensional phosphoamino acid analysis

1. Prepare the samples as above but resuspend the sample in the first-dimension loading buffer.
2. Cut a 10 cm × 10 cm piece of a cellulose TLC plate.
3. Load 1 μl of the sample; let it dry; load on top 1 μl of ddH$_2$O containing 8 μg of phosphoserine, 8 μg of phosphothreonine, and 8 μg phosphotyrosine.
4. Wet the plate in the first-dimension buffer, allowing the buffer to wick into the spotted sample evenly from all directions simultaneously (as described above).
5. Electrophorese 20 min at 1500 V (amino acids run to the positive pole). Remove the plate from the apparatus and let it dry.
6. Wet the plate as above in the second-dimension buffer. The sample should be arranged in a line: allow the buffer to wick into the sample appropriately.

7. Electrophorese 18 min at 1300 V. Remove from the apparatus and dry. Visualize the standards with 0.2% ninhydrin in acetone; dry, and expose.

Solutions

- *1-D-phosphoamino acid loading buffer*: 0.8 mg ml^{-1} phosphoserine, 0.8 mg ml^{-1} phosphothreonine, 0.8 mg ml^{-1} phosphotyrosine, and 0.001% xylene cyanol FF, orange G, and acid fuchsin.
- *1-D electrophoresis buffer*: pyridine–glacial acetic acid–water, 5:50:945 (v/v/v).
- *2-D electrophoresis buffer (first dimension)*: water–glacial acetic acid–88% formic acid, 800:150:50 (v/v/v), pH 1.9
- *2-D electrophoresis buffer (second dimension)*: water glacial acetic acid–pyridine, 945:50:5 (v/v/v), pH 3.5

Trouble shooting

- Cooling unit should be tested in advance: if not working well, it may burn the TLE. If the signal is low, try loading more sample and use a phosphoImager to expose the assay.
- HCl may react in the presence of alkali: make sure that samples containing alkali are not being dried in the speed vacuum at the same time.

Protein kinase overexpression

General strategies

Induction of protein kinases upon stimulation is done in the lymphoid cell that expresses it physiologically. However, for enzymological purposes, overexpression of the kinase is required for analyzing substrate specificity and carrying out structure/function analysis. Below we include the protocols related to lymphoid protein kinase overexpression in bacteria, baculovirus, and lymphoid cells.

Bacterial expression

Commentary

Bacterial expression is a fast and inexpensive method for protein expression. However, if active enzyme has to be obtained, several precautions must be taken. Gex vectors have been used with success in the expression of soluble and active lymphoid protein kinases. The protocols for the use of Gex vectors for protein production and purification have been described previously (Smith and Johnson 1988). Below we include some modifications for the expression of soluble and active bacterial protein kinases.

Materials

E. coli X90
E. coli JM109
LB and ampicillin
Triton X-100 lysis buffer
Tip sonicator

GSD

Glycerol

IPTG (isopropyl β-D-thiogalactoside)

Glutathione–agarose beads.

Procedures

Modification of the procedure for the expression of protein-kinases using Gex vectors

1. Subclone the cDNA encoding the kinase under study in the selected Gex vector. Use *E. coli* JM109 bacteria for the cloning process.
2. To induce the protein production, transform *E. coli* X90 (Carrera 1993) and grow three different colonies overnight at 37°C (10 ml of LB plus ampicillin).
3. Dilute 1 ml aliquots 1:10 in LB plus ampicillin and grow for 1 h (shaking at 37°C). Add 0, 0.1, and 1 mM IPTG to a series of aliquots. Incubate 3 h shaking at 37°C.
4. Extract the enzyme from the bacteria: spin the bacterial culture and resuspend it in 500 µl Triton X-100 lysis buffer (see recipe above) including phosphatase and protease inhibitors. Sonicate the cells in Triton X-100 buffer at 4°C using a tip sonicator (try two rounds of sonication with 10 s pulses, 5 min apart, at a duty cycle of 70% and output control at medium) (See Note a). If the signal is low, try to load more sample and use a phosphoImager to expose the assay.
5. Spin at $13 \times 10^3 g$ for 30 min at 4°C. Supernatant contains the soluble enzyme and the pellet contains the insoluble fraction. To solubilize the enzyme present in the pellet, boil it in 500 µl GSD (see Note b).

6. The fraction containing the soluble enzyme can be analyzed immediately or stored at –80°C in aliquots in the presence of 10% glycerol.

Notes
(a) Sonication should be optimized in each individual apparatus. An alternative protocol (Crews et al 1991) for protein solubilization without sonication can also be tried in parallel.
(b) The sample contains genomic DNA that makes manipulation difficult. To fragment genomic DNA, boil and freeze the sample approximately five times until viscosity decreases.

Solubility test for protein kinases expressed in bacteria

1. Take 20 µl of the Triton X-100 extract (soluble fraction), add 10 µl GSD, and boil 2 min. Load on to SDS-PAGE. Take 20 µl of the GSD containing the Triton X-100 insoluble fraction.
2. Resolve the samples by SDS-PAGE and analyze by Coomassie blue staining. From the different conditions tested, the optimal conditions will yield maximum solubility (approximately 90% of the protein in the soluble fraction) in the different colonies.

Activity test for protein kinases expressed in bacteria

1. Mix in 1.5 ml tubes 100 µl of the different Triton X-100 extracts containing the soluble enzyme with 10 µl of glutathione–agarose beads. Rock the tubes 15 min at 4°C.
2. Spin at low speed (100g) in a microfuge; discard the supernatant.
3. Wash the precipitates twice in lysis buffer and four times in 0.5 M LiCl, 50 mM Tris pH 7.5. To wash, add 1 ml

of the washing buffer, vortex, spin in a microfuge at low speed, and discard the supernatant.

4. Wash the immunoprecipitate with 50 mM Tris pH 7.5.

5. Enzyme may be tested by kinase assay directly.

6. Enzyme may also be released from the GST and the beads (start in this case from 500 ml of bacterial culture, and avoid the presence of protease inhibitors). Resuspend the beads in 150 mM NaCl, 2.5 mM CaCl$_2$, and add 1:500 thrombin:fusion protein (1 μg of human plasma thrombin for approximately 500 μg of fusion protein). Incubate 1 h at 25°C. Thrombin present in the reaction (only 0.2% of the mix) can be separated by chromatography. Collect the supernatant and add 1 mM PMSF; dilute 50–100 times in PBS and concentrate by dialysis using Centricon filters.

7. Elution from the beads of only the fusion protein can be done using 1 volume of 50 mM Tris–HCl pH 8.0, 5 mM reduced glutathione (check that final pH is 7.5). Incubate 2 min at 4°C; collect the supernatant. Repeat this step. Dilute 50–100 times in PBS and concentrate by dialysis using Centricon filters. Alternatively, treat directly with thrombin to release only the enzyme (Note a).

8. Carry out the kinase assay as above.

9. After the kinase reaction, boil the samples in GSD. Prepare samples containing protein concentration controls (samples with 50 ng, 100 ng, 200 ng, and 500 ng BSA boiled in GSD) and resolve them in parallel by SDS-PAGE (Laemmli 1970).

10. Stain the gel using Coomassie blue, dry, expose, quantify moles of phosphate transferred into exogenous substrate per mole of enzyme (concentration estimated in comparison with DNA standards), and choose the conditions for enzyme preparation that yield the highest specific activity (Note b).

11. The optimal conditions can be applied for larger-scale preparations of purified enzyme and the resulting aliquots stored for subsequent analysis at –80°C in 50 mM Tris, pH 7.5, 10% glycerol, 0.005% Triton X-100.

Notes
(a) At least for pp56^{lck} both bacterial Gex-protein kinase and the free protein kinase displayed similar specific activities (Carrera et al 1993).
(b) If possible, the purified bacterial enzyme should be compared with enzyme purified from eukaryotic cells and should display similar specific activity. Samples are useful for western blot enzymatic determinations and for substrate specificity analysis.

Trouble shooting

- To avoid protein kinase denaturation, keep the enzyme at 4°C and store with 10% glycerol at –80°C; thaw only once.
- Strong induction conditions often result in the agglutination–denaturation of the fusion protein in bacterial inclusion bodies. Avoid strong induction conditions. To make them mild, try lower temperatures, low IPTG concentration (or no IPTG), short incubation times for induction, and slower conditions for growing the bacterial culture.
- If, upon purification, the sample is contaminated with bacterial proteins, increase the washes and include two washes with 0.1% SDS–Tris 50 mM pH 7.5.
- If, during the washing protocol, the glutathione beads attach to the Eppendorf tube, add 0.005% Triton X-100 to the washing buffers.

Baculoviral expression

The baculoviral expression system utilizes a helper-independent virus (*Autographa californica* nuclear polyhedrosis virus) which can be grown to high titers in cells adapted to spinner culture. The transduced protein can represent up to 1% of the total protein in infected cells, being a good source for enzyme purification and kinetic studies. The system also facilitates the study of interactions between two proteins by coinfection with the appropriate second virus. Finally, this system represents a eukaryotic expression system and thus utilizes many of the lymphoid protein modification, processing, and transport mechanisms. The protocols for the use of baculoviral techniques were described and detailed by Summers and Smith (1987). The use of the new generation of baculoviral vectors (commercially available) that allow selection of recombinants using a reporter gene is recommended.

Overexpression on lymphoid cells

Materials

Electroporation cuvettes
Ultracentrifuge
CsCl
RPMI–10%FCS
G418

Electroporation of lymphoid cell lines for transient and stable expression of protein kinases

1. Subclone the cDNA encoding the lymphoid kinase in an appropriate expression vector (pEF BOS is recommended for transient expression, (Mizushima and Nagata 1990) and pSFFV-Neo is recommended for stable expression (Fuhlgrigge 1988)).
2. Make a large preparation of DNA by cesium chloride gradient. Resuspend pure DNA in TE. Quantify DNA concentration.
3. Prepare 10^7 cells per sample, resuspended in 1 ml of RPMI–10% FCS at room temperature.
4. Put $50\,\mu g$ of the cDNA encoding the protein in a sterile Eppendorf tube.
5. Add 1 ml of RPMI containing 10^7 cells to the Eppendorf with the cDNA. Mix well.
6. Transfer to the electroporation cuvette (0.4 cm electrode, gap 50).
7. For the first experiment electroporate at $960\,\mu Fa$ and different voltages, for example, 260, 280, 300, 310, 320, 330, and 340 V, to optimize the best transfection conditions for a given cell. Immediately after electroporation, transfer to 5 ml of RPMI–10%FCS. Incubate 24 h at 37°C, 5%CO_2.
8. For transient expression (see Note a), collect the cells that are ready for analysis of the expression of the electroporated protein kinase (see methods above).
9. For stable expression, collect the cells and resuspend them at 10^6 cells ml^{-1} in RPMI containing the appropriate concentration of G418 (the minimum dose of G418 that kills 100% of the cells in 7–10 days, previously optimized). Place 0.5 ml per well in 24-well plates (2 ml capacity). Add 0.5 ml of RPMI–10%FCS containing the appropriate concentration of G418 every 5 days. Colonies should grow in 2–4 weeks.

Note
(a) Transient expression can be used to co-electroporate the cDNA encoding your protein and a reporter

cDNA; this has been used successfully for CAT assays (Woodrow 1993).

Trouble shooting

- Electroporation efficiency decreases if cells are not 100% viable (check the culture; it should be in exponential growth). Electroporation efficiency also decreases at low DNA concentrations, if small amounts of specific DNA must be used complement it with carrier (irrelevant) DNA.
- If cell viability upon electroporation is low, the DNA may still have bacterial contaminants (make a second CsCl gradient), or may contain CsCl (wash the pellet of DNA after CsCl gradient with 70% cold EtOH).

References

Carrera AC, Sanchez-Madrid F, Lopez-Botet M, Benabeu C, O de Landazuri M (1987) Involvement of the Cd4 molecule in a post-activation event on T-cell proliferation. Eur J Immunol 17: 179–186.

Carrera AC, Alexandrov K, Roberts TM (1993) The conserved Lys of the catalytic domain of protein kinases is actively involved in the phosphate transfer reaction and not required for anchoring ATP Proc Natl Acad Sci USA 90: 442–446.

Casnellie JE (1991) Assay for protein kinases using peptides with basis residues for phosphocellulose binding. Methods Enzymol 200: 115–120.

Cooper JA (1990) The src family of protein kinases. In Kemp B, Alewood PF, eds. Peptides and Protein Phosphorylation, CRC Press, Boca Raton, p. 85.

Crews CM, Alessandrini AA, Erickson RL (1991) Mouse Erk-1 gene product is a serine/threonine protein kinase that has the potential to phosphorylate Tyr. Proc Natl Acad Sci USA 88: 8854–8859.

Eisenthal R, Cornish-Bowden A (1974) The direct linear plot. Biochem J 139: 715–720.

Fraker PJ, Speck JC (1978) Protein and cell membrane iodinations with a sparingly soluble chloroamine. Biochem Biophys Res Commun 80: 849–857.

Fuhlgrigge RC, Fine SM, Unanue ER, Chaplin DD (1988) Expression of membrane IL-1 by fibroblast transfected with murine pro-interleukin-1α cDNA. Proc Natl Acad Sci USA 85: 5649–5653.

Laemmli UK (1970) Cleavage of structural proteins during the assembly of the head of bacteriophage T4. Nature 227: 680–685.

Lineweaver H, Burk D (1934) The determination of enzyme dissociation constants. J Am Chem Soc 56: 658–670.

Luo K, Hurley TR, Sefton BM (1990) Transfer of proteins to membranes facilitates both cyanogen bromide cleavage and two-dimensional proteolytic mapping. Oncogene 5: 921–923.

Mizushima S, Nagata S (1990) pEF-BOS, a powerful mammalian vector. Nucleic Acids Res 18: 5322–5326.

Morrison DK, Kaplan DR, Rapp U, Roberts TM (1988) Signal transduction from membrane to cytoplasm: growth factors and membrane-bound oncogene product increase Raf-I phosphorylation and associated protein kinase activity. Proc Natl Acad Sci USA 85: 8855–8859.

Smith DB, Johnson KS (1988) Single step purification of polypeptides expressed in Escherichia coli as fusions with glutathione S transferase. Gene 67: 31–40.

Strugill TW, Ray LB, Erikson E, Maller JL (1988) Insulin-stimulated MAP-2 kinase phosphorylates and activates ribosomal S6 kinase II. Nature 334: 715–718.

Summers MD, Smith GE (1987) A manual of methods for baculovirus vectors and cloned insect cell culture procedures. Texas Agricultural Experimental Station Bulletin 1555, Texas A.M. University College Station, TX.

Woodrow MA, Rayter S, Downward J, Cantrell DA (1993) p21ras function is important for T-cell antigen receptor and protein kinase C regulation of nuclear factor of activated T-cells. J Immunol 150: 3853–3861.

Whole mount *in situ* hybridization of mouse embryos

14.6

José Luis de la Pompa[1,2]
Vincent Aguirre[3]
Tak W. Mak[1,2]
José Carlos Gutiérrez-Ramos[3]

[1]*The Amgen Institute,*
[2]*Ontario Cancer Institute, University of Toronto, Ontario, Canada*
[3]*The Center for Blood Research, Inc. and the Department of Genetics, Harvard Medical School, Boston, Massachusetts, USA*

TABLE OF CONTENTS

Immunology Methods Manual
ISBN 0–12–442712–X

Introduction

The analysis of the spatiotemporal expression pattern of potentially developmentally relevant genes is one of the first steps in understanding their role. Whole mount *in situ* hybridization is a rapid and convenient method, introduced in *Drosophila* by Tautz and Pfeifle (1989); its potential is not restricted to strict developmental studies, it may also be used to analyze the expression of molecules involved in the commitment and differentiation of the immune system (Barcena et al 1991).

Hematopoietic cell development begins in the mouse embryo at approximately day 7 of gestation and is thought to originate in the yolk sac and proceed to the fetal liver before becoming resident in the adult bone marrow. The first visible differentiated hematopoietic cells in the conceptus are of the erythroid lineage, which appear in the 7.5 day postcoitum (dpc) yolk sac (Russell and Bernstein 1966) and later at 9 dpc in the fetal liver (Johnson and Jones 1973). Multilineage erythroid/myeloid progenitors have also been found in 8 dpc yolk sac (Moore and Metcalf 1970). Recently, the aorta–gonad–mesonephros (AGM) region of the embryo has been demonstrated to be a novel site of hematopoiesis (7–10 dpc) as well, containing multilineage progenitors including those of the lymphoid lineage (Medvinsky et al 1993; Godin et al 1993). The studies of hematopoietic progenitor activity in post-implantation and early midgestation embryos are sparse, either because the techniques for the identification of such cells *in vivo* are technically difficult or because such activity is limited. If these limitations can be suitably overcome, the molecular characterization of these progenitor cells *in situ* could provide information about their location and migration patterns, the regulation of the genes they express, and possibly upstream regulatory genes whose expression could be integral for their commitment and differentiation.

Over time the initial procedure of whole mount hybridization has been nicely adapted by a number of laboratories to other experimental systems such as *Xenopus* (Hemmati-Brivanlou et al 1990), chick (Goulding et al 1993), and mouse (Conlon and Rossant 1992; Wilkinson and Nieto 1993; Conlon and Herrmann 1993). In mouse embryos, the procedure can be applied ideally between 7.5 and 10.5 dpc. At later gestational stages, owing to problems with penetrance reagents, it is more convenient to initially dissect the organ of interest (i.e., fetal liver or fetal thymus) and subsequently perform *in situ* analysis.

Briefly, the method involves the following steps: (1) production of a nucleic acid probe labeled with a uridinylate moiety conjugated to the steroid digoxigenin (Boehringer Mannheim); (2) fixation and preparation of embryo or tissue; (3) hybridization of probe to embryo or tissue, and washing to remove unhybridized probe; and (4) visualization of probe by an immunohistochemical reaction using an antidigoxigenin antibody coupled to alkaline phosphatase (Boehringer Mannheim).

The study of RNA distribution in the whole embryo or organ has obvious advantages over procedures using sectioned material. The most noteworthy is that the overall pattern of hybridization is immediately apparent and small groups of cells that express the gene of interest can easily be identified. We present a nonradioactive whole mount *in situ* hybridization method that we have successfully used for post-implantation mouse embryos and organs.

Materials

Care must be taken to avoid degradation of cellular RNA. To prevent RNAase contamination, we currently treat solutions with 0.1% diethylpyrocarbonate (DEPC) and autoclave them. In addition, all glassware required prior to hybridization is baked.

General reagents

Solutions marked with an asterisk (*) should be RNAase-free. Protein-free solutions should be autoclaved after treatment with DEPC. Protein-containing solutions should be made with DEPC-treated water. Unless otherwise stated, solutions are stored at room temperature. In specific cases, the company and/or catalog number of the product is provided.

- BCIP:
  5-bromo-4-chloro-3-indolyl-phosphate (X-phosphate) 50 mg ml^{-1} solution in dimethylformamide (Boehringer Mannheim cat no. 1383 221). Store at −20°C.
- 10% Blocking reagent* (Boehringer Mannheim cat no. 1096176). Store at −20°C.
- 10% BSA (Sigma). Store at −20°C.
- 10% CHAPS* (Sigma)
- 0.25 M DTT* (Stratagene, Boehringer Mannheim). Store at −20°C.
- 0.4 M EDTA (pH 8.0)*
- Formamide* (Gibco, Stratagene). Store at −20°C.
- 200 mg ml^{-1} glycine*. Store at −20°C.
- 50 mg ml^{-1} heparin*. Store at −20°C.
- 30% hydrogen peroxide (Sigma). Store at 4°C.
- 4 M LiCl*
- 1 M MgCl$_2$*
- NBT: 4-nitroblue tetrazolium chloride. 100 mg ml^{-1} solution in dimethylformamide, 70% (v/v) (Boehringer Mannheim cat. no. 1383 213). Store at −20°C.
- Normal sheep serum (Cedar Lane, Gibco). Store at 4°C.
- NTMT: 100 mM NaCl, 100 mM Tris–HCl pH 9.5, 50 mM MgCl$_2$, 0.1% Tween 20. Prepare fresh just before use.
- 10× Nucleotide mix* (Boehringer Mannheim): 10 mM ATP, 10 mM CTP, 10 mM GTP, 6.5 mM UTP, 3.5 mM DIG-UTP in Tris–HCl pH 7.5.
- 10× PBS*
- 20 mg ml^{-1} proteinase K*. Store at −20°C.
- 20× SSC, pH 7.0*
- 100 mM spermidine* (Sigma). Store at −20°C.
- 10% Tween 20* (Sigma).
- 10× Transcription buffer* (Boehringer Mannheim): 0.4 mM Tris–HCl, pH 8.0; 60 mM MgCl$_2$; 100 mM DTT; 20 mM spermidine. Store at −20°C.
- 10 mg ml^{-1} yeast tRNA* (Sigma cat no. R-9001). Store at −20°C.
- 2 M Tris–HCl, pH 8.0*
- 2 M Tris–HCl, pH 9.5.

Fixatives, aldehydes, and solvents

- Paraformaldehyde (Sigma). Store at 4°C.
- 25% glutaraldehyde (Sigma). Store in aliquots at −20°C.
- Ethanol and methanol (HPLC grade)
- Xylene (histological grade).

Methods

Synthesis of digoxigenin-labeled RNA probes

Single-stranded RNA probes labeled with digoxigenin–UTP are synthesized from DNA template as indicated by the manufacturer (Boehringer Mannheim).

1. Mix the following reagents in the following order at room temperature:

Linearized DNA template ($1\,\mu g\,ul^{-1}$)	$1\,\mu l$
Sterile distilled water (DEPC-treated!)	$11\,\mu l$
10× Transcription buffer	$2\,\mu l$
0.25 M DTT	$1\,\mu l$
10× Nucleotide mix	$2\,\mu l$
Placental ribonuclease inhibitor ($40\,U\,\mu l^{-1}$) (Note a)	$1\,\mu l$
RNA polymerase ($20\,U\,\mu l^{-1}$) (Note a)	$2\,\mu l$

2. Incubate at 37°C for 2 h.
3. Stop the reaction by adding $2\,\mu l$ of 0.4 M EDTA (pH 8.0).
4. The quality of the transcription can be checked by taking a $1\,\mu l$ aliquot and running it on a 1% agarose gel. An RNA band approximately 10 times more intense than the template DNA should be seen, indicating that about $10\,\mu g$ of probe has been synthesized.
5. Precipitate probe by adding $2.5\,\mu l$ of 4 M LiCl and $75\,\mu l$ of 100% ethanol, and incubate at –20°C for 2 h or at –70°C for 30 min. Spin in a microcentrifuge at 4°C for 30 min. Resuspend the pellet in $25\,\mu l$ of water, add $2.5\,\mu l$ of 4 M LiCl and $75\,\mu l$ of 100% ethanol and precipitate again.

Spin again, and wash the pellet twice with 70% ethanol, and air dry.

6. Dissolve the pellet in $200\,\mu l$ of hybridization solution (Note b) at 37°C for 10 min. Store at –20°C. Assuming that $10\,\mu g$ of RNA has been obtained from each μg of template used, the approximate probe concentration would be $0.05\,\mu g\,\mu l^{-1}$.

Notes
(a) From, e.g., Stratagene or Boehringer Mannheim.
(b) Hybridization solution*: 50% formamide, 5× SSC, 2% blocking reagent, 0.1% Tween 20, 0.5% CHAPS, $50\,\mu g\,ml^{-1}$ yeast tRNA, 5 mM EDTA (pH 8.0), $50\,\mu g\,ml^{-1}$ heparin.

Fixation and pretreatment of embryos

1. Dissect out the embryos in PBT (PBS with 0.1% Tween 20). In embryos older than 8.5 dpc, punch a hole in the anterior neuropore, to avoid the trapping of reagents.
2. Fix groups of 10 embryos (8.5 dpc) in 15 ml of freshly prepared 4% paraformaldehyde (see Note a) in PBS, overnight at 4°C with rocking.
3. Wash the embryos twice for 10 min each in PBT at 4°C.
4. Wash for 10 min each with 25%, 50%, 75% methanol–PBT, then twice with 100% methanol at 4°C. Embryos can be stored at this stage at –20°C for 2–3 weeks.
5. Rehydrate by taking the embryos though the methanol series in reverse, 10 min each at room temperature, and

then wash three times with PBT.

6. Bleach the embryos with 6% hydrogen peroxide in PBT for 30 min at room temperature with rocking. Wash five times with PBT.

7. Transfer the embryos to 2 ml screwcap tubes (e.g., Nalgene freezing vials) Treat with 20 μg ml^{-1} proteinase K in PBT for 4 min at room temperature with rocking (for 8.5 dpc embryos).

8. Wash with freshly prepared 2 mg ml^{-1} glycine in PBT and then twice with PBT for 5 min each. Special care must be taken because embryos are very fragile at this stage until they are refixed.

9. Refix the embryos with 0.2% glutaraldehyde–4% paraformaldehyde in PBS at room temperature for 20 min with rocking.

10. Wash twice with PBT.

11. Add 2 ml of hybridization solution and incubate at 65°C with rocking, for at least 3 h. Embryos can be stored in this mix at –20°C for weeks.

Note

(a) 4% Paraformaldehyde: Dissolve 4 g paraformaldehyde in 60 ml calcium and magnesium-free PBS* at 65°C, filtered through 0.45 μm filter unit, adjust to 100 ml and cool on ice. Use freshly prepared.

0.5–2 μg ml^{-1}.

2. Incubate at 65°C overnight.

3. Wash the embryos with 2 ml volumes of the following solutions, 5 min each at 65°C: 100% hybridization solution; 75% hybridization solution (25% 2× SSC); 50% hybridization solution (50% 2× SSC); 25% hybridization solution (75% 2× SSC).

4. Wash twice for 30 min each with 2× SSC, 0.1% CHAPS at 65°C.

5. Wash with 2× SSC, 0.1% CHAPS, twice for 10 min at room temperature.

6. Wash with 0.2 × SSC, 0.1% CHAPS, twice for 30 min at 55°C.

7. Wash with PBS twice for 10 min at room temperature. Special care must be taken because the embryos become sticky in PBS and tend to attach to the pipette tip.

8. Wash with PBT for 5 min at room temperature.

Antibody incubation and immunohistochemical detection of probe

Antibody incubation and washings are carried out with gentle rocking.

Hybridization and post-hybridization washes of embryos

All the following hybridization and washing steps are performed with gentle rocking.

1. Replace the hybridization solution (see above) with 2 ml of fresh hybridization solution, and add probe to

1. Block embryos with 10% heat-inactivated goat serum (see Note a), 1% BSA in PBT, for 3 h at room temperature.

2. During this time, preabsorb the alkaline phosphatase-coupled antidigoxigenin antibody by diluting it to 1:2000 in 10% heat-inactivated goat serum, 1% BSA, and heat-inactivated embryo powder (Note b) in PBT. Rock gently at 4°C for 3 h. Spin at full speed in microfuge at 4°C for 10 min. Take the supernatant.

3. Remove the 10% serum from the

embryos, replace with the preabsorbed antibody and incubate overnight at 4°C.

4. Wash at least five times with 0.1% BSA in PBT for 1 h each at room temperature. Embryos older than 10 dpc should be washed in 10 ml volumes.
5. Wash twice with PBT for 30 min each at room temperature.
6. Wash at least three times with NTMT, 10 min each. A precipitate forms when embryos are transferred from the PBT to the NTMT; they should be washed until the NTMT solution is clear.
7. Start immunohistochemical reaction by incubating embryos in NTMT containing 4.5 µl NBT and 3.5 µl BCIP per ml. Protect embryos from light and rock them for the first 10 min. Subsequently, stand the embryos in the dark and check periodically until color develops to the desired extent.
8. Stop color reaction, washing with PBT three times, 10 min each. Fix the stained embryos with 4% paraformaldehyde in PBS overnight at 4°C with rocking.
9. For observation under a dissecting microscope, dehydrate the embryos through 25%, 50%, 75% methanol–PBT and 100% methanol, 10 min each. The dehydration intensifies the purple reaction products to dark blue. Rehydrate by taking the embryos through the reverse methanol series to PBT. Clear the embryos by passing them into 50%:50% glycerol–PBT. In this solution, embryos are ready for observation and photography.

Note
(a) Heat-inactivated goat serum: A 20% solution of goat serum in PBT is heated at 70°C for 30 min before use.
(b) Heat-inactivated embryo powder: A few mg of mouse embryo powder in 0.5 ml of PBT are heated at 70°C for

30 min before use.

Embryo powder preparation (Harlow and Lane 1988)

1. Homogenize 13.5-dpc embryos in a minimum volume of ice-cold PBS.
2. Add 4 volumes of ice-cold acetone and mix vigorously. Keep on ice for 30 min.
3. Centrifuge at 10,000g for 10 min, remove the supernatant and wash the pellet with ice-cold acetone and spin again.
4. Transfer the pellet to a piece of filter paper; spread out the pellet, rub it into a fine powder, and allow it to air dry at room temperature. Store at 4°C.

Clearing and embedding of embryos

Although the reaction product using NBT–BCIP is soluble in organic solvents such as ethanol and xylene, after overnight fixation in 4% paraformaldehyde the embryos can be quickly dehydrated and cleared for paraffin embedding.

1. Transfer the embryos to a glass container (e.g., scintillation vial) and dehydrate them through 25%, 50%, 75%, and 100% ethanol, 15 min each with gentle rocking.
2. Clear embryos in xylene, three times for 10 min each, with rocking. Embed in a 50%:50% xylene–wax (Paraplast) mix, 15 min, and three changes of paraffin wax (Paraplast) at 60°C,

60 min each. Place embryos into a plastic mold, orient them with a warmed needle under a dissecting scope, and allow the wax to set.

3. 6–10 μm sections are dewaxed through two changes of xylene, 10 min each, and mounted in Permount or Entellan (Merck).

In addition, the use of horseradish

peroxidase (HRP)-linked antidigoxigenin antibody may be considered. Although probably not as sensitive as alkaline phosphatase, it gives a brown or black reaction product that is stable to clearing agents.

Embryonic expression of CD34: an illustrative example

We have chosen CD34 as an example of an immunologically relevant gene whose expression pattern in the early embryo should provide valuable information about the processes involved.

The cell surface antigen CD34 is selectively expressed on myeloid and lymphoid progenitor cells as well as hematopoietic stem cells (Shaper and Civin 1984). This stage-specific expression suggests a potential regulatory function for CD34 during hematopoiesis. CD34 is also expressed on vascular endothelial cells and appears to be the cell surface ligand for the L-selectin present on the surface of leukocytes (Baumhueter et al 1993). Thus, the interest in studying the pattern of expression of CD34 in the early embryo is twofold. First, it might identify the first hematopoietic progenitors as they are generated and determined from mesodermal cells during embryonic life. Second, its expression on endothelial cells or their progeni-

tors might reveal the establishment of leukocyte–endothelial interactions necessary for the colonization of lymphoid organs by progenitor cells. The study of CD34 expression by whole mount *in situ* hybridization should aid in the examination of two chief processes that occur in the mid-gestation embryo: hematopoiesis and angiogenesis.

Whole mount *in situ* hybridization analysis of CD34 expression in 9.5 dpc mouse embryos shows a strong signal in endothelial cells and the yolk sac. Stainings of the 9.5 dpc yolk sac show bright staining of cells with flattened morphology consistent with blood vessel endothelium while only random hematopoietic cells and surrounding mesenchymal cells show signal (not shown). The 9.5 dpc embryo body shows a very specific pattern of CD34 RNA staining in blood vessels and periaortic areas (Fig. 14.6.1).

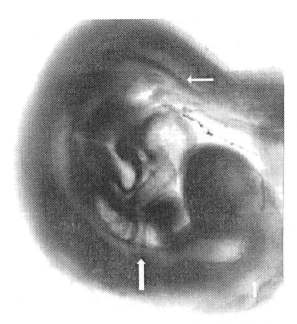

Figure 14.6.1 Whole mount *in situ* hybridization analysis of CD34 expression at 9.5 dpc of murine gestation. A 20–25 somite embryo is viewed laterally. Staining is seen lining the left first branchial arch as it leaves the ventral aortic sac and extends to the dorsal aorta (long horizontal arrow). Signal is also seen caudally in the dorsal aorta (vertical arrow) and rostrally at the base of the developing neural tube (short horizontal arrow).

Comments and trouble shooting

Levels of expression of immunologically relevant genes

Some developmentally regulated genes shown to have key roles in early mouse embryonic embryogenesis have been detected easily by whole mount *in situ* hybridization owing to their high level of expression in a spatially restricted domain. It is conceivable, however, that the detection of the expression of some of the immunologically relevant genes to be assayed by the readers of this manual (e.g., cytokines, cytokine receptors, adhesion molecules and their ligands, and so on) may be hampered by low expression levels of the corresponding genes. Thus, it seems reasonable first to evaluate expression levels using a highly sensitive technique such as quantitative RT-PCR (see Section 5). This could help in the determination of whole mount *in situ* hybridization as a suitable technique.

Determination of specific versus nonspecific signal can be difficult for genes expressed in a low, albeit detectable, pattern, even if that expression is ubiquitous. Hence, the evaluation of the signal will depend largely on the experience acquired with that particular probe and its corresponding sense negative control.

Manipulation of yolk sacs

As described in the Introduction, the yolk sac is a major site of hematopoietic activity in the early embryo and, therefore, likely to be analyzed frequently. Manipulation of dissected yolk sacs is not an easy task owing to their fragility. To preserve this fragile tissue, we have observed that it is better leave the entire amnion attached to the embryo proper throughout the entire procedure. Prior to visualization, the yolk sac may be dissected away for isolated analysis. However, presence of the entire amnion presents a problem because of the protrusion of substantial amounts of membrane from the embryo proper. Therefore, it is best to process such specimens in individual containers from dissection through visualization to avoid entanglement and possible damage.

Technical tips that might help to increase the signal/noise ratio

- Fixation: We have not found differences in signal or background when varying fixation time of embryos from 2 h to overnight.
- The proteinase K treatment is essential and its length depends on the age of the embryos and may vary for each batch of proteinase. For example, for 8.5 and 9.5 dpc embryos we have achieved good results with digestion times of 3 and

5 min, respectively.
- After fixation of proteinase K-treated embryos, it is not essential to treat embryos with the reducing agent sodium borohydride (Conlon and Hermann 1993).
- The detergent CHAPS was initially introduced by Harland (1991) to prevent nonspecific stickiness of the probe via the digoxigenin moiety and we have obtained good results using it.
- Hybridization temperatures of 55–65°C yield good results, although, in general, 65°C gives a clearer background and still a strong signal.
- Ribonuclease treatment is omitted provided that the probe is very specific. It seems to decrease specific signal and make the embryos sticky.
- The use of 2 mM levamisole in the NTMT washes to inhibit endogenous phosphatases does not seem to affect background and is omitted.

Other recommendations

Bleaching of embryos in hydrogen peroxide is reduced to a minimum and, in certain cases, may be unnecessary. After fixation and washing, we have stored embryos in methanol at –20°C (Conlon and Hermann 1993) or processed them into hybridization solution for storing at –20°C as well (Wilkinson and Nieto 1993) without detecting major signal differences. They may be stored for 3–4 weeks.

Acknowledgments

We thank Ron Conlon for introducing us to the mouse whole mount *in situ* technique and to Angela Nieto for her useful advice. Jose C. Gutiérrez-Ramos is the Amy C. Potter fellow.

References

Barcena A, Sanchez MJ, de la Pompa JL, Toribio ML, Kroemer G, Martinez-A C (1991) Involvement of the interleukin 4 pathway in the generation of functional gamma/delta T cells from human intrathymic pro-T cells. Proc Natl Acad Sci USA 88: 7689–7693.

Baumhueter S, Singer MS, Henzel W (1993) Binding of L-selectin to the vascular submucin CD34. Science 262: 436–438.

Conlon RA, Rossant J (1992) Exogenous retinoic acid rapidly induces anterior ectopic expression of murine *Hox-2* gene *in vivo*. Development 116: 357–358.

Conlon RA, Hermann BG (1993) Detection of messenger RNA by *in situ* hybridization to postimplantation embryo whole mounts. *Methods Enzymol* 225: 373–383.

Godin IE, Garcia-Porrero JA, Coutinho A, DieterlenLievre F, Marcos MAR (1993) Para-aortic splanchnopleura from early mouse embryos contains B1a cell progenitors. Nature 364: 67–70.

Goulding MD, Lumsden A, Gruss P (1993) Signals from the notochord and floor plate regulate the region-specific expression of two *Pax* genes in the developing spinal cord. Development 117: 1001–1016.

Harland RM (1991) *In situ* hybridization: an improved wholemount method for *Xenopus* embryos. Methods Cell Biol 36: 685–695.

Harlow E, Lane D (1988) Antibodies: A Laboratory Manual. Cold Spring Harbor Laboratory Press, Cold Spring Harbor, NY.

Hemmati-Brivanlou A, Frank D, Bolce MB, Brown BD, Sive HL, Harland RM (1990) Localization of specific mRNAs in *Xenopus* embryos by whole mount *in situ* hybridization. Development 110: 325–330.

Johnson GR, Jones RO (1973) Differentiation of the mammalian hepatic primordium *in vitro*. I. Morphogenesis and the onset of hematopoiesis. J Embryol Exp Morphol 30: 83–96.

Medvinsky AL, Samoylina NL, Muller AM, Dzierzak EA (1993) An early pre-liver intra-embryonic source of CFU-S in the developing mouse. Nature 364: 64–67.

Moore MAS, Metcalf D (1970) Ontogeny of the haematopoietic system: yolk sac origin of *in vivo* and *in vitro* colony forming cells in the developing mouse embryo. Br J Haematol 18: 279–296.

Russell ES, Bernstein SE (1966) Blood and blood formation. In Green EL, ed. Biology of the Laboratory Mouse, 2nd edn. McGraw-Hill, New York, pp. 351–372.

Shaper JH, Civin CL (1984) Antigenic analysis of johnson hematopoiesis. III. A hematopoietic progenitor cell surface antigen defined by a monoclonal antibody raised against KG-a cells. J Immunol 133: 157–165.

Tautz D, Pfeifle C (1989) A non-radioactive *in situ* hybridization method for the localization of specific RNAs in embryos reveals transcriptional control of the segmentation gene *hunchback*. Chromosoma 98: 81–85.

Wilkinson DG, Nieto MA (1993) Detection of messenger RNA by *in situ* hybridization to tissue sections and whole mounts. Methods Enzymol 225: 361–373.

Measurement of mRNA for cytokine and cytokine receptor genes by quantitative PCR

14.7

Gui-quan Jia
José Carlos Gutiérrez-Ramos

The Center for Blood Research, Inc. and the Department of Genetics, Harvard Medical School, Boston, Massachusetts, USA

TABLE OF CONTENTS

Immunology Methods Manual
ISBN 0–12–442712–X

Introduction

The development and functioning of leukocytes are controlled by a large number of cytokines (reviewed by Arai et al 1990; Nicola 1989). These mediators are expressed by many different cell types and can exert restricted and pleiotropic effects on various cell lineages.

Detection and quantification of cytokines and their receptors at the protein level could be hampered by low expression levels of their corresponding genes as well as lack of monoclonal antibodies (mAbs) for each cytokine or receptor subunit. In addition, the sensitivity of assays such as flow cytometry or various binding assays using receptor-specific antibodies or the ligand for the receptor itself is low (Chirmule et al 1991). A suitable alternative for such quantification is analysis of the expression of these genes at the mRNA level. However, standard approaches such as Northern blot and RNAase protection also suffer from lack of sensitivity under some circumstances. In addition, it is difficult to compare the results from different experiments and different research groups because mRNA levels are measured in relative terms. The ability to detect low levels of expression of a cytokine or its receptors is critical given that some of these molecules may require only a few ligand molecules to exert full biological function on a target cell.

This chapter describes a very sensitive and quantitative method for the measurement of mRNA molecules per microgram of RNA or per cell based on competitive PCR. This method should, however, be generally applicable to any new PCR-based system.

General strategies

Reverse transcription (RT)-PCR is an extremely powerful method for mRNA analysis. However, it can be difficult to obtain quantitative information with this technique. This is because the two sequential enzymatic steps involved in this method, the synthesis of a cDNA from an RNA template and the amplification of the cDNA by the polymerase chain reaction, are subject to variables such as enzyme activity, affinity for the substrate, etc. which can considerably affect amounts of product, resulting in artificial differences in the final measurements. The coamplification of a heterologous mRNA species (e.g. β-actin) with the mRNA from the target gene can control for these differences to some degree. However, because RNA levels are quantified in relative terms, it is still difficult to compare results obtained from different experiments. To overcome these limitations, the use of homologous internal DNA and RNA standards has become common practice. This strategy allows the control of variations in amplification efficiencies and determination of the absolute value of mRNA molecules for a family of genes. In this method, an exogenous RNA or DNA standard is added to the target sample and amplified simultaneously in a single PCR reaction mixture. This makes it possible to calculate the absolute level of target mRNA or cDNA present in the original sample. This method, named competitive or quantitative PCR, although first described by Wang and coworkers (Wang et al 1989), has been improved and applied to the measurement of several gene families such as cytokines, oncogenes, etc. (Bouaboula et al 1992; Kaashoek et al 1991; Scheuermann and Bauer 1993).

Competitive PCR takes place in a truly competitive fashion because the standard and target sequence(s) compete for the same primers and, therefore, for amplification. It circumvents many of the disadvantages of the other quantitative methods, allowing accurate quantification of cellular mRNA levels and measurements of relative changes in amount of mRNA.

In this approach, illustrated in Fig. 14.7.1, a synthetic RNA standard shares primer binding sites with the target cellular RNA, but has different internal sequences or 'stuffer'. A series of dilution of the standard RNA made *in vitro* are added to a constant amount of sample RNA

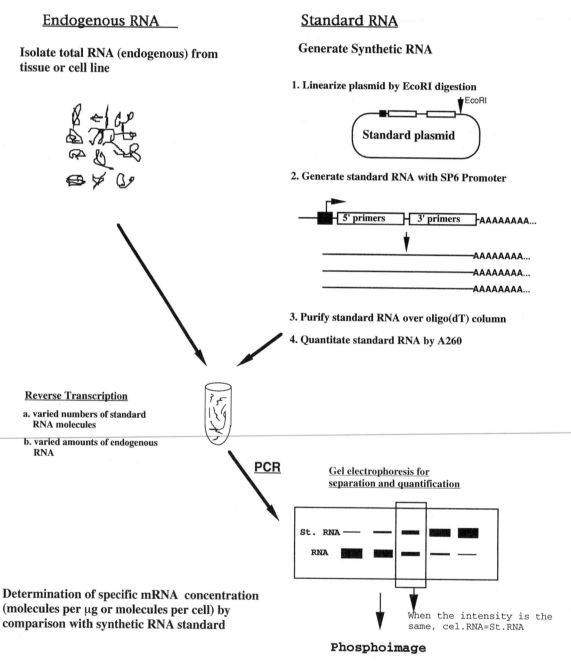

Endogenous RNA

Isolate total RNA (endogenous) from tissue or cell line

Standard RNA

Generate Synthetic RNA

1. Linearize plasmid by EcoRI digestion

Standard plasmid

2. Generate standard RNA with SP6 Promoter

5' primers 3' primers AAAAAAAA...
AAAAAAAA...
AAAAAAAA...
AAAAAAAA...

3. Purify standard RNA over oligo(dT) column

4. Quantitate standard RNA by A260

Reverse Transcription

a. varied numbers of standard RNA molecules

b. varied amounts of endogenous RNA

PCR

Gel electrophoresis for separation and quantification

St. RNA

RNA

Determination of specific mRNA concentration (molecules per μg or molecules per cell) by comparison with synthetic RNA standard

When the intensity is the same, cel.RNA=St.RNA

Phosphoimage

Figure 14.7.1 Schematic representation of PCR-based mRNA quantification procedure.

and reverse transcribed. These reverse transcription materials are used for a series of PCR reactions. Following PCR, the amplification products are analyzed by gel electrophoresis and the amounts of products generated by the standard and the target are determined for each individual reaction. The size difference between standard and cellular RNA can be distinguished by gel electrophoresis (shown in Fig. 14.7.1). The concentration of target RNA is measured by titrating the cellular RNA versus different concentrations of standard RNA. Careful kinetic analysis to ensure identical amplification efficiency for cellular and standard RNA is critical for accurate measurement.

Generation of standard RNA

The internal standards that can be used for quantitative PCR are either identical in sequence and different in length to the target RNA (homologous standards) or completely different in sequence but sharing the target motifs for the amplification primers (heterologous standards). Heterologous standards may be preferable since, when homologous competitive fragments are used as standards, it is possible that heteroduplexes are formed between sequences in the standard and the target RNA.

Heterologous standards can be constructed simply and efficiently by assembling, in the same vector, all the primer pairs to be used in the PCR experiments. In this way, primer pairs not used in a given reaction function as a 'stuffer' in the heterologous competitive standards (Bouaboula et al 1992; Scheuermann and Bauer 1993; Legoux et al 1992). We will use the generation of a cytokine receptor standard RNA vector as an example of how to design and construct a template for *in vitro* synthesis of standard RNA.

Construction of the PCR standard plasmid pSPCR1

The plasmid pSPCR1 contains target sequences for PCR primer pairs that are specific for 22 different cytokine receptor mRNAs. The structure of pSPCR1 is depicted in Fig. 14.7.2. All the primer sequences are located between an SP6 promoter and a poly(A) stretch so that the pSPCR1 plasmid can be used as a template for *in vitro* transcription and subsequent generation of synthetic standard RNA.

Design of oligonucleotide primers

The following issues should be considered in designing an oligonucleotide primer.

1. Twenty-nucleotide sequences present in different exons of a known mouse cytokine receptor gene seem to be a good choice. They should have a moderate G+C content (40–60%) and should not have an obvious secondary structure.

   Note: Twenty nucleotides is long enough for a primer. To discriminate the PCR products that have arisen from RNA and DNA, it is necessary to design the primer pairs in such a way that they are located in different exons.

2. The chosen target sequences for the primers should be selected so as not to contain complementary sequences, especially at the 3' ends. Usage of the appropriate software for oligo analysis

such as the Oligo4.0 program (National Biosciences, Inc.) would help in the design.

3. The difference in the sizes of the amplified fragments using standard RNA and endogenous RNA as templates should be around 100–150 bp.

    Note: The difference in size between standard and endogenous RNA should be easily distinguishable by electrophoresis, while being as small as possible to eliminate concern about differences in electrophoretic behavior due to size differences.

4. To avoid the inclusion of sequences in the vector that do not prime efficiently, or give unexpected patterns following the PCR reaction, it is better to test the primer pairs with RNA preparations known to contain the mRNA species of interest before they are included in the vector that will be used as a template for standard RNA synthesis.

Assembling of primer pairs into a template

pSPCR1 is constructed by sequential PCR reactions with overlapping primers (Scheuer-mann and Bauer 1993) (Fig. 14.7.3). The 5'

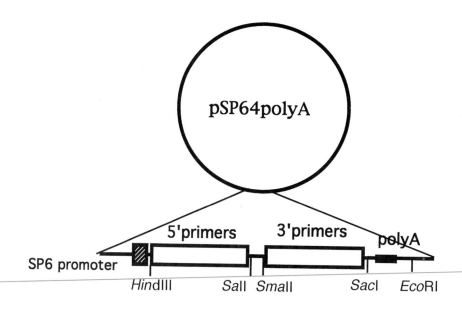

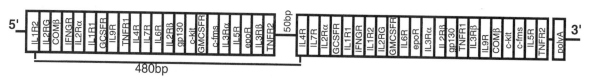

Figure 14.7.2 Structure of pSPCR1 plasmid used to generate the standard RNA. Plasmid pSPCR1 is constructed using overlapping synthetic oligonucleotides as described. The plasmid includes two 480 bp blocks of synthetic DNA which are integrated by an array of twenty-two 20-mers corresponding to the 5' and 3' PCR primer pairs specific for twenty-two different cytokine receptors and subunits. Distances, in base pairs, are measured from the start of SP6 polymerase transcription. The complete list of the genes included in pSPCR1 is shown.

primer and 3' primer blocks are generated independently using six overlapping 100-mer oligonucleotides which are used sequentially as templates for five steps of amplification.

Materials

- 10× PCR buffer II: 500 mM KCl, 100 mM Tris–HCl pH 8.3. Keep at –20°C; thaw completely and mix well before using.
- 10 mM dNTP: 10 mM dATP, 10 mM dTTP, 10 mM dGTP, 10 mM dCTP. Keep at –20°C; thaw completely and mix well before use.
- 25 mM MgCl₂; keep at –20°C; thaw completely and mix well before use.

Procedure

1. Twelve 100-mer oligonucleotides containing all primer pairs in the designed order are synthesized automatically using a DNA synthesizer.
2. Mix the following reactants on ice (50 µl final):

10× PCR buffer II	5 µl
10 mM dNTP × 4	1 µl ×4
25 mM MgCl₂	3 µl
Taq polymerase	2 units
H₂O to	50 µl (minus volume of oligonucleotides)

3. Add 10 pmol of each of the two 100-mer oligonucleotides (A and B in Fig. 14.7.3) to the above mixture and

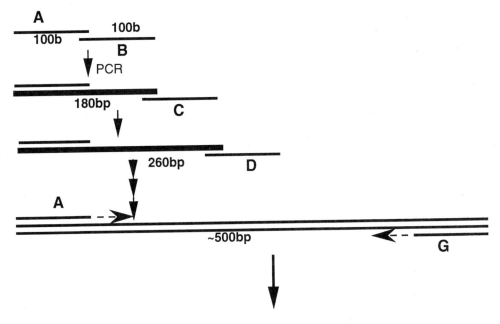

5' primers Block Synthesis

Isolate fragment and clone into pCRII

Figure 14.7.3 Construction of the inserts containing aligned target sequences for PCR primers. The synthetic insert is constructed using overlapping oligonucleotides following the protocol detailed in the text.

overlay the reaction mixture with mineral oil.

Note: Oligos A and B contain 20 complementary nucleotides at their 3' ends.

4. Denature for 5 min at 94°C.
5. Cycle twice on a Perkin Elmer DNA Thermal Cycler with the following parameters: denaturation at 94°C × 30 s, annealing at 45°C × 30 s, and extension at 72°C × 30 s.
6. Cycle 8 times with the following parameters: denaturation at 94°C × 30 s, annealing at 65°C × 30 s, and extension at 72°C × 30 s.

Note: After these 8 cycles, a 180 bp product with the combined sequence of Oligos A and B should have been generated.

7. Repeat step 2.
8. Add 10 pmol of oligonucleotide A and oligonucleotide C. These will be the PCR primers for step 7.

Note: Once again, the 20 nucleotides at the 3' end of oligonucleotide C are identical to the 20 nucleotides at the 5' end of oligonucleotide B.

9. Add 1 μl of PCR product from step 6 as template to the reaction mixture in step 8 and overlay the reaction mixture with mineral oil.
10. Repeat steps 4 and 5.
11. Repeat step 6 but with 5 cycles.

Note: This step results in an extension of 80 bp in the amplification product.

12. As performed in steps 8 and 9, 1 μl of the PCR product generated in the previous step is used as a template for the next PCR step in combination with oligonucleotides A and D (Fig. 14.7.3). When the last overlapping oligonucleotide is used, 20 PCR cycles should be performed to generate enough final product for the subcloning.

Note: The products from each PCR step should be analyzed by electrophoresis on a 1.5% agarose gel followed by ethidium bromide staining. If the background is high, the product should be purified from the preparative agarose gel by electroelution and used for the subsequent PCR reaction.

Cloning of the standard fragment into the vector

5' and 3' fragments containing the final primer blocks from the last step of PCR described above are purified from the agarose gel and subcloned into pCRII (Invitrogen Co., San Diego, CA, USA) then into pSP64poly(A) vector following general strategies.

1. Products (25 μl) from the last PCR step are analyzed by electrophoresis on a 1.2% agarose gel stained with ethidium bromide. The expected band (480 bp) is cut from the gel.

Note: There usually is a smear around the expected band. The band should be cut as sharply as possible to avoid the cloning of partial products.

2. DNA is purified from the gel slice and cloned into pCRII following a routine protocol described elsewhere in this manual.
3. Ten independent clones are picked; the size of their insert is verified by restriction enzyme analysis and they are sequenced.
4. After sequencing, the clone containing the least number of errors is chosen. Those errors due to mutations occuring in the middle of the primer sequence have to be corrected by site-directed mutagenesis and sequenced again to verify mutation reversion.

Note: These point mutations are presumably generated by errors generated by the *Taq* polymerase amplification. If the mutations are located in the 5' or 3' of the target sequence for a given primer, one can sometimes shorten the length of the primers in such a way that the mutation is not included within the annealing area and therefore does not affect the efficiency of amplification. If this does not work, one has to repair those mutations as well by site-directed mutagenesis.

5. Once the primer blocks have been corrected, they are cloned into pSP64polyA using unique restriction sites that were engineered at the ends of each primer block (shown in Fig. 14.7.2).

Synthesis, purification and quantification of pSPCR1 standard RNA

The synthetic standard RNA is generated *in vitro* by transcription driven from the SP6 promoter. Subsequently, the RNA is purified with oligo(dT) beads, taking advantage of the synthetic standard RNA carrying poly(A) at the 3' end.

Note: Reagents and solutions used throughout RNA procedures should be prepared using standard methods for handling RNA.

Materials

- 10× buffer (*in vitro* transcription): 0.4 M Tris-Cl, pH 7.5, 0.1 M MgCl$_2$, 50 mM DTT, 0.5 mg ml^{-1} BSA
- 10 mM NTPs: ATP, CTP, GTP, UTP

Note: NTPs (nucleoside triphosphates) are used for *in vitro* transcription, not dNTPs (deoxynucleotide triphosphates)

- SP6 RNA polymerase (Boehringer Mannheim, Germany)
- DNAase I, RNAase-free (Boehringer Mannheim, Germany)
- High-salt buffer (HSB): 400 mM LiCl, 10 mM Tris–HCl pH 8, 1 mM EDTA
- Low-salt buffer (LSB): 10 mM Tris–HCl, pH 8, 1 mM EDTA
- 3 M sodium acetate pH 4.8
- Oligo(dT)–cellulose beads (Pharmacia Biotech)

Procedure

1. To generate a template for *in vitro* RNA synthesis, the plasmid pSPCR1 is linearized by digestion with *Eco*RI (Fig. 14.7.1).
2. DNA is then routinely extracted with phenol–chloroform and EtOH-precipitated.
3. 2 μg of linearized plasmid DNA is used as template by adding it to the following mixture (20 μl final):

10 mM NTP × 4 (ATP, CTP, GTP, UTP)	2 μl × 4
10× buffer	2 μl
SP6 polymerase	20 units
H$_2$O to	20 μl

Incubate for 60 min at 37°C.

4. Add 20 units of DNAase I (RNAase-free) to the reaction mixture and incubate for 15 min at 37°C.
5. 2 μl of the product is analyzed by agarose gel electrophoresis to ensure that the standard RNA is of the correct size (1041 nucleotides).
6. The rest of the product is mixed with 0.2 ml of oligo(dT)–cellulose beads in HSB and incubated for 30 min on ice.

Note: Since the beads are stored in a different buffer that is not suited for the reaction, they should be pelleted by centrifugation and resuspended in the same volume of HSB immediately before use.

7. The slurry is then loaded into a 1 ml syringe-column.
8. The column is washed with 10 ml of HSB.
9. Four 100 μl aliquots of LSB are loaded subsequently over the column to elute the poly(A) RNA into a 1.5 ml Eppendorf tube.
10. The purified synthetic RNA is precipitated by adding 40 μl of 3 M sodium acetate (pH 4.8) and 1 ml of 100% ethanol and storage at –20°C overnight.
11. The RNA is then recovered by centrifugation at maximum speed in an Eppendorf centrifuge for 30 min at 4°C, rinsed with 80% ethanol, and resuspended in 15 μl of DEPC-treated water.

Note: To obtain an accurate measurement of the standard RNA, purification of poly(A) + RNA is performed to ensure that the *in vitro*-synthesized standard RNA is full length and free of plasmid DNA contamination.

12. The concentration of the standard RNA is measured as the absorbance at 260 nm and the number of RNA molecules is estimated according to the size of the standard RNA.

The length of the synthetic RNA molecules obtained using the plasmid pSPCR1 as a template is 1041 bp and therefore 1 μg of synthetic RNA is equivalent to 1.76×10^{12} molecules.

Preparation of cellular RNA

Successful RNA extraction depends on the rapid and thorough inactivation of RNAases. Various methods have been developed to this end, all of which employ conditions that are of necessity deleterious to protein activity.

Note: Reagents and solutions used for RNA procedure should be prepared using standard methods for handling RNA.

Small scale

For RNA extraction involving a small number of cells (fewer than 10^4 cells), or from single cells, RNA can be isolated by magnetic beads linked to oligo(dT)$_{20}$.

Large scale

For large-scale reactions, total RNA is isolated by the modified acid phenol procedure (Chomczynski and Sacchi 1987). This method works very well when RNA must be extracted from frozen tissue or cell pellets (see p. 1203).

Materials

- BioMag Oligo(dT)$_{20}$ (PerSeptive Diagnostics, Inc. Cambridge, MA, USA)
- Guanidine thiocyanate solution (GTS) without 0.5%(w/v) sarcosyl
- Binding buffer: 20 mM Tris (pH 8.0), 0.5 M NaCl, 0.1% sodium azide
- Wash buffer: 7 mM Tris (pH 8.0), 0.17 M NaCl, 0.1% sodium azide

- Magnetic separation unit

Procedure

1. Cell pellets (a known amount of cells) are resuspended in 20 µl of the guanidine thiocyanate solution and lysed by vigorous vortexing.
2. 40 µl of the binding buffer is added and incubated at 65°C for 5 min.
3. While the sample mixture is being heated, 5 µl of completely suspended Oligo(dT)$_{20}$ beads is aliquoted into a microcentrifuge tube.
4. This aliquot of beads is subjected to magnetic separation for 30 s using the units provided by the vendor and the supernatant is discarded. The beads are washed once with 10 µl of the binding buffer and left as a wet cake.
5. Once the sample mixture has been heated for 5 min, it is centrifuged at 14,000g for 5 min.
6. The supernatant obtained in step 5 is transferred into the tube containing the washed Oligo(dT)$_{20}$ beads from step 4. The mixture is then incubated at room temperature for 3 min.
7. The beads are recovered from the mixture by magnetic separation using the unit provided by the vendor.
8. The beads with the attached RNA are washed twice with 20 µl of the washing buffer.
9. Washed RNA is eluted twice from the beads with 10 µl of DEPC-treated water at 56°C for 3 min.
10. RNA is then concentrated by Speedvac.

Titration analysis

This entails measurement of the expression of the target gene in a particular cell type. The amount of endogenous mRNA for a particular gene is titrated versus known amounts of the synthetic RNA standard.

Reverse transcription

First-strand cDNAs are generated by reverse transcription according to the manufacturer's protocol (RT-PCR kit, Perkin Elmer Co., Norwalk, CT, USA).

Materials

- 25 mM MgCl$_2$
- 10× PCR Buffer II, dNTPs ×4 (see 'Assembling of Primer Pairs into a Template')
- RNAase inhibitor (Perkin Elmer)
- Reverse transcriptase (Perkin Elmer)
- Oligo(dT)$_{16}$ (Perkin Elmer)

Procedure

1. Prepare a master mix for reverse transcription using the Perkin Elmer RT-PCR kit on ice.

Component	Volume	Final
25 mM MgCl$_2$	4 µl	5 mM
10× PCR buffer II	2 µl	1×
dNTPs ×4	2 µl ×4	1 mM each
RNAase inhibitor	1 µl	1 U µl^{-1}
Reverse transcriptase	1 µl	2.5 U µl^{-1}
Oligo(dT)$_{16}$	1 µl	2.5 µM

2. Prepare five different dilutions (4-fold) of the synthetic standard RNA.

Note: For 2 µl of total cellular RNA,

dilutions should start at 10^8 molecules of standard RNA. When using the RNA prepared from a known number of cells, the starting point for the dilutions of the standard RNA should be predetermined in calibration experiments. In our experiments, dilutions typically start at 10^5 molecules of standard RNA with the cellular RNA from 1000 cells. This figure will vary depending on the cells used and the level of expression of the gene studied. The exact dilutions should be modified accordingly.

3. Mix 2 μg of the total cellular RNA (or the RNA obtained directly from a known amount of cells) with five different dilutions (4-fold) of the standard RNA. Add DEPC-treated water to the tubes to bring the final volume to 3 μl.

   Note: If the RNA is obtained from a known amount of cells, do not let their number be greater than 10^5.

4. Heat the mixtures at 65°C for 2 min then cool on ice immediately.
5. Add 17 μl of the master mix for reverse transcription to each RNA mixture and overlay with 20 μl of mineral oil.
6. Incubate this mixture at 25°C for 10 min, then at 37°C for 20 min and finally at 42°C for another 20 min.
7. Incubate at 99°C for 5 min, then cool at 4°C for 5 min.
8. Dilute the reaction mixtures from 2 μg of total cellular RNA to 100 μl (1:5).

   Note: Reaction mixtures set up with RNA from a known number of cells should not, in principle be diluted unless, with experience, one realizes that the gene(s) to be studied are expressed at such high levels that they can not be quantified without previous dilution.

The reverse transcription (RT) products can be analyzed immediately or stored at –20°C until the PCR analysis is performed.

Polymerase chain reaction

Materials

See 'Assembling of Primer Pairs into a Template'.

Procedure

1. Take 1 μl of the RT products from each of the cellular/standard RNA dilutions and put into five different PCR tubes.
2. Prepare PCR master mix as follows:

Component	Volume (10 μl)	Final
10× PCR buffer II	1 μl	1×
25 mM MgCl$_2$	0.6 μl	1.5 mM
dNTPs ×4	0.2 μl	200 μM
5' primer (specific)	0.4 μl	(0.2 μM)
3' primer (specific)	0.4 μl	(0.2 μM)
[α-^{32}P]dATP	0.2 μCi	
Taq polymerase	1.0 unit	1 U/10 μl
H$_2$O	to 9 μl	

Note: The reaction volume can be varied proportionally and the isotope is added for quantitative detection. For this purpose, other materials used for quantification can be substituted for [α-^{32}P]ATP (e.g., biotin-labeled nucleotides, digoxigenin-labeled nucleotides, etc.).

3. Mix the PCR master mix and the RT product(s) (which will be used as template for the PCR) and overlay with mineral oil.
4. Carry out the PCR reaction on a Perkin Elmer/Cetus thermal cycler. Denature samples at 94°C for 3 min and then subject them to 30 cycles of

denaturation at 94°C × 30 s, annealing at 50°C–60°C for 30 s, and extension at 72°C × 30 s.

Note: The annealing temperature should be determined empirically or by the Oligo program. The optimal annealing temperature for most primers around 20 nucleotides in length with moderate G+C content ranges between 45°C and 60°C. The total number of PCR cycles to be used depends on the total amount of cDNA used and on the level of expression of the gene being assayed. For example, if RNA from only around 50 cells is being used, the number of PCR cycles should go up to 55.

The PCR products are analyzed directly or can be stored at 4°C until analysis.

Results readout

The most straightforward approach for quantification of the PCR products derived from the standard and cellular RNA is measurement of the incorporation of labeled nucleotides or labeled primers into the PCR products.

Materials

- 4–6% polyacrylamide gel
- 1× TBE buffer
- Electrophoresis equipment
- Phosphorimaging screen

Procedure

1. 1 μl of 10× loading buffer is added to each PCR reaction mixture.
2. 5 μl of each PCR reaction mixture from the five different dilution tubes is electrophoresed in a 4–6% polyacrylamide gel in 1× TBE buffer. An appropriate marker should be loaded in parallel.

   Note: The concentration of polyacrylamide depends on the size of the amplified product.

3. After electrophoresis, the gel is stained with 0.5% ethidium bromide in 1× TBE for 15 min at room temperature.

4. A photograph of the gel adjacent to a fluorescent ruler should be taken under UV to confirm the size of bands (from both the standard and cellular templates).

   Note: If a radioactive marker is used in the electrophoresis, steps 3 and 4 can be omitted.

5. The gel is dried in a vacuum drier for 30 min at 80°C (similarly to the process for a sequencing gel).
6. The dry gel is then exposed on a phosphorimaging screen overnight at room temperature.
7. The specificity of the radioactive PCR products is established by comparing the size of the amplified products to the expected cDNA bands on the stained gel.
8. The quantification of specific bands is achieved by the phosphorimager system in combination with Imagequant Software (Bio-Rad, Hercules, CA, USA).

   Note: Compatible software can also be used.

9. The amount of radioactivity contained within each of the two bands (corresponding to standard and

cellular RNAs, respectively) as obtained from the analysis of the screen, is plotted on a logarithmic scale against the known amount of standard RNA molecules (Fig. 14.7.4a).

Note: These three values should be plotted for each experimental tube. Thus each tube has one x value (the known amount of standard molecules in that experimental tube), and two y values (the radioactivity signal of the standard band (y_1) and that of the cellular band (y_2) obtained from

phosphorimaging analysis of the two bands from that experimental tube. The amount of cellular RNA molecules is directly interpolated from the crossover point of the two curves (Fig. 14.7.4).

To correct for any variation in RNA content and cDNA synthesis in the different preparations, each sample should be normalized on the basis of its β_2-microglobulin or β-actin RNA contents (see 'Comments and Trouble Shooting' later).

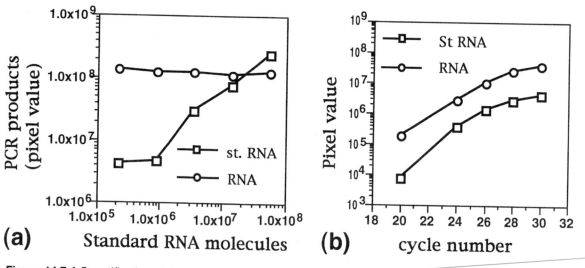

Figure 14.7.4 Quantification of cytokine receptor mRNA. (a) The radioactive signal of specific bands derived from standard and cellular cDNA was quantified and the amount of PCR product (pixel value) was plotted against the amount of standard RNA included in the cDNA reaction mix. The amount of cellular RNA is directly inferred from the crossover point of the two curves. (b) Kinetic analysis of simultaneous amplification to confirm the accuracy of the measurement. The radioactive values of PCR products were plotted on a logarithmic scale against the number of amplification cycles.

Kinetic analysis

This is the most critical step of the whole quantification procedure. To insure the accuracy of the estimated number of cellular RNA molecules as determined in the preceding subsection, verification of similar amplification

efficiencies for the standard and cellular RNA has to be performed.

Since heterologous competitor standard which has a different sequence than that of the cellular target RNA molecules (except for the primer sequence) has been used, there could be a difference in the amplification efficiencies of standard and cellular target RNAs. To insure that there is a similar amplification efficiency, kinetic experiments should be carried out at various cycle numbers (from 20 to 30 cycles). These kinetic experiments are performed with fixed amounts of cellular and standard RNA that correspond to the experimental point at which the concentrations of standard and cellular RNA were closest to the theoretical intersection between the two curves in the titration plot. In this case, several duplicate tubes are subjected to a different number of cycles. The same protocol described for the titration analysis is followed for the kinetic analysis. After the indicated number of cycles, the different samples are processed as indicated above. The phosphoimaging values obtained from the cellular and standard bands are plotted on a logarithmic scale against the number of amplification cycles (Fig. 14.7.4b).

Note: The number of cycles used in the kinetic analysis depends on the number of cycles used in the titration analysis.

The slopes of the two curves generated in the kinetic plot (corresponding to the standard and cellular RNAs) should be parallel. If this is the case, the amplification efficiency is similar for both templates, and the amount of cellular RNA can be interpolated directly from the titration plot (or alternatively from the kinetic analysis plot). If the two curves are not parallel, the conditions for the PCR should be changed.

Note: At this stage of the procedure, the concentrations of both templates (cellular and standard) must be adjusted until they become equal.

Comments and trouble shooting

Considerations on RNA preparation

There are many methods of RNA extraction routinely used for large-scale RNA preparation that yield very low amounts of RNA when used with a small number of cells (fewer than 10^6 cells). When the amount of sample is limited, this poor recovery becomes very important. In some experimental protocols, cell lysates (Alard et al 1993) are directly used for RT-PCR. However, we have observed that the results obtained from such experiments are not always reliable and can yield high backgrounds. This is probably due an RNA population mixed with proteins, genomic DNA, and semidegraded RNA. Thus, we recommend the microbeads protocol for the purification of RNA. The use of oligo(dT) beads for purification of poly(A) + RNA from cell lysates has further advantages: (1) stronger denaturation chemicals can be used in the RNA preparation to destroy RNAases completely and protect RNA from degradation; (2) most cellular proteins and genomic DNAs can be removed from the RNA sample. Conveniently, the RNA still attached to the beads can be used directly for reverse transcription and also for PCR amplification. In our experience, there are no significant differences in the amplification efficiencies when using the RNA template eluted from the beads or noneluted. Using this method, we have been able to study cytokine gene expression in RNA samples derived from 20 to 10^4 cells.

An alternative approach for the study of mRNA expression in small populations of cells could be based on RT-PCR amplification of mRNA supported on membranes as reported by Ruiz and Bok (1993). The positive band on a Northern blot was cut out, and the piece of membrane with the fixed RNA was used as template for RT-PCR. In this case, the portion of the membrane containing the RNA source can be repeatedly used for RT-PCR and therefore the expression of different genes can be measured from the same small amount of RNA. We have used a similar method to that described above which is based on fixing the cellular and standard RNAs on a small piece of nylon membrane. The membrane is then subjected to competitive RT-PCR. Unfortunately, our experience has been that the RT-PCR amplification efficiencies of both templates differed between experiments with the same membrane pieces and within the same experiment between different membrane pieces. Experiments are underway to solve these technical problems, which are probably related to saturation of the membranes or with even distribution of the template on the membranes.

Polymerase chain reaction

We have developed some technical tips over time in our laboratory. Performing the RT-PCR reaction in small volumes has the advantage of savings in the amount of template RNA to be used for each reaction (which can be critical when the amount of sample is limited or when there is a large battery of genes to be measured), as well as in reagent costs. Since only $10\,\mu l$ of PCR product is used for the quantification, we reduced the reaction volumes to $10\,\mu l$ in the PCR amplification (see also 'Polymerase Chain Reaction'). Incubation times should be kept as short as possible to reduce overall cycling time and to minimize the risk of non-specific amplification. Denaturing and annealing times of 30 s should be adequate. Extension times of 30 s for targets that are less than 1 kilobase long are also adequate (in a thin-walled tube). The number of cycles required depends on the abundance of target RNA and of the efficiency of the PCR (generally from 30 to 60 cycles).

PCR product quantification

The most straightforward approach is to measure the amount of product by measuring the incorporation of labeled nucleotides or labeled primers into the PCR products generated and resolving them by gel electrophoresis. Although direct, the use of labeled nucleotides in PCR

can be problematic. Trace amounts of unincorporated label often remain in the electrophoretic gel as the product bands migrate, resulting in a 'trail' of label throughout the lane. This sometimes makes it difficult to measure the amount of incorporated label. In this respect, electrophoresis with polyacrylamide gel is a better choice since the background level is 20-fold higher with agarose gel than with polyacrylamide gel electrophoresis (Bouaboula et al 1992). On the other hand, the agarose gel is more difficult to dry than is polyacrylamide. Because only one isotope is incorporated per molecule of PCR product in labeled primers method, the method involving incorporation of labeled nucleotides is much more sensitive than with labeled primers. In theory, the different numbers of A nucleotide sites in standard and cellular RNA should be taken into consideration. However, in our experience, we have not observed significant differences related to different A nucleotide sites in standard and cellular RNA, probably because of the relatively large sizes of the amplified fragments (200–800 bp), and because of the large numbers of molecules in the PCR amplifications. It is important to increase the sensitivity of detection for the measurement of low gene expression from limited numbers of cells.

To control for variations due to errors in the spectrophotometric measurement of RNA amounts in the different samples, the PCR results obtained for the expression of the different genes studied should be normalized to the expression of β-actin in the same set of samples. These normalization experiments can be performed with several housekeeping genes (for example β-actin or β_2-microglobulin), which have been reported to have highly stable expression and have been used widely as calibration genes in Northern blot and PCR. However, there are several reports of problems with use of β-actin as a control for adequate RNA preparation and cDNA synthesis (Taylor and Heasman 1994; Cross et al 1994). We have found similar problems with this gene in our calibration experiments, suggesting that β-actin should not generally be used as a control for RT-PCR reactions. This is especially true for RT-PCR from very small amounts of RNA. In this regard, we found that β_2-microglobulin is better as a control for quantity and quality of RNA and cDNA transcription for RT-PCR experiments.

References

Alard P, Lantz O, Sebagh M, Calvo CF, Charpentier B (1993) A versatile ELISA-PCR assay for mRNA quantification from a few cells. Biotechniques 15: 730.

Arai K, Nishida J, Hayashida K, et al (1990) Cytokines: coordinate regulation of immune and inflammatory responses. Annu Rev Biochem 59: 783.

Bouaboula M, Legoux P, Pessegue B, et al (1992) Standardization of mRNA titration using a polymerase chain reaction method involving co-amplification with a multispecific internal control. J Biol Chem 267: 21830–21838.

Chirmule N, Oyaizu N, Kalyanaraman VS, Pahwa S (1991) Misinterpretation of results of cytokine bioassays. J Immunol Methods 137: 141–144.

Chomczynski P, Sacchi N (1987) Single-step method of RNA isolation by acid guanidinium thiocyanate–phenol–chloroform extraction. Anal Biochem 162: 156–159.

Cross NCP, Lin F, Goldman JM (1994) Appropriate controls for reverse transcription polymerase chain reaction (RT-PCR). Br J Haematol 87: 218.

Kaashoek JGJ, Mout R, Falkenburg JHF, Landegent JE (1991) Cytokine production by the bladder carcinoma cell line 5637: rapid analysis of mRNA expression levels using a cDNA-PCR procedure. Lymphokine Cytokine Res. 10: 231–235.

Legoux P, Minty C, Delpech B, Minty AJ, Shire D (1992) Simultaneous quantitation of cytokine mRNAs in interleukin-1 beta stimulated U373 human astrocytoma cells by a polymerisation chain reaction method involving coamplification with an internal multi-specific control. Eur Cytokine Netw 6: 553.

Nicola NA (1989) Hematopoietic cell growth factors and their receptors. Annu Rev Biochem 58: 45–77.

Ruiz A, Bok D (1993) Direct RT-PCR amplification of mRNA supported on membranes. Biotechniques 15: 882.

Scheuermann RH, Bauer SR (1993) Polymerase chain reaction-based mRNA quantification using an internal standard: analysis of oncogene expression. Methods Enzymol 218: 446–473.

Taylor JJ, Heasman PA (1994) Control genes for reverse transcription/polymerase chain reaction (RT-PCR). Br J Haematol 86: 444–445.

Wang AM, Doyle MV, Mark DF (1989) Quantitation of mRNA by the polymerase chain reaction. Proc Natl Acad Sci USA 86: 9717–9721.

Generation of immunological chimeras at fetal stages of lymphocyte development by *in utero* transplantation

14.8

Aliki-Anna Nichogiannopoulou
José-Carlos Gutiérrez-Ramos

The Center for Blood Research, Inc. and the Department of Genetics, Harvard Medical School, Boston, Massachusetts, USA

TABLE OF CONTENTS

Immunology Methods Manual
ISBN 0–12–442712–X

Introduction

The generation of hematopoietic chimeras has provided a powerful tool for the study of the development and maturation of the immune system. Syngeneic, congeneic, or allogeneic bone marrow has been the source of hematopoietic stem cells for the generation of such chimeras. Typically, the transplanted population is depleted of all mature T cells in an attempt to avoid graft-versus-host-disease (GVHD). Further engraftment failure rates can be decreased by conditioning the host with lethal doses of irradiation and/or chemotherapy. The conditioning regimen ablates all mature and progenitor hematopoietic cells potentially able to recognize and reject the allograft. Alternatively, immunodeficient mouse strains can be used as recipients. The severe combined immunodeficient (SCID) mouse (Bosma et al 1983) and the recombinase activating gene-1 or -2 (*RAG-1* or *RAG-2*) deficient mouse (Mombaerts et al 1992; Shinkai et al 1992), which lack B and T lymphocytes, have both been used as recipients for bone marrow transplantation. However, immunodeficient mouse strains with profound lymphoid deficiencies can still mount nonspecific immune responses through their macrophages, NK cells, and granulocytes. Thus, sublethal conditioning remains necessary prior to transplantation. Conditioning regimens affect all cycling cells and often result in severe side-effects, such as gastrointestinal toxicity, which may cause death prior to engraftment.

An additional limitation of adult transplantation is the inherent inability to study the influence of embryonic events. At least one immunodeficiency has been reported which seems to result from a developmental block during fetal thymic development. This immunodeficiency could not be corrected by bone marrow transplantation in adult mice, presumably owing to irreversible degeneration of the thymic microenvironment during fetal life (Hollaender et al 1995).

These considerations call for a fetal transplantation protocol that can evade the immunological defenses of the recipient without the use of toxic conditioning and that can also permit engraftment at a time before irreversible degeneration of the lymphopoietic compartment occurs. A fetal transplantation protocol would also enable the engraftment of donor cells during the earliest stages of hematopoiesis. The latter consideration is relevant to the development of the 'nonconventional' lymphoid lineages, i.e., the CD5$^+$ B cells and the Vγ3 Thy1$^+$ dendritic epidermal T cells. The development of these lineages is reportedly restricted to the fetal stages of lymphopoietic development. Adult transplantation experiments with either irradiated or immunodeficient recipients have shown that the adult lymphopoietic microenvironment does not support differentiation of bone marrow stem cells into these 'embryonic' lymphocyte lineages. Accordingly, reconstitution of these lineages requires the interaction of the transplanted cells with the fetal microenvironment.

We have developed a protocol for the engraftment of hematopoietic donor cells into the fetal liver of immunodeficient mouse embryos at early stages of fetal lymphocyte development. The fetal liver is the main hematopoietic organ in the developing mouse embryo from the time of hepatic vascularization at day 10 until birth at day 21, when hematopoiesis shifts to the bone marrow, where it remains through adulthood (Gilmour 1942; Sachs et al 1990). For that reason, the fetal liver between embryonic days 13 and 17 was chosen as the target for our transplantation protocol.

The protocol allows for engraftment of a donor cell population in the fetal liver of a developing mouse embryo. The immunological immaturity of the recipient permits allogeneic transplantation without conditioning. Also, the engrafted cells and their progeny can interact with the recipient hematopoietic system at an early point in development, allowing for the pattern of lymphocyte ontogeny to be dissected.

Injection of cells through the uterine wall into the embryo is the route chosen to obtain fetal hematopoietic engraftment. Two alternative forms of *in utero* microinjection have previously been described: injection of cells into the placental circulation of day 11 embryos (Fleischman and Mintz 1979, 1984), and injection of cells or viruses into the yolk sac or in the amniotic cavity of day 7–9 embryos (Weismann et al 1978; Jaenisch 1985). These methods have not been employed widely, largely because of the lack of precision in the placement of the transplanted cells, limitations on the degree of engraftment, and the high procedure-related fetal mortality.

The following protocol permits the reproducible delivery of cells into the primary site of fetal hematopoiesis with minimal mortality.

Materials and methods

Mice

Success of the *in utero* fetal transplantation protocol depends largely on the mouse strain used as well as on overall health of the colony. We found it necessary to breed the *RAG-2* mutation, originally in C57BL/6 129 background, into the FVB background to improve breeding behavior and litter size.

All animals are kept in microisolator cages in our SPF facility and provided with sterilized food and water. The light and background noise cycle is kept constant throughout breeding of the colony and during postoperative recovery of pregnant female mice.

Male breeders were kept individually in microisolator cages. To establish timed pregnancies, one or two female mice are placed with a male in a cage overnight. The morning of vaginal plug detection is defined as embryonic day 1. Mice are not allowed to mate for more than 12–14 h to ensure accuracy in the determination of embryonic age. Embryonic age is confirmed morphologically at the time of surgery. Average gestation is 21 days.

Anesthesia

Careful choice of anesthetic is essential. Some anesthetic agents have been reported to result in a high frequency of spontaneous abortions after surgery. We successfully use a mixture of xylazine and ketamine to anesthetize pregnant mice for the duration of surgery (approximately 1 h).

- Ketaset: ketamine HCl 100 mg ml^{-1} (Aveco)
- Rompun: xylazine 20 mg ml^{-1} (Miles)
- Ether: anhydrous (Aldrich)

1. Prepare the anesthetic by mixing Ketaset 1:1 with Rompun.
2. Inject 60 μl of the mixture with a 1 ml insulin syringe (28G$\frac{1}{2}$) intraperitoneally into each pregnant mouse (60 μl/25 g body weight).

Note: Ether can be administered should the mouse start to wake up before the operation is completed. A 50 ml tube containing ether-soaked cotton wool can be placed over the head of the mouse for a few seconds. Ether administration often correlates with an increased rate of spontaneous abortion and should be used only when absolutely necessary.

Microscope and light source

A binocular dissecting stereomicroscope providing total magnification between ×2 and ×20, such as Nikon SMZ-10 or SMZ-U, or Zeiss Stemi SV 6 or SV 11 without understage illumination should be used.

If the dissecting microscope has a built-in light source, this should remain off throughout the surgical procedure, to prevent overheating of the animal and desiccation of the uterus. A fiber-optic light source should be used instead.

Surgical equipment

Most surgical instruments listed have been purchased from Roboz, Inc. (Rockville, MD, USA), but other suppliers with similar instruments can be used.

- Micro dissecting forceps with smooth jaws
- Micro dissecting forceps with longitudinal serrations
- Atraumatic forceps
- 50 mm-long bulldog clamps
- Micro dissecting scissors
- Surgical suture (5.0 silk, 16 mm curved tapered needle, silicone treated)
- 9 mm wound clips
- Cotton swabs
- Animal support platform: A slab of

Styrofoam with two additional slabs glued on top to create a trough is used. If the width of the trough is narrow enough, no additional restrain or support is needed.

- Lavage solution: Lactated Ringer's solution (LR) containing $100\,U\,ml^{-1}$ penicillin and $100\,\mu g\,ml^{-1}$ streptomycin is used for lavage. The solution is prewarmed to 37°C before use.
- Microinjection apparatus: Cell suspensions are injected with an automated Narishige IM-200 (New York, London) pneumatic microinjector supplied with a foot switch.

Microinjection needles

The needles for microinjection are pulled from borosilicate glass capillaries (World Precision Instruments, Sarasota, FL, USA) with an outer diameter of 1.0 mm and an inner diameter of 0.58–0.60 mm. The needles are prepared by pulling the glass capillaries on a mechanical pipette puller (Flaming/brown Micropipette Puller, Model P-87, Sutter Instruments). Instrument settings should allow for pulling a pipette with an internal tip diameter of 20–30 μm. The needles are beveled on a stone beveler (K.T. brown type Micropipette Beveler, Model BV-10, Sutter Instruments) to acquire a bevel of 50 μm. Needles are stored in 150 mm Petri dishes on strips of autoclaved modeling clay.

Note: The microinjection needles are designed to be solid enough for puncturing the uterine wall, have a wide enough inner diameter to avoid clogging, yet be small and delicate enough to minimize embryonic damage.

Injection procedure

Preparation of cells to be injected

Cells to be injected must be prepared as a single cell suspension in protein-free IMDM. Since the injected volume has to be kept between 0.5 and 1.0 µl, cells must be concentrated. However concentrations higher than 1×10^5 cells µl^{-1} (1×10^8 cells ml^{-1}) are not recommended, as clogging of the microinjection needles may occur. Cells must be kept on ice to prevent aggregation.

Surgical procedure

1. Anesthetize a pregnant mouse as described above and place on its back on the support platform. Swab the abdomen with 75% alcohol and make a ventral midline incision. A 2 cm-long incision is usually sufficient. *Note:* We always try to keep the incision as short as possible without compromising uterine accessibility, as this minimizes surgical closure time and hence the dose of anesthesia required.
2. Carefully separate the connective tissue underneath the skin from the dermis and attach the cutaneous margins to bulldog clamps for retraction. *Note:* Separation of the skin from the underlying connective tissue allows maximal stretching of the skin, keeping the incision length short.
3. Carefully lift the peritoneal wall, avoiding the underlying bowel, and incise along the midline, avoiding blood vessels. Thoroughly lavage the peritoneal cavity with LR until all bleeding (if any) ceases. *Note:* During the entire procedure keep the uterine horns and the peritoneal cavity of the mouse moist by repeated lavages with LR.
4. Gently, with atraumatic forceps pull out both uterine horns and count the decidual swellings in each horn.
5. Starting from the ovarian end of the right uterine horn, each embryo is supported with a cotton swab and transilluminated with a fiber-optic light source to make the fetal liver visible (Plate 3). The fetal liver at embryonic day 14 is a dark red structure, occupying a large proportion of the embryo's peritoneal cavity, and should be readily visualized. If necessary, the embryo can be viewed under the stereoscope to facilitate orientation.
6. With the microinjection needle loaded and attached to an automated microinjector, penetrate the uterine wall at a 45° angle to a depth of 1 to 2 mm and inject 0.5–1 µl of cell suspension. Each embryo should be manipulated separately. *Note:* Do not inject the embryo most proximal to the cervix on each side. In our experience this increases viability of the whole litter and decreases postoperative abortion rate. Minimizing handling of the uterus also minimizes trauma and increases survival rates.
7. After injecting the last embryo, thoroughly rinse both uterine horns with LR and place them back in the peritoneal cavity.
8. Lavage extensively with LR and distend the peritoneal cavity with LR before surgical closure. *Note:* This significantly increases survival, probably by preventing maternal dehydration during recovery from anesthesia as well as preventing infection.
9. Using 5.0 silk suture, close the peritoneum with a continuous locking stitch. Close the skin with the same type of suture, but using straight

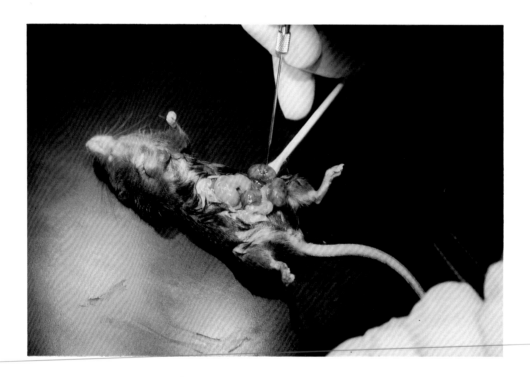

Plate 3 An anesthetized day 14 pregnant mouse with the left uterine horn exposed. The first embryo from the ovarian end is supported by a cotton swab and injected into the fetal liver, between front and hind limbs.

stitches instead. They are sturdier than the continuous locking stitch and will prevent evisceration. Wrap the mouse in a piece of paper tissue and place it on its back in the cage. Allow recovery

from anesthesia under a heating lamp. Place some mouse food soaked in water on the floor of the cage for the recovery period.

Comments

This protocol overcomes some of the limitations inherent in the *in utero* transplantation methodologies described by others. Fleischman and Mintz (1979, 1984; Fleischman et al 1982) describe the use of the W/W mutant anemic mice with an endogenous stem cell defect due to a mutation in c-kit (Orr et al 1990) as recipient embryos. Allogeneic normal fetal liver cells were microinjected into the blood vessels of the fetal placenta on day 11 of gestation. The injected cells reach the fetal circulatory system via efferent blood vessels and establish hematopoiesis, marking the newborn recipients with allogeneic hematopoietic cells. Through the transplacental injection route, the donor cells are placed directly into the circulation, assume their usual route of migration, and thus have direct access to normal sites of differentiation. The degree of engraftment appeared to be directly correlated with the severity of the anemia, with the most anemic recipients showing the highest engraftment rates. Severely anemic W/W mice, however, are nonfertile (Geissler et al 1981) and

must be mated as wild type heterozygotes, giving rise to only 25% homozygous mutants, i.e., engraftable progeny.

In utero microinjection can also be accomplished by injecting cells into the amniotic cavity between days 8.75 and 9.25 of gestation. The cells deposited in the amniotic cavity enter and eventually colonize the embryo (Jaenisch 1985). The success of this protocol is largely dependent on the migratory nature of the injected cells, which is directly correlated to their capacity to engraft the recipient (Weismann et al 1978). Therefore this route of injection is more suitable for neural-crest cells, primordial germ cells, and yolk sac blood–island cells.

Retroviral suspensions (Jaenisch 1980; Walsh and Cepko 1988) and virus-producing cells (Lemischka 1986) have also been introduced into mouse embryos in this manner, facilitating *in situ* infection at very early stages and subsequent lineage analysis of the progeny.

References

Bosma GC, Custer RP, Bosma MJ (1983) A severe combined immunodeficiency mutation in the mouse. Nature 301: 527.

Fleischman RA, Mintz B (1979) Prevention of genetic anemias in mice by microinjection of normal hematopoietic stem cells into the fetal placenta. Proc Natl Acad Sci USA 76: 5736.

Fleischman RA, Mintz B (1984) Development of adult bone marrow stem cells in H-2-compatible and -incompatible mouse fetuses. J Exp Med 159: 731.

Fleischman RA, Custer RP, Mintz B (1982) Totipotent stem cells: normal self-renewal and differentiation after transplantation between mouse fetuses. Cell 30: 351.

Geissler EN, McFarland EC, Russel ES (1981) Analysis of pleiotropism at the dominant white spotting (W) locus of the house mouse: a description of ten new W alleles. Genetics 97: 337.

Gilmour JR (1942) Normal hemopoiesis in intrauterine and neonatal life. J Pathol 52: 25.

Hollaender GA, Wang B, Nichogiannopoulou A, et al (1995) Developmental control point in induction of thymic cortex regulated by a subpopulation of prothymocytes. Nature 373: 350.

Jaenisch R (1980) Retroviruses and embryogenesis: microinjection of Moloney leukemia virus into midgestation mouse embryos. Cell 19: 181.

Jaenisch R (1985) Mammalian neural crest cells participate in normal embryonic development on microinjection into post-implantation mouse embryos. Nature 318: 181.

Lemischka IR, Raulet DH, Mulligan RC (1986) Developmental potential and dynamic behavior of hematopoietic stem cells. Cell 45: 917.

Mombaerts P, Iacomini J, Johnson RS, Herrup K, Tonegawa S, Papaioannou V (1992) RAG-1 deficient mice have no mature B and T lymphocytes. Cell 68: 869.

Orr UA, Avivi A, Zimmer Y, Givol D, Yarden Y, Lonai P (1990) Developmental expression of c-kit, a proto-oncogene encoded by the W locus. Development 109: 911.

Sachs L, Abraham NG, Wiedermann CJ, Levine AS, Konwalinka G (eds) (1990) Molecular Biology of Hematopoiesis. Intercept, England.

Shinkai Y, Rathburn G, Lam K-P, et al (1992) RAG-2 deficient mice lack mature lymphocytes owing to inability to initiate V(D)J recombination. Cell 68: 855.

Walsh C, Cepko CL (1988) Clonally-related cortical cells show several migration patterns. Science 241: 1342.

Weismann IL, Papaioannou VE, Gardner RL (1978) In: Differentiation of Normal and Neoplastic Hematopoietic Cells, Cold Spring Harbor Laboratory Press, Cold Spring Harbor, NY, p. 33.

Outlook

14.9

Carlos Martinez-A.

Centro Nacional de Biotecnología, Universidad Autonóma, Madrid, Spain

Improvement of methodologies as well as development of new methodologies will help to increase our appreciation of their applications and to solve the technical problems that remain. For example, whole-mount *in situ* hybridization is now being used to detect apoptotic cell death during lymphoid development. The development of new nonradioactive labeling systems and detection substrates will allow the detection of the expression of different molecules (e.g., a given cytokine and its receptor) simultaneously in the same tissue or organ. In the future, a combination of *in situ* hybridization and PCR amplification will surely increase the sensitivity of this technique.

None of the methods described for detecting apoptosis allows for unequivocal distinction between apoptosis and necrosis, given that apoptosis is typically followed by secondary necrosis. To discriminate these types of cell death, it is imperative either to assess cellular morphology by electron microscopy or to determine the electrophoretic pattern of DNA fragmentation. The development of techniques that clearly will distinguish these processes in a simple way is therefore imperative.

Retroviral gene transfer has great potential and this methodology will become more popular with the advent of new generations of vectors such as those containing internal ribosomal entry sites (IRES) to insure expression of the transduced sequences and helper-free high-titer packaging cell lines. The majority of the protocols designed to generate high chimerism in the receptor animals usually introduce a selective step to enrich the transduced populations. Most of them are based on the antibiotic resistance conferred by the provirus. This practice could eliminate populations in which the expression of the reporter gene would be under the threshold necessary for survival. This inconvenience might be ameliorated, or at least controlled, using vectors that bear marker genes encoding proteins expressed in the cell surface; then the selection may be carried out by cell sorting. The combination in the same vector of these marker genes and tags recognizable by PCR would be an extremely powerful tool for lineage analysis. Finally, a deeper understanding of the stem cell biology could permit the description of new growth factors critical for allowing efficient *in vitro* maintenance and transduction.

Biology has focused in the last 10 years on the functional characterization of proteins by genetic analysis. Although the gain or loss of function clearly establishes the involvement of a given gene in a particular function, reality often demonstrates the presence of redundancy. Attention has lately focused on the

development of *in vitro* biochemical techniques aiming at the characterization of the fine structure–function relationship of a given protein. Enzymes constitute perfect candidates since their catalytic activities can be tested *in vitro*. The present section includes the latest developments in the analysis of protein kinase catalytic activities, effector preferences, and regulation that will complement genetic analysis and crystallographic studies. In this regard, expectation is centered on the development of techniques for analyzing the structure of purified proteins or complexes with their regulators and/or effectors. In addition, detailed knowledge of how enzymes mutate in different subdomains will enable the preparation of mutated genes or new proteins that will retain only some of their original characteristics.

Index

Index